Lecture Notes in Mechanical Engineering

r further volumes:
://www.springer.com/series/11236

Giorgio Dalpiaz · Riccardo Rubini
Gianluca D'Elia · Marco Cocconcelli
Fakher Chaari · Radoslaw Zimroz
Walter Bartelmus · Mohamed Haddar
Editors

Advances in Condition Monitoring of Machinery in Non-Stationary Operations

Proceedings of the Third International Conference on Condition Monitoring of Machinery in Non-Stationary Operations CMMNO 2013

Springer

Editors
Giorgio Dalpiaz
Gianluca D'Elia
Engineering Department
University of Ferrara
Ferrara
Italy

Fakher Chaari
Mohamed Haddar
Mechanical Engineering Department
National School of Engineers of Sfax
Sfax
Tunisia

Riccardo Rubini
Marco Cocconcelli
Department of Science and Engineering Methods
University of Modena and Reggio Emilia
Reggio Emilia
Italy

Radoslaw Zimroz
Walter Bartelmus
Diagnostics and Vibro-Acoustics Science Laboratory, Institute of Mining Engineering
Wrocław University of Technology
Wrocław
Poland

ISSN 2195-4356 ISSN 2195-4364 (electronic)
ISBN 978-3-642-39347-1 ISBN 978-3-642-39348-8 (eBook)
DOI 10.1007/978-3-642-39348-8
Springer Heidelberg New York Dordrecht London

Library of Congress Control Number: 2013948862

Printed on acid-free paper

Springer is part of Springer Science+Business Media (www.springer.com)

Preface

Third edition of the International Conference on Condition Monitoring of Machinery in Non-Stationary Operations (CMMNO13) was held in Ferrara, Italy. This yearly event merges an international community of researchers who met—in 2011 in Wrocław (Poland) and in 2012 in Hammamet (Tunisia)—to discuss issues of diagnostics of rotating machines operating in complex motion and/or load conditions.

The growing interest of the industrial world on the topics covered by the CMMNO13 involves the fields of packaging, automotive, agricultural, mining, processing, and wind machines in addition to that of the systems for data acquisition. The participation of speakers and visitors from industry makes the event an opportunity for immediate assessment of the potential applications of advanced methodologies for the signal analysis.

As a matter of fact, signals acquired from machines often contain contributions from several different components as well as noise. Therefore, the major challenge of condition monitoring is to point out the signal content that is related to the state of the monitored component. This is before all else demanding when the machines operate in non-stationary conditions.

The book is the collection of the CMMNO13 Proceedings, and it is divided into the following parts, namely:

Part I: Keynote Speeches
Part II: Rolling Bearing Diagnostics
Part III: Modelling of Dynamics and Fault in Gear Systems
Part IV: Signal Processing for Machine Condition Monitoring
Part V: Experimental and Numerical Modeling of Machine Dynamics
Part VI: Mechanical Systems Diagnostics

Part I collects some of the speeches given at CMMNO13 by: Prof. Nicolò Bachschmid, Prof. Cécile Capdessus and Prof. Diego Galar.

The topic of the Part II and III is the vibration analysis for the diagnosis of faults in Gearbox and Bearings. De facto, these components play a pivotal role in the rotating machine scenario. Even if the study on incipient failure detection of gearboxes and bearings started over two decades ago, there is still a great need to implement new algorithms for fault diagnostics especially when the machine operates in nonstationary conditions.

Part IV shows that modern condition monitoring extensively requires advanced signal processing techniques. This part embraces complex techniques as well as the use of the "Instantaneous Angular Speed" and the "Empirical Mode Decomposition".

Part V addresses the expected steps for success of condition monitoring methods: test rig development and the numerical modeling of machine dynamics. As a matter of fact, the real behavior of machines can be only studied with numerical models experimentally validated.

In the last parts of the book, a miscellaneous of particular case studies involving cutting tools, excavators, rolling mills, and water distribution networks are presented. The papers presented in this part give the solution to individual applications of monitoring and diagnostics of such complex machines. The presented test cases come directly from industry needs and observation of industrial real problems.

Contents

Part V Experimental and Numerical Modeling of Machine Dynamics

Part I
Keynote Speeches

Dynamical Behavior of Rotating Machinery in Non-Stationary Conditions: Simulation and Experimental Results

Nicolò Bachschmid and Steven Chatterton

Abstract Condition monitoring of rotating machinery is generally performed in *stationary* conditions or quasi-stationary conditions. Assuming linearity of the system its dynamic behaviour can be simulated in the frequency domain. Simulated results are then compared to measured results and the comparison allows to apply model based diagnostic procedures. As soon as the system presents strong non-linearity, simulation must be performed in the time domain, including also iterative procedures, which may become cumbersome. When the dynamic behaviour of linear or non-linear systems in non-stationary conditions is simulated, the time domain integration must be necessarily used. Accuracy of simulated results gets weaker; comparison with measured results for diagnostic purposes becomes difficult or ineffective; model based diagnostic approach seems not applicable. Monitoring of machines in "strong" *non-stationary* conditions is generally performed only by means of accurate signal analysis, without modelling the machine or the process. In this paper the simulation of some typical behaviour of rotating machines in weak or strong non-stationary conditions in the time domain is presented and discussed. Some experimental results are also presented and compared to simulations. Further examples of systems with strong non-linearity, that in stationary conditions exhibit non-stationary vibrations, are also given.

Keywords Blade Flutter Instability · Blade Vibration · Condition Monitoring · Friction Contact Model · Non-Stationary Conditions · Rotating Machinery · Rotor Rubbing · Steam Whirl Instability

---

N. Bachschmid (✉) · S. Chatterton
Department of Mechanical Engineering, Politecnico di Milano, Via G. La Masa 1 20156 Milan, Italy
e-mail: Nicolo.Bachschmid@polimi.it

G. Dalpiaz et al. (eds.), *Advances in Condition Monitoring of Machinery in Non-Stationary Operations*, Lecture Notes in Mechanical Engineering,
DOI: 10.1007/978-3-642-39348-8_1,

1 Introduction

Continuous condition monitoring of rotating machinery by means of vibration measurements is a common praxis in all kind of industrial plants. The data collected and the comparison with reference data allow to perform the surveillance of the machine, to detect malfunctioning conditions and to trigger alarms or shut downs. In intelligent monitoring systems these data can be processed in order to enable the use of automatic diagnostic procedures. Among these the model based diagnostic procedure, that allows to compare measured results with calculated ones obtained by means of models of the machine and of the malfunctions, seems to be the most promising procedure. In stationary conditions this procedure proved to be highly effective when data are available in many different quasi-stationary (or steady state) operating conditions, typically during the long lasting run-down transients of turbo-groups which are considered a sequence of steady state operating conditions at different rotational speeds. Linearity of the system must be assumed, simulations are made in the frequency domain, and the model based diagnostic procedure is then developed in the frequency domain. As soon as the system becomes non-linear, although being quasi-stationary, this approach is not anymore accurate or even not applicable, depending on the strength of the non-linearity. The simulation of highly non-linear systems must be effected necessarily in the time domain. For smaller or weaker non-linearity also iterative procedures in the frequency domain and the harmonic balance approach can be used.

When the condition of the machine is non-stationary, time domain simulation is generally needed (instead of frequency domain) in order to enable the model to follow the changing condition of the machine.

Non-stationary conditions may be generated by:

1. Rapidly changing the rotational speed or the frequency of excitation
2. Changing the excitation amplitude and frequency
3. Changing system parameters
4. Instability
5. Systems with strong non-linearity (which with stationary excitation exhibit non-stationary motion).

Some of these changes in non-stationary condition may be slow, some others are rapid.

When changes are slow and the system is linear, the approximation of considering discrete steps in time in which the condition is considered steady or stationary (and the vibrations are then calculated in the frequency domain), can be used, which is enormously convenient for computer time. The model based diagnostic approach in the frequency domain can then be used.

Conversely, when changes are rapid, time domain integration must be necessarily used.

The simulation in the time domain of systems modeled by means of 3D finite elements with rather refined mesh, and still more when some non-linearity is present (like friction contacts for instance), may become so cumbersome that simulations are not anymore affordable. Modal reduction can then be used for the linear part of the system for strongly reducing the degrees of freedom: with few degrees of freedom time domain integration is then affordable.

One of the problems in monitoring machines in non-stationary conditions is certainly the definition of a reference situation. The actual transient conditions of the machine should be identical to the reference conditions (e.g. same speed and thermal transient, same initial conditions and so on), to allow a comparison and define the deviation from normal behavior and therefore the malfunctioning condition. Otherwise only an accurate model can predict the behavior in different transient conditions, which could be a very difficult task.

A series of examples of simulations (sometimes compared to experimental results) will be given and discussed in the following.

2 Simulation of Transient Behavior of Rotating Machinery Driven by Electric Motors

In this example the start up of a three-phase electrical motor driving a centrifugal pump under no load, modeled with finite beam elements (one torsional degree of freedom per node) and concentrated mass and moments of inertia of the impeller, is simulated. The start up torque time history is given by the electric motor manufacturer. Figure 1 shows the driving torque (bottom diagram) and the results obtained as rotation, velocity or speed, acceleration as typical example of Runge–Kutta time integration.

Of course also torsional deflections are calculated in each element of the finite element model, from relative rotations between the element nodes, from which maximum stresses can be evaluated and compared to the material strength. Such kind of calculation is always required for shaft lines in which electrical machines (generators or motors) are present. Generally the worst conditions are the torsional vibrations excited by short circuits, where the torque may have peaks higher even by one order of magnitude with respect to rated torque. When a torsional natural frequency is excited during this transient severe stresses may appear in some section of the shaft line and shaft strength must be checked. Fortunately this is a transient event and material fatigue may not be considered.

3 Simulation of Transient Behavior of Vibrations of Blades During Steam Turbine Run-up Transient

This example is taken from [1]. The analyzed blade row is the last stage (conventionally called L-0) of a low pressure (LP) steam turbine, composed by 12

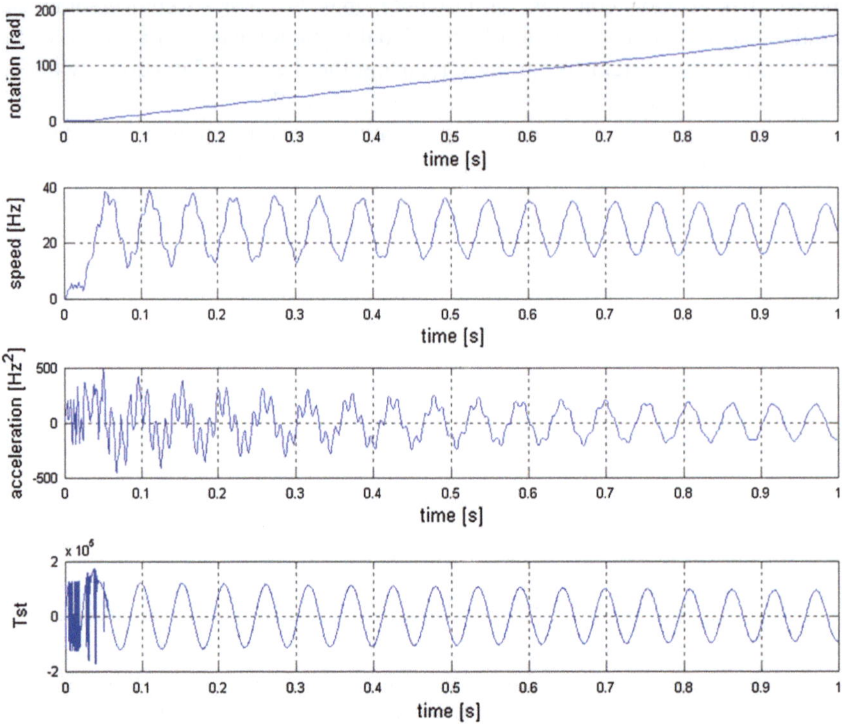

Fig. 1 Start up transient of electric motor driven pump evaluated in one node of the f.e. model

packs of 10 blades each. The blades are grouped in packs of 10 blades by welded shrouds on the tip of the blades and snubbers or lashing wires at an intermediate length of the blade. The first vibrating mode of the pack, which corresponds to an in-phase tangential motion of the blades, has a numerical frequency of 115 Hz. This frequency could be excited in resonance with an engine order 3 at a rotational speed of 2,300 rpm during the run-up of the turbine. This is exactly what happened during the run-up of the machine on the plant, as can be seen from the experimental vibrations reported in Fig. 2. The excitation comes from a non uniform static pressure distribution around the rotating blade row which has apparently a rather strong 3rd harmonic component in the Fourier expansion of the steam pressure around the circumference.

Figure 2 shows the vibration time histories of the first blade of each of the 12 groups as function of the rotational speed during the speed rise between 2,207 and 2,440 rpm which occurred in 7s (with an acceleration of 3.48 rad/sec^2). For the shaft lateral vibration measurement the acceleration is sufficiently low to consider the transient as a sequence of steady state conditions at different rotational speeds. Records were obtained by means of a tip-timing measurement system. Colors indicate phase of vibration with respect to the 1xrev. reference, and show typical phase rotation by passing resonance. Natural frequencies, damping and even

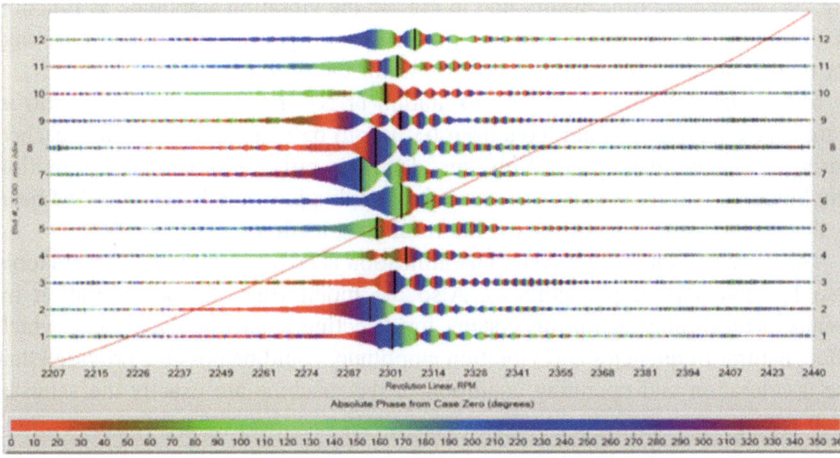

Fig. 2 12 blade pack vibration time histories passing a resonant frequency

excitation seem to be slightly different from pack to pack, due to manufacturing and assembling tolerances, and to some irregularity in the excitation. Being the excitation a 3rd engine order (with a 3xrev. exciting frequency) its acceleration is 10.45 rad/s^2 (three times the rotational acceleration). The effect of this rather high acceleration is clearly visible in the graph: the maximum peak is followed by a series of smaller peaks due to a beating phenomenon between natural and exciting frequency. From this experimental record it is possible to identify both the excitation strength and the modal damping of the blade pack, which are both unknown.

A one d.o.f. modal model of the pack has been build: the stiffness of the pack for this tangential deflection has been defined by applying static tangential unit loads to a finite element model of the blade pack and evaluating the deflections. The modal mass has been deduced from known natural frequency. The modal damping has been defined by comparing the time histories calculated with different values of damping ratios, and unit excitation, to the experimental curves: best fitting defines the modal damping. These time histories are obtained with the usual Runge–Kutta time integration. Being the system linear the time history shapes are independent from intensity of excitation. These curves are shown in Fig. 3: in

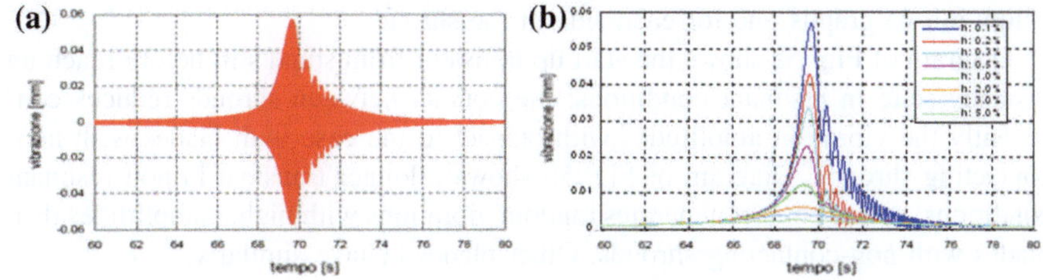

Fig. 3 **a** Vibration time history; **b** Vibration amplitude time histories for different damping ratios

Fig. 3a one generic time history and in Fig. 3b the vibration amplitude as function of time for different modal damping ratios. Resulting best fitting has been found for modal damping between 0.2 and 0.3 % for the different packs.

Maximum amplitude with unit excitation (1 N) and 0.2 % damping ratio in stationary conditions would have been 63 μm, to be compared to 42 μm reached during transient excitation. With this damping ratio and measured vibration amplitude of 2 mm then also the excitation amplitude of the blade pack has been evaluated by proportionality: its value resulted something less than 50 N.

This example shows that system identification is possible with very simple tools also for non-stationary conditions. Also the monitoring of blade vibration by means of tip timing measuring system, for detecting cracks in blade roots, which affect natural frequencies and vibration amplitude, could be possible provided that the acceleration and operating condition of the machine are exactly the same in each run-up transient.

4 Simulation of Blade Rows with Non-Linear Contact Conditions: Single Blade Modeled as 1 d.o.f. System

This example is taken from [2]. A complete integrally shrouded blade row with gaps between shrouds has been modeled by means of 1 d.o.f. per blade modal model representing its first vibration mode, and with non-linear spring and dampers for simulating the contact force between shrouds. Strong model reduction is strictly necessary for having acceptable computer time in calculating blade row behavior. Some gap is allowed between shrouds, harmonic excitation close to resonance is applied and resulting vibrations are calculated. The real blade mounted on a test block and its f.e.m. model showing its first vibration mode are represented in Fig. 4b, and the circular reduced model of the complete row composed of 120 blades is shown in Fig. 4a.

Due to non-linearity time step integration has to be used, and vibrations can result with some chaos, as it is shown in Fig. 5, where vibration time histories of a single blade are represented in two different excitation conditions. The red curve indicates the behavior when shrouds are never in contact (in case of high gap between shrouds) and black curve indicates the resulting vibrations when shrouds get in contact. The contact state (0 open and 1 closed) is indicated by the blue line below in two graphs one for each side of the shroud.

Diagram of Fig. 5a shows the start up transient from stand still before reaching a steady state in resonant conditions: the contact between shrouds reduces consistently the vibration amplitudes with respect to the case with blades with non-contacting shrouds. Diagram of Fig. 5b shows a longer transient in non resonant conditions, where contact generates random vibrations with higher amplitudes than blades with non-contacting shrouds. Other blades behave similarly.

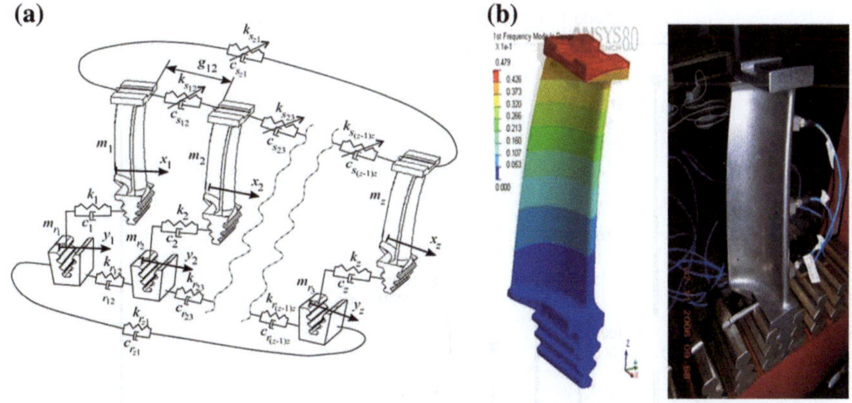

Fig. 4 Integrally shrouded blade row with non-linear contact conditions between shrouds

This is an example of a system in stationary conditions, with a stationary excitation, that behaves randomly which is not exactly a stationary condition, due to a high contact non-linearity. Monitoring blade vibrations in these conditions (which are stationary random) would be rather difficult, despite the fact that in the vibration spectra as that one shown in Fig. 6, some lines are repetitive and could be used as reference.

The contact model is obviously rather rough. More realistic models require to define in the contact area parts which are in sticking contact, parts which are in a sliding or micro-sliding contact, and parts which are separated, which are function of friction coefficient and of contact normal and tangential stiffness, which in turn depend on penetration depth. A more realistic calculation is presented in the following.

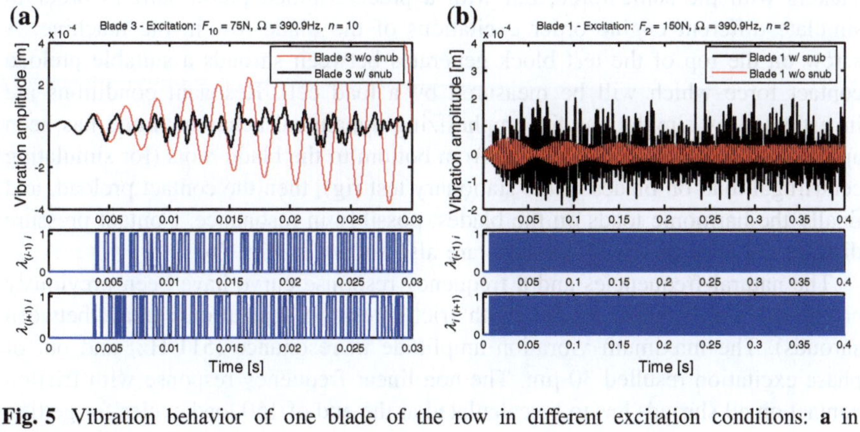

Fig. 5 Vibration behavior of one blade of the row in different excitation conditions: **a** in resonance; **b** out of resonance

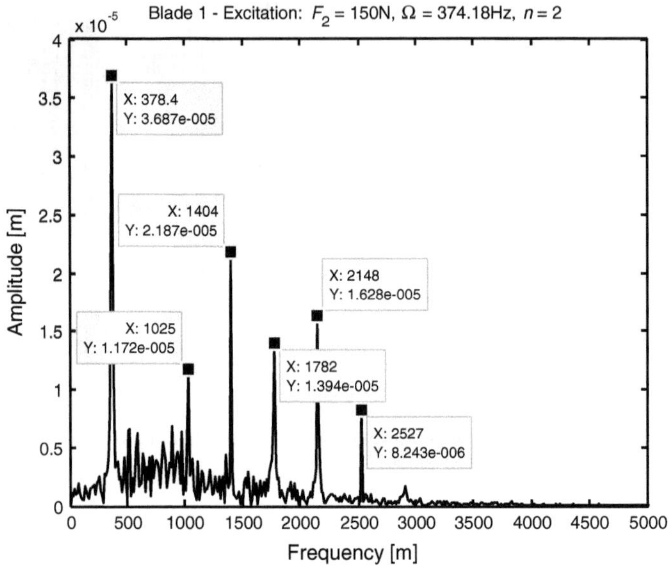

Fig. 6 Vibration spectrum of blade 1 of the row excited slightly out of resonance

5 Simulation of Groups of Blades with Non-Linear Friction Contact Conditions: Full Finite Element Models for Blades and Shrouds

Different contact conditions with different contact forces between shrouds of a group of blades mounted on a test block (represented in Fig. 7a) have been simulated. The complete analysis is described in detail in [3] In the experimental tests, the two internal blades (blades 2 and 3) will be excited independently by two shakers with the same force, but with a predetermined phase shift in order to simulate different engine order excitations of the blade row in the machine. A screw on the top of the test block generates between shrouds a suitable preload contact force which will be measured by a load cell. Resonant conditions are investigated by simulations for emphasizing dynamical behavior. Load has been applied in three steps: first the load from bottom to the blade roots (for simulating centrifugal load on blades in the stationary test rig), then the contact preload, and finally the harmonic loads on the blades, possibly in resonance. Contact pressure distribution after the two load steps are also shown in Fig. 7b.

The natural frequencies and a frequency response curve have been previously calculated with a linear model (with frictionless no separation contact between shrouds). The maximum vibration amplitude in resonance (511 Hz) and out of phase excitation resulted 30 μm. The non-linear frequency response with friction contact on all shrouds has to be calculated at the end of 150 load cycles for getting a steady state solution. Therefore the computation of the steady state response of

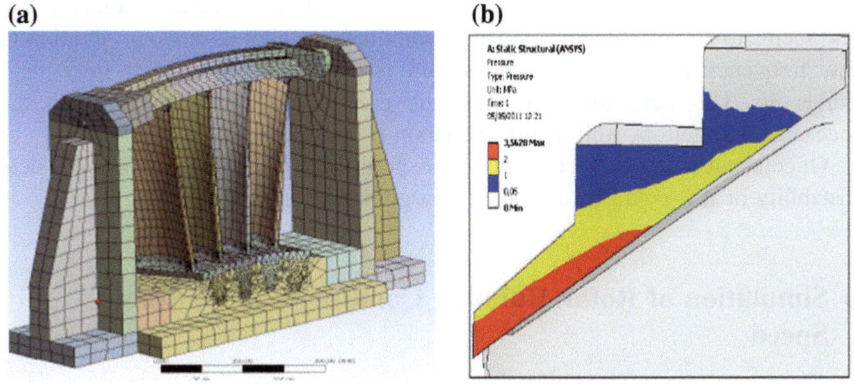

Fig. 7 **a** F.e.m. model of the test rig with the group of blades **b** resulting shroud contact pressure distribution

the system in correspondence of one exciting frequency requires up to a maximum of 150 h with a cluster of 32 processors. The results are stored only for the last load step (when steady state condition has been reached) and may require up to 90 GByte. Several hours are required to download these results on a workstation.

Some results are shown in Fig. 8a: vibration amplitude has been reduced dramatically due to the increased stiffness in sticking contact areas, and to friction in micro-slipping contact in the remaining contact areas: only 1.8 μm (Fig. 8a), with a phase delay of 45° between the two excitations applied to the blades, was the maximum amplitude compared to the frictionless result of 30 μm. Vibrations are sinusoidal like the excitation force, apparently no non-linear behavior is visible.

But when contact preload is reduced (Fig. 8b) then non-linear behavior appears, vibration is periodical but non sinusoidal, slipping is enhanced and vibration amplitude reduces consistently. And when friction coefficient is reduced from 0.2 to 0.1 (Fig. 8c) vibration remains periodical, amplitude as well as slipping increases. These behaviors are steady state.

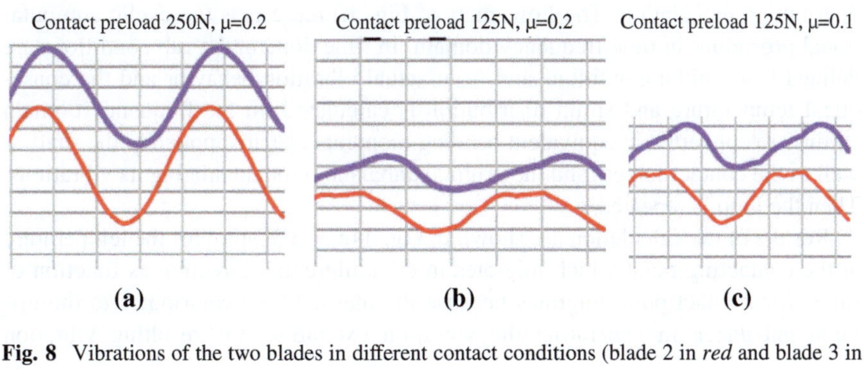

Fig. 8 Vibrations of the two blades in different contact conditions (blade 2 in *red* and blade 3 in *blue*)

Reducing further the friction coefficient to 0.01 vibration becomes non periodic, and seems to be unstable but limited in amplitude. A kind of random vibration in low frequency range (sub-harmonic with respect to the excitation frequency) builds up. This behavior is an unsteady condition, despite the fact that the mechanical system as well as the excitation is stationary.

Of course it must be checked (best by experiments) if this behavior is a physical instability or if it could be attributed to a numerical instability.

6 Simulation of Rotor Rubbing Conditions at Constant Speed

An interesting example of non-stationary thermal state of a shaft due to rubbing which generates non-stationary vibrations concerns the so called spiral vibrations. This behavior is described in detail in [4]. The heat which develops during rubbing of a rotating shaft against a stator, generates a bow of the shaft which evolves in time, and the changing bow generates vibrations. This is a typical example of a rotating shaft in non-stationary conditions. This behavior should therefore be studied in the time domain. But vibrations are 1xrev., at rotational speed frequency of 50 Hz, and should be calculated in time domain with time steps of a fraction (e.g. 1/100) of the rotation period (let's therefore consider a suitable time step of 0.0002 s). Heat distribution due to the friction force evolves slowly, the rotor bow shows appreciable changes in time periods of minutes, depending on shaft diameter. A suitable time step for evaluating the thermal state of the rotor could be 1 min. Using a time step of 0.0002 s for calculating a transient of hours would be a cumbersome exercise. Therefore we can consider the rotor state in each minute of the thermal transient as stationary for the vibrations, and calculate the changes in vibration of the shaft in the frequency domain. This has been done in the simulation of the so-called spiral vibrations of a rotor. The designation comes from the fact that the measured vibration vector moves on a spiral path in a polar plot.

Figure 9 shows the model of the shaft, beam elements for calculating shaft vibrations and a refined mesh in radial, tangential and axial directions for heat distribution calculation. The flow chart of Fig. 10 represents the cyclic computational procedure in time/frequency domain. In time domain the rub conditions are defined from initial conditions and from actual vibration behavior and the consequent temperature and strain distribution is calculated. In the frequency domain strains are converted in equivalent bending moments that are applied to the shaft as well as the contact forces (and the original unbalance) for calculating its vibrations. Then the loop is repeated.

Results of the calculation are shown in Fig. 11a as a 3D plot for the temperature at the contacting point which migrates in circumferential direction as function of time. The contact point migrates because the thermal bow contributes to the original unbalance in generating the vibration excitation: the resulting vibration

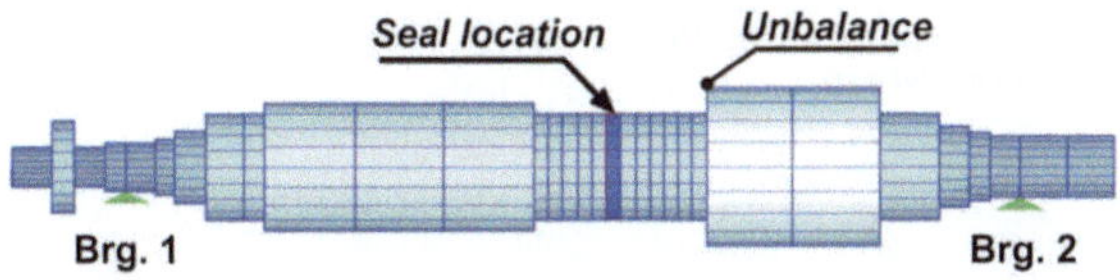

Fig. 9 Model of the shaft of a HP-IP steam turbine rubbing at seal location

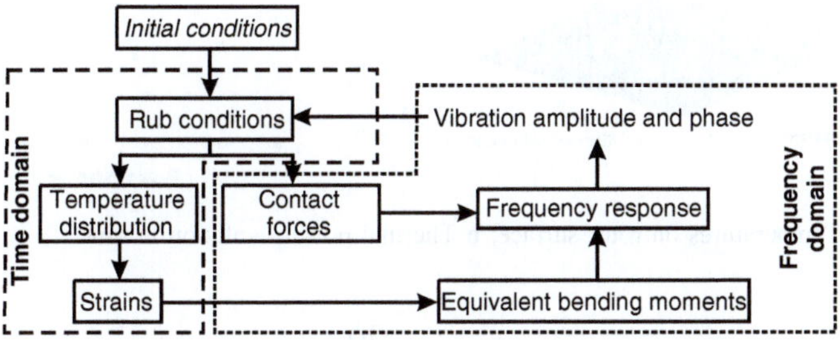

Fig. 10 Flow chart of the computation method in time/frequency domain

vector changes direction as well as the contact point. Figure 11b shows the thermal power introduced in the rotor due to friction as function of time steps.

Figure 12a then shows the equivalent bending moments at a certain time instant as calculated from thermal strains, which simulate the bow. Finally Fig. 12b shows the polar plot of the spiral vibration (amplitude and phase) that could be measured in correspondence of the bearings of the shaft, at the rotational speed of 3,300 rpm. In this case it is an unstable spiral, because vibrations are increasing. But also stable spiral vibrations can occur, when amplitude of vibration decreases and finally the contact is lost. After that when heat is again uniformly distributed the unbalance will force the shaft to rub again and the cycle repeats. This occurs at a rotational speed of 3,500 rpm, as it is shown in Fig. 13a.

Finally also an experimental result compared to the simulated one is represented in Fig. 13b. The rubbing occurred close to a bearing in correspondence of a oil seal ring in a 50 MW turbo-group.

7 Experimental Evidence of Full Annular Rubs During a Speed Transient, and Model Based Identification of the Generated Changing Bow

This is an example of a turbo-group (schematically represented in Fig. 14) in which the HP-IP rotor experienced a full annular rub during a run-down transient approaching its 1st natural lateral frequency: therefore we have a non-stationary speed and a non-stationary thermal state. The complete case history is described in

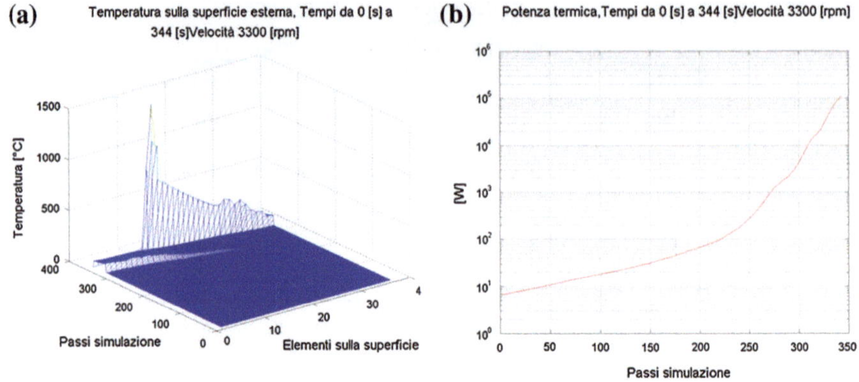

Fig. 11 **a** Temperatures on rotor surface; **b** Thermal power evolution in time

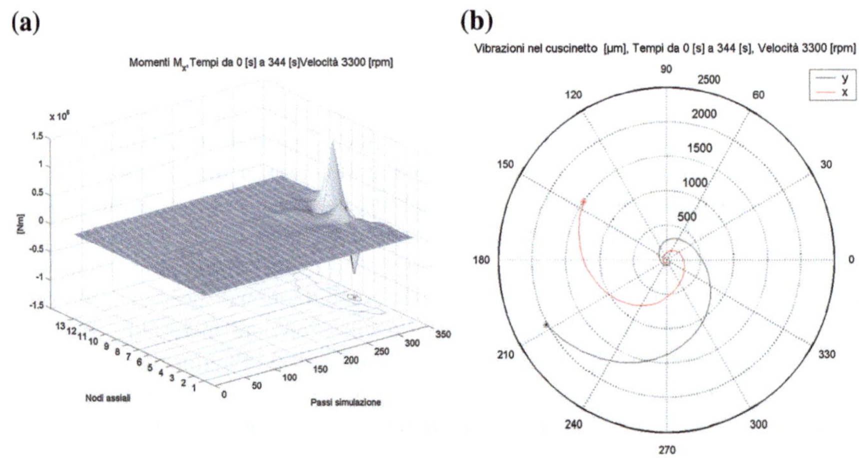

Fig. 12 **a** Equivalent bending moments simulating bow; **b** Unstable spiral vibration (3,300 rpm)

detail in [5]. The run-down transient occurs, as usual for big turbo-groups, with a sufficient low deceleration so that the transient can be considered as a sequence of steady state conditions. As in previous example also here the thermal transient is rather slow: the generated bow is slowly changing, resulting vibrations are changing slowly and this behavior can be analyzed by a sequence of steps with different speed and heat conditions, in which vibrations can be simulated in the frequency domain.

Measured vibrations in the two bearings of the rubbing turbine are reported in the two Bode diagrams of Fig. 15. The run-down had been interrupted as vibration amplitude reached alarm levels close to the shaft first critical speed, rotational speed was increased to get out of the critical speed and trying to relax the thermal stresses and straighten the shaft. Since this attempt was unsuccessful, the run-down

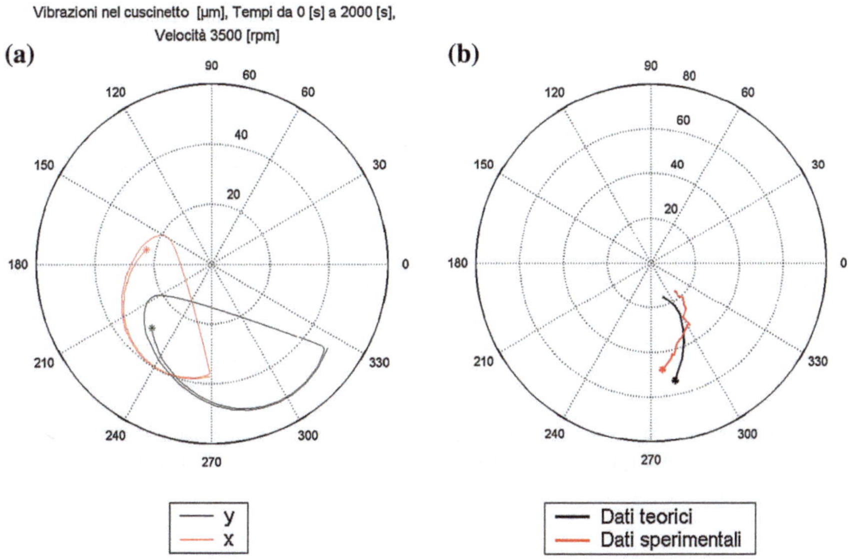

Fig. 13 **a** Stable cyclic spiral vibrations; **b** Measured and simulated spiral vibrations

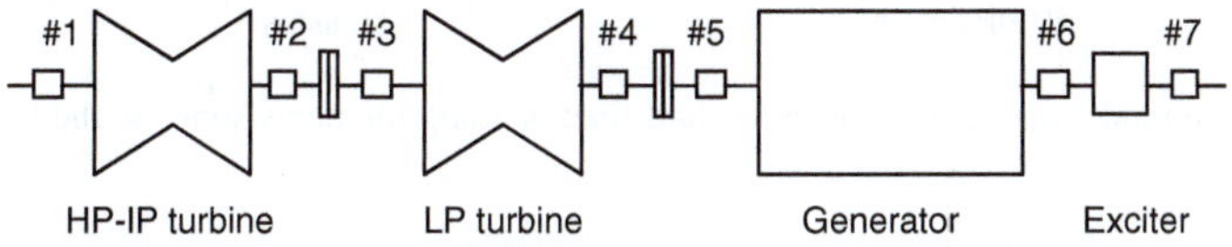

Fig. 14 Turbo-group with a rub in HP-IP turbine rotor

was completed in rubbing conditions for a consistent part of the run-down. At minimum speed a consistent thermal bow remained as can be seen from residual amplitude at 500 rpm in bearing 2. That a severe full annular rub occurred can be deduced also from phase changes: analyzing amplitude and phase of the 1xrev vibration amplitudes measured in the two bearings in between 1500 and 1800 rpm, we have at same rotational speed completely different vibration vectors (with different amplitude and different phase). This is due to the changing thermal bow generated by the rubbing friction.

In order to verify where the rubbing occurred with the aid of the beam element model of the machine, from vibrations in the bearings during the run-down from 1,800 rpm the amount and the location of the rub and the developing bow were identified.

Figure 16 shows the finite element model of the shaft line (composed of the steam turbines only) and the model used to represent the thermal bow by equal and opposite bending moments.

Figure 17a shows how the most probable position of the rub was identified from best fitting of calculated vibrations to measured ones, by means of the evaluation

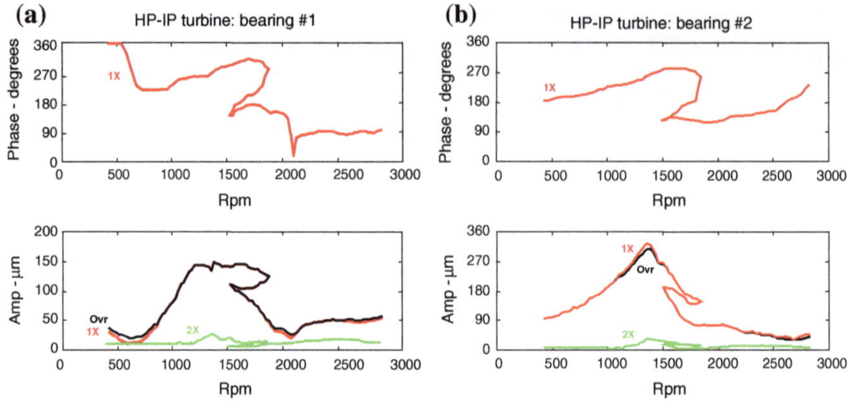

Fig. 15 Interrupted run-down transient of a rubbing steam turbine rotor: stopping at 1,500 rpm, reaccelerating till 1,800 rpm, and completing the run-down: **a** bearing 1; **b** bearing 2

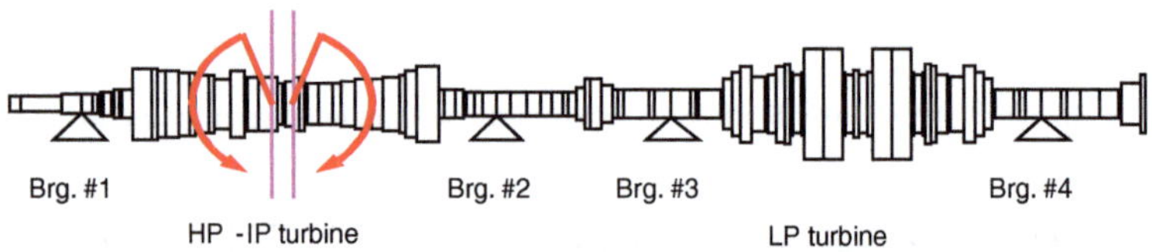

Fig. 16 Fem model of the steam turbine shaft line: bending moments simulate the local thermal bow

of the residues (kind of squared error as vector difference between measured and calculated vibrations) in the model based identification procedure. This has been done in the frequency domain considering stationary conditions at different rotational speed steps. Minimum value of residue identified the most probable position of the rubbing section.

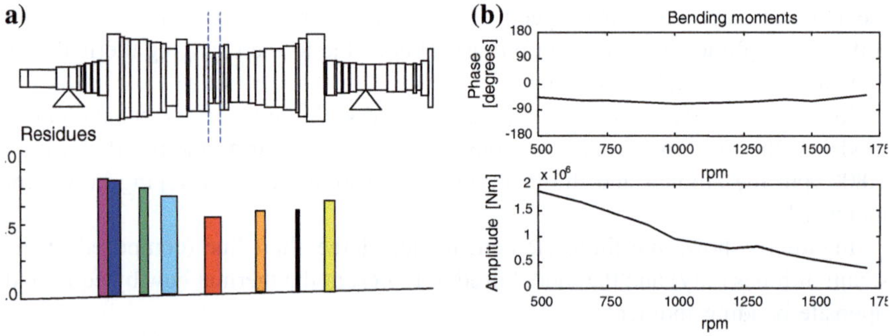

Fig. 17 **a** Most probable location of the rub; **b** Most probable bending moment versus speed

Figure 17b shows the value and phase of the identified bending moment during the run-down transient. Phase is almost constant showing that the contact was not migrating around the shaft circumference, amplitude is increasing showing that the shaft remained in contact with the stator during the whole run-down transient, increasing the developed heat in the same position.

This is again an example where a slowly changing non-stationary condition has been successfully modeled with different steps of stationary conditions, which allowed to remain in the frequency domain, and to use the model based identification procedure in the frequency domain.

8 Experimental Evidence of a Steam Whirl Instability

Instability is a typical strongly non-stationary condition of a rotating machine, which must be overcome rapidly for avoiding failures. Vibrations increase rather suddenly as will be shown in the following case history of a steam whirl instability that occurred in the HP rotor of a 460 MW steam turbine, as described in detail in [6].

These vibrations can be simulated in the frequency domain, taking account of negative damping in the eigen-value of the model of the machine, which defines the rate of the vibration amplitude increase.

Figure 18 represents the sketch of the turbo-group, and Fig. 19 shows a waterfall diagram of the vibration spectra measured in bearing 1 at different time instants before, during and after the event.

In the first part of the diagram the speed (3,000 rpm) is constant, the power (not shown) is increasing. 1xrev. vibration component at 50 Hz is rather low, but a small sub-harmonic component at 32 Hz (first critical speed of the shaft) is present which could indicate an arising instability. At hour 20.42 since power has further increased, suddenly the sub-harmonic component amplitude increases and reaches the shut down level in around 30 s. As the steam inlet valves have been closed, the sub-harmonic component disappears and the speed of the machine decreases. This is the typical symptom of a steam whirl instability, which occurs close to the maximum power (and steam flow) of a turbine. Regarding the modeling the excitation comes from cross coupled stiffness coefficients which develop in blade rows as function of power and clearances, and from steam bearing effect in labyrinth seals as function of the swirl motion. These effects must be taken into account as well as the damping effects of the oil film bearings, for analyzing stability of the machine. The evaluation of these effects are affected by some uncertainties, therefore the design of the machine resulted stable: that means that

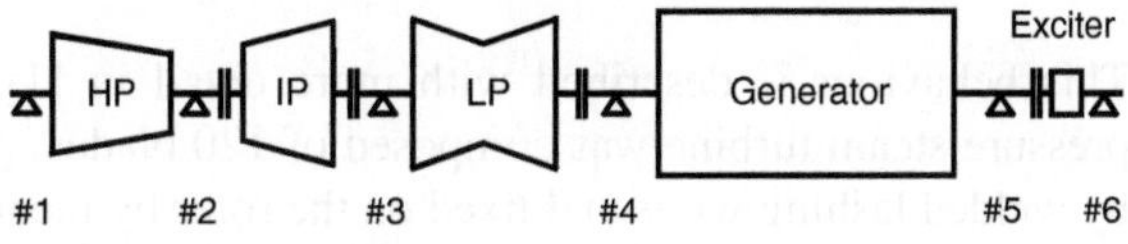

Fig. 18 Sketch of the 460 MW turbo-group where the HP rotor experienced a steam whirl instability

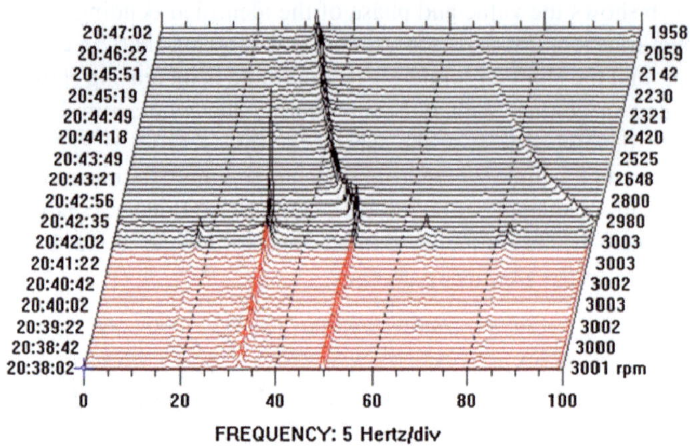

Fig. 19 Waterfall diagram of the vibration spectra measured in bearing 1 of HP rotor during load increase

Fig. 20 Vibration spectrum in bearing 1 just before shut-down

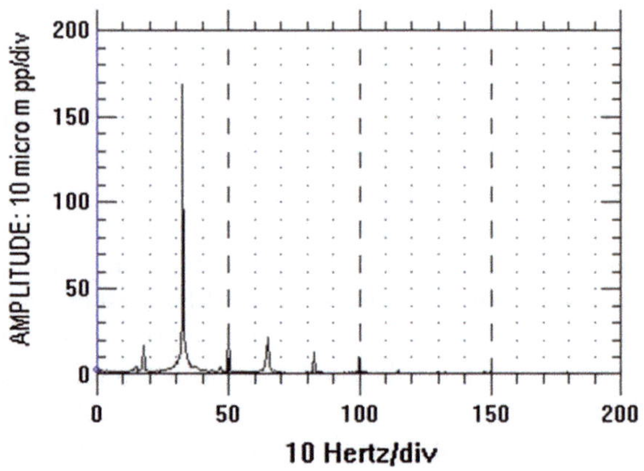

the real part of the first eigenvalue of the model of the machine was negative. The measured behavior showed instead a positive value of the real part of the eigenvalue, which was also rather high. Figure 20 shows with more detail one of the spectra recorded during the event.

9 Blade Flutter Instability

This behaviour is described with more detail in [1]. The last stage of the low pressure steam turbine was composed of 120 blades, grouped in packs of 8 blades by welded lashing wires and fixed on the rotor by means of fir tree side entry roots.

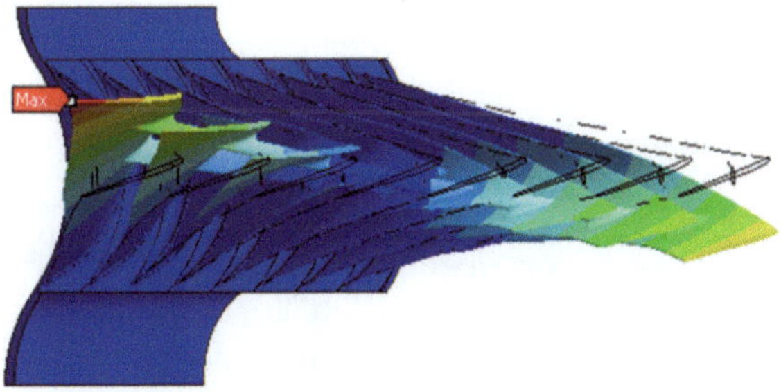

Fig. 21 3rd vibration mode (225 Hz) of the blade pack as calculated from f.e. model modal analysis

During commissioning of the turbine the machine was operated in off design conditions where the long last stage blades may experience high vibrations. An inspection after this period revealed cracks in blade roots and failures in lashing wires. Taking into account also the centrifugal field, the modal analysis allowed to identify the following natural frequencies and associated modes: the 1st (tangential) mode at 115 Hz, the 2nd (axial) mode at 185 Hz and the 3rd (so called X-mode) at 225 Hz. This last mode is shown in Fig. 21 and is the vibration mode that was strongly excited as the cracks and lashing wire failures mostly developed in the first and last blade of the packs. The machine was repaired and equipped with a tip timing blade vibration measuring system. Figure 22 shows the deflections of one blade (blade no. 53) during a speed transient from about 2,700 to 2,890 rpm.

The blade is deflected due to centrifugal force and is vibrating at a frequency of 217 Hz, which is close to one of the natural blade pack frequencies (which resulted to be at 225 Hz at rated speed from calculation). The corresponding pack mode is called X-mode. Resonant conditions could occur at 2,700 rpm with the 5th EO excitation but did not appear. High vibrations are measured instead at a rotational speed of 2,770 rpm, but are not related to the rotational speed (excitation is at a non-integer engine order of 4.7), frequency of vibration is constant and not changing with speed as occurs for the vibrations due to some engine order excitation (that are marked with blue lines in Fig. 22). Vibrations growth and decay depend on some other operating condition parameter, not on speed. Therefore these vibrations must be due to some fluid-dynamic instability. Maximum apparent vibration peak-to-peak amplitude in tangential direction was 14.1 mm, which corresponds to 4.7 mm in the X-mode vibration direction. The fatigue stresses corresponding to similar vibration amplitudes could have caused cracks and failures. Colours which indicate phase of vibration with respect to the 1x rev. reference are randomly distributed around maximum vibration peak, indicating a non-synchronous vibration, where the phase is random. Figure 23 shows the distribution of blade vibrationamplitudes over the blade row composed of 120 blades. The vibration amplitude distribution shows clearly that maximum amplitude exists each 8–9 blades: the first and the last blade of the group of 8 blades are vibrating with maximum amplitude (typical for the X-mode shown in Fig. 21). These blades were also those where cracks close to the roots and lashing wire failures had been

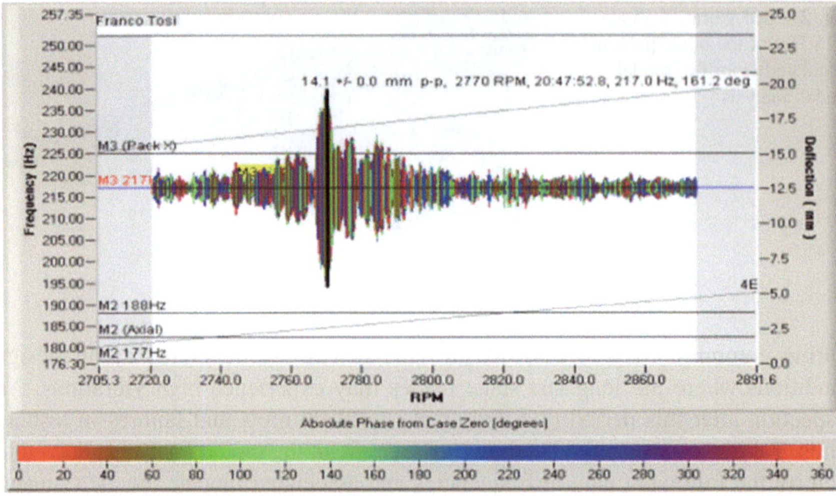

Fig. 22 Transient unstable vibrations of blades during a speed transient of the steam turbine

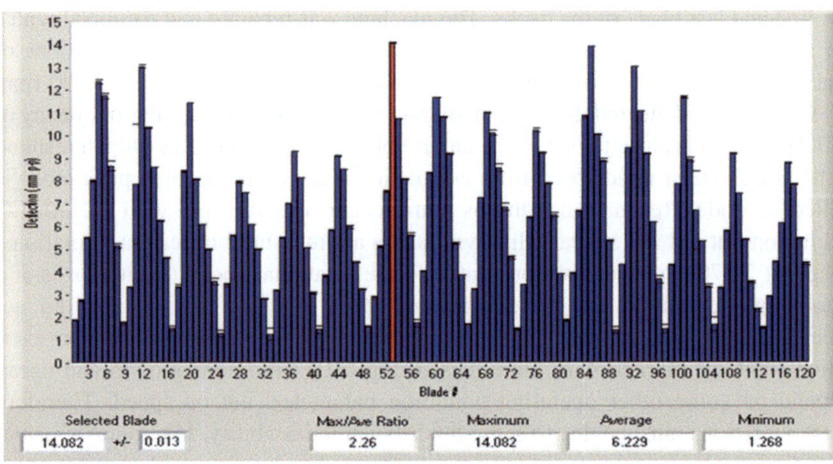

Fig. 23 Vibration amplitude distribution along the blade row during instability

found. In the row of 120 blades there are 15 packs of 8 blades each, therefore 15 peaks and 15 nodes are found in the vibration distribution around the row.

During this non-stationary condition of the machine, small variation in output power and condenser pressure cause apparently the onset of this instability, the evolution over time of the vibrations, with its growth and decays that are extremely difficult to be modeled and cannot be simulated or predicted with sufficient accuracy.

10 Conclusions

The simulation of typical vibration behavior of rotating machines in "weak" or "strong" non-stationary conditions is presented and discussed. The behavior of slowly changing systems (weak non-stationary condition) can be simulated in a mixed approach using both time domain and frequency domain approach. This is the case for instance when the vibration behavior of rotors during thermal transients are simulated. Some experimental results are also presented and compared to simulations. Further examples of systems with strong non-linearity, that in stationary conditions exhibit non-stationary vibrations, are also given. Finally two strong non-stationary instability conditions (cases 8 and 9) are presented of rotating shafts and blade rows: the first one can be simulated rather easily with a linear model, the latter is very difficult to be simulated.

References

1. SanvitoM, Pesatori E, Bachschmid N, Chatterton S (2012) Analysis of LP steam turbine blade vibrations: experimental results and numerical simulations, 10th International conference on vibrations in rotating machinery IMechE, London, pp 11–13
2. Bachschmid N, Pennacchi P, Pesatori E, Turozzi G (2007) On the "snubbing" mechanism for reducing blade vibration, proceedings of the ASME international design engineering technical conferences and computers and information in engineering conference, DETC2007, 1261–1270
3. Ferrante M, Pesatori E, Bachschmid N (2012) Simulation of the dynamic behavior of a group of blades with friction contacts, 10th international conference on vibrations in rotating machinery IMechE, London, 11–13
4. Bachschmid N, Pennacchi P, Vania A (2000) Modeling of spiral vibrations due to rub in real rotors, 7th international conference on vibrations in rotating machinery IMechE, C576/082
5. Vania A, Bachschmid N, Pennacchi P (2001) Analysis of light rotor-to-stator contacts in large turbine-generator units, 4th International conference on acoustical and vibratory surveillance methods and diagnostic techniques, Compiègne, France
6. Bachschmid N, Pennacchi P, Vania A (2007) Some results in steam whirl analysis, 12th IFToMM world congress, Besançon, France, 18–21

10 Conclusions

The simulation of typical vibration behavior of rotating machinery ("weak" and "strong" non-stationary [illegible]

[illegible]

Speed Transform, a New Time-Varying Frequency Analysis Technique

Cécile Capdessus, Edgard Sekko and Jérôme Antoni

Abstract Due to the periodical motions of most machinery in steady state operation, many diagnosis techniques are based on frequency analysis. This is often performed through Fourier transform. Some extensions of these techniques to the more general case of non stationary operation have been proposed. They are based on signal processing advances such as time–frequency representations and adaptive filtering. The technique proposed in this paper is based on the observation that, when under non stationary operation, the vibrations of a machine are still tightly related to the speed variations. It is thus suggested to decompose the vibration signal over a set of time-varying frequency sine waves synchronized with the speed variations, instead of fixed frequency sine waves. This set of time-varying frequency sine waves is shown to be an orthonormal basis of the subspace it spans in the case of linear frequency variations. An insight to the improvement such decomposition can provide for spectral analysis, cyclostationary analysis and time–frequency representation is given. Some application examples are presented over both simulated signals and real-life signals.

Keywords Vibration analysis · Non stationary operation · Time-varying frequency sine-waves · Decomposition over an orthonormal basis

C. Capdessus (✉) · E. Sekko
Laboratoire PRISME, 21 Rue Loigny-la-Bataille 28000 Chartres, France
e-mail: Cecile.capdessus@univ-orleans.fr

E. Sekko
e-mail: Edgard.Sekko@univ-orleans.fr

J. Antoni
Laboratory of Vibrations and Acoustics, University of Lyon (INSA),
69621 Villeurbanne Cedex, France
e-mail: jerome.antoni@insa-lyon.fr

G. Dalpiaz et al. (eds.), *Advances in Condition Monitoring of Machinery in Non-Stationary Operations*, Lecture Notes in Mechanical Engineering,
DOI: 10.1007/978-3-642-39348-8_2, © Springer-Verlag Berlin Heidelberg 2014

1 Introduction

In rotating machinery, under varying rotation speed, the vibration signal is non-stationary and its statistic characteristics vary with time. For such a signal, conventional general harmonic analysis, performed through Fast Fourier Transform (FFT), doesn't provide accurate information for the spectrum of the signal, due to the non-constant speed. In order to overcome the limitations of this technique, appropriate methods have been dedicated to varying speed cases. The best known methods are Order Tracking analysis by short-time Fourier transform (STFT) [1], windowed Fourier transform (WFT) or angular resampling [2]. These last years, new methods such as Vold Kalman filter [3] or Gabor Order Tracking approach [4] have been introduced. The main characteristic of these methods is that they are based on resampling scheme or sampled STFT. However in many cases, all previous techniques have limited resolution or show a number of gaps [5]. For the STFT, depending on the window used, analysis quality may be affected. Regarding the angular resampling of the vibration signal, its drawback is that the resonance frequencies of the rotating machine are disturbed by the process.

In order to avoid the previous shortcomings, we propose a new technique called Speed Transform (ST) when the speed varies linearly. This new approach consists in expanding the vibration signal into a series of elementary oscillatory functions, whose frequencies depend on the variation of the speed. The main advantage of the new approach is that it adapts to the vibration signal and components so that the resonance frequencies are preserved.

The paper is organized as follows. In Sect. 2, ST is presented in both theoretical and practical implementation aspects. In Sect. 3, ST is applied to simulated data and in Sect. 4 it is applied to real-life data. Our conclusion is presented in Sect. 5.

2 Speed Transform, Theory and Practical Implementation

Applying Fourier analysis to a signal consists in decomposing that signal over a basis of elementary oscillatory functions. The definition of the Fourier transform of a signal $s(t)$ is well known and is given by:

$$S(f) = \int_{-\infty}^{+\infty} s(t)\, e^{-2\pi jft} dt \tag{1}$$

where the Fourier transform of the signal $s(t)$ is denoted by $S(f)$, t stands for time and f for frequency. When actually applied to vibration signals, it usually comes to a slightly different tool that can be described by:

$$S(f) = \frac{1}{2T} \int_{-T}^{+T} s(t)\, e^{-2\pi jft} dt \tag{2}$$

It thus consists in evaluating the closeness of the signal $s(t)$ to a set of elementary oscillatory functions $e^{2\pi jft}$ over a finite time interval T, or in other terms, evaluating a mean contribution of each of these $e^{2\pi jft}$ functions to the signal over that interval. In what follows, this practical tool will be considered as Fourier transform.

The tool described by Eq. (2) exhibits a nice asymptotic property. Let us evaluate the closeness of two functions of the basis $e^{2\pi jf_1 t}$ and $e^{2\pi jf_2 t}$ when the time interval tends to infinity, i.e.:

$$\lim_{T\to+\infty} \frac{1}{2T} \int_{-T}^{+T} e^{-2\pi jf_1 t}\, e^{+2\pi jf_2 t}\, dt = \left[\begin{array}{c} 0\ for\ f_1 2 \\ 1\ for\ f_1 = f_2 \end{array} \right. \tag{3}$$

This is equal to zero for $f_1 \neq f_2$ (cross terms) and equal to one for $f_1 \neq f_2$ (auto terms). This property allows to evaluate the contribution of any $e^{2\pi jft}$ function to the signal by applying Eq. (2), provided that the time interval is long enough to ensure that the cross-terms vanish.

This tool is thus well fitted to the physical nature of vibration signals under stationary operation. Indeed, the periodical movements of machinery generate periodical components within the temporal moments of the signals. Due to the nice property described by Eq. (3) these periodical components produce spectral lines through Fourier analysis.

Under non stationary operation, Fourier analysis keeps its nice mathematical properties but does not fit any more to the physical model of the data. Indeed, many components of the vibration signal follow the speed variations, so that they are not any more periodic. All the components that are tied to the rotation frequency of the machine spread over a frequency band. Retrieving their amplitude or even detecting them through Fourier analysis thus becomes difficult.

This is why we propose to use a basis of elementary oscillatory functions whose frequencies follow the speed variations. This would allow preserving the main advantage of frequency analysis, which is to fit to the physical model of the data. But what about the mathematical properties of the decomposition over such a time-varying basis? More precisely, do a set of $e^{2\pi j \int_0^t \Delta f(u)\, du}$ functions, whose frequencies $n\Delta f(t)$ follow the speed variations, still exhibit the property described by Eq. (3)? We showed (see appendix) that provided the variations of $n\Delta f(t)$ are linear versus time, that mathematical property still holds.

Let us give some specifications for this new analysis tool that will be called speed transform:

- first, the frequencies of the basis functions should all be proportional to the speed variations,
- their frequencies must always be smaller than the Nyquist frequency,
- they should be equi-spaced over the analysis band, in terms of proportion of the rotation speed, rather than in terms of frequency band,
- they should be equi-spaced with such step $\Delta f(t)$ that no component tied to the rotation speed could be missed in the analysis interval.

In order to satisfy these specifications, let us first define some expressions: $r(t)$ will stand for the rotation speed in Hertz and $T = NT_e$ for the duration of the vibration signal, with N the total number of samples and f_e the sampling period. The different parameters used to generate the basis are shown on Fig. 1. The rotation frequency (in green) and the basis functions frequencies (blue) are represented versus the number of samples in reduced frequency.

The example is given for a $N = 64$ samples base, with speed variations in reduced frequency given by $r(t) = 0.1 + 0.001 * t$. The green plot represents the rotation frequency variation $r(t)$ and the red one represents the time-varying frequency resolution $\Delta f(t)$ of the speed transform. It is proportional to $r(t)$ and the proportionality coefficient is an integer value k such that $r(t) = k\Delta f(t)$ and $max\{\Delta f(t)\} \leq \frac{1}{T}$ in order to have a sufficient resolution not to miss any component. The basis B is composed of $\frac{N}{2} + 1 = 33$ oscillatory functions $b_n(t)$ calculated as follows:

$$B = \left[b_n(t) = e^{2\pi jn\int_0^t \Delta f(u)\,du}, 0 \leq n \leq N/2 \right] \tag{4}$$

On Fig. 2 is displayed the modulus of the basis correlation matrix. As expected from theory, it is a diagonal matrix with unitary diagonal terms. The non diagonal terms should be zero but some side effects appear due to the fact that the basis functions are finite length. These side effects can be minimized by applying an apodisation window, such as Hamming window, as shown on Fig. 3.

As usual, the shape of the main lobe and side-lobes depends on the apodisation window. The width of the main lobe and the maximum speed resolution $max\{\Delta f(t)\}$ are both related to the inverse of the length of the signal. This ensures that one of the calculated samples at least will be located on the main lobe. This prevents any component whose variations are not a integer multiples of $\Delta f(t)$ from not being detected. In order to improve the amplitude estimation of these components some interpolation can be applied, as in classical spectral analysis.

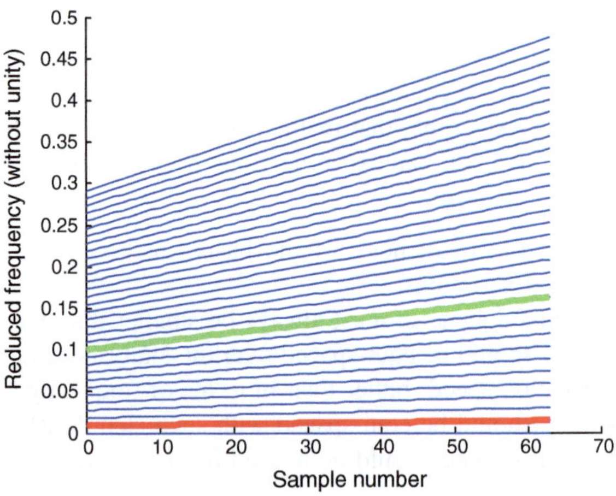

Fig. 1 Frequencies of the basis functions

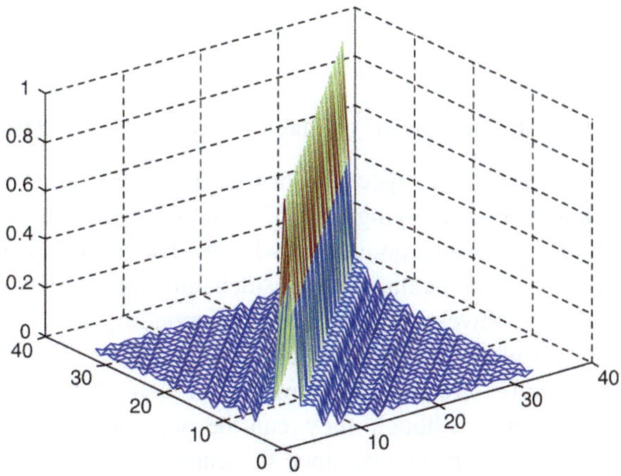

Fig. 2 Modulus of the correlation matrix of the basis

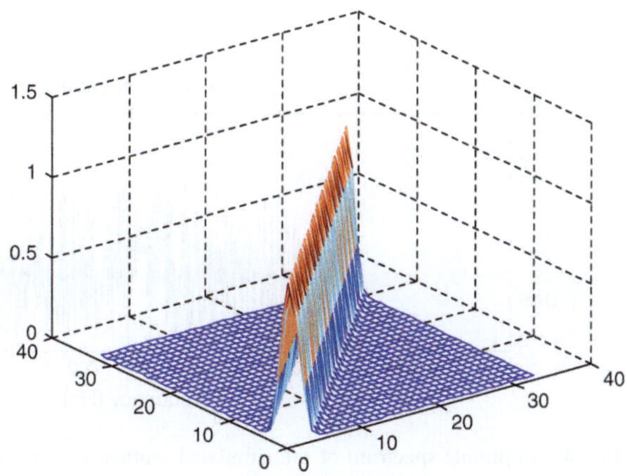

Fig. 3 Modulus of the correlation matrix of the basis computed with apodisation

3 Application to Simulated Data

We first applied this technique to the analysis of simulated toothed gearing vibrations. The features of the simulated experiment are the following ones:

- the two wheels are respectively 20 and 22 toothed ones,
- the rotation frequency of the 20 toothed wheel is variable and given by $f_{20}(t) = 16 + 3t$, the gearing frequency by $f_{eng}(t) = 20f_{20}(t)$ and the 22 toothed wheel frequency by $f_{22}(t) = f_{eng}(t)/22$.
- the vibration signal is given by $s(t) = s_{eng}(t)(s_{20}(t) + s_{22}(t))$ with:
- $s_{eng}(t) = \cos\left(2\pi\int_0^t f_{eng}(u)du(t)t + \varphi_{eng}\right)$ where φ_{eng} is a random phase

- $s_{20}(t) = \sum_{m=1}^{8} \cos\left(2\pi m \int_0^t f_{20}(u)du + \varphi_{20,m}\right)$ with $\varphi_{20,m}$ random (resp. $s_{22}(t)$, $f_{22}(t)$ and $\varphi_{22,m}$).
- the total number of available samples is $N = 20000$, i. e. a 1s time interval.

The amplitude spectrum of this simulated signal is displayed on Fig. 4. It was calculated by applying a 65,536 sample Fourier transform over the whole signal, with a Hamming apodisation window. Due to the speed variation, the energy of the gearing fundamental and its sidebands spread over several frequency channels. Only the lowest frequencies, corresponding to the inferior sidebands of the 36 toothed wheel modulation, can be separated. This comes from the fact that their frequencies are varying slower than that of the higher frequency components. Nevertheless, though they can be separated, the amplitude displayed on the spectrum is erroneous, since the energy of each sideband is spread over several neighboring frequency channels.

This problem cannot be totally overcome by the use of time–frequency distributions. Two spectrograms were calculated with different frequency resolutions. The spectrogram displayed on Fig. 5 was computed over slices of 1,024 samples,

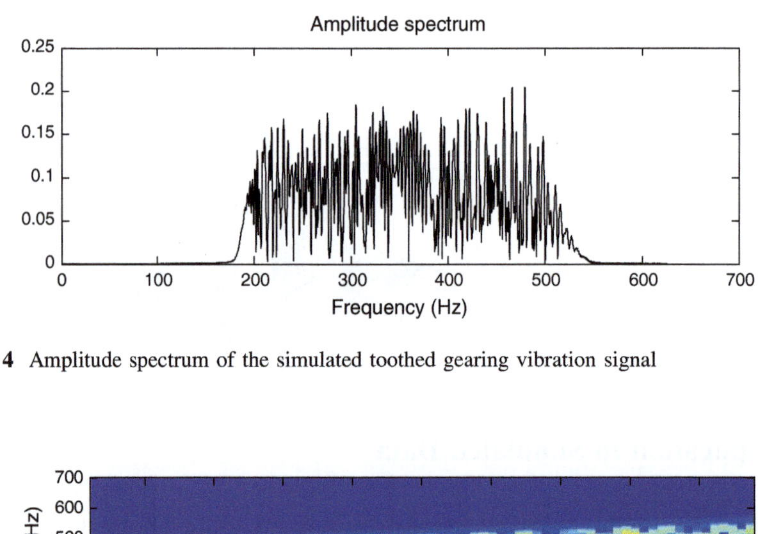

Fig. 4 Amplitude spectrum of the simulated toothed gearing vibration signal

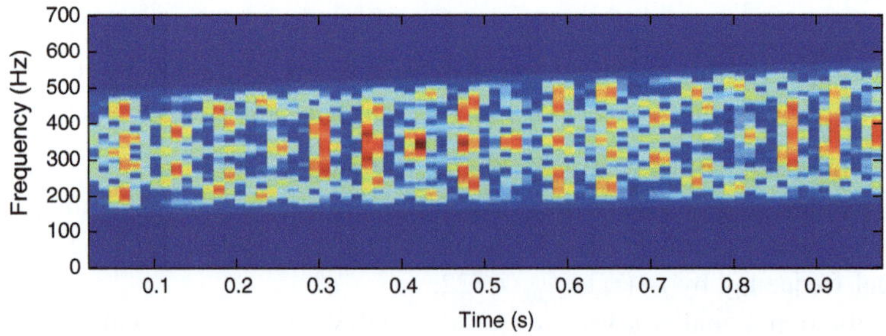

Fig. 5 Spectrogram of the vibration signal computed over 1024 sample slices

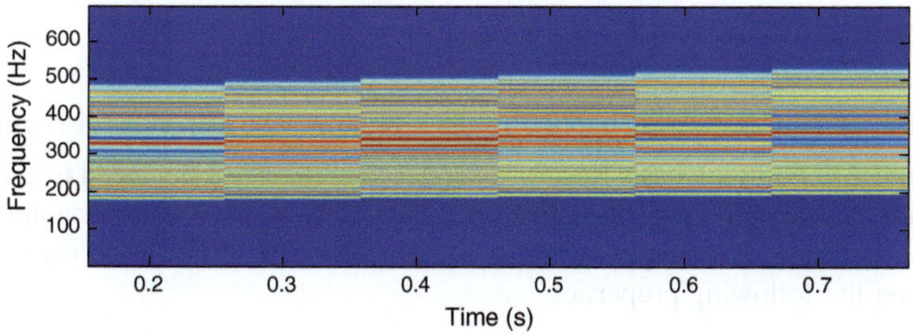

Fig. 6 Spectrogram of the vibration signal computed over 8192 sample slices

Hamming windowing and a¾ of a slice overlap. The frequency resolution is thus $\Delta f = 19.53\,Hz$ and the temporal resolution $\Delta t = 0.05\,s$, which would be enough to follow the temporal variations of the speed but does not allow separating the sidebands. Whereas the second spectrogram (Fig. 6), calculated the same way but over 8,192 sample slices, not only fails to follow the temporal variations ($\Delta t = 0.4\,s$) but cannot either separate the sidebands, despite a better frequency resolution ($\Delta f = 2.44\,Hz$), because the sidebands spread over neighboring frequency channels.

Whereas the speed transform of the signal, displayed on Fig. 7, succeeds in separating all the sidebands and estimating their amplitude. Instead of being represented versus frequency, it is represented versus the proportion of the speed signal that was used to build the speed transform basis. Here the speed signal was that of the 20 toothed wheel, so that all the sidebands corresponding to that wheel appear at harmonic positions of the speed signal. The transform was computed over the whole signal with Hamming windowing and interpolated by 8.

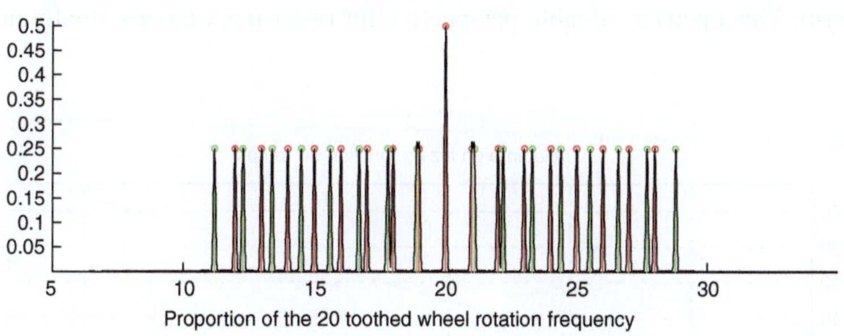

Fig. 7 Speed transform (in *black*) displayed versus the proportion of the 20 toothed wheel rotation frequency. *Green lines* are the true 22 toothed sidebands and *red* ones the true gearing fundamental and true 20 toothed sidebands

4 Application to Real-Life Data

Here the speed transform is applied to the vibration signal of a diesel engine.

In order to build the basis functions $b_n(t) = e^{2\pi jn\int_0^t \frac{r(u)}{K}du}$ we first need to estimate from the tachometer signal the function $g(t) = 2\pi\int_0^t r(u)\,du$, i.e. the angular position of the shaft, if we suppose that it is zero for $t = 0$. This can be easily done based on the following properties:

- $g(t)$ must be an order 2 polynomial
- It must be such that $sin(g(t)) = 0$ at each tick of the tachometer signal

A curve is built by associating an integer multiple of 2π to each tick time and a curve fitting procedure allows finding the order 2 polynomial fitted to that curve. The coefficients of this polynomial are then used to compute the value of $g(t)$ at any time t.

The speed transform has first been computed on the time interval between 30.6 and 32 s, where the rotation frequency is stationary and equal to 75 Hz (see Fig. 8), in order to compare the result to Fourier transform.

The amplitude spectrum and the speed transform calculated on that interval are displayed on Figs. 9 and 10. This confirms that the usual Fourier transform is actually a particular case of the proposed speed transform. Note however a slight improvement of the latter as compared to the former due to the compensation of small speed fluctuations.

The technique has then been applied between 25 and 30 s, i.e., on a time interval where the speed is linearly increasing. The amplitude spectrum and speed transform are displayed on Figs. 11, 12 and 13. The capability of the speed transform to adjust to linear speed variations is clearly demonstrated, even in the high frequency range. Whereas one can only distinguish the H2 harmonic (firing frequency of the engine) in the Fourier transform, all multiples of the H1 (crankshaft rotation) and H1/2 (thermodynamic cycle) are visible in the speed transform. This opens a valuable perspective for order tracking at virtually no cost

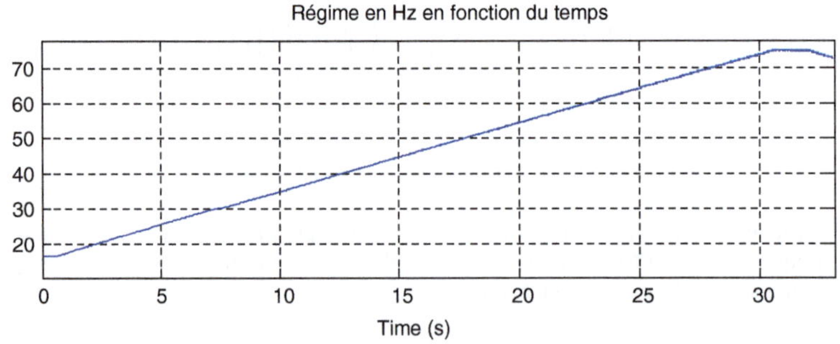

Fig. 8 The rotation speed of the diesel engine estimated from the tachometer signal

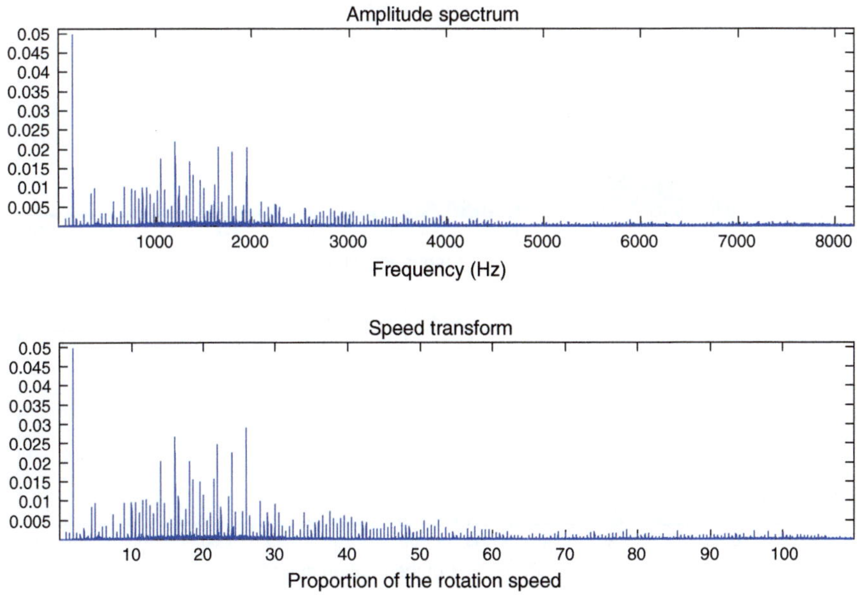

Fig. 9 The Amplitude spectrum and speed transform calculated between 30.6 and 32 s

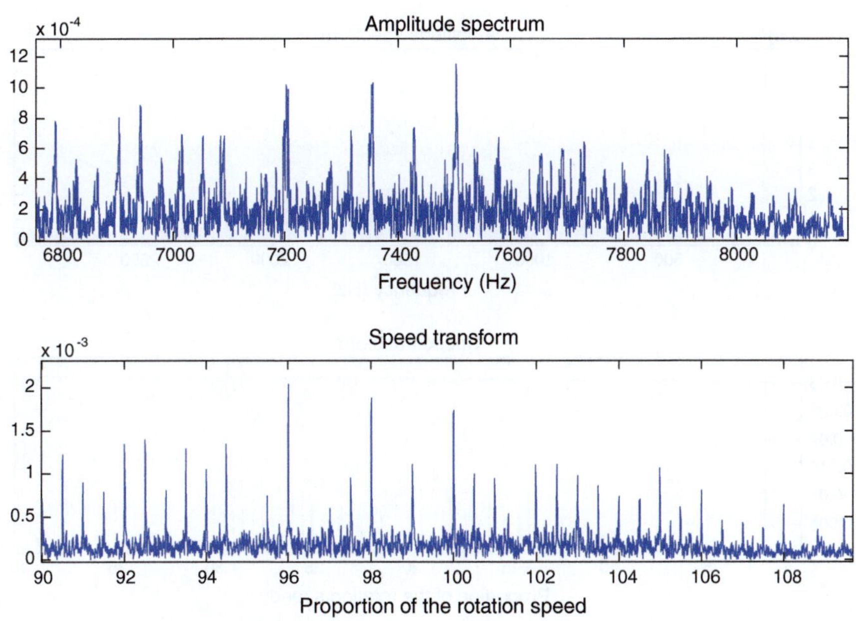

Fig. 10 Zoom on the high frequencies

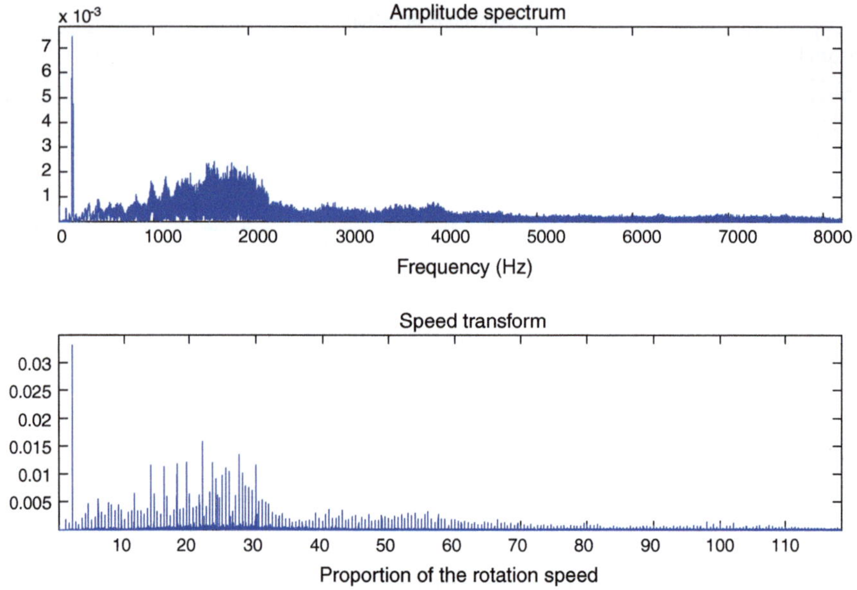

Fig. 11 The Amplitude spectrum and speed transform calculated between 25 s and 30 s

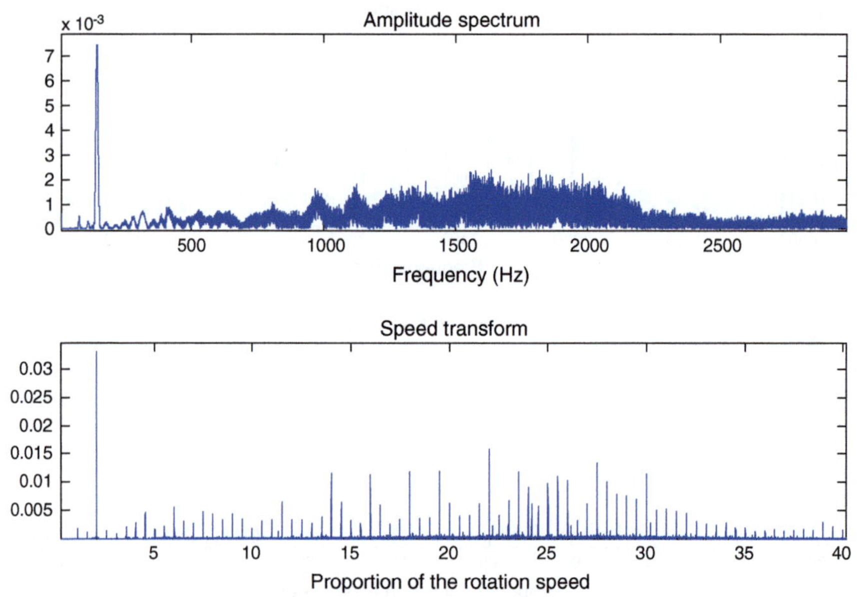

Fig. 12 Zoom on the low frequencies

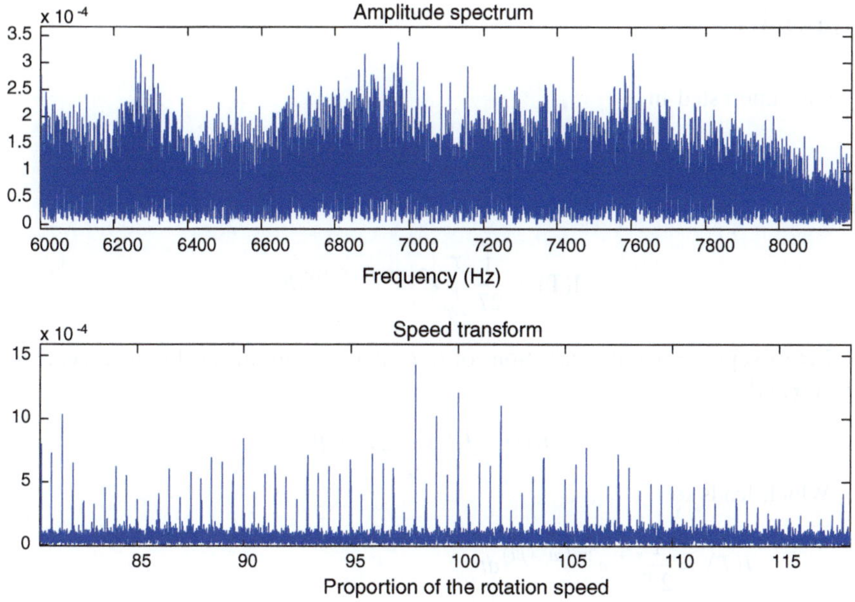

Fig. 13 Zoom on the high frequencies

as compared to other techniques, such as those based on time-varying filters or angular resampling preconditioning.

5 Conclusion

Fourier analysis plays a prominent role in the vibration analysis of rotating machines. Strictly speaking, it applies only to the situation where the machine is rotating at exactly constant speed. The extension of Fourier analysis to non-stationary operation is a current and active field of research. Whereas valuable solutions exist, for instance based on angular resampling, this paper proposed a new speed transform which inherits all the properties of the Fourier transform—in particular the orthonormality of its functional basis—when applied to linear speed variations. This not only has the advantage of simplicity, but it also returns properly scaled results. The speed transform has potential in several applications, and in particular in order tracking. Due to its resemblance to Fourier analysis, it opens a bunch of signal processing possibilities dedicated to non-stationary signals, such as non-stationary demodulation, non-stationary envelope analysis, cyclo-non-stationary analysis, etc. A short-time version of the speed transform is also conceivable to track arbitrary speed variations that can be approximated as piece-wise linear.

Appendix

We are interested in:

$$\lim_{T\to\infty} I(T)$$

where:

$$\mathrm{I(T)} = \frac{1}{2T}\int_{-T}^{T} e^{2\pi j(f_1(t)-f_2(t))t}dt$$

Let us suppose that the variations of $f_1(t)$ and $f_2(t)$ are linear. In this case, there exist $\alpha\, et\, \beta$ such that:

$$f_1(t) - f_2(t) = 2\alpha t + \beta$$

Which leads to:

$$I(T) = \frac{1}{2T}\int_{-T}^{T} e^{2\pi j\left(\alpha t^2+\beta t\right)}dt$$

$$I(T) = \frac{1}{2T}\int_{-T}^{T} e^{2\pi j\left(\alpha\left(t+\frac{\beta}{2\alpha}\right)^2-\frac{\beta^2}{4\alpha}\right)}dt = \frac{1}{2T}e^{-2\pi j\frac{\beta^2}{4\alpha}}\int_{-T}^{T} e^{2\pi j\left(\alpha\left(t+\frac{\beta}{2\alpha}\right)^2\right)}dt$$

The variable is changed from t to x:

$$\left|\begin{array}{ll} x & = t+\frac{\beta}{2\alpha} \\ dt & = dx \\ -T & +\frac{\beta}{2\alpha}\leq x\leq T+\frac{\beta}{2\alpha} \end{array}\right.$$

Which leads to:

$$I(T) = \frac{1}{2T}e^{-2\pi j\frac{\beta^2}{4\alpha}}\int_{-T+\frac{\beta}{2\alpha}}^{T+\frac{\beta}{2\alpha}} e^{2\pi j\alpha x^2}dx$$

The integral can be decomposed into three parts:

$$\int_{-T+\frac{\beta}{2\alpha}}^{T+\frac{\beta}{2\alpha}} e^{2\pi j\alpha x^2}dx = I_1 + I_2 + I_3$$

with:

$$\left| \begin{array}{l} I_1 = \int_{-T+\frac{\beta}{2\alpha}}^{-T} e^{2\pi j\alpha x^2} dx \\ I_2 = \int_{-T}^{-T} e^{2\pi j\alpha x^2} dx \\ I_3 = \int_{T}^{T+\frac{\beta}{2\alpha}} e^{2\pi j\alpha x^2} dx \end{array} \right.$$

The integrals I_1 et I_3 are finite so that the only problem is to calculate:

$$\lim_{T\to\infty} \frac{1}{2T} e^{-2\pi j\frac{\beta^2}{4\alpha}} I_2(T)$$

Let $\mathrm{J(T)} = \int_{-T}^{T} e^{2\pi j\alpha t^2} dt$

$$\lim_{\mathrm{T}\to+\infty} \mathrm{J(T)} = J_\infty = \int_{-\infty}^{+\infty} e^{2\pi j\alpha t^2} dt$$

Let us change variable t to: $u = \sqrt{2\pi\alpha}\, t$
The expression becomes:

$$J_\infty = \frac{1}{\sqrt{2\pi\alpha}} \int_{-\infty}^{+\infty} e^{ju^2} du = \frac{1}{\sqrt{2\pi\alpha}} 2 \int_{0}^{+\infty} e^{ju^2} du$$

Which is proportional to the well known Fresnel integral: $\int_{0}^{+\infty} e^{ju^2} du = \frac{\sqrt{\pi}}{2} e^{j\frac{\pi}{4}}$

So that $J_\infty = \frac{1}{\sqrt{2\alpha}} e^{j\frac{\pi}{4}}$
This proves that $\lim_{T\to\infty} I(T) = 0$ whenever $f_1(t) - f_2(t) \neq 0$
If $f_1(t) - f_2(t) = 0$, it is easy to show that $I(T) = 1$ for any value of T.

References

1. Meltzer G, Ivanov YY (2003) Fault detection in gear drives with non-stationary rotational speed—part II: the time frequency approach, mechanical systems and signal processing, 17, pp 1033–1047
2. André H, Daher Z, Antoni J (2010) Comparison between angular sampling and angular resampling methods applied on the vibration monitoring of a gear meshing in non stationary conditions, ISMA
3. Pan MC, Lin YF (2006) Further exploration of Vold Kalman filtering order tracking with shaft speed information-(I) theoretical part, numerical implementation and parameter investigations. Mech Syst Sign Process 20(5):1134–1154
4. Hui S, Wei J, Qian S (2003) Discrete gabor expansion for order tracking, will appear in IEEE trans of instrumentation and measurements
5. Bandhopadhyay DK, Griffiths D (1995) Methods for analyzing order spectra, SAE paper 951273

SMART: Integrating Human Safety Risk Assessment with Asset Integrity

Diego Galar, Peter Sandborn, Uday Kumar and Carl-Anders Johansson

Abstract Maintenance activities are commonly organized into scheduled and unscheduled actions. Scheduled maintenance is undertaken during pre-programmed inspections. Maintenance operations try to minimize the risk of deterioration based on a priori knowledge of failure mechanisms and their timing. However, in complex systems it is not always possible to schedule maintenance actions to mitigate all undesired effects, and SMART systems, which monitor selected parameters, propose actions to correct any deviation in normal behavior. Maintenance decisions must be made on the basis of accepted risk. Performed or not performed scheduled tasks as well as deferred corrective actions can have positive or negative consequences for the company, technicians and machines. These three risks should be properly assessed and prioritized as a function of the goals to be achieved. This paper focuses on how best practices in risk assessment for human safety can be successfully transferred to risk assessment for asset integrity.

Keywords SMART · Risk · Diagnosis · Prognosis · Maintenance

1 Introduction

Terms such as 'hazard', 'integrity', 'reliability', 'risk', 'robustness' and 'survivability' have become part of our daily vocabulary. For instance, the derogatory term unreliable is used to describe the undependable behavior of an individual or

D. Galar (✉) · U. Kumar · C.-A. Johansson
Lulea University of Technology, Luleå, Sweden
e-mail: diego.galar@ltu.se

P. Sandborn
CALCE, Center for Advanced Life Cycle Engineering, University of Maryland, Maryland, USA
e-mail: sandborn@umd.edu

G. Dalpiaz et al. (eds.), *Advances in Condition Monitoring of Machinery in Non-Stationary Operations*, Lecture Notes in Mechanical Engineering,
DOI: 10.1007/978-3-642-39348-8_3,

an item, whereas the cautionary term risk is used to warn of possible exposure to an adverse consequence. There is an increasing effort to sharpen the notions of risk and reliability and to quantify them, as quantification is required for normative decision making, especially for decisions pertaining to human safety and well-being. The coupling of quantified measures of risk with normative decision making is called risk management.

Risk management is the underlying concept of the currently popular SMART devices and SMART approaches. Manufacturing environments typically include large machines with many moving parts, such as factory robots, welding machines and the like, which can pose a safety risk to factory workers who work in close proximity to them. An assessment of the level of risk of harm to operators using a machine or process entails the collection of qualitative and quantitative data with regard to potentially hazardous situations. Added to this, failed machines result in work slow downs and stoppages that cost companies money due to lost production and maintenance costs. This paper focuses on how best practices in risk assessment for human safety can be successfully transferred to risk assessment for asset integrity.

International standard ISO 14121 [25]—'Safety of machinery—Principles for risk assessment'—defines 'risk' as comprising two factors: consequences (or severity of injury) of a hazardous situation; and the probability of occurrence of the hazardous situation. The probability of occurrence is further divided into two constituent parts: frequency and duration of exposure to a hazard. Such factors can be generically applied to all types of machinery to estimate the level of risk.

A risk assessment is routinely performed to determine various risk factors that might be encountered in the machines. Such assessment includes determining the types of risk and the degrees of risk associated with each type. To maximize the effectiveness of the risk assessment and risk estimation, it is essential that accurate and detailed data are available upon which judgments concerning the basic factors above can be made. Risk assessments are used to implement risk reduction measures and to reduce risk to acceptable levels.

Safety related control systems are commonly integrated into machinery to reduce risk. In general, the standard approach to risk assessment is to mitigate all the factors that can be anticipated. Often the factors are embedded in the usage and training documentation supplied with the machine. But at all these stages, some aspects are typically unknown and some decisions have to be based on assumptions about future machine usage characteristics. If the actual machine usage characteristics differ from those assumed, the risk assessment becomes invalid. Such unanticipated variations in the risk factors can arise due to both internal and external conditions. Despite all the aforementioned variations, it is not uncommon for a manufacturing operation to be working with an original risk assessment that is no longer valid. This can result in unanticipated hazards to machine operators, leading to worker injuries and unexpected machine servicing, accompanied by production line delays. Such failures defeat the purpose of the initial risk assessment.

Traditionally, machines are equipped with safety controls that mitigate potential risks associated with the operation of that machine.

Figure 1 shows several devices commonly used to monitor critical parameters to assure a safe environment. These parameters include: vibration, temperature, load and other parameters that are used as condition monitoring indicators for the health assessment of the machine. This is so important that most electronics manufacturers have separate catalogues of products for condition monitoring or safety assurance. Safety issues were strongly regulated by international organizations, like ISO, IEC etc., during the 1980s and 1990s. Acceptable vibration, noise and/or temperature values for the human body are now commonly accepted by manufacturers, end users, etc. In other words, risk assessment and technology have been moving forward for years.

The type of safety related control system depends on the risk assessment. If the machine usage characteristics are such that the risk assessment becomes invalid, it could mean that the safety-related control system does not provide the required type of functionality or level of integrity. Protective decisions are based on a priori risk assessment and when certain limits are reached, alarms will be activated and safety actions will be performed. Paradoxically, there is a difference between the risk assessment for the protection of humans (operators and maintainers), which is extremely regulated, and the 'protection' of a productive asset. The production required to meet customer expectations, is an additional aspect that requires risk assessment as well.

Risk assessment is dynamic for all involved players. Variations in machine usage that can modify the original risk assessment, include increased frequency or duration of machine operation, higher power levels, improper maintenance, processing of unexpected materials, unexpected environmental conditions, use by persons with competency levels lower than expected, occurrence of mechanical and software-related system faults, and non-standard system components or operational modifications. These changing scenarios convert the risk assessment into an iterative process that should be conducted at all phases of a machine's design, manufacture, commissioning and operation, with the results handed from one stage to the next, so as to take into account variations in risk factors. These

Fig. 1 Safety/Condition Monitoring devices for vibration and temperature control

risks can potentially affect the three players: humans, assets and businesses. However, only human effects are regulated. Note that there have been recent attempts like PAS55 [1], or the upcoming ISO 55000 [22–24] (inspired by PAS55) where the asset will be recognized as a key stakeholder that needs to be taken care of in terms of risk (Fig. 2).

For this risk assessment, the overwhelming amount of data is an issue. In the past, risk assessments have typically been performed manually during installation or maintenance of the machine components to determine safety barriers for persons or recommended preventive programs to keep the asset in good condition. Today, it is a prohibitively labor-intensive process to conduct ongoing manual risk assessments of all machines in a factory setting. Therefore, there is a need to overcome the aforementioned deficiencies associated with conventional systems and devices by integrating the IT capabilities to perform these assessments with accuracy in dynamic and changing environments.

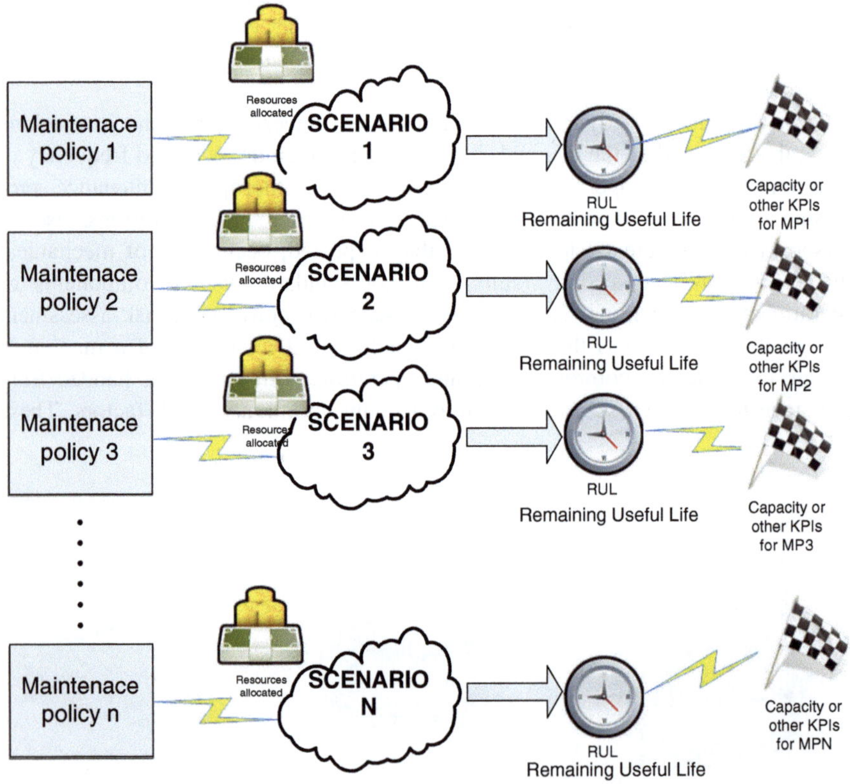

Fig. 2 Risk assessment as an iterative process for optimum maintenance decision making

2 Risk Assessment for Maintenance Actions

Risk assessments are the basis of a risk based maintenance approach. The ‘S’ for Safety in RAMS terminology has been a powerful driver in the transfer of risks from people to machinery.

Well-known science fiction author Isaac Asimov [2] created the following three laws of robotics:

1. A robot may not injure a human being or, through inaction, allow a human being to come to harm.
2. A robot must obey the orders given to it by human beings, except where such orders would conflict with the First Law.
3. A robot must protect its own existence as long as such protection does not conflict with the First or Second Laws.

Reading these statements carefully, one sees that Asimov has prioritized the behavior of machines and created the basis for risk assessment. In summary, risk assessment should be primarily performed to protect human integrity, since maintainers or operators work closely with productive assets. However, once the requirements to protect humans are fulfilled, the asset must protect itself. This pinpoints the concept of SMART devices as protection for people and self-protection of the machinery.

SMART devices must perform evaluations of operational risks to aid maintenance decisions concerning assets under certain conditions. These devices will always prioritize the safety of humans but will adapt when the operating condition modifies the scenario and the machine’s risk is changed.

The risk assessment will condition the real maintenance planning for the asset according to the risks the humans want to accept for themselves and for the health of the machinery. Risk tolerance is up to the individual to accept as a function of economic, social or environmental conditions. For example, in vehicle health monitoring, most direct interventions (not taking shop repairs into account) consist of corrective actions aimed at replacing or repairing failing components or components in the course of degradation, scheduled periodic inspections, and preventive actions such as replacement of equipment with a limited useful life.

These consequences will have risk implications for both human integrity and asset integrity. Figure 3 shows the dilemma of a SMART maintenance system in a railway context. On the one hand, the rolling stock must reach a certain destination. On the other hand, its condition may pose a threat to the integrity of the vehicle, to the people travelling on it and, finally, to the business. The maintainer must prioritize the performed action as a function of the acceptable risk, and sacrifice either the vehicle or the business goals, since human safety cannot be compromised.

The criteria used to evaluate risk-based decisions are moving from a purely economic perspective to the consideration of social and environmental factors to protect humans and assets [3]. This requires that asset maintenance practices must

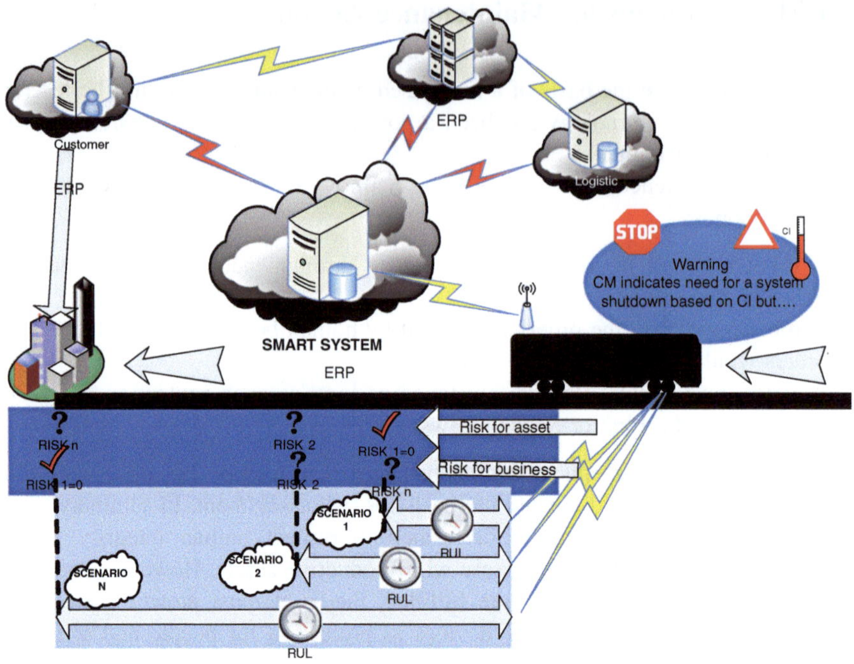

Fig. 3 SMART maintenance system for a railway

include financial performance as well as environmental protection and societal impact; [4].

The most important steps in the procedure of risk assessment are summarized in Fig. 4, and the role of models and methods for reliability evaluation and maintenance is emphasized. The concept of SMART maintenance proposes a plant maintenance management based on Risk Based Maintenance (RBM) as an extension of the well-known and establish Risk Based Inspection (RBI) for risky and hazardous industries like oil and gas (API 580 [15] and 581 [16]). This integrated approach to risk assessment and maintenance planning seeks to balance acceptable risks of failure with mission accomplishment and improved safety and health, reduce costs by repairing or replacing deteriorating equipment at the optimal time and eliminate ineffective inspections or interventions.

2.1 SMART Maintenance Assessment

SMART maintenance proposes models and methods for supporting managers and practitioners in planning and executing maintenance actions in the framework of risk-based maintenance, as all stakeholders accept that maintenance decisions involve the acceptance of certain intrinsic risks. SMART devices must have

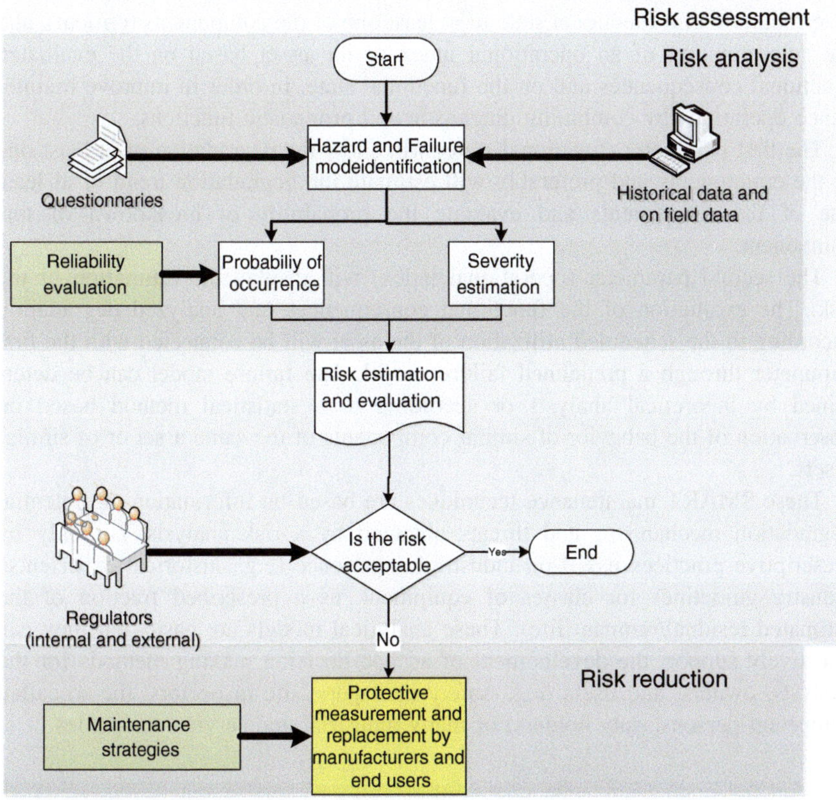

Fig. 4 Flowchart of risk assessment

embedded methods for aiding in the decision making for maintenance operations on assets and individual components. This assessment method is the core of SMART's decision support mechanism and must comprise the following steps:

- Analysis of the degradation of the components;
- Evaluation of the functional consequences, for the asset, of the analyzed degradation according to the scheduled utilization of the machine; and
- Determination of an operational plan for the asset according to the evaluated functional consequences.

The method to assess this risk should make it possible to optimize the maintenance operations on an asset according to its utilization by determining an operational risk, or in other words, a risk of disturbance of normal utilization. The operational risk is based on a prognosis of failure of at least one component, this prognosis being determined by analysis of degradation of the component based on measurement and historical data.

The assessment should comprise the calculation of two relevant parameters: the determination of a functional state of at least one of the components (critical), and the determination of an operational index of the asset based on the evaluated functional consequences and on the functional state, in order to improve maintenance operations by combining diagnostic and prognostic functions.

The first parameter (functional state) will show the degradation of at least one of the components and preferably will estimate the degradation trend of at least one of the components and evaluate the probability of breakdown of this component.

The second parameter (operational index) will display the estimation of the risk. The evaluation of the functional consequences and analyzed degradation according to the scheduled utilization of the asset will be connected with the first parameter through a predefined failure model. The failure model can be determined by theoretical analysis or according to a statistical method based on observation of the behavior of similar components of the same asset or of similar assets.

These SMART maintenance techniques are based on information on potential degradation mechanisms and threats obtained by a risk analysis, not only by prescriptive practices based on industrial experience (e.g., historical experience, industry guidelines for classes of equipment, as a prescribed fraction of the estimated residual/remnant life). These analytical models are basic, but they can effectively support the development of ad hoc decision making methods for the analysts, owners, and users (e.g., safety managers, site inspectors, the so-called competent persons, duty holders) of many industrial and service companies.

3 SMART Interface Standards for the Condition Monitoring of Machines

Predictive maintenance uses many sensors, including temperature, pressure, and vibration sensors. In the past, personnel have used portable instruments to periodically measure a machine's condition. Recently, continuous or real-time monitoring of machinery has been made possible by mounting sensors on the machine and using a data acquisition system and a host computer to collect the data.

Unfortunately, the application of permanently installed sensors on all machines in a manufacturing plant is very costly. However, standardized smart sensor interfaces can potentially drive down the equipment price and enhance its functionality and features. Sensor-to-network interfacing is an important issue in the machine maintenance and asset management industry for mission-critical and safety–critical applications. Unexpected machine failures are not only costly to correct but may take human lives, making safety for humans the real driver of standardization. The high costs of machinery and maintenance software management systems call for a high degree of operational reliability and machine uptime. Despite the technological advancements of recent decades, state-of-the-art machine condition monitoring and diagnostic systems have yet to be widely accepted and deployed on the factory floor. This is attributed mainly to the lack of standardized sensor interfaces for data acquisition, the great variety of ad hoc algorithms for information processing, data interchange, fault diagnosis, and machine remaining life prognosis, and the lack of standards for a plug-and-play sensor for interoperability.

A number of interface standards, including IEEE 1451.1 [18], 1451.2 [17], 1451.3 [19], and 1451.4 [20], are being drafted for networking smart transducers. There are also a number of standardized digital interfaces and communication protocols as well as networked smart transducer models. The IEEE 1451 standards provide for self-describing measurement and control devices, allowing one to choose best-in-class products from preferred vendors for true plug-and-play capability. The IEEE 1451 set of standards defines a Transducer Electronic Data Sheet (TEDS), electronic interfaces, and wired/wireless connection of analog and digital communications, creating an ideal modular distributed architecture.

Many of these technologies are being developed by the armed forces all over the world, especially the U.S. Navy. These technologies will eventually migrate to industrial, commercial, and consumer markets for condition-based maintenance (CBM) of machines. Machinery Information Management Open Systems Alliance (MIMOSA [26]), a non-profit industry trade association, aimed to establish an open architecture and a set of protocols for exchanging complex sensor information between CBM systems. It focused on the development of a Common Relational Information Schema (CRIS) protocol, a common language for transferring and exchanging machine condition monitoring and assessment data such as pressure and temperature between a client's proprietary database and another remotely located user, using common data communication conventions such as Java, XML, and Distributed Component Object Model (DCOM). Another related effort, Open System Architecture for Condition Based Maintenance (OSA-CBM) program (MIMOSA OSA-CBM v3.2 standard), developed an open architecture and standard for distributed CBM software components. MIMOSA/CRIS has been adopted by OSA-CBM as the core infrastructure for distributed machine maintenance information communication.

However, neither the MIMOSA nor the OSA-CBM specifications defines a sensor interface and communication protocols for acquiring sensor data. OSA-CBM recommended interfaces between its program structure and other published standards from the IEEE and International Organization for Standardization (ISO). Thus, the IEEE 1451 smart transducer interface standard seemed to be able to play a key role in completing the process from the acquisition of data at the sensor level to the transfer of the sensor information to the enterprise level, where MIMOSA and OSA-CBM are set up to manage the information. The approach of integrating the IEEE 1451 standard with the MIMOSA and OSA-CBM specifications was investigated. It was found to be feasible to establish a link between the IEEE 1451 standard and MIMOSA and OSA-CBM architecture.

In summary, the distributed measurement control concept with networked sensors and actuators would work quite well with distributed control, remote condition monitoring and conditioned-based maintenance of machinery. Taking this approach in advanced system design will reduce the total-life-cycle cost of a system in terms of modular sensor system design, use of commercially off-the-shelf sensors and actuators, ease of maintenance and upgrades by simply 'plug-and-playing' components. Finally, with the use of standardized components and interfaces, interoperability is no longer an issue.

The IEEE 1451 smart transducer interface standards are defined to allow manufacturers to build transducers of various performance capabilities that are interoperable within a networking system. The IEEE 1451 standards have provided the common interface and enabling technology for the connectivity of transducers to microprocessors, control and field networks, and data acquisition and instrumentation systems. The standardized transducer interface specified by the standards allows the self-description of sensors. It also provides a standardized mechanism to facilitate the 'plug-and-play' of sensors to networks. The network-independent smart transducer model defined by IEEE 1451.1 and common command, transducer interfaces, and functionality defined by IEEE 1451.0 will allow sensor manufacturers to support multiple networks and protocols; therefore, transducer-to-network interoperability is on the horizon. The expanding Internet market has created an opportunity for sensor and network manufacturers to exploit web-based and smart sensor technologies.

The advancement of wireless technology will revolutionize the use of sensors and actuators. The emerging Bluetooth and Wi-Fi technologies have promised to free the wires from the sensors and put billions of wireless devices into operation, particularly in the field of condition monitoring and conditioned-based maintenance of machinery and equipment; see Ramamurthy et al. [5]. The sensor community is carefully evaluating these wireless connectivity schemes in an effort to find a better way to ease the connection and integration of sensors into the application domain. Wireless sensor connectivity will continue to be explored until it is proven that wireless connection is as good as and as secure as a wired connection.

Advanced and high performance microprocessor technology will enhance the performance of DAQ modules and reduce their size and cost. The next generation of data acquisition systems, networkable, integrated sensing devices, will fuse sensors and wireless technology into a single module. The modular wireless DAQ will be embedded with various algorithms (Fast Fourier Transform, Wavelet Transform, close-loop control, filtering, etc.), selectable remotely by the host to perform a specific operation and monitor the health of an asset to predict its remaining life. These low cost, wireless, modular, sophisticated, networkable sensing devices based on the IEEE 1451 standards will be deployed easily by place and play. Users stand to benefit from these innovations and new applications.

4 Diagnosis and Prognosis as Maintenance DSS Enablers

Diagnostics is an intrinsic and valuable element of asset health assessment. Its applications and contributions to the quality of maintenance span all aspects of asset management. Beyond the value to individual machines, diagnostics also contribute to the improvement of production systems and the development of new technologies in various sectors. Diagnostic tests serve a role in the measurement and tracking of standards and quality of maintenance by maintainers and end users, thereby making an essential and multifunctional contribution to machine health. In addition, the protection of machine health relies on diagnostics to detect non-visible damage that reduces the Remaining Useful Life (RUL) of the asset.

Diagnostic information is indispensable for decision-making by maintainers, end users, manufacturers, even regulators. With such information, decisions can be made for asset specific and population wide treatments, measures, procedures, and services. Diagnostics provides key and sometimes critical information at multiple junctures along the life cycle, from risk assessment and early diagnosis, to machine follow-up and breakdown management to minimize shutdowns and maximize availability. The principal uses of diagnostics include diagnosis, primary risk assessment (i.e., predictive and early problem identification), prognosis, selection of corrective action if needed, selected condition monitoring and deterioration management. Diagnostics provide maintainers with information essential to making appropriate decisions. In diagnosis, one or multiple tests are used, typically in combination with machine history and technician experience, to identify a particular existing problem or condition. Some tests or test combinations may identify the existence of multiple degradations and machine health problems in addition to the primary diagnosis, providing information that can inform selection among alternative actions or adjusting a selected treatment.

Figure 5a shows a vibration velocity spectrum of a centrifugal pump with numerous pathologies. Unfortunately, due to an aging effect, rotating machines develop multiple problems simultaneously; all are visible in spectral analysis. The real challenge is the correct identification of the most relevant and current problem since incorrect identification can produce reduced mitigation. Problems like imbalance, misalignment, looseness, oil problems or loose bolts are merged and displayed simultaneously. By way of contrast, (Fig. 5b) shows one spectrum with one clear defect; this seldom happens, but if it does, identification of the problem is immediate and accurate.

Diagnostics can detect incipient problems or determine which machines are at increased risk for developing certain types of deterioration. Determination of increased risk may allow maintainers to take measures to prevent or reduce the risk of developing an undesired condition, including increased condition monitoring, operation changes and preventive interventions.

Detection of emerging deterioration before symptoms appear or at an early symptomatic stage allows significant opportunities for early prevention and treatment. Accurate and early detection and identification enable assessment of

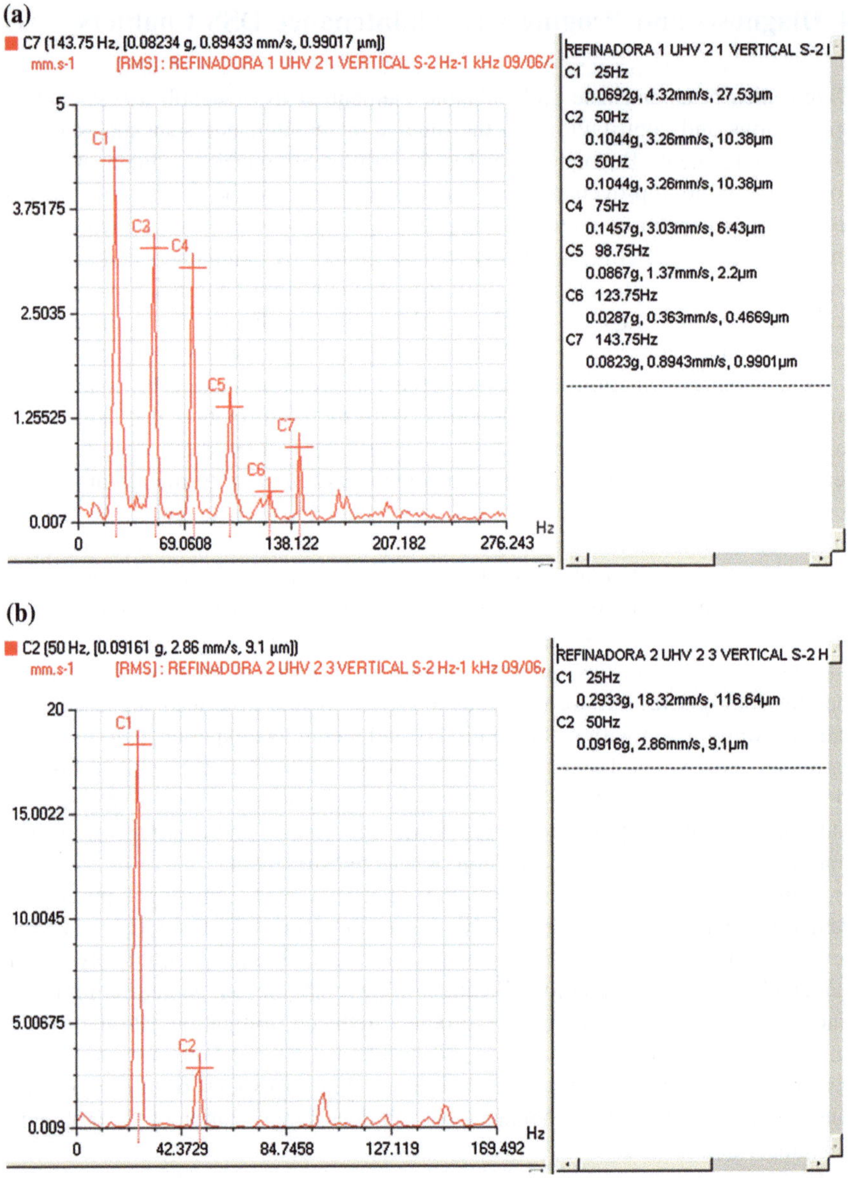

Fig. 5 **(a)** Spectrum of an asset with multiple pathologies, **(b)** Spectrum of an asset with one identified problem

machine health status that can translate into fewer shutdowns, decreased life-cycle costs, improved operation, and reduced maintenance costs. Diagnostics are evolving continually to enable more sensitive and specific detection of failure mechanisms at earlier stages via measurement of multiple parameters. These diagnostics offer new opportunities for timely disease prevention and treatment.

The 'profile' (manufacturer, environment, operations) or other predispositions of a machine may influence its response to a maintenance action. Emerging maintenance technologies use information about asset variability to allow targeted treatment selection tailored to individual needs. Context diagnostics are profile-based diagnostic tests used to determine the individual benefits or harms of taking certain actions. The knowledge of targeted treatments can allow the maintainer to avoid potentially harmful or ineffective maintenance actions, resulting in improved machine health outcomes and cost savings as a result of more effective decision-making. Context based decisions are contributing to an ideological shift within the maintenance community from a 'one size fits all' approach to 'right maintenance actions for the right machine'. Databases that compile and present such information are becoming available for scientists to study and maintainers to see how profile variations may relate to actions outcomes; see Candell et al. [6]. As the use of context data becomes more integrated into maintenance practice guidelines, electronic records and decision support systems will increasingly include context-based policies in routine treatment decisions. The increasing use of context technologies has great potential to yield better treatment selection and failure management strategies.

Diagnostic tests may also be used to assess the degree of damage progression or severity and the likelihood of recovery or risk of future adverse outcomes. The prognostic information is frequently used to inform treatment decisions tailored to individual machine health status and needs. Prognostic assessment can include testing for certain co-existing problems (e.g., unbalance, misalignment, wear, oil problems). The presence of co-existing failure mechanisms may inform necessary alterations in maintenance options and operation regimen. Certain chronic failures require continuous monitoring to avoid serious consequences or maintenance complications, so screening for emerging resistance to maintenance actions or co-occurring failures is essential. When used for these purposes, diagnostics are instrumental in helping maintainers manage complex, or later-stage machine health problems or conditions. Effective failure monitoring and management is often linked to reduced maintenance utilization, lower maintenance costs, and improved operation.

SMART maintenance systems based on Real Time Condition Monitoring and optimum Decision Support Systems allow maintainers to conduct rapid diagnostic in situ tests, rather than sending data to centralized laboratories. Such rapid diagnostics provide technicians with information on machine health status and maintenance options during maintenance inspections or scheduled actions. This immediate responsiveness reduces delays in effective decision-making; it also allows rapid responses to critical situations such as sudden breakdowns caused by the fast degradation of mechanisms and to routine and non-critical situations. As a result, it has the potential to reduce maintenance department costs.

Consumer expectations for diagnostics, such as rapid results, increased automation, simpler operation, and enhanced portability continue to drive the development of SMART devices. As time goes on, SMART diagnostics will play a significant role in maintenance decision-making, particularly in areas where rapid and accurate responses are closely tied to outcomes, such as the diagnosis of an imminent failure.

Diagnostics of large populations of assets are used to detect problems in individuals and to track population-level outbreaks. of the numerous applications for population diagnosis include global degradation, transmission of problems, and identification of common failure mechanisms or resistance to maintenance (preventive or corrective) actions to determine future risks. As novel diagnostics continue to emerge, machine health threats can be characterized and contained more quickly and efficiently, affecting fewer individuals and improving health management options. Use of diagnostics for these applications informs appropriate maintenance decisions and containment efforts to reduce the spread of the failure mechanisms among the population. Diagnostics development in this area has focused on rapid and accurate results, as well as portable, easy-to-use instruments. Technological advances increase flexibility and responsiveness to changing maintenance needs. Newly developed rapid detection diagnostics may help decrease the time between introduction of a new failure mechanism and detection, enabling faster and more effective maintenance response. Many of these emerging diagnostics are being adapted for field use in emergency situations, ideally allowing containment efforts through self-maintenance actions (via actuator) to begin before maintenance crews come into the picture.

Emerging technologies, such as tribotronics, are useful for associating various products with oil status or machine health. As new markers are validated, and as the significance of various combinations of markers are better understood, these technologies are being adapted rapidly for a range of diagnostic applications. One expanding trend is multiplexing, which involves conducting tests for more than one marker in the same test sample. This testing paradigm is also being developed in array formats, where multiple multiplex tests can be performed on the same platform or chip.

As diagnostics become increasingly integrated and capable of generating vast amounts of data, analytical advances and ease of interpretation will facilitate adoption and diffusion of these technologies into routine maintenance practice. For example, interpretation of a tribology marker assay that includes several hundred tests may be too complex for use in general maintenance practice without software or information processing capabilities to assist with analysis and presentation of diagnostic results. As products that identify many hundreds or thousands of markers, sophisticated analytical tools will be necessary to decipher the relationships between collected and processed information and the predisposition to initiate the failure mechanism.

Advances in electronic maintenance records and decision support software will assist maintainers to extract meaning from increasingly complex diagnostic results. Computerized systems are currently assisting with the processing of certain tests, and similar systems will decrease the diagnostic interpretation time, allowing more rapid translation into appropriate prevention or treatment efforts.

Diagnostics play a profound role in the life cycle of an individual, the general population, and the environment. The significance of diagnostics is anticipated to increase as it evolves with other interventions and with machine health information technology. Cloud computing interfaces are anticipated to better link the profiles of machines, technicians, end users, manufacturers and regulators; see Galar et al. [7]. The new diagnostic interfaces have the potential to redefine, on a global scale, the relationship among participants through timely and relevant information available at the 'fingerprint' level; see Wandt et al. [8]. The convergence of diagnostics and machine health information technology is driving the development of more rapid, accurate and high throughput diagnostic information. In some instances, future diagnostic devices will need to incorporate advanced information technology with greater integration capabilities, ease of use, and compatibility with other instruments or information resources (e.g., electronic records or databases). Despite current advances, portable diagnostic devices, for example, instruments or field use devices, still lack features and capabilities to further enhance the ability for maintainers and end users to capture, interpret, and use diagnostic information with greater speed, precision, and context. Conventional diagnostics provide minimal information such as a numerical value to a maintainer but minimal contextual information as to the potential relevance of the result at a higher plane of observation. The majority of conventional diagnostics also fall short of providing rapid real-time assay results which is an essential element for time critical diagnostic decision-making.

The recent technology development using SMART devices has enabled rapid and highly sensitive diagnostics to be used for a plethora of tests. The capabilities of these diagnostic technology platforms, when combined with advances in computing, telecommunications, and satellite technologies, will enhance the capacity and the potential of diagnostics, setting the stage for a paradigm shift in the generation and management of machine health information and services, Karim et al. [9].

5 Decision-Making Strategy: The Mitigation of the Risk

Fault detection and diagnosis, in general, are based on variables measured by instruments and observed variables and states by human operators. The automatic processing of measured variables for fault detection requires analytical process knowledge, and the evaluation of observed variables requires human expert knowledge or heuristic knowledge. Therefore, fault detection and diagnosis can be considered within a knowledge-based approach, [10, 11, 12]. Figure 6 shows an overall scheme of the proposed architecture for SMART maintenance applications.

In the first phase, the actual sensing elements must be deployed together with the power supply, integrated computing and the wireless communication with the surroundings. There are common condition monitoring techniques like vibrations, ultrasound, oil analysis or thermography with increasingly less expensive

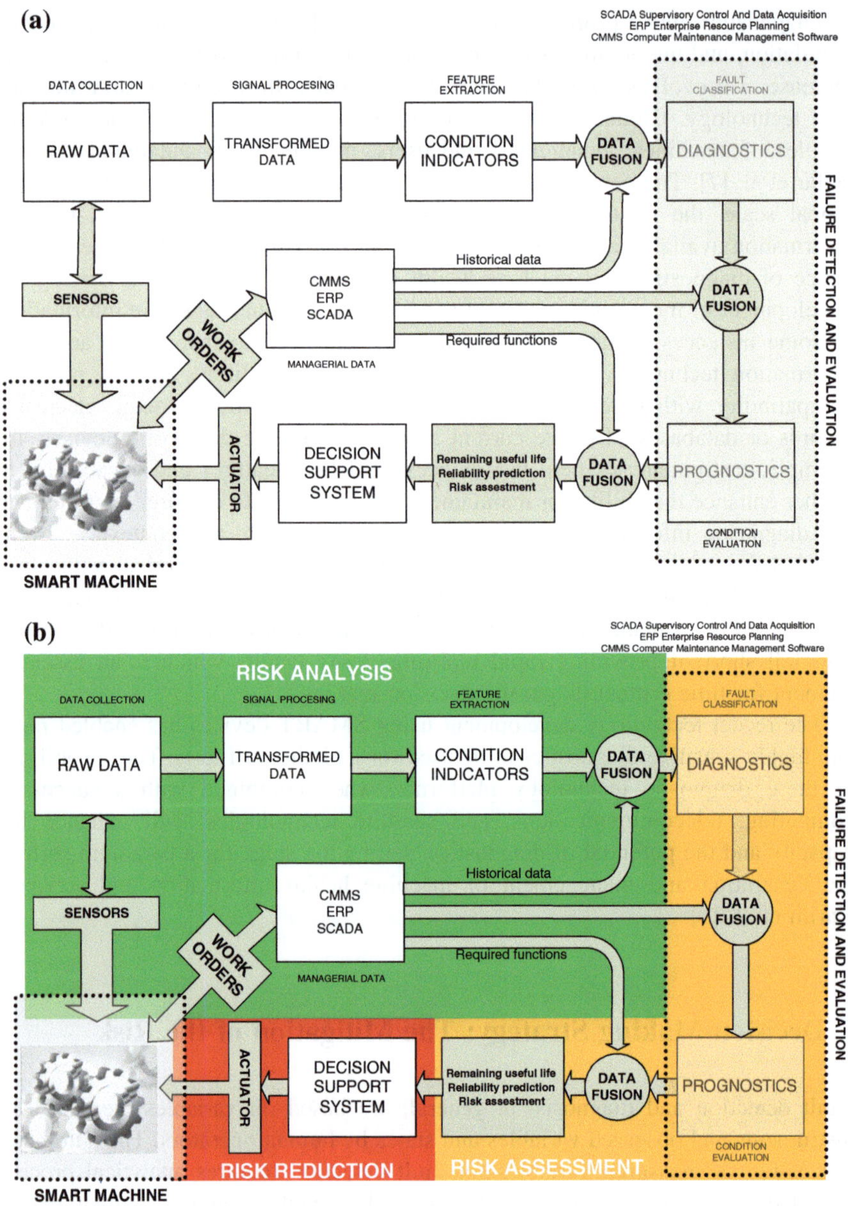

Fig. 6 **(a)** Proposed SMART architecture, **(b)** Risk stages in SMART architecture from identification to mitigation

acquisition systems, but besides these parameters, others provided by passive electrical sensing (inductive, resistive, and capacitive), active electrical sensing (electromagnetic fields, piezoelectric) can also reflect the condition of some parts of the machine. These sensing techniques should be used to measure the operating conditions (speed, load, temperature) in order to configure the fingerprints. The empirical relationships between sensor output and actual physical condition must be established but there are multiple ways to do it.

Information extracted from the sensors can be manipulated in different ways. Some relationships between variables' values and potential consequences with mitigating actions can result from personnel experience—mostly maintenance crew and operators who know the behavior of the machine. This information is valuable, and there are many efforts to retain this knowledge within companies due to the aging of maintenance crews all over the world. However, the relationship between acquired data and existing faults for further decision making requires the systematization and automation of diagnosis and prognosis with optimum maintenance decision making.

Models can be data driven or physical based; see Galar et al. [7]. Data driven models use a black box approach where acquired data are processed and a number of features are calculated to find those with peculiar behavior, anticipating anomalies and malfunctions of the asset. Physical models use physics of the failure to mathematically model the failure mechanism, first to identify it, and then to predict how long it will last.

According to Galar et al. [13], the information fusion of such relevant sources, models and maintenance experience can provide a powerful tool because the weaknesses of both methods are negated.

The hybrid models proposed in the SMART maintenance approach consider the analytical knowledge about the process to produce quantifiable, analytical information. To do this, data processing based on measured process variables is performed to generate the characteristic values:

- Limit value checking of direct, measurable signals. The characteristic values are the violated signal tolerances; these are usually supplied by experienced technicians and operators.
- Signal analysis of directly measurable signals using signal models like correlation functions, frequency spectra, autoregressive moving average (ARMA) or the characteristic values, e.g., variances, amplitudes, frequencies or model parameters. These techniques, together with AI tools like ANN, SVM or LSM, constitute the basis of data driven methods for diagnosis and prognosis
- Process analysis using mathematical process models together with parameter estimation, state estimation and parity equation methods. The characteristic values are parameters, state variables or residuals. Information provided by the process and properly modeled constitutes the foundation of physics based methodologies.

Features such as physically defined process coefficients or special filtered or transformed residuals are extracted and then compared to the normal features of the non-faulty process. To this end, methods of change detection and classification are applied. The SMART architecture does not neglect the potential benefits of either data or physics driven methods. The resulting changes (discrepancies) in measured signals, signal models or process models are considered as analytic symptoms.

The real challenge is the comparison itself and the concept of discrepancy. Maintainers and end users usually wonder how big the increment of a variable should be to consider it a real threat. This consideration is the first step of the decision making process; it determines if risk is really present or if this is a false alarm. Figure 7 shows a taxonomy of analytical fault-detection methods.

5.1 Simple Thresholds

Some features have been proposed as single-shot diagnostic variables with an absolute threshold; therefore, they do not required historical data to determine trending. For example, kurtosis is said to indicate a damaged bearing for any value significantly higher than 3, say 4.5. Realistically, such variables are too simple to offer single-shot diagnosis under most circumstances.

Typically, a threshold has to be set in a case-by-case manner. It can be set absolutely or relatively to, for example, the maximum of an exemplar signal (or the average of a number of signals). An absolute maximum can be set for machine life threatening events such as collision or spindle bearing seizure, or a threshold can be set at, say, 200 % of the maximum of an exemplar signal. Usually, there is more than one threshold for different purposes (see Fig. 8), such as tool fracture or a missing tool. Response time is scheduled according to the seriousness and urgency of an event. In some cases, immediate action is necessary while, in others, a slower response is more appropriate. When the exemplar is the previous cycle, the threshold is no longer fixed and is sometimes called a floating threshold. Figure 8 shows this behavior in the dynamic thresholding of static thresholds as a function of operation and maintenance issues where aging effects or changes in the process may require different levels of actuation. A wrong static threshold is the

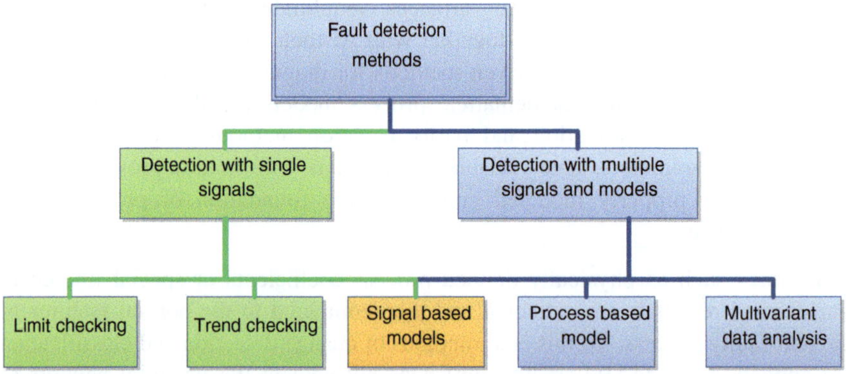

Fig. 7 Analytical fault detection methods

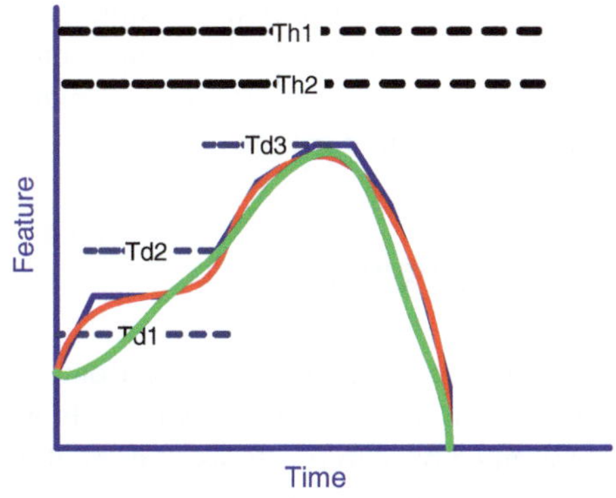

Fig. 8 Ordinary and time dependent thresholds in part signature

one proposed by ISO 10816 [21]to characterize the severity of the vibration. This severity does not depend on the age of the machine and conditions in the operation, so the standard is equally applicable to new assets and to machines close to the disposal phase. A dynamic threshold would solve these issues. In fact, one of the main functions of SMART maintenance practices is the proposal of dynamic thresholds based on contextual conditions.

5.2 Time/Position-Dependent Thresholds

As mentioned above, due to the highly non stationary nature of most manufacturing process cycles and their associated health indicators, a constant threshold is frequently too limiting, as shown in Fig. 8. Unfortunately, these constant thresholds are the most common in the standards and in manufacturers' recommendations. Because of their simplicity and independency of maintenance and operation, they are usually too conservative. The optimum way to consider the aging process or changing environments is to use finite duration thresholds which can be placed in or across different time or position segments of a cycle to enable more refined monitoring, as shown in Fig. 8 with thresholds Td1, Td2 and Td3.

5.3 Statistical Process Control

When several variables are measured by sensors, absolute values may not reflect the right information. Instead, statistical values are considered and the control chart can be used directly. When indirect sensing is involved, an indirect method must be used. In the case of displacement measurement using accelerometer, a double integration process is needed, and the thresholds must take this process into consideration. Gong et al. [14] proposed a two-step method integrating an online sensor and a control chart for such an application. A manufacturing process is monitored by a sensor and a sample of parts is inspected whenever a warning is issued by the sensor. If the sample mean is outside the predetermined control limits, the process is suspended to conduct a diagnosis to determine the cause for possible corrective actions.

5.4 Part Signature

A part signature can be produced from a single observation but is more frequently produced from repeated observations of a cycle by averaging. The signature can be represented by step functions or more flexible parametric or nonparametric curves through curve fitting or nonparametric profiling such as principal curves.

Envelopes can then be placed over and under the signature to monitor process stage or faults. Timing variations between the part signature and the current cycle must be considered to minimize missed detection or false alarms. More sophisticated classifications of the local or global deviations from the part signature have also been developed. With this method, a signature with deviations can be fixed as a pattern to check how different the acquired signal is from the previous signals which produced the signature. Unfortunately, this averaging has a low pass effect, minimizing the transitions and making the part signature smoother. This smooth average result can compromise the impartiality of the decision system when one spurious peak occurs.

5.5 Waveform Recognition

Another criterion to identify an anomaly is the recognition of an abnormal waveform. This is especially relevant in non-stationary conditions, i.e. the opposite of part signature or thresholding, when transients happen and the waveform of the transient corresponds to different environmental or operating conditions. For example, in some tool monitoring systems, a number of force waveform patterns of tool breakage and chipping are stored in the system. When a carbide tool breaks, the cutting force suddenly rises and then drops to zero, while the force mostly just drops to zero when breakage occurs in ceramic tools. The system continuously matches the incoming signal to the stored waveform patterns. Waveform recognition could be devised not to be sensitive to variables such as cutting conditions and material properties.

5.6 Pattern Recognition

With the features extracted by signal processing, one can take advantage of a number of pattern classification methods, such as linear discriminant function, fuzzy logic, neural net, fuzzy neural net, decision tree, support vector machine, etc., to monitor the process stage or machine/process faults. Typically, exemplars are made available through empirical means, and features providing the most discriminant power are identified through statistical analyses. A classifier is then 'trained' to optimally classify the exemplars in terms of some objective functions. These functions measure the 'distance' between elements to detect outliers and perform the clustering in the functions of different criteria. Figure 9 shows a typical classifier using Support Vector Machine where the 'distance' between elements depends on a previously selected kernel function. Classifiers are very sensitive to these configuration parameters; they occasionally handle outliers and anomalies in a less than optimum way.

In multi-sensor configurations, sensor fusion is frequently done in a similar way. Due to the proliferation of feature extraction and classification tools, this is an area where much research has been done, sometimes at the cost of more fundamental research from which more insights could have been gained. The community will benefit greatly if further research is directed at the development of theoretical guidelines and systematic utilization instead of just trying different combinations of the tools made available by the AI community.

5.7 *Severity Estimator: The Prognosis Step*

Last but not least is the prognostication of the failure. Once a fault is detected, it is usually desirable to determine its severity. Severity estimation must answer the 'when' question, since 'what' and 'why' have already been answered by diagnosis. For example, it is desirable to track tool wear or a gear tooth crack to know when one has to change the tool or perform a maintenance action. A severity estimator is a mathematical representation of the relationship between some features and the actual severity. It is best if an analytical model can be established, as this will minimize the need for empirical data over an actual failure trajectory. When an analytical model is unavailable or infeasible, a black box model calibrated by

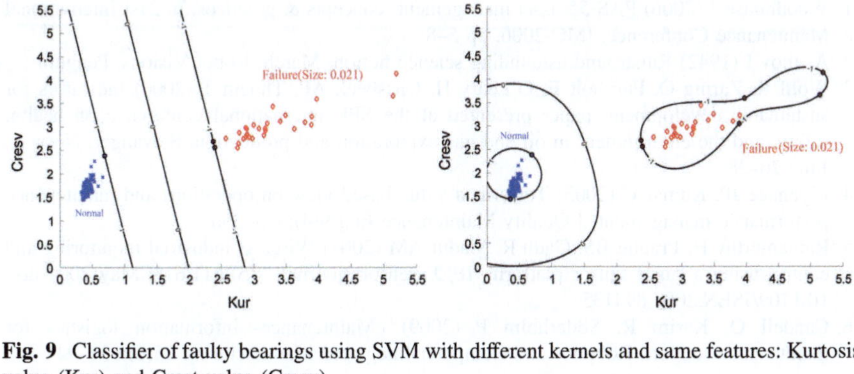

Fig. 9 Classifier of faulty bearings using SVM with different kernels and same features: Kurtosis value (Kur) and Crest value (Cresv)

empirical run-to-fail data can be used; however, these models are not generally appreciated in the industry due to their lack of transparency for decision making, as technicians must trust in the black box outputs. Finally, discrete-time time-invariant Markov models have been used to describe a machine's deterioration trajectory.

6 Conclusion

SMART maintenance approaches will become feasible solutions to maintenance problems and replace the traditional concept of condition monitoring. Risk assessment of maintenance decisions based on RUL estimations is a promising strategy to evaluate the various possible consequences of performed maintenance actions or deferred maintenance. This methodology relies on collected information, creation of operational scenarios and potential deviations from expected behavior. Underlying technologies like sensors and algorithms must be properly connected, however, preferably according to existing standards if any.

At this point, these risk assessment policies are seldom considered in maintenance decisions, and most existing policies focus on protecting human beings and the environment. It is time to redirect safety issues towards asset integrity and the accomplishment of business goals other than human safety.

A number of technologies are already available; the current lack of standardization is a problem that will hopefully be sorted out in the upcoming ISO 55000.

Acknowledgments This work is partially supported by SKF-UTC. The authors gratefully acknowledge the helpful comments and suggestions from the Advanced Condition Monitoring Center in LTU and associated partners.

References

1. Woodhouse J (2006) PAS-55-asset management: concepts & practices. In 21st International Maintenance Conference, IMC-2006, pp 5–8
2. Asimov I (1942) Runaround astounding science fiction, March, Robot Visions. Penguin
3. Wolff R, Zaring O, Furuholt E, O'Leary H, Gaasbeek AP, Thoem T (2000) Indicators for sustainable development, paper presented at the SPE international conference on health, safety, and the environment in oil and gas exploration and production, Stavanger, Norway, June 26–28
4. Liyanage JP, Kumar U (2003) Towards a value based view on operations and maintenance performance management. J Quality Maintenance Eng 9(4):333–350
5. Ramamurthy H, Prabhu BS, Gadh R, Madni AM (2007) Wireless industrial monitoring and control using a smart sensor platform, IEEE sensors journal, 7(5):611,618, May 2007 doi: 10.1109/JSEN.2007.894135
6. Candell O, Karim R, Söderholm P (2009) eMaintenance—information logistics for maintenance support. Robotics and Computer-Integrated Manufacturing 25(6):937–944

7. Galar D, Palo M, Van Horenbeek A, Pintelon L (2012) Integration of disparate data sources to perform maintenance prognosis and optimal decision making. Insight-non-destructive testing and condition monitoring, 54(8):440–445(6)
8. Wandt K, Karim R, Galar D (2012).Context adapted prognostics and diagnostics. International conference on condition monitoring and machinery failure prevention technologies, London, UK
9. Karim R, Söderholm P, Candell O (2009) Development of ICT-based maintenance support services. J Quality Maintenance Eng 15(2):127–150
10. Rasmussen J (1993) Diagnostic reasoning in action. IEEE trans on system, man and cybernetics, 23(4):981–991
11. Struss P, Malik A, Sachenbacher M (1996) Qualitative modeling is the key to automated diagnosis. In 13th IFAC World Congress, San Francisco, CA, USA
12. Isermann R (1994) Integration of fault-detection and diagnosis methods. In Proceedings IFAC symposium on fault detection, supervision and safety for technical processes (SAFEPROCESS), pp 597–609, Espoo, Finland
13. Galar D, Gustafson A, Tormos B, Berges L (2012) Maintenance decision making based on different types of data fusion. Podejmowanie decyzji eksploatacyjnych w oparciu o fuzję różnego typu danych. Eksploatacja i Niezawodnosc, Maintenance and Reliability 14(2), 135–144
14. Gong L, Jwo W, Tang K (1997) Using on-line sensors in statistical process control. Manage Sci 43(7):1017–1029
15. API Publication 581 (2000) Risk based inspection, base resource document 1st (edn)
16. API Publication 580 (2002) Recommended practice for risk based inspection 1st (edn)
17. IEEE Std 1451.2-1997. IEEE Standard for a smart transducer interface for sensors and actuators–transducer to microprocessor communication protocols and transducer electronic data sheet (TEDS) Formats, instrumentation and measurement society, TC-9, institute of electrical and electronic engineers, New York. 10016–5997, SH94566
18. IEEE Std 1451.1-1999 (2000) IEEE standard for a smart transducer interface for sensors and actuators–network capable application processor information model, instrumentation and measurement society, TC-9, institute of electrical and electronic engineers, New York. 10016–5997, SH94767
19. IEEE Std 1451.3-2003 (2004) IEEE standard for a smart transducer interface for sensors and actuators–digital communication protocols and transducer electronic data sheet (TEDS) formats for distributed multidrop systems, instrumentation and measurement society, TC-9, institute of electrical and electronic engineers, New York. 10016–5997, SH95174
20. IEEE Std 1451.4-2004 (2004) IEEE standard for a smart transducer interface for sensors and actuators–mixed-mode communication protocols and transducer electronic data sheet (TEDS) formats, instrumentation and measurement society, TC-9, institute of electrical and electronic engineers, New York. 10016–5997, SH95225
21. ISO 10816-1, Mechanical vibration. Evaluation of machine vibration by measurements on non-rotating parts. Parts 1–6
22. ISO (2012a) ISO/CD 55000.2 Asset management—overview, principles and terminology, ISO/TC 251/WG 1
23. ISO (2012b) ISO/CD 55001.2 Asset management—management systems—requirements, ISO/TC 251/WG 2
24. ISO (2012c) ISO/CD 55002.2 Asset management—management systems—guidelines for the application of ISO 55001, ISO/TC 251/WG 2
25. ISO 14121 (1999) Safety of machinery-Principles of risk assessment. International Organisation for standardisation. ISO 14121-1:2007 Safety of machinery—risk assessment– part 1: principles, 1st (edn)
26. MIMOSA Website. http://www.mimosa.org/

Part II
Rolling Bearing Diagnostics

Incipient Fault Detection in Bearings Through the use of WPT Energy and Neural Networks

Maria Jesus Gomez, Cristina Castejon and Juan Carlos Garcia-Prada

Abstract Bearings are one of the more widely used elements in rotating machinery, reason why they have focused the attention of many researches in the last decades. The aim is to obtain a methodology that allows a reliable diagnosis of this kind of elements without dismounting them from the machine, and detecting the failure in incipient stages before a critical failure occurs. This manuscript develops and improvement of a technique showed in [1] of automated diagnosis of bearings through vibration signals, using the coefficients of the Multirresolution Analysis (MRA) and Multilayer Perceptron (MLP) neural network (NN). Data were obtained from a quasi-real industrial machine, where bearings were supporting axial and radial loads while rotating at different speeds. This technique offered very good results when diagnosing healthy and faulty bearings, nevertheless the reliability decreased when distinguishing between different kinds of failures. The novel technique showed in the present work, increases the success rates obtained using the same data: not only allows detecting early faults but also their location with high accuracy. The methodology exposed in this work is based on the use of the relative energy of the Wavelet Packets Transform (WPT), and NN, concretely, the RBF.

Keywords Early bearing diagnosis · Condition monitoring · WPT · Relative energy · Neural networks · RBF

M. J. Gomez (✉) · C. Castejon · J. C. Garcia-Prada
MAQLAB Group, Mechanical Department, Universidad Carlos III de Madrid, Av. de la Universidad, 30 28911 Madrid, Spain
e-mail: mjggarci@ing.uc3m.es

C. Castejon
e-mail: castejon@ing.uc3m.es

G. Dalpiaz et al. (eds.), *Advances in Condition Monitoring of Machinery in Non-Stationary Operations*, Lecture Notes in Mechanical Engineering,
DOI: 10.1007/978-3-642-39348-8_4,

1 Introduction

The aim of condition monitoring is to detect failures in rotating machines before a critical damage occurs. This kind of maintenance has a lot of advantages, because it makes no necessary the dismounting of a machine to check the status of its elements. Besides, the probability of detecting a failure before it becomes critical increases, avoiding losses and making the operations safer. By these reasons, automation of fault diagnosis in industrial processes has been the aim of many researchers in the last decades.

Concretely, rolling bearings are one of the more widely used elements in rotating machinery, and its failure is one of the foremost causes of breakdowns in this kind of machines. Bearings are fundamental elements in the support subsystem, which hold great part of the static and dynamic loads, reason why they have high risks of failure. Most of the researches related to bearing fault diagnosis agrees with the use of vibration signals, due to they contain valuable information about failures [2, 3], however Acoustic Emission (AE) have been also appropriately used with accuracy to diagnose bearings, as in the case of [4].

Based on the use of this kind of signals, most authors classifies the techniques to diagnose bearings in three approaches: time domain based on statistical parameters [5], frequency domain analysis [6], and time–frequency analysis such as Wavelet Transform (WT) [1, 7] and Hilbert-Huang Transform (HHT) [8].

Diagnosis based on time domain statistical parameters has shown low effectiveness when it is applied to incipient faults or when the system is exposed to low loads, as pointed in [9]. By this reason, the use of time domain statistical parameters as unique way to extract features is not common.

The analysis of the frequency domain is the most classical approach to detect failures in rotating machinery, and concretely the Envelope Analysis is the more popular fault diagnosis method of rolling bearing. Envelope analysis means exploiting cyclostationary of second order (CS2) that appears when bearing defects exist [10]. However, this classical tool is seriously affected by the noise, especially in early fault stage. Some studies have been carried out to improve the results of this method, as for example in [11]. In other cases, to solve this problem, the envelope analysis has been combined with other techniques as the Wavelet Transform (WT), as in the case of [12]. Another tool usually applied to examine the frequency domain of the signals is the Empirical Mode Decomposition (EMD), used to obtain Intrinsic Mode Functions (IMFs), as shown in [13].

The Hilbert-Huang Transform (HHT) is a time–frequency analysis technique based in the EMD. The HHT offers high reliability, as in the case of [14].

The same way as HHT, the Wavelet Transform (WT) also offers information both in time and frequency domain, providing the proper treatment both for stationary and for non-stationary signals. WT gives also a multi resolution analysis, so it is especially useful to diagnosis of defects [15]. With this purpose, WT have been widely used and not only for bearings, but also for general rotating machinery as in [16], for gears as shown in [17], for shafts [18], and for structural elements as in [19].

However, the use of the WT is a complex task due to the great diversity of critical parameters which must be chosen, such as the mother wavelet and the decomposition level. On the other hand, until few time ago, the WT had another bigger disadvantage: the incapability for decomposing the high frequency bands trough the Multi resolution Analysis (MRA). Wavelet Packets Transform (WPT) constitutes an improvement of the MRA [20], due to the ability to decompose all the frequency bands. Thus, applications of WPT are highly increasing, and nowadays is the most used technique to treat signals in many fields, as in the case of speech recognition [21], denoising [22], and treatment of electrocardiographs [23], among others.

WPT coefficients can be used directly as features, as they content reliable information about failures [24]. However, other information related to the WPT coefficients can be also used as features, as has been demonstrated in [25], where statistical parameters are calculated, and in [26], where the energy of the WPT is successfully used as crack indicator.

In a diagnosis procedure, after features extraction, an intelligent classification system is also needed. A lot of intelligent classification systems have been developed and used for monitoring systems, as fuzzy classifiers, used in [27, 28], genetic algorithms [18], and the most used, the Support Vector Machines [29, 30] and Neural Networks (NN) [31].

In [1], an algorithm was developed to diagnose four conditions of ball bearings: healthy, inner race fault, outer race fault, and ball fault. The data were obtained at three different rotating speeds: 10, 20 and 30 Hz. The algorithm was based in the use of the MRA coefficients, after selecting of the optimal frequency band (the one where the coefficients presented larger differences between health bearing and the faulty conditions). This coefficients were used to train a Multilayer Perceptron (MLP) NN. With this methodology, high success rates were achieved, obtaining no false alarms, and distinguishing reasonably, for the speeds of 20 and 30 Hz, the healthy bearings from the faulty. However, the MLPs generated had problems to distinguish between different kinds of faults.

The aim of this work is to improve the results obtained from the analysis carried out in [1] working with the same data. The energy of the coefficients of the improved technique WPT will be used to feed, in this case, a Radial Base Function (RBF) neural network.

2 Experimental Setup

The vibration signals were obtained from a rig developed by the UNED mechanical department. FAG 7206 B single ball bearings were tested at three different rotation speeds set to 10, 20 and 30 Hz, and controlled by a photo tachometer. The rig is shown in Fig. 1.

In Fig. 1 the first elements observed, starting on the right hand-side, are axial and radial pneumatic cylinders, which apply loads of 2.5 and 3 bars respectively.

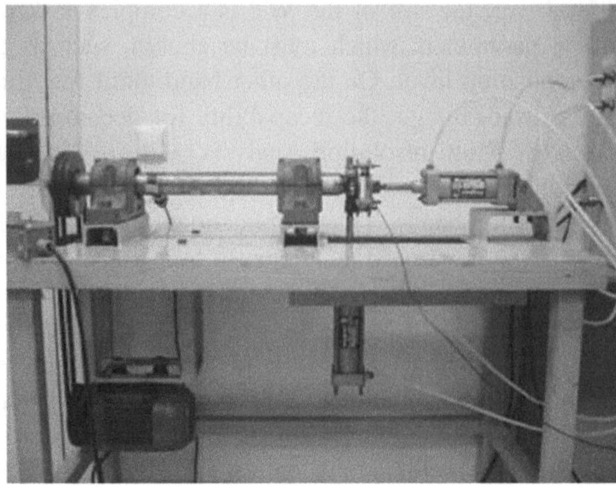

Fig. 1 Bench bank used for the measurements. UNED lab

Following, the bearings assembly can be seen. A transmission pulley is directly connected to the motor by a V-belt.

The measurement chain is composed by a B&K 4383 accelerometer, a B&K NEXUS amplifier and a DAS-1200 Keithley acquisition card. The sampling rate was set to 5,000 Hz, and all the acquired signals had 5,120 points.

The tests were carried out with healthy bearings. Later several faults were induced to the bearings to carry out the tests, including inner race fault, outer race fault, and ball fault. A pit 2 mm long was artificially induced in the inner or outer race by an electric pen. In the case of the rolling ball, multiple slots in the surface were performed to simulate the flacking phenomenon.

Finally, 284 signals are obtained: 196 signals for each rotation speed, and 49 for each fault condition.

3 Wavelet Packets Transform

Wavelet Transform (WT) is specially efficient to carry out local analysis of non stationary signals. It obtains correlation coefficients between a signal and a mother wavelet function selected. When WT is applied in a discrete way, called Discrete Wavelet Transform (DWT), the signal is decomposed in information of approximation and detail with recursive filters low and high pass. WPT consists on the application of the DWT in a recursive way, until a decomposition level selected, according to the scheme shown in Fig. 2.
where $W(k,j)$ represents the coefficients of the signal in each packet. k is de decomposition level and j is the position of the packet within the decomposition level. Then, each correlation vector $W(k,j)$ has the structure of the Eq. (1):

$$W(k,j) = \{w_1(k,j), \ldots, w_N(k,j)\} = \{w_i(k,j)\} \quad (1)$$

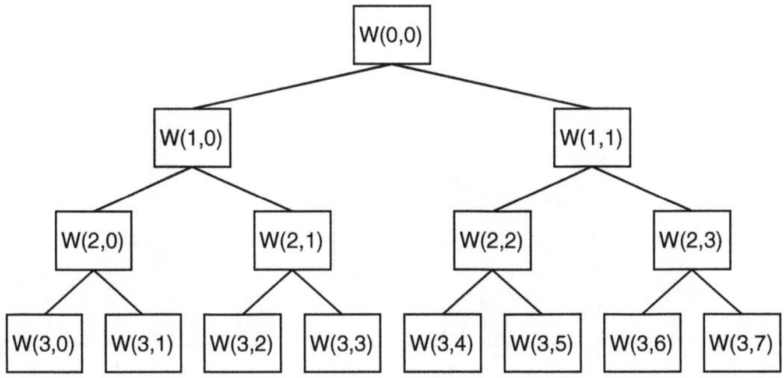

Fig. 2 WPT analysis, procedure of decomposition in approximation and detail information through low pass filters and high pass filters, until decomposition level 3

where i is the position of the coefficient within its packet.

3.1 Energy of the WPT Coefficients for Feature Extraction

The concept of energy used in the WPT analysis is very close to the Fourier Theory [24]. The energy of the packets is obtained from the sum of all the squares of the coefficients of each packet, according to Eq. (2):

$$Ek,j = \sum_i \{w_i(k,j)\}^2 \tag{2}$$

The relative energy, as a normalized parameter proposed in [26], is calculated as shown in Eq. (3):

$$Ek,j_{rel} = \frac{Ek,j}{E_t} \tag{3}$$

where E_t is the total energy of the signal, calculated as the sum of all the energies of the packets.

3.2 Features Extraction

Using the definition of the energy of the packets described above, the transformations are carried out. The mother wavelet used is the Daubechies 6 (DB6), due to its effectiveness in this area has been already proved in previous related works [1, 15, 16].

The decomposition level has been set to 3. This level was chosen because the better classification results were obtained with this value. The patterns extracted

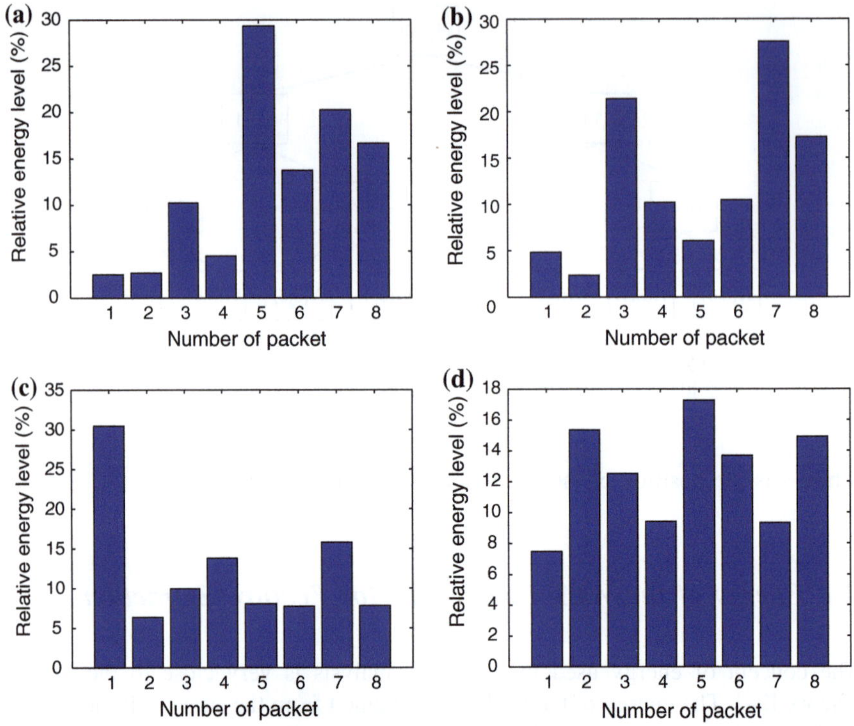

Fig. 3 WPT relative energies (%) at decomposition level 3 with mother wavelet DB6. **a** Healthy bearing. **b** Inner race fault bearing. **c** Outer race fault bearing. **d** Ball fault bearing

then, are vectors of 8 elements, which seems to be a proper number of inputs for the NN. The decomposition level determines the frequency resolution offered by each packet, that in this case is 312.5 Hz.

At each rotation speed, the features of all the conditions of fault are extracted. An example of the results obtained is shown in Fig. 3.

4 Classification System

The architecture of NN used as intelligent classification system is the Radial Base Function (RBF), because it has offered better results than the MLP and the Probabilistic (PNN) in previous related works [15]. RBFs are constituted by three layers of neurons, one of input, one or more hidden and one of output.

RBF architecture has a lot of advantages such as fast training and easy optimization. This is due to the low number of design parameters that must be decided by the designer, where the more critical are the number of neurons in the hidden layer and the activation function.

A critical parameter of the activation function is the spread. The spread is a constant that means the critical distance between the input and the weight vector. When this distance is reached, the output gets a value lower than a threshold.

The optimization of RBF parameters is carried out by a process examining the number of neurons of the hidden layer, and the success rate versus the spread. The value of spread that minimizes the number of neurons in the hidden layer (to reduce the computational cost) and maximizes the success rate is chosen.

5 Results and Discussion

After training several NN to optimize the design parameters, a total number of three RBFs were chosen, one for each rotation speed. Each NN is fed with 49 features of length 8 by each condition at every speed. The number of outputs of each NN is 4, one for each condition. The characteristics of the trainings are presented in Table 1.

During training process, the algorithm actualizes weight vectors between layers until the sum squared error (SSE) falls beneath an error goal (set to 0.2) or a maximum number of neurons in the hidden layer has been reached (700 neurons).

Success rates obtained at each speed are presented in Table 2, where the best results exposed in [1] are also presented in order to make a comparison.

As can be observed, success rates have been increased in the present work. The previous work used the coefficients of a narrow frequency band as features, while in this work uses the energies of the whole signal (specifically the 8 packets generated

Table 1 Design parameters of the RBF at the three rotation speeds

	RBF10		RBF20		RBF30	
2*Data Distribution	Training	75 %	Training	75 %	Training	75 %
	Test	25 %	Test	25 %	Test	25 %
Spread value	0.6		1.3		1.4	
Goal	0.2		0.2		0.2	
Number of neurons in the hidden layer	30		19		7	
Number of inputs at the pattern	8		8		8	
Number of outputs	4		4		4	

Table 2 Comparison of the success rates at different speeds with the obtained results in previous related work with the same data

	Current work		Previous related work [1]	
Speed (Hz)	Hidden neurons	Network accuracy (%)	Hidden neurons	Network accuracy (%)
10	30	91.38	30	85.71
20	19	91.78	30	81.63
30	7	92.58	30	77.04

at decomposition level 3) are selected. The number of inputs has been reduced from 18 coefficients to 8 levels of energy; however the information is related to a wider frequency band. The improvement can be assigned to other effects that occur in the machine when a fault appears, and a narrower band cannot detect.

The improvement of the success rates can also be assigned to the use of WPT in place of MRA, and to the use of the RBF architecture as a substitute for MLP.

From Table 2, it can be stated that the rotation speed of 30 Hz offers the best results both in terms of success rate (with a 92.58 %), and of computational cost, giving the lower number of neurons in the hidden layer with respect to the speeds of 10 and 20 Hz.

In Fig. 4 the partial results of the classification are shown.

The increasing of the success rates in this work allows a considerably better discrimination between the kind of fault with respect to the previous work [1], especially in the case of 30 Hz. Regarding the discrimination between health and faulty bearing, results are also better at 30 Hz, where zero false alarms can be found. The probability of not detecting a faulty bearing has been significantly reduced at this speed.

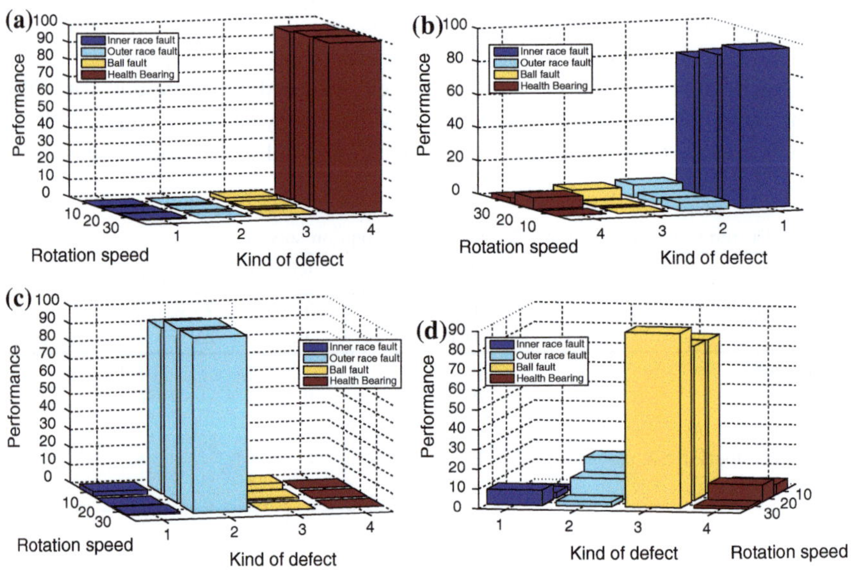

Fig. 4 Performance (%) of the three different NNs. **a** Health bearing classification. **b** Inner race fault classification. **c** Outer race fault classification. **d** Outer race fault classification

6 Conclusions

With base on the data obtained in the previous related work [1], a new analysis has been carried out to diagnose early faults in ball bearings at three different locations. The methodology has been changed: instead of using the coefficients of a specific frequency band with a MRA analysis, the WPT relative energies of the whole signal have been used. Besides, the architecture of the NN is the RBF, instead of the MLP.

The results obtained in the previous work have been improved. The success rates when distinguishing healthy from faulty bearings have been increased, however the better improvements have been achieved when discriminating between different kinds of fault, where the previous methodology had a serious lack. In the present work, the RBFs can detect with high accuracy the location of the fault at the three speeds.

It can be stated that both the diagnosis and the computational cost of the NN, are improved when the rotation speed increases, so the better solution found in this work is the NN trained with the data obtained at 30 Hz.

Acknowledgments The authors wish to thank to the Mechanical Department of UNED and Omar Lara for their assistance with the experimental part.

References

1. Castejon C, Lara O, García-Prada J (2010) Automated diagnosis of rolling bearings using mra and neural networks. Mech Syst Sig Process 24:289–299
2. Du Q, Yang S (2007) Application of the EMD method in the vibration analysis of ball bearings. Mech Syst Sig Process 21:2634–2644
3. Hu Q, He Z, Zhang Z, Zi Y (2007) Fault diagnosis of rotating machinery based on improved wavelet package transform and SVMs ensemble. Mech Syst Sig Process 21:668–705
4. Amarnath M, Sugumaran V, Kumar H (2013) Exploiting sound signals for fault diagnosis of bearings using decision tree. Measur J Int Measur Confederation 46(3):1250–1256
5. Muruganatham B, Sanjith M, Krishnakumar B, Murty SS (2013) Roller element bearing fault diagnosis using singular spectrum analysis. Mech Syst Sig Proccess 35:150–166
6. Li B, Chow M, Tipsuwan Y, Hung J (2000) Neural-network based motor rolling bearing fault diagnosis. IEEE Trans Industr Electron 47:1060–1069
7. Al-Raheem K, Abdul-Karem W (2010) Rolling bearing fault diagnostics using artificial neural networks based on laplace wavelet analysis. Int J Eng Sci Technol 2:278–290
8. Rai V, Mohanty A (2007) Bearing fault diagnosis using fft of intrinsic mode functions in hilbert-huang transform. Mech Syst Sig Process 21:2607–2615
9. Keller J (2003) Vibration monitoring of UH-60A main transmission planetary carrier fault. U.S. ARMY AMCOM. Aviation Engineering Directorate
10. Randall R, Antoni J, Chobsaard S (2000) A comparison of cyclostationary and envelope analysis in the diagnostics of rolling element bearings. In: ICASSP'00 Proceedings of the Acoustics, Speech, and Signal Processing, pp. 3882–3885
11. Yujie G, Jingyu L, Jie L, Zhanhui L, Wentao Z (2013) A method for improving envelope spectrum symptom of fault rolling bearing based on the auto-correlation acceleration signal. Appl Mech Mater 275–277:856–864

12. Barakat M, Badaoui ME, Guillet F (2013) Hard competitive growing neural network for the diagnosis of small bearing faults. Mech Syst Sig Process. doi:10.1016/j.ymssp.2012.11.002. URL http://dx.doi.org/10.1016/j.ymssp.2012.11.002
13. Zhang Y, Zuo H, Bai F (2013) Classification of fault location and performance degradation of a roller bearing. Meas: J Int Meas Confederation 46:1178–1189
14. Jiang L, Li B, Li X (2013) An imporved HHT method and its application in fault diagnosis of roller bearing. Appl Mech Mater 273:264–268
15. Lara O, Castejon C, Garcia-Prada J (2006) Bearing fault diagnosis based on neural network classification and wavelet transform. WSEAS Trans Sig Process 2(10):1371–1378
16. Adewusi SA (2001) Wavelet analysis of vibration signals of an overhang rotor with a propagating transverse crack. J Sound Vib 5:777–793
17. Baydar N, Ball A (2003) Detection of gear failures via vibration and acoustic signals using wavelet transform. Mech Syst Sig Process 17:787–804
18. Xiang J, Zhong Y, Chen X, He Z (2008) Crack detection in a shaft by combination of wavelet-based elements and genetic algorithm. Int J Solids Struct 17:4782–4795
19. Douka E, Loutridis S, Trochidis A (2003) Crack identification in beams using wavelet analysis. Int J Solids Struct 40:3557–3569
20. Liu B, Ling S (1997) Machinery diagnostic based on wavelet packets. J Vib Control 3:5–17
21. Vignolo L, Milone D, Rufiner H (2011) Genetic wavelet packets for speech recognition. Expert Syst Appl
22. Mercorelli P (2013) A denoising procedure using wavelet packets for instantaneous detection of pantograph oscillations. Mech Syst Sig Process 35(1–2):137–149
23. Tinati M, Mozaffary B (2006) A wavelet packets approach to electrocardiograph baseline drift cancellation. Int J Biomed Imaging, pp. 1–9
24. Hu Q, He Z, Zhang Z, Zi Y (2007) Fault diagnosis of rotating machinery based on improved wavelet package transform and SVMs ensemble. Mech Syst Sig Process 21:688–705
25. Shen C, Wang D, Kong F, Tse P (2013) Fault diagnosis of rotating machinery based on the statistical parameters of wavelet packet paving and a generic support vector regressive classifier. Meas J Int Meas Confederation 46:1551–1564
26. Feng Y, Schlindwein F (2009) Normalized wavelet packets quantifiers for condition monitoring. Mech Syst Sig Process 23:712–723
27. Lou X, Loparo K (2004) Bearing fault diagnosis based on wavelet transform and fuzzy inference. Mech Syst Sig Process, pp. 1077–1095
28. Sugumaran V, Ramachandran K (2011) Fault diagnosis of roller bearing using fuzzy classifier and histogram features with focus on automatic rule learning. Expert Syst Appl 38:4901–4907
29. Yang J, Zhang Y, Zhu Y (2007) Intelligent fault diagnosis of rolling element bearing based on svms and fractal dimension. Mech Syst Sig Process 21(5):2012–2024
30. Yuan S, Chu F (2007) Fault diagnostics based on particle swarm optimization and support vector machines. Mech Syst Sig Process 21–4:1787–1798
31. Bin G, Gao J, Li X, Dhillon B (2012) Early fault diagnosis of rotating machinery based on wavelet packets-empirical mode decomposition feature extraction and neural network. Mech Syst Sig Process 27:696–711

Bearing Fault Detection Using Beamforming Technique and Artificial Neural Networks

Walace de Souza Pacheco and Fernando A. N. C. Pinto

Abstract The importance of predictive maintenance optimization has been recognized over the past decades. A relevant aspect in the process of machinery noise control is the proper identification of noise sources. Microphone-array-based methods are known as alternatives for noise source identification in machines. In this work, the "Beamforming" technique is used to visualize the directionality pattern of the noise emitted by a rotating machine and a study is presented to compare the performance of machine condition detection using different architectures of classifiers based on Artificial Neural Networks. Sound maps from a rotating machine are used as inputs to classifiers for two-class (normal or fault) recognition. The classifier is trained with a subset of the experimental data for known machine conditions and is tested using the remaining data set. The procedure is illustrated using data from experimental sound maps of a rotating machine. The effectiveness of the classifiers and the network training is improved through the use of the Karhunen-Loève transform on the sound map data.

Keywords Acoustics · Source identification · Predictive maintenance · Artificial neural network · Karhunen-Loève transform

W. de Souza Pacheco (✉) · F. A. N. C. Pinto
Department of Mechanical Engineering, Universidade Federal do Rio de Janeiro, C.P. 68503, Rio de Janeiro 21941-972, Brazil
e-mail: walacepacheco@gmail.com

F. A. N. C. Pinto
e-mail: fcpinto@ufrj.br

G. Dalpiaz et al. (eds.), *Advances in Condition Monitoring of Machinery in Non-Stationary Operations*, Lecture Notes in Mechanical Engineering,
DOI: 10.1007/978-3-642-39348-8_5, © Springer-Verlag Berlin Heidelberg 2014

1 Introduction

Condition monitoring and diagnostics are widely used in almost every field of industry with applications in automation, quality control and predictive maintenance. The use of vibration signals is quite common in the field of condition monitoring of rotating machinery and the use of acoustic measurements is a promising way of complementing predictive maintenance systems. In this work, the data analyzed are derived from monitoring of a experimental test rig. Sound maps have been generated in order to observe the operating conditions of rotating machinery through spatial patterns of sound emission. The conventional Beamforming technique in the frequency domain is used to generate the sound maps.

Artificial neural networks (ANNs) have been applied in problems of pattern recognition, dynamic system identification, problems of pattern classification, fitting function, process control, time series forecasting [1]. In the present work, it will be applied in diagnosis of machine conditions and automated detection treating these as classification problems based on learning pattern from empirical data modeling.

A study is presented to compare the performance of bearing fault detection using different architectures of classifiers based on Artificial Neural Networks, namely, Multilayer Perceptron Neural Networks (MLP). A microphone array with twelve elements acquires time-domain acoustical signals from a rotating machine with normal and defective bearings. These signals are processed in order to obtain sound maps. By comparing the sound maps of a machine running in normal and faulty conditions, detection of faults like bearing defects is possible.

2 Sound Map Data

Figure 1 shows a rotating machine consisting of electric motor, mating gears, unbalanced disks and rolling bearings and an array circular with twelve microphones spaced by an angle of 30° at a radius r = 130 mm which was mounted 500 mm above from machine. Separate measurements were obtained for two conditions, one with normal bearings and the other with a defective bearing on the bearing block indicated in Fig. 1.

The signals of twelve microphones in the circular array were acquired with a PXI-unit from National Instruments. The acquisition hardware can be seen in Fig. 2. The signals were obtained under laboratory conditions, with low background noise. Shaft revolution was sensed by an optical sensor giving one pulse for each turn of the shaft. This signal was also connected to the data acquisition system. The machine was set up to rotate at 1,920 rpm. Measurements were obtained at a sampling rate of 20,000 samples/s. In the present work, these time-

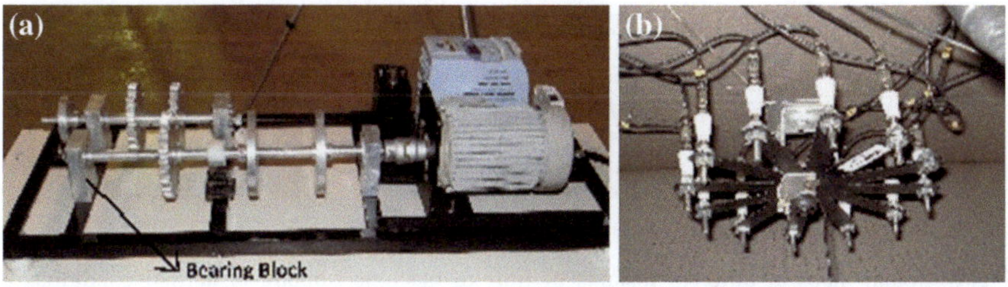

Fig. 1 Equipment and array used in the experiment. **a** Test-rig. **b** Array circular with 12 microphones

domain data were processed to create the sound maps through a conventional Beamforming algorithm in the frequency domain, implemented in Labview.

According to Brandstein and Ward [2], Beamforming is a method of mapping noise sources by assessing sound pressure levels based upon the direction from which they originate. The idea behind the technique is that a coherent sound, in the form of a plane wave, coming from a specified direction and being received by different microphones will lead to similar signals that are delayed on time based on the different travel paths [3]. Beamforming allows a "real-time" analysis of the incident sound, since it can picture a snapshot of the waves reaching the array. Of course this snapshot is restricted by the amount of time necessary to acquire the samples needed for the FFT-algorithm being used.

The sound maps generated by an array of microphones installed near the experimental test rig, are analyzed in the 1/3 octaves band centered about 1,250, 1,600, 2,000, 2,500, 3,150, 4,000, 5,000, 6,300 and 8,000 Hz. The different directionality patterns of the sound maps are related to machine running in normal and faulty conditions, this variation can be observed in Figs. 3 and 4 respectively, in the 1/3 octave band centered about 6,300 Hz.

Two sets of experimental data, with normal and defective bearings, were considered. For each set, two signal groups consisting about 200,000 samples were

Fig. 2 Experimental setup and Acquisition hardware

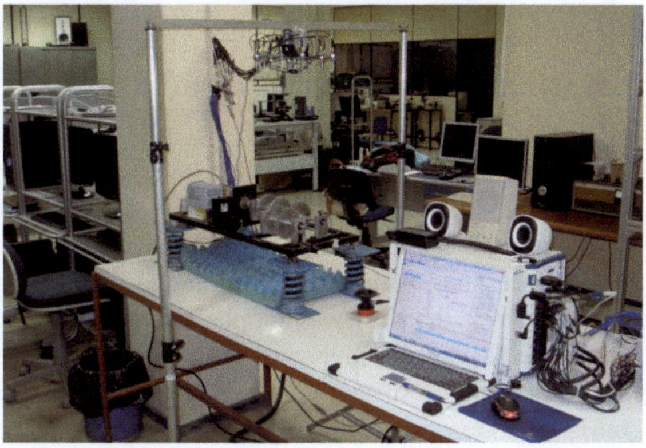

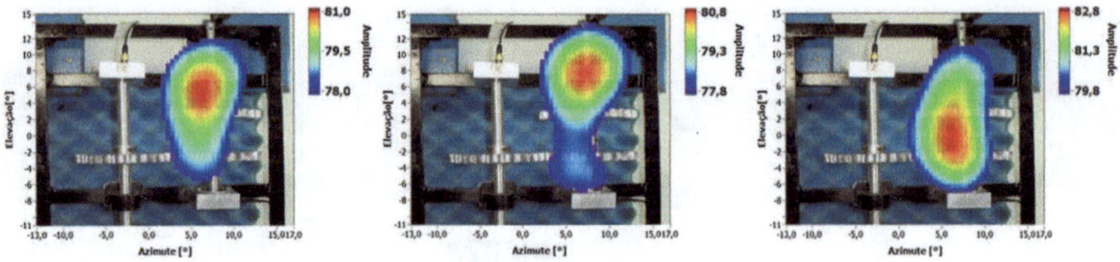

Fig. 3 Sound maps at different instants of time to machine running in normal conditions at 6,300 Hz

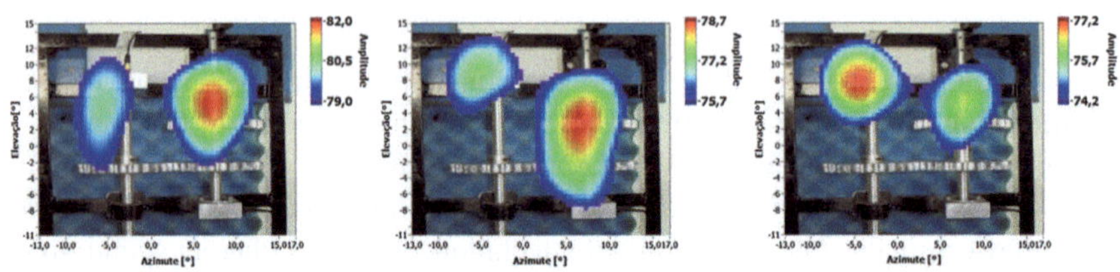

Fig. 4 Sound maps at different instants of time to machine running in fault condition at 6,300 Hz

obtained using the microphones monitoring the machine condition. These samples were divided into 140 bins of 2,500 samples each with 50 % overlapping. These bins were further processed, in Labview, to create one set of sound maps for each 1/3 octave band.

The total set is a matrix 1,080 × 140 × 2 consisting of 1,080 sound pressure levels, 120 for each 1/3 octave band, associated with each direction of the sound map, i.e., each row of the matrix represents a sound map orientation, 10 × 12 pixels for a specific 1/3 octave band. The columns represent the time series of the sound maps, 140, and two bearing conditions, normal and defective. These data are used as inputs to the classifier. The classifier is trained with a subset of the experimental data for the known machine conditions and is tested using the remaining set of data.

3 Karhunen-Loève Transform

The Karhunen-Loève transform (KLT), also called Principal Component Analysis or Hotelling Transform, is a well-known statistical method for feature extraction, data compression and so far it has been broadly used in a large series of signal and image processing, pattern recognition and data analysis applications.

In this work, the KLT was adopted for dimensionality reduction. Using the KLT technique, basically a higher dimensional data space can be transformed onto a lower dimensional space [4]. The sound map data used for classification purpose consist of 280 vectors with 1,080 dimensions, 140 vectors for each of the 2 classes (normal and faulty). Initially all 1,080 dimensions were used for training. After that was successfully implemented, the KLT technique was used to reduce the number of dimensions from 1,080 to 49, according the Broken Stick test [5].

Essentially, the KLT uses the covariance matrix obtained from the input data vectors. By looking at the covariance matrix of the data vectors, one can determine which of the dimensions are highly variable across the data space. This leading to a standard Eigenvalue problem. The KLT technique thus transforms the existing data space into a new data space. The variance of the resulting, transformed data set can be controlled along with the reduction in the dimension of the set, i.e., the number of components in the data vector is reduced. At the same time, most of the information contained in the original data set is retained. With KLT, new features obtained are a linear combination of the original ones.

4 Artificial Neural Network

Artificial Neural Networks (ANNs) were introduced as models of biological neural networks. An ANN is a system composed of interconnected processing elements, called neurons, which are arranged in layers. Each neuron is responsible for mapping linear and nonlinear data input and output, mainly determined by its activation function [1].

Among different kinds of ANNs, multilayer perceptron (MLP) neural networks are quite popular and will be used in this work. MLPs consist of an input layer of source nodes, one or more hidden layers of computation nodes or 'neurons' and an output layer [Haikin [6]]. Figure 5 shows an example of a ANN architecture that is widely used in practical applications.

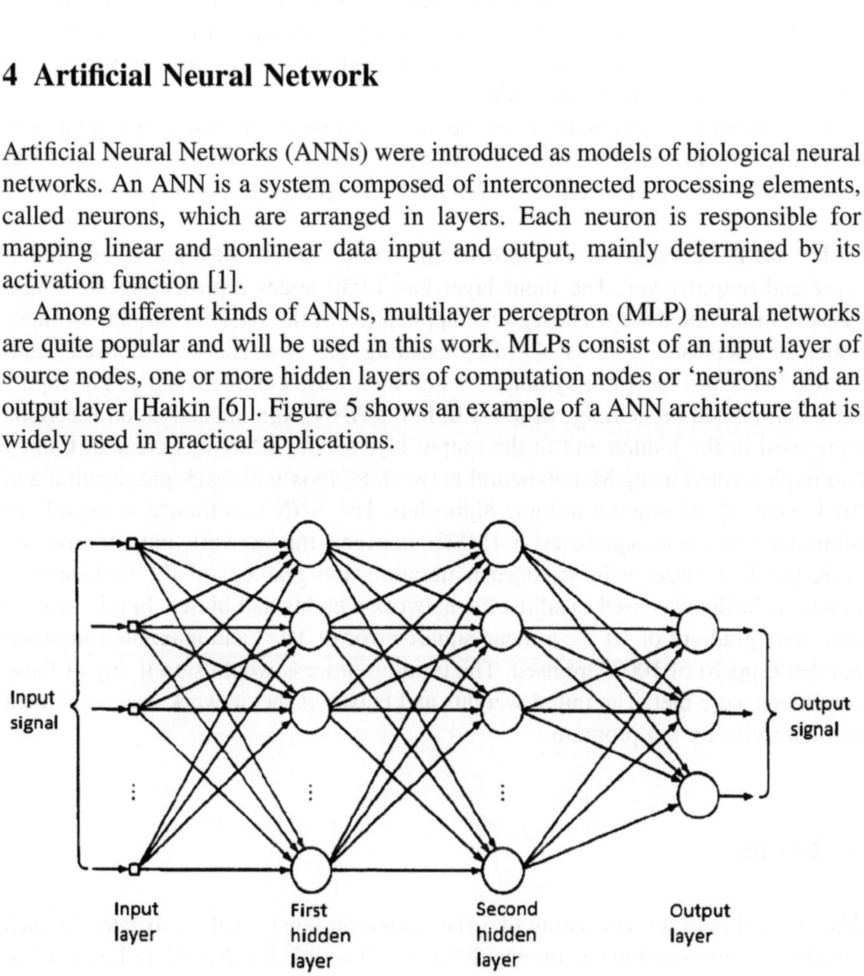

Fig. 5 Architecture of a multilayer perceptron (MLP) neural networks [6]

The neuron can be represented mathematically through the following expressions:

$$v_k = \sum_{j=1}^{m} w_{kj} x_j \tag{1}$$

and

$$y_k = \varphi(v_k + b_k) \tag{2}$$

where, $x_1, x_2, \ldots, x_m$ are the input signals; $w_{k1}, w_{k2}, \ldots, w_{km}$ are the synaptic weights; v_k is the output of linear combiner; $\varphi(.)$ is the activation function and y_k is the output signal of neuron. The model neuron also includes a bias externally applied, represented by b_k. The bias has the effect of increasing or decreasing the net inflow of the activation function, depending on whether it is positive or negative, respectively. The parameters weights and biases together, constitute the adaptive parameters in the network.

The number of nodes in the input and the output layers depend on the number of input and output variables respectively. The number of hidden layers and the number of nodes in each hidden layer affect the generalization capability of the network.

The MLP Neural Network, used in this work, consists of input layer, hidden layer and output layer. The input layer has 1,080 nodes representing the sound maps in the original way. Through the application of the KLT the number of input nodes is decreased to only 49. Representing the two classes, "normal" and "failed" bearings the 2 binary output nodes are always complementary. The inputs were normalized in the range of 0.0–1.0. In the ANN, sigmoid activation functions were used in the hidden and in the output layers. The ANN was created, trained and implemented using Matlab neural network toolbox with back-propagation and the Levenberg–Marquardt training algorithm. The ANN was trained iteratively to minimize the mean square error (MSE) between the network outputs and the corresponding target values. At each iteration, the gradient of the performance function (MSE) was used to adjust the network weights and biases. In this work, a minimum gradient of 10^{-10}, a mean square error of 10^{-6} and maximum iteration number (epoch) of 100 were used. The training process would stop if any of these conditions were met. The initial weights and biases of the network were generated automatically by the program.

5 Results

The original data set, consisting of 1,080 sound pressure levels associated to each direction of the sound maps, in nine different 1/3 octave bands, split in form of 140 bins, for each bearing condition, were divided into two subsets. The first 84 bins of each signal, or 60 % of bins, were used for training the ANNs and the rest were used for validation. The target value of output node was set 1 and 0 for normal and failed bearings, respectively.

The KLT technique was used to reduce the number of dimensions from 1,080 to 49, drastically reducing the number of input nodes.

The classification results are presented to see the effects of the Karhunen-Loève transform for the diagnosis of machine condition using MLPs. To evaluate the effect of the architecture of the neural network on the performance of classifiers, neural networks of different architectures with one or two hidden layers were tested for each classifier. These architectures were chosen on the basis of training trials. Table 1 and 2 show a summary of the best results obtained with the training of six types of architectures of MLP.

The neural network architecture 49 × 4 × 4 × 2 (49 neurons in the input layer, 4 neurons in the first hidden layer, 4 neurons in the second hidden layer and 2 neurons in output layer) showed the lowest success rate of the patterns presented to it, i.e., the network has recognized only 91.43 %. The architecture with better ability to generalize the network was 49 × 19 × 2 with success rate of 98.93 %. In general, it is noted in Tables 1 and 2 that the other architectures were also able to classify with efficiency above 91.43 %. It should be noted that the increase in the number of layers or neurons does not necessary leads to improved performance.

The training time needed for the different architectures depends on the amount of neurons used. With the dimension reduction achieved by the KLT the corresponding times are much lower than those for the original set. For the same highest success rate, of 98.93 %, the training time is reduced from 80.1 to 2.98 s. The smallest time of only 0.64 s still allows a success rate of 98.57 %.

Table 1 Results of success rate of different MLP architectures without dimension reduction

Architecture	Training time (s)	Success rate (%)
1,080 × 3 × 2	28.7	97.86
1,080 × 5 × 2	80.1	98.93
1,080 × 9 × 2	413.5	95
1,080 × 4 × 3 × 2	59.8	98.93
1,080 × 3 × 6 × 2	18.35	93.21
1,080 × 6 × 5 × 2	187.6	98.57

Table 2 Results of success rate of different MLP architectures with dimension reduction

Architecture	Training time (s)	Success rate (%)
49 × 6 × 2	0.64	98.57
49 × 18 × 2	2.24	95.36
49 × 19 × 2	2.98	98.93
49 × 3 × 3 × 2	0.88	98.57
49 × 4 × 4 × 2	0.78	91.43
49 × 7 × 3 × 2	0.93	98.21

Even with data with dimension reduced by KLT the failure introduced in the experimental test rig could be recognized. The results show that the MLP can be used satisfactorily as an alternative technique for fault diagnosis and classification, using maps sound in conjunction with KLT.

6 Conclusions

In this work the performance of a bearing fault detection system is investigated. The use of sound maps is proposed in order to correlate possible failures with their spatial locations. The great amount of data from these maps, in different frequency ranges, poses a difficulty for the training of ANN based classifiers.

The application of the KLT allowed the use of an approximation with fewer dimensions than the original data set. It simplifies the database of the sound maps and greatly improves the training of the ANN.

The recognition of patterns on the sound maps can also be used to detect the incipient failures of the machine components, through the on-line monitoring system, reducing the possibility of catastrophic damage and the machine down time.

Further work includes the application of the classifiers in a real gas turbine to the failure prediction in the framework of its maintenance system.

References

1. Bishop CM (2006) Pattern recognition and machine learning. Springer, New York
2. Brandstein M, Ward D (2001) Microphone arrays: signal processing techniques and applications. Springer, New York (Digital Signal Processing)
3. Christensen JJ, Hald J (2004) Beamforming. Technical review No. 1, Brüel & Kjær sound and vibration measurement A/S
4. Pacheco WS, Pinto FANC (2011) Application of the Karhunen Loève transformation on spatial sound mappings of a gas turbine. In: Proceedings of 18th international congress on sound and vibration, Rio de Janeiro, Brazil
5. Jackson DA (1993) Stopping rules in principal components analysis: a comparison of heuristical and statistical approaches. Ecol Soc Am 74(8):2204–2214
6. Haykin S (1999) Neural networks, 2nd edn. Prentice-Hall, Upper Saddle

HOS Analysis of Measured Vibration Data on Rotating Machines with Different Simulated Faults

Akilu Yunusa-Kaltungo, Jyoti K. Sinha and Keri Elbhbah

Abstract Vibration-based condition monitoring (VCM) has gained tremendous successes in the detection and differentiation of faults associated with rotating machines, through the installation of various numbers of vibration transducers at individual bearing pedestals of the monitored machine. This chapter however exposes the future potentials of the use of the higher order spectra (HOS) i.e., the bispectrum and the trispectrum for rotating machines faults diagnosis (FD). The aim of this is to achieve a significant reduction in the number of vibration transducers required at each bearing pedestal, without necessarily compromising valuable information required for the diagnosis. Four cases (healthy, shaft misalignment, cracked shaft and shaft rub) were simulated on an experimental rig with two rigidly coupled shafts supported by four ball bearings. Only four accelerometers (one at each bearing pedestal) were used for this study. The HOS results were compared for the different conditions of the rig. The observations and findings are presented in the chapter.

Keywords Rotating machines · Condition monitoring · Spectrum · Higher order spectrum · Bispectrum · Trispectrum

---

A. Yunusa-Kaltungo (✉) · J. K. Sinha · K. Elbhbah
School of Mechanical, Aerospace and Civil Engineering, University of Manchester, Oxford Road, Manchester M13 9PL, UK
e-mail: Akilu.kaltungo@postgrad.manchester.ac.uk

J. K. Sinha
e-mail: jyoti.sinha@manchester.ac.uk

K. Elbhbah
e-mail: keri.elbhbah@postgrad.manchester.ac.uk

G. Dalpiaz et al. (eds.), *Advances in Condition Monitoring of Machinery in Non-Stationary Operations*, Lecture Notes in Mechanical Engineering,
DOI: 10.1007/978-3-642-39348-8_6, © Springer-Verlag Berlin Heidelberg 2014

1 Introduction

The impacts of machine failures with respect to safety, environment, profit and market share losses are increasingly becoming enormous by the minute [1]. A huge proportion of the operations in most manufacturing as well as service providing industries are dependent on rotating machines, which forms the basis for a continuous search for tools and techniques that will effectively enhance the early detection of incipient failures in these machines [2].

Vibration-based condition monitoring (VCM) over the years has been effectively used for the diagnosis of faults in rotating machines, with one of its most established and successful diagnostic techniques based on spectrum and other signal analysis. The use of spectrum analysis for fault diagnosis in rotating machines has been principally done through the examination of the presence of different harmonics and sub-harmonics of the machine's rotational speed in the vibration spectrum, and this was clearly elaborated in the studies by Goldman and Muszynka [3]. Similarly, studies by Sinha [4] provided tangible details about the concepts and applications of VCM. Despite the tremendous successes that have been achieved by these conventional techniques, their requirements for numerous vibration transducers (accelerometers and/or proximity probes) at each bearing location could be overwhelming. If a large rotating machine with an appreciable number of bearings (such as some turbo-generators) is to be monitored, the number of vibration transducers needed, the data to be analyzed and the time will be enormous, which may in turn complicate the entire process of fault diagnosis. Hence, capitalizing on the recent advances achieved in the area of computational sciences may enhance the feasibility of tremendously reducing the transducer requirements at each bearing location without necessarily compromising any of the information needed for fault diagnosis.

Emerging studies have revealed the capabilities of higher order spectra (HOS), namely the bispectrum and trispectrum [5] for the diagnosis of various faults related to different rotating machines [6–11]. The greatest strength of HOS is in the fact that it achieves a combination of the various frequency components present in a signal; thereby providing the relationships between the harmonics and sub-harmonics responses of the running speed of the rotating machine, through the aid of one-point measurement per bearing [6, 12]. Through this, there exists a great possibility of reducing the amount of transducers required per bearing, during machinery vibration measurements. Therefore, this study simulates four different cases (healthy, misalignment, crack and rub) on a medium scale experimental rig; where two rigidly coupled shafts were supported on a relatively stiff support through the aid of 4 ball bearings. The vibration experiments have been conducted using 4 accelerometers for measurement (one installed at each bearing pedestal) in the horizontal direction. Hence, the spectrum, bispectrum and trispectrum have been computed. The chapter provides the results as well as the potentials of combining bispectrum and trispectrum in vibration-based fault detection and analysis in rotating machines.

2 HOS

Bispectrum [6, 13–15] and trispectrum [5, 8] are the two types of HOS [10, 11] used in the present study. The computational approaches used for the spectrum and the HOS (bispectrum and trispectrum), are briefly discussed here.

2.1 *Spectrum*

The power spectrum of a time domain signal $x(t)$, is calculated by the discrete Fourier transform (DFT) of the signal as follows;

$$S_{xx}(f_k) = E[X(f_k)X^*(f_k)], \quad k = 1, 2, 3, \ldots, N \tag{1}$$

where $S_{xx}(f_k)$ is the power density, $X(f_k)$ and $X^*(f_k)$ are respectively the DFT and its complex conjugate at frequency (f_k) for the considered time domain signal$x(t)$. N is the number of frequency points while the mathematical operator E [.] denotes the mean.

2.2 *Bispectrum*

The bispectrum on the other hand is the double Fourier transform of the third-order moment of a time signal $x(t)$, which is computed by the DFT as [14, 15];

$$B_{xxx}(f_l, f_m) = E[X(f_l)X(f_m)X^*(f_l + f_m)], \quad l + \mathrm{m} \leq N \tag{2}$$

The bispectrum gives the coupling between the frequencies at f_l, f_m and$f_l + f_m$ for the considered time domain signal $x(t)$. Assuming that the frequencies f_l andf_m denote the pth and qth harmonics of the running speed of a rotating machine respectively, then the bispectrum (B_{xxx}) component could also be written as B_{pq} [6].

2.3 *Trispectrum*

Similarly, the trispectrum represents the triple Fourier transform of the fourth-order moment of a time signal, which is computed thus [5, 8];

$$T_{xxxx}(f_l, f_m, f_n) = E[X(f_l)X(f_m)X(f_n)X^*(f_l + f_m + f_n)], \quad l + \mathrm{m} + n \leq N \tag{3}$$

Just as in the case of the bispectrum, if f_l, f_m and f_n respectively denote the *p*th, *q*th and *r*th harmonics of the running speed of a rotating machine, then the trispectrum could also be represented by T_{pqr} [5].

3 Experimental Set-up

The photographic representation of the experimental rig is shown on Fig. 1, which is situated in the Dynamics Laboratory of the University of Manchester. The rig principally consists of two rigidly coupled steel shafts of uniform diameters (20 mm) but varying lengths (1 and 0.5 m respectively), which were supported by four ball bearings mounted on relatively stiff pedestals (just as indicated by Fig. 1). The 1 m shaft is connected to the electric motor via a flexible coupling. There are three balance steel discs of dimensions 125 mm (OD) $\times$ 15 mm (thickness), with two of the discs fitted on the long shaft (first disc is 30 mm from the drive motor and the second is 19 mm from the second bearing) and the third on the shorter shaft (21 mm from both bearings 3 and 4) as shown on Fig. 1 [6].

4 Simulation of Faults

The following four cases (healthy, misalignment, cracked shaft and shaft rub) have been simulated in the experimental rig and vibration data have been collected at a constant rotational speed of 2,040 RPM (34 Hz), which corresponds to half of the first natural frequency of the rig. For all four cases, only four accelerometers (one at each bearing pedestal in the horizontal direction, due to reduced stiffness in this direction) were used for the collection of the vibration responses. All vibration data were recorded on to a PC through the aid of a 16-channel, 16-bit Data Acquisition Card (NI 6229), for subsequent signal processing using the MatLab code. Further details about the simulated faults are also available in Elbhbah and Sinha [6].

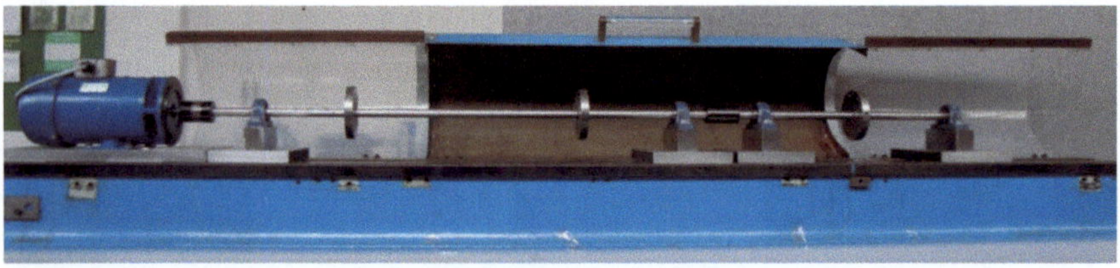

Fig. 1 Photographic representation of the experimental rig

5 Data Analysis and Results

The measured vibration data at 4 bearings have been analyzed using spectrum, bispectrum and trispectrum signal processing techniques. The FFT for all four cases (healthy, misalignment, crack and rub) were calculated using an 80 % overlap with a Hanning window, frequency resolution (df) of 0.6104 Hz, a sampling frequency (f_s) of 5,000 Hz and number of data points, $N = 8{,}192$.

It was observed in the earlier study that the bispectrum analysis shows much better diagnostic features than spectrum analysis [6]. Now, the trispectrum analysis is also introduced and then both bispectrum and trispectrum are combined for the observation of the best possible diagnostic features for different faults. The merit of HOS is that they merge different frequency components of a signal, so that the coupling between harmonics and sub-harmonics responses may generate peculiar characteristics at several running speeds that could enhance fault diagnosis. The measured vibration data from this experiment were also used to compute both the bispectrum and the trispectrum. Some of the HOS plots from two bearings (1 and 3) at a running speed of 2,040 RPM (34 Hz) are shown on Figs. 2 and 3. In the bispectrum, B_{11} signifies the coupling between 1x (twice) and 2x components of the spectrum; B_{12} signifies coupling between 1x, 2x and 3x components in the spectrum; B_{13} signifies coupling between 1x, 3x and 4x components in the spectrum; B_{22} signifies the coupling between 2x (twice) and 4x components in the spectrum; B_{ss} signifies coupling between 0.5x (sub-harmonic components twice) and 1x; and B_{s1} signifies coupling between 0.5x, 1x and 1.5x components in the spectrum. Similarly, in the trispectrum, T_{111} signifies coupling between 1x (thrice) and 3x components in the spectrum; $T_{112} = T_{211} = T_{121}$ signifies coupling between 1x (twice), 2x and 4x components in the spectrum; T_{sss} signifies coupling between sub-harmonic components 0.5x (thrice) and 1.5x components in the spectrum; $T_{ss1} = T_{1ss} = T_{s1s}$ signifies coupling between sub-harmonic components 0.5x (twice), 1x and 2x components in the spectrum. It is evident that both the bispectrum and trispectrum plots shown in the Figs. 2 and 3 provided distinctions between the four simulated cases (healthy, misalignment, crack and rub).

The research study by Elbhbah and Sinha [6] only made a comparison between the spectrum and the bispectrum, and concluded that the appearance of components B_{11} and B_{12} $(= B_{21})$ in the healthy case could be as a result of some residual rotor unbalance and little misalignment between bearings 2 and 3. Furthermore, the healthy case peaks were of significantly lower magnitudes when compared to either the misalignment or crack cases. In addition to B_{11} and B_{12} $(= B_{21})$, the misalignment case also possessed a B_{22} component. Although the crack case contained similar components as the misalignment case, it possessed an additional $B_{13} = B_{31}$ components, which were of higher magnitudes than those of the misalignment case. The rub case displayed an entirely different feature from the other three cases, owing to the fact that most of the rotor's unbalance energy have been converted to sub-harmonic responses, which was responsible for the cluster of peaks around B_{ss} and B_{s1} $(= B_{1s})$ [6].

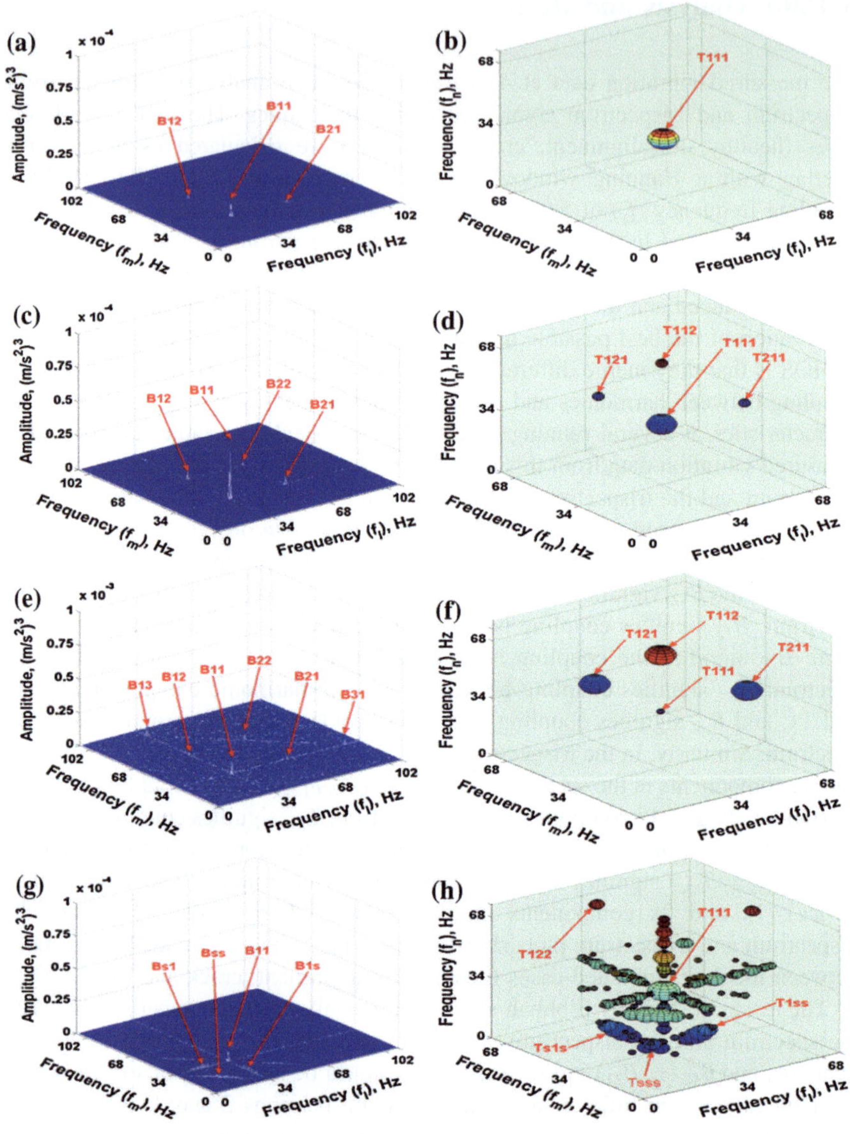

Fig. 2 Typical amplitude bispectra **a**, **c**, **e** and **g** and trispectra **b**, **d**, **f** and **h** for bearing 1; **a**, **b** (healthy), **c**, **d** (misalignment), **e**, **f** (crack) and **g**, **h** (shaft rub)

In the current study however, the trispectrum has been calculated (using Eq. 3) and included. For the trispectrum plots, the healthy case at both bearings (1 and 3) only possessed the T_{111} component, which was also adjudged to be due to residual misalignment and unbalance between bearings 2 and 3. It must be mentioned that the trispectrum responses were quite consistent across all bearings.

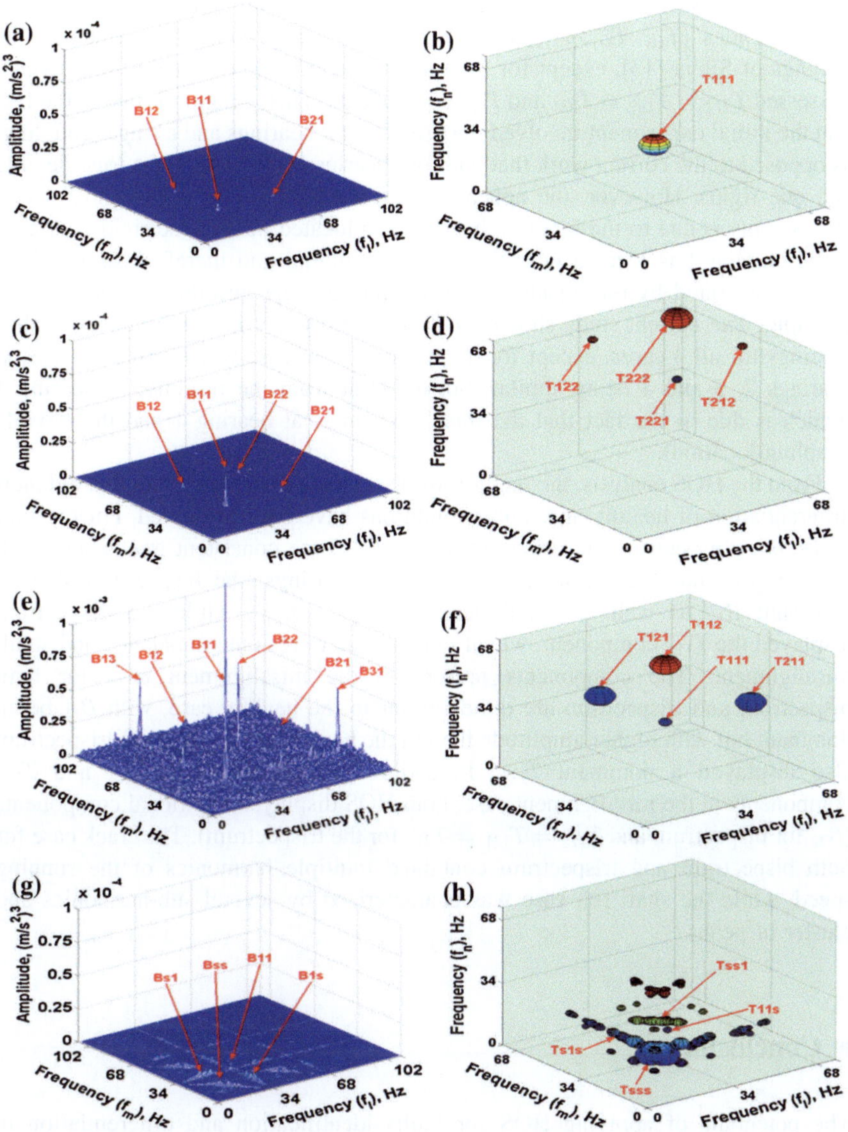

Fig. 3 Typical amplitude bispectra **a**, **c**, **e** and **g** and trispectra **b**, **d**, **f** and **h** for bearing 3; **a**, **b** (healthy), **c**, **d** (misalignment), **e**, **f** (crack) and **g**, **h** (shaft rub)

The misalignment case at bearing 1possessed components $T_{112} = T_{121} = T_{211}$ which were small and equal in size, and a large T_{111} component. Misalignment at bearing 3 showed small $T_{212} = T_{122} = T_{221}$ and a large T_{222}. The crack case possessed a response that was somewhat a reverse of the misalignment case in size, but similar in components (i.e., $T_{112} = T_{121} = T_{211}$ were large and the T_{111}

was small). Just as in the case of the bispectrum, the rub case displayed several sub-harmonics (T_{sss}, T_{ss1}, T_{s1s}, T_{1ss}, *etc.*). These observations conformed to the findings of Sinha [13], except for the misalignment response at bearing 3, which possessed $T_{212} = T_{122} = T_{221}$ and T_{222} components. This variation is due to the fact that the initial experiment involved the use of just 2 bearings and a single coupling, as opposed to the current work that had four bearings and 2 couplings (one flexible and one rigid). However, the appearance of the $T_{212} = T_{122} = T_{221}$ and $T_{222}T_{212}$ components is due to the fact that bearing 3 is located next to the rigid coupling, while bearing 1 is located near the flexible coupling and therefore some of the energy generated by the misalignment at bearing 1 are absorbed by the flexible coupling. The current study showed a strong consistency in the responses at all 4 bearings for all 4 cases, except for the misalignment case which had responses at bearings 2, 3 and 4 being similar, but different from the response at bearing 1 (which is due to the fact that the misalignment is at bearing 1 and the flexible coupling location).

From the HOS analysis, the features of bispectrum and trispectrum that aid their differentiation of healthy and faulty conditions have been presented. For the four cases, both bispectrum and trispectrum showed some consistent trends across all bearing pedestals. In the healthy case at all bearings, the bispectrum showed dominant B_{11} as well as small peaks of $B_{12}\,(= B_{21})$, while the trispectrum displayed the T_{111} component, which may be due to residual unbalance and small misalignment. The components present in the misalignment case for both bispectrum and trispectrum are quite similar to the healthy case, with B_{11} being dominant but with higher amplitude than in the healthy case, while the trispectrum also displayed a dominant T_{111}. In addition to the dominant B_{11} and T_{111} components in the misalignment case, both HOS displayed additional components (B_{22} for bispectrum and $T_{112} = T_{121} = T_{211}$ for the trispectrum). The crack case for both bispectrum and trispectrum contained multiple harmonics of the running speed, while the shaft rub case was characterized by several sub-harmonics and cluster of peaks.

6 Conclusion

The potentials of applying HOS for faults identification and differentiation in rotating machines have been explored with the experimental simulation of four cases (healthy, misalignment, crack and shaft rub) on an experimental rig. It was noticed that the HOS provided a clear distinction between healthy and faulty conditions, and also indicates the possibilities of identifying different faults using a certain combinations of bispectrum and trispectrum components. *For instance, the presence and high magnitudes of bispectrum components B_{11}, B_{12} and B_{22} could be combined with trispectrum components T_{111} and T_{112} for the identification of crack, while sub-harmonic components B_{ss} and T_{sss} are indicators of shaft rub.* Hence, the current study highlights the possibilities of eliminating the use of multiple sensors in orthogonal directions in VCM. However, the consistency of the fault classification and identification needs to be further established by simulating more faults of varying sizes and locations on different rotating rigs.

Acknowledgments Akilu Yunusa-Kaltungo wishes to thank the Petroleum Technology Development Fund (Federal Government of Nigeria) for sponsoring his PhD study.

References

1. Renwick JT, Babson PE (1985) Vibration analysis—a proven technique as a predictive maintenance tool. IEEE Trans Ind Appl IA-21(2):324–332
2. Yunusa-Kaltungo A, Sinha JK (2012) Condition monitoring: a simple and practical approach. LAP Lambert Academic Publishing GmbH & Co, Saarbrücken (dissertation converted into a book). ISBN 978-3-8473-7807-3109
3. Goldman P, Muszynska A (1999) Application of full spectrum to rotating machinery diagnostics. Orbit 20(1):17–21 (Bently Nevada Corporation, USA, 1st Quarter)
4. Sinha JK (2002) Health monitoring techniques for rotating machinery. Ph.D. thesis. University of Wales, (Swansea University), Swansea, UK
5. Collis WB, White PR, Hammond JK (1998) Higher-order spectra: the bispectrum and trispectrum. Mech Syst Sig Proc 12(3):375–394
6. Elbhbah K, Sinha JK (2012) Bispectrum: a tool for distinguishing different faults in rotating machine. In: Proceedings of ASME Turbo Expo 2012
7. Jiang L, Liu Y, Li X, Tang S (2011) Using bispectra distribution as a feature for rotating machinery fault diagnosis. Measurement 44:1284–1292
8. Sinha JK (2007) Higher order spectra for crack and misalignment identification in shaft of a rotating machine. Struct Health Monit Int J 6(4):325–334
9. Montero FEH, Medina OC (2008) The application of bispectrum on diagnosis of rolling element bearings: a theoretical approach. Mech Syst Sig Proces 22(3):588–596
10. Howard IM (1997) Higher-order spectral techniques for machine vibration condition monitoring. Proc Inst Mech Eng Part G: J Aerosp Eng 211(4):211–219
11. Wang CC, Jamestoo GP (2002) Rotating machine fault detection based on HOS and artificial neural networks. J Intell Manuf 13(4):283–293
12. Elbhbah K, Sinha JK (2012) A Composite vibration spectrum for a machine for vibration based condition monitoring. In: Proceedings for the ASME 2012 international design engineering technical conferences and computers and information in engineering conference, 12–15 August, Chicago, Illinois, USA
13. Elbhbah K, Sinha JK (2010) Bispectrum for fault diagnosis in rotating machines. In: Proceedings of international congress on sound and vibration, Cairo
14. Sinha JK, Elbhbah K (2013) A future possibility of vibration based condition monitoring of rotating machines. Mech Syst Sig Proc 34(1–2):231–240
15. Sinha JK (2009) Higher order coherences for fatigue crack detection. Eng Struct 31(2009):534–538

Signal Complexity and Gaussian Process Models Approach for Bearing Remaining Useful Life Estimation

Pavle Boškoski, Matej Gašperin and Dejan Petelin

Abstract Standard bearing fault detection features are shown to be ineffective for estimating bearings remaining useful life (RUL). In this paper we propose a new approach estimating bearing RUL based on features describing the statistical complexity of the envelope of the generated vibrations and a set of Gaussian process (GP) models. The proposed approach is shown to be sensitive to incipient condition deterioration which allows timely and sufficiently accurate estimates of the RUL. The proposed approach was evaluated on the data set comprising of 17 bearing runs with natural fault evolution.

Keywords Bearing prognostics · Statistical signal complexity · Gaussian process models

1 Introduction

Several surveys show that bearing faults represent the most common cause for failure of mechanical drives [1, 2]. As a result, a plethora of methods have been developed addressing the problems of bearing fault detection and prognostics. Most of the available methods rely on a well-established feature set, based on characteristic bearing fault frequencies. However, these features are shown to be ineffective for estimating bearing's remaining useful life (RUL) [3]. Addressing the problem of bearing fault prognostics, in this paper we propose a combination

---

P. Boškoski (✉) · M. Gašperin · D. Petelin
Jožef Stefan Institute, Jamova cesta 39, SI-1000 Ljubljana, Slovenia
e-mail: pavle.boskoski@ijs.si

M. Gašperin
Faculty of Electrical Engineering/RICE, University of West Bohemia, Plzen, Czech Republic

G. Dalpiaz et al. (eds.), *Advances in Condition Monitoring of Machinery in Non-Stationary Operations*, Lecture Notes in Mechanical Engineering,
DOI: 10.1007/978-3-642-39348-8_7,

of new features based on the statistical complexity of the envelope of bearing's vibrations and Gaussian process (GP) models for estimating bearing's RUL.

Majority of the available approaches describe the relationship between the defect growth and the trend of some statistical characteristic of the generated vibrations like energy, peak-to-peak values, RMS, kurtosis, crest factor etc. [3–5]. Usually these values are calculated from the generated vibrations filtered on specific frequency bands. The ratios of these features from various frequency bands are employed for estimating the bearings RUL. The effectiveness of these ratios can be explained through the relation between the time evolution of the excited bearing's natural frequency and the deterioration of the bearing's RUL [6–8]. Based on this assumption, bearing's RUL was estimated using approaches such as: tracking the evolution of the vibration energy using hidden Markov models [9] or tracking the increase of the dimensional exponents of the generated vibrations [10].

Following these two approaches, we propose a set of features that quantify the statistical complexity of the generated vibrations. The concept of statistical complexity is readily applied for analysis of EEG signals [11, 12]. In the context of bearing prognostics, any change in the bearing surface can be treated as a source of additional signal components with complex dynamics, hence increasing the statistical complexity of the generated vibrations. Our results show that the evolution of the statistical complexity of the generated vibrations can be directly related to the bearings RUL. Additionally, the process for calculating the statistical complexity requires no prior information about the operating conditions and no previous knowledge about the physical characteristics of the monitored drive [13, 14].

Using the Rényi entropy based statistical complexity, the bearing's RUL was estimated using GP models, which are probabilistic, non-parametric models. GP models search for relationships among the measured data rather than approximating the modelled system by fitting the parameters of the selected basis functions, which is common for other black-box identification approaches. The predictions of GP models are represented by a normal distribution. Because of their properties GP models are especially suitable for modelling when data is unreliable, noisy or missing. Their uses and properties for modelling are reviewed in [15]. In this paper the GP models are used for two purposes: filtering noisy features and estimating the RUL.

The proposed approach consists of four main steps. In first step three features, in-depth described in Sects. 2 and 3, are extracted from the acquired vibrations. The process of the numerical estimation of these features is presented in Sect. 4. In second step these features are filtered using GP models, described in Sect. 5. Afterwards, GP models are used for the estimation of RUL values based on filtered features. The final RUL estimation is obtained by fusion of all estimated RUL values. The evaluation of the approached is presented in Sect. 6.

2 Signal Complexity

The definitions of the statistical complexity of a signal varies depending of the context. In context of signals one can define two extremes: periodic and purely random signals. Both cases belong to the class of low complexity signals: the former due to its repetitive pattern and the latter due to its compact statistical description [16, 17]. Consequently, "complex" signals should be located somewhere in between, a typical candidates are signals generated by a system with chaotic behaviour.

For a random signal, generated by a random source with probability distribution $\mathscr{P}$, the statistical complexity $\mathscr{C}(\mathscr{P})$ can be assessed through the information carried by the generated signal [11]. The statistical complexity provides a link between the entropy of the source $H(\mathscr{P})$ and the "distance" between the probability distribution $\mathscr{P}$ and the uniform distribution $\mathscr{P}_e$ as [11]:

$$\mathscr{C}(\mathscr{P}) = Q_0 D_\alpha^w(\mathscr{P}, \mathscr{P}_e) H_\alpha(\mathscr{P}), \tag{1}$$

where $\mathscr{P}_e$ is the uniform distribution and Q_0 is a normalisation constant so that $Q_0 D_\alpha^w(\mathscr{P}, \mathscr{P}_e) \in [0, 1]$. The values $H_\alpha(\mathscr{P})$ and $D_\alpha^w(\mathscr{P}, \mathscr{P}_e)$ are the Rényi entropy and Jensen-Rényi divergence respectively, and are defined as [18, 19]:

$$H_\alpha(\mathscr{P}) = \frac{1}{1-\alpha} \ln \sum_{p \in \mathscr{P}} p^\alpha(x), \; a \geq 0 \, a \neq 1 \tag{2}$$

$$D_\alpha^w(\mathscr{P}, \mathscr{Q}) = H_\alpha(w\mathscr{P} + (1-w)\mathscr{Q}) - \{wH_\alpha(\mathscr{P}) + (1-w)H_\alpha(\mathscr{Q})\}. \tag{3}$$

The statistical complexity $\mathscr{C}(\mathscr{P})$ is usually plotted versus the entropy $H_\alpha(\mathscr{P})$ [11]. Such a plot always covers a specific pre-defined area, as shown in Fig. 1.

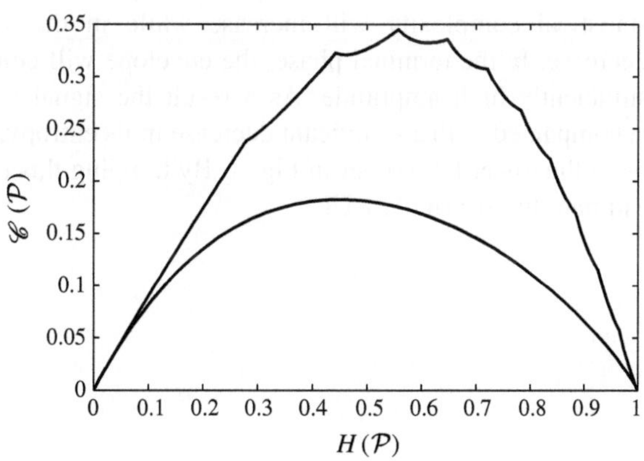

Fig. 1 Signal's statistical complexity area. The predefined shape of the plot outlines the possible time evolution of the signal's complexity. By trending the evolution of the statistical complexity within the predefined area one can perform the prognostics task

3 Complexity of Bearing Vibrations

Healthy bearings produce negligible vibrations. However, in the case of surface damage, vibrations are generated by rolling elements passing across the damaged site on the surface. Each time this happens, impact between the passing ball and the damaged site triggers a system impulse response $s(t)$. The time of occurrence of these impulse responses as well as their amplitudes should be considered as purely random processes. Consequently, the vibrations generated by damaged bearings can be modeled as [20]:

$$y(t) = \sum_{i=-\infty}^{+\infty} A_i s(t - t_i) + n(t), \tag{4}$$

where A_i is the impulse of force that excites the entire structure and t_i is the time of its occurrence. The final component $n(t)$ defines an additive random component that contains all non-modeled vibrations as well as environmental disturbances.

Generally the impulse response $s(t)$ is influenced by the transmission path from the point of impact to the measurement point [21]. As the position of the damaged spot on the bearing surface rotates the transmission path changes in time. However, regardless of its true form, $s(t)$ is charaterised by its high-frequency signature. Since this is the only characteristic relevant for our analysis, we will adopt the model (4) as sufficiently accurate one.

Evolution of the statistical complexity of the generated bearing vibrations The main diagnostic information regarding bearing faults are the time moments t_i in (4). Therefore, the usual approach is to analyze the envelope of the generated vibrations. In our case, we look for any changes in the statistical characteristics of the envelope [13].

In the case of healthy bearings, due to the lack of impacts, the envelope of the generated vibrations will be without any visible structure. Therefore, the envelope will have low complexity but high entropy, i.e. such a signal would be positioned in the lower right corner in Fig. 1. The occurrence of a surface fault will introduce some "structure" in the envelope of the generated vibrations. Consequently, its statistical complexity will increase while in the same time the entropy will decrease. In the terminal phase, the envelope will contain impulse responses with sufficiently high amplitude. As a result the signal complexity will sharply drop accompanied with a significant decrease in its entropy, hence the final position will be in the lower left corner in Fig. 1. By trending this evolution, one will be able to estimate the bearing's RUL.

4 Wavelet Based Estimation of the Statistical Complexity of the Signal Envelope

The first step in the calculation of the statistical complexity is the estimation of the PDF $\mathscr{P}$ of the envelope of the generated vibrations. Due to the link between the signal's envelope and its instantaneous power [22], the PDF is estimated through the energy distribution of the wavelet packet transform (WPT) coefficients [23]. WPT is described by a binary tree structure, as shown in Fig. 2. Each node in WPT tree with depth d_M is marked as (d,n), where depth $d = \{1,2,\ldots,d_M\}$ and $n = \{1,2,\ldots,2^d\}$ stands for the number of the node at depth d. The wavelet coefficients, in the set of terminal nodes T, contain all information regarding the analysed signal.

Each of the n nodes at level d contains N_d wavelet coefficients $W_{d,n,t}$ $t = 0,\ldots,N_d - 1$, $N_d = 2^{-d}N_s$, N_s is the sample length of the signal [24]. Using these coefficients, the portion of the signal's energy $E_{d,n}$ for each node (d,n) reads [25]:

$$E_{d,n} = \sum_{t=0}^{N_d-1} \left\| W_{d,n,t} \right\|^2 \tag{5}$$

and total signal's energy becomes:

$$E_{tot} = \sum_{\substack{t=0 \\ d,n\in T}}^{N_d-1} \left\| W_{d,n,t} \right\|^2 = \sum_{d,n\in T} E_{d,n}. \tag{6}$$

The set $\mathscr{P}^{d,n}$ expresses the contribution of each wavelet coefficient to the energy of the signal within the terminal node (d,n):

$$\mathscr{P}^{d,n} = \left\{ p_t^{d,n} = \frac{\left\| W_{d,n,t} \right\|^2}{E_{d,n}}, t = 0,\cdots,N_d - 1 \right\}. \tag{7}$$

A similar set can be defined for the contribution of the energy of each terminal node $(d,n) \in T$ in the total energy of the signal E_{tot}:

$$\mathscr{P}^{T} = \left\{ p_{d,n} = \frac{E_{d,n}}{E_{tot}},\ d,n \in T \right\}. \tag{8}$$

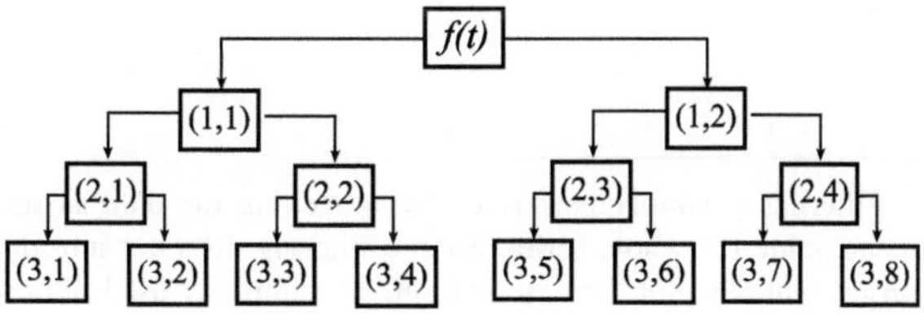

Fig. 2 Example of a full WPT tree with depth $d_M = 3$

The elements contained in both sets $\mathscr{P}^{d,n}$ and $\mathscr{P}^T$ can be treated as realisation of a random process. Based on these realisations one can estimate the corresponding probability distributions and calculate their entropies and statistical complexity according to relations (1–3).

Condition monitoring based on the statistical characteristics of the sets $\mathscr{P}^{d,n}$ and $\mathscr{P}^T$. The values of the selected features (1–3) are calculated on short non-overlapping windows, where the initial values are regarded as reference ones. As time evolves, the presence of faults will alter the envelope PDF for particular node, hence changing the feature values. As a result, RUL can be estimated by tracking the evolution of their values. It is important to stress that the window length is usually very short so that one can assume that within its duration the operating condition is almost constant. If speed varies mildly the distribution pattern will not change much as shifted harmonics will remain within the frequency band associated to a particular node. If changes in the operating speed are severe, it might happen that the frequency content from one node moves to the adjacent node, thus fooling entirely the diagnostic reasoning. On the other hand, mild variations in load normally have no significant impact on the frequency distribution pattern.

5 Gaussian Process Models

Features (1–3) based on $\mathscr{P}^{d,n}$ and $\mathscr{P}^T$ are quite noisy. Therefore we filter them using GP models.[1] Afterwards based on these filtered features, GP models are used for estimating RUL.

A Gaussian process is a collection of random variables which have a joint multivariate Gaussian distribution. Assuming a relationship of the form $y = f(\mathbf{x})$ between input $\mathbf{x}$ and output y, we have $y_1, \ldots, y_N \sim N(0, \mathbf{K})$, where $\mathbf{K}_{pq} = \mathrm{Cov}(y_p, y_q) = C(\mathbf{x}_p, \mathbf{x}_q)$ gives the covariance between output points corresponding to input points $\mathbf{x}_p$ and $\mathbf{x}_q$. Thus, the mean $m(\mathbf{x})$ and the covariance function $C(\mathbf{x}_p, \mathbf{x}_q)$ fully specify the Gaussian process.

The value of covariance function $C(\mathbf{x}_p, \mathbf{x}_q)$ expresses the correlation between the individual outputs $f(\mathbf{x}_p)$ and $f(\mathbf{x}_q)$ with respect to inputs $\mathbf{x}_p$ and $\mathbf{x}_q$. It should be noted that the covariance function $C(\times, \times)$ can be any function that generates a positive semi-definite covariance matrix. Most commonly used covariance function is a composition of the square exponential covariance function and the constant covariance function presuming white noise:

[1] Filtering is simply performed by modelling the data as time-series and then estimating the mean value for whole series. Such a filtering does not introduce any additional lag in the time series, which is not the case with other commonly used filtering methods.

$$C(\mathbf{x}_p, \mathbf{x}_q) = v_1 \exp\left[-\frac{1}{2}\sum_{d=1}^{D} w_d (x_{dp} - x_{dq})^2\right] + d_{pq} v_0, \tag{9}$$

where w_d, v_1 and v_0 are the hyperparameters of the covariance function, D is the input dimension, and $\delta_{pq} = 1$ if $p = q$ and 0 otherwise. Hyperparameters can be written as a vector $\Theta = [w_1, \ldots, w_D, v_1, v_0]^T$. Hyperparameters w_d indicate the importance of individual inputs. If w_d is zero or near zero, it means that the inputs in dimension d contain little information and could possibly be neglected.

To accurately reflect the correlations presented in the data, the hyperparameter values of the covariance function need to be optimized. Due to the probabilistic nature of the GP models, instead of minimizing the model error, the probability of the model is maximized.

Consider a set of N D-dimensional input vectors $\mathbf{X} = [\mathbf{x}_1, \mathbf{x}_2, \ldots, \mathbf{x}_N]^T$ and a vector of output data $\mathbf{y} = [y_1, y_2, \ldots, y_N]$. Based on the data $(\mathbf{X}, \mathbf{y})$, and given a new input vector $\mathbf{x}^*$, we wish to find the predictive distribution of the corresponding output y^*. Based on training set $\mathbf{X}$, a covariance matrix $\mathbf{K}$ of size $N \times N$ is determined. The overall problem of learning unknown hyperparameters q from data corresponds to the predictive distribution $p(y^*|\mathbf{y}, \mathbf{X}, \mathbf{x}^*)$ of the new target y, given the training data $(\mathbf{y}, \mathbf{X})$ and a new input $\mathbf{x}^*$. In order to calculate this posterior distribution, a prior distribution over the hyperparameters $p(\Theta|\mathbf{y}, \mathbf{X})$ can first be defined, followed by the integration of the model over the hyperparameters

$$p(y^*|\mathbf{y}, \mathbf{X}, \mathbf{x}^*) = \int p(y^*|\Theta, \mathbf{y}, \mathbf{X}, \mathbf{x}^*) p(\Theta|\mathbf{y}, \mathbf{X}) d\Theta. \tag{10}$$

The computation of such integrals can be difficult due to the intractable nature of the non-linear functions. Therefore the general practice for estimating hyperparameter values is minimising the following negative log-likelihood function:

$$\mathscr{L}(\Theta) = -\frac{1}{2}\log(|\mathbf{K}|) - \frac{1}{2}\mathbf{y}^T\mathbf{K}^{-1}\mathbf{y} - \frac{N}{2}\log(2\pi). \tag{11}$$

GP models can be easily utilised for regression calculation. Based on training set $\mathbf{X}$, a covariance matrix $\mathbf{K}$ of size $N \times N$ is calculated. The aim is to find the distribution of the corresponding output y^* for some new input vector $\mathbf{x}^* = [x_1(N+1), x_2(N+1), \ldots, x_D(N+1)]$. The predictive distribution of the output for a new test input has normal probability distribution with mean and variance

$$\mu(y^*) = \mathbf{k}(\mathbf{x}^*)^T\mathbf{K}^{-1}\mathbf{y}, \tag{12}$$

$$\sigma^2(y^*) = \kappa(\mathbf{x}^*) - \mathbf{k}(\mathbf{x}^*)^T\mathbf{K}^{-1}\mathbf{k}(\mathbf{x}^*), \tag{13}$$

where $\mathbf{k}(\mathbf{x}^*) = [C(\mathbf{x}_1, \mathbf{x}^*), \ldots, C(\mathbf{x}_N, \mathbf{x}^*)]^T$ is the $N \times 1$ vector of covariances between the test and training cases, and $k(x^*) = C(\mathbf{x}^*, \mathbf{x}^*)$ is the covariance between the test input itself.

As can be seen from (13), the GP model, in addition to mean value, also provides information about the confidence in prediction by the variance. Usually, the confidence of the prediction is depicted with $2s$ interval which is about 95 % confidence interval. This confidence region can be seen in the example in Fig. 3 as a grey band. It highlights areas of the input space where the prediction quality is poor, due to the lack of data or noisy data, by indicating a wider confidence band around the predicted mean.

6 Results

The proposed approach was evaluated on the data set for the IEEE PHM 2012 Data Challenge [26]. Provided data consist of three batches, each corresponding to to different speed and load conditions. The generated vibrations were sampled with 22 for duration of 100, repeated every 5 min. The experiments were stopped when the RMS value of the generated vibrations surpassed 20. The available vibration signals were analysed using WP tree with depth $d_M = 4$, which results into 16 terminal nodes. All features are filtered using GP models. The filtered statistical complexity $\mathscr{C}(\mathscr{P})$ for one particular node is shown in Fig. 4.

The time evolution of the statistical complexity $\mathscr{C}(\mathscr{P})$ has similar shape as the theoretical one. In time the statistical complexity evolves from the area of low complexity and high entropy towards the area of low complexity and low entropy. During this evolution the value of $\mathscr{C}(\mathscr{P})$ passes through the apex of the predefined area shown in Fig. 1.

Prediction results Using experimental runs from the first and the third batch as training set, 16 GP models were defined, one for each of the 16 WP nodes. Each GP model describes the most probable evolution of the three features $D_\alpha^w(\mathscr{P}^{d,n}, \mathscr{P}_e)$, $H_\alpha(\mathscr{P}^{d,n})$ and $H_\alpha(\mathscr{P}^T)$ in respect to the RUL normed in the interval $[0, 1]$. At each time moment, we calculate the likelihood of the bearings RUL based on the estimated GP model estimates.

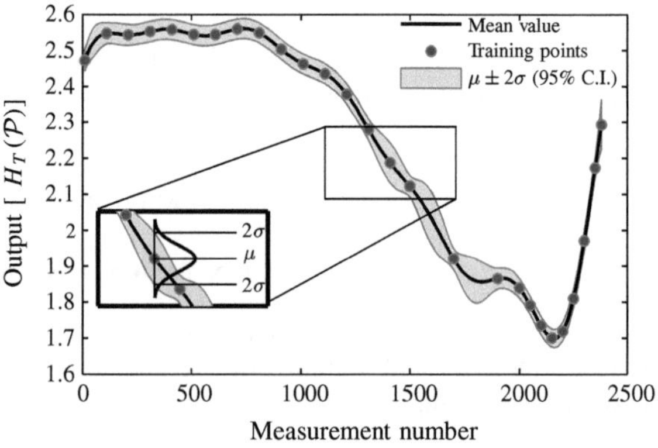

Fig. 3 Modelling with GP models: in addition to mean value (prediction), we obtain a 95 % confidence region for the underlying function f (shown in grey)

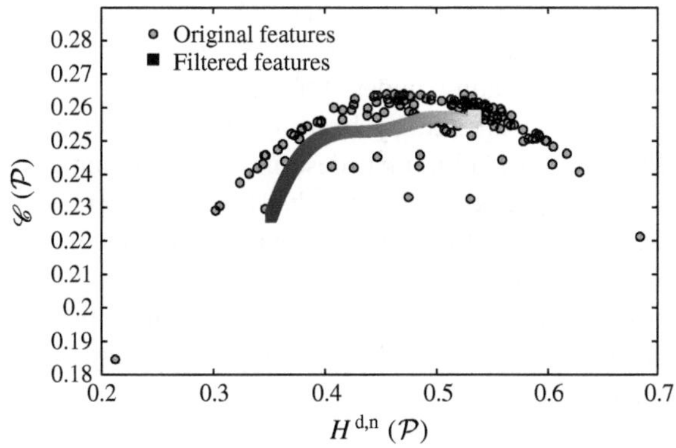

Fig. 4 Evolution of envelope complexity where shading describes the time. In the beginning the generated vibrations have low complexity $\mathscr{C}(\mathscr{P})$ and high entropy $H^{3,3}(\mathscr{P})$. In time, the statistical complexity slightly increases, where as the value of $H^{3,3}(\mathscr{P})$ decreases. At the end, both $\mathscr{C}(\mathscr{P})$ and $H^{3,3}(\mathscr{P})$ significantly decrease

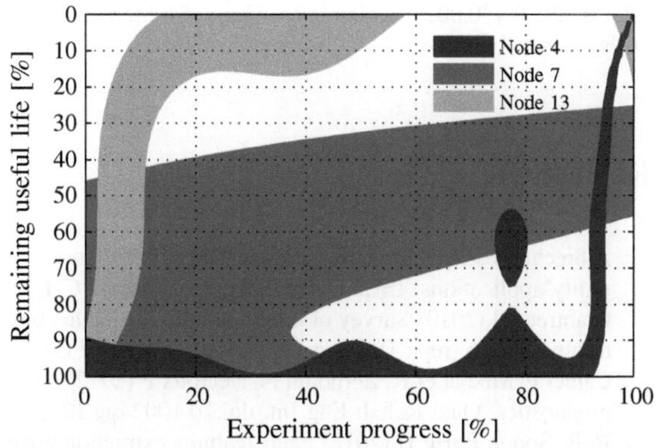

Fig. 5 RUL estimates for different WP nodes. Initial signs of deterioration are firstly detected by the high frequency nodes, like the 13 node. Low frequency nodes become sensitive towards the very end of the experiment. A proper RUL estimate should be based on the information carried by these ratios

These likelihood estimates for one bearing run are shown in Fig. 5, which shows that RUL sensitivity differs among the WP nodes. For instance, high frequency nodes, such as node 13, give first indications that the bearing has reached the end of its useful life. Conversely, RUL estimates of low frequency nodes, for instance node 4, are over-optimistic during the majority of the experiment. The actual bearing condition becomes visible only towards the end of the experiment. Such an observation leads to a conclusion that the first signs of bearing condition deterioration become visible in the high frequency parts of the signal. As the condition deteriorates sensitivity shifts towards features extracted from the lower frequency bands.

7 Conclusions

The combination of statistical complexity features coupled with Gaussian process models provides a suitable solution for estimating beating RUL. The proposed approach is generally applicable, as it requires no prior knowledge neither about the bearing physical characteristics nor about the bearing's operating condition. Therefore, the generated GP models were calculated using features extracted from bearings operating under different operating conditions and were evaluated on vibrations generated by bearings that operated under previously "unseen" conditions. These GP models describe the evolution of the selected features in respect to the bearing RUL. The results show that decrease in the bearing condition shifts the sensitivity of the features, making the features extracted from high frequency bands sensitive to initial damage and features from low frequency bands sensitive to severe damage. Consequently, this relation between condition deterioration and frequency dependent feature sensitivity can be employed for estimating the bearing RUL.

Acknowledgments We like to acknowledge the support of the Slovenian Research Agency through the Research Programme P2-0001, the Research Project L2-4160 and project EXLIZ CZ.1.07/2.3.00/30.0013, which is co-financed by the European Social Fund and the state budget of the Czech Republic.

References

1. Albrecht PF, Appiarius JC, Shrama DK (1986) Assessment of the reliability of motors in utility applications. IEEE Trans Energy Convers **EC-1**:39–46
2. Crabtree CJ (2010) Survey of commercially available condition monitoring systems for wind turbines. Tech. rep., Durham University, School of Engineering and Computing Science
3. Camci F, Medjaher K, Zerhouni N, Nectoux P (2012) Feature evaluation for effective bearing prognostics. Qual Reliab Eng Int. doi:10.1002/qre.1396
4. Li R, Sopon P, He D (2012) Fault features extraction for bearing prognostics. J Intell Manuf 23:313–321. doi:10.1007/s10845-009-0353-z
5. Lybeck N, Marble S, Morton B (2007) Validating prognostic algorithms: a case study using comprehensive bearing fault data. In: aerospace conference, 2007 IEEE, pp 1–9
6. Qiu J, Seth BB, Liang SY, Zhang C (2002) Damage mechanics approach for bearing lifetime prognostics. Mech Syst Signal Process 16(5):817–829
7. Randall RB (2011) The challenge of prognostics of rolling element bearings. In: wind turbine condition monitoring workshop
8. Wang W (2008) Autoregressive model-based diagnostics for gears and bearings. Insight 50(5):1–5
9. Ocak H, Loparo KA, Discenzo FM (2007) Online tracking of bearing wear using wavelet packet decomposition and probabilistic modeling: a method for bearing prognostics. J Sound Vib 302(4–5):951–961
10. Janjarasjitt S, Ocak H, Loparo K (2008) Bearing condition diagnosis and prognosis using applied nonlinear dynamical analysis of machine vibration signal. J Sound Vib 317(1–2):112–126
11. Kowalski AM, Martin MT, Plastino A, Rosso OA, Casas M (2011) Distances in probability space and the statistical complexity setup. Entropy 13(6):1055–1075
12. Martin M, Plastino A, Rosso O (2006) Generalized statistical complexity measures: geometrical and analytical properties. Phys A 369(2):439–462
13. Boškoski P, Juričić Đ (2012) Fault detection of mechanical drives under variable operating conditions based on wavelet packet rényi entropy signatures. Mech Syst Signal Process 31:369–381

14. Boškoski P, Juričić Đ (2012) Rényi entropy based statistical complexity analysis for gear fault prognostics under variable load. In: condition monitoring of machinery in non-stationary operations. Springer Berlin Heidelberg, pp 25–32
15. Rasmussen CE, Williams CKI (2006) Gaussian processes for machine learning. MIT Press
16. Adami C (2002) What is complexity? BioEssays 24(12):1085–1094
17. Crutchfield JP, Young K (1989) Inferring statistical complexity. Phys Rev Lett 63(2):105–108
18. Basseville M (2010) Divergence measures for statistical data processing. Tech. rep., IRISA
19. Rényi A (1960) On measures of information and entropy. In: 4th Berkeley symposium on mathematics, statistics and probability
20. Randall RB, Antoni J, Chobsaard S (2001) The relationship between spectral correlation and envelope analysis in the diagnostics of bearing faults and other cyclostationary machine signals. Mech Syst Signal Process 15:945–962
21. Antoni J, Randall RB (2003) A stochastic model for simulation and diagnostics of rolling element bearings with localized faults. J Vib Acoust 125(3):282–289
22. Antoni J (2009) Cyclostationarity by examples. Mech Syst Signal Process 23:987–1036
23. Mallat S (2008) A wavelet tour of signal processing, 3rd edn. Academic Press, Burlington
24. Percival DB, Walden AT (2000) Wavelet methods for time series analysis. Cambridge University Press, Cambridge
25. Blanco S, Figliola A, Quiroga RQ, Rosso OA, Serrano E (1998) Time-frequency analysis of electroencephalogram series iii. Wavelet packets and information cost function. Phys Rev E 57(1):932–940
26. Nectoux P, Gouriveau R, Medjaher K, Ramasso E, Morello B, Zerhouni N, Varnier C (2012) Pronostia: an experimental platform for bearings accelerated life test. In: IEEE international conference on prognostics and health management. Denver, CO, USA

Estimating Rolling Element Bearing Stiffness Under Different Operational Conditions Through Modal Analysis

William Jacobs, Rene Boonen, Paul Sas and David Moens

Abstract This paper presents a novel test rig, developed to analyse the behaviour of rolling element bearings subjected to highly varying loads. The design is optimised to measure the bearing behaviour, free from dynamics of the surrounding structure. In the current study, the test rig is used to evaluate the stiffness of a deep groove ball bearing under different operational conditions. The bearing behaviour is measured using the modal analysis technique. Then, an analytical model of the test structure is fitted on the data to estimate the bearing stiffness. The stiffness estimation is validated using a dummy bearing with a known stiffness. Finally, the stiffness of a mounted ball bearing is estimated. The paper evaluates the effect of a radial static load on the bearing stiffness. Stationary and operational conditions are compared as well. A clear difference between the stiffness of a rotating and non-rotating bearing is observed.

Keywords Rolling element bearing stiffness · Modal analysis

1 Introduction

In open literature, little information on the characteristics of rolling element bearings under dynamic conditions is available. This is probably due to the difficulty of performing a sufficiently accurate measurement and the multitude of

W. Jacobs (✉) · R. Boonen · P. Sas
Department of Mechanical Engineering, KU Leuven,
Celestijnenlaan 300B, 3001 Heverlee, Belgium
e-mail: william.jacobs@mech.kuleuven.be

D. Moens
Campus De Nayer, Department of Applied Engineering, Lessius Mechelen,
Jan De Nayerlaan 5, 2860 Sint-Katelijne-Waver, Belgium
e-mail: david.moens@mech.kuleuven.be

D. Moens
Department of Mechanical Engineering, KU Leuven, Leuven, Belgium

G. Dalpiaz et al. (eds.), *Advances in Condition Monitoring of Machinery in Non-Stationary Operations*, Lecture Notes in Mechanical Engineering,
DOI: 10.1007/978-3-642-39348-8_8,

parameters influencing the results [1]. Some information is available on the dynamics of machine tool spindles, as part of research conducted in the seventies. An overview is given in [2]. Special rigs were built for excitation and response measurement of the rotor. It was concluded that the dynamics of the bearings are mainly influenced by their preload, speed, lubricant and clearance.

Recently, a novel rolling element bearing test rig was developed at KU Leuven. The test rig is able to apply a fully controlled multi-axial static and dynamic load on the bearing. Also, different types and sizes of bearings can be mounted. As the load acting on the bearing is well-known, load identification techniques can be developed and validated. Furthermore, the influence of dynamic excitations on the lifetime of bearings can be investigated. Finally, bearing models can be validated using this test rig. In the current study, the relation between an external dynamic load and the bearing response is analysed. Also, the bearing stiffness is experimentally determined for different load and speed conditions.

This paper first introduces the test rig in Sect. 2. A full review of the test rig can be found in an earlier publication of the authors [3]. In order to understand the bearing response, the dynamics of the surrounding structure should be known. Sect. 3 of the paper therefore details about the test rig dynamics. In Sect. 4, an analytical model to estimate the bearing stiffness from modal measurements is introduced. The stiffness measurement is experimentally validated in Sect. 5, using a dummy bearing with a known stiffness. In the last section, the stiffness of a deep groove ball bearing is analysed for different operational conditions.

2 Test Rig Design

The main concept of the test rig is outlined in Fig. 1. An electric motor drives a shaft through a flexible coupling. The shaft is supported by two bearings, forming a rigid spindle. At the end of the shaft, a third bearing is mounted. This is the test bearing. The load is directly applied on the stationary outer ring of the test bearing.

The rig makes it possible to test bearings of different types, such as deep groove ball bearings and tapered roller bearings, and bearings of different sizes. Using a clamping mechanism called collet chuck, the shaft can be adjusted to fit different bearings. The test bearing is mounted on a small auxiliary shaft, adapted to its bore

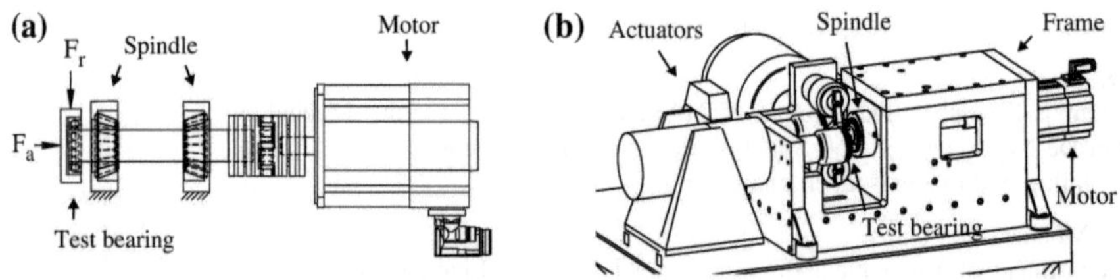

Fig. 1 Concept **a** and overview **b** of the test rig

diameter. After inserting this auxiliary shaft into the main shaft of the spindle, the locknut is tightened forming a stiff connection between both shafts. Also, using an intermediate adaptor sleeve in between the bearing and the housing, different bearings can be fitted in the housing.

The load imposed on the bearing is controlled in the radial and axial direction, independent of each other. Furthermore, the load has a static and dynamic component in both directions. In this way, it is possible to simulate different real-life situations, where i.e. gear meshing forces are acting on the bearing. Air springs apply a static force up to 10 kN, while electrodynamic shakers generate a dynamic force with an amplitude up to 1 kN and a frequency up to 500 Hz, in each direction. Figure 2 gives an overview of the actuator configuration. The static load is generated by four air springs, transferring their force to the bearing using an arm on the housing. Two air springs control the axial force ($F_{a,st}$) and two air springs control the radial force ($F_{r,st}$). The dynamic load is directly introduced on the bearing housing through the stingers of the shakers: one stinger for the axial direction ($F_{a,d}$) and one stinger for the radial direction ($F_{r,d}$).

3 Test Rig Dynamics

In order to enable a correct interpretation of the bearing measurements, the dynamics of the test rig should be analysed first. Both the housing of the test bearing and the frame of the test rig are solid structures, designed to keep the resonances of the rig outside the range of the bearing excitation up to 500 Hz. According to finite element (FE) calculations, the first flexible mode of the assembly housing and sleeve occurs at a resonance frequency of 695 Hz. The frame shows a first flexible mode at 663 Hz. It is a closed and rigid structure, mounted on four bushings. The bushings dynamically decouple the frame from the

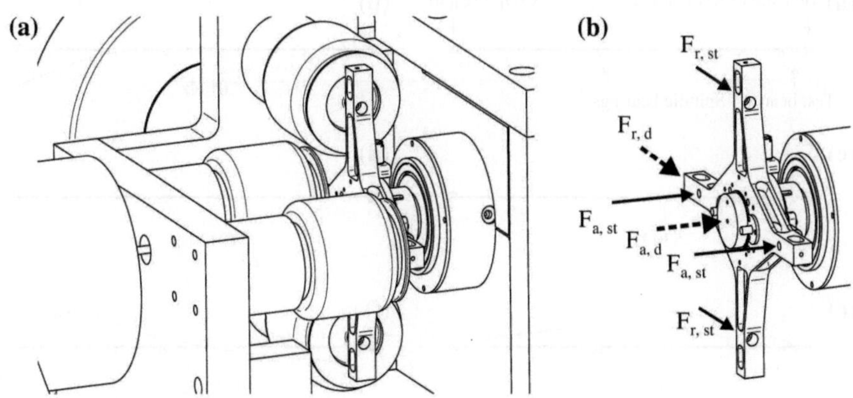

Fig. 2 Actuators **a** and their corresponding force vectors **b**

environment. The six rigid body modes of the test rig moving on its bushings are all located between 4 and 16.4 Hz. An earlier publication of the authors [4] details about the dynamics of the bearing housing and the frame, including modal analyses to validate the FE calculations.

The spindle of the test rig consists of a solid shaft, supported by two tapered roller bearings in a cylindrical housing. The tapered roller bearings are mounted in O-configuration and axially preloaded to increase the bending stiffness of the spindle. During design, a one-dimensional model of the stepped shaft is used to evaluate its dynamics. The stiffness of the bearings depends on the axial preload, and is estimated based on [6]. A radial stiffness of 510×10^6 N/m, an axial stiffness of 263×10^6 N/m and a tilt stiffness of 160×10^3 Nm/rad is used. The shaft is connected to the motor through a flexible coupling, decoupling both parts dynamically. Only the mass of the vibrating part of the coupling should be taken into account. It is modeled as a point mass. The test bearing housing is modeled as a point mass as well, as its dynamics were analysed separately. In between the bearing housing and spindle shaft, a stiffness element is added. It represents the test bearing. The value of this stiffness is set to 35×10^6 N/m, an average value of the possible test bearing stiffness's. The FE model and the first five modes are shown in Fig. 3. The third mode, at 1,012 Hz, is the rigid body mode of the shaft in which the shaft axially translates.

In order to understand the test bearing movement, the frequency response function (FRF) between an input force on the bearing housing and the displacement of the test bearing is analysed. Figure 4 shows the displacement of the housing mass (*solid line*) and the displacement of the bearing seat on the front of the shaft (*dashed line*). At 530 Hz, a strong displacement of the housing mass moving on the test bearing stiffness is observed. This resonance corresponds to the second mode of Fig. 3. It is not a spindle mode, as it only appears due to the connection with the test bearing and the housing mass. The other modes of Fig. 3

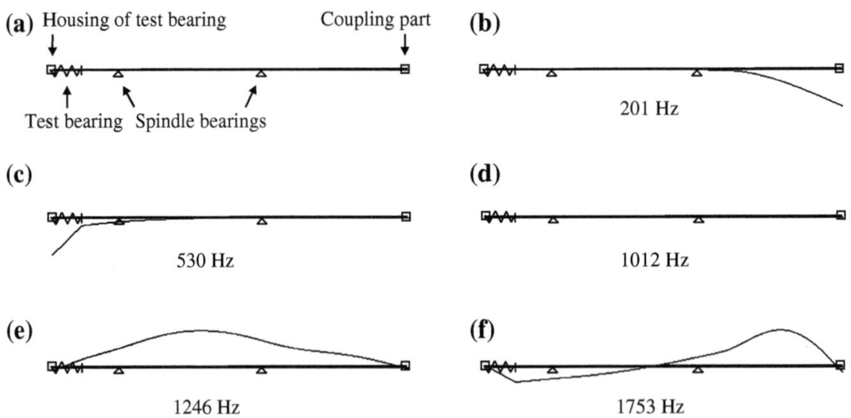

Fig. 3 Model **a** and first five modes **b–f** of the spindle and test bearing assembly

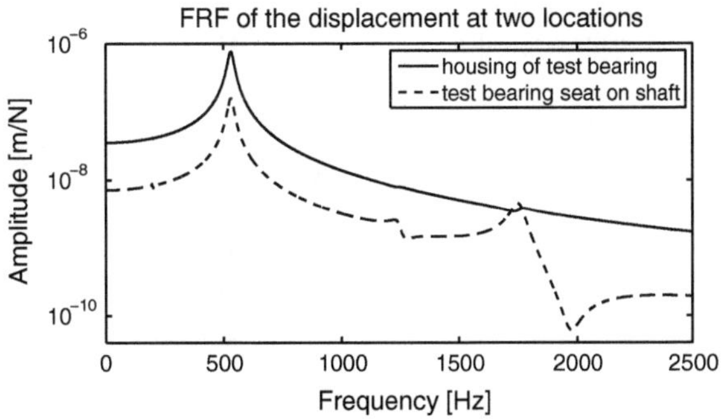

Fig. 4 FRF of the displacement at two locations of the FE model

are spindle modes. Clear radial motion of the test bearing seat on the shaft first appears at 1,753 Hz. Also recalling the third mode of Fig. 3, axial motion of the test bearing seat first appears at 1,012 Hz. Therefore, it is concluded that the bearing can be excited up to 500 Hz, without any influence of the spindle dynamics.

To validate the spindle dynamics, the modes of the shaft were experimentally measured. A shaker load is introduced on the test bearing housing, while 3 accelerometers measure the response of the shaft. As the spindle shaft is mounted in a cylindrical housing, it was only possible to install accelerometers on the front and the back of the shaft. Therefore, a full modal analysis could not be performed. Figure 5 shows the sum of FRFs of the 3 accelerometer signals. To analyse the effect of the air spring load on the spindle dynamics, the measurement was repeated for different load levels of the air springs. The FRFs are highly influenced by structural resonances of the different test rig components above 700 Hz. Also, below 150 Hz, tilt modes of the test bearing influence the FRFs. Nevertheless, the first and fifth spindle mode of the simulation, having a high response at the sensor locations, could be identified. The first mode appears at 287 Hz. The fifth mode was found at 1,899 Hz for 500 N radial load, 1,908 Hz for 1,000 N radial load and 1,915 Hz for 1,500 N radial load. As the radial load acting on the tapered roller bearings, introduced by the air springs, is very small compared to the axial preload, the stiffness of the spindle is only little affected by the air pressure. The location of this fifth mode is used in the next section to improve the estimation of the test bearing stiffness.

4 Analytic Model to Estimate the Bearing Stiffness

The FE model of the previous section shows a clear rigid body motion of the test bearing housing moving on the bearing stiffness. The frequency of this rigid body mode is mainly determined by the stiffness of the test bearing. Therefore, identification of the mode is used to estimate the bearing stiffness. In the current study,

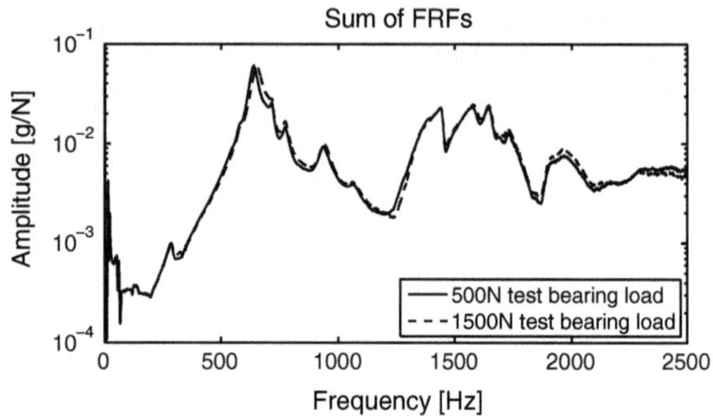

Fig. 5 Sum of FRFs of modal analysis spindle

a deep groove ball bearing 6302 is mounted. Five accelerometers are used to measure the motion of the test bearing housing. A full review of this measurement is given in [4]. In order to estimate the bearing stiffness, a model is fitted on the measured FRFs. For example an analytical model or a multi-body model can be used. Here, the radial bearing vibrations is described by the analytical model of Fig. 6. The parameters of the model are summarized in Table 1.

Each mass has one degree of freedom, namely a displacement in the vertical direction. Also, a vertical force acts on m_H, representing the shaker force. Since the housing of the test bearing has its first flexible mode at 695 Hz, it is considered as a single mass below this frequency. The value of this mass m_H is known from CAD. The spindle is represented by a single spring k_S and mass m_S. This is valid for excitations up to 1,908 Hz. Above this frequency, flexible modes of the spindle influence the results. The frame is considered as a single mass m_F for excitations up to 660 Hz, its first flexible mode. Both k_B and k_{AS} are derived from data provided by the manufacturer of the bushings and air springs. In order to determine the values of the combined stiffness k_{TB+MS}, the bending stiffness k_S and the masses m_S and m_F, the model is fitted on measured FRFs. The stiffness k_{TB+MS} is the combination of k_{TB} and k_{MS} in series. The bending stiffness k_S is considered constant for different test bearing loads, as the measured bending mode of the spindle is only little affected by this load. The masses m_S and m_F are the equivalent

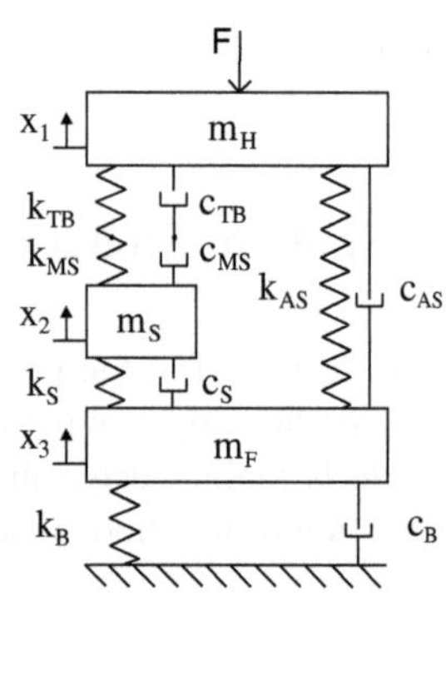

Fig. 6 Analytic model to estimate the bearing stiffness

Table 1 Parameters of the 3DOF model

Parameter	Description	Value
m_H	Mass of the test bearing housing	2.566 kg
m_S	Equivalent mass of the spindle	2.8 kg
m_F	Equivalent mass of the frame	205 kg
k_{TB}	Stiffness of the test bearing	To be determined
k_{MS}	Stiffness of the mounting system	114×10^6 N/m
k_{AS}	Combined stiffness of the air springs	650×10^3 N/m
k_S	Bending stiffness of the spindle	348×10^6 N/m
k_B	Combined stiffness of the bushings	180×10^3 N/m

masses of both the spindle and the frame, concentrated at the location of the test bearing. The simulated system has three natural frequencies, tuned according to their value as experimentally measured:

- 4.7 Hz: the motion of m_F moving on the bushing stiffness k_B.
- 626 Hz: the motion of m_H moving on the stiffness k_{TB+MS}.
- 1,908 Hz: the first bending mode of the spindle shaft exciting the bearing.

The stiffness of the modular mounting system of the test bearing, as described in Sect. 2, is combined into k_{MS}. It comprises the stiffness of the bearing housing, the intermediate sleeve, the auxiliary shaft and the collet chuck. All the connections, the line contact between the housing and the sleeve, the sleeve and the bearing, the bearing and the shaft and the chuck connection between the auxiliary shaft and spindle shaft are included as well. The stiffness k_{MS} is measured experimentally using a solid bearing, a steel disk with the same dimensions as the test bearing. Since the stiffness of the steel disk is at least a factor 10 higher than k_{MS}, k_{TB+MS} approximates k_{MS}. Therefore, k_{MS} is identified using the same modal analysis as applied to determine the test bearing stiffness, now with the steel disk mounted in the test rig and neglecting k_{TB} in the model.

Figure 7 compares the measured *(black lines)* and calculated *(grey lines)* FRFs when the test bearing is inserted. Below 150 Hz, the measured FRFs are influenced by the tilt modes of the bearing, which are not modeled. Above 700 Hz, the flexible modes of the surrounding structure appear. However, in between 150 and 700 Hz, the correspondence between the model and the measurements is good. It can be noted that the axial translational mode of the bearing influences the measurement around 300 Hz, an effect which is not observed in the simulated FRFs.

5 Validation of the Stiffness Estimation

To validate the stiffness estimation, a dummy bearing with a known stiffness is mounted in the test rig. The design of the dummy bearing is shown in Fig. 8a. It consists of an inner ring, outer ring, and a flexible structure in between.

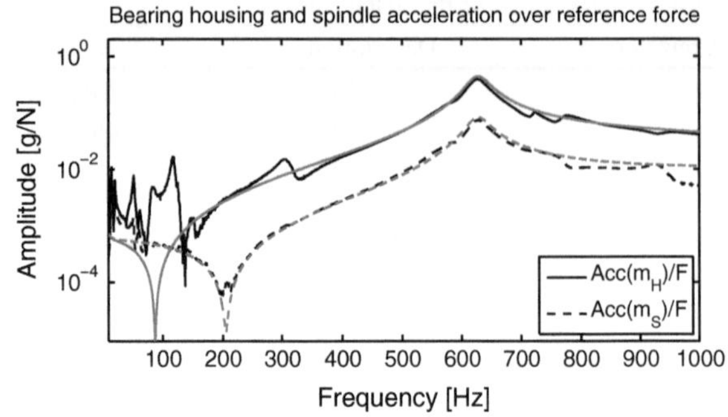

Fig. 7 Comparison between measurements and 3DOF model

The intermediate structure contains 4 leaf springs, resulting in a stiffness comparable to a normal test bearing stiffness. The stiffness is mainly determined by the thickness of the leaf springs, and accurately defined using the 2D FE model of Fig. 8b. The dummy bearing was manufactured through electric discharge machining (EDM) of a Sverker steel disk, obtaining a precision on its dimension of 0.02 mm. To take into account the variability on the geometry and the material properties, the stiffness was determined for two different models:

- Model 1: the thicknesses of the leaf springs are 0.02 mm smaller than their nominal values, the Young's modulus of the material is 200 GPa. This model serves as a lower bound for the actual stiffness, the calculated stiffness is 31.14×10^6 N/m.
- Model 2: the thicknesses of the leaf springs are 0.02 mm bigger than their nominal values, the Young's modulus of the material is 210 GPa. This model serves as an upper bound for the actual stiffness, the calculated stiffness is 34.12×10^6 N/m.

The stiffness of the dummy bearing is measured at different load levels of the air springs. The measurement is repeated to check the repeatability. The results are given in Fig. 9. The stiffness increases slightly as the load level increases. This, most likely due to an increase of stiffness in the line contacts of the mounting system, an effect which is not incorporated in k_{MS}. The grey areas indicate the

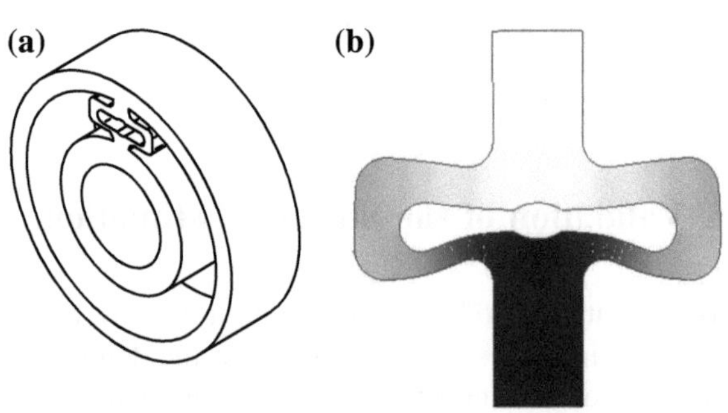

Fig. 8 Design **a** and deformation **b** of dummy bearing

upper and lower limit of the stiffness, given by the FE model. In conclusion, the estimated stiffness is within the limits of the model. Also, the repeatability of the estimation is good.

6 Bearing Stiffness Under Different Operational Conditions

In this final section of the paper, the measurements of the dynamic bearing response are combined with the 3DOF model to estimate the bearing stiffness of the mounted test bearing in different operational conditions. The influence of the radial static load on the bearing is analysed. Also, the influence of the bearing speed is investigated.

As the radial static load on a bearing increases, the stiffness of the bearing in the direction of the static load increases. More rolling elements transfer the load through the bearing and the contact surface between the rolling elements and the rings increases. Also, the stiffness of all the line contacts in the system increases. Table 2 summarizes the estimated bearing stiffness k_{TB} for different radial loads.

Next, the influence of the bearing speed is analysed. Tests at the same and different speeds are performed, and the bearing stiffness k_{TB} is estimated. The results are summarized in Table 3. It is concluded that the speed has no significant influence on the stiffness. This conclusion corresponds to the observations of [1, 5]. The spread on the stiffness estimation at different speeds is very small, and equal to the spread on the estimation at the same speeds.

Finally, the bearing stiffness of a non-rotating bearing is estimated. Table 4 shows the stiffness for three different positions of the shaft. When changing the position of the shaft, the configuration of the rolling elements in the loaded zone of the bearing changes. Therefore, a slight change in stiffness can be expected. More importantly, the stiffness of the rotating bearing is significantly lower than the stiffness of the non-rotating bearing. A mean decrease of 8.8 % is noted when the bearing is put in operation. Future research will analyse the effect on the bearing

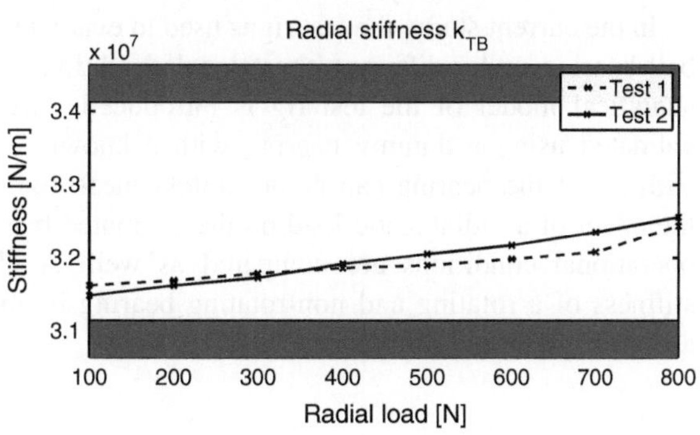

Fig. 9 Measured radial stiffness of the dummy bearing

Table 2 Influence of radial load on bearing stiffness (at 600 RPM)

Load (N)	Resonance (Hz)	Stiffness (N/m)
500	551.7	46.1×10^6
1,000	570.1	51.3×10^6
1,500	586.9	56.6×10^6

Table 3 Influence of speed on bearing stiffness (at 1,000 N radial load)

Speed (RPM)	Resonance (Hz)	Stiffness (N/m)
600	572.9	52.1×10^6
600	570.1	51.3×10^6
1,200	570.6	51.4×10^6
1,200	572.1	51.9×10^6

Table 4 Influence of shaft position on bearing stiffness (at 0 RPM and 1,000 N radial load)

Position	Resonance (Hz)	Stiffness (N/m)
1	583.5	55.5×10^6
2	586.4	56.4×10^6
3	588.0	57.0×10^6

stiffness of both the increased temperature and the formation of the lubricant film, when the bearing is put in rotation.

7 Conclusion

An innovative and versatile bearing test rig has been developed. The test rig allows easy adjustment to mount different types and sizes of rolling element bearings. The bearings can be preloaded up to 10 kN, and excited up to 500 Hz. In this way, a wide range of bearings can be tested in real-life conditions. The test rig is developed to analyse the behaviour of rolling element bearings subjected to highly varying loads. The paper shows that the design is optimised to measure the bearing behaviour, free from dynamics of the surrounding structure.

In the current study, the test rig is used to evaluate the stiffness of a deep groove ball bearing under different operational conditions. To estimate the stiffness, an analytical model of the test rig is introduced. The stiffness estimation is then validated using a dummy bearing with a known stiffness. It is shown that the stiffness of the bearing can be accurately measured. Finally, the paper evaluates the effect of a radial static load on the estimated bearing stiffness. Stationary and operational conditions are compared as well. A clear difference between the stiffness of a rotating and non-rotating bearing is observed. Future research will aim to explain this difference.

Acknowledgments This research is funded by a Ph.D. grant of the Agency for Innovation by Science and Technology (IWT). Part of this work was performed through the support of the IWT SBO-project Prognostics for Optimal Maintenance.

References

1. Kraus J, Blech J, Braun S (1987) In situ determination of rolling bearing stiffness and damping by modal analysis. J Vib Acoust Stress Reliab Design 109:235
2. Stone B (1982) The state of the art in the measurement of the stiffness and damping of rolling element bearings. CIRP Annals-Manuf Technol 31(2):529–538
3. Jacobs W, Boonen R, Sas P, Moens D (2012) The effect of external dynamic loads on the lifetime of rolling element bearings: accurate measurement of the bearing behaviour. J Phys: Conf Series 364(1):
4. Jacobs W, Boonen R, Sas P, Moens D (2012) Measuring the rigid body behaviour of a deep groove ball bearing setup. Proceedings of ISMA2012-USD2012, pp 715–726
5. Guo Y, Parker R (2012) Stiffness matrix calculation of rolling element bearings using a finite element/contact mechanics model. Mechanism and Machine Theory, pp 32–45
6. Brändlein J, Eschmann P, Hasbargen L, Weigand K (1999) Ball and Roller Bearings: Theory, Design and Application, Wiley, UK

Acknowledgements. This research is funded by a Ph.D. grant of the Agency for Innovation by Science and Technology (IWT). Part of this work was performed through the support of the IWT SBO-project Prognostics for Optimal Maintenance.

Parametric Analysis Focused on Non-linear Forces in Oil-film Journal Bearings

Andrea Vania, Paolo Pennacchi and Steven Chatterton

Abstract Many investigation methods used to identify the most common faults in rotating machines do not consider the non-linear behaviour of oil-film journal bearings with an adequate care. This chapter shows the results of a parametric analysis performed to study the sensitivity of non-linear effects in the oil-film forces to changes of some parameters of the synchronous (1X) filtered orbit of the journal. This study is focused on the influence on non-linear forces caused by changes of the maximum amplitude and circularity of the journal orbit as well as by changes of the inclination angle of the major principal axis of the 1X elliptical orbit. Moreover, also the effects of the shaft rotational speed, bearing load and the average journal position have been taken into account. A procedure to perform this sensitivity analysis for different types of journal bearing is described. Then, the results obtained by the analysis of the behaviour of a two-lobe elliptical oil-film journal bearing are shown and discussed.

Keywords Oil-film journal bearings · Non-linear dynamics · Rotating machines · Diagnostics

1 Introduction

It is well known that the oil-film forces in sleeve journal bearings are considerably non-linear. The importance of this non-linearity depends on many factors like: shaft rotational speed, journal average position, amplitude and shape of the journal orbit, oil-film temperature. The effects of the non-linear component of the oil-film forces on the shaft vibration can be rather limited as they are often smoothed by the inertia forces of the shaft. However, the occurrence of very high non-linear

A. Vania (✉) · P. Pennacchi · S. Chatterton
Department of Mechanical Engineering, Politecnico di Milano, Via G. La Masa 1, 20156 Milan, Italy
e-mail: andrea.vania@polimi.it

G. Dalpiaz et al. (eds.), *Advances in Condition Monitoring of Machinery in Non-Stationary Operations*, Lecture Notes in Mechanical Engineering,
DOI: 10.1007/978-3-642-39348-8_9,

oil-film forces can cause not negligible super-synchronous vibrations of the rotating machine. With regard to diagnostic purposes, this phenomenon can cause false fault identifications as the presence of shaft super-synchronous vibrations is often ascribed to dangerous faults like: rotor-to-stator rubs, shaft-crack propagations, severe machine misalignments.

In fact, many fault symptom analysis techniques do not give an adequate importance to the non-linear effects in the oil-film forces. Conversely, serious and common primary faults like, for instance, a high unbalance and a high shaft thermal bow, can cause unexpected super-synchronous harmonic components in the frequency spectrum of shaft vibrations. This can occur when large journal orbits generate considerable non-linear effects in the oil-film forces.

Some sleeve journal bearings are more sensitive to a non-linear behavior depending on their geometrical characteristics. The sensitivity of the oil-film forces of journal bearings to non-linear effects has been investigated, by means of a model-based method, by evaluating the oil-film forces caused by 1X filtered journal orbits having different amplitudes and shapes, that is by orbits caused only by synchronous vibrations. This approach makes the results of this study independent from the mechanical characteristics of the shaft and the characteristics of the rotor-system excitations.

This paper shows the results of a parametric analysis performed to study the sensitivity of the non-linear effects in the oil-film forces to changes of some parameters of the 1X orbit of the journal. More in particular, the influence on non-linear forces caused by changes of the orbit maximum amplitude and degree of circularity Orbit Shape Factor (OSF), as well as by changes of the inclination angle of the major principal axis of the elliptical orbit, has been considered in this study. Moreover, also the effects of the shaft rotational speed and the average journal position have been taken into account. A procedure to perform this sensitivity analysis for different types of journal bearings is described. Then, the results obtained by the analysis of the behaviour of a two-lobe elliptical oil-film journal bearing are shown and discussed.

2 Investigation Method

It is important to specify that the goal of this investigation is to study the sensitivity to non-linear dynamic effects in the oil-film forces of journal bearings, not to simulate the non-linear vibrations of the journal. Therefore, contrary to other usual methods [1, 2], the oil-film forces caused by only synchronous vibrations were considered. Let us denote by x and y the horizontal and vertical displacements of the journal with respect to the bearing centre. Then, the synchronous vibrations of the journal, which rotate with the angular velocity Ω, can be expressed as:

$$\begin{aligned} x(t) &= X\cos(\Omega t + \varphi) \\ y(t) &= Y\cos(\Omega t + \beta) \end{aligned} \tag{1}$$

As said above, the method used in this study to investigate the oil-film force does not require to define the time-history of the excitations that cause the 1X vibrations expressed by Eq. (1). In this way, the presence of not null super-synchronous oil-film forces is a symptom of non-linear effects.

The vibrations expressed by Eq. (1) cause the journal to describe an elliptical orbit having a major and minor principal axes $\bar{a}_{1X}$ and $\bar{b}_{1X}$, respectively. The parameter $\bar{a}_{1X}$ is the maximum vibration amplitude of the journal. Then, let us denote by θ the inclination angle of the major principal axis with respect to the horizontal axis x. In the case of null vibration amplitudes, the components of the journal position, x_0 and y_0, are determined by the static bearing load, W, whose horizontal and vertical components are W_x and W_y, respectively. The degree of circularity of the journal orbit, which is often elliptical, can be expressed by the ratio OSF between the minor and major principal axes of the orbit.

For each given set of values of Ω, W_x and W_y, the parameters $\bar{a}_{1X}$, OSF and θ can be varied inside suitable ranges of interest. For each given couple of the parameters OSF and θ, the changes of the maximum vibration amplitude $\bar{a}_{1X}$ cause changes of the orbit dimension without modifying the shape factor. For each couple of the parameters $\bar{a}_{1X}$ and θ, the changes of the OSF cause variations of the flatness of the elliptical orbit. In the end, for each couple of parameters $\bar{a}_{1X}$ and OSF, the changes of the angle θ cause variations of the direction along which the maximum amplitude of the journal vibration occurs.

In general, the increase of the parameter $\bar{a}_{1X}$ causes an increase of the non-linear effects in the oil-film forces. However, the importance of these effects can be highly influenced also by the orientation of the major principal axis of the orbit and by the orbit flatness.

For a given bearing load the minimum thickness h_{min} of the oil-film depends on the shaft rotational speed. Low values of h_{min} can cause a considerable increase of the non-linear effects in the oil-film forces. The lowest instantaneous value of the minimum thickness of the oil-film that occurs during a complete revolution of the shaft is affected by the parameters $\bar{a}_{1X}$, OSF and θ.

For each set of these parameters the oil-film forces were evaluated, over an entire orbit, for N samples (with $N = 256$) equally spaced in the time. For each analysis, the average position of the journal was varied by applying an iterative technique in order to obtain mean values of the horizontal and vertical oil-film forces that equilibrate the corresponding components of the bearing load. Then, the harmonic content of the oil-film forces was evaluated.

Being the journal vibration synchronous (1X), the presence of not null super-synchronous oil-film forces is a symptom of non-linear effects. In general, the amplitude of the 2X harmonic component is the highest in the super-synchronous oil-film forces.

It is important to consider that, in general, the maximum 1X and 2X oil-film forces do not occur in the same direction. Moreover, the direction of the maximum nX oil-film force does not necessarily coincide with that of the maximum vibration amplitude. Therefore, in order to make independent the results of this study from the directions along which the oil-film forces were estimated, the maximum and minimum amplitude, $\bar{F}_{a_{n\mathrm{X}}}$ and $\bar{F}_{b_{n\mathrm{X}}}$, of the nX oil-film forces were evaluated.

The oil-film forces can be obtained on the basis of the evaluation of the oil-film pressure, p (s, z), on the bearing: where s is the coordinate in the circumferential direction of the i-th bearing lobe and z is the coordinate in the axial direction. For finite length bearings, the pressure distribution $p(s, z)$ is given by the following Reynolds equation [3–5]:

$$\frac{\partial}{\partial s}\left(h^3\frac{\partial p}{\partial s}\right)+\frac{\partial}{\partial z}\left(h^3\frac{\partial p}{\partial z}\right)=6\,\mu\left(U\frac{\partial h}{\partial s}+2\frac{\partial h}{\partial t}\right) \tag{2}$$

where μ is the lubricant viscosity, h (s, z) is the oil-film thickness, and U is the circumferential velocity of the journal. The oil-film forces were evaluated by solving the Reynolds Eq. (2), numerically, by means of a finite difference method [3–5]. With regards to this, classical boundary conditions were used: that is null pressure in the inlet and outlet grooves, as well as null lateral pressure. Moreover, the following classical Reynolds conditions were used in the cavitation regions:

$$\frac{\partial p}{\partial s}=0 \quad p=0 \tag{3}$$

3 Case Study

This chapter shows the results obtained by the analysis of the behaviour of an elliptical oil-film journal bearing (Fig. 1). The main geometrical characteristics of this bearing are reported in Table 1. Owing to the short length of the chapter only some of the results obtained for a shaft rotational speed of 1,000 rpm are shown.

In order to consider dimensionless quantities the amplitude of the major and minor elliptical orbits were divided by the assembled radial clearance C_b of the bearing. That is:

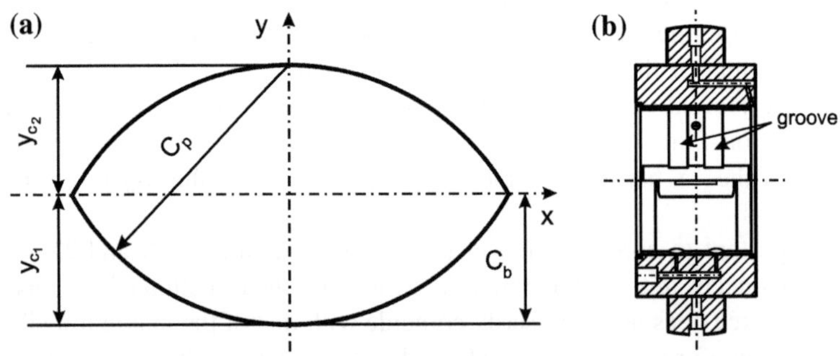

Fig. 1 **a** Configuration of an elliptical oil-film journal bearing. **b** Section view of a real bearing

Table 1 Bearing characteristics

	Symbols	
Diameter	D	457 mm
Length	L	254 mm
Machined clearance	C_p	0.608 mm
Assembled clearance	C_b	0.304 mm
Lower lobe width	$\Delta\theta$	160°
Pre-load factor	m_p	0.5
Oil viscosity	μ	26.57 cSt
Bearing load (vertical)	W	200 kN

$$a_{1X} = \bar{a}_{1X}/C_b \, b_{1X} = \bar{b}_{1X}/C_b \quad (4)$$

A vertical bearing load of 200 kN was considered in this study. The amplitude of the parameter a_{1X} was varied from 0.05 to 0.4 while the OSF was varied from 0.1 to 0.5. In the end, the inclination angle θ of the major principal axis a_{1X} was varied from 5° to 55°.

For each case study, the oil-film forces were evaluated over a complete shaft revolution. Then, the harmonic content of these forces was computed. Afterwards, the maximum amplitude of the 1X and 2X oil-film forces generated during a complete orbit, $\bar{F}_{a_{1X}}$ and $\bar{F}_{a_{2X}}$, were evaluated.

Figure 2 shows some 1X elliptical orbits that have been obtained by changing only the maximum vibration amplitude a_{1X}. The clearance profile shown in Fig. 2 delimits the area of the bearing inside which the journal position can be contained without causing any contacts between shaft and Babbit metal. These orbits do not have the same centre because the increase of the orbit dimension causes changes also of the mean value of the horizontal and vertical oil-film forces that must

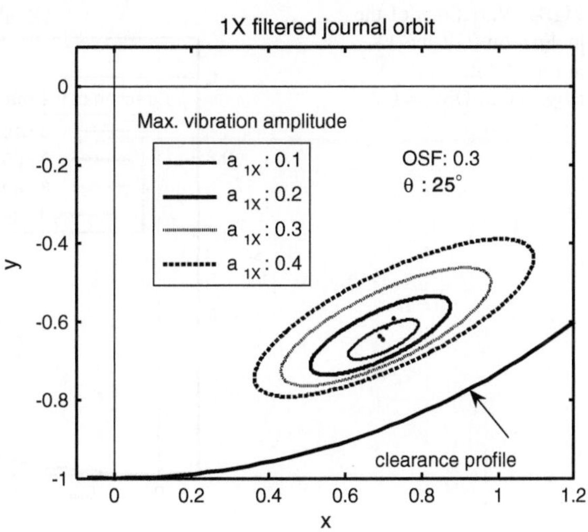

Fig. 2 Variations of the maximum amplitude of the 1X journal orbits (OSF = 0.3, $\theta = 25°$)

equilibrate the assigned bearing load. However, all these orbits have the same shape factor and the same orientation.

Figure 3 shows some 1X elliptical orbits that have been obtained, for a given couple of values of the parameters a_{1X} and θ, by changing the OSF.

Also in this case, the journal orbits do not have the same centre, owing to the above mentioned reasons. In the end, Fig. 4 shows some 1X elliptical orbits that have been obtained, for a given couple of values of the parameters a_{1X} and OSF, by changing the inclination θ of the major principal axis.

In order to manage dimensionless quantities, the maximum amplitude of the 1X and 2X oil-film forces was divided by that of the bearing load W. That is:

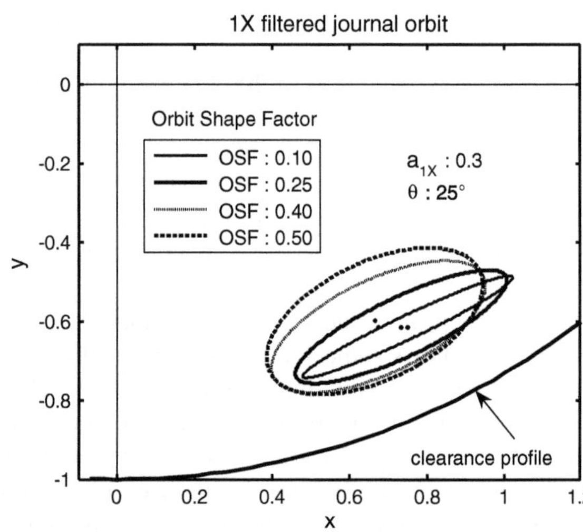

Fig. 3 Variations of the shape factor of the 1X journal orbits ($a_{1X} = 0.3$, $q = 25°$)

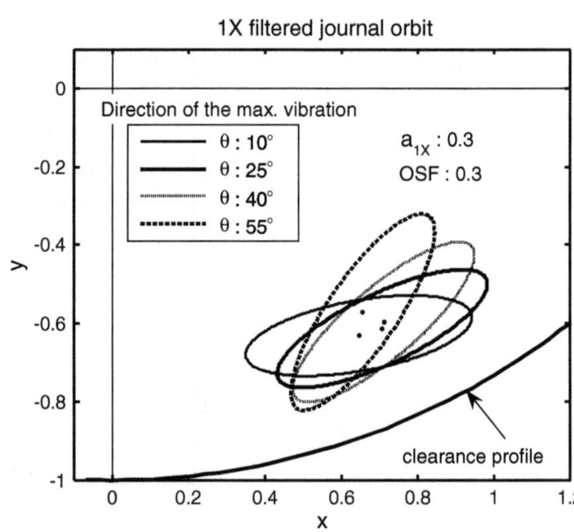

Fig. 4 Variations of the inclination θ of the 1X journal orbits ($a_{1X} = 0.3$, OSF $= 0.3$)

$$F_{a_{1X}} = \bar{F}_{a_{1X}}/W \quad F_{a_{2X}} = \bar{F}_{a_{2X}}/W \tag{5}$$

Figures 5 and 6 show the influence of the parameters a_{1X} and OSF on the amplitude of the 1X and 2X oil-film forces, $F_{a_{1X}}$ and $F_{a_{2X}}$, respectively.

These results were obtained for a value of the angle θ equal to 20°. The results illustrated in Fig. 5 show that, especially for the highest values of a_{1X}, the decrease of the orbit flatness causes an important increase of the 1X oil-film forces.

The results illustrated in Fig. 6 show that increasing values of a_{1X} cause an increase of the sensitivity of the 2X oil-film forces to the OSF.

Moreover, it is possible to find a limited range of values of the OSF parameter inside which the 2X oil-film forces, that is the non-linear effects, are minimised. Figures 7 and 8 show the influence of the parameters OSF and θ on the amplitude of the 1X and 2X oil-film forces, $F_{a_{1X}}$ and $F_{a_{2X}}$, respectively.

These results were obtained for a value of the parameter a_{1X} equal to 0.3. The results illustrated in Figs. 7 and 8 show that the direction of the maximum

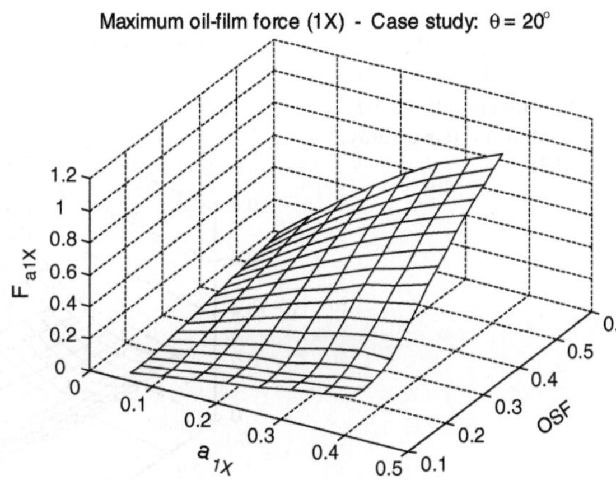

Fig. 5 Influence of the parameters a_{1X} and OSF on the maximum magnitude of the 1X oil-film force (case study: $\theta = 20°$)

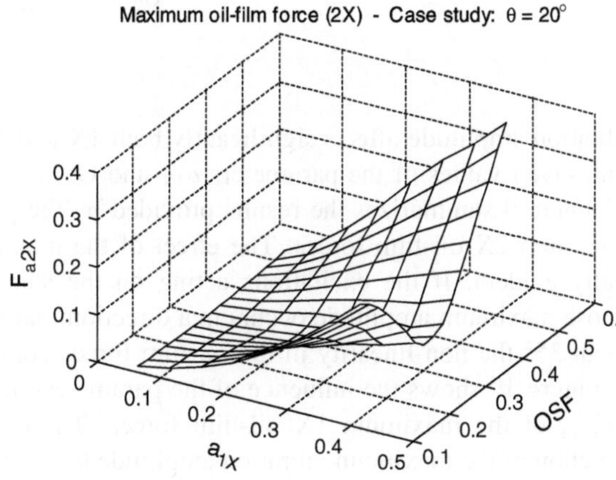

Fig. 6 Influence of the parameters a_{1X} and OSF on the maximum magnitude of the 2X oil-film force (case study: $\theta = 20°$)

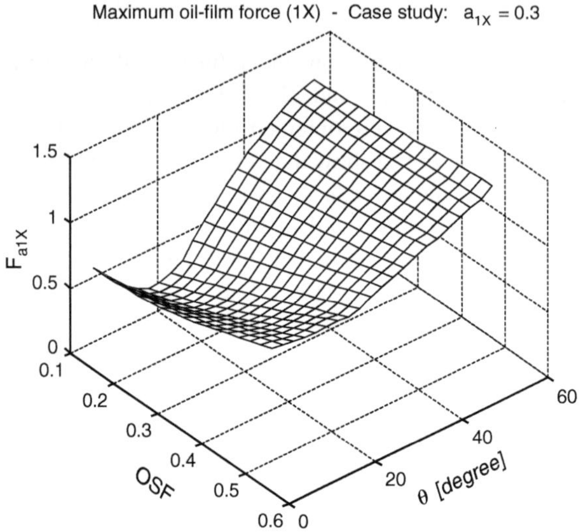

Fig. 7 Influence of the parameters OSF and θ on the maximum magnitude of the 1X oil-film force (case study: a_{1X}= 0.3)

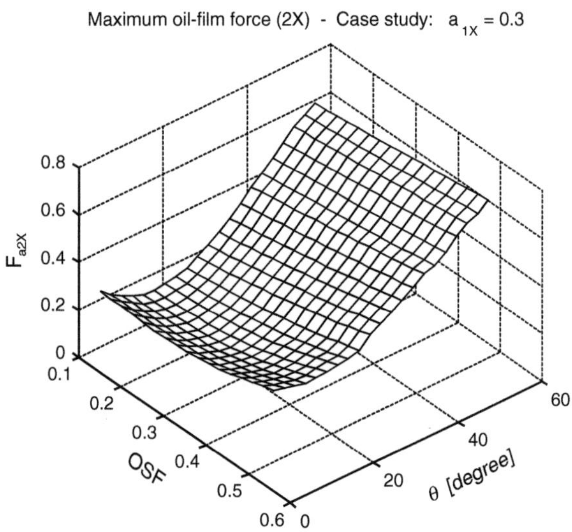

Fig. 8 Influence of the parameters OSF and θ on the maximum magnitude of the 2X oil-film force (case study: $a_{1X} = 0.3$)

vibration amplitude affects significantly both 1X and 2X oil-film forces associated with given values of the parameters a_{1X} and OSF.

Figure 9 summarizes the results provided by the parametric analysis about the maximum 2X oil-film forces. The effect of the maximum vibration amplitude is really evident. If the excitations acting on the shaft cause large journal orbits whose maximum amplitude occurs in a direction that forms an angle θ higher than about 35° the non-linearity in the oil film forces considerably increases.

Figure 10 shows the influence of the parameters a_{1X} and OSF on the direction, $\beta F_{a_{1X}}$, of the maximum 1X oil-film forces. These results were obtained for a direction of the maximum vibration amplitude $\theta = 30°$. Although the amplitude of

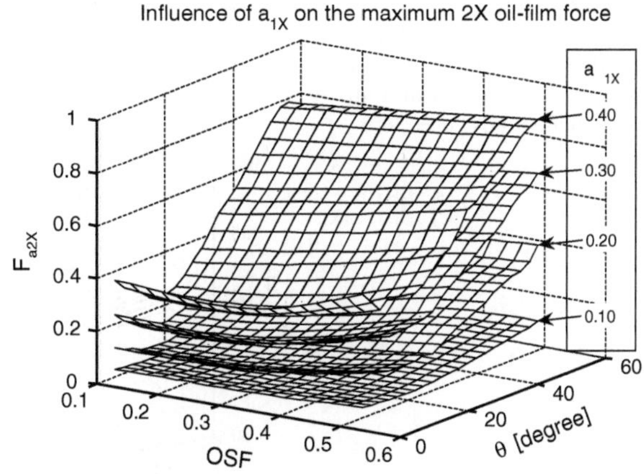

Fig. 9 Influence of the parameters a_{1X}, OSF and θ on the maximum magnitude of the 2X oil-film force

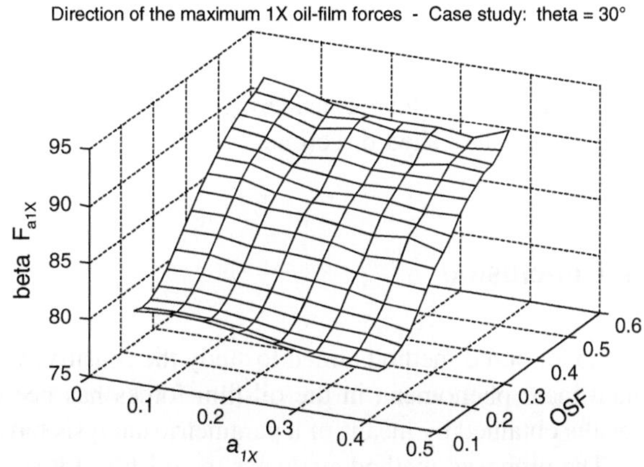

Fig. 10 Influence of the parameters a_{1X} and OSF on the direction of the maximum magnitude of the 1X oil-film force (case study: $\theta = 30°$)

the horizontal 1X vibration of the journal is higher than the corresponding vertical one, the direction of the maximum magnitude of the 1X oil-film forces is nearly vertical. This is a consequence of the considerable anisotropy of the oil-film stiffness that characterized the elliptical journal bearings.

Similarly, Fig. 11 shows the influence of the parameters a_{1X} and OSF on the direction, $\beta F_{a_{2X}}$, of the maximum 2X oil-film forces. Also in this case study, these results were obtained for a direction of the maximum vibration amplitude $\theta = 30°$. Owing to the shape and mechanical characteristics of the oil-film that is generated when the journal moves along the lower arc of the orbit, the direction of the maximum magnitude of the 2X oil-film forces, caused by the non-linear effects, is nearly vertical.

These results show that the most important non-linear effects, here represented by the maximum magnitude of the 2X oil-film forces, not necessarily occur in the same direction along which the highest journal vibration occurs. This is the reason for which, in this investigation, the authors decided to take into account both shape and amplitude of the 1X journal orbit to evaluate the harmonic content of the oil-

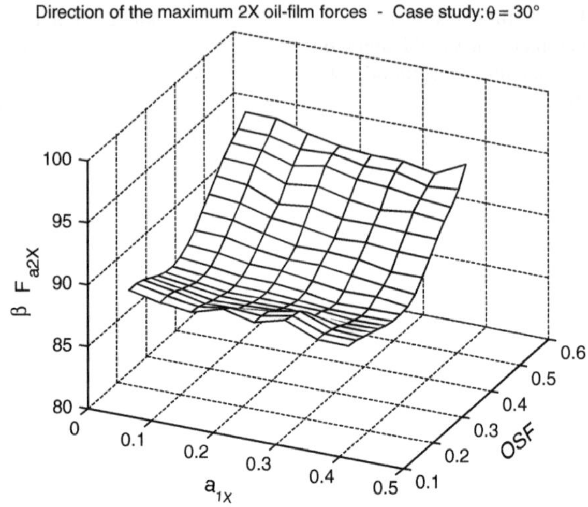

Fig. 11 Influence of the parameters a_{1X} and OSF on the direction of the maximum magnitude of the 2X oil-film force (case study: $\theta = 30°$)

film forces rather than considering the maximum vibration level evaluated in a single pre-established direction.

4 Conclusion

A model-based method aimed to study the sensitivity of sleeve journal bearings to non-linear phenomena in the oil-film forces has been shown in the chapter. The results obtained by means of a parametric analysis have been shown and discussed.

The proposed method evaluates the oil-film forces caused only by synchronous vibrations, that is by 1X elliptical orbits. The influence of the basic parameters that define these orbits on the synchronous and super-synchronous oil-film forces have been investigated. The results of this parametric analysis have been shown and discussed.

The method proposed by the authors to evaluate the sensitivity of a fluid-film journal bearing to give rise to non-linear effects in the lubricant forces has shown to be able to point out the contribution of single factors to the non-linear phenomena. The results provided by this investigation method can be very useful for diagnostic purposes and for an optimization of the bearing design aimed at improving its performances.

References

1. Vania A, Pennacchi P, Chatterton S (2012) Analysis of the sensitivity to non-linear effects in the oil-film forces of journal bearings. In: Proceedings of 10th IMECHE international conference on vibrations in rotating machinery, London, UK, C1326-037
2. Bachschmid N, Pizzigoni B, Tanzi E (2000) On the 2xrev—vibration components in rotating machinery excited by journal ovalization and oil-film non-linearity. In: Proceedings of 7th IMECHE international conference on vibrations in rotating machinery, University of Notthingam, Nottingham, UK, C576-083, pp 449–458
3. Hori Y (2006) Hydrodynamic lubrication. Springer, Tokyo
4. Stachowiak GW, Batchelor AW (2005) Engineering tribology. Butterworth Heinermann, Burington, USA
5. Szeri AZ (1998) Fluid film lubrication—theory and design. Cambridge University Press, Cambridge

Diagnostic of Rolling Element Bearings with Envelope Analysis in Non-Stationary Conditions

Pietro Borghesani, Roberto Ricci, Steven Chatterton and Paolo Pennacchi

Abstract In the field of rolling element bearing diagnostics, envelope analysis has gained in the last years a leading role among the different digital signal processing techniques. The original constraint of constant operating speed has been relaxed thanks to the combination of this technique with the computed order tracking, able to resample signals at constant angular increments. In this way, the field of application of this technique has been extended to cases in which small speed fluctuations occur, maintaining high effectiveness and efficiency. In order to make this algorithm suitable to all industrial applications, the constraint on speed has to be removed completely. In fact, in many applications, the coincidence of high bearing loads, and therefore high diagnostic capability, with acceleration-deceleration phases represents a further incentive in this direction. This chapter presents a procedure for the application of envelope analysis to speed transients. The effect of load variation on the proposed technique will be also qualitatively addressed.

Keyword Bearing diagnostics · Envelope analysis · Computed order tracking · Jitter · Squared envelope spectrum

1 Introduction

The history of models of vibration signals coming from damaged rolling element bearings starts with the famous and successful study of McFadden [1]. He presented the vibration signal as amplitude modulated Dirac comb in convolution with the transfer function of the signal. The Dirac comb, representing the impacts of the rolling elements on the damaged race, was modelled as deterministically periodic, with frequency depending on the location of the defect (e.g. Ball Pass

P. Borghesani · R. Ricci · S. Chatterton · P. Pennacchi (✉)
Department of Mechanical Engineering, Politecnico di Milano,
Via G. La Masa 1 20156 Milan, Italy
e-mail: paolo.pennacchi@polimi.it

G. Dalpiaz et al. (eds.), *Advances in Condition Monitoring of Machinery in Non-Stationary Operations*, Lecture Notes in Mechanical Engineering,
DOI: 10.1007/978-3-642-39348-8_10, © Springer-Verlag Berlin Heidelberg 2014

Frequency Inner for an inner ring damaged bearing). This model has been refined and developed in the works of Randall et al. [2, 3], introducing the important concept of *jitter*, i.e. a stochastic delay/anticipation of each impact with respect to the average periodic behaviour, and consequently identifying the strong 2nd order cyclostationary nature of this signal. A further step forward is due to Antoni et al. [4], finally proposing a domain change for the description of the phenomenon, from the time to the angular domain. This allowed extending the analysis of damaged bearings to the cases with speed fluctuations around an average rotational speed. To perform the digital signal processing of such signal, a coupling of envelope analysis and computed order tracking (COT) was proposed. Envelope analysis has gained in the last decades a leading role in the diagnostics of rolling element bearings, owing to its simplicity and low computational effort. In particular, the squared envelope spectrum (SES), belonging to the family of cyclostationary analysis tools [5], is prevailing against equally effective indicators, thanks to the possibility of developing sound statistical tests for the presence of damage symptoms.

The coupling with COT techniques, developed for other applications [6, 7, 8] has relaxed the SES constraints on the constancy of speed, allowing its implementation in many industrial applications where uncontrolled fluctuations occur around an average rotational frequency of the bearing shafts. The traditional approach is to apply the envelope analysis to an angular resampled vibration signal [9]. The effectiveness of such algorithms has been proved thoroughly in the case of small speed fluctuations [10], however, the need for more robust, automatic algorithms, suitable for different operating conditions, is pushing for a more general relaxation of the constraint on speed constancy.

In particular, it happens often that load on the bearings is high only when the machine is either accelerating or braking, as for instance it happens in the traction system of a train. As demonstrated in [10], on the one hand high bearing load increases the power of the vibration signals components indicating a fault, making its symptoms clearer, on the other hand it reduces their sensitivity to deviations of the characteristic bearing frequencies.

The aim of this chapter is therefore to describe a procedure, based on SES and COT, for the analysis of bearing vibration signals in highly variable operating conditions. The chapter will start introducing some modifications to the aforementioned models, in order to understand the advantages of the newly proposed technique Finally, in order to verify the sensitivity of this method to the typical problems of real vibrations [11], an experimental test will be presented.

2 Modelling a Damaged Bearing Signal: Domain Definition

The last two models described in the introduction show a single difference: the domain in which the signals are defined. In particular the model developed by Randall et al. [2] describes the vibration signal $x(t)$ in the time domain:

$$x(t) = \sum_{i=-\infty}^{+\infty} A_i s(t - iT - \tau_i) + n(t) \tag{1}$$

The so modeled signal, except for the environmental noise $n(t)$, is represented by a repetition of the impulse response of the structure $s(t)$, triggered and A_i-modulated by the ith impact of the rolling elements with the fault. The train of impacts is characterized by an average expected period T and a variable deviation (delay/anticipation) τ_i, called *jitter*, different for each ith impact.

On the contrary Antoni et al. [4] propose a full angular domain definition of the same quantities:

$$x(\theta) = \sum_{i=-\infty}^{+\infty} A_i s(\theta - i\Theta - \psi_i) + n(\theta) \tag{2}$$

where Θ is the average angular period of two consecutive impulses and ψ_i the equivalent jitter in the angular domain.

Even if it might seem that the last formulation completely releases the model from the rotational speed of the machine, actually both the models are valid only in case of almost constant speed. If on the one hand a variable speed will clearly invalidate the first model, introducing a bias in the time domain periodicity of the impulse train, on the other hand the same variation of speed will distort the replicas of the angular domain defined transfer function $s(\theta)$ in the second model.

In case of significant speed variations, the physical dependence of the transfer function on time (and not on the angle) leads to the necessity of defining different $s_i(\theta)$ for each impact, since the relationship $\theta = \theta(t)$ varies with the speed.

A more suitable signal model for highly variable speed is thus the following:

$$x(t) = \sum_{i=-\infty}^{+\infty} A_i s(t - T_i) + n(t) \quad \text{with} \quad T_i = \sum_{k=1}^{i} \frac{\Theta}{\bar{\omega}_k} + \psi_i \tag{3}$$

where $\bar{\omega}_k$ is the average angular speed in between the $(k-1)$th and the kth impact. Therefore, the impulse train is effectively dominated by an average angular period Θ, while the impulse response is defined in the time domain, where it keeps the same form for each impact.

3 Combining SES and COT in Case of Variable Speed

A simple efficient algorithm to obtain a squared envelope spectrum (SES) is based on the process on Fig. 1a, i.e. the *traditional squared envelope spectrum* (T-SES) procedure. It combines the COT and SES in sequence, in particular transforming the raw vibration signal in angular domain, and then obtains the SES by band-pass filtering and squaring the resampled signal.

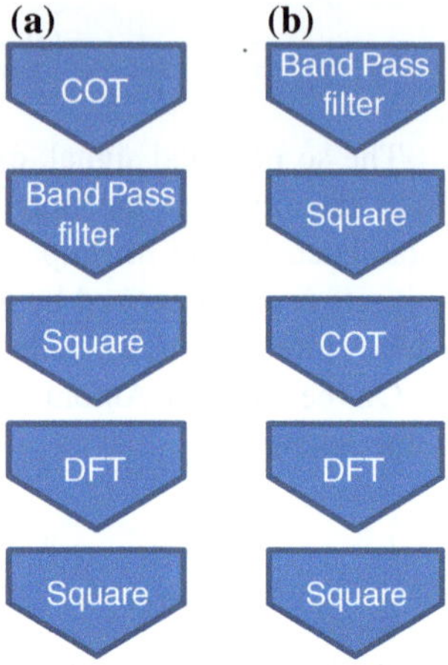

Fig. 1 **a** T-SES procedure, valid in case of small fluctuations of operating speed. **b** RS-SES procedure, for highly variable speed (transients)

However, if applied to harsh transients, the quality of the results can decrease drastically. This is a consequence of the fact that this procedure relies on the signal model described in Eq. (2), which is not anymore representative of the real behavior of the system in strongly variable speed conditions. A more suitable solution can be obtained on the basis of the signal model of Eq. (3), which considers the impulse response function periodic in the time domain and, on the contrary, defines the impulse train in the angular domain. Therefore the authors propose that the demodulation is performed directly in the time domain, to later apply COT on the so obtained envelope. This procedure, defined as *reversed sequence squared envelope spectrum procedure* (RS-SES) is synthesized in Fig. 1b. A potential drawback of the technique is the absence of a step of synchronous averaging for pre-whitening, which can be in any case substituted by alternative techniques like cepstrum pre-whitening [12]. The indexes obtained by means of RS.SES are, in terms of statistical properties, equivalent to the traditional ones, and can indergo to same statistical tests [13].

4 The Effect of Load

The speed is not the only factor affecting the dynamics of a damaged rolling element bearing. As already shown in [14], a variation of load on the bearings often induces a deviation of the bearing characteristic frequencies. In particular, for low loads, the bearing characteristic frequencies measured experimentally can differ significantly from the corresponding values calculated analytically on the basis of the geometry of the bearing. As highlighted in Table 1, the link with the fundamental train frequency (*FTF*) drives the behavior of all the bearing characteristic frequency, under the hypothesis, verified experimentally, that *FTF* is always lower than its theoretical value.

Table 1 Deviations of bearing characteristic frequencies in case of outer ring fixed on the support and rotating inner ring

Characteristic frequency	Acronym	Dependence on FTF	Expected deviation
Fundamental train frequency	*FTF*	*FTF*	Lower
Ball spin frequency	*BSF*	$\left(1+\frac{PD}{BD}\right)FTF$	Lower
Ball pass frequency outer	*BPFO*	$NB \cdot FTF$	Lower
Ball pass frequency inner	*BPFI*	$NB(1 - FTF)$	Higher

A powerful tool for a first assessment of the significance of the deviation phenomenon on a bearing vibration signal is the *reversed sequence envelope spectrogram* (RS-ESg). Its calculation starts by following the same procedure of RS-SES until angular-resampled envelope signal is obtained. Thus, RS-ESg is finally obtained by computing the spectrogram of this angular resampled quantity. RS-ESg is a 2 variables function of order and rotation, indicating the order content of the signal at different rotations of the shaft.

5 Experimental Validation

Figure 2 shows the test-rig used for the experimental validation and consisting of a main shaft, driven by an AC motor controlled by means of an inverter and coupled to a spiral bevel gearbox with orthogonal axis. The gearbox output shaft is connected to a rotating shaft supported by two roller bearings: the bearing under test is located in the middle of the rotating shaft and it is preloaded by means of a simple mechanism based on spring acting on the bearing. A screw governs the

Fig. 2 Test-rig layout

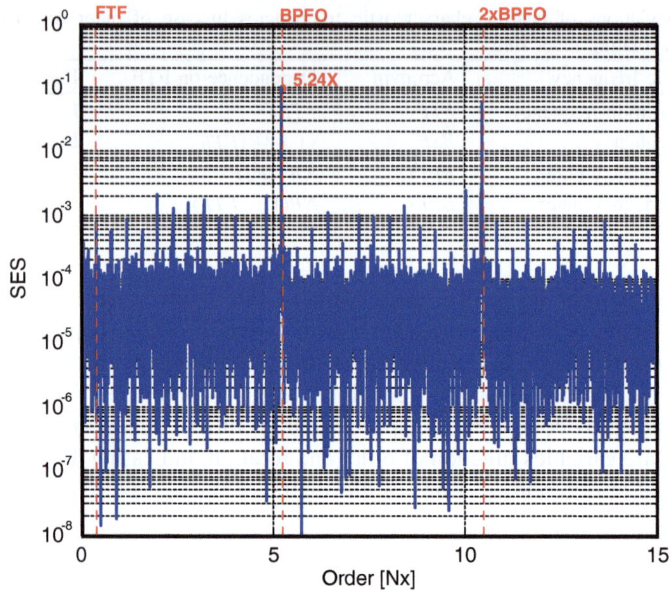

Fig. 3 RS-SES in case of high load acceleration ramp test

compression of the spring, allowing the setting of the load on the bearing. Another AC motor, similar to the first one, is acting as a brake for power recovery.

The cylindrical roller bearing SKF NJ 206 ECP under test had been damaged with a spall on the outer ring, thus, given the bearing geometrical data, a theoretical BPFO with a frequency of 5.24 times (5.24×) the bearing shaft frequency is expected.

In the first test, a constant acceleration has been imposed to the electric motor so that the bearing shaft reaches 1,000 rpm in approximately 40 s, while bearing load is maintained high and approximately equal to 500 N.

The results of the analyses, performed following the newly proposed procedure, are presented in Fig. 3, where the RS-SES is shown. The actual value of the BPFO is constant during the test and very close to the expected theoretical value, resulting in high and concentrated peaks in the SES. The upper half of the available frequency range has been chosen as demodulation band (6,250–12,500 Hz), accordingly to the typical values for bearing diagnostics and considering the spectrum of the signal.

The newly proposed procedure is, in this case, by far more suitable, since it would be difficult to choose a single demodulation band for the whole angular resampled signal for the T-SES, as a consequence of the wide range of operating speeds obtained during the test. This is clearly demonstrated by the two spectrograms of Fig. 4a and b, respectively related to the original signal and the angular resampled signal. As it is evident in Fig. 4a, the resonance of the test rig around 10 kHz is excited during almost all the test, thus it represents a good pass-band for the demodulation filter in the RS-SES procedure. On the contrary, the same

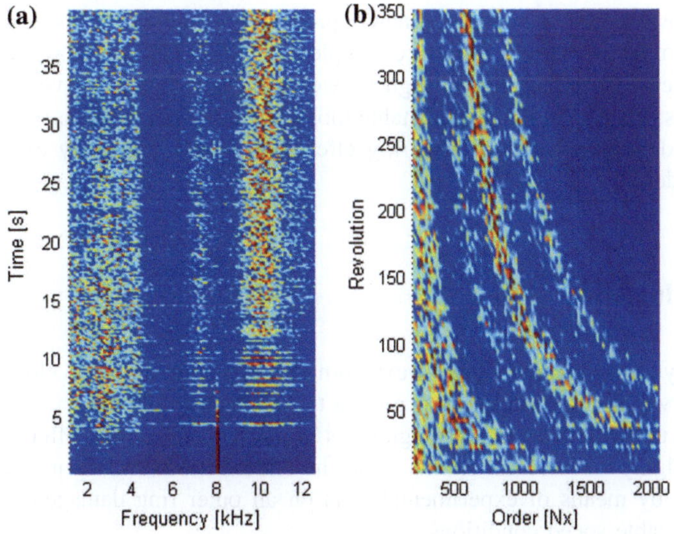

Fig. 4 Spectrograms of the vibration signal: **a** in time–frequency domain; **b** after COT in angular-order domain

Fig. 5 SR-ESg in case of low load acceleration ramp

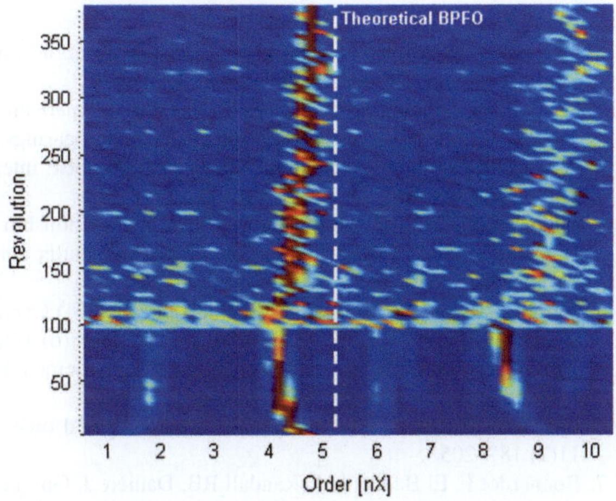

resonance is no more constant in the angular resampled signal (Fig. 4b), where it is impossible to choose a proper demodulation band to apply the T-SES.

The second test presented in this chapter has been obtained with lower load (200 N). In this case, the jitters from the theoretical BPFO become more important and more variable, as shown in the SR-ESg of Fig. 5. The variability of the BPFO is probably due to a higher sensibility to small fluctuations of load, as a consequence of the lower average load applied [14], making the roller more prone to slip on the races.

As shown in this example, the newly proposed procedure allows detecting the presence of a fault also in this more complex condition of variable speed and low load. When load variability is high as well, and particularly in very low load conditions, RS-ESg is the most suitable tool for bearing diagnostics, since the RS-SES could be affected by a *smearing* effect due to the variability of the actual BPFO order.

6 Conclusions

The newly proposed procedure for extending the applicability of envelope analysis to highly variable speed cases has proven to be effective. This is a crucial step in the effort to introduce bearing diagnostics in the industrial field, often characterized by highly variable conditions. Also the effect of load has been qualitatively described by means of experimental tests on an outer ring damaged bearing in highly variable speed conditions.

References

1. McFadden PD, Smith JD (1984) Model for the vibration produced by a single point defect in a rolling element bearing. J Sound Vib 96(1):69–82
2. Randall RB, Antoni J, Chobsaard S (2000) A comparison of cyclostationary and envelope analysis in the diagnostics of rolling element bearings, acoustics, speech, and signal processing, 2000, vol 6. In: Proceedings of 2000 IEEE international conference on ICASSP '00, pp 3882–3885
3. Randall RB, Antoni J, Chobsaard S (2001) The relationship between spectral correlation and envelope analysis in the diagnostics of bearing faults and other cyclostationary machine signals. Mech Syst Sig Proces 15(5):945–962
4. Antoni J, Bonnardot F, Raad A, El Badaoui M (2004) Cyclostationary modelling of rotating machine vibration signals. Mech Syst Sig Proces 18(6):1285–1314
5. Antoni J (2007) Cyclic spectral analysis of rolling-element bearing signals: facts and fictions. J Sound Vib 304(3–5):497–529
6. Fyfe KR, Munck EDS (1997) Analysis of computed order tracking. Mech Syst Sig Proces 11(2):187–205
7. Bonnardot F, El Badaoui M, Randall RB, Danière J, Guillet F (2005) Use of the acceleration signal of a gearbox in order to perform angular resampling (with limited speed fluctuation). Mech Syst Sig Proces 19:766–785
8. Borghesani P, Pennacchi P, Chatterton S, Ricci R (2013) The velocity synchronous discrete Fourier transform for order tracking in the field of rotating machinery. Mech Syst Sig Proces http://dx.doi.org/10.1016/j.ymssp.2013.03.026
9. Randall RB, Antoni J (2011) Rolling element bearing diagnostics—a tutorial. Mech Syst Sig Proces 25(2):485–520
10. Randall RB, Sawalhi N, Coats M (2011) A comparison of methods for separation of deterministic and random signals. Int J Condition Monit 1(1):11–19
11. Borghesani P, Pennacchi P, Randall RB, Ricci R (2012) Order tracking for discrete-random separation in variable speed conditions. Mech Syst Sig Proces 30:1–22

12. Borghesani P, Pennacchi P, Randall RB, Sawalhi N, Ricci R (2013) Application of cepstrum pre-whitening for the diagnosis of bearing faults in variable speed conditions. Mech Syst Sig Proces 36(2):370–384
13. Borghesani P, Pennacchi P, Ricci R, Chatterton S (2013) Testing second order cyclostationarity in the squared envelope spectrum of non-white vibration signals. Mech Syst Sig Proces http://dx.doi.org/10.1016/j.ymssp.2013.05.012
14. Pennacchi P, Borghesani P, Chatterton S, Ricci R (2011) An experimental based assessment of the deviation of the bearing characteristic frequencies. In: Proceedings of 6th international conference acoustical and vibratory surveillance methods and diagnostic techniques, Compiegne, France, 24–26 Oct 2011

Bearing Fault Identification using Watershed-Based Thresholding Method

H. Fandiño-Toro, O. Cardona-Morales, J. Garcia-Alvarez and G. Castellanos-Dominguez

Abstract In this work, a novel thresholding method is proposed to improve the accuracy in segmentation process on thermal images. Characteristics of the thermal distribution around convex Regions of Interest (ROI) are the core of this method, used as input markers for a segmentation process based on watershed transform. This method based on data variability reduces the classification error by about 10 % and reduces the number of features by about 80 % from the set of 360 elements. Moreover, the proposed method provides some tracks for fault localization, demonstrated for a bearing unbalance test rig.

Keywords Fault identification · Thresholding · Region of interest · Thermal images

1 Introduction

Bearing faults, like motor shaft misalignment, cause damages on rotating machine parts such as couplings, bearings, engine components and loads, among others [1]. Therefore, the main goal of machine maintenance is to avoid the unexpected

H. Fandiño-Toro (✉) · O. Cardona-Morales · J. Garcia-Alvarez · G. Castellanos-Dominguez
Signal Processing and Recognition Group, Universidad Nacional de Colombia,
Km 9 way to airport, Campus la Nubia, Manizales, Republic of Colombia
e-mail: hfandinot@unal.edu.co

O. Cardona-Morales
e-mail: ocardonam@unal.edu.co

J. Garcia-Alvarez
e-mail: jcgarciaa@unal.edu.co

G. Castellanos-Dominguez
e-mail: cgcastellanosd@unal.edu.co

G. Dalpiaz et al. (eds.), *Advances in Condition Monitoring of Machinery in Non-Stationary Operations*, Lecture Notes in Mechanical Engineering,
DOI: 10.1007/978-3-642-39348-8_11, © Springer-Verlag Berlin Heidelberg 2014

machine damages, using two maintenance strategies: corrective and preventive. For the former one, the machine operator repairs or replaces the damaged part as soon as the fault is present. In addition, the latter one suggests the frequent inspection on the machine, assessing the machine part deterioration degree. Hence, an adequate preventive maintenance would be accomplished by providing adequate information of some Region of Interest (ROI) on thermal image, where the hot spots would be located. However, due to the characteristics of thermal images, conventional methods of ROI segmentation are not adequate [2], giving this segmentation to be manually accomplished by the human expert.

In this work, the proposed multi-level thresholding method improves the ROI segmentation method for thermal images, thus giving local information for machine fault identification. Based on the local variability of the pixel distribution of thermal images, the method gives the following features: first, the threshold adjustment is invariant of the statistical distribution model estimated from the image; secondly, the thresholding method fits with segmentation techniques for images exhibiting variability around local maximal values.

2 Proposed Thresholding Method

Variations measured on thermal image are limited by the geometry of each machine part. Thus, thermal changes occur at low contrast areas, bounded by smooth edges. Therefore, ROI segmentation is adequate to characterize several machine parts, depending on the thermal pattern exhibited in operation. In this case, the Watershed Transform is a commonly segmentation technique, where the topological gradient image generates borders, namely watershed lines, which define the contour of a segmented object on the image.

Proposed method consists on a modified iterative extraction of disjoint intervals, based on the pixel distribution in the image, generating the quantized image with the weighted average of the values belonging each interval, and then assigned to represent each element of image [3]: noting $\boldsymbol{X}$ as the original thermal image, k as a threshold parameter, and $[T_1, T_2]$ as the interval calculated for some iteration, following algorithm describes the proposed method:

Step 1. The first interval (at the iteration $n = 0$) is defined as $[T_a(0), T_b(0)] = [\mathrm{a}, \mathrm{b}]$, where $a = min\ (\boldsymbol{X})$ and $b = max\ (\boldsymbol{X})$;

Step 2. For each interval $[T_a(n), T_b(n)]$ at the iteration n, the weighted mean μ and standard deviation are calculated a [j]

Step 3. Thresholds for the next iteration n + 1 are calculated as: $T_a(n+1) = \mu - k\sigma$ and $T_b(n+1) = \lfloor \mu + k\sigma \rfloor$, where $\lfloor . \rfloor$ denotes floor operation;

Step 4. A new interval is created as: $[T_a(n+1),\ T_b(n+1)] = [T_a(n+1)+1,\ T_b(n+1)-1]$, the increment/decrement of the interval values by 1 avoids interval overlapping;

Step 5. Pixel values belonging inside remaining intervals $T_{Ra} = [T_a(n),\ T_a(n+1)]$ and $T_{Rb} = [T_b(n),\ T_b(n+1)]$ are represented in quantized image with the respective values $\mu(T_{Ra})$ and $\mu(T_{Rb})$;

Step 6. If $T_a(n) - T_b(n) \geq 2$, update $n = n + 1$, then go to Step 2.

Hence, the proposed method suggests that those pixel sets containing a high-grade of variability belong to a convex ROI. Moreover, parameter k is unknown. An inadequate choice of this parameter can lead to inaccurate quantization of the image, therefore a value of k is empirically estimated to get several intervals related with convex *ROIs*.

Another problem concerns with objects not belonging to the fault identification process, like cables and sensors, appearing as segmented *ROIs*. To address this problem, the segmentation process uses a region–based masking, eliminating those connected regions being less than z pixels in the image. This assumes that the objects inside the ROI of size less than z pixels are not relevant to the identification process. Thus, following algorithm describes the segmentation process:

Step 1. The proposed thresholding algorithm processes the thermal image, giving quantized masks. The first high–valued quantized interval corresponds to *ROI*–candidate areas, generating the binarized image by assigning ones at the *ROI*–candidate areas, and zero otherwise.

Step 2. Watershed Transform segments the ROI on thermal image, using as local minimal those values at the center inside of the watershed lines. It gives a number of watershed regions.

Step 3. Comparison between statistical mode of the values of each watershed region, and the statistical mode of same region in binarized image. If the latter value is equal to zero, the watershed region is discarded. Thus, labeling of the remaining watershed regions is *ROIm*, $m = 1, \ldots, M$, where M is the number of remaining *ROIs*.

3 Feature Extraction for Fault Identification

If the thermal camera detects some isotropic heat propagation, then an isotropic gradient operator (like Gaussian) is capable to identify a possible fault, by estimating the probability that a pixel belongs to a relevant *ROI* [4]. However, faults can appear as anisotropic contour regions; then, an anisotropic operator would obtain information about the value and direction in which heat spreads. In this case, the usage of Sobel and Prewitt gradient operators allow detecting changes of heat propagation magnitude and direction in image for fault identification purposes.

Thus, the gradient operators T_x and T_y are calculated for the directions x and y of the thermal image $\boldsymbol{X}$, respectively, as

$$T_x = \begin{bmatrix} -1 & 0 & +1 \\ -2 & 0 & +2 \\ -1 & 0 & +1 \end{bmatrix}; \quad T_y = \begin{bmatrix} -1 & -2 & -1 \\ 0 & 0 & 0 \\ +1 & +2 & +1 \end{bmatrix} \tag{1}$$

$$T_x = \begin{bmatrix} -1 & 0 & +1 \\ -1 & 0 & +1 \\ -1 & 0 & +1 \end{bmatrix}; \quad T_y = \begin{bmatrix} -1 & -1 & -1 \\ 0 & 0 & 0 \\ +1 & +1 & +1 \end{bmatrix} \tag{2}$$

where Eqs. (1) and (2) refer to the Sobel and Prewitt operators, respectively. Thus, the convolution between image and either vertical T_x or horizontal T_y gradient operator gives the gradient vectors $G_x = T_x * \boldsymbol{X}$ and $G_y = T_y * \boldsymbol{X}$, respectively. Finally, Eq. (3) extracts the gradient direction feature α at each pixel position (x, y) as:

$$\alpha(x, y) = \tan^{-1}\left(\frac{G_y}{G_x}\right) \tag{3}$$

4 Experimental Set-Up

Evaluation of the proposed method consists on its implementation on a rotating machine fault identification system. Specifically, the method belongs to the ROI segmentation process on the thermal image used for fault identification. The following steps are part of the evaluation process: (1) processing of acquired thermal image by the proposed method, giving a quantized image under a selected parameter k; (2) mapping of the quantized image, giving the watershed lines within the *ROI* segmentation algorithm; (3) gradient feature extraction from segmented *ROIs*; (4) calculation of classification error rate for different machine conditions, using parameter k as variable and gradient values as inputs.

4.1 Test Rig and Image Database

Figure 1 shows the testing rotating machine of the fault identification system, consisting of a three-phase induction motor, with a rigid coupling between its shaft and another shaft containing 2 drilling wheels. Insertion of weights of arbitrary mass in one of these wheels induces the following two unbalance types: the first one, by inserting weights in the drilling wheel closer to 12 cm from the dock, being labeled as *First Wheel Unbalance* (FWU); the second one, called *Second Wheel Unbalance* (SWU), by inserting weights into the farther bearing at 43.5 cm

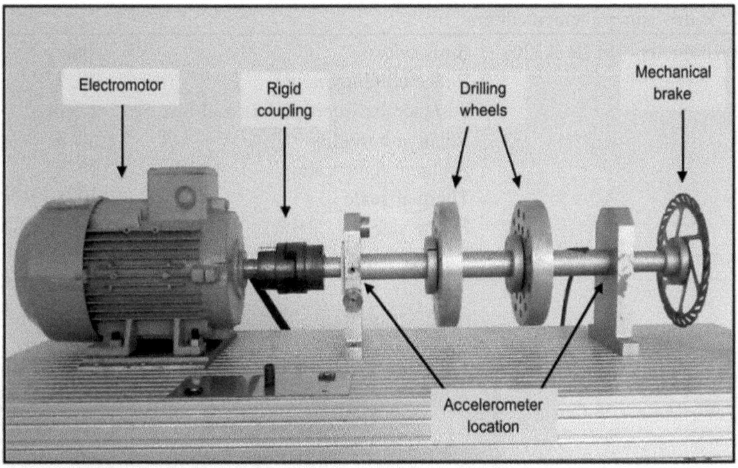

Fig. 1 Test rig image

from the coupling. Machine operation recording lasts 2 h since its startup. Thus, there is an observation for each different operating condition: Normal, FWU and SWU.

Thermal image is the result of the decomposition of recorded video frames into a YUV-space color image sequence. At a video recording rate of 1 frame per 60 s, and noting that the first and second hour of machine operation is an adequate observation time (when the motor achieves a stable temperature), the total number of recorded images per observation is then 60. Moreover, each required image comes from the Y intensity plane of each element of YUV-space sequence, since this plane is a projection of the thermal chroma values. Table 1 shows other video recording parameters.

By observation of the operation conditions, a preliminary segmentation of a ROI corresponding to the first and second bearing aims to improve the fault identification process [5], because thermal variations in this *ROI* are directly related to the mechanical effects due to unbalance. Thus, the *ROI* labeled as *ROI*12 in Fig. 2 is the database image, used for the segmentation algorithm.

4.2 Thresholding and Segmentation

The k-parameter adjustment assures a finite number of disjoint intervals related with convex *ROIs*. So, at a determined value k, the high–valued intervals keep the maximal information of thermal distribution. In this case, values of $k = 1, 2, 3$ are previously assigned. Besides, a region–based masking is implemented after the thresholding process to prevent including of non-relevant objects for the identification process, such as the sensor and cable objects placed along the machine.

Table 1 Video camera specifications

Camera parameters (FLIR A320)	Emissivity	0.82
	Reflected temperature	20 °C
	Distance between camera and test rig	1.5 m
	Relative humidity	50 %
	Ambient temperature	20 °C
	Thermal scale	10–50 °C
	Frame size	640 × 480 pixels
Video acquisition parameters	Video format	MPEG-2

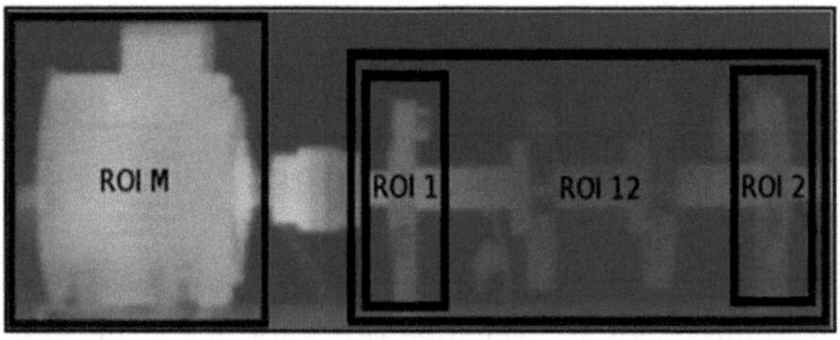

Fig. 2 Some preliminary ROIs

By observation of non-relevant objects in image, an assumed ROI size threshold is z = 200 pixels. This value would vary for another experiments and test rigs. The coordinates of pixels belonging selected ROIs are part of the data required for the feature extraction process.

4.3 Feature Extraction and Classification Error Rate

The Edge Direction Histogram (EDH), proposed in (6), determines the relevant phase values α used for fault identification, for each segmented ROI M: (1) Eq. (3) extracts the gradient direction $\alpha$for each point, using either Prewitt or Sobel operator; (2) rounding of each direction, up to the nearest integer value, gives the direction vector $\vec{p} = \{\alpha(x,y)\} \in N$, for each coordinate $(x,y) \in N$; (3) Principal Component Analysis (PCA),based on histogram of direction vector $\vec{p}$, gives the-relevance weight for each direction.

After sorting of the relevant directions, evaluation of classification performance requires 10 cross-validation trials (70 % training and 30 % validation images, sorted randomly), by comparing the classification error between trials using all directions and trials using the most relevant directions. Noting that location is the main characteristic for considered fault conditions (FWU and SWU), the evaluation for each selected ROI M would give a track for not only fault identification, but also fault localization, defined as Localized Fault Identification (*LFI*) [2]. The number of relevant directions α are found when the classification error value is less than 10 %.

5 Results

5.1 Thresholding and Segmentation

Figure 3 shows the mean and standard deviation of quantized values for the three most high–valued intervals, for each considered operation condition, calculated by proposed thresholding method, using $k = 1, 2, 3$. In this case, the optimal value k' is found at the maximal distance between the mean value of first and second interval. This condition implies that the method provides adequate thresholds for the segmentation process, allowing this an accurate ROI segmentation [6]. As result, the selected value is $k' = 1$.

Moreover, the mean number of iterations achieved in Table 2 measures the proposed threshold method performance, for considered parameter values $k' = 1, 2, 3$. In this case, the lower number of iterations is found at $k = k' = 1$, for the three considered conditions.

Figure 4 shows an example of the calculated intervals per iteration, using the proposed thresholding method at k' In this case, the test image corresponds to the 3600 s interval, under FWU condition; thresholding process stops at 4 iterations, giving a total of 8 intervals. For the sake of illustration, interval labels are by magnitude level order. As result, the interval number 1 gathers the highest thermal values, related with high variability.

Thereafter, results on segmentation process (Fig. 5) describes the following: (1) quantization of thermal database image, by the proposed thresholding method using the parameter k', gives the quantized image of Fig. 5a, where the highest

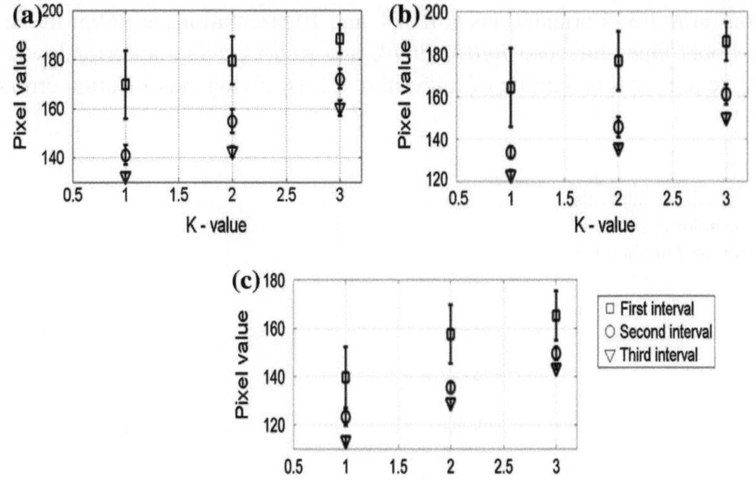

Fig. 3 Intervals with higher pixels for the considered operating conditions. **a** FWU. **b** SWU. **c** Normal

Table 2 Thresholing iterations for different k values

Operating condition	K value		
	1	2	3
FWU	4	7	13
SWU	4	7	17
Normal	4	7	18

interval (in white) fits with convex ROIs; (2) the masks belonging to the first interval composes the binarized image shown in Fig. 5b; (3) the watershed transform segments the thermal image, using region borders as watershed lines shown in Fig. 5c; (4) the region–based masking eliminates regions not belonging to identification processes; in this case, the regions belonging to an accelerometer sensor and a power cable disappear from segmented image (Fig. 5d); (5) finally, the labels for each segmented ROI masks appears in Fig. 5e as $ROIm, m = 1, \ldots, 5$.

5.2 Feature Extraction and Classification Error Rate

Direction feature extraction, for each labeled *ROIm* and for the database image *ROI*12, gives correspondent sample vectors required as inputs of classification process. Figure 6 shows the results on the comparison between classification error of each *ROIm* and error of *ROI*12. For the Prewitt operator, *ROI*3, *ROI*4 and *ROI*5 achieve an error lower than *ROI*12, requiring less number of relevant α values. Moreover, the classification error obtains less than 10 % using around of 8 orientations at *ROI*3, 4 orientations at *ROI*4, and 10 orientations at *ROI*5. In the case of the Sobel Operator, also *ROI*3, *ROI*4, and *ROI*5 achieve an error lower than *ROI*12, requiring less number of relevant α values, giving classification errors less

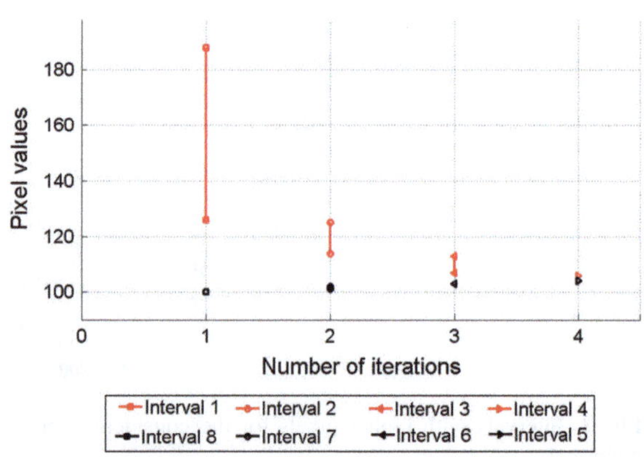

Fig. 4 Resulting intervals after thresholding process using proposed method k = 1

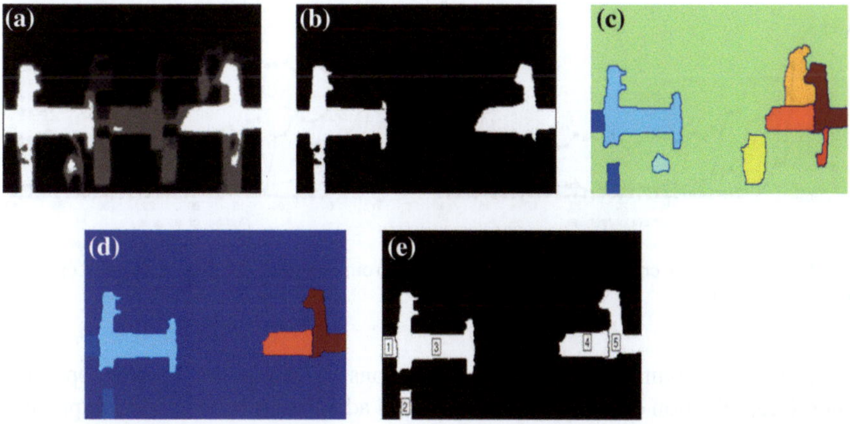

Fig. 5 ROI12 thermal image segmentation process. **a** Quantized image. **b** Mask for ROIs. **c** Preliminary Watershed segmentation. **d** Masks for ROIs after false object extraction. **e** Labeled ROIs after segmentation

than 10 % using around of 8 orientations at *ROI*3, 3 orientations at *ROI*4, and 8 orientations at *ROI*5.

6 Discussion

Given results of Fig. 4 indicate that the proposed thresholding method is adequate for fault identification, by observing the calculated intervals for different operation conditions. For example, using adjust parameter $k = 1$, a thermal variation between 120 and 140 of the quantized value of the first interval means that the machine is in Normal condition, whereas a thermal variation above 140 of the quantized value, at the same interval, would be identified as a fault.

From given results of Fig. 5, the proposed method gathers the most measured variations into one interval, using the fewest number of iterations at $k = 1$. In this case, the first interval achieves the most variable values, at the first iteration. Since the thresholding process is not enough for adequate fault localization, results of Fig. 5a and b give a relationship between the most variable interval and convex *ROI*, relating both thresholding and segmentation processes. In this case, the highest interval fits with convex ROIs of greater value variability. Using the given results of Fig. 5c–e, the proposed method helps to the watershed segmentation process with the following: firstly, avoiding the over–segmentation; secondly, eliminating those regions not belonging for the identification process.

Results given from Fig. 6 show that the number of relevant elements of the direction gradient feature α is around 10. Although the minimal number of relevant directions correspond to *ROI*4 using the Sobel operator, the *ROI*4 and *ROI*5 can

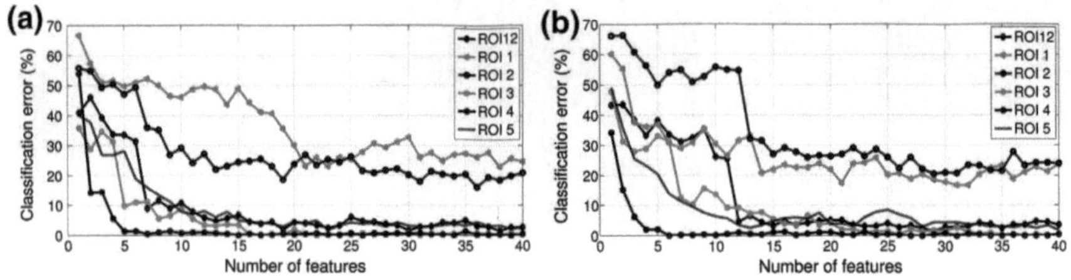

Fig. 6 Classification error using gradient features from segmented ROIs. **a** Prewitt operator. **b** Sobel operator

achieve a less number of relevant directions α, using the Prewitt operator. Therefore, selection of latter operator gives an adequate feature extraction process. Hence, the proposed thresholding method, with parameter $\kappa = 1$, gives segmented *ROIs* reducing the bearing fault classification error by about 10 % and reducing the number of relevant features by about 80 %, in comparison with the image *ROI*12. Finally, using results shown in Fig. 6, for both operators, *ROI*4 would be the key region for fault localization.

7 Conclusion

In this work, a proposed multi–level thresholding method improves the segmentation process of convex Regions of interest (*ROI*) for image–based fault identification and location systems.

Method evaluation uses a bearing fault identification system, requiring as input database a thermal image acquired on related test rig, the watershed transform as segmentation process, and the orientation gradientα as fault identification feature. Results provide the following highlights: (1) the method provides an adequate measure for fault identification, calculating an interval related with convex ROIs, being candidates to enclose a fault zone; (2) reduction of feature elements is suggested, to increase the fault classification rate, giving the Prewitt operator as adequate for feature extraction; (3) for this test rig, segmented *ROI*3, *ROI*4 and *ROI*5 are candidate regions for fault location.

Following works will be under consideration: firstly, a comparison between the quantized value of the proposed thresholding method and other measures, such as the statistical measures proposed in [7] for fault identification systems using thermal images; secondly, the usage of another feature, like Local Binary Patterns, to improve the accuracy of fault identification and localization system; finally, the use of method on other test rigs related with machine faults, acquiring more information about the fault phenomena.

Acknowledgments The authors acknowledge to Colciencias and the Universidad Nacional de Colombia for the financial support of the research projects "*Sistema autónomo de monitoreo de vibraciones para diagnóstico de fallas no estacionarias*" (with code 1101-521-28792).

References

1. Hoppler R, Errath R (2007) Motor bearings, not must a piece of metal. In: Proceedings of IEEE cement industry technical conference record, pp 214–233
2. Fandiño-Toro H, García-Álvarez J, Castellanos-Domínguez G (2011) Performance evaluation of measures for the thermographic detection of motor faults by mass unbalance. In: Proceedings of the international conference surveillance 6
3. Arora S, Acharya J, Verma A, Panigrahi P (2008) Multilevel thresholding for image segmentation through a fast statistical recursive algorithm. Pattern Recogn Lett 29(2):119–125
4. Ortiz-Jaramillo B, Fandiño-Toro H, Benitez-Restrepo H, Orjuela-Vargas S, Castellanos-Domínguez G, Philips W (2012) Multi-resolution analysis for region of interest extraction in thermographic, non-destructive evaluation. In: Proc SPIE 8295:82951
5. Widodo A, Satrijo D, Huda M, Lim G, Yang B (2011) Application of self organizing map for intelligent machinefault diagnostics based on infrared thermography images. In: Proceedings of IEEE 2011 sixth international conference on bio-inspired computing: theories and applications (BIC-TA), pp 123–128
6. Gao X, Xiao B, Tao D, Li X (2008) Image categorization: graph edit distance + edge direction histogram. Pattern Recogn 41(10):3179–3191
7. Younus A, Yang B (2009) Wavelet coefficient of thermal image analysis for machine fault diagnosis. In: Proceedings of the international conference on mechanical engineering, ICME09, pp 1–6

Envelope Cepstrum Based Method for Rolling Bearing Diagnostics

Milena Martarelli, Paolo Chiariotti and Enrico Primo Tomasini

Abstract The task of identifying a faulty roller element bearing has been so far faced through the use of envelope analysis. As it is well known the main issue linked to such approach is related to the definition of the optimal band-pass filter which can enhance the defect characteristics when the vibration signal is affected by severe noise. The Kurtogram has overcome this limit by letting the optimal band-pass filter be selected in a semi-automatic way, that is by exploiting the potentials of the Spectral Kurtosis. This paper aims at presenting an alternative algorithm which is able to cope with faults characterised by an impulsive-periodic nature. It is well known that faults characterised by periodic-impulsive nature are identifiable by means of cepstral analysis while damages inducing modulation effects are usually assessed via envelope processing. The presented algorithm combine two instruments, since it is based on the Fourier spectrum of the cepstrum squared envelope. Such spectrum allows to isolate the modulation effect by centring the modulating frequency around the DC component. In this paper the algorithm is applied to both synthesized data reproducing typical damaged rolling bearing signals and experimental data. Results achieved by exploiting the proposed algorithm are compared to the ones obtained by applying conventional envelope analysis based on Spectral Kurtosis.

Keywords Rolling bearing diagnostics · Envelop analysis · Cepstrum

---

M. Martarelli (✉)
Università e-Campus, Novedrate, Italy
e-mail: m.martarelli@univpm.it

P. Chiariotti · E. P. Tomasini
Department of Industrial Engineering and Mathematical Sciences (DIISM), Università Politecnica delle Marche, Ancona, Italy
e-mail: p.chiariotti@univpm.it

E. P. Tomasini
e-mail: ep.tomasini@univpm.it

G. Dalpiaz et al. (eds.), *Advances in Condition Monitoring of Machinery in Non-Stationary Operations*, Lecture Notes in Mechanical Engineering,
DOI: 10.1007/978-3-642-39348-8_12,

1 Introduction

Quality control of rotating machinery is often based on the analysis of the machine vibro-acoustic signature. Indeed, the potential defect induces an anomalous vibration that becomes noise. By monitoring the trend of that signature is possible to discover damages at the very initial stage; by checking the variations of the frequency components the presence of a specific defect can be diagnosed. Typical systems where such diagnostic approach is applied are rolling element bearings. When the rolling element strikes a local fault on either the outer or inner race, the structural high-frequency resonances of the system, between the bearing and the measurement location, are excited. McFadden also showed in [1] that the series of these bursts is further amplitude modulated by both the load borne by the rolling element and the change in position of the fault, when the latter is moving, with respect to the fixed measurement location.

The roller element bearing diagnostics is generally assessed monitoring its vibration signature. As described in [2], however, it has to be recognised that the characteristics which prove a bearing diagnostic technique to be efficient depend on its ability in detecting and enhancing the impulsiveness of the signals. Several signal processing techniques have been presented to the scientific community during the last decades, ranging from linear prediction models to Minimum Entropy Deconvolution, as reported in [3]. One the most popular approach relies on the envelope analysis [4]. The main issue related to envelope analysis depends on the necessity of band-pass filtering the raw signal in order to enhance the bearing signature. The optimal frequency band of this filter is sometimes tough to identify. For such reason Antoni [5, 6] proposed the use of the Kurtogram as a semi-automatic procedure for the detection of the optimal band-pass filter. The Kurtogram, especially in its Fast Kurtogram version [7], improves and speeds up the selection of the optimal band-pass filter by the use of the Spectral Kurtosis. Within the last years the technique has been become the reference technique for rolling element bearing diagnostics.

This paper aims at reviewing the use of the Amplitude Cepstrum, firstly proposed by Randall [8], for the identification of those kind of defects. The algorithm consists in calculating the Fourier transform of the Squared Envelope of the Cepstrum (hereafter named FSEC), which corresponds to the Fourier Transform of the Squared Amplitude Cepstrum. The advantage of using this kind of approach grounds on the possibility to link both the ability of the cepstrum in evidencing echoes and periodic impacts and the capability of the envelope to extract modulations.

The approach is tested on both simulated and experimental data. The usefulness of a combined use of FSEC and wavelet denoising is also shown.

2 Diagnostics Methodology

Spectral analysis is sometimes not the best instrument to highlight modulation phenomena. Indeed when low frequency phase and amplitude modulations are present the spectrum exhibits a large number of closely spaced peaks; moreover, phase modulation induces the presence of several sidebands, which contribute in enriching the spectral content as well. The spectral content is even richer if several modulation groups are present. Those characteristics make spectral analysis quite though.

The cepstrum analysis represents a valid technique to tackle such issue, since it is capable of enhancing the periodic structures of a signal; moreover, if compared to the autocorrelation function, the use of the logarithm allows the weaker components be analysed. In this paper power cepstrum has been adopted following (1) [8]:

$$C_P(\tau) = F^{-1}\{\ln[S_{xx}(f)]\} \tag{1}$$

with:

F^{-1} the inverse Fourier Transform

$\ln[S_{xx}(f)]$ the zero-padded logarithm of the single-sided autopower spectrum of the time signal.

A powerful method to enhance modulations is the signal envelope, which is the modulus of the corresponding analytic signal [8], following (2):

$$env[C_P(\tau)] = \sqrt{\mathrm{Re}^2[C_P(t)] + H\{\mathrm{Re}[C_P(t)]\}^2} \tag{2}$$

where H represent the Hilbert transform.

The Fourier transform of the Squared Envelope of the Cepstrum (FSEC) is equal, for the cepstrum definition, to the autocorrelation of the logarithmic spectrum, but with the advantage of not requiring the use of correlation algorithms.

$$FSEC\,(f) = F\{env^2[C_P(\tau)]\} \tag{3}$$

As it will be shown in the following such tool is able to evidence modulation phenomena induced by periodic impulses generated by localized defects.

However, it is well known, when performing a vibration analysis on a mechanical system, that acquired signals are commonly spoiled by noise. A useful approach aiming at enhancing the capabilities of the FSEC when working on real data consists in pre-processing these data applying some denoising techniques. Among these approaches, a really useful denoising technique is represented by wavelet filtering [9, 10]. The main advantage in using wavelets consists in working with non-constant time–frequency resolutions (that is one of the main differences with respect to the standard Short Time Fourier Transform approach) therefore better balancing between the two domains.

3 Analysis of Results from Synthetic Data

The main objective of a diagnostic algorithm is that of detecting a fault signal in the presence of a strong background noise. That is quite typical when the main task is for instance the identification of faulty bearings in complex machinery. The understanding of the capabilities of a technique results easier if tests are performed on a synthetic signal, since noise can be added in a controlled way. According to Antoni [6] many incipient faults in rotating machinery produces a series of repetitive transient forces, which therefore excite the system structural resonances. That holding, the noise-free signal of a faulty bearing can be synthesized according to (4)

$$x(t) = \sum_k X_k h(t - \tau_k) \tag{4}$$

where *h(t)* is the impulse response excited by a single impact and where $\{X_k\}_{k \in Z}$ and $\{\tau_k\}_{k \in Z}$ are sequences of random variables which respectively take into account for possible random amplitudes and random occurrences of the impacts (which therefore describes the slips of the rolling element bearings). Moreover, the difference $\tau_{k+1} - \tau_k$ is a non-negative random variable whose mean specifies the average rate of repetition of the impacts. That holds for any index *k*.

Gaussian random noise *n(t)* can be further added to the signal in (4) in order to test the performance of the algorithm in presence of a strong background noise, following (5).

$$y(t) = x(t) + n(t) \tag{5}$$

Figure 1 shows the synthesised signal *y(t)* representing the vibration signature of a faulty rolling element bearing (outer race fault—f_{defect}: 52 Hz). The presence of impacts is attenuated by the addition of Gaussian noise which lowers the Signal to Noise Ratio to 5 dB.

The envelope analysis performed filtering the original signal with the optimal filter obtained from the SK represents the reference technique in assessing the

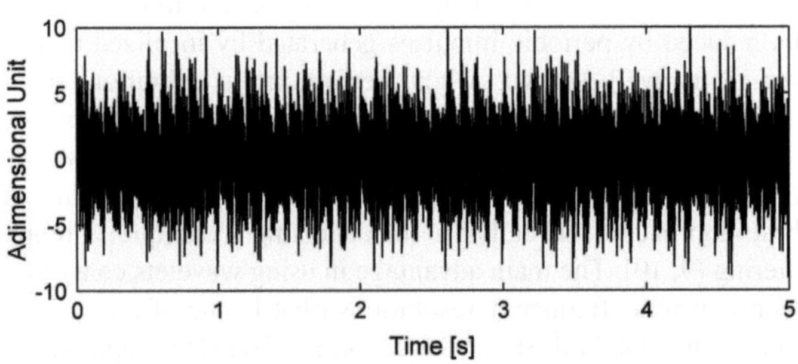

Fig. 1 Synthetized signal representing a faulty rolling element bearing (outer race fault)—5 dB SNR

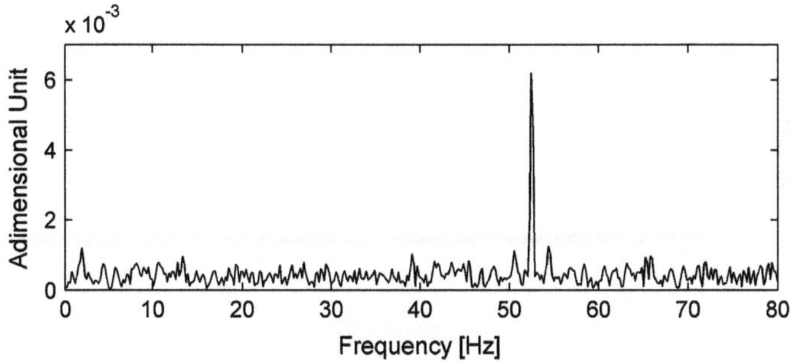

Fig. 2 Envelope spectrum of the synthetized signal filtered through the SK approach

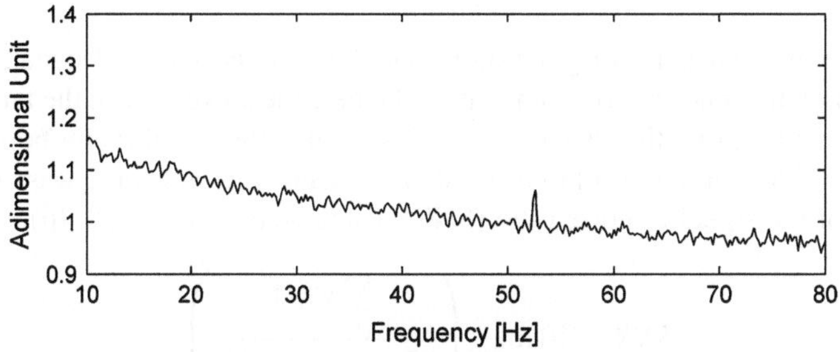

Fig. 3 FSEC of the synthetized signal *y* (*t*)

presence of a faulty rolling element bearing. Figure 2 reports the envelope spectrum of the signal *y(t)* after the Wiener filtering step obtained from the SK calculation. The Weiner filter is identified through the Fast Kurtogram approach [7] (optimal band-pass filter central frequency: 1,171 Hz; optimal band-pass filter bandwidth: 781 Hz). The defect characteristic frequency is clearly well identified by the SK approach.

The application of the FSEC algorithm on *y(t)* results in the spectrum shown in Fig. 3. The defect characteristic frequency (f_{defect}: 52 Hz) is again well identified by the algorithm, which gives as good performances as the SK approach. The FSEC performances can be further enhanced if a proper Wavelet Denoising (WD) pre-processing step is performed on the raw signal *y(t)*, as clearly demonstrated by Fig. 4. The applied denoising method is based on soft-thresholding [11] and minimax principle [12] to estimate the threshold which is rescaled on the basis of the noise variance calculated from the wavelet coefficients at the first level. The mother-wavelet Db11 at level 2 is exploited to denoise the raw signal.

As it is well known, to successfully perform a denoising pre-processing the "mother wavelet" should be chosen in such a way to better approximate and capture the transient spikes of the original signal. The choice of the mother wavelet can be based on eyeball inspection of the spikes of signal, or it can be selected based

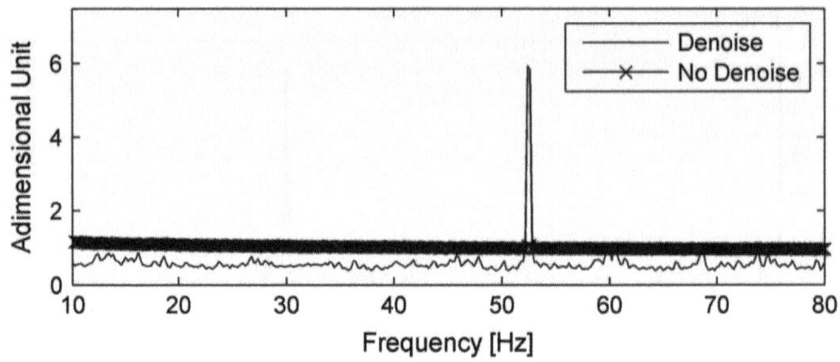

Fig. 4 FSEC of the synthetized signal *y* (*t*)—comparison of the FSEC results with/without the wavelet denoising pre-processing step

on correlation between the original signal and the wavelet-denoised signal, or based on the cumulative energy over some interval where the spikes due to the fault occur. We chose the Db11 mother-wavelet since it gave the higher values for both the two approaches. The optimal level to be used in the denoising procedure was chosen on the basis of the signal to noise ratio (SNR), calculated according to (6),

$$SNR = 10\log_{10}\left(\frac{\sum_t y^2(t)}{\sum_t (\bar{y}(t) - y_d(t))^2}\right) \quad (6)$$

where $y(t)$ is the original signal, $y_d(t)$ is the denoised signal and $\bar{y}(t)$ refers to the mean value of $y(t)$.

Figure 5 shows the SNR values evaluated on the signals obtained from the WD-FSEC and SK approaches. It can be seen that the WD-FSEC approach produces a 5 dB increment in the SNR.

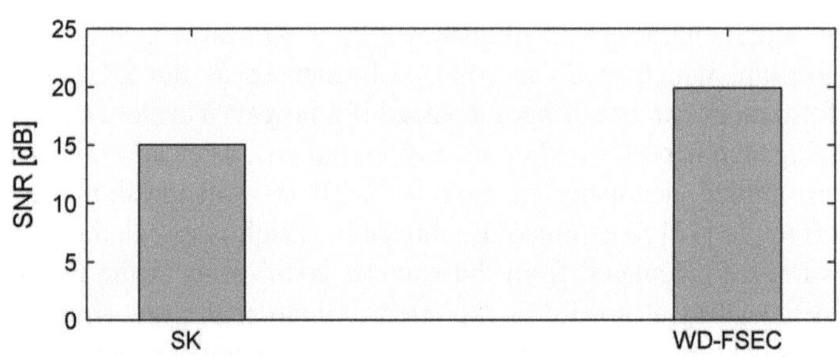

Fig. 5 SNR comparison of SK-based envelope spectrum of the vibration acceleration signal versus FSEC

4 Analysis of Results from Experimental Data

The FSEC algorithm was applied to experimental data measured on a test-bench installing an SKF 1205ETN9 self-aligning ball bearing with a fault in the outer race. The motor was driving the system at a stationary speed of 560 rpm: this makes defect frequency equal to 49 Hz.

The vibration signature of the rolling bearing was measured by an accelerometer mounted on the bearing case. Data were acquired with a sampling rate of 51,200 Samples/s. A scheme of the measurement set-up is shown in Fig. 6. The acceleration time history of the acquired signal is given in Fig. 7.

The envelope spectrum of the signal filtered using the filter identified by the spectral kurtosis is shown in Fig. 8, where the peak at the defect frequency (49 Hz) is clear. The optimal band-pass filter (central frequency: 10,400 Hz; filter bandwidth: 540 Hz) was found exploiting the Fast Kurtogram technique.

One of the main requirement of the FSEC approach relates to the presence of well-structured periodic impacts on the signal. Whenever the noise masks these impacts FSEC suffers. For this reason it is important, in order to usefully exploit such technique, to perform, as pre-processing, a wavelet denoising operation.

Figure 9 shows the FSEC results after the cleaning operation performed through wavelet denoising (Sym5 mother-wavelet at level 2). The peak at the characteristic frequency of the defect is even more pronounced (if compared to the

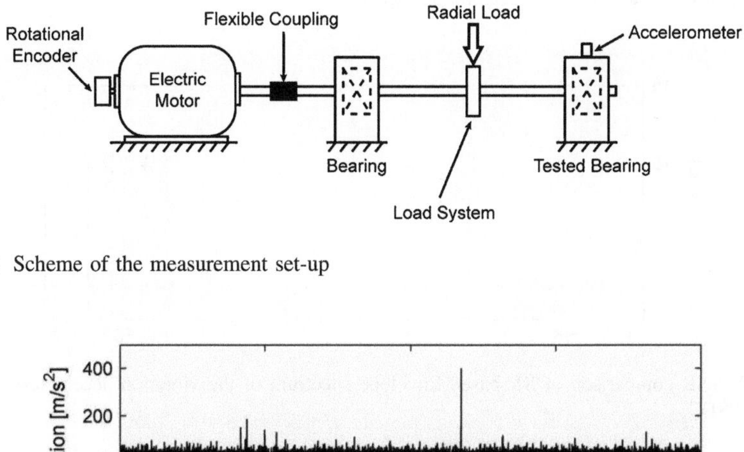

Fig. 6 Scheme of the measurement set-up

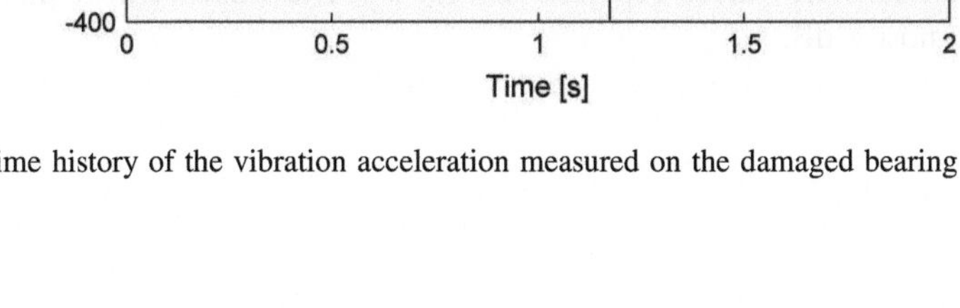

Fig. 7 Time history of the vibration acceleration measured on the damaged bearing

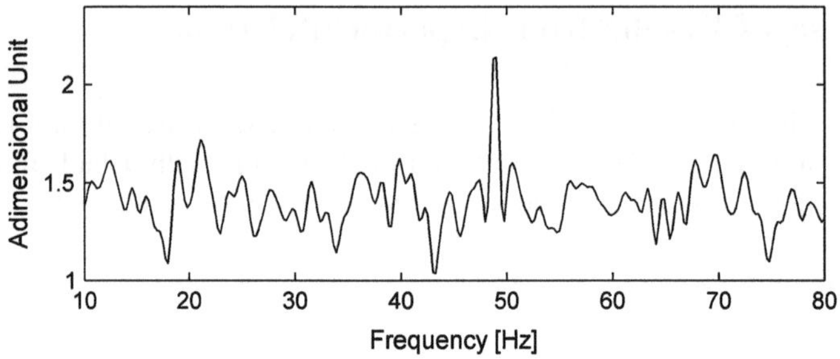

Fig. 8 Envelope spectrum of the vibration acceleration signal filtered with the filter identified by the Fast Kurtogram

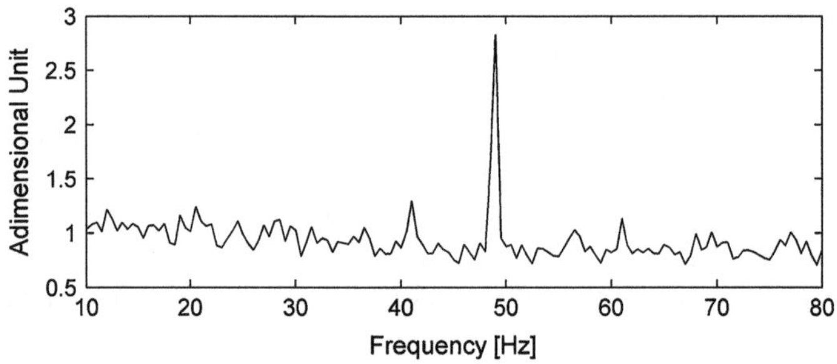

Fig. 9 FSEC of the vibration acceleration signal

Fig. 10 SNR comparison of SK-based Envelope spectrum of the vibration acceleration signal versus FSEC

noise pedestal) with respect to the one appearing in Fig. 8. This is even clearer if looking at the SNR values reported in Fig. 10. The improvement in the SNR reaches almost 8 dB.

5 Conclusion

This paper aimed at reviewing the use of the envelope of the cepstrum of a signal (also known as amplitude cepstrum) for the detection of faulty roller element bearing. This approach can be considered an interesting alternative to the classic Fast-Kurtogram-based envelope analysis, which is in reason considered the reference technique in bearing diagnostics. The algorithm has been tested on both simulated and real vibration signals and results obtained have been compared to those got from Fast-Kurtogram-based envelope analysis thus showing the capabilities of the FSEC approach. It has been recalled that the cepstrum-based technique works properly if well-structured impacts are present on the raw signal. Whenever the noise masks these phenomena (containing the true information) the algorithm can still be used so long as a proper denoising pre-processing procedure is also performed, for instance wavelet denoising. In such case, the FSEC approach performs even better than the SK approach, since the FSEC results in a better SNR. The slightly higher computational complexity of the combined use of wavelet denoising and FSEC is therefore repaid by the improvement in the identification of the defect.

Acknowledgments This work has been carried out in the framework of the COST Action TU 1105 "NVH analysis techniques for design and optimization of HYBRID and ELECTRIC vehicles".

References

1. McFadden PD, Smith JD (1984) Model for the vibration produced by a single point defect in a rolling element bearing. J Sound Vib 96(1):69–82
2. Randall RB (2010) Vibration-based condition monitoring. Wiley, New York
3. Randall RB, Antoni J (2011) Rolling element bearing diagnostics-A tutorial. Mech Syst Sig Process 25(2):485–520
4. Darlow MS, Badgley RH, Hogg GW (1977) Application of high frequency resonance techniques for bearing diagnostics in helicopter gearboxes. Technical Report, USA Army Air Mobility Research and Development Laboratory, pp 74–77
5. Antoni J, Randall RB (2006) The spectral kurtosis: a useful tool for characterising nonstationary signals. Mech Syst Sig Process 20(2):282–307
6. Antoni J, Randall RB (2006) The spectral kurtosis: application to the vibratory surveillance and diagnostics of rotating machines. Mech Syst Sig Process 20(2):308–331
7. Antoni J (2006) Fast computation of the kurtogram for the detection of transient faults. Mech Syst Sig Process 21(1):108–124
8. Randall RB, Hee J (1981) Cepstrum analysis, Bruel and Kjaer Technical Review, No. 3
9. Rioul O, Vetterli M (1981) Wavelets and signal processing. IEEE Signal Processing Magazine
10. Rao R, Bopardikar A (1998) Wavelet transforms, introduction to theory and applications. Addison-Wesley, Reading
11. Donoho D (1995) De-noising by soft-thresholding. IEEE Trans Inf Theory 41(3):613–627
12. Rosas-Orea MCE, Hernandez-Diaz M et al. (2005) A comparative simulation study of wavelet based denoising algorithms. Proceedings of the 15th international conference on electronics, communications and computers

Condition Monitoring of Rotating Machines Using Vibration and Bearing Temperature Measurements

Adrian D. Nembhard, Jyoti K. Sinha, A. J. Pinkerton and K. Elbhbah

Abstract Conventional vibration-based condition monitoring (VCM) of rotating machines with a multiple bearing system, such as Turbo-generator (TG) sets, is data intensive. Since a number of sensors are required at each bearing location, the task of diagnosing faults on such systems may be impossible for even an experienced analyst. Hence, the current study aims to develop a simplified fault diagnosis (FD) method that uses just a single vibration and a single temperature sensor on each bearing. Initial trials on an experimental rotating rig indicate that supplementing vibration data with temperature measurements gave improved FD when compared with FD using vibration data alone. Observations made from the initial trials are presented in this paper.

Keywords Vibration monitoring · Condition monitoring · Rotating machinery · Fault diagnosis · Principal component analysis

A. D. Nembhard (✉) · J. K. Sinha · K. Elbhbah
School of Mechanical, Aerospace and Civil Engineering (MACE),
The University of Manchester, Manchester M13 9PL, UK
e-mail: Adrian.nembhard@postgrad.manchester.ac.uk

J. K. Sinha
e-mail: Jyoti.Sinha@manchester.ac.uk

K. Elbhbah
e-mail: Keri.Elbhbah@postgrad.manchester.ac.uk

A. J. Pinkerton
Department of Engineering, Lancaster University, Bailrigg, Lancaster LA1 4YR, UK
e-mail: aj.pinkerton@lancaster.ac.uk

G. Dalpiaz et al. (eds.), *Advances in Condition Monitoring of Machinery in Non-Stationary Operations*, Lecture Notes in Mechanical Engineering,
DOI: 10.1007/978-3-642-39348-8_13,

1 Introduction

Conventional vibration-based condition monitoring (VCM) is a mature technique for fault diagnosis of rotating machines, however it is a relatively involved process that mandates judgment and expertise from a trained analyst. The limitations of the method are further compounded when considering VCM of a rotating machine with multiple bearings, say a multi-stage Turbo-generator (TG) set. In such case, a number of vibration sensors and measurements are required at each bearing location [1] which results in the acquisition of large volumes of data. The method then becomes a computational burden during data processing and the task of diagnosing faults on these systems may be daunting, if not impossible, for even an experienced analyst. Thus a more simple but robust technique is usually required and would be well appreciated by the relevant industries.

Recent studies [2, 3] have suggested two VCM methods that require significantly reduced number of vibration sensors. Elnady et al. [2] proposed the use of the on-shaft vibration (OSV) measurement technique that requires special arrangement of the measurement instrumentation. Sinha and Elbhbah [3] used just a single vibration sensor on each bearing and then data fusion for diagnosis, which is slightly computationally involved. The present study aims to keep both data acquisition and processing simple and develop a diagnosis technique that uses fewer sensors while preserving moderate computational load. With the wide availability of temperature monitoring systems on rotating machines in industries and studies confirming the sensitivity of temperature to rotating machinery faults [4–8], an opportunity exists to integrate temperature and vibration data for effective fault diagnosis. Hence, a simplified fault diagnosis (FD) method is proposed that uses just a single vibration and a single temperature sensor on each bearing. The temperature measurement is expected to compensate for the reduction in vibration sensors while replacing the need for advanced and complex signal processing of the vibration data in the fault diagnosis process.

The proposed vibration and the temperature measurements are made on an experimental rotating rig with two coupled rotors supported through 4 ball bearings. Different faults are simulated in the rig and both vibration and temperature measurements are collected and analysed in Sect. 2. Earlier studies used Principal Component Analysis (PCA) as a tool for the diagnosis in rotating machines [2, 9, 10], so this method is applied in the present study. Results from the analysis are presented and discussed in Sect. 3 and finally conclusions are drawn in Sect. 4.

2 Procedure

2.1 Experimental Set Up

Figure 1 shows a photograph of the experimental rig used for the experiment [6]. The set up consists of two 20 mm nominal diameter dissimilar length (100 and

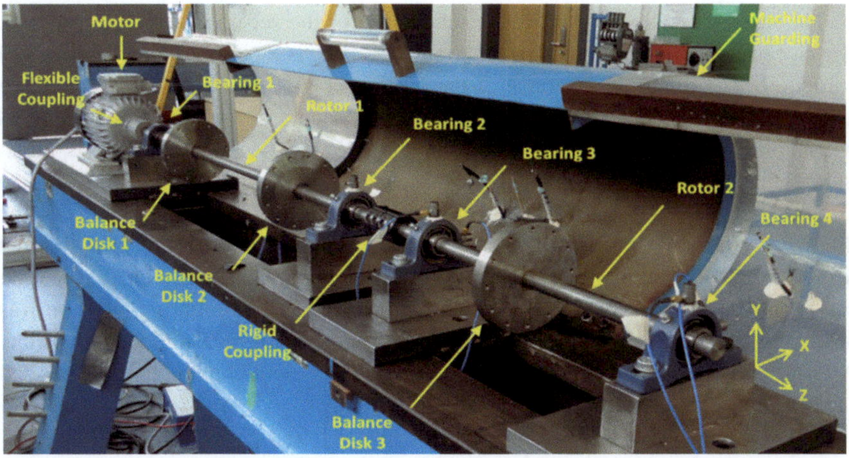

Fig. 1 Experimental rig mechanical layout

50 mm) rigidly coupled rotors that are supported by four grease lubricated ball bearings. These are secured atop flexible steel pedestals that are bolted to a large lathe bed secured to the concrete flooring. Machined sections accommodating balancing disks are mounted on each rotor. System drive is provided by a 0.75 kW, 3 Phase, 3,000 rpm motor that is mated to the rotor-bearing system via a semi-flexible coupling.

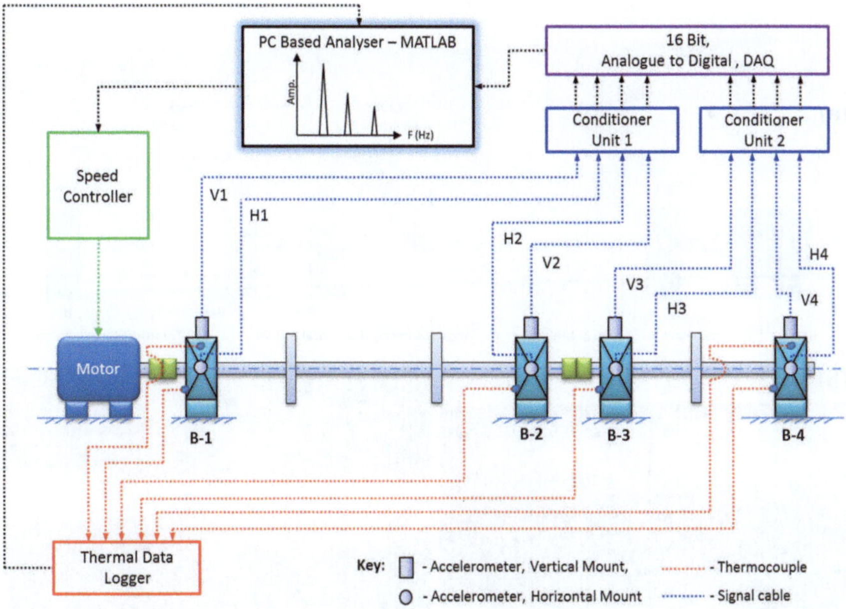

Fig. 2 Schematic of software and instrumentation

The instrumentation and software schematic for the set-up is depicted in Fig. 2. Rig speed is varied with a speed controller that is operated from a personal computer. The dynamic response of the system is then measured with 100 mV/g accelerometers. Each bearing location has two accelerometers that are mounted with adhesive in mutually perpendicular directions. The vibration data is transmitted through two four channel signal conditioners to a 16 Bit Analogue to Digital (A/D) Data Acquisition System. Data logging software then stores the digitized vibration data on the personal computer. To measure the thermal response of the system, K-type thermocouples are attached between the bearing casing and outside of the outer race of each bearing. This mounting position was used to get the most immediate and accurate temperature measurements possible. All four temperature readings were captured with an eight channel data logger and saved to the personal computer for later analysis.

2.2 Simulated Faults and Experiments

Experiments were performed at 2,400 rpm (40 Hz). Data for the healthy machine condition was first collected in order to establish baseline conditions for the rig. Data was then collected for three fault conditions; cracked rotor, rotor rub and coupling misalignment. The cracked rotor condition was simulated with the crack in three different positions, as shown in Fig. 3a. In each case a "breathing crack" [7], with a depth of 20 % shaft diameter was created (see Fig. 3b, c).

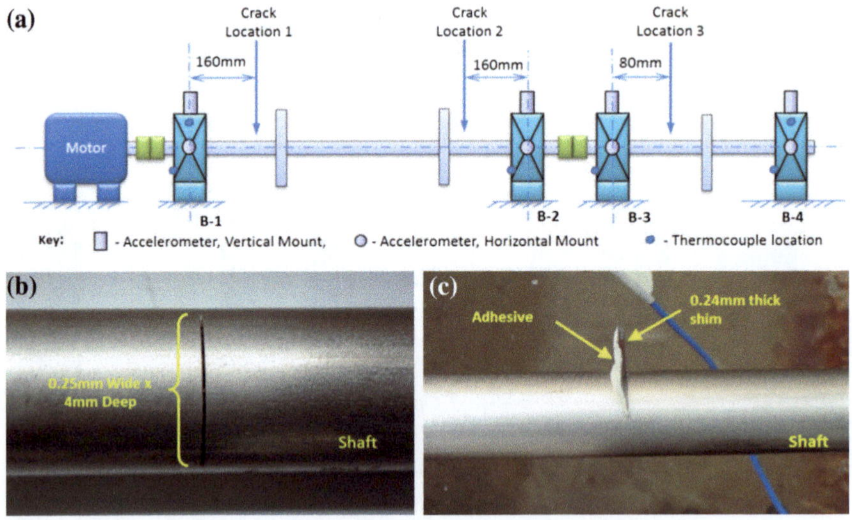

Fig. 3 Details of crack simulation: (a) locations tested, (b) crack dimensions and (c) shim details

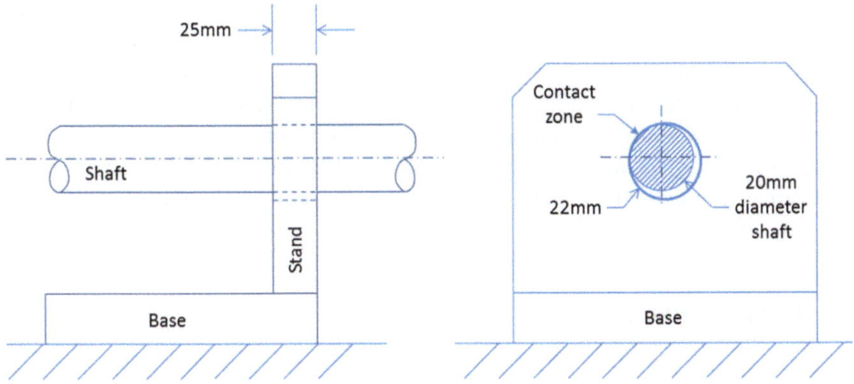

Fig. 4 Details of apparatus used for rub simulation

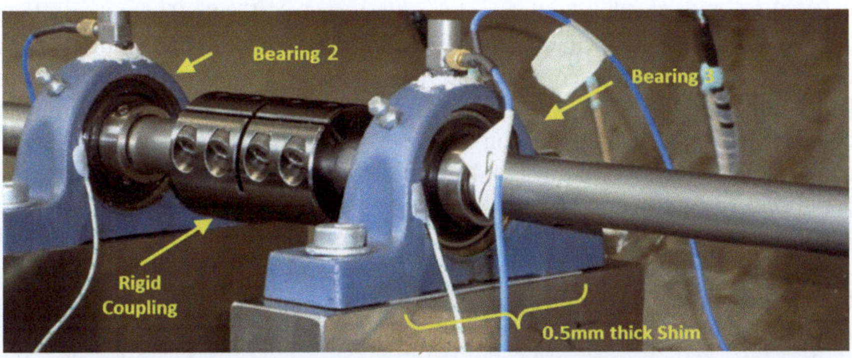

Fig. 5 Details of misalignment simulation at bearing 3

Table 1 Summary of experimental procedure used. One fault scenario was simulated at a time

No	Code[a]	Scenario name	Data collection method
1	Healthy	Healthy	On starting system from rest, bearing temperatures were recorded for a period of ten minutes at 5 s intervals. Vibration data, at a sampling frequency of 10 kHz, was collected at the ten minute point for a total sample time of one minute. After each experiment the system was allowed to cool to ambient temperature before configuring the rig for a different scenario
2	Cr Nr1	Crack near Bearing 1	
3	Cr Nr2	Crack near Bearing 2	
4	Cr Nr3	Crack near Bearing 3	
5	Mlign	Misalignment	
6	Rb Nr4	Rub near Bearing 4	

[a] Same nomenclature is used in figures throughout rest of paper

Rub was simulated by a Perspex apparatus consisting of a base bolted to the lathe bed (at 115 mm from bearing 4) and a stand (see Fig. 4). Misalignment was the final scenario tested in order to minimise the effect of any residual misalignment that could be retained in the system after testing. A steel shim was installed under Bearing 3 housing to induce angular misalignment in the y–z plane across the rigid coupling located between Bearing 2 and Bearing 3 (see Fig. 5). Table 1 summarizes the experiments and data collection procedure employed.

2.3 Data Analysis

For each machine condition, the acceleration spectrum was generated from data measured in the horizontal radial direction between 5 Hz and 1 kHz. Each spectrum was analysed to gain insight to the condition being simulated.

Principal Component Analysis [11] was then performed; first with vibration data only and then with vibration and temperature data. Firstly, for each machine condition, the measured vibration data was segmented into 20 observations. Each observation was used to compute one time domain feature (root mean square [rms]) and three frequency domain features (amplitudes of 1x, 2x and 3x harmonic components). Each bearing location was treated as a separate set of features. Hence a total of 16 features (4 bearings $\times$ 4 features) were computed. These were used to populate a feature matrix, X_v, for loading to the PCA algorithm. Each scenario simulated was treated as a different set of observations; hence matrix X_v would have 16 rows (features) and 120 columns (6 scenario x 20 observations). Before computing of Principal Components (PCs), each element in X_v was converted to zero mean and unit variance [2, 11].

Secondly, temperature measurements were processed to add to the vibration data. Since temperatures were recorded during machine run up, it was necessary to do an extrapolation to obtain steady state bearing temperatures. For this a simple thermodynamic model of the bearing and a curve fitting process were used. Assuming the majority of heat loss from a bearing was via conduction to the steel pedestals, which could have acted as a large heat sink, the bearing temperature change values would be as shown in Eq. (1).

$$\Delta T = (T_{max} - T_o)e^{-At} \quad (1)$$

(where ΔT is temperature increase, T_{max} is steady state temperature, T_o is ambient temperature, A is an arbitrary variable and t is time).

As shown in Fig. 6a, unknown variables in the equation for each condition were adjusted until the model (dashed lines) matched the warm up curves (solid lines). These variables were then used to generate full steady state curves (see Fig. 6b) for each fault condition. Full interrogation of the acceleration spectra was also done to

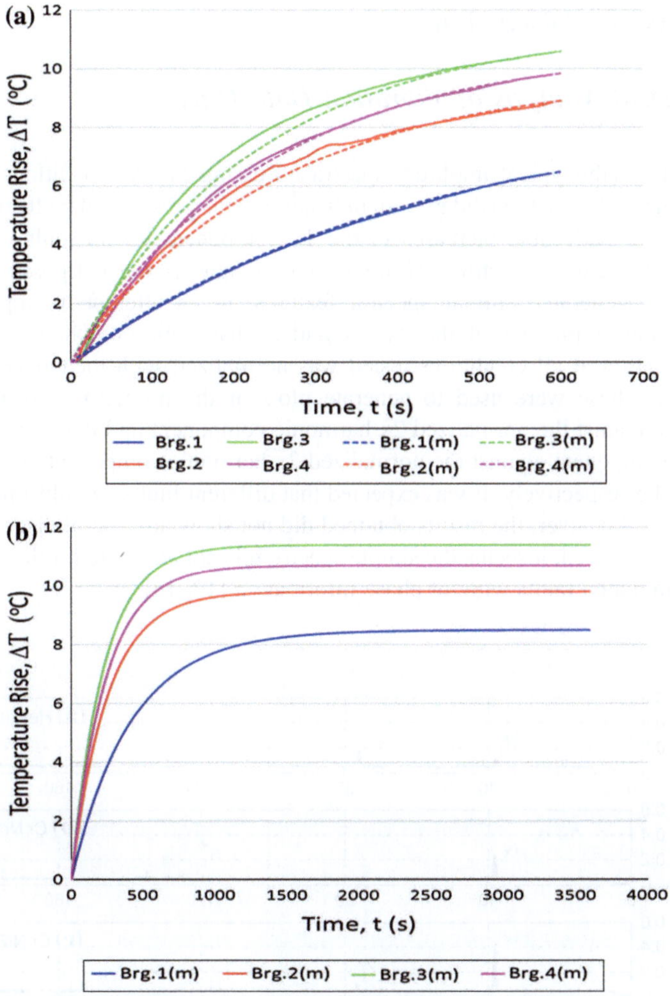

Fig. 6 Sample showing progression of temperature trends in *curve* fitting process: **a** transient curve fitting and **b** generation of steady state curves

ascertain if any bearing fault was present; as this would have affected the temperature measurements. The steady state temperatures calculated for each bearing were included as an additional feature, which resulted in a 20 × 120 feature matrix, X_{t+v}. The process of normalizing elements in X_{t+v} was repeated and PCs were calculated.

3 Results and Discussion

3.1 Spectral Analysis of Vibration Data Only

Figure 7 shows the typical amplitude spectra at Bearing 2 for all conditions tested. It can be seen that the method did give an indication of the presence of fault conditions, as there were noticeable increases in the 1x component for all fault conditions relative to the healthy spectrum. However, fault diagnosis was not possible as different faults generated similar spectral features. In an attempt to improve the diagnosis, the amplitudes of the 4x, 3x and 2x harmonic components for each bearing location at all conditions tested was normalized with their respective 1x component. These were used to generate plots of the normalized 3x harmonic component against the normalized 2x harmonic component and the normalized 4x harmonic component against the normalized 2x harmonic component, as shown in Fig. 8a and b respectively. It was expected that different faults would form clusters in each plots, however, the results obtained did not show any useful discriminative feature. As expected, it seems the simple spectrum is not adequate for the diagnosis of these simulated faults without phase information [12, 13].

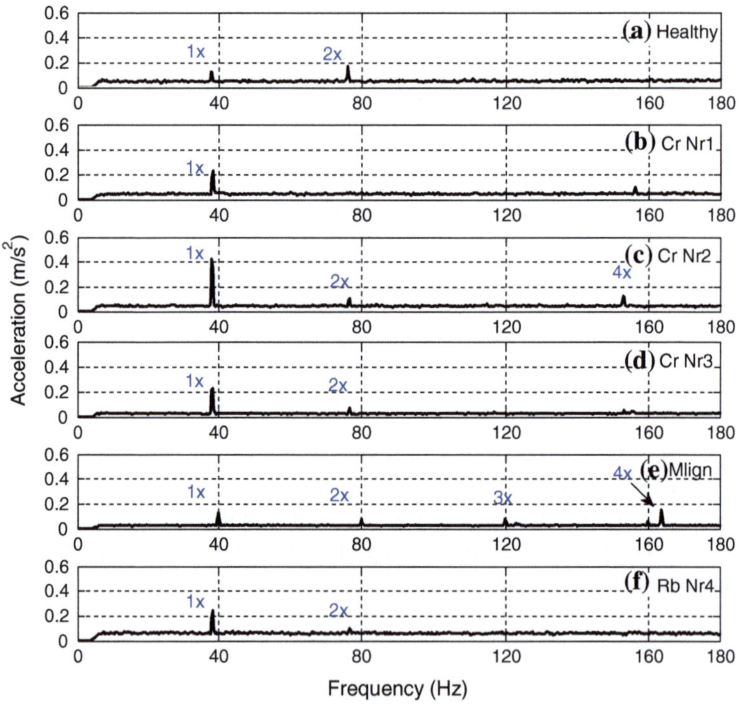

Fig. 7 Typical amplitude spectra at bearing 2 for all conditions tested

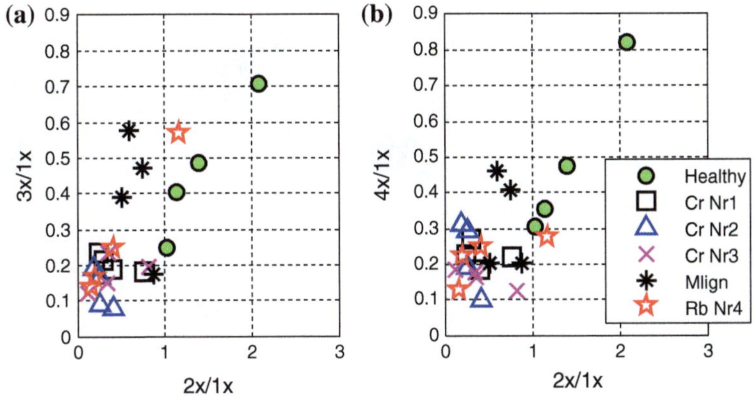

Fig. 8 Comparison of spectral features for all conditions tested at 40 Hz: **a** 3x/1x Vs 2x/1x **b** 4x/1x Vs 2x/1x

3.2 PCA with Vibration Data Only

Figure 9 shows a two dimensional (2D) PC plot which correlates the 2nd PC (PC2) against the 1st PC (PC1). Healthy condition did not occupy a defined space as there was some overlap with it and the Rub near bearing 4 data points. It was also observed that there was overlap between Misalignment and Crack near bearing 1 and 3 data. This is analogous to results obtained in the spectrum analysis (see Fig. 7) where it was not possible to distinguish cracked rotor from misalignment. It therefore appears that PCA with vibration data alone is no more useful than spectrum analysis in this case.

3.3 PCA with Vibration and Temperature Data

Figure 10 shows the 2D PC plot that was produced. It can be seen that the addition of temperature fully separates healthy from all faulty data. The overlap between Misalignment and Crack near bearing 1 and 3 that was noticed in PCA with vibration data only (in Fig. 9) does not exist in this plot. Therefore the addition of temperature puts each fault condition in a clearly defined region, which may be useful for fault diagnosis. Additionally, when compared to the simple spectrum in Fig. 7c; where Crack near bearing 2 was observed to be the most severe fault (possessing the largest 1x component and with noticeable increase in 2x harmonic component with respect to the healthy spectrum), a similar observation is made here, as the said condition has the greatest separation from the healthy case. Therefore, in addition to fault classification, it seems this method may be able to provide useful information on fault severity. Further experimentation would be required to verify this.

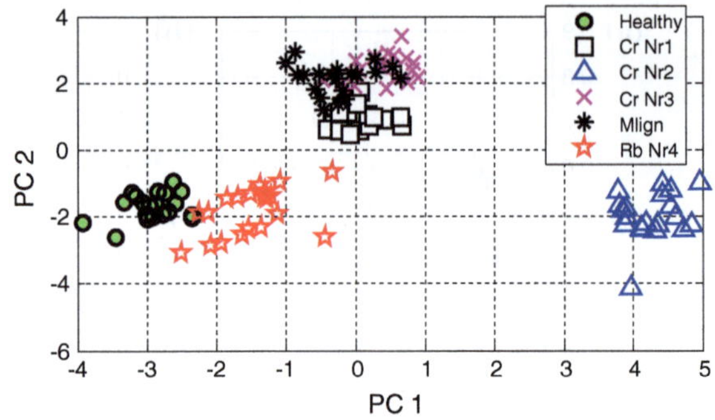

Fig. 9 Feature extraction by PCA without Temperature—40 Hz

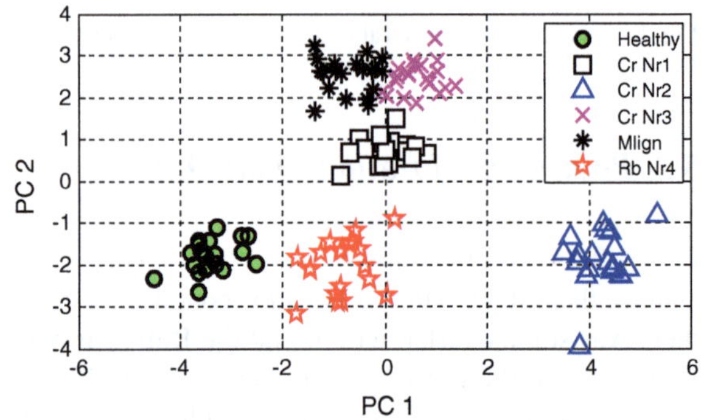

Fig. 10 Feature extraction by PCA, with Temperature—40 Hz

4 Conclusion

A fault diagnosis technique for rotating machinery rotor-related faults is proposed using a single vibration sensor together with a simple temperature sensor on each bearing. Initial trials show that supplementing vibration data with temperature measurements gives better separation of healthy and faulty data; allowing clearer diagnosis of faults when compared with vibration data alone. The proposed method also seems to be simple and non-intrusive and thus have the potential for future application. However, further experimentation on different rigs with different faults is required to fully explore the potential and validate the usefulness of the method.

Acknowledgments A. D. Nembhard wishes to sincerely thank the Commonwealth Scholarship and Fellowship Plan for affording him the PhD Commonwealth Scholarship to the UK from Jamaica to pursue the current research.

References

1. British Standards Institution (2009) BS ISO 10816-1:1995+A1:2009 Mechanical vibration—Evaluation of machine vibration by measurement of on non-rotating parts—Part 1: general guidelines. BSI, Milton Keynes
2. Elnady ME, Sinha JK, Oyadiji SO (2012) Condition monitoring of rotating machines using on-shaft vibration measurement. In: IMechE, 9th international conference on vibrations in rotating machinery. Springer, London, UK 11–13 Sep 2012
3. Sinha JK, Elbhbah K (2013) A future possibility of vibration based condition monitoring of rotating machines. Mech Syst Sig Process 34(1–2):231–240
4. Craig M, Harvey TJ, Wood RJK, Masuda K, Kawabata M, Powrie HEG (2009) Advanced condition monitoring of tapered roller bearings, Part 1. Tribol Int 42(11–12):1846–1856
5. Gaberson HA (1996) Rotating machinery energy loss due to misalignment. Energy conversion engineering conference 1996, IECEC 96, proceedings of the 31st intersociety, vol 3. pp 1809–1812
6. Nembhard AD (2011) Comparison of thermal and vibration analyses for the condition monitoring of rotating machines. MSc. The University of Manchester
7. Sabnavis G, Kirk RG, Kasarda M, Quinn D (2004) Cracked shaft detection and diagnostics: a literature review. Shock Vib Dig 36(4):287–296
8. Yong-Wei T, Jian-Gang Y (2011) Research on vibration induced by the coupled heat and force due to rotor-to-stator rub. J Vib Control 17(4):549–566
9. Li W, Shi T, Liao G, Yang S (2003) Feature extraction and classification of gear faults using principal component analysis. J Qual Maintenance Eng 9(2):132
10. Liying J, Xinxin F, JianguoC, Zhonghai L (2012) Fault detection of rolling element bearing based on principal component analysis. In: Proceedings of the 2012 24th Chinese control and decision conference (CCDC). IEEE, New Jersey, USA, 23–25 May 2012
11. Jolliffe IT (2002) Principal component analysis, 2nd edn. Springer, New York
12. SinhaJK (2002) Health monitoring techniques for rotating machinery. Ph.D. University of Wales Swansea (Swansea University)
13. Sinou JJ (2009) Experimental response and vibrational characteristics of a slotted rotor. Commun Nonlinear Sci Numer Simul 14(7):3179–3194

A Comparative Analysis of Detecting Bearing Fault, Using Infrared Thermography, Vibration Analysis and Air-Borne Sound

Nikolaos G. Athanasopoulos and Pantelis N. Botsaris

Abstract A comparative analysis has been performed as an effort to obtain a better idea of how the fault is appearing over a rolling bearing. In this paper are presented the results of this analysis between three methods of detecting faults on bearings: infrared thermography, vibration analysis and air-borne sound. Those methods are applied on a specific rolling bearing and developed on an experimental set up. The conducted experiment depicted that this comparison is feasible as the results of each method are relevant.

Keywords Rolling bearing · Fault detection · Vibration · Thermography · Air-borne sound · Condition monitoring · Signal processing · Fast Fourier transform

1 Introduction

Ball and rolling element bearings are perhaps the most widely used components in industrial machinery. They are used to support load and allow relative motion inherent in the mechanism to take place. Different methods are used for the detection and diagnosis of bearing defects; they may be broadly classified as vibration and acoustic measurements, temperature measurements and wear debris analysis [1].

N. G. Athanasopoulos (✉) · P. N. Botsaris
Democritus University of Thrace, School of Engineering, Department of Production Engineering and Management, Laboratory of Mechanical Design, 12 Vas. Sofias, Xanthi, Thrace-Hellas, Greece
e-mail: nikoatha@gmail.com

P. N. Botsaris
e-mail: panmpots@pme.duth.gr

G. Dalpiaz et al. (eds.), *Advances in Condition Monitoring of Machinery in Non-Stationary Operations*, Lecture Notes in Mechanical Engineering,
DOI: 10.1007/978-3-642-39348-8_14,

The temperature is an important value measured to present the bearing fault. Measurements of the temperature are used in parallel with other parameters like vibration measurements for the detection of fault. The temperature of the bearings and the lubricant is an indication of the state of machine. Increased temperature in outer rings bearings warns for the initiation of the damage.

As regards the temperature of the lubricant is usually from 10 to 20 °C lower than the bearing. Most lubricants for each 15 °C increase in temperature over 70 °C, have their life halved (or even more), occurring negative effect on the life of the bearing [2]. A recently used method to capture the temperature of the whole bearing is the infrared thermography. The infrared thermography is the technique that allows (from the radiation emitted by a scene, of adapted equipments and techniques of mastery of the measure situation) to obtain the spatial and temporal distribution of the temperatures of this observed scene. In this study, the measuring of temperature based on infrared thermography allows us to detect the presence of abnormally warm zones on the surface of the bearing [3]. Problems with bearings are usually found by comparing the surface temperatures of similar bearings working under similar conditions. Overheating conditions appear as "hot spots", within an infrared image as usually found by comparing similar equipment [4].

Another method of detecting the fault is through analyzing the mechanical vibrations and transforming them into the Frequency-domain. Frequency-domain or spectral analysis of the vibration signal is perhaps the most widely used approach of bearing defect detection. The advent of modern fast Fourier transform (FFT) analyzers has made the job of obtaining narrowband spectra easier and more efficient. Both low and high frequency ranges of the vibration spectrum are of great interest, in assessing the condition of the bearing [5]. The interaction of defects in rolling element bearings produces pulses of very short duration whenever the defect strikes or is struck owing to the rotational motion of the system. These pulses excite the natural frequencies of bearing elements and housing structures, resulting in an increase in the vibrational energy at these high frequencies. The resonant frequencies of the individual bearing elements can be calculated theoretically. Each bearing element has a characteristic rotational frequency. With a defect on a particular bearing element, an increase in vibrational energy at this element's rotational frequency may occur. These characteristic defect frequencies can be calculated from kinematic considerations i.e., the geometry of the bearing and its rotational speed [6].

Apart from the measurement of mechanical vibration that caused by the machine, it is also measured the airborne acoustic ("air-borne sound"). Acoustic noise from mechanical vibrations produced within a machine from metallic components, lubricants, other moving objects or materials, from entrapped air or steam in the inner part of the machine. This noise in the air around the engine, called Air born sound. Measurement of acoustic noise can be used for the detection of defects in rolling element bearings [7].

2 Materials and Methods

2.1 Experimental Set-up

The instrumentation and experimental setup that was used for the needs of the present work is presented in Fig. 1.

The ball bearing that was used for charging was a common deep groove ball bearing with single mounting (type 6206 by Nachi). It was adapted in an axis and inside the load device (Fig. 1). The accelerometer was placed on the top of the load device. The bearing charged with a continuous radial load equal to 48,54 KN. The sensor signals were of altered tendency and with suitable conditioning circuitry leveled to the host computer, for storing and further processing with the LabView (National Instrument) program.

2.2 Applying Load on the Bearing

The load device for the applying of charge of the bearing was studied for both radial and axial stress. For radial stress, the surface growing tension is $\sigma_1 = 57.6\ \mathrm{N/mm^2}$, while the yield point of steel St 37-2 is $R_e = 210\ \mathrm{N/mm^2}$ (safety factor on the surface deformation = 3.6), while for axial the surface stress in the cap of the structure is equal to $\sigma_2 = 34.3\ \mathrm{N/mm^2}$ ($E_2 = 728\ \mathrm{mm^2}$) (safety factor on the surface deformation in the cap = 6.1) [2]. After studying the

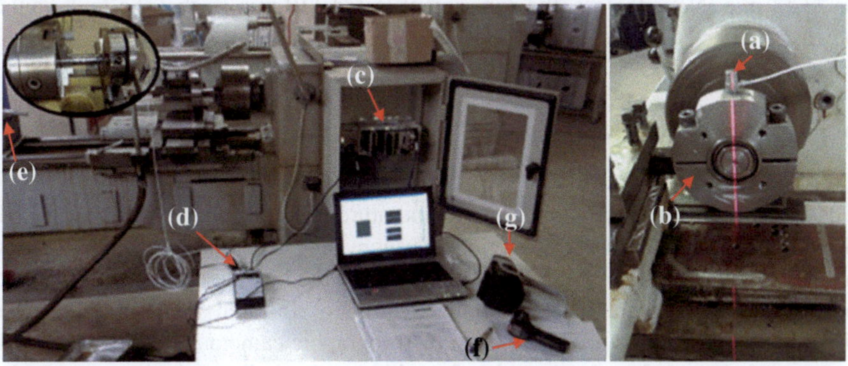

Fig. 1 Experimental setup: **a** One accelerometer (for vibration on the radial axis) (type 8702 B25 by Kistler), **b** A load device for the measurement of charge (constructed by a previous work in our laboratory (Fig.2) [2]), **c** A DAQ card for acquired data (type CompactDAQ—9172 by national instruments and the module NI 9233). **d** The coupling circuit for the accelerometer. **e** A microphone for the measurements of the acoustic emissions (Voltcraft Digital Schallpegelmass-gerat 329). **f** An Infrared thermometer (type KIRAY 100 by KIMO). **g** A portable thermal imager (type IVN 780-P by IMPAC)

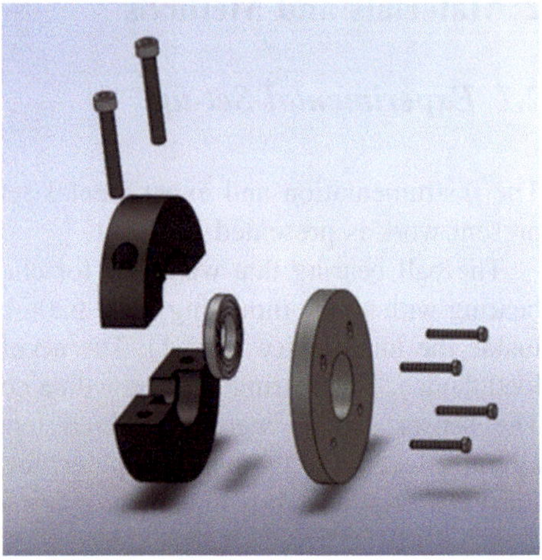

Fig. 2 Steel construction (Stall 37-2), with shear modulus G equal to 80 KN/mm^2

strain-machine, there had to be calculated the load of the bearings, caused by the screw, and therefore had to be calculated the torque from the torque wrench that were used on the two radial screws, by using the equations of Mann :

Friction by Mann: $M_{an} = M_G + M_A = F \times r_2 \times \tan(\alpha + \rho\prime) + F_{\mu A} \times r_A = 49.99$ Nm

where $\mathbf{M_G}$, is the friction torque of the thread and $\mathbf{M_A}$ is the torque from the torque wrench, for F = 24.27 kN, $\rho\prime = 9$, $r_2 = d_2/2$, $\alpha = 2, 9^o$, $\mu_A = 0, 11$, $r_a = 0.7$ d, $d_2 = 10, 863$ mm.

The basic rating life of the bearing 6206 for radial load equals to 48.54 KN according to ISO 281:1990 is: $L_{10} = \left(\frac{C}{P}\right)^p = \mathbf{0.0731}$ or $L_{10h} = \frac{10^6}{60\,n} L_{10} = 1.219$ h or 73.14 min. Nevertheless, the experiment lasted 110 min so that the fault could be easier detected.

2.3 Temperature and Infrared Thermography

The measuring of the rolling bearing temperature has been done with the infrared thermometer (Fig. 3). After analyzing the increasing temperature, further details had to be added in order to specify which parts of the bearing exactly did the fault occurred. At this point were used the method of Infrared thermography. The thermographs were acquired by a fixed thermographic camera and were processed and recorded every 20 min. A study of machine tool spindle to obtain an assembly presenting a limited temperature increase at high speed monitored the fatigue life of bearings [8]. It showed by model and by experiment that a temperature rise between 16 and 34 °C was to be expected in normal working conditions. In the study of J.-S. and K.-W Chen, for bearing load for high speed spindle, it was clearly showed that the temperature of the bearing depend on the load and on the rotational speed [6].

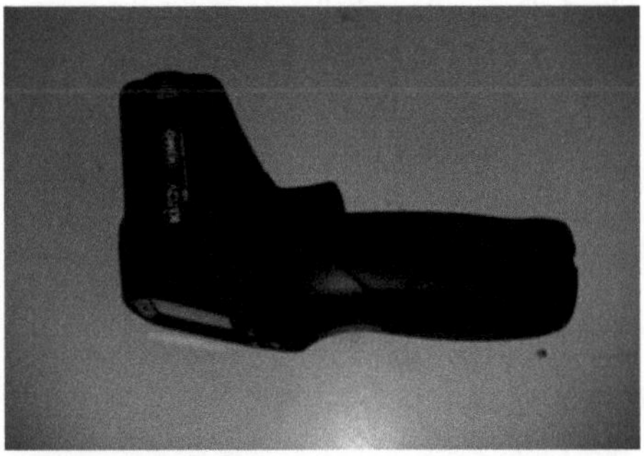

Fig. 3 Infrared thermometer

2.4 Vibrations and Characteristic Frequencies

It is marked as mentioned before that, the vibrations width is an indicator of the vibration power and depends from the speed (Rpm) and the load (N) of the bearing. For a particular bearing geometry, inner raceway, outer raceway and rolling element faults generate vibration spectra with unique frequency components. It is these unique frequency components and their magnitudes that make it possible to determine the condition of the bearing. The bearing defect frequencies are linear functions of the running speed of the motor. Outer race and Inner race frequencies are also linear functions of the number of balls in the bearing. For a bearing, with the outer ring stationary, bearing key frequencies are calculated as follows (1–4) [9]:

$$\text{Fundamental Train Frequency (FTF)}: \quad \mathrm{f} = \frac{f_s}{2\cdot}\left(1 - \frac{\mathrm{B_d}}{\mathrm{P_d}}\cos\phi\right) \tag{1}$$

$$\text{Inner Race Frequency (IR)}: \quad \mathrm{f_i} = \mathrm{Z}(f_s - FTF) \tag{2}$$

$$\text{Ball spin frequency (BS)}: \quad \mathrm{f_b} = \frac{\mathrm{P_d\ f_s}}{\mathrm{B_d}2\cdot}\left[1 - \left(\frac{\mathrm{B_d}}{\mathrm{P_d}}\right)^2 \cos\phi\right] \tag{3}$$

$$\text{Outer Race Frequency (OR)}: \quad \mathrm{f_o} = \mathrm{Z}(FTF) \tag{4}$$

where fs, Bd, Pd, φ and Z are the revolutions per second of the IR or the shaft, the ball diameter, the pitch diameter, the contact angle and the number of balls, respectively (Fig. 4). The contact angle for the ball bearings carrying no thrust load is assumed to be zero.

By the above equations it is calculated the defected frequencies for the bearing of this experiment. While knowing that is turning in the machine-tool of system with speed of 1,000 Rpm with $\cos\varphi = 1$, Bd = 10.40 mm, Pd = 47, 25 mm and Z = 9, the defected frequencies are: FTF = 58.49 Hz, IR = 91, 51 Hz, BS = 36, 03 Hz, and OR = 6.5 Hz.

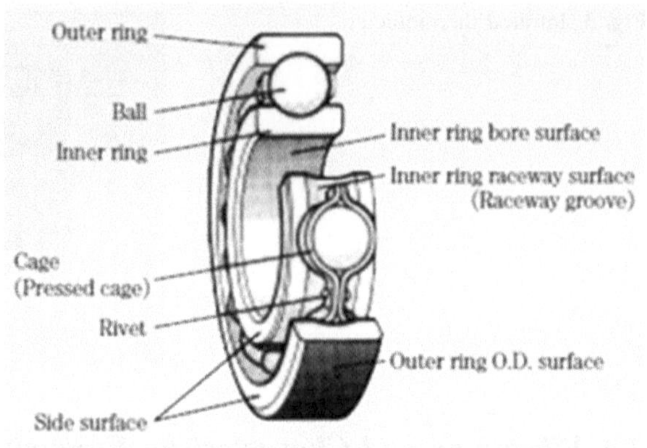

Fig. 4 Deep groove ball bearing

The ball bearing, as all the machine elements have a characteristic natural frequency vibration, while at the same time the frequencies with which their components are vibrating can be calculated, as the exterior ring, the internal ring or their balls. With the Fourier Transformation there can be distinguished the frequencies of ball bearing's elements and conceive if there is any damage in the particular component, so that the fault of the bearing can be averted. The principal advantage of the method is that the repetitive nature of the vibration signals is clearly displaced as peaks in the frequency spectrum at the frequency where the repetition takes place [10]. The FFT was applied on the signal acquired by LabView on the software environment of MatLab.

2.5 Air-Borne Sound

In this paper the air-borne sound were measured by the portable microphone (Fig. 5).

The microphone is a sensor that converts sound pressure into an electrical signal. Of great importance is how to load the microphone right. It should be placed always in the same place as the previous measurements so that the results

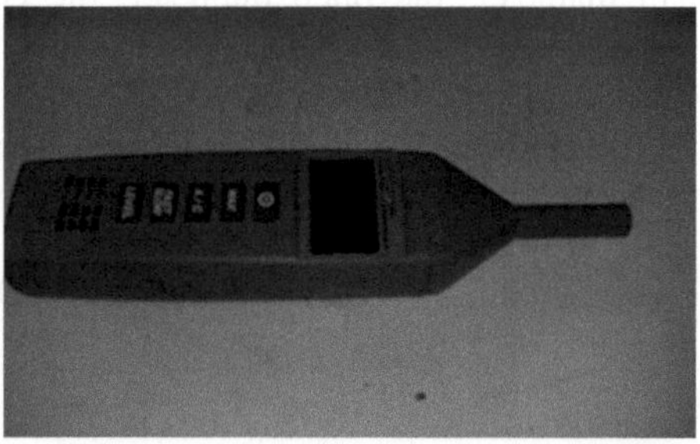

Fig. 5 Portable microphone

are comparable. The microphone should be placed at a distance greater than twice the length of the machine but not too far away from it as the received signal will permeate other sounds.

3 Results and Discussion

3.1 Temperature and Infrared Thermography

In Fig. 6, it is presented the evolution of temperature over time measuring by the portable thermometer.

It is shown that at the beginning of the operation (the first 20 min) there was a high increase of the temperature. A steady low increasing temperature followed and at the end of the experiment a further increase had occurred depicting the fault of the bearing. Those temperatures were above the limit of normal working conditions of the bearing that is around 60 °C [11].

In the experiments that were conducted for the purpose of this paper there where depicted faults in two parts of the bearing, by the portable thermal imager. On the Fig. 7, it is shown that there was a hot spot with max temperature of 69.5 °C on one part of the outer ring, while the rest parts temperature were about 26 °C below.

The other part that an increased temperature was occurred is on the rolling elements (the balls), shown by Fig. 8.

At the Fig. 8 the numbers (1–9), represent the nine balls of the bearing, and at the ball number 3 there was an increased temperature, about 5 °C more than the others, declining a fault at this ball.

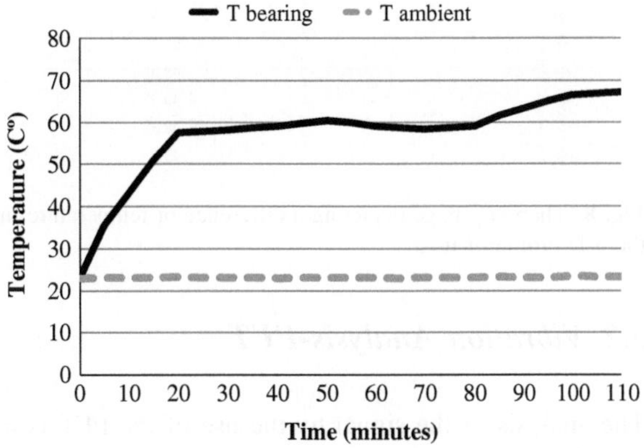

Fig. 6 The evolution of temperature over time diagram

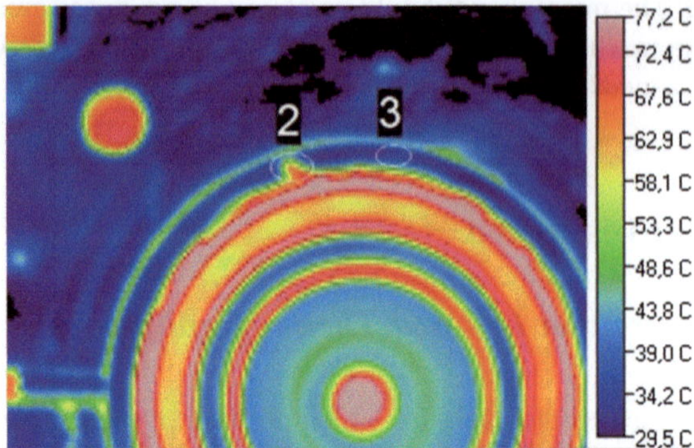

ROI	ROI Label	Min C	Max C	Avg C	Emissivity	Ambient C	Delta C
	2	33,5	69,5	50,9	0,90	20,7	
	3	31,6	43,3	35,4	0,90	20,7	

Fig. 7 Thermal image depicting a hot spot on the outer raceway

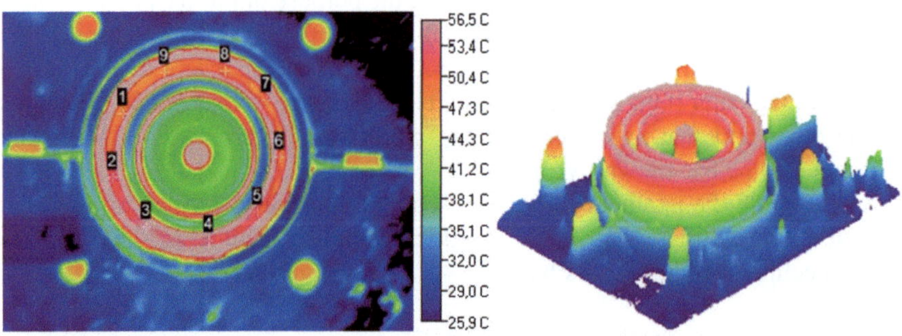

ROI	ROI Label	Min C	Max C	Avg C	Emissivity	Ambient C	Delta C
+	1		49,7		0,90	19,4	
+	2		52,3		0,90	19,4	
+	3		55,9		0,90	19,4	
+	4		53,4		0,90	19,4	
+	5		51,3		0,90	19,4	
+	6		49,9		0,90	19,4	
+	7		49,9		0,90	19,4	
+	8		49,5		0,90	19,4	
+	9		49,3		0,90	19,4	

Fig. 8 Thermal image depicting a difference of temperature in the balls of the bearing and also the 3-D profile of it

3.2 *Vibration Analysis-FFT*

The analysis of the signal by the use of the FFT is presented on the Fig. 9.

It is depicted the appearance of fault in various part of the rolling element, defined by the appearance of the characteristic frequencies and their harmonics.

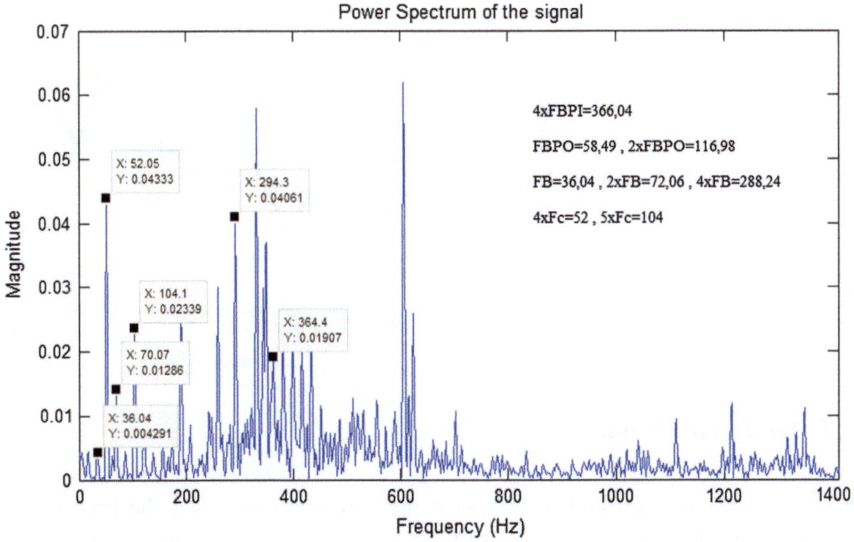

Fig. 9 The extracted plot from MatLab with the characteristics frequencies

More precisely, on the plot it is visible the first second and forth harmonics of the FB (FB = frequency when one point on a rolling element is damaged), declining fault on the rolling elements (the balls). Also the first and second harmonics of the FBPO (FBPO = frequency when one point on outer raceway is damaged), showing the fault on the outer raceway of the bearing. Some fault can also be seen at the parts of the cage (Fc) (fourth and fifth harmonics) and at the inner raceway (fourth harmonic).

3.3 Cost Analysis

At this point, a comparison of the cost between thermography and vibration analysis was also vital in order to compare those two methods on a more large scale view. The Table 1 depicts the total cost of each method respectively for the equipment they occurred. It is shown that the method of thermography is cheaper than the one of the vibration analysis.

Table 1 Total cost of vibration analysis and thermography equipment

Method	Used equipment					Total cost
Vibration analysis	Accelerometer	Software	Coupling circuit	DAQ—9172	Work hours	7,000–9,000 euro
Thermography	Thermal imager	Software	Work hours			5,000–7,000 euro

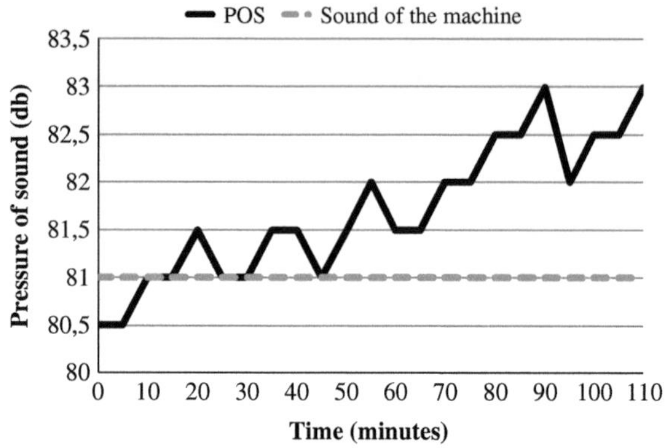

Fig. 10 The sound pressure evolution over time during the experiment

3.4 Air-Borne Sound

In the Fig. 10, it is presented the sound pressure evolution over the time of the conducted experiment. It is shown that over the time there was an increase of the sound pressure depicting extensive use of the machine as well increased stressing of the bearing element.

4 Conclusions

In this paper, a comparative analysis had been used, by applying three methods. The results had shown that the data acquired by the analyzed thermal images, the signal processing from the accelerometers and the rising levels of the air born sound are all relevant between them. The vibration analysis is the most common and precisely method of detecting fault on bearings but still it's a method requiring well trained staff and greater time and money for the acquisition of the results. So the results are depicting that even if the vibration analysis is the most common method of detecting fault on bearings, it still can be replaced by the use of other methods less accurate but less difficult to operate. Further testing under variable and/or axial loads is under investigation by the present research team before a final conclusion can be made. The finite element analysis will be also examined, as promising method of detecting the fault on bearings.

References

1. Tandon N, Choudhury A (1999) A review of vibration and acoustic measurement methods for the detection of defects in rolling element bearings
2. Rodopoulos K (2009) Design, study and construction of module for stress on bearings. Thesis presented on DUTH University

3. Maldague XPV (2001) Theory and practice of infrared technology for nondestructive testing. Wiley, New York
4. Leemans V, Destain M, Kilundu B, Dehombreux P (2011) Evaluation of the performance of infrared thermography for on-line condition monitoring of rotating machines
5. Mouroutsos SG, Chatzisavvas I (2009) Study and construction of an apparatus that automatically monitors vibration and wears in radial ball bearings which are loaded in radial direction. 2009 International conference on signal processing systems, IEEE computer society, pp 292–296
6. Chen J-S, Chen K-W (2005) Bearing load analysis and control of a motorized high speed spindle. Mach Tools Manuf 45(12–13):1487–1493
7. Tandon N, Choudhury A (1999) A review of vibration and acoustic measurement methods for the detection of defects in rolling element bearings. Tribol Int 32:8
8. Jiang A, Mao H (2010) Investigation of variable optimum preload for a machine tool spindle. Int J Mach Tools Manuf 50(1):19–28
9. Ocak Hasan, Loparo KA (2004) Estimation of the running speed and bearing defect frequencies of an induction motor from vibration data. Int J Mech Syst Sig Process 18:515–533
10. Botsaris PN, Koulouriotis DE (2007) A preliminary estimation of analysis methods of vibration signals at fault diagnosis in ball bearing. 4th International conference on NDT, Chania
11. Detweiler WH (2011) Common causes and cures for roller bearing overheating. SKF USA Inc., King of Prussia, PA

Monitoring Lathe Tool's Wear Condition by Acoustic Emission Technology

Tobias Pinner, Hermann Sommer Obando, Georg Moeser and Wolfgang Burger

Abstract Manufacturing of mechanical components with a large lot size is usually an automated process. The knowledge of the in situ *condition* of the tool can help to tailor its exchange cycles in order to reduce costs by using it to its full capacity, without compromising workpiece quality. We aim to monitor the *condition of a lathe's tool* cutting tip by use of *acoustic emission* (*AE*) *technology* in order to increase its service life. Unlike previous research approaches that verified the potential of AE to monitor tool wear, we concentrate on identifying and overcoming application challenges of automated tool wear control that impede a broad and economic use of tool condition monitoring (TCM) in industry.

Keywords Tool wear · Condition monitoring · Acoustic emission · Lathe tool cutting tip · Cooling lubricant influence · Sensor requirements

1 Introduction

Manufacturing of mechanical components with a large lot size is predominantly accomplished by automated tool machines. The manufacturing procedure is programmed and stored into a control device before starting the subsequent batch production of the component. This period is characterized by the continuous unattended execution of predefined manufacturing steps, using various tools depending on the type of manufacturing plant.

The production quality is influenced by different parameters that can be optimized by the manufacturer (like drilling speed, feed, tool material). Usually, the quality lowers continuously with the amount of manufactured components because

T. Pinner (✉) · H. Sommer Obando · G. Moeser · W. Burger
Karlsruhe Institute of Technology (KIT), IPEK: Institute of Product Engineering, Karlsruhe, Germany
e-mail: tobias.pinner@kit.edu

G. Dalpiaz et al. (eds.), *Advances in Condition Monitoring of Machinery in Non-Stationary Operations*, Lecture Notes in Mechanical Engineering,
DOI: 10.1007/978-3-642-39348-8_15, © Springer-Verlag Berlin Heidelberg 2014

of the tool's wear and it can worsen abruptly due to tool defects (e.g. drill breakage or lathe tool rupture). In order to keep the fabrication tolerances within the acceptable deviation, the manufacturing tools are replaced regularly. Due to inconstancy of materials, tool faults and the variation of manufacturing parameters, the time-based changing of tools can lead to a premature or belated exchange. The late tool exchange leads to useless production output as the fabrication tolerances and quality can no longer meet the requirements. This outweighs the costs for a premature exchange and thus, the time schedules usually tend to arrange for an early tool exchange.

The knowledge of the in situ condition of workpiece quality and tool wear can help to *tailor tool exchange cycles* in order to reduce costs by using the tool to its full capacity. Furthermore, sudden tool failure can be recognized and the subsequent output of unusable components can be anticipated and prevented. In this paper, we present requirements for a sensor system to monitor a lathe tool's condition in terms of wear. After presenting the state of the art solutions for process monitoring of manufacturing machines (with a focus on lathes), we summarize *boundary conditions* and define *requirements* for monitoring a lathe tool's condition. We also present the *results* of our first measurements on a CNC-lathe and how they affect the system requirements. The paper addresses the question, which requirements arise from the objective of monitoring a lathe tool's wear and how they can be identified and sharpened.

Acoustic Emission (AE) Analysis turned out to be of advantage for our needs and have been used for similar purposes, like condition monitoring of hydrodynamic journal bearings [2, 3, 9]. Due to this fact, we accomplished ground experiments with SAE-Sensors attached to the lathe tool. These experiments can be regarded as one source for identifying and sharpening requirements for monitoring a lathe-tool's wear condition.

The paper is structured as follows: Sect. 2 deals with a more detailed specification of the *initial objectives*, based on the presentation of the lathing process and machine layout. Section 3 comprises the state of the art presentation and depicts our *methodological approach* to concretise the system of objectives and to specify fundamental design decisions. Within Sect. 4 we present first *experimental results* from Acoustic Emission measurements on a lathe-tool. Section 5 presents the derivation of *requirements*, based on the measurements of Sect. 4 and on other considerations. Section 6 depicts first ideas for a technical realization and *summarizes* the output of this paper.

2 System Description and Objective Definition

2.1 Monitoring of the Machining Process

Machining is one of the most important manufacturing processes. In order to remove material from a workpiece it can be applied to metallic as well as non-metallic materials either [5]. Economic pressure drove the area of manufacturing into developments that raise the overall efficiency. This results in high machining speeds and increasing loads on machines and tools, accompanying an increased wear. There is a need for great effort in *process monitoring and control*, diagnostics and maintenance in order to avoid machine malfunctions and machine breakdown. Machine downtime must be minimized and process stability must be maximized during manned and unmanned production to reach the goal of high productivity [14]. Dornfeld [6] summarized: "These objectives are to ensure safety, prevention of damage to the machine, prevention of rejected workpieces, prevention of idle time on the machine and optimal use of resources." Automated surveillance of manufacturing machines is known to be a central factor to minimize downtime, lower production costs and increase output quality.

In many cases the desired measure cannot be monitored directly. This applies e.g. for the tool wear or the output quality that both could easily be examined manually after manufacturing but not during the machining process itself. Because of this lack of measurability, other measurable quantities linked to the desired value (e.g. the output quality or the tool wear) are used by a mathematical model. In our research, we concentrate on a number of CNC turning centres with milling capabilities. Every turning centre features two independent spindles that have a coaxial adjustment, and thus they can exchange the workpiece for a conjoint machining. Every spindle has an overhead tool revolver that contains one or more lathe tools or drills.

2.2 Definition of Objectives

We aim to monitor the condition of a lathe's tool cutting tip as an *indicator* for the output quality of the workpiece. With help of this condition monitoring the service life shall be extended up to the maximum admissible. Hereby, average operating life should be extended by more than 10 %. Furthermore, unpredictable tool failures shall be recognized in order to halt the operation for tool replacement. Potential necessary sensor or controller elements should work autonomously for at least three months without any need for service (e.g. battery change) and their costs are targeted to be kept below 50 % of those for current monitoring solutions.

3 State of the Art and Our Approach

Machining monitoring methods concerning tool wear have been under scientific examination for over 25 years. Shiraishi [11] compares different methods of tool wear detection in a paper of 1988. As is established in measurement technology, he also distinguishes between direct methods that measure the occurring tool wear as measurand and indirect measures that use indicators to deduce the correlating tool wear. He summarizes at that time indirect methods turned out to be of advantage, as reliability of direct methods of wear measurement posed problems in reliability. Back then, force detection and acoustic emission (AE) methods seemed promising for industrial use.

Nowadays, commercially available CNC-Machining monitoring systems base on a multitude of different sensors that can be integrated into manufacturing machines. The range of sensor technologies reaches from force measurement and vibration monitoring to AE-sensors. Usually, measures are proved to be within a predefined (previously adjusted) scope and a notification is sent if there are any discrepancies. Hence, tool wear in terms of breakage can be securely identified. Nevertheless, no low-cost single-sensor application could be found that permits a reliable and continuous flank wear monitoring. Abellan-Nebot and Romero Subirón [1] give a review of machining monitoring systems basing on indirect measurement methods and thus using artificial intelligence (AI) process models. Widespread sensor systems for monitoring applications are dynamometers for force and torque measurement to identify cutting forces, accelerometers for vibrations on cutting tools, AE sensors for high frequency acoustic emission during the cutting process and current sensors for measuring the motor current as an indicator for motor torque that is, in turn, proportional to the cutting force [1]. Teti, Jemielniak, O'Donnell and Dornfeld give another profund review of methods for monitoring machining operations in [13]. They summarize different scopes of monitoring and referring sensor solutions. For the monitoring of tool condition (i.e. tool wear, breakage, geometry, temperature) they found the application of AE and vibrations measurement as well as cutting force, hydraulic pressure, motor power, optical and sound sensors plus the combination of them. For the surveillance of surface integrity, cutting force, vibrations and temperature are used as well. Additionally, the spindle motion displacement had been used. The main reason for lacking commercial availability of scientifically examined monitoring solutions remains the complex and non-universal use of evaluation methods [13].

In our work, we try to find a technical solution for a sensor system that integrates seamlessly within existing manufacturing environments without needing an individual setup. Our approach bases upon the finding that AE signals serve as an adequate source for the identification of lathe tool wear [4] and enable the differentiation between various effects concerning tool behavior, like tool and chip breakage (burst type signal, see [11]) or elastic deformation of material (continuous type, see [8]). Additionally to the six key issues from [1] that deal with sensor selection and signal processing, the overall measurement system has to be considered. Based on process considerations of lathe operations, fundamental requirements are derived. Additionally to the findings of research groups on tool wear analysis by acoustic emission [1, 7, 10, 13], we conduct further reference studies with AE-sensors in order to validate existing correlations between AE-signal and tool wear condition. Furthermore, the influence of environmental factors (like cooling lubricant, EMC, temperature) will be analyzed. First findings of these examinations are presented below. To reach our objectives we develop a tailored AE-sensor that suits the identified requirements at the lowest possible cost. This sensor will be evaluated based on reference measurements.

4 Experimental Results and Boundary Conditions

In order to determine the system requirements we designed a set of experiments to be performed on-site monitoring a CNC turning center. They regarded the cutting process as well as the machine environment influence on SAE measurements. In this paper we describe the execution and present the results of these experiments.

The measurement chain employed consists of four main elements. The *SEA-sensor* is a KISTLER sensor type 8152B1. It is a broadband acoustic emission sensor suited for the frequency range from 50 to 400 kHz. Outside of this range, sensor sensitivity, thus measurement reliability, drops. Along with the actual vibrating piezoelectric element is an impedance converter that works as a pre-amplifier. However, the high-frequency *amplifier* KISTLER 5125B1 is in charge of both the main amplification and feeding the preamplifier. Additionally, it contains a high- and a low-pass filter that are set to 500 and 50 kHz respectively. The signal is then sent through a 16-Bit *A/D-converter* from the type NI USB-6251 with a sample rate of 1.25 MHz. Measurement data processing, storage and display are handled by a MATLAB framework developed at the institute.

4.1 Experimental Setup

In our experiments we measured the acoustic emissions during the machining of different workpieces with cutting tips showing different grades of wear. For a machine equipped with each a barely used cutting tip and later on a discarded one, we measured emissions during the machining of aluminum, steel and brass parts. The machining of each of the pieces took around four seconds and was conducted using common cutting parameters depending on the material. The sensor bases on a piezoelectric transducer that converts the mechanical acoustic wave into a proportional voltage signal. From this signal, the AE-power of specific frequency bands can be derived, which is specified as a positive squared voltage value.

4.1.1 Effect of Cooling Lubricant Application and Idle Running

In order to determine whether the cooling lubricant application has an influence on the AE-signal, we performed a measurement of its application without any machining taking place at the time. A time sample of 1 ms was analyzed for the time before the application of cooling lubricant, thus the idle running, as well as a sample during the application of lubricant, as illustrated in Fig. 1. The idle running

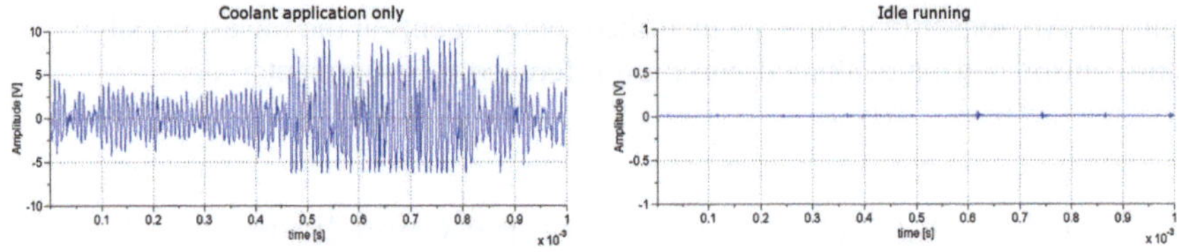

Fig. 1 Raw AE-signals of active cooling lubricant application (*left*) and idle running (*right*)

of the lathe appears to have no effect on AE-measurements. However, the mere application of cooling lubricant, even when no machining is taking place, produces significant signals, comparable to those that occur while machining aluminum. Early spectral analysis of the signal has shown the frequency range of this influence to be too wide to be neutralized through a band-stop filter without also affecting important machining data. This effect makes it hard to ensure that the measured AE actually corresponds to the machining itself. However, Sturm and Förster [12] state the possibility of recognizing the part of the signal relevant to the machining process performing an impulse density variation analysis on the raw signal, where changes in the relative incidence of signal values exceeding a certain threshold is analyzed.

4.1.2 Workpiece Material Influence

The before mentioned effect of the cooling lubricant application was present during the machining of aluminum and steel parts but not during the lubricant-free machining of brass parts. Figure 2 (left) shows a representative sample of the time signal (1 ms) measured during the machining of aluminum parts. The upper part of the figure shows the results when machining took place using a barely used cutting tip. The lower graph depicts the results when using a worn out cutting tip. The latter shows how worn out tips tend to cause interruptions, in which the amplitude of the signal is at least very close to zero, much more often than new tips do. Using a new tip causes the total amount of samples with a voltage lower than 0.2 V to drop from 60.01 %. A worn out one leads to a value of 53.81 %. The results for the machining of brass parts are depicted in Fig. 2 (right). The signal interruptions now appear to occur more often, when a barely used cutting tip is equipped for machining of the parts. This observation is supported by the fact that for the measurements on brass with a worn out cutting tip 80.08 % of all samples showed a voltage lower than 0.2 V, while only 75.04 % of samples using a new cutting tip did so.

The third material studied was stainless steel (X5CrNi18-10). The corresponding graphs are contained in Fig. 3. The presence of more interruptions regardless of the cutting tip in use is evident. However, they seem even more present when using a worn out cutting tip. The percentage of sample values below

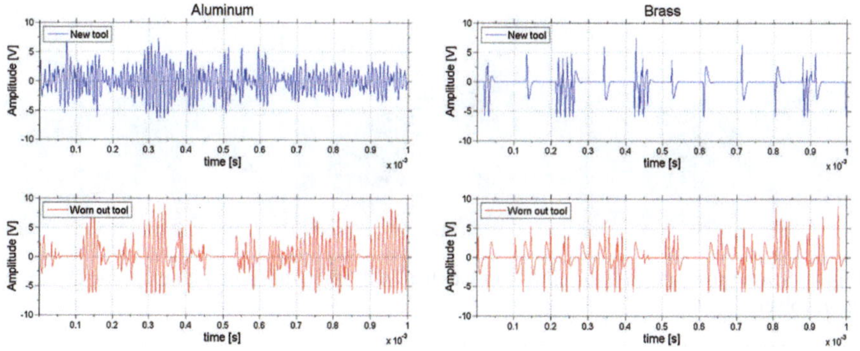

Fig. 2 Sample of the time signal (1 ms) measured during the machining of aluminum parts (*left*) and during the machining of brass parts (*right*)

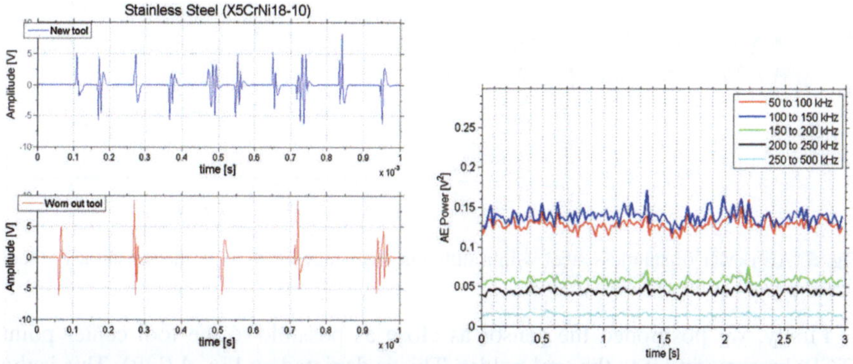

Fig. 3 Sample of the time signal (1 ms) measured during the machining of stainless steel parts (X5CrNi18-10) (*left*) and AE-power of stainless steel within different frequency bands with the predominant sensor positioning (*right*)

0.2 V for machining with a barely used cutting tip is 83.96 %, while a worn out one caused it to raise to 93.05 %.

Further analysis of time signals and their spectrums are pending, as well as a statistical analysis of the machining of higher lot sizes. Furthermore, the effect of the application of cooling lubricant requires further investigation as it is the most predominant source of undesired AE.

4.1.3 Sensor Positioning Influence

We placed the sensor at three different positions on the machine and repeated and measured the acoustic emission of the same cutting process for each of them. In order to determine whether the signal received by the sensor is adequate for our purposes, we analyzed the AE-power within a set of different frequency bands during the machining of stainless steel, and generated graphs for three seconds of the process.

Fig. 4 Sensor positions (*left* to *right*): Close to the TCP, at the tool post with magnetic clamp, at the revolver

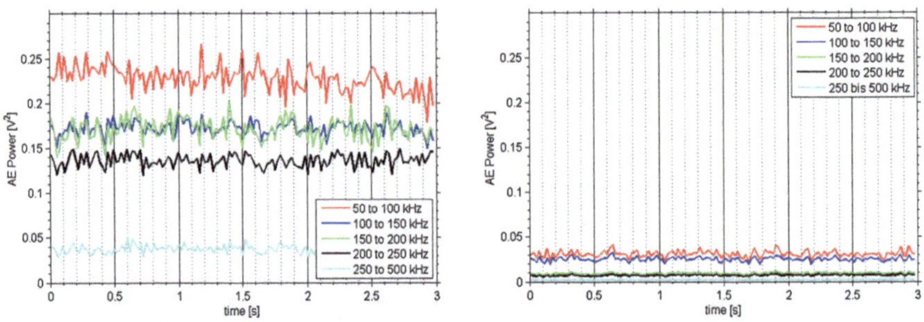

Fig. 5 AE-power of stainless steel within different frequency bands with the sensor at the tool post (*left*) and at the revolver (*right*)

Firstly, we positioned the sensor as close as possible to the tool center point (TCP) by screwing it to the tool holder. This is depicted on Fig. 4 (left). This is the same position the sensor was in, when all of the previously presented results were measured. The resulting mean over 151 samples of the AE-power for different frequency ranges is illustrated in Fig. 3 (right). Secondly, we attached the sensor to the tool post using a magnetic clamp. The location of the sensor and the clamp can be taken from Fig. 4 (middle). The positioning of the sensor further away from the cutting process but closer to the tool post's cooling-lubricant nozzle led to an AE power increase across all frequency ranges. Only the frequency range between 100 and 150 kHz remained relatively unaffected, as illustrated in Fig. 5 (left). Lastly, we used the magnetic clamp to position the AE-sensor even further away of the TCP, at the body of the machine's revolver (see Fig. 4, right). As expected, Fig. 5 (right) depicts how the power substantially dropped across all frequencies.

5 Result: Derivation of Requirements

The sensor system to be developed should be applicable to a broad *variety* of different lathes and machine arrangements. This refers to different amounts of lathe tools, spindles and the overall number of lathes that are integrated into the automated manufacturing process of a specific company. Physical environment and operating conditions have to be consistent for the fail-safe application of identified analysis algorithms. This can be achieved by attaching the sensor element to the "*common denominator*" of different lathes. By nature, the operational contact appears to be the point of highest reproducibility: The contact is formed only by the workpiece and the cutting tip that is identical on every other lathe. Already the tool holder may be found in different materials and sizes, although it is standardized to three specific shapes. The farther we move away from the operational contact, the more differences characterize a certain lathe.

The cutting tip itself is not suitable for a sensor integration due to its small size and its single-use concept. The tool holder features an universal shape and an adequate narrow distance to the operational contact. In order to be all-purpose by means of different lathe layouts, *the sensor element has to be integrated into the tool holder without altering its outer shape*. Every shape that differs from the standardized tool holder design may conflict with any parts of the lathe, like the spindle or the revolver. According to our measurements regarding sensor positioning and state of the art findings, signal quality suffers from a high amount of surfaces between the signal's source and the measuring point. Additionally, protecting the sensor from direct exposure to the cooling lubricant application may reduce its influence on measurements. As we aim to integrate the sensor into the tool holder, the signal transmitting surfaces are of low count anyway.

The requirement to integrate the sensor element into the tool holder, causes other requirements to arise: The *sensor must submit its data without being tethered to the surrounding environment*, as the revolver is multi-turn and would destroy a data wire. This implicates that the acquired raw-data must be preprocessed and prepared for a wireless transmission. A transmitter unit has to be integrated as well as a power supply to *provide electrical energy for the preprocessing and submission* of data. To fully take advantage of an automated tool wear monitoring, the sensing unit must be able to act autarkical for an adequate period of time. We consider a service life of at least three months to be practical (Table 1).

The transmitted data must be received by a *processing unit* in order to analyze the individual tool condition. Basically, one processing unit can handle one or multiple lathe tools. We consider *one processing unit per machine* to form a reasonable architecture of data processing. One unit per tool seems to be impractical due to the high cost and because the continuous data analysis of every tool is not necessary—a machine of two spindles can only have two tools under operation at the same time. On the other hand, one unit per manufacturing hall necessitates a high range of wireless communication for each sensor, resulting in a high energy demand. Keeping transmitter and receiver close-by with a predefined distance allows the reduction of transmitting power. Several machines can be smoothly *controlled* via *a central user interface* that is installed additionally. Alternatively, every single lathe could be equipped with an interface that visualizes the need for a tool change.

The presented experiments show a significant dependency between sensor signal and workpiece material, as well as a high influence of cooling lubricant

Table 1 Derived Requirements

	Requirement	Description
1	Holder integration	The sensor element must be integrated into the tool holder
2	Data submission	The data submission of the sensor unit must be wireless
3	Data preprocessing	Prior to the submission, the sensor data has to be prepared
4	Power supply	There must be an integrated power supply that drives the unit for at least three months
5	Processing architecture	One processing unit per lathe executes the analysis algorithms and can either forward or display the results
6	Tailored analysis	Different materials require different analysis algorithms
7	Manufacturing data acquisition	Manufacturing parameters (e.g. material, cooling lubricant) must be acquired

application on the measurement of AE, in addition to the expected influence of the condition of the cutting tip. Therefore, a *distinction between different materials under operation and manufacturing states might be required, in order to use tailored analysis algorithms*. Nevertheless, the distinction could potentially be made using the sensor signal only and a universal analysis algorithm. This would simplify the integration of the monitoring system into existing machines in comparison to acquiring this information from the lathe's manufacturing data. Overall, every manufacturing parameter that turns out to be of relevance should be extracted from the acoustic emission signal only.

6 Conclusion and Outlook

While the potential of AE technology for tool condition monitoring is well-known, we see the challenge in turning scientific results into a reliably working and economic sensor system. Considering those questions regarding sensor positioning and the effect of idle running of the lathe are satisfactorily answered, we came to three main conclusions regarding further experimental endeavors. The first is the need for additional analysis of the raw measurement data, i.e. the time signals, with the state of the art methods presented in Sect. 3. Furthermore, the disturbing effect on AE measurements of the application of cooling lubricant is to be either reduced or neutralized so that tool wear detection is unaffected by it. This could potentially be achieved either by an adequate signal post processing and/or the new sensor positioning suggested in Sect. 5. Finally, data regarding experiments considering the whole service life of a cutting tip are necessary. This should be combined with the examination of bad cutting parameters' influence towards the AE signal. Finding regularities between AE signal and cutting parameters holds potential for an extended use of the sensor for identifying bad cutting conditions. Taking the presented conclusions into consideration, a first tool wear detection solution for the presented purposes seems feasible in the near future.

References

1. Abellan-Nebot JV, Romero Subirón F (2009) A review of machining monitoring systems based on artificial intelligence process models. Int J Adv Manuf Technol 47(1–4):237–257
2. Albers A, Nguyen T, Burger W (2012) Energy-efficient hydrodynamic journal bearings by means of condition monitoring and lubrication flow control. Int J Condition Monit 2(1):4
3. Albers A, Nguyen T, Burger W (2012) Reducing power loss of hydrodynamic journal bearing by means of condition monitoring and automatic lubrication flow control. In: Proceeding TAE
4. Bhuiyan MSH, Choudhury SI, Nukman Y (2012) Tool condition monitoring using acoustic emission and vibration signature in turning. In: Proceedings of the world congress on engineering 2012 vol III, Newswood Ltd. : International Association of Engineers, London, U.K., vol III
5. Davim JP (ed) (2008) Machining: fundamentals and recent advances. SpringerLink, Bücher, Springer London, London
6. Dornfeld D (1992) Application of acoustic emission techniques in manufacturing. NDT E Int 25(6):259–269
7. Kamarthi SV, Kumara SRT, Cohen PH (2000) Flank wear estimation in turning through wavelet representation of acoustic emission signals. J Manuf Sci Eng 122(1):12
8. Moriwaki T, Tobito M (1990) A new approach to automatic detection of life of coated tool based on acoustic emission measurement. J Eng Ind 112(3):212
9. Nguyen T, Gold S, Weber J, Albers A (2012) Acoustic emission analysis in the early stage of heavy duty hydrodynamic: hydrostatic hybrid pod drive bearing development. In: Proceeding of the 9th international conference on condition monitoring and machinery failure prevention technologies CM2012/MFPT2012, London
10. Pontuale G, Farrelly FA, Petri A, Pitolli L (2003) A statistical analysis of acoustic emission signals for tool condition monitoring (TCM). Acoust Res Lett Online 4(1):13
11. Shiraishi M (1988) Scope of in-process measurement, monitoring and control techniques in machining processes: part 1: in-process techniques for tools. Precis Eng 10(4):179–189
12. Sturm A, Förster R (1990) Maschinen- und Anlagendiagnostik für die zustandsbezogene Instandhaltung. B.G. Teubner, Stuttgart
13. Teti R, Jemielniak K, O'Donnell G, Dornfeld D (2010) Advanced monitoring of machining operations. CIRP Ann Manuf Technol 59(2):717–739
14. Weck M, Brecher C (2006) Werkzeugmaschinen 3: Mechatronische Systeme, Vorschubantriebe, Prozessdiagnose. Springer, Berlin [u.a.]

Part III
Modelling of Dynamics and Fault in Gear Systems

Joint Power-Speed Representation of Vibration Features. Application to Wind Turbine Planetary Gearbox

Jacek Urbanek, Marcin Strączkiewicz and Tomasz Barszcz

Abstract Wind turbine condition monitoring is essential task in the process of maintaining machine operation at the optimal level. It is related to ensuring the profitability of investment and the provision of security in the environment of the turbine. However, operational conditions of turbine associated with non-stationary nature of the stimulus which is the wind, impede the correct diagnosis of the machine. In addition, a multitude of parameters adversely affects the clarity of predictions and setting alarm thresholds. In the article, the authors evaluate the impact of generator output power and rotational speed on selected vibration-based feature value. The study was performed for wind turbine planetary gearbox during fault development of the ring. It was possible due to historical data consisting peak-to-peak (P2P) values together with corresponding values of rotational speed and generator output power. For the purpose of the experiment the method that bases on calculation of arithmetic mean of the data in the segments corresponding to the chosen ranges of both rotational speed and generator output power is presented. Results are given in the form of three-dimensional charts, which allow assessing the impact of parameters on the studied feature. The paper shows that for machinery operating under varying regime proposed representation might serve as a valuable method for fault detection. Additionally, authors highlight the importance of analysis of vibration-based features as a function of two variables (rotational speed and power/load).

Keywords Wind turbines · Vibration analysis · Non-stationary operation · Planetary gearbox

---

J. Urbanek (✉) · M. Strączkiewicz · T. Barszcz
AGH University of Science and Technology, 30059 Kraków, Poland
e-mail: urbanek@agh.edu.pl

G. Dalpiaz et al. (eds.), *Advances in Condition Monitoring of Machinery in Non-Stationary Operations*, Lecture Notes in Mechanical Engineering,
DOI: 10.1007/978-3-642-39348-8_16,

1 Introduction

Over the last few years wind power is becoming an important sector of the energy industry. Therefore more attention is paid to the aspect of operation maintenance of the wind power generating machinery [1–4]. It does not only enable to limit the possible breakdown cost and time of repair, but in addition it provides higher productivity of the machinery [5].

Because of the fact that wind turbines operate under non-stationary wind behavior [6], thus the analysis of vibration-based diagnostic features might often lead to inconsistent conclusions. It has been shown that vibration-based features strongly depends on the load of monitored machinery. Therefore, it has been proposed in [7, 8] to use feature-load representation to present distribution of features against the operating conditions. As has been presented it provides better effectiveness in classification of data than simple statistical feature processing.

The issue of fault detection of machinery operating in non-stationary conditions has been widely investigated in the recent time. Many diagnostic techniques has been employed for this issue, including wavelets, the order analysis, adaptive filters and exploiting cyclo stationarity of vibrations [9–11], etc. The offline processing for multidimensional features has been investigated in [12, 13] and may include data processing using Principal Component Analysis, data projection technique, outliers analysis, etc. However, due to the economical and practical reasons building of advanced feature extracting module for online processing is not justified and is expected to be simplified. Such difficulty occurs not only for wind turbines, but also in many fields where rotating machinery is employed, e.g. mining, marine or aviation industries. It is important to remember that condition monitoring on this field has not yet found complete solutions and still needs improvement.

Therefore, the main aim of this paper is to present the analysis of the influence of operation parameters on the vibration-based features on the example of peak-to-peak (P2P) as it is in authors' opinion one of the most general analysis. Commonly used industrial wind turbine is chosen as a test object for presented investigation. For the purpose of the experiment, evolution of P2P estimator represented as a two dimensional function of rotational speed and generator output power is investigated during time period in which the wear of planetary gearbox has developed.

The paper is organized as follows. After the introductory part the investigated wind turbine design is briefly described along with placement of sensors used for the research. Next, the methodology of the investigation is explained. The results of the research are described in the further section and followed by the conclusions.

2 Investigated Wind Turbine

For the purpose of the investigation, commonly used turbine with nominal power of 1,500 kW was selected. The turbine has two nominal states of operation which depend on the wind speed—except for the nominal operational mode that corresponds to 1,800 RPM of generator rotational speed it also is enabled to operate on, so called, the "low" operational state that corresponds to 1,050 RPM of generator rotational speed.

In Fig. 1 one may find a kinematic system of the analysed wind turbine. The main rotor is driven by three blades and supported by the main bearing. It passes the torque to the planetary gearbox. Second bearing supporting the rotor is incorporated into the gearbox. The planetary gear has three planets, which are driven by the planet carrier. The planets transmit the torque to the sun gear, in the same time increasing the rotational speed. The sun shaft passes the torque from the planetary gear to the two-stage parallel gear. The parallel gear contains three shafts: the slow shaft clutched to the sun shaft, the intermediate shaft and the high speed shaft, which drives the generator. The generator produces AC current of a varying frequency. This current is converted first into DC and then into AC current of frequency equal to the grid frequency. Electric transformations are performed by the controller at the base of the tower. The drive-train multiplies the rotational speed from about 25 RPM on the main rotor to about 1,800 RPM at the generator.

Typical requirement for wind turbine drive-train condition monitoring systems is to measure vibrations with six measurement channels that each covers separate area of drive-train [14]. Namely: main bearing, planetary gearbox, 1st stage of parallel gearbox, 2nd stage of parallel gearbox, front of the generator and back of the generator

Additional required measurement covers rotational speed in order to asses acquired vibration data to proper operational state. In majority of condition monitoring systems dedicated for wind turbines value of generated power is measured as well as a supporting feature.

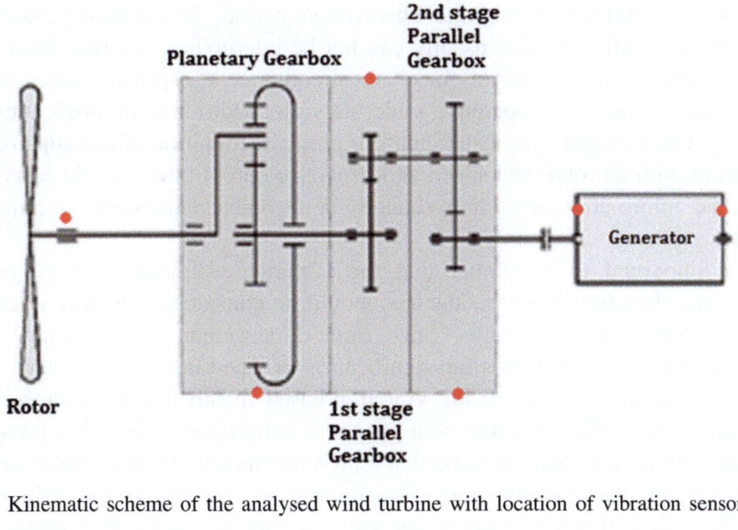

Fig. 1 Kinematic scheme of the analysed wind turbine with location of vibration sensors (*red dots*)

Presented study is performed for the planetary gear of wind turbine that suffered extended wear of the ring. Data used in described investigation was obtained from industrial condition monitoring system operating on commercial wind-farm. Presented measurements were carried out on the described turbine for over ten months (from 25th June 2009 to 16th April 2010), which resulted in 29 000 data points for each parameter (P2P, rotational speed and generator output power).

3 Proposed Representation of Vibration-Based Features

The main objective of this paper is to present selected vibration-based feature; namely, P2P as a two dimensional function of power and rotational speed. It is the authors' belief that such representation will provide better understanding of the influence of those parameters on measured vibration signals. In commonly used industrial condition monitoring systems dedicated for wind turbines vibration-based features are segregated according to rotational speed. Figure 2 presents P2P of vibration signals measured on planetary gearbox casing for around 10 months (from 25th June 2009 to 16th April 2010). Red color represents data obtained during operation in "high" operational state ($\sim$1,800RPM), while blue color corresponds to "low" operational state ($\sim$ 1,050RPM). Please note that horizontal axis represent subsequent samples number not the date time. Such representation is used in order to simplify the description of proposed representation method. According, to a commercial company that provided the data for described experiment, monitored turbine suffered extensive wear of the ring of planetary gearbox. Presented data covers time period in which mentioned fault has developed.

In Fig. 2, for both operational states, insignificant increase of observed feature (P2P) can be noticed during the observation period. Unfortunately, such low increase of vibration feature usually can not be interpreted as a manifestation of development of the fault of any mechanical component, especially when dealing with complex machinery operating under varying conditions. Although, presented data might be considered as insufficient for proper recognition of investigated fault its relation to both rotational speed and generator output power might carry some additional information regarding changes of technical conditions of monitored object.

It is important to remember that the vibration estimators depend on two parameters, therefore those parameters should be considered together when analyzing vibration-based features. Such three-dimensional representation should allow observers to evaluate the significance of operational parameters. Since provided database consist, beside vibration-based features (P2P), corresponding rotational speed values together with generator output power there is a possibility to define two-dimensional function that represents the relation between those three parameters. Figure 3 presents three dimensional representation of described data. It can be noticed that not whole power-speed surface is covered by corresponding

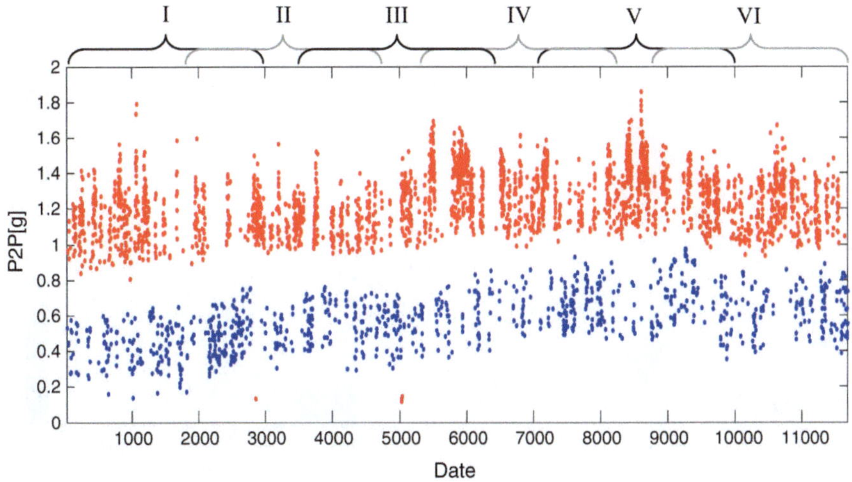

Fig. 2 Evolution of P2P values from 25th June 2009 to 16th April 2010. *Red* "high" operational state, *blue* "low" operational state. Brackets at the *top* of the figure covers data sets used for experiment

P2P values. It is caused by operational character of observed wind turbine. When wind turbine operates with relatively low rotational speed it can not generate significant power. Therefore, there are no vibration-based features in database that correspond to low rotational speed and high power values at the same time. Presented situation is similar when dealing with high rotational speed values but low power. However, range of generator output power is much wider when turbine operates with high rotational speed than when it operates with low speed. It is due to the fact that wind turbines often use generated power to feed processes related to their own operation (e.g. control mechanisms).

As shown in Fig. 3 representation of vibration-based features on power-speed plane results in three-dimensional "cloud of scattered points". Such representation might be relatively difficult for visual examination not to mention automatic data processing algorithms. In order to present scattered data as a three-dimensional surface that could be clear to analyze the authors wish to propose the method that bases on calculation of arithmetic mean of the data in the segments corresponding to chosen ranges of both rotational speed and generator output power. Mean value of preferred feature calculated for chosen power-speed segment is given by:

$$Feat_{\Delta RPM}^{\Delta P}(P, RPM) = \frac{1}{n}\sum_{n=1}^{n} Feat_{RPM'}^{P'}(n), \tag{1}$$

$$P' \subset \{P - \Delta P/2,\ P + \Delta P/2\}, \tag{2}$$

$$RPM' \subset \{RPM - \Delta RPM/2,\ RPM + \Delta RPM/2\}. \tag{3}$$

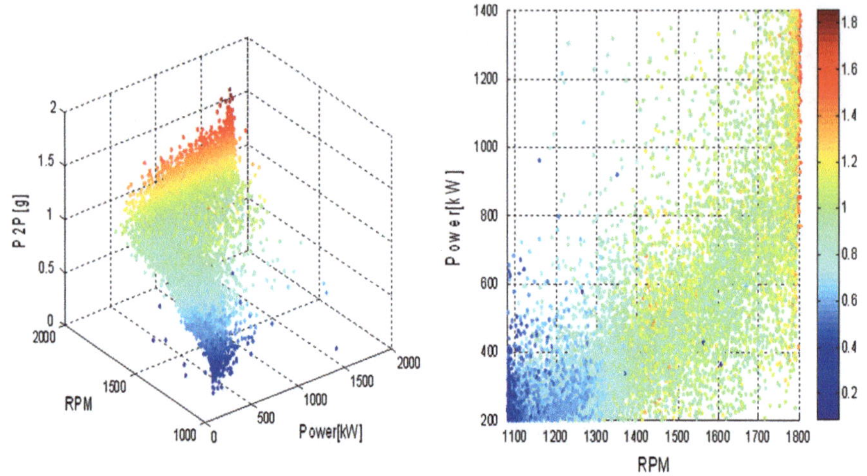

Fig. 3 P2P values represented as a two dimensional function of generator output power and rotational speed. *Left* three-dimensional plot, *right* colormap

where P stands for Power. ΔP and ΔRPM are range-widths of power and rotational speed respectively $Feat_{RPM'}^{P'}(n)$ is the nth vibration-based feature sample corresponding to the range of power P' and the range of rotational speed RPM'.

Figure 4 presents P2P values represented as a three-dimensional surface after application of formula given by Eq. (1). For following presentation first 3,000 samples of P2P feature, power and rotational speed, from database was used (marked by roman I in Fig. 2). Selected parameter ΔP was equal to 200 kW while $\Delta RPM$was equal to 100 RPM.

Figure 4 presents average P2P value in relation to both: rotational speed and power. It can be seen that resulted surface does not cover whole range of power for each value of rotational speed. It is caused by previously mentioned lack of vibration-based feature data that corresponds to those parameters. Additionally, strong dependency between rotational speed and P2P might be seen in the range

Fig. 4 P2P values represented as a two-dimensional function of generator output power and rotational speed

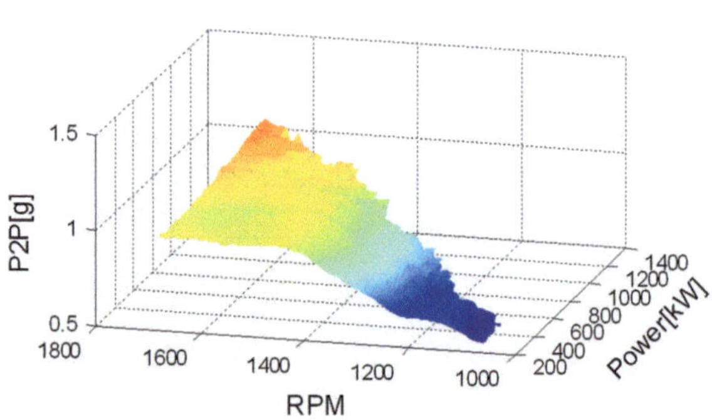

from 1,000 to 1,400 RPM. Above that value of rotational speed its relation to P2P value becomes less significant. For each value of rotational speed relation of observed feature value to power might be described as almost linear. When power values increase P2P values increase as well. Please mind that given example was calculated for planetary gearbox in proper technical condition. The next chapter presents the investigation of the influence of technical condition of studied object on proposed power-speed characteristics.

4 Results of the Experiment

Figure 5 contains proposed power-speed characteristics calculated for data described in Sect. 3. In order to investigate the influence of technical condition of planetary gearbox on proposed representation vibration-based feature data was divided into six consecutive sub-sections. Selected sub-sections are circumscribed by brackets in Fig. 1 and marked by roman numbering. Each sub-section contains

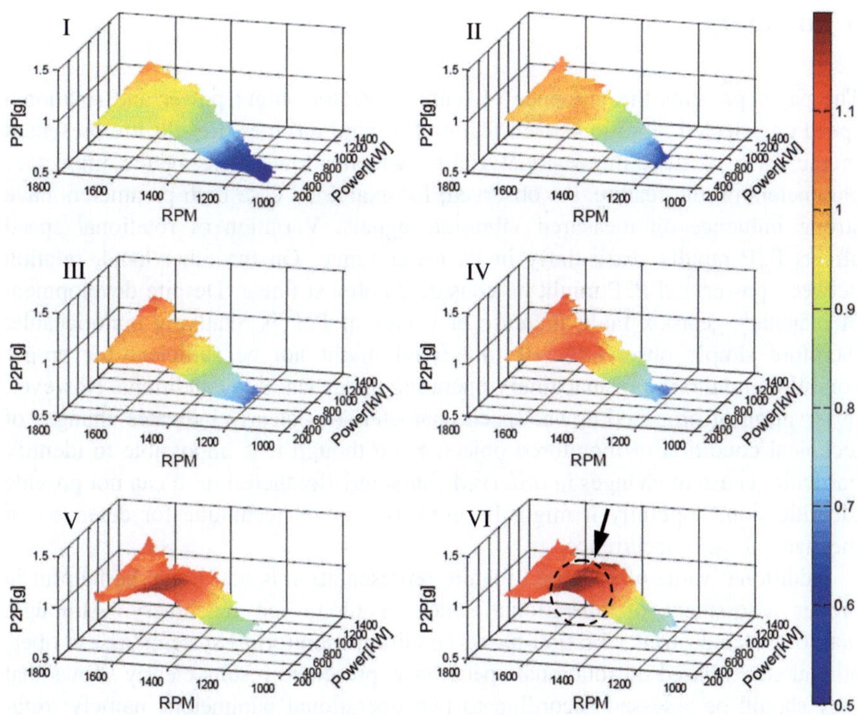

Fig. 5 P2P values represented as a two-dimensional function of generator output power and rotational speed for subsequent data segments. Roman numbers represents data segments marked in Fig. 2

3,000 samples of selected feature values (P2P) with corresponding values of power and rotational speed. As seen in Fig. 2 each sub-sections overlapped each other by approximately 40 %. As well as for results presented in Fig. 4, ΔP was equal to 200 kW while ΔRPM was equal to 100 RPM. Please note that all plots in Fig. 4 have the same scale.

Even though evolution of P2P values presented in Fig. 2 does not indicate clear the change of technical condition of observed planetary gearbox, power-speed characteristics show recognizable changes. Significant increase of P2P values might be seen around 1,400 RPM, especially for lower values of generator output power (marked in Fig. 5 VI). Additionally, slight increase of all P2P values can be noticed within observation period. It is the authors' belief that such irrelevant increase of feature values is caused by relatively inconsiderable influence of developed fault on measured vibration signal. Presented results confirms general idea of data representation proposed by Bartelmus and Zimroz in Ref. [7]. It is worth mentioning that subsequent representations of presented characteristic show that mentioned changes have evolutionary character; therefore, they might indicate development of planetary gearbox fault.

5 Summary

The paper presents the influence of both: generator output power and rotational speed on selected vibration-based feature for wind turbines. Results are presented in the form of three-dimensional charts, which allow assessing the impact of parameters on the feature. As observed, for examined case both parameters have strong influence on measured vibration signals. Variation of rotational speed affects P2P rapidly particularly in its lower range. On the other hand, relation between power and P2P might be considered almost linear. Despite development of planetary gearbox fault, increase of values of P2P is relatively unnoticeable; therefore simple observation of time-trend might not be sufficient for proper condition monitoring of machinery operating under variable conditions. However, observation of proposed power-speed characteristics clearly indicates changes of technical condition of monitored object. Even though it is impossible to identify particular cause of changes in observed characteristic; therefore; it can not provide identification, hopefully it might be used as a novel technique for detection of mechanical fault occurrence.

Additional value of proposed feature representation is that it can be helpful in proper assessment of operational states of monitored machinery. Although, majority of industrial vibration-based condition monitoring systems assess operational states based on rotational speed only, presented results clearly shows that they should be assessed according to two operational parameters, namely: rotational speed and power/load. Vibration features assessment without considering the influence of varying load might often lead to erroneous conclusions regarding technical condition of machinery in non-stationary operations.

In general, it is the authors belief that representation of selected diagnostic feature as a function of two main operational parameters (rotational speed and power/load) can give comprehensive information about dynamic character of observed machinery. It can be a valuable source of knowledge not only about the influence of operational parameters on selected diagnostic feature but also about object resonances or the character of its operation.

References

1. Urbanek J, Barszcz T, Uhl T (2012) Comparison of advanced fault detection methods for rolling bearings fault in wind turbines. Metrol Meas Syst XIX(4):715–726
2. Barszcz T, Randall RB (2009) Application of spectral kurtosis for detection of a tooth crack in the planetary gear of a wind turbine. Mech Syst Sig Process 23:1352–1365
3. Kusiak A, Li W (2011) The prediction and diagnosis of wind turbine faults. Renew Energy 36(1):16–23
4. Zimroz R, Bartkowiak A (2011) Investigation on spectral structure of gearbox vibration signals by principal component analysis for condition monitoring purposes. J Phys Conf Ser 305(1):012075
5. Żółtowski B (2004) Cempel Cz. (red.), Inżynieria diagnostyki maszyn, Instytut Technologii Eksploatacji PIB, Warszawa
6. Barszcz T, Bielecki A, Wójcik M (2010) ART-type artificial neural networks applications for classification of operational states in wind turbines. Lecture Notes in Computer Science 6114 LNAI PART 2 (2010), pp 11–18
7. Bartelmus W, Zimroz R (2009) A new feature for monitoring the condition of gearboxes in non-stationary operation conditions. Mech Syst Sig Proc 23(5):1528–1534
8. Bartelmus W, Chaari F, Zimroz R, Haddar M (2010) Modelling of gearbox dynamics under time varying non-stationary operation for distributed fault detection and diagnosis. Eur J Mech Solids 29(4):637–646
9. Antoni J (2009) Cyclostationarity by examples. Mech Syst Sig Process 23(4):987–1036
10. Makowski R et al. (2011) Adaptive bearings vibration modelling for diagnosis. Lecture Notes in Computer Science (in Lecture Notes in Artificial Intelligence). LNAI, vol 6943, pp 248–259
11. Urbanek J, Antoni J, Barszcz T (2012) Detection of signal component modulations using modulation intensity distribution. Mech Syst Sig Process 28:399–413
12. Worden K, Staszewski WJ, Hensman JJ (2010) Natural computing for mechanical systems research: a tutorial overview. Mech Syst Sig Process 25(1):4–111
13. Bartkowiak A, Zimroz R (2012) Data dimension reduction and visualization of multidimensional data with application to gearbox diagnostics data: comparison of several methods. Solid State Phenomena 180:177–184 doi:10.4028/www.scientific.net/SSP.180.177
14. Gellermann T (2003) Requirements for condition monitoring systems for wind turbines. AZT Expertentage vol 10, 11 Nov 2003, Allianz 2003

Parallel Autoregressive Modeling as a Tool for Diagnosing Localized Gear Tooth Faults

Paweł Rzeszuciński and James R. Ottewill

Abstract One of the standard approaches widely used in the field of localized gear tooth fault diagnosis is the creation of residual signals i.e. signals obtained after removing the deterministic frequency components from a Time Synchronously Averaged vibration signals. Most of the time these components are removed based on the knowledge of the characteristic gearbox frequencies. Sometimes however such information is not available. AR modeling, a type of time series modeling, has been found to be capable of faithfully estimating the deterministic content of the signal allowing meaningful residual signals to be created. An improvement to the classic AR modeling approach is proposed in this text. The method is applied to experimental data taken from a gearbox in both healthy and faulty condition. The improvement derived from the new method is quantified through a comparison with results obtained by applying Time Synchronous Averaging and the classic AR modeling method to the experimental data.

Keywords Condition monitoring · Autoregressive modeling · Residual signal · Localized tooth faults

1 Introduction

Autoregressive modeling, where the current value of a signal can be estimated based on the weighted sum of its past values, is the most common type of time series modeling [1]. In the mathematical terms the word "autoregression" informs

P. Rzeszuciński (✉) · J. R. Ottewill
ABB Corporate Research Center, Ul. Starowiślna 13A, 31-038 Kraków, Poland
e-mail: pawel.rzeszucinski@pl.abb.com

J. R. Ottewill
e-mail: james.ottewill@pl.abb.com

G. Dalpiaz et al. (eds.), *Advances in Condition Monitoring of Machinery in Non-Stationary Operations*, Lecture Notes in Mechanical Engineering,
DOI: 10.1007/978-3-642-39348-8_17,

that values of a given variable appear in an equation as a function of itself. In such an arrangement the future values can be estimated by regressing the current values on historical values of itself (hence the term *auto*) [2]. Used alone or in conjunction with other signal processing techniques, it has been successfully applied in the field of gearbox condition monitoring for the purpose of detecting localized gear tooth faults [3–6]. The method has been found to be capable of precisely estimating signals that contain clear, distinct peaks in their frequency spectra. Because gear vibrations mainly comprise harmonic sinusoids, with the most predominant components relating to frequencies originating from the meshing of the gears and the rotation of the shafts, AR models are capable of faithfully reconstructing the deterministic part of a healthy gear signal [3]. Applying AR modeling to gear vibration signals allows any changes within the signal that originate from non-deterministic, transient events to be observed. This is achieved by creating residual vibration signals through the subtraction of the waveform obtained using the AR modeling method from the original signal. The expected residual signal for a healthy gear is comprised of normally distributed noise [7]. In the case of a localized gear tooth fault being present in one of the teeth, the residual signal contains only information generated by the impulsive nature of contact during the meshing action between the two gears [8]. In this paper, an improvement to the classic AR modeling method for diagnosing localized gear tooth faults, based on the creation of residual signals using a combination of AR models of different order, is proposed.

2 Theoretical Background

Any autoregressive dataset may be estimated by using a finite number of its historical values. For a signal $x(n)$ the AR process can be defined as per Eq. 1 [5]:

$$\hat{x}(n) = \sum_{k=1}^{p} a(k)x(n-k) \tag{1}$$

where $\hat{x}(n)$ is the estimated n-th value of the dataset, $a(k)$ is the k-th coefficient of the AR model and p is the order of the AR model. It can be seen that the n-th sample can be estimated as the sum of the weighted previous samples. The AR model coefficients are most commonly derived by means of the second-order statistical characteristics of the signal. This is usually embodied in the form of the autocorrelation matrix. The AR process satisfies the Yule-Walker equations, hence the AR model coefficients can be found efficiently by solving them. Those equations can be represented in a matrix form as per Eq. 2 [5] where $r_{xx}(k)$ is the autocorrelation sequence of signal $x(n)$. The equations can be solved using a number of different approaches, the most common including Levinson-Durbin recursion or the Burg method. A number of different tools for solving the Yule-Walker equations are implemented in Matlab software, and the mathematical details can be found in e.g. [9, 10].

$$\begin{bmatrix} r_{xx}[0] & r_{xx}[-1] & \cdots & r_{xx}[p-1] \\ r_{xx}[1] & r_{xx}[0] & \cdots & r_{xx}[-p+2] \\ \vdots & \vdots & \ddots & \vdots \\ r_{xx}[p-1] & r_{xx}[p-2] & \cdots & r_{xx}[0] \end{bmatrix} \begin{bmatrix} a[1] \\ a[2] \\ \vdots \\ a[p] \end{bmatrix} = - \begin{bmatrix} r_{xx}[1] \\ r_{xx}[2] \\ \vdots \\ r_{xx}[p] \end{bmatrix} \tag{2}$$

3 Application to Gear Fault Diagnosis

Performing diagnosis with the use of the AR model begins with estimation of the parameters of the model, which are derived from vibration signal generated by the monitored gears operating under healthy conditions. Once established, the coefficients of the model can be used as a filter to perform signal estimation based on linear prediction as per Eq. 1. Such an estimated signal has the same phase as the original signal hence by subtracting one from another a residual signal can be obtained. Since the filter coefficients are derived based on a healthy signal, the components of the current vibration that is considered to have originated from normal machine operations will be faithfully represented, and so filtered out during the creation of the residual signal. The resulting residual signal is expected to be randomly distributed and represents the AR model prediction error [3].

On the other hand, as the localized gear tooth fault develops, short duration excitations modulate the vibration signal resulting in shaft speed sidebands that appear around the GMF and its harmonics. This, in turn, changes the amplitude distribution of the signal [11, 12]. The spacing of the sidebands is equal to the relevant shaft speed—the driving or the driven—depending on which shaft the faulty gear is attached to [8]. In addition to the sidebands appearing as a result of the amplitude modulation of the signal, the tooth fault impulses excite a broad spectrum of frequencies adding to the amount of baseline noise in the signal. As a fault becomes more severe a greater number nondeterministic frequency components appear in the spectrum. The AR model, derived from a healthy signal, is not capable of faithfully estimating such changes and, as a result, new features start to appear in the residual signal.

4 Optimal AR Model Order

In AR modeling, the current value of the signal is estimated based on the weighted sum of the past values of the signal. It is therefore important to establish the optimal amount of past samples to be included in the estimation of the value of the current sample (i.e. establish the order of the model). The order value has an influence on the shape of the estimated power spectrum. Using too many samples increases the uncertainty of the estimation and may cause some spurious spectral peaks to appear in the estimated signal, whereas too few samples are likely to cause inaccurate peak detection, hence lead to poor sample value prediction [5, 13]. For this reason some a priori knowledge is essential to estimate the optimal order of the model. There are many algorithms that are capable of tackling similar optimization problems, however for gear fault diagnosis [4] suggested using solutions derived by Akaike: the Final Prediction Error (FPE) or Akaike Information Criterion (AIC). Both methods find the optimal relation between the changes in the variance of the signal and the amount of AR model estimation error.

The algorithms can be implemented as per Eqs. 3 and 4 respectively [4]:

$$\mathrm{FPE}(k) = \frac{N+k}{N-k}\rho_k \tag{3}$$

$$\mathrm{AIC}(k) = Nln(\rho_k) + 2k \tag{4}$$

where N is the length of the analyzed signal, k is the order of the AR model and ρ_k is the AR model estimation error power. Applying any of the techniques on the vibrations generated by a healthy gearbox leads to optimization of the AR model order.

Once the optimal AR model length is established, the appropriate filter coefficients can be obtained by solving the Yule-Walker equations as per Eq. 2. In this text the solution to the equations, together with the AR model error power have been established with the use of the *aryule* Matlab function. The resultant filter coefficients are then used to estimate the samples of the currently analyzed signal. Since, as mentioned before, the estimated signal is in phase with its original version, therefore the step of subtracting the estimated and the original signal is a straightforward operation [1]. Eventually the residual signal can be analyzed for any signatures of a localized gear tooth fault.

5 Parallel Autoregressive Modeling

The use of the AR model for creation of residual signals can be further improved with the proposed extension to the method—Parallel Autoregressive modeling (PAR). In the proposed method two residual signals are derived in parallel based on the same vibration signal. One residual signal is derived from a signal estimate created with the use of the optimal order of the AR model, whereas the second signal uses a slightly higher model order. The output of the PAR is a residual signal that is created as a result of subtraction of the two initially estimated residual signals. The reasoning behind this approach is that by estimating the signal based on two different, yet very similar, models the process of creation of the final residual signal will lead to even greater decrease of the amplitude of the "healthy" part of the signal, while retaining clear presence of the faulty signatures.

This is due to the fact that two estimations created using a filter of slightly different model orders can both faithfully estimate the deterministic part of the signal, while still being unable to reliably represent the transient components appearing in the signal as a result of the incipient gear tooth fault. Even though some differences in the estimation of the signal will be present for both residual signals, they will be much greater in the case of the impulsive part of the estimation. Essentially the resultant residual signal will have decreased amplitude in the healthy region, while the section with the transient event will remain relatively high. This is likely to give an earlier indication of tooth fault (through greater separation between the healthy and faulty parts of the signal) and lead to an improved process of automated fault detection with the use of specific Condition Indicators e.g. kurtosis.

6 Results

Data used to validate the results were generated on an experimental rig in the Diagnostics Laboratory of the ABB Corporate Research Centre in Kraków, Poland. A schematic of the system is shown in Fig. 1. The system is comprised of a 1.5 kW induction motor operating at a speed of 1,400 RPM which was driving a two-stage, parallel shaft, helical reduction gearbox. The system was loaded by a 6 kW DC motor with nominal speed of 1,500 RPM which, in turn, was connected with a permanent magnet motor. The last part of the system was used purely for the purpose of obtaining an encoder signal which was permanently mounted to its shaft. The encoder signals were later used to derive the Time Synchronous Averages of the vibration signals. After 30 tests for a nominally healthy gearbox

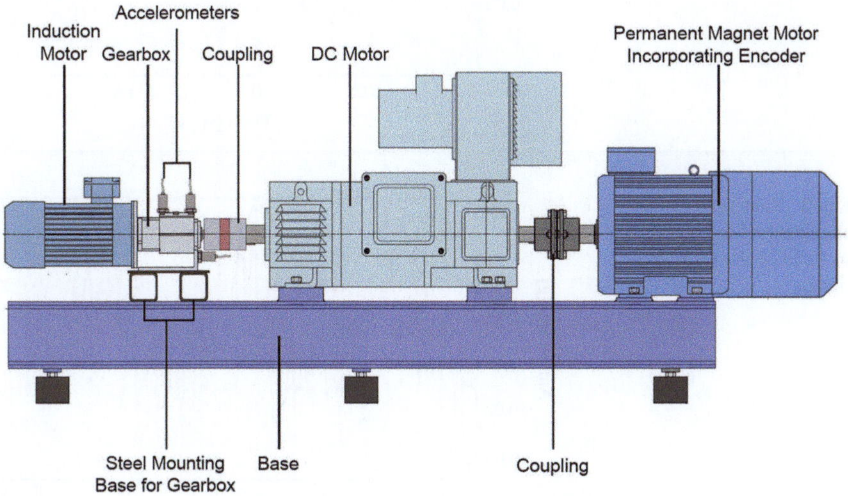

Fig. 1 Schematic of the experimental test rig. Repeated from [14]

condition were recorded, the gearbox had a pit artificially seeded to one of the teeth on the 2nd stage output gear. The vibration generated by this gear was subject to the analysis in this text. Vibration signals were recorded for 26.88 s at the rate of 4,444 samples per second with the use of SKF CMSS 2110 accelerometers. A much more detailed description of the test rig and the process of seeding the fault can be found in [14].

Figure 2 shows two residual signals created during the implementation of the PAR model. The first signal is created from a 68th order model, which was computed by both order selection criteria (Eqs. 3 and 4) as being the optimal for the estimation of the Time Synchronously Averaged original vibration signal. The second signal is estimated with an 88th order model. Both presented waveforms contain a lot of similarities, especially in the healthy part of the signal. Magnified views on the healthy and the faulty parts of the residual signals show that the healthy part has been estimated virtually identically (Fig. 3) however there are clear differences in the part containing the signatures of the localized gear tooth fault (Fig. 4). Subtraction of the two residual signals (Fig. 5) leads to creation of a signal where the healthy section has significantly lower amplitude (compared with the initial residual signal) and, as a result, the presence of the impulse is much clearer. Consequently the kurtosis values derived on the final residual signal are

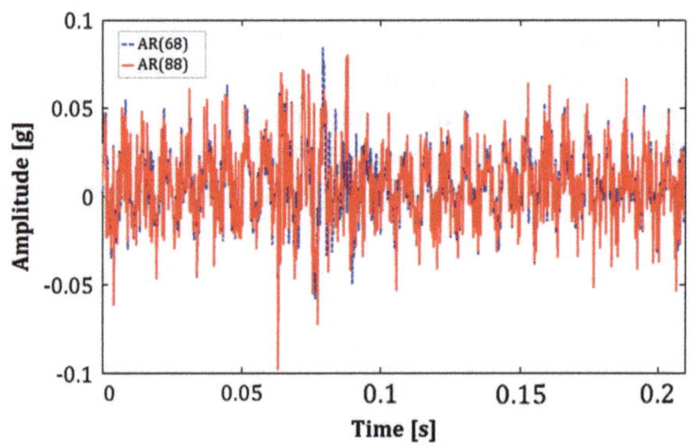

Fig. 2 Two residual signals created based on the PAR model

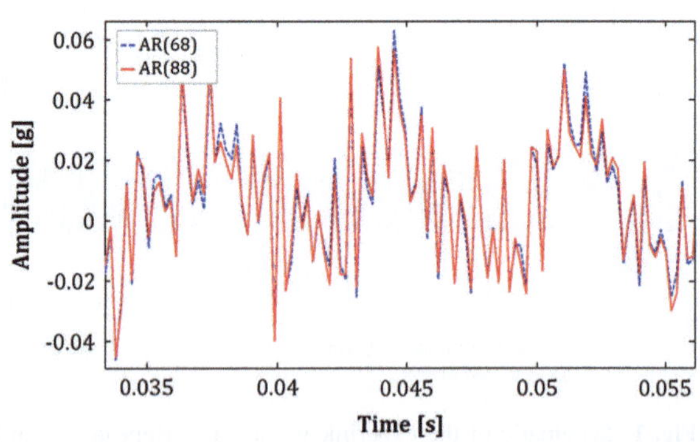

Fig. 3 Near perfect alignment of the healthy sections of the residual signals

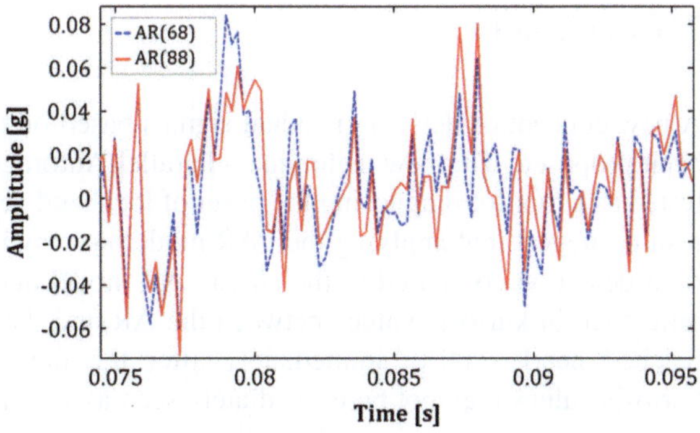

Fig. 4 Differences of the sections of the residual signals that contain signatures of tooth fault

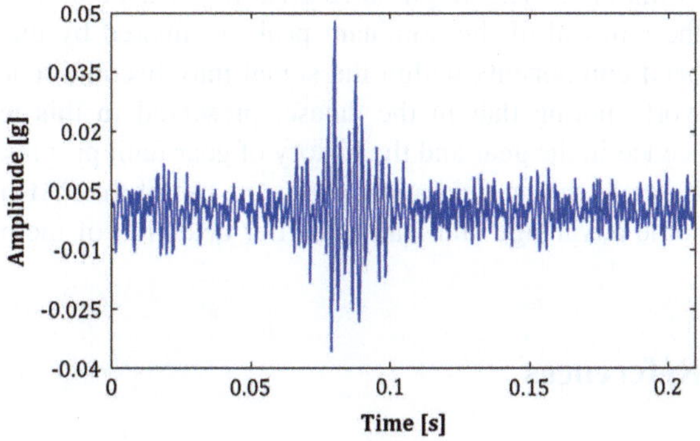

Fig. 5 Residual signal created as an output from the PAR model

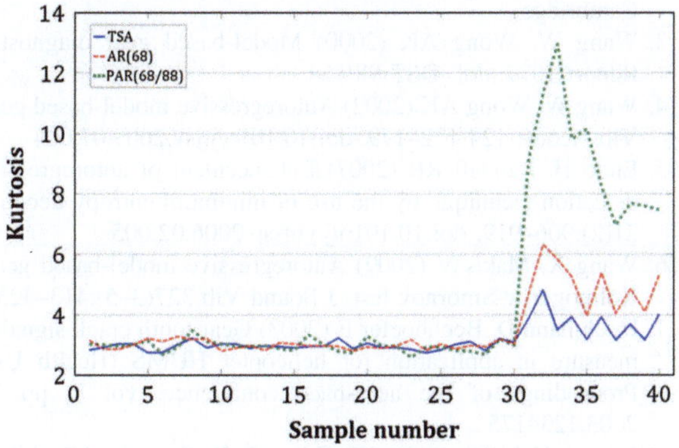

Fig. 6 Kurtosis values for the TSA, AR based residual signal and PAR based residual signal

greater when compared with the TSA and the classic AR based residual signal (Fig. 6). It may be observed that the impulsive content within the residual signal obtained with the use of the proposed method contained much clearer fault indication, which led to stronger reaction form the kurtosis indicator.

7 Conclusions

A new concept of deriving residual signals based on the classic AR modeling has been proposed. The new technique—Parallel Autoregressive modeling—has been tested as a tool for diagnosing the onset of localized gear tooth faults. The obtained results suggest that applying the PAR modeling may increase the chances of early fault detection compared to the classic AR model and the TSA signal alone. The difference in kurtosis values between the AR and the PAR based residual signals reached nearly 200 % immediately after the fault was seeded. The obtained kurtosis values may not be immediately seen as very advantageous in a laboratory environment however it might indicate increased capabilities of detecting localized gear tooth faults in their early stages through higher sensitivity to the presence of impulses. This might have great importance in noisy conditions where, through the removal of the dominant peaks estimated by the model, all transient, broadband components within the signal may become much easier to detect. It is also worth noting that in the dataset presented in this text the fault was artificially seeded in the gear and the history of gear fault progress could not be observed. It is believed that in field applications the use of the PAR model estimation might give some advantage and lead to earlier detection of incipient faults.

References

1. Matignon R (2005) Neural network modeling using Sas enterprise miner. AuthorHouse
2. Barreto H, Howland FM (2005) Introductory econometrics. Cambridge University Press, Cambridge
3. Wang W, Wong AK (2000) Model-based gear diagnostic techniques, DSTO Technical Report 1079. doi: DST 98/164
4. Wang W, Wong AK (2002) Autoregressive model-based gear fault diagnosis. Trans ASME J Vib Acoust 124:172–179. doi:10.1016/j.jsv.2009.07.004
5. Endo H, Randall RB (2007) Enhancement of autoregressive model based gear tooth fault detection technique by the use of minimum entropy deconvolution. Mech Syst Sig Process 21(2):906–919. doi:10.1016/j.ymssp.2006.02.005
6. Wang X, Makis V (2009) Autoregressive model-based gear shaft fault diagnosis using the Kolmogorov-Smornov test. J Sound Vib 327(3–5):413–423. doi:10.1016/j.jsv.2009.07.004
7. Hochmann D, Bechhoefer E (2003) Gear tooth crack signals and their detection via the FM4 measure in application for helicopter HUMS (Health Usage and Management System). Proceedings of the aerospace conference, vol 7. pp 3313–3326. doi:10.1109/AERO.2003.1234175
8. Rzeszuciński PJ, Sinha JK, Edwards R, Starr A, Allen B (2012) Amplitude of probability density function (APDF) of vibration response as a robust tool for gearbox diagnosis. Strain 48(6):510–516. doi:10.1111/j.1475-1305.2012.00849.x
9. Percival DB, Wladen AT (1993) Spectral analysis for physical applications: multitaper and conventional univariate techniques. Cambridge University Press, Cambridge
10. Broersen PMT (2006) Automatic autocorrelation and spectral analysis. Springer, Berlin
11. McFadden PD (1985) Low frequency vibration generated by gear tooth impacts, nondestructive testing (NDT). International 18(5):279–282. doi:10.1016/0308-9126(85)90008-2

12. Combet F, Gelman L (2011) Novel adaptation of the demodulation technology for gear damage detection to variable amplitudes of mesh harmonics. Mech Syst Sig Process 25:839–845. doi:10.1016/j.ymssp.2010.07.008
13. Sörnmo L, Laguna P (2005) Bioelectrical signal processing in cardiac and neurological applications. Academic Press, London
14. Ottewill JR, Orkisz M (2013) Condition monitoring of gearboxes using synchronously averaged electric motor signals. doi:10.1016/j.ymssp.2013.01.008

Modulation Sidebands of Planetary Gear Set

M. Karray, F. Chaari, A. Fernandez Del Rincon, F. Viadero and M. Haddar

Abstract In this paper a torsional model of planetary gear set test bench is developed. This bench is composed by two identical planetary gears connected by a common shaft. A tri axial accelerometer is mounted in both rings. The mechanism leading to modulation sidebands is modelled. Time histories are characterized by a periodic fluctuation. Spectra showed sidebands around the mesh frequency and its harmonics. Simulation is achieved to demonstrate amplitude modulation and rich sidebands that agree well with the experimental results.

Keywords Planetary gears · Torsional model · Amplitude modulation · Sidebands

M. Karray (✉) · F. Chaari · M. Haddar
Dynamics of Mechanical Systems Research Unit, National School of Engineers of Sfax, BP1173, 3038 Sfax, Tunisia
e-mail: maha-karray@hotmail.fr

F. Chaari
e-mail: fakher.chaari@gmail.com

M. Haddar
e-mail: mohamed.haddar@enis.rnu.tn

A. Fernandez Del Rincon · F. Viadero
Department of Structural and Mechanical Engineering, University of Cantabria, Cantabria, Spain
e-mail: fernandra@unican.es

F. Viadero
e-mail: viaderof@unican.es

G. Dalpiaz et al. (eds.), *Advances in Condition Monitoring of Machinery in Non-Stationary Operations*, Lecture Notes in Mechanical Engineering,
DOI: 10.1007/978-3-642-39348-8_18, © Springer-Verlag Berlin Heidelberg 2014

1 Introduction

Planetary gear sets are widely used in many applications due to advantages over parallel shaft arrangement such as high power density and large reduction in a small volume [1]. Despite their distinguishing advantages, noise and vibration remain key concerns in these applications. Mesh and bearing dynamic forces are the primary sources of such behavior. Two approaches were adopted model based and experimental analysis. Major analytical studies proposed lumped-parameters models to predict free and forced vibration characteristics of planetary gear sets where the gears are rigid bodies interconnected by springs representing teeth in mesh and support bearings. The modeling of the dynamic behavior of planetary gear have been initiated by Cunliffe et al. [2] who developed a model with a 13° of freedom to analyze the frequencies and mode shapes with single fixed carrier. Kahraman [3] established an analytical model including variable mesh stiffness. He extended this model to a 3-dimensional model for the case of helical gears and determined the influence of mesh phasing between planets on the dynamic response [4]. Using a 3D model Kahraman and Blankenship [5] determined the distribution of loads on the planets. Kahraman [6] reduced the model to a purely torsional model to find the natural frequencies and the corresponding vibration modes. Parker [7] rigorously proved the effectiveness of the choice of mesh phasing between planets to reduce vibration due to the process of meshing.

There are few experimental works on planetary gear vibrations to understand the complex dynamics of planetary gears. Hidaka et al. [8–14] published a series of reports, they studied some important issues such as load distribution, effect of different meshing-phase among sun/ring-planet meshes, etc. [15] investigated with experimental and theoretical approaches the influence of several system level factors in gear stress. Few experimental researches were dedicated to two stage planetary gear and this can be explained by the complexity of this transmission.

The study presented in this paper is expected to provide the understanding of the dynamic behavior of two stages planetary gear and the origin of amplitude modulation in vibration signals.

2 Dynamic Model of Planetary Gear Set

The planetary gear dynamic model presented in this paper is inspired from a planetary gear bench set developed at the University of Cantabria of Spain (Fig. 1).

This model is based on the one developed by [16]. Only rotational motions of the gear bodies are considered and translational degrees of freedom in that model are eliminated.

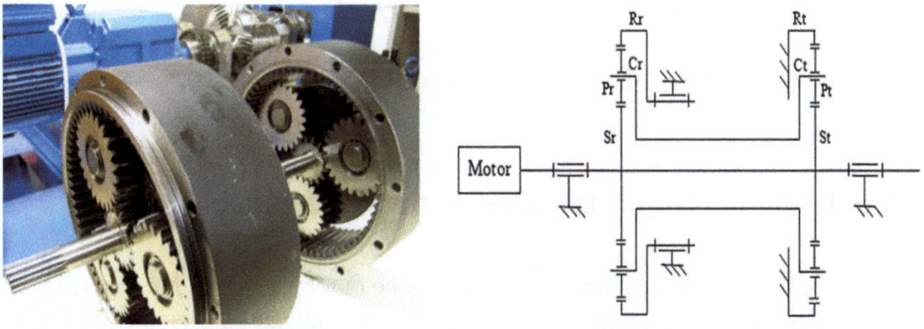

Fig. 1 Planetary gear set rig

The gear bodies are assumed rigid with moment of inertia I_c, I_r, I_s, I_p of respectively the carrier, the ring, the sun and the planets. The planetary gear set have 3 planets.

The sun-planet and the ring-planet tooth meshes are modeled as linear spring k_{sn}, k_{rn}, n = 1…N. In our case we consider that all sun-planet mean mesh stiffness are equal $k_{sn} = k_{sp}$ and all ring-planet mesh stiffness are equal $k_{rn} = k_{rp}$. The planets are assumed identical and equally spaced.

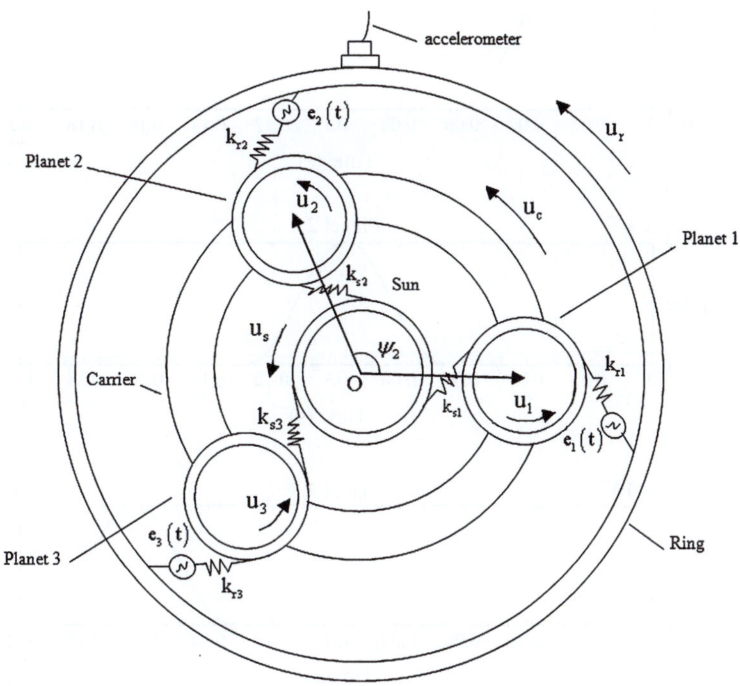

Fig. 2 Planetary gear model

The system's equation of motion is given by: $M\ddot{q} + K(t)q = F$.

With q is the vector of degrees of freedom, M is the inertia mass, K(t) is the stiffness matrix and F is the external applied forces.

3 Modeling of Amplitude Modulation Phenomenon

In this model an accelerometer is mounted on the outer surface of the ring (Fig. 2).

The measured acceleration for the accelerometer indicates a periodic fluctuation variation in vibration amplitudes as planets pass through this transducer location. This fact causes the apparition of an amplitude modulation of vibration in time histories which results in amplitude modulation with sidebands around mesh frequencies and harmonics.

For a complete revolution of the carrier, the accelerometer experiences the disturbances from all 3 planets in sequence. The influence of every planet on the transducer is assumed for duration of $T_c/3$ with T_c is the rotational period of the carrier. According to this assumption, when the planet i approaches to the

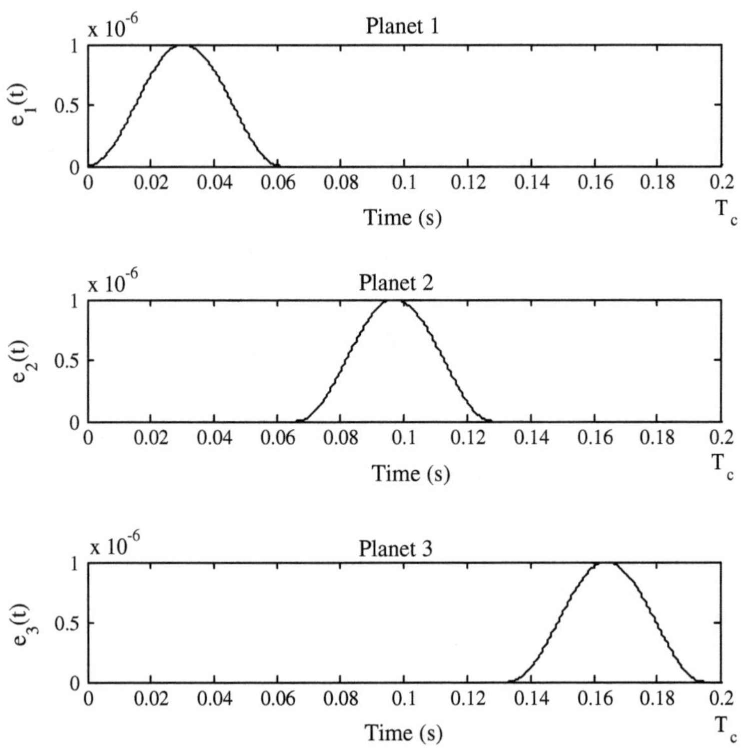

Fig. 3 Hanning function e(t)

transducer location its influence in terms of vibration amplitude will increase for the first $T_c/6$ to reach to its maximum level when the planet *i* is at the transducer location, then it diminish to zero at the end of the next $T_c/6$ time period. This will be followed by planet $i + 1$ that will dominate the response of the transducer for the next $T_c/3$, and so on.

This phenomenon can be modeled by a three Hanning function $e_1(t)$, $e_2(t)$ and $e_3(t)$ as presented in Fig. 3. These functions will be introduced as additional displacement on the different lines of action on the ring-planets meshes as shown in Fig. 2.

4 Numerical Simulation

The characteristics of the gear system are giving in Table 1. We choose 1,500 rpm as rotational speed of the motor, so that the speed of the carrier is about 300 rpm and the mesh frequency is 325 Hz.

Figure 4 shows the time varying acceleration on the ring without taking into account the influence of transducer location. The corresponding spectrum is presented in Fig. 5. The mesh frequency and its harmonics are well observed.

Now the influence of the position of the accelerometer is introduced in the model which will cause an additional force $F_1(t)$ to be added to the initial static force F. Figure 6 shows the time varying acceleration on the same ring, it is well observed a periodic fluctuation of the signal. The corresponding spectrum shown in Fig. 7 reveal significant sidebands around the gear mesh harmonic frequency.

Table 1 Planetary gear characteristics

	Carrier	Ring	Sun	Planet
Number of teeth	–	65	16	24
Moment of inertia (kg m^2)	0.0021	0.697	0.0003	0.002
Base diameter (mm)	57.55	249.38	61.38	92.08
Mesh stiffness (N/m)	$k_{sp} = 2.75110^8$		$k_{rp} = 4.127210^8$	

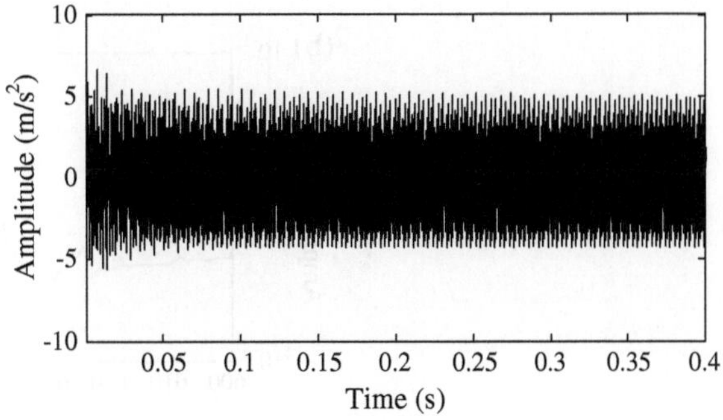

Fig. 4 Evolution of the acceleration on the ring without taking into account the influence of transducer location

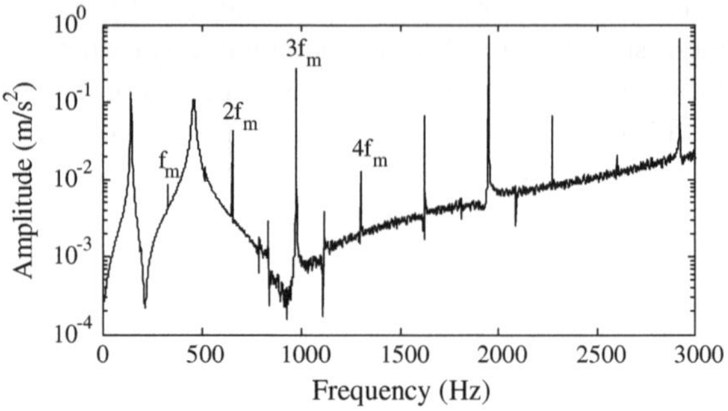

Fig. 5 The spectrum of the acceleration on the ring without taking into account the influence of transducer location

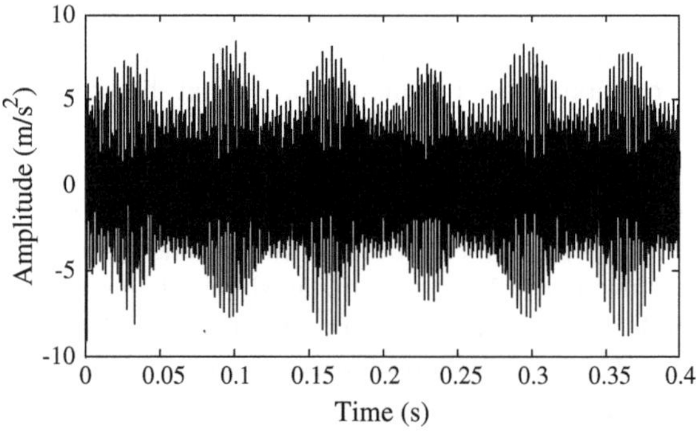

Fig. 6 Evolution of the acceleration on the ring with taking into account the influence of transducer location

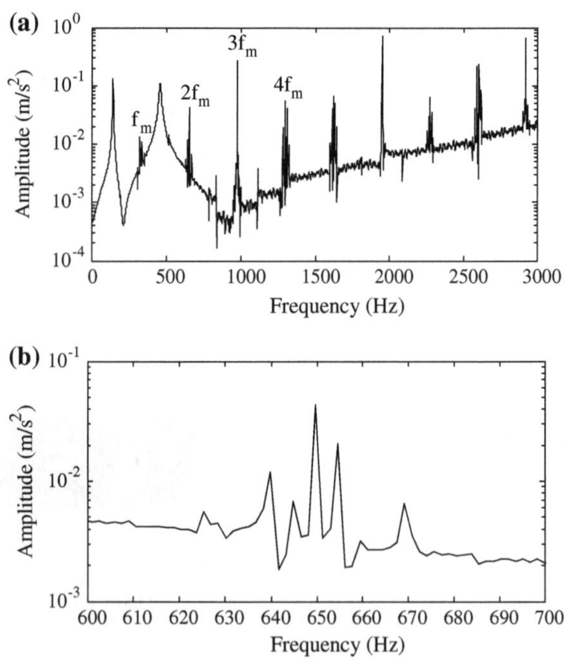

Fig. 7 The spectrum of the acceleration on the ring with taking into account the influence of transducer location

The sidebands are found to be at $nf_m \pm 3mf_c$ where f_c is the carrier rotational frequency and (n, m) are integer.

In order to validate the results an experimental test is made on the planetary gear test bench with a carrier speed of 300 rpm.

Figure 8 shows the time evolution of the acceleration measured on the ring. An apparent amplitude modulation is observed. As a result the spectra of this time history presented in Fig. 9 exhibit significant number of sidebands around the gear mesh harmonics.

The results obtained with numerical simulation are confirmed experimentally.

Fig. 8 Evolution of the acceleration on the ring

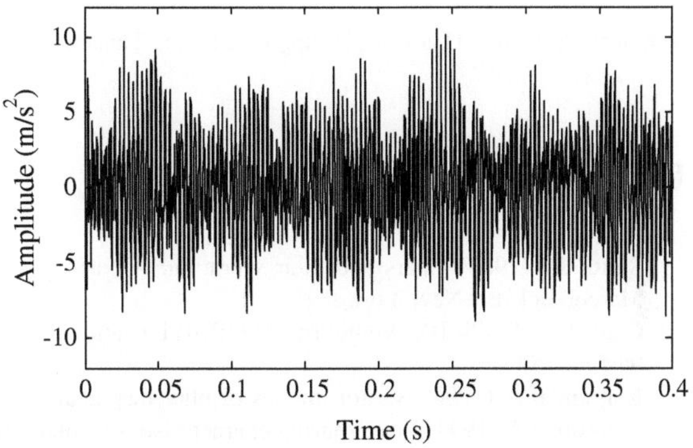

Fig. 9 The spectrum of the acceleration on the ring

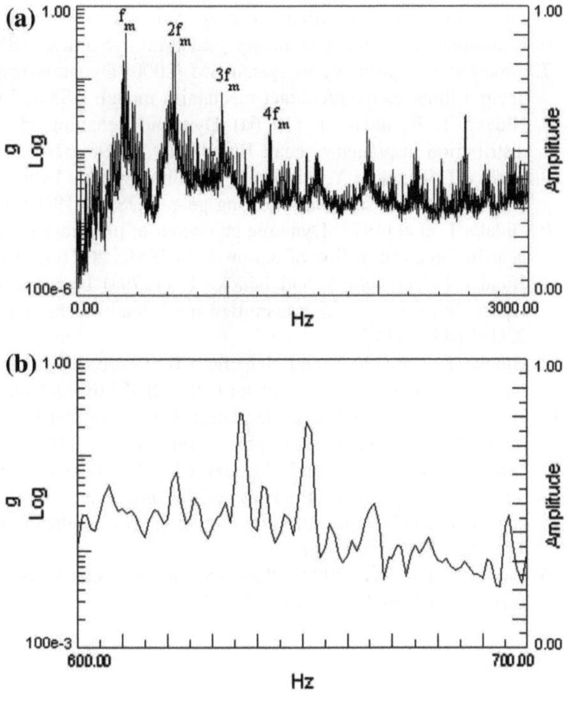

5 Conclusion

In this work modulation sideband of planetary gear set is investigated analytically and experimentally. A torsional model is developed. Then modulation sidebands are taken into account by a suitable modelling. An experimental investigation shows the validity of numerical results. The Time varying acceleration of the ring is characterized by an amplitude modulation that results in significant number of sidebands around the gear meshes harmonics. This result reveals one of the sources of amplitude modulation. Practically, position of vibration transducer is very important when diagnosing a planetary gear transmission.

Acknowledgments This work is supported by the Tunisian-Spanish Joint Project no A1/037038/11.

References

1. Smith JD (1983) Gears and their vibration. A basic approach to understanding gear noise. Marcel Dekker, New York
2. Cunliffe F, Smith JD, Welbourn DB (1974) Dynamic tooth loads in epicyclic gears. J Eng Ind 94:578–584
3. Kahraman A (1994) Natural modes of planetary gear trains. J Sound Vib 173:125–130
4. Kahraman A (1994) Load sharing characteristics of planetary transmissions. Mech Mach The 29:1151–1165
5. Kahraman A, Blankenship GW (1994) Planet mesh phasing in epicyclic gear sets. Proceedings of international gearing conference. Newcastle, UK, p 99–104
6. Kahraman A (1994) A planetary gear train dynamics. ASME J Mech Des. 116(3):713–720
7. Parker RG, Agashe V, Vijayakar SM (2000) Dynamic response of a planetary gear system using a finite element/contact mechanics model. ASME J Mech Des 122(3):305–311
8. Hidaka T, Terauchi Y (1976a) Dynamic behavior of planetary gear, 1st report. Load distribution in planetary gear. Bull JSME 19:690–678
9. Hidaka T, Terauchi Y, Ishioka K (1976b) Dynamic behaviour of planetary gear, 2nd report. Displacement of sun gear and ring gear. B JSME 19(138):1563–1570
10. Hidaka T, et al (1977) Dynamic behaviour of planetary gear, 3rd report. Displacement of ring gear in direction of line of action. Bull JSME 20:1663–1672
11. Hidaka T, Terauchi Y and Ishioka K (1979a) Dynamic behaviour of planetary gear, 4th report. Influence of the transmitted tooth load on the dynamic increment load. Bull JSME 22(168):877–884
12. Hidaka T, Terauchi Y and Nagamura K (1979b) Dynamic behaviour of planetary gear, 5th report. Dynamic increment of torque. Bull JSME 22(169):1017–1025
13. Hidaka T, Terauchi Y and Nagarnura K (1979c) Dynamic behaviour of planetary gear, 6th report. Influence of meshing-phase. Bull JSME 22(169):1026–1033
14. Hidaka T, Terauchi Y and Nagamura K (1979d) Dynamic behaviour of planetary gear, 7th report. Influence of the thickness of the ring gear. Bull JSME 22(170):1142–1149
15. Ligata H (2007) Impact of system-level factors on planetary gear set behaviour. Dissertation, The Ohio State University
16. Lin J, Parker RG (2002) Planetary gear parametric instability caused by mesh stiffness variation. J Sound Vib 249:129–145

A Novel Method of Gearbox Health Vibration Monitoring Using Empirical Mode Decomposition

Jacek Dybała and Adam Gałęzia

Abstract Nowadays, many industrial types of machinery rely on different types of gears to transmit rotational torque. Gearbox faults are one of the major reasons for breakdown of industrial machinery. Therefore, gearbox diagnosing is one of the most important topics in machine condition monitoring. A number of signal processing techniques are described for the vibrodiagnostics of gearboxes, but there are also different limitations for vibration based gear diagnostic methods. For some specific requirements (e.g. time-triggered signal acquisition), not all of described techniques can be always applied in industrial reality. This paper introduces a novel, easy to use method of gearbox health vibromonitoring based on Empirical Mode Decomposition (*EMD*) and a time-domain analysis of vibration signal parts. Six sets of data collected from gearboxes are used to validate the proposed method. The experimental results demonstrate that the gear tooth defect can be detected and evaluated at an early stage of development when both Empirical Mode Decomposition and statistical analysis technique are used.

Keywords Vibration condition monitoring · Gearbox diagnostics · Empirical mode decomposition (EMD) · Time-domain signal analysis

J. Dybała (✉) · A. Gałęzia
Warsaw University of Technology, Institute of Vehicles,
ul. Narbutta 84, 02-524 Warsaw, Poland
e-mail: jdybala@simr.pw.edu.pl

A. Gałęzia
e-mail: agalezia@simr.pw.edu.pl

G. Dalpiaz et al. (eds.), *Advances in Condition Monitoring of Machinery in Non-Stationary Operations*, Lecture Notes in Mechanical Engineering,
DOI: 10.1007/978-3-642-39348-8_19, © Springer-Verlag Berlin Heidelberg 2014

1 Introduction

Gearboxes are widely used in industrial machinery. Gearbox diagnostics is important for guaranteeing machine safety and production efficiency, its damage can cause the breakdown of a machine and lead to serious consequences. Condition monitoring of the gearbox allows tracking the development of its technical degradation. Monitoring of gearbox degradation helps to prevent unpredictable downtimes and serious failures.

Gearbox is a complex vibration system whose components interact generating complex vibration signal. The analysis of machine vibration signals is one of the most important means used for condition monitoring of gearboxes. The vibration signal of a gearbox carries information about the fault in the gears and fault detection is possible through analyzing the vibration signal using different signal processing techniques. Many researchers have proposed various methods for vibrodiagnosing gearboxes. For fault diagnosis of gearboxes, researchers have proposed statistical analysis [1], time domain averaging [2], time–frequency analysis [3], cyclo-stationary analysis [4] and some special analyses based on intermodulation phenomena [5], the local meshing plane concept [6] and the pattern recognition approach [7]. These papers have considerably enriched the knowledge of gearbox condition monitoring, but gearbox diagnostics, in general, is still a challenging problem.

One of the most important issues in condition monitoring of gearboxes is to detect a fault as early as possible, and alert operator before it develops into a catastrophic damage. Generally one can distinguish two main degradation processes of gears: degradation of contact surface and tooth breaks [8]. Degradation of contact surface, such as pitting, evolves slowly in time causing that contact surfaces degrade in an increasing number of teeth deteriorating meshing. It is difficult to detect this at an early stage of degradation. Tooth breaks can be driven by a sudden increase of load or can be caused by fatigue. If gears are constantly working under loads which are slightly above designed values, fatigue-driven break can occur. It is crucial to detect a break at an early stage of its development as failure proceeds very rapidly.

Extraction of diagnostically useful features from vibration signal often requires signal enhancement techniques such as: signal filtering (classical, adaptive and optimal filters), modeling (cyclo-stationary, stochastic) or decomposition (wavelets) [9–15]. Recently, a novel decomposition method called Empirical Mode Decomposition (*EMD*) has been proposed [16]. Recent publications on *EMD* [17–21] show its effective application in many diagnostic tasks.

The following paper investigates the *EMD*-based approach for gearbox diagnostics. By using *EMD* a raw vibration signal is decomposed into the first Intrinsic Mode Function (*IMF*) and the final residue after the extraction of the first empirical mode. Then, a novel index called Health Index (*HI*) is proposed to describe the condition of a gearbox. Health Index is based on kurtosis of the first empirical mode and kurtosis of the final residue. To validate the proposed method, six sets of data collected from gearboxes are used. Results of the analysis show that the proposed method is effective in condition monitoring of gearboxes, especially in the detection of gear tooth faults.

2 The Proposed Method

The values of many statistical parameters (e.g. kurtosis) of vibration signal are known to increase when approaching the end of a gearbox life. However, operating conditions (e.g. load) as well as construction solutions (e.g. stiffness of gear shafts) cause that in certain operational cases the values of statistical parameters have different levels and may vary significantly. This creates the need for an universal indicator, which would be used for gearbox condition monitoring in different construction and maintenance situations. Such an indicator would indicate that a gear tooth is damaged and gearbox destruction can occur in short time.

In this section, we propose a Health Index to monitor gearbox condition. Health Index proposed herein is based on Empirical Mode Decomposition and a time-domain analysis of vibration signal parts. Empirical Mode Decomposition (*EMD*) decomposes a signal into Intrinsic Mode Functions (*IMF*s) which represent different oscillation modes embedded in the signal. Based on the *EMD* algorithm, any signal $x(t)$ can be reconstructed by a linear superposition of empirical modes:

$$x(t) = \sum_{i=1}^{n} c_i(t) + r_n(t) \tag{1}$$

where $c_i(t)$ is i-th empirical mode and $r_n(t)$ is the final residue after the extraction of n empirical modes. The successive modes include signal components from different frequency bands ranging from high to low frequency. Therefore, EMD corresponds to a non-stationary and nonlinear filtering [22].

Each empirical mode $c_i(t)$, called Intrinsic Mode Function (*IMF*), fulfills the following two conditions [16]: (1) in the whole empirical mode, the number of mode local extremes and the number of mode zero-crossings are equal or differ at most by one and (2) at any point, the local average of upper and lower envelope is zero. The algorithm for the extraction of *IMF*s from original signal $x(t)$ is called sifting process and it consists of the following steps [23]:

Step 1: Define the residue as $r_0(t) = x(t)$.
Step 2: Identify all the local extremes (maxima and minima) of $x(t)$.
Step 3: Connect all the local maxima (respectively minima) with a line known as the empirically determined upper envelope $E_{max}(t)$ [respectively the lower envelope $E_{min}(t)$].
Step 4: Construct the mean of empirically determined upper and lower envelope $m(t) = 0.5 \cdot (E_{min}(t) + E_{max}(t))$.

Step 5: Define the detail (proto-*IMF*) as $d(t) = x(t) - m(t)$ and replace $x(t)$ by $d(t)$.
Step 6: Repeat steps 2 ÷ 5 until $d(t)$ meets the *IMF* conditions and the stoppage criterion of the sifting process is fulfilled, then derive *i*-th *IMF* from $d(t)$ and replace $x(t)$ by $r_i(t) = r_{i-1}(t) - d(t)$.
Step 7: If the stoppage criterion of the signal's decomposition is fulfilled then finish the decomposition process; otherwise, go to step 2.

The second *IMF* condition is too rigid to use, so it is necessary to change it to implement the *EMD*. The local average of upper and lower envelope must be close to zero according to some criterion. The evaluation of how small the amplitude of the local average is may be done in comparison with the amplitude of the corresponding mode. In [24] authors introduce a new criterion based on the local mode amplitude $a(t) = 0.5 \cdot (E_{max}(t) - E_{min}(t))$ and the evaluation function $\sigma(t) = |m(t)/a(t)|$. In this paper, $d(t)$ meets the second *IMF* condition, when $max(\sigma(t)) < \theta$ (the coefficient θ was equal to 0.5).

A critical part of the *EMD* procedure is the stoppage criteria of the sifting process and decomposition process. The stoppage criterion of the sifting process determines the point when sifting is complete and a new *IMF* has been found. The stoppage criterion of the decomposition process determines how many components will be extracted from the signal.

A detailed description of Empirical Mode Decomposition together with the sifting and decomposition stoppage criteria is provided in [23].

In the proposed approach, a raw vibration signal $x(t)$ is decomposed into the first empirical mode $c_1(t)$ and the final residue $r_1(t)$ using *EMD*:

$$x(t) = c_1(t) + r_1(t) \tag{2}$$

The value of Health Index (*HI*) is calculated for a gearbox vibration signal in the following way:

$$HI = \frac{K_{c1} - K_{r1}}{K_x} \tag{3}$$

where K_x is the kurtosis value of the raw vibration signal $x(t)$, K_{c1} is the kurtosis value of the first empirical mode $c_1(t)$ and K_{r1} is the kurtosis value of the final residue $r_1(t)$. In case an early fault occurs, *HI* should be higher than the threshold which equals 0. The fact that HI is higher than 0 indicates that Kc1 > Kr1.

3 Case Study with Real Vibration Data

In this paper, six data sets of vibration signals collected from a diagnostics test stand are used to validate the proposed *HI*. Vibration signals from experiments of a fatigue break of the gear tooth were recorded on back-to-back tester (Fig. 1). This test stand allows for performing accelerated durability tests of toothed wheels.

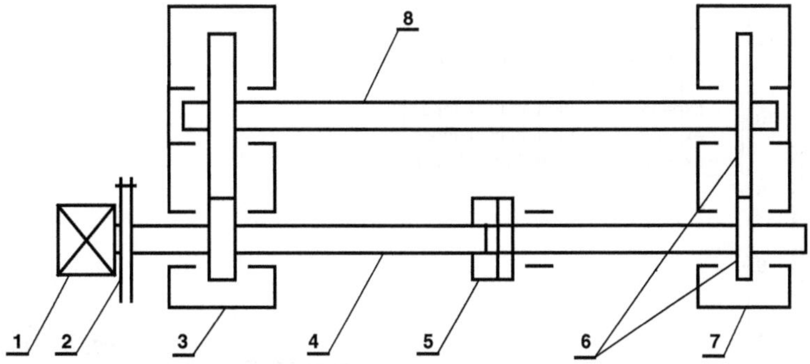

Fig. 1 Diagram of a test stand: **1** engine, **2** clutch, **3** closing gearbox, **4** coupling shaft, **5** stretching clutch, **6** examined wheels, **7** examined gearbox, **8** shaft

The test stand consists of two gearboxes driven by an electric motor. The shaft between the pinions is divided, which enables to tilt one of its parts versus the other and thus to introduce relevant meshing forces. To enable the measurement of the torsional moment introduced in shaft, tensometers were placed on it. Helical gears were mounted in the closing gearbox which demonstrated bigger durability than the spur gears mounted in the examined gearbox. Due to this, during all experiments, always the examined toothed gear was damaged. The toothed wheels were made of 20H2N4A steel, carbon-coated and hardened to 60 HRC. Each test of accelerated fatigue break of the gear tooth was performed until complete tooth break. During the research 3 experiments were performed with low stiffness shafts and 3 with high stiffness shafts. Each experiment was of a different duration but our analysis was limited to the last 600 s before the complete break of the gear tooth.

The test stand was characterized by the following parameters: gearbox load 1,200 $\div$ 1,500 Nm; motor speed 1,450 $\div$ 1,500 rpm; gear ratio of both gearboxes 1.296; module of toothed wheels 4 mm; number of pinion teeth 27; number of wheel teeth 35; distance of wheel axis 125 mm.

A triple-axle Bruel and Kjaer 4504A accelerometer was installed in the upper part of the examined toothed gear's body, over the place where the bearing of pinions shaft is mounted. Vibration signals were recorded continuously with sampling frequency 20,000 Hz. In the examined gearbox an induction sensor was mounted which caused test stand shutdown upon tooth breaking. The momentary speed of pinion shaft was also registered. Vibration signal measured in vertical direction has been chosen for further analysis, as more informative, because this direction was more parallel to the line of teeth contact.

It is worth pointing out that the crack at the base of the tooth was not initiated artificially (e.g. by cutting it) and was only related with fatigue. In order not to disturb the process of fatigue-related breaking of a tooth, the test was carried out continuously from the start till the complete break of a gear tooth.

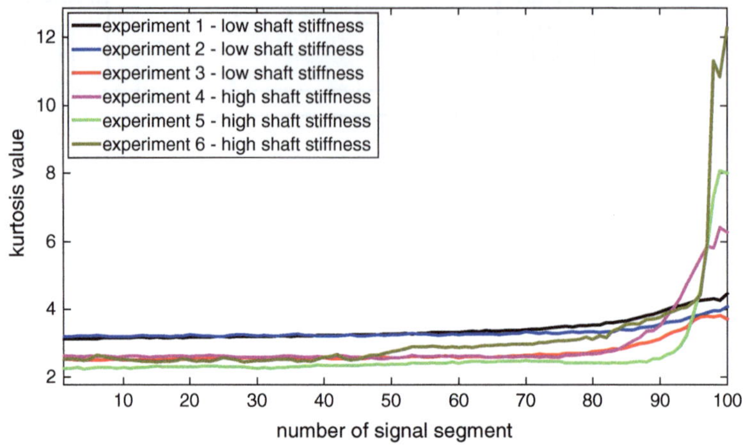

Fig. 2 Kurtosis values of raw vibration signals from 6 experiments

The values of kurtosis of raw vibration signals from 6 experiments are presented in Fig. 2. Presented values of parameter were calculated from the last 100 succeeding signal segments, each lasting for 6 s. For each case, the break of the gear tooth occurred on the 100th signal segment. High variability of kurtoses of machine vibration signals makes it difficult to infer about current condition of gears (Fig. 2).

To investigate the usability of the proposed indicator, *HI* values were calculated from the above mentioned 100 succeeding signal segments before the tooth break.

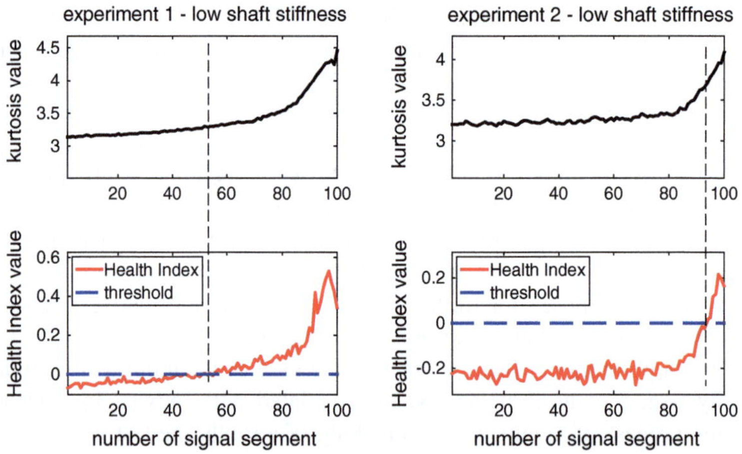

Fig. 3 Kurtosis values of raw vibration signals (*black*) and corresponding *HI* (*red*) for experiments 1 and 2

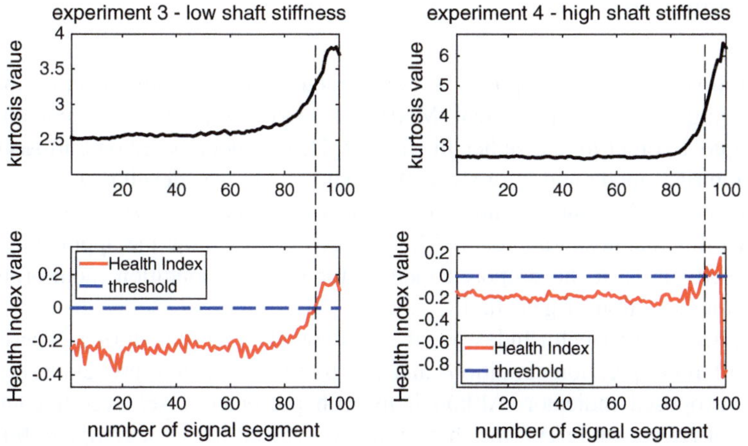

Fig. 4 Kurtosis values of raw vibration signals (*black*) and corresponding *HI* (*red*) for experiments 3 and 4

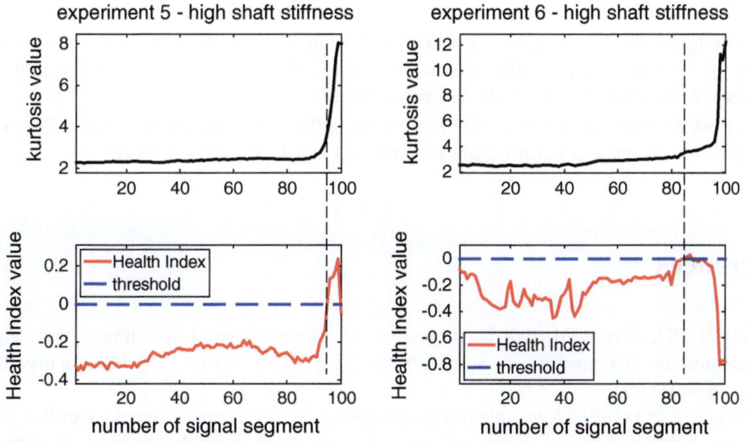

Fig. 5 Kurtosis values of raw vibration signals (*black*) and corresponding *HI* (*red*) for experiments 5 and 6

Figures 3, 4, 5 present kurtosis values of raw signal and corresponding *HI* values for each experiment.

The research results show that the proposed Health Index, with fault detection condition defined as the first occurrence of $HI > 0$, demonstrates good performance in the detection of a fatigue break of the gear tooth. In each presented case, despite different operation conditions (e.g. load) as well as construction solutions (e.g. stiffness of gear shafts), *HI* exceed threshold, which allows for detecting the development of a fatigue break of the gear tooth.

4 Conclusion

The following paper proposes a novel method of gearbox health vibration monitoring based on Empirical Mode Decomposition and statistical analysis technique. In order to realize health warning, Health Index based on kurtosis of the first empirical mode and kurtosis of the final residue was developed.

Six sets of lifetime vibration data collected from gearboxes were used to validate this diagnostic approach. The research results show the applicability and the effectiveness of the proposed method for gearbox condition monitoring and automatic detection of gear fault.

Application of Health Index to vibration signals, recorded during experiments with different operation conditions and construction solutions, proves universality of the proposed indicator. Although in each presented case, overall course of values of *HI* was different, the threshold was crossed before a complete break of the gear tooth giving the operator enough time to stop the machine and avoid a catastrophic damage.

The analysis results illustrate that the proposed method performs successfully in detecting gear failure and has a big potential in engineering applications.

Acknowledgments The authors thank Robert Gumiński (Warsaw University of Technology, Institute of Vehicles) for providing measurement data. The authors would also like to thank the reviewer for his valuable comments and suggestions.

This work has been carried out in the framework of the COST Action TU 1105 "NVH analysis techniques for design and optimization of HYBRID and ELECTRIC vehicles".

References

1. Samuel PD, Pines DJ (2005) A review of vibration-based techniques for helicopter transmission diagnostics. J Sound Vibr 282(1–2):475–508. http://dx.doi.org/10.1016/j.jsv.2004.02.058
2. McFadden PD (1991) A technique for calculating the time domain averages of the vibration of the individual planet gears and the sun gear in an epicyclic gearbox. J Sound Vibr 144(1):163–172. http://dx.doi.org/10.1016/0022-460X(91)90739-7
3. Williams WJ, Zalubas EJ (2000) Helicopter transmission fault detection via time-frequency, scale and spectral methods. Mech Syst Sig Process 14(4):545–559. http://dx.doi.org/10.1006/mssp.2000.1296
4. Zimroz R, Bartelmus W (2009) Gearbox condition estimation using cyclo-stationary properties of vibration signal. Key Eng Mat 413:471–478. http://dx.doi.org/10.4028/www.scientific.net/KEM.413-414.471
5. Radkowski S (2008) Non-linearity and intermodulation phenomena tracking as a method for detecting early stages of gear failures. Insight: Nondestr Test Condition Monit 50(8):419–422. http://dx.doi.org/10.1784/insi.2008.50.8.419
6. Mączak J (2012) Local meshing plane analysis as a source of information about the gear quality. Mech Syst Sig Process. http://dx.doi.org/10.1016/j.ymssp.2012.09.012
7. Dybała J (2012) Vibro diagnostics of gearboxes using NBV-based classifier: a pattern recognition approach. Mech Syst Sig Process. http://dx.doi.org/10.1016/j.ymssp.2012.08.021

8. Randall RB (2011) Vibration-based condition monitoring: industrial, aerospace and automotive applications. Wiley, New York. ISBN 978-0-470-74785-8
9. Antoni J (2009) Cyclostationarity by examples. Mech Syst Sig Process 23(4):987–1036. http://dx.doi.org/10.1016/j.ymssp.2008.10.010
10. Barszcz T (2009) Decomposition of vibration signals into deterministic and nondeterministic components and its capabilities of fault detection and identification. Int J Appl Math Comput Sci 19(2):327–335. http://dx.doi.org/10.2478/v10006-009-0028-0
11. Combet F, Gelman L (2009) Optimal filtering of gear signals for early damage detection based on the spectral kurtosis. Mech Syst Sig Process 23(3):652–668. http://dx.doi.org/10.1016/j.ymssp.2008.08.002
12. Lin J, Zuo M (2003) Gearbox fault diagnosis using adaptive wavelet filter. Mech Syst Sig Process 17(6):1259–1269. http://dx.doi.org/10.1006/mssp.2002.1507
13. Makowski R, Zimroz R (2011) Adaptive bearings vibration modelling for diagnosis. Adapt Intell Syst, Lect Notes Comput Sci 6943:248–259. http://dx.doi.org/10.1007/978-3-642-23857-4_26
14. Makowski R, Zimroz R (2012) Application of schur filtering for local damage detection in gearboxes. In: Fakhfakh T et al. (ed) Condition monitoring of machinery in non-stationary operations. Proceedings of the second international conference "condition monitoring of machinery in non-stationary operations" CMMNO'2012, part 3. Springer, Berlin, ISBN (Print): 978-3-642-28767-1. e-ISBN (online): 978-3-642-28768-8. http://dx.doi.org/10.1007/978-3-642-28768-8_32, pp 301–308
15. Zimroz R, Bartelmus W (2012) Application of adaptive filtering for weak impulsive signal recovery for bearings local damage detection in complex mining mechanical systems working under condition of varying load. Solid State Phenom 180:250–257. http://dx.doi.org/10.4028/www.scientific.net/SSP.180.250
16. Huang NE, Shen Z, Long SR, Wu MC, Shih HH, Zheng Q, Yen NC, Tung CC, Liu HH (1998) The empirical mode decomposition and the Hilbert spectrum for nonlinear and non-stationary time series analysis. Proceedings of the royal society of London, series A: mathematical, physical and engineering sciences 454(1971):903–995. http://dx.doi.org/10.1098/rspa.1998.0193
17. Dong H, Qi K, Chen X, Zi Y, He Z, Li B (2009) Sifting process of EMD and its application in rolling element bearing fault diagnosis. J Mech Sci Technol 23(8):2000–2007. http://dx.doi.org/10.1007/s12206-009-0438-9
18. Gao Q, Duan C, Fan H, Meng Q (2008) Rotating machine fault diagnosis using empirical mode decomposition. Mech Syst Sig Process 22(5):1072–1081. http://dx.doi.org/10.1016/j.ymssp.2007.10.003
19. Liu B, Riemenschneider S, Xu Y (2006) Gearbox fault diagnosis using empirical mode decomposition and Hilbert spectrum. Mech Syst Sig Process 20(3):718–734. http://dx.doi.org/10.1016/j.ymssp.2005.02.003
20. Loutridis SJ (2004) Damage detection in gear systems using empirical mode decomposition. Eng Struct 26(12):1833–1841. http://dx.doi.org/10.1016/j.engstruct.2004.07.007
21. Yu D, Cheng J, Yang Y (2005) Application of EMD method and Hilbert spectrum to the fault diagnosis of roller bearings. Mech Syst Sig Process 19(2):259–270. http://dx.doi.org/10.1016/S0888-3270(03)00099-2
22. Feldman M (2011) Hilbert transform applications in mechanical vibration. Wiley, New York. ISBN 978-0-470-97827-6
23. Dybała J, Zimroz R (2012) Application of empirical mode decomposition for impulsive signal extraction to detect bearing damage: industrial case study. In: Fakhfakh T et al. (ed) Condition monitoring of machinery in non-stationary operations. Proceedings of the second international conference "condition monitoring of machinery in non-stationary operations" CMMNO'2012, part 3. Springer, Berlin. ISBN (Print): 978-3-642-28767-1. e-ISBN (online): 978-3-642-28768-8. http://dx.doi.org/10.1007/978-3-642-28768-8_27, pp 257–266

24. Rilling G, Flandrin P, Gonçalvès P (2003) On empirical mode decomposition and its algorithms. Proceedings of the 6th IEEE-EURASIP workshop on nonlinear signal and image processing (NSIP-03), 8–11 June. Grado, Italy. http://perso.ens-lyon.fr/patrick.flandrin/NSIP03.pdf. Accessed 12 Jan 2013

Artificial Immune Systems for Data Classification in Planetary Gearboxes Condition Monitoring

Edyta Brzychczy, Piotr Lipiński, Radoslaw Zimroz and Patryk Filipiak

Abstract In the paper a problem of diagnostic data classification is discussed. The classic condition monitoring approach requires two examples of machines: one in a good and one in a bad condition. From the industrial perspective such a requirement is often very difficult to fulfill, especially in the case of machines with an unique design. To overcome it, we proposed to use the Artificial Immune System (AIS) based approach to classify multidimensional diagnostic data. AIS allows to recognize a change of the machine condition based on a training phase using the dataset related to a good condition. To validate the proposed procedure and assess efficiency of the condition recognition, an extra data set from another machine (of the same type) in a bad condition was used. In the paper several key issues related to the selection of parameters have been discussed.

Keywords Planetary gearbox · Condition monitoring · Non-stationary operations · Multidimensional symptom space · Artificial immune systems · Classification

E. Brzychczy (✉)
Faculty of Mining and Geoengineering, AGH University of Science and Technology, Cracow, Poland
e-mail: brzych3@agh.edu.pl

P. Lipiński · P. Filipiak
Institute of Computer Science, University of Wroclaw, Wroclaw, Poland
e-mail: lipinski@ii.uni.wroc.pl

R. Zimroz
Diagnostics and Vibro-Acoustics Science Laboratory, Wroclaw University of Technology, Wroclaw, Poland
e-mail: radoslaw.zimroz@pwr.wroc.p

G. Dalpiaz et al. (eds.), *Advances in Condition Monitoring of Machinery in Non-Stationary Operations*, Lecture Notes in Mechanical Engineering,
DOI: 10.1007/978-3-642-39348-8_20,

1 Introduction

In this paper, we propose an efficient method of planetary gearbox vibration based data classification based on biologically-inspired computing i.e. artificial immune systems. Planetary gearbox is a crucial element of energy processing mechanical systems. Requirements related to high-transmitted power, high transmission ratio and small dimensions/weight cause that planetary gearboxes are complex and used in advanced systems (helicopters, wind turbines, mining machines, etc.) so there is a need to assure their reliability during operation. Feature classification problem in condition monitoring of planetary gearboxes working in non-stationary conditions was the subject of many research and publications [1, 2, 3, 4, 5], in which two class data were taking into account. In practice sometimes we have no possibility to gain data according to bad condition of machines, we have only good condition indicators, so classification process is difficult or impossible to provide [6, 7]. In such case we can use biologically-based algorithms i.e. artificial immune systems [8–14]. The artificial immune system uses the negative selection algorithm (NSA), resembling the positive and negative selection process of maturation of some cells in the nature, which aims at detection of abnormal data samples by a number of trained pattern detectors, where these pattern detectors are usually built on the basis of normal data samples without using any abnormal data samples. In contrary to many popular classification techniques, such as decision trees or neural networks, the artificial immune system approach does not require negative data samples (i.e. feature vectors describing the planetary gearboxes in bad conditions) in the training data set, because the negative selection algorithm creates the pattern detectors discovering relations and similarities among positive data samples without comparing to negative data samples. Such an approach has proved its efficiency in many practical applications, ranging from simple classification methods to complex real-time fault detection systems.

2 Data Description. Problem Definition

In this section, we will discuss the problem and available data obtained during experimental work. Due to some redundancy, obtained data were subjected to further processing in order to eliminate correlation and minimize dimension of data used for classification. As it was said the purpose of this paper is to provide an efficient procedure for condition monitoring of complex machine i.e. multistage planetary gearbox used in one of the biggest machine working in the industry, namely bucket wheel excavator, see Fig. 1. The problem of diagnosis this kind of heavy machinery systems have been referred previously by Bartelmus and Zimroz [4]. The novel solution proposed here is that in order to monitor the condition we

Fig. 1 Example of investigated machine

don't need bad condition data set that is usually used in order to perform the data classification (to establish threshold value) and finally the condition recognition. In this paper we used spectral based data (obtained from spectral analysis of vibrations acquired from planetary gearbox) in order to estimate wear of geared wheels. Please note that is it not related to localized damage, the most exploited subject for gearboxes. In next sections we will provide short explanation how input data have been formed.

2.1 Feature Extraction

In the proposed approach, the condition of the planetary gearbox used in the bucket wheel excavator id described by 15 dimensional feature vectors, extracted earlier from the original vibration signal via spectral analysis (SA) (Figs. 2, 3). So, as an input data we have used matrix of diagnostic data MDD with 15 columns related to 15 spectral components (where pp1 is amplitude of the planetary mesh frequency PMF, pp2 is amplitude of second harmonic of the PMF, ...pp15 is amplitude of 15th harmonic of the PMF). A life time $\theta_G = 10{,}000$ h has been estimated for this machine, we will consider it in the paper as machine in good condition. Final dimension of the diagnostic features matrix was 951×15. For two data sets prepared based on these data we trained and tested our immune system. We were lucky to acquire data also from second machine with lifetime $\theta_B = 20{,}000$ h (designated to major overhaul and called as machine in bad condition). Dimension of the diagnostic features matrix for bad condition case was $\theta_B = 1{,}232 \times 15$. These data were used for validation only, they were not considered during training of the system.

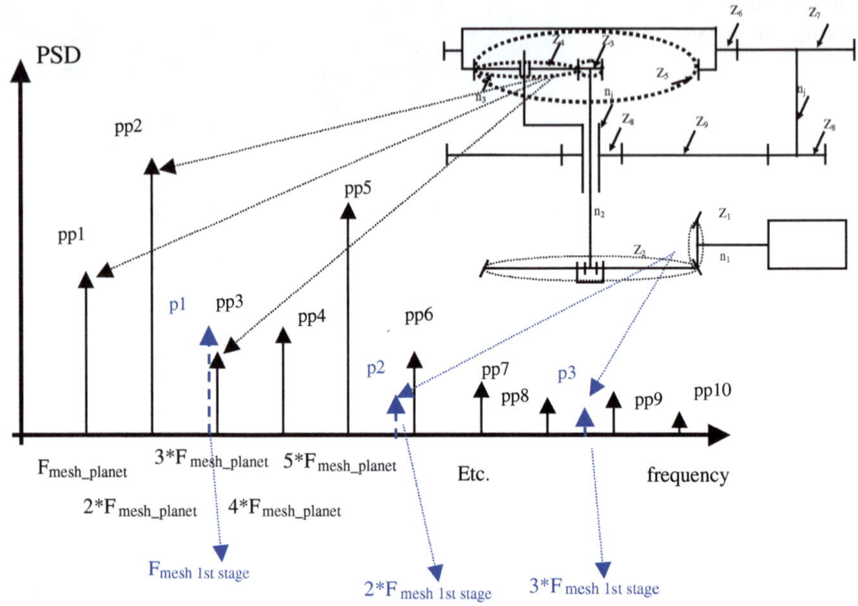

Fig. 2 An idea of physical explanation of spectral based feature extraction

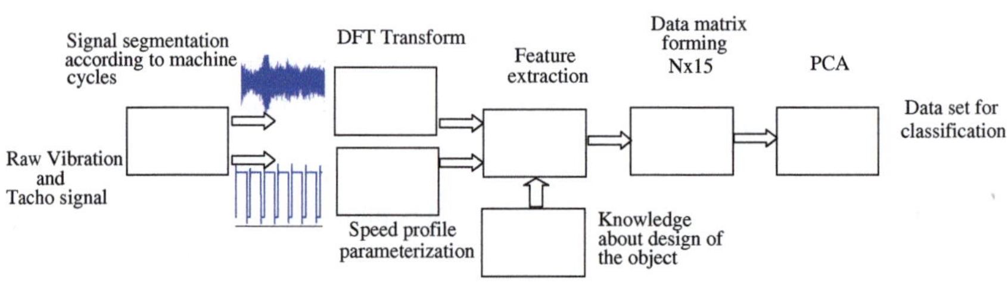

Fig. 3 A scheme of raw signals processing for feature extraction

2.2 Data Preprocessing Using Principal Component Analysis

Principal Component Analysis (PCA) is a frequently used analytic method for pre-processing or visualization of high dimensional data, features extraction, dimensionality reduction, lossy data compression, etc. [15]. There exist a number of formal definitions of PCA, each of which leads to the same algorithm described below [16]. A commonly used definition states that PCA is the orthogonal projection of the data onto a lower dimensional linear space, known as the principal subspace, such that the variance of the projected data is maximized. It is worth noticing that such maximization is equivalent to minimization of projection error.

Let $\{x_n\}$ be a set of observations where $x_n \in R^d$ $(d > 1)$ for each n = 1,..., N. Let S be a covariance matrix of $\{x_n\}$, calculated as follows

$$S = \frac{1}{N}\sum_{n=1}^{N}(x_n - \bar{x})(x_n - \bar{x})^T,$$

where

$$\bar{x} = \frac{1}{N}\sum_{n=1}^{N}x_n$$

is a mean value of samples.

It is proved (e.g. in Bishop [16] that eigenvectors $\{v_1, v_2, \ldots, v_n\}$ of S (with their corresponding eigenvalues $\lambda_1, \lambda_2, \ldots, \lambda_n$ such that $\lambda_1 \geq \lambda_2 \geq \ldots \geq \lambda_n$) form a basis of Euclidean space R^d, arranged in such a manner that first m < d eigenvectors (i.e. $v_1, v_2, \ldots, v_m$) constitute m principal components of given samples. It means that samples $\{x_n\}$, when mapped with transformation matrix $[v_1, v_2, \ldots, v_n]$, can be reduced to the first m coefficients assuring the maximal variance of projected data. In this paper PCA was used for the purpose of visualization of classification results. Due to high redundancy in investigated data [17, 18] classification results could be clearly visualized in 2-dimensional space of first two principal components. It should be noted that further research provided by Bartkowiak and Zimroz [19] proved that dependency between these data in fact slightly nonlinear and require more than just 2 PC but it is neglected here.

3 One Class Classification: A Short Review

A typical classification procedure tries to split a given set of samples into groups of elements that are similar in some sense (e.g. ones sharing the same features or located in the same subspace). In such cases, a classifier is usually trained with a portion of known data. After a successful training process, it is able to classify previously unknown samples by recognizing similarities to data analyzed before. One-class classification, introduced in [20], is a very specific classification problem because of both its simplification and complexification compared to the procedure described above. The aim is to distinguish the elements, given a set of samples, that belong to one class from the remaining ones. However, only samples from this one class can be used in a training period and no prior knowledge about the elements from outside the class is available. Such problem occurs frequently in real life situations. For instance, in anomaly detection like frauds, medical conditions or device malfunctions [21]. A commonly used approach in such cases is an application of Artificial Immune System that is proposed in this paper.

4 Artificial Immune Systems. A Theoretical Background

Artificial Immune Systems (AIS) are relatively new branch of Natural Computing. Their key feature is an algorithm based on the behavior of living organisms immune system. The Brunet [22] publication is considered one of the first in this area. The design process of AIS includes the three following steps:

- Selecting a representation of system elements (i.e. mainly antigens and antibodies),
- Defining a set of mechanisms for evaluating an influence of each component of immune system and a system itself on the environment (usually by choosing an appropriate distance metric),
- Adaptation procedure.

Elements of a system can be represented by many data structures that address the requirements of a given problem. Choosing an appropriate data representation is the key factor of AIS because the distinction between self and non-self molecules and/or cells is strictly based on it. Most frequently used representation is a sequence of attributes stored in a vector of real values, integers, binary data or predefined symbols.

Let $Ab = < Ab_1, Ab_2, \ldots, Ab_L >$ (where $L > 0$) represent an antibody and $Ag = <Ag_1, Ag_2, \ldots, Ag_L >$ represent an antigen. Hence, their mutual correspondence can be evaluated with some similarity or complementarity measure that allows for distinction between *self* and *non-self* elements.

In the case of real-valued vector based representation, commonly used metrics are [23]:

1. Euclidean:

$$D(Ab, Ag) = \sqrt{\sum_{i=1}^{L} (Ab_i - Ag_i)^2}$$

2. Minkowski:

$$D(Ab, Ag) = \left[\sum_{i=1}^{L} |Ab_i - Ag_i|^L\right]^{\frac{1}{L}}$$

3. Manhattan:

$$D(Ab, Ag) = \sum_{i=1}^{L} |Ab_i - Ag_i|$$

4. Maximum:

$$D(Ab, Ag) = \max_{i=1,\ldots,L} |Ab_i - Ag_i|$$

In the case of binary data representation, a Hamming distance can be used instead:

$$D(Ab,\, Ag) = \sum_{i=1}^{L} \delta_i, \quad where: \delta_i = \begin{cases} 1\, for\, Ab_i \neq Ag_i \\ 0\, otherwise \end{cases}$$

Since Hamming distance measures a complementarity between two binary vectors, it can be formulated equivalently as a subtraction L-D(Ab, Ag).

Let $\varepsilon > 0$ be the similarity threshold. If D(Ab, Ag) $< \varepsilon$ then molecule Ab is recognized as *self*. Otherwise, Ab is considered *non-self* and, as a consequence, an immune response is initiated.

Basic mechanisms modeled in AIS are:

- Clonal selection,
- Negative selection.

Clonal selection mechanism imitates the behavior of B-cells during activation and pathogen lysis. While intruder invasion is recognized immune cells are stimulated to select the matching antigen and reproduce it. A reproduction process is based on mitosis which is the reason why this sort of immune response is referred to as clonal selection or clonal expansion. Each case of reproduction, including B-cells reproduction, results in possible errors (called *mutations*). One of distinctive features of immune system is the fact that the probability of mutation is inversely proportional to the similarity between antibody and antigen that attacks the parent cell. Thus, if the similarity is relatively high then the probability of mutation is low. Otherwise, in the case of low similarity, the mutation rate is very high (so called *hypermutation*).

It is worth mentioning that a presence of mutation can be very useful since it helps for more precise adaptation of detectors to invading antigens. As a result, such well-adapted detectors are partially released into plasma and partially stored in the immune memory then circulating in organism as a part of adaptive immune system.

Computational model of this process, named Clonal Selection Algorithm Castro and Zuben [24] and its applications, are widely used in AIS dedicated to data analysis [25], optimization problems [26] and many others.

Negative selection mechanism is a key part of T-cells selection observed in a thymus. If some cell or molecule known as *self* is recognized and matched by an immature T cell, then this T-cell is automatically destroyed and, as a result, removed from the spectrum of T-cells. Only remaining lymphocytes, i.e. those that doesn't match any self cells, are able to get outside thymus and become a part of

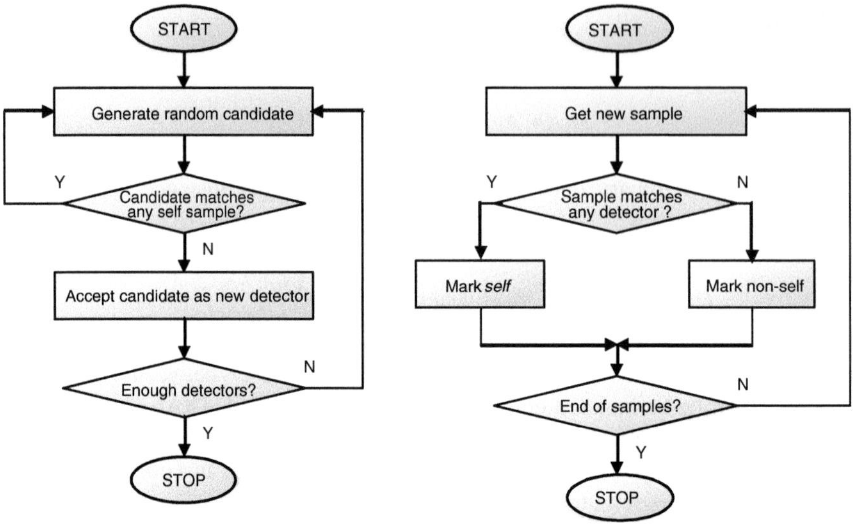

Fig. 4 Negative selection algorithm [23]: learning process (*left diagram*) and classification process (*right diagram*)

immune system. Such sort of selection can take place in living organisms because a thymus is surrounded by a selective barrier that allows only host organism molecules for passing through. As a result, T-cell learns to tolerate self cells by not responding to self molecules present in thymus.

An algorithmic model of immune system selection based on T-cells called Negative Selection Algorithm (NSA) was introduced in [27]. Numerous further improvements were proposed in data representation [28], detectors generation [29] and matching rules [30].

AIS models exploiting negative selection mechanism are commonly used for detection of novelties [13] or anomalies [31], e.g. frauds, medical conditions or device malfunctions [21].

Figure 4 depicts a negative selection algorithm introduced in [23] that was used for experiments in this paper. A learning process (presented on the left diagram) consists of a loop of consecutive generating random candidates for antibodies. Each new candidate is exposed to the set of training data. Note, that only *self* samples are stored in the training set, so if a candidate matches any sample from this set, it is automatically erased, otherwise it is accepted as a new detector and included in the set of antibodies. The distinction between matching and not matching candidates is determined by the distance between this candidate and a sample (using a predefined metrics) as it was described before in this section. Clearly, the described process guarantees that after the training period all samples in the training set will be recognized by the obtained AIS as *self*. A classification process (presented on the right diagram) works very similar. New samples (of previously unknown data) are exposed to the set of antibodies. If any antibody

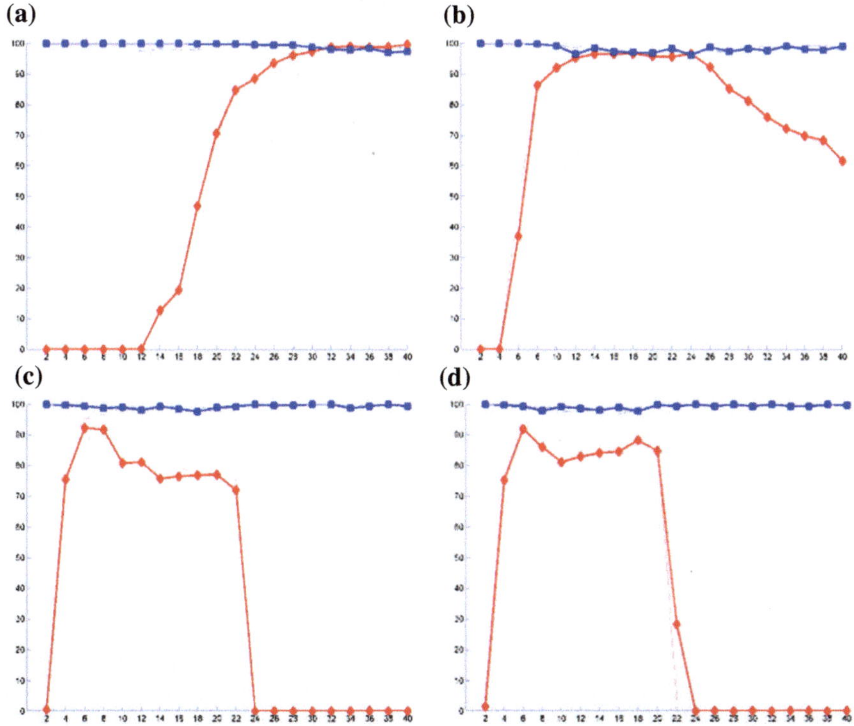

Fig. 5 Plots of averaged true positive (*blue lines*) and averaged true negative (*red line*) results out of 20 runs of negative selection algorithm using the four distance measures: (**a**) Manhattan, (**b**) Euclidean, (**c**) Minkowski, (**d**) Max; x-axis represent distance thresholds ε (even numbers from 2 to 40); y-axis represent percent of matches. Plots corresponding to 30,000 antibodies are printed with *bold lines*

matches such sample, it is automatically marked as *non-self* antigen, otherwise as *self*.

1. The proposed classification procedure

Two sets of data were available for the experiments—951 samples of good condition data and 1,232 samples of bad condition data. The 2/3 of randomly selected good condition data (i.e. 637 samples) were used for the learning purpose. Classification of the remaining 1/3 (314 samples) of good condition data and all bad condition data (1,232 samples) was the subject of a validation process.

A classification result was considered *true positive* if a sample of good condition data was recognized as good. Similarly, a result was considered *true negative* if a sample of bad condition data was recognized as bad. Similarity thresholds $\varepsilon > 0$ for the distance D(Ab, Ag) between antigen $Ag \in R^L$ and antibody $Ab \in R^L$ used in the experiment were even numbers from 2 to 40. Sets of 10, 20, 30, 40 and 50 thousands of antibodies were trained and tested for each run of the algorithm.

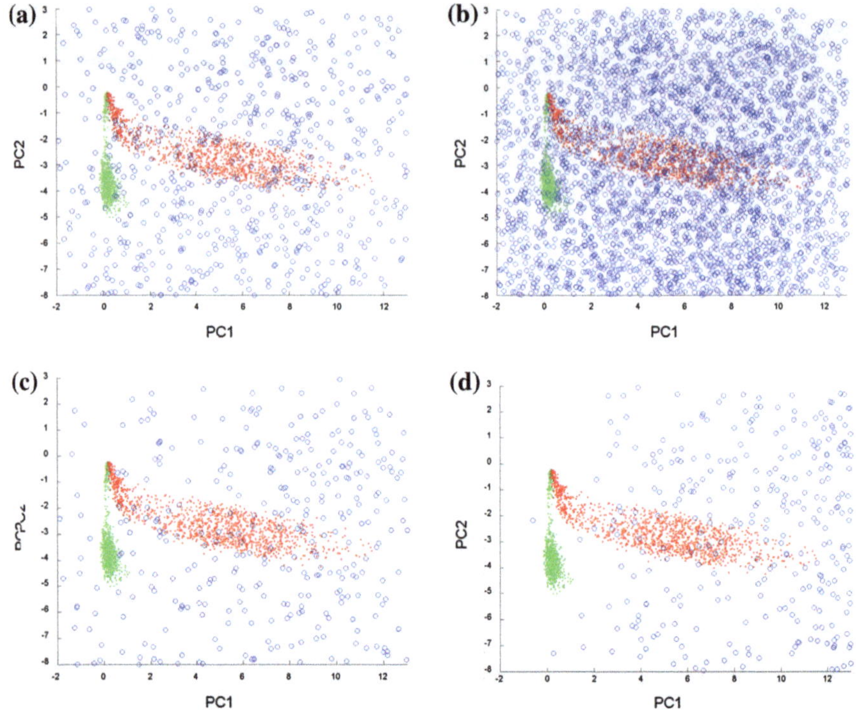

Fig. 6 Plots of first two principal components of good data (*green dots*) and bad data (*red dots*). *Blue circles* represent locations of detectors with: (**a**) optimal settings, (**b**) surplus of detectors, (**c**) deficiency of detectors, (**d**) oversized distance

Figure 5 presents the results of calculations. Plots of averaged true positive are drawn with blue lines and averaged true negative with red lines. Values corresponding to 30,000 antibodies are printed in bold.

The correspondence between the value of *true positive* and *true negative* is evident. In the case of Manhattan distance, the greater value of similarity threshold $\in$, the greater percentage of negative. In the cases of remaining distance measures, setting some threshold values strictly outperforms both lower and greater values. Optimal settings are in the ranges: 12–26 (for Euclidean distance), 6–8 (Minkowski distance) and 8–18 (Max distance). Clearly, best results were obtained using Manhattan and Euclidean distance. Choosing the number of antibodies also plays a very important role. Too small number of antibodies leads to considerably long distances between them. As a result, the search space is not sufficiently covered with detectors causing high risk of misclassification.

On the other hand, an increment of antibodies leads to grow of computational cost. The experiment revealed that the optimal number of detectors is about 30,000 since further increment of antibodies did not lead to significant improvement in performance. As it is clearly seen, a key factor of AIS is the selection of learning parameters. When a threshold ε is too small, only good condition data can be recognized which is completely useless. On the contrary, choosing great values of threshold leads to misclassification of some negative samples, which are close to the positive samples (c.f. Fig. 6d).

Another important configuration issue is the number of detectors. A large number of detectors may lead to overfitting problem and possibility of detecting unknown good condition samples as bad condition ones (c.f. Fig. 6b). On the other hand, a small number of detectors may lead to some "gaps" in the space and accepting negative samples from these gaps (c.f. Fig. 6c).

5 Conclusion

In paper a proposal of procedure of condition monitoring of the complex planetary gearbox is shown. Due to unique type of design of these types of machines, in many situations there is no chance to perform two class classification as it has been already done by [4, 18] or [7, 17] in their previous works. Recent development in the field of pattern recognitions, especially in biologically-inspired computing, motivated us to apply AIS approach for the purpose of diagnosis of planetary gearbox working under non-stationary operations. It has been done for the very first time in that context. We successfully applied AIS to classification of planetary gearboxes condition monitoring in good condition, and next thanks to possessed data from the gearbox of the same type just before replacement, we have validated of our system for bad condition data. During validation process classification error at the level below 5 % was obtained. The obtained result is very good. One should note that this classification error cannot be straightly compared with the classification error of other techniques (such as Neural Network, Decision Trees, etc.), because AIS use only positive samples in the learning process. Lack of "bad condition" data makes impossible to use NN or DT. In practice, the classification with AIS allows to construct monitoring systems without observations of negative cases or with only a few such observations. Further development may improve the proposed approach with regard to optimal dimension of detectors set and parameterization of affinity measure.

Acknowledgments This work is partially supported by the statutory grant No. S20096 (R. Zimroz).

References

1. Stander CJ, Heyns PS, Schoombie W (2002) Using vibration monitoring for local fault detection on gears operating under fluctuating load conditions. Mech Syst Sig Process 16(6):1005–1024
2. Samuel PD, Pines DJ (2005) A review of vibration-based techniques for helicopter transmission diagnostics. J Sound Vib 282(1–2):475–508

3. Cempel C, Tabaszewski M (2007) Multidimensional condition monitoring of machines in nonstationary operation. Mech Syst Sig Proces 21(3):1233–1241
4. Bartelmus W, Zimroz R (2009) A new feature for monitoring the condition of gearboxes in nonstationary operating systems. Mech Syst Sig Proces 23(5):1528–1534
5. Zimroz R, Bartkowiak A (2011) Investigation on spectral structure of gearbox vibration signals by principal component analysis for condition monitoring purposes. J Phys: Conf Series 305(1):012075
6. Bartkowiak A (2010) Anomaly, novelty, one-class classification: a short introduction. In: Ajith A et al (ed) Proceedings of computer information systems and industrial management applications 2010, vol 1–6 Ostrava. (IEEE Xplore doi:10.1109/CISIM.2010.5643699)
7. Bartkowiak A, Zimroz R (2011) Outliers analysis and one class classification approach for planetary gearbox diagnosis. J Phys: Conf Series 305(1):012031
8. Cen J, Shao L, Zhang Q, Xu B (2009) Fault diagnosis research of rotating machinery based on artificial immunity. Coal Mine Machin, vol 3
9. Clifton DA, Bannister PR, Tarassenko LA (2006) Application of an intuitive novelty metric for jet engine condition monitoring. In: Ali M, Dapoigny R (eds) IEA/AIE LNAI 4031, pp 1149–1158
10. Clifton DA, Clifton LA, Bannister PR (2008) Automated novelty detection in industrial systems. Stud Computat Intell (SCI) 116:269–296. (www.springerlink.com)
11. Montechiesi L, Cocconcelli M, Rubini R (2012) Artificial immune system for condition monitoring based on Euclidean distance minimization. Proceedings of the 2nd international conference on condition monitoring of machinery in non-stationary operations. Hammamet, Tunisia, 26–28 Mar
12. Montechiesi L, Cocconcelli M, Rubini R (2012) Application of the artificial immune systems for bearings diagnostics in servomotors. Proceedings of the international conference on noise and vibration engineering (ISMA2012). Leuven, Belgium
13. Surace C, Worden K (2010) Novelty detection in a changing environment: a negative selection approach. Mech Syst Sig Process 24:1114–1128
14. Worden K, Staszewski WJ, Hensman JJ (2011) Natural computing for mechanical systems research: a tutorial overview. Mech Syst Sig Process 25:4–111
15. Jolliffe IT (2002) Principal component analysis, 2nd edn. Springer, New York
16. Bishop CM (2007) Pattern recognition and machine learning. Springer, New York
17. Bartkowiak A, Zimroz R (2012) Data dimension reduction and visualization with application to multidimensional gearbox diagnostics data: comparison of several methods. Diffus Defect Data Pt B: Solid State Phenom 180:177–184
18. Zimroz R, Bartkowiak A (2012) Two simple multivariate procedures for monitoring planetary gearboxes in non-stationary operating conditions. Mech Syst Signal Process (in press). http://dx.doi.org/10.1016/j.ymssp.2012.03.022
19. Bartkowiak A, Zimroz R (2012b) Curvilinear dimensionality reduction of data for gearbox condition monitoring. Przeglad Elektrotechniczny 88(10B):268–271
20. Moya M, Hush D (1996) Network constraints and multi- objective optimization for one-class classification. Neural Netw 9(3):463–474
21. Chandola V, Banerjee A, Kumar V (2009) Anomaly detection: a survey. ACM Comput Surv 41(3):15
22. Burnet F (1974) Clonal selection and after. Theoretical immunology. Marcel Dekker Inc., New York
23. Dasgupta D, Zhou J (2007) Revisiting negative selection algorithms. Evol Comput 15(2):223–251
24. Castro de LN, Zuben von FJ (2000) The clonal selection algorithm with engineering applications, GECCO 2000
25. Hunt JE, Cooke DE (2002) Learning using an artificial immune system. J Netw Comput Appl 19(2):189–212
26. Castro de LN, Timmis JI (2002) An artificial immune network for multimodal function optimization, CEC 2002

27. Forrest S, Perelson AS, Allen L, Cherukuri R (1994) Discrimination in a computer. In: Proceedings of IEEE symposium on research in security and privacy, pp 202–212
28. Gonzales F, Dasgupta D, Nino LF (2003b) A randomized real-value negative selection algorithm, ICARIS 2003
29. Ayara M, Timmis J, Lemos de R, Castro de I, Duncan R (2002) Negative selection: how to generate detectors, ICARIS 2002
30. Gonzales F, Dasgupta D, Gomez J (2003) The effect of binary rules in negative selection, GECCO 2003
31. Dasgupta D, Yu S, Majumdar N (2005) MILA - Multi-Level Immune Learning Algorithm and its application to Anomaly Detection. Soft Comput J 9(3):172–184 (Online Publication was in December 2003)

Gearbox Condition Monitoring Procedures

Walter Bartelmus and Radoslaw Zimroz

Abstract There is a need to treat a gearbox as a subsystem which consist of several elements like gears, bearings and shafts incorporated into a box. The gearbox is incorporated into a system of a drive, for example an electric motor, and a driven machine. When a system is in operation the mention elements interacts each other. When preparing condition method one has for disposal results of the research directed to evaluation of isolated faults like a tooth crack, tooth breakage, pitting, scuffing, misalignment and so on. It is taken in advance that only one of the mentioned faults occurs in the system. The research is done at the condition of constant load or constant rotation speed. There is also done research under condition of different constant levels of the load and rotation speed. The scenario of degradation process of gearboxes is that only one fault occurs and developed in the system. In real gearbox systems many different scenarios of degradation process may occur. The presented paper will show gearbox condition monitoring procedures, which is equivalent to different scenarios. There are given some steps, which ought be considered when gearbox condition monitoring procedures are developed. Having in mined these steps one can control the gearbox degradation process extending a gearbox live and reducing maintenance cost, creating, what is called, the failure prevention technology. For presenting this paper stimulates us also the papers presented in the MSSP "Special Issue" on the "Condition monitoring of machines in non-stationary operations" where some authors have tendency of treating the machine as the collection of separate not connected elements in the same way as it was stated before.

Keywords Gearbox · Condition monitoring · Design factors · Operation factors

W. Bartelmus (✉) · R. Zimroz
Diagnostics and Vibro-Acoustics Science Laboratory, Wroclaw University of Technology, Pl Teatralny 2, 50-051 Wroclaw, Poland
e-mail: walter.bartelmus@pwr.wroc.pl

R. Zimroz
e-mail: radoslaw.zimroz@pwr.wroc.pl

G. Dalpiaz et al. (eds.), *Advances in Condition Monitoring of Machinery in Non-Stationary Operations*, Lecture Notes in Mechanical Engineering,
DOI: 10.1007/978-3-642-39348-8_21,

1 Introduction

Vast literature on condition monitoring for example [1–3] do not gives the diagnostics procedures for gearbox condition monitoring as one think a gearbox as a system, which consists of many elements like: bearings, gears, shafts and there is interaction between these elements, when the system is in operation. For developing the suitable gearbox condition monitoring procedures there is a need to develop some steps:

- First step, one ought to understand the relation between factors having influence diagnostic signals and condition of gearboxes as it is given in [4] and [5].
- Second step, one ought to know information on relation between the factors, which have influence on vibration diagnostic signal and diagnostic signal presentation in the form of spectra as it is given in papers [6] and [7].
- Third step, one ought to understand the different form of gearbox degradation processes as it is given [8–16].

On the bases of all the mentioned papers there is possibility do develop the gearbox condition monitoring procedures.

2 Relation Between Factors Having Influence Diagnostic Signals and Condition of Gearboxes

The factors [5] are divided as it is given in Fig. 1. The primary factors are divided as the design and production technology.

The secondary factors are divided by operation and change of condition. The design factors can be determined on a design specification presented in the form of design drawings where all details of diagnosed object are presented. The production technology factors are resulted from production that means, parts machining and their assembling. They are imbedded into the product and they reveal itself during object/machine operation. Because the operation condition change during operation so they should be taken into consideration. The design,

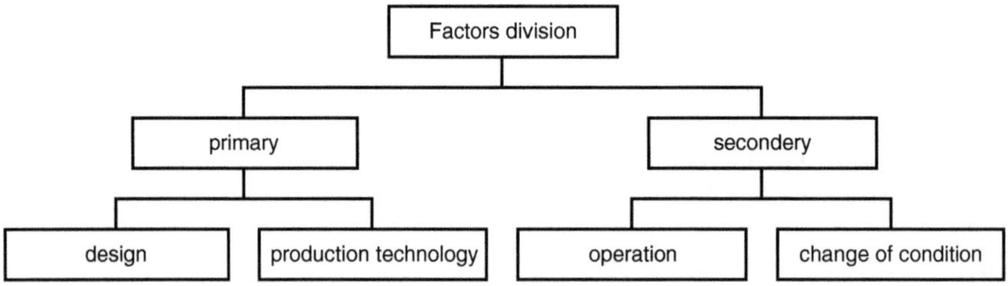

Fig. 1 General division of factors influencing vibration signals

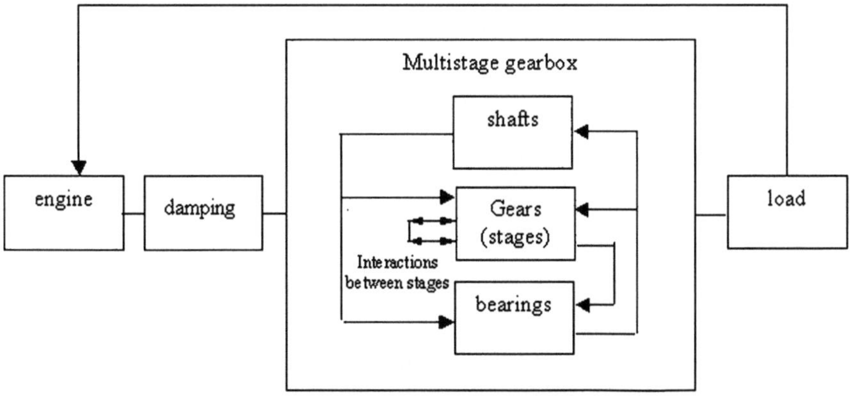

Fig. 2 Interaction between components in multistage gearbox system and external components to the gearbox such as: electric motor/engine, damping coupling, external load

production technology factors determine the reference machine condition. In this paper the reference machine condition is determined by the load susceptibility characteristics (LSC) [11] and [12]. It is the load trending of diagnostic features at the current condition for a given stage of a gearbox. The LSC is the result of regression analysis, presented later. The operation factors are determined by outer load variation, which may be connected with excavation process in mining industry or varying wind power in a wind turbine. The change of condition of gearboxes is determined by all forms of faults, which occur during gearbox operation. In the Fig. 2 [10] there is given the scheme of the system where an interaction between components in multistage gearbox system and external components to the gearbox such as: electric motor/engine, damping coupling, external load are presented. The presented system operates in certain environment (temperature, humidity, dustiness), which has influence on degradation—change of a gearbox condition. Having in mined all the mentioned factors one can control the gearbox degradation process [8, 13, 14] extending a gearbox live and reducing maintenance cost creating, what is called, the failure prevention technology.

3 Diagnostic Signal Presentation in the Form of Spectra

Figure 3 shows two driving systems for the bucked wheel of a bucked wheel excavator. In both systems there is used a planetary gearbox. In the system given in Fig. 3a the planetary gearbox is characterised by standstill rim (gear z_3) and rotating sun (gear z_1). The planetary gearbox in system Fig. 3b the rim (gear z_5) and the sun (gear z_3) are rotating. Equivalently their ratios are given in (1) and (2).

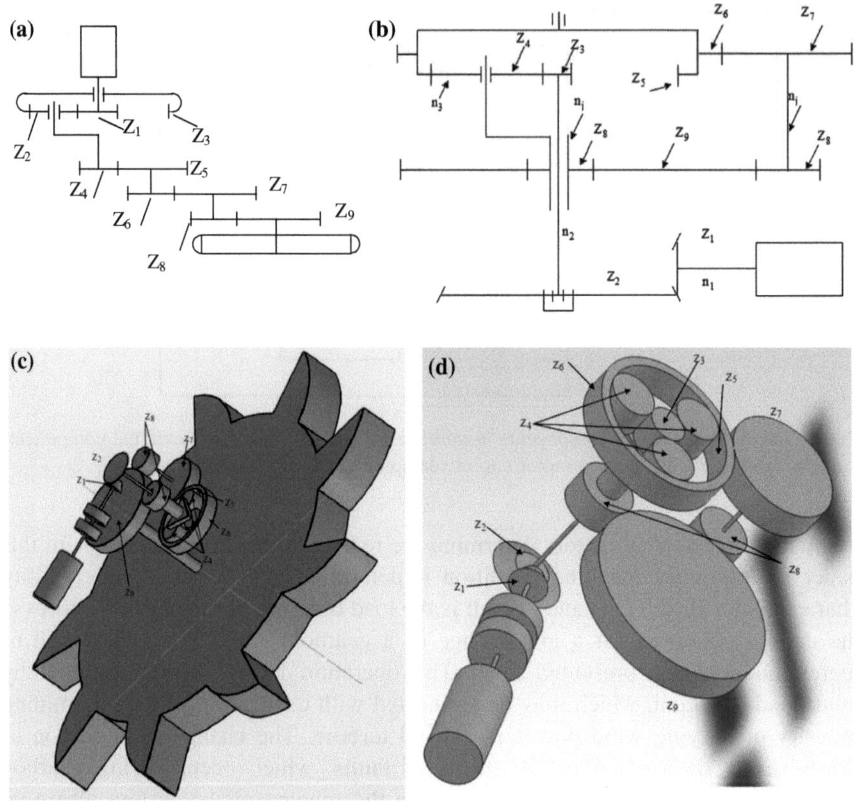

Fig. 3 Two schemes (**a**) and (**b**) of different designs of driving systems, (**c**) and (**d**) more details for driving system (**b**), in (**d**) bucked wheel is neglected

$$u_p = 1 + \frac{z_3}{z_1} \tag{1}$$

$$u_p = 1 + \frac{z_5}{z_3} + \frac{z_5 z_7}{z_3 z_6} \tag{2}$$

These two systems are presented for comparison to show the difference in the design of the system and the difference in the ratios of the planetary gearboxes. Further consideration is directed to the system given in Fig. 3b. For planetary gearbox condition monitoring there is a need to calculate a meshing frequency, more about the meshing frequencies calculations look into [6] and [7]. According the principle given there the meshing frequency for the planetary gearbox incorporating into the system Fig. 3b is given by the statement (3). More details of the system presenting in Fig. 3b is given in Fig. 3c and Fig. 3d, so the meshing frequency for the planetary gearbox in system Fig. 3b is

$$f_{34} = f_{45} = \frac{n_{2j} z_3}{60} = \frac{(n_2 - n_j) z_3}{60} \tag{3}$$

where:

n_{2j} related speed rotation of a shaft which rotates with the speed rotation n_2 (RPM),

n_2 absolute speed of the second shaft (RPM),

n_j arm/carrier speed rotation (RPM),

z_3 number of teeth in gear 3.

To use the above statement one ought have more statements which are connected with the gearbox system given in Fig. 3b. The complete ratio of the system is

$$u_c = u_s u_p u_w \tag{4}$$

The bevel stage ratio equals to

$$u_s = \frac{z_2}{z_1} \tag{5}$$

The planetary gearbox ratio equals to as given in (2). The cylindrical stage gear ratio equals to

$$u_w = \frac{z_8}{z_9} \tag{6}$$

The arm/carrier speed rotation

$$n_j = \frac{n_2}{u_p} \tag{7}$$

For the gearbox condition monitoring there also needed other frequencies connected with gearing local faults [6] and [7] and roller elements bearings faults.

4 Different Form of Gearbox Degradation Processes

As it is given in [15] when considering a gearbox failure one should take into consideration primary and secondary misalignment of shafts which are in a gearbox system. The presented paper is mainly concentrated on inner gearbox shafts and gears misalignment (IGSGM). The IGSGM can be evaluated during gearbox operation on the base of vibration signal and operation load of a gearbox. On the base of vibration and load characteristic presented in two dimension space the regression line can be evaluated. This regression line is called the susceptibility characteristic (SC) [11], load trending. The primary cause of misalignment can be assessed at the beginning of a gearbox operation by SC, load trending. The primary cause of misalignment evaluated by SC is the measure of gearbox quality. The inner secondary cause of misalignment is developed during a gearbox operation in the process of bearing wear. In the paper [11] it is discussed the issue of the cause of a gearbox secondary misalignment. It is described the case of planetary gearbox condition evaluation where is giving the evidence that the frictional wear of bearings cause IGSGM. The IGSGM is evaluated on the base of SC and bearing frictional wear is measured after the planetary gearbox dismantling and tooth gearing faults as surface micro cracking has been noticed.

Some discussions on gearbox root cause analysis is given in [8]. Here is given a case describing the root cause analysis (RCA) of a bevel gears, which shows that the reason of gear condition change is IGSGM. The RCA is based on the knowledge gained from experiment presented in [11]. The bevel gear (Fig. 4) shows developed scuffing (there is more on surface distress in [16]) and a tooth gear fatigue development, and heat tints. The heat tints show that the surface bulk temperature would raise to about 250 °C. It was also inferred that the tooth breakage is the fourth cause a tooth gear degradation process. The primary cause is misalignment, second scuffing, third fatigue crack (details in Fig. 5), fourth brittle tooth breakage. Figure 6 shows a scheme of fatigue fracture area with beach marks and brittle fracture area. These two considered cases shows that the main cause of gearbox failure is IGSGM. At the end ought to be underlined that primary cause of misalignment, which comes for example from misaligned motor and gearbox shaft position should be eliminated before putting system into operation according the technologies given in [17].

To understand the increase of bulk temperature of gear teeth some computer simulation study is given here, more theory is given in [4]. The reason of increased temperature is the power lost during teeth friction.

Fig. 4 Heavily effected/ destructed gear teeth, and colours heat tint

Fig. 5 View of damaged gear, fatigue fracture area with beach marks, brittle fracture area

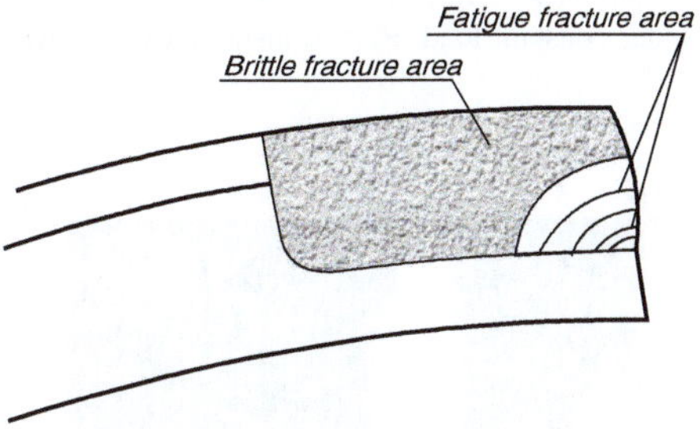

Fig. 6 Scheme of fatigue fracture area with beach marks and brittle fracture area

The friction power is equal to a product of the slipping velocity and the friction force. The slipping velocity can be calculated as an absolute value of the difference between the tangent velocities to the gear teeth.

$$v_p = v_{t1} - v_{t2} \tag{8}$$

The instantaneous friction lost power is

$$N_{lost} = v_p \mu P_n \tag{9}$$

Plots of instantaneous friction loss are shown in Figs. 7 and 8 for different friction coefficient values: $\mu = 0.02$ and 0.1.

Instantaneous efficiency value is defined as

$$\eta = \frac{N_s - N_{lost}}{N_s} \tag{10}$$

where: N_s—the motor's power, W;

N_{lost}—the power of the losses due to friction in the meshing, W.

Plots of the above parameter when the gear is properly lubricated and the frictional resistance is defined by friction coefficient $\mu = 0.02$, Fig. 7. For comparison, plots of the power of the losses and efficiency plots for $\mu = 0.1$ are shown in Fig. 8. As it follows from Fig. 7, efficiency is in a range of 0.098–1 and at $\mu = 0.1$ it ranges from 0.98 to 1 (Fig. 8). A comparison of Figs. 7 and 8 shows that for steady running (period 4) at m varying from 0.02 to 0.1, i.e. at a fivefold

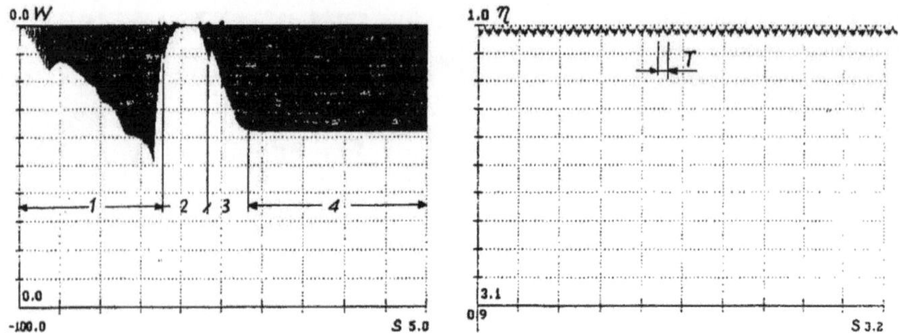

Fig. 7 Plot of power of system losses due to friction in meshing for $\mu = 0.02$: 1 ÷ 4, different periods of system run: *1* acceleration of the system, *2* free run, *3* loading of the system, *4* run under constant load. Plot of instantaneous efficiency values for $\mu = 0.02$

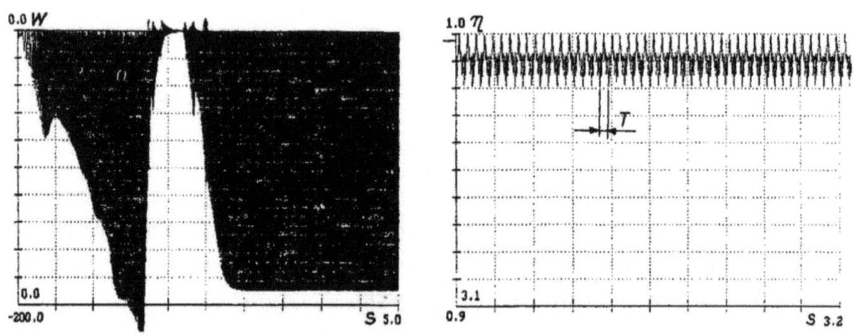

Fig. 8 Plot of power of system losses due to friction in meshing for $\mu = 0.02$: 1 ÷ 4, different periods of system run: *1* acceleration of the system, *2* free run, *3* loading of the system, *4* run under constant load. Plot of instantaneous efficiency values for $\mu = 0.1$

increase in the value of the friction factor, the maximum instantaneous losses are respectively $Nl_{lost} = 38$ and 195 W and hence the loss ratio is 195/38 = 5. Thus the fivefold increase in the friction factor corresponds to the fivefold increase in the power of the frictional losses. For now has been given comparison when a gear run in perfect condition $\mu = 0.02$ and at oil dry friction $\mu = 0.1$. It is dry friction for fine roughness condition. When gear surface is in the scuffing condition Fig. 4 the friction coefficient may arise to $\mu = 0.4$ even more. In the case of $\mu = 0.4$ using linear relation which govern the above consideration an instantaneous efficiency (10) $\eta = 0.245$. This support the idea that surface bulk temperature assessed on heat tints Fig. 4 can be high, and in considered case is about 250 °C.

5 Load Susceptibility Characteristics as a Measure of Gearbox Condition

Load susceptibility characteristics according to [11] and [12] are the measure of gearbox condition. The load susceptibility characteristics can be expressed by linier relation [11]

$$y = ax + b \tag{11}$$

where y—the value of the signal feature, x—the operating conditions (instantaneous speed in this case) indicator, a, b—the parameters to be determined.

Here is described the case of planetary gearbox condition evaluation where is giving the evidence that the frictional wear of bearings cause IGSGM. The IGSGM is evaluated on the base of SC and bearing wear is measured after the planetary gearbox (Fig. 3b) dismantling and tooth gearing faults as surface micro cracking has been noticed. In a paper [11] the term load susceptibility (LS) or SC is introduced. The load susceptibility is given by the regression characteristics in Fig. 9b as diagnostic features for planetary gearbox as a function of rotation speed RPM; for a gearbox in good ("o" dots) and bad condition ("x" dots). In this case, presentation of susceptibility characteristics for an electric motor is based on a linear relationship between the transmitted moment and rotation speed. It means that one may use the load susceptibility characteristics as the function of a load or function of a rotation speed RPM as it is given in Fig. 9b. In the presented cases of the load susceptibilities the characteristics are interpreted as follows. The case for a good condition of gearbox shows that planetary gearbox behaves as a linear system under increasing load, that means with increasing load the system elements deflection increases, causing linear increase of diagnostic feature, which is dynamic inter teeth force related. In the case of bad condition as result of frictional

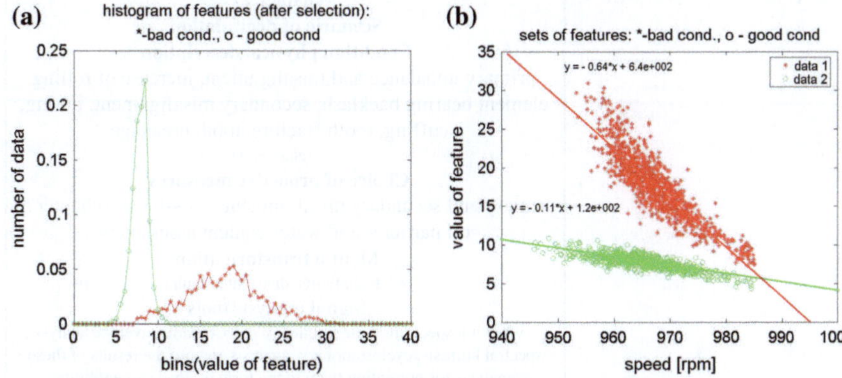

Fig. 9 **a** Data distribution of measured diagnostic parameters. **b** Load susceptibility characteristics as diagnostic features for planetary gearbox as a function of rotation speed RPM; for a gearbox in good (*o dots*) and bad condition (*x dots*)

wear of bearings the gear mesh under the condition of an IGSGM what gives a linear increase of the gear cooperation error and linear increase of inter teeth force, which cause linear increase of a vibration acceleration signal as it is presented by linear regression line Fig. 9b in a case of the bad gear condition. Figure 9b also shows very good separation of data for the good and bad condition. Better than is given in Fig. 9a when the data distribution functions overlap each other.

6 Proactive and Failure Prevention Technology and Concussions

In Fig. 10 there is given modified from [18] basic scheme of elements of vibration gearbox diagnostic method and elements of failure prevention technology.

In the scheme Fig. 10 there are considered several elements as:

Factors influencing vibration signal, which are divided as is given in the introduction into four groups namely: design, production technology, operation, condition change.

Interaction of gearbox elements as is given in Fig. 2 and environment influence should be considered.

On the basis of knowledge on a gearbox degradation process [8, 13–16], which comes from the factors analysis, interaction of gearbox elements, environment influence different degradation scenarios can be described.

Fig. 10 Modified basic scheme of elements of vibration gearbox diagnostic method and elements of failure prevention technology

Factors influencing vibration
design, production technology, operation, condition change
(chapter 2)
Interaction of gearbox elements, environment influence
(chapter 2)
Scenario of degradation
Condition physical description
primary unbalance and misalignment, increase of rolling element bearing backlash, secondary misalignment. pitting, scuffing, tooth fracture, tooth breakage
(chapter 4)
Choice of proactive measures
primary and secondary misalignment assessment, (chapter 5)
oil particles and water content assessment
Mental transformation
local faults, distributed faults
Signal analysis tools
vibration spectrum, time frequency spectrogram, envelope analysis, spectral kurtosis, cyclostationary, analysis and use the results of these analysis for presenting them in the form of load susceptibility characteristics
(chapter 3 and 5)
Inferring
gear condition

Taking also into consideration knowledge on a gearbox degradation process [8, 13–16] choice of proactive measures ought be used as like primary and secondary misalignment assessment, oil particles and water content assessment. For the secondary misalignment the load susceptibility characteristics should be used.

In degradation scenarios there one ought to take into consideration order of possible events, which describe: primary unbalance and misalignment, increase of rolling element bearing backlash and secondary misalignment. Teeth or tooth pitting, scuffing, fracture, breakage as it is described in the Chap. 4.

On the base of primary misalignment evaluation proactive measure should be under taken eliminating some possible misalignment according to [17].

On the base of secondary misalignment evaluation proactive measure should be under taken, which goes to rolling element bearings replacement.

On the base of oil particles and water content assessment oil should be purified or replaced to reduce a gearbox elements wear.

For signal analysis should be used: vibration spectrum, time frequency spectrogram, envelope analysis, spectral kurtosis, cyclo-stationary analysis and used the results of these analysis for presenting them in the form of, load susceptibility characteristics.

Inferring on gear condition should be made during a gearbox operation from its beginning and on the base of different signal analysis make suggestions on use proactive measures, to reduce cost of maintenance.

Acknowledgments This paper was financially supported by Polish State Committee for Scientific research 2010–2013 as research project NN 504147838.

References

1. Neville W, Sachs PE (2007) Practical plant failure analysis. Taylor & Francis, United Kingdom
2. Hatangadi AA (2005) Plant and machinery failure prevention. McGraw-Hill, New York
3. Gridkhor P, Scheffer C (2004) Practical machinery vibration analysis and predictive maintenance. Elsevier, Netherlands
4. Bartelmus W (2006) Condition monitoring of open cast mining machinery. Oficyna Wydawnicza Politechniki Wrocławskiej Wrocław
5. Bartelmus W (2009) Influence of design, production technology, operation and change of condition factors on gearbox vibration diagnostic signals. Proceedings of resent developments in acoustic, noise and vibration. The 16th international congress on sound and vibration, Kraków, Poland, 5–9 July
6. Bartelmus W, Zimroz R (2011) Vibration spectra characteristics frequencies for condition monitoring of mine machinery compound and complex gearboxes. Scientific papers of the institute of mining of the Wroclaw University of Technology, mining and geology XVI, Wroclaw, pp 17–34
7. Bartelmus W (2011) Fundamentals for condition monitoring and diagnostics for driving bucket wheel system with overload mechanism of bucket wheel excavator. Scientific papers of the institute of mining of the Wroclaw University of Technology, mining and geology XVI Wroclaw, pp 5–16

8. Bartelmus W (2008) Root cause and vibration signal analysis for gearbox condition monitoring. Insight Nondest Test Condition Monit 50(4):195–201
9. Bartelmus W (1992) Vibration condition monitoring of gearboxes. Mach Vibr 1992(1):178–189
10. Bartelmus W, Zimroz R (2009) Vibration condition monitoring of planetary gearbox under varying external load. Mech Syst Sig Process 23:246–257
11. Bartelmus W, Zimroz R (2009) A new feature for monitoring the condition of gearboxes in non-stationary operation conditions. Mech Syst Sig Process 23:1528–1534
12. Bartelmus W, Chaari F, Zimroz R, Haddar M (2010) Modelling of gearbox dynamics under time varying non-stationary operation for distributed fault detection and diagnosis. European Journal of Mechanics—A/Solids 29:637–646
13. Bartelmus W, Zimroz R (2009) Gearbox condition degradation models. Proceedings of conference on condition monitoring and machine failure prevention technology. Dublin, Ireland, 22–25 June
14. Bartelmus W (2011) Gearbox damage process. Conference series in 9th international conference on damage assessment of structures, DAMAS (Electronic document) 2011. Oxford, UK, 11–13 July. J Phys 305(1):9
15. Bartelmus W (2012) Secondary misalignment as the main cause of gearbox failure. International conference on condition monitoring, London, United Kingdom, 12–14 June
16. Morales-Espejel GE (2012) Understanding and preventing surface distress. Gear Solutions, 29–38 July
17. Wowk V (2000) Machine vibration; alignment. McGraw-Hill, New York
18. Bartelmus W, Zimroz R (2011) The gearbox vibration load susceptibility as the measure of its condition: approaches and procedures for diagnostic relation discovery. International conference on condition monitoring, Cardiff, United Kingdom, June

Vibration Monitoring of Winch Epicyclic Gearboxes Using Cyclostationarity and Autoregressive Signal Model

Bassel Assaad and Mario Eltabach

Abstract This paper proposes a model-based technique using a combination of cyclostationary and autoregressive signal modelling in order to detect wear in a multistage planetary gear of lifting cranes. The first-order cyclostationarity is exploited by the analysis of the Time Synchronous Average part (TSA) of the angular resampled vibration signal. Then an autoregressive model (AR) is applied to the TSA part in order to extract a residual signal containing pertinent fault signatures. The paper also explores the efficiency of a number of methods commonly used in vibration monitoring. Condition monitoring indicators are then extracted from different treated signals. In the experimental part, all these techniques are applied to a test bench data of a lifting winch. The goal is to trend the evolution of the extracted features during the test. This study reveals that the proposed procedure using this combination enhances the ability to detect and diagnose mechanical wear of winch planetary gears.

Keywords Condition monitoring · Time synchronous averaging · Cyclostationarity · Autoregressive modeling · Epicyclic gearboxes · Winches

1 Introduction

The epicyclic (planetary) gear system refers to a compound gear system with planet gears between a centre sun gear and an outer ring gear. Its compact design makes it well suited for lifting machines, ranging from driving motors to driven machines,

---

B. Assaad · M. Eltabach (✉)
Centre Technique des Industries Mécaniques, CETIM, 52 av Félix Louat,
BP 80067, 60304 Senlis Cedex, France
e-mail: mario.eltabach@cetim.fr

B. Assaad
e-mail: bassel.assaad@cetim.fr

G. Dalpiaz et al. (eds.), *Advances in Condition Monitoring of Machinery in Non-Stationary Operations*, Lecture Notes in Mechanical Engineering,
DOI: 10.1007/978-3-642-39348-8_22,

with reduced speed. These planetary gears are prone to several defects that reduce their performance and give rise to adverse effects such as noise, vibration and cracks. Therefore, it is crucial to identify the various faults emanating from these types of gearboxes to avoid any unscheduled. The practical Condition Monitoring (CM) based on vibration analysis has historically been the technique of choice for machinery maintenance and fault diagnosis. It includes some specialized techniques such as Time Synchronous Averaging (TSA) [1, 2], time–frequency analysis [3, 4], spectral kurtosis [5], mathematical modelling [6–8], etc. It has been shown that the TSA technique is important for the extraction of the vibration signal features. Nevertheless, the TSA alone might not be sufficient in isolating the damage in its infancy. Sometimes, it is preferable to combine the TSA with other techniques such as the autoregressive filtering in order to improve the Detection ability of gear faults. Since the residue signal (difference between the filtered signal and its original part) contains pertinent faults signatures, what we propose is to apply and explore the pertinence of many usual statistical and power indicators to this last signal (residue AR signal). This paper proposes many condition indicators (CIs) extracted from the residual, the difference and the residue signals.

2 Vibration Induced by Epicyclic Gears

The epicyclic gear train vibrations are fundamentally different from those of the fixed-axis gear train. They are difficult to analyse because of two reasons. First, there are multiple planet gears producing similar vibrations. The second reason is the relative motion of planet gears to the transducer located on the gearbox housing. Indeed, the vibration transmission paths from the gear mesh point to the transducer vary with time. Since rotating machines involve cyclic patterns, the transmission of the vibration signal is periodic, and so its spectrum contains components at the fundamental gear mesh frequencies (GMF) and their harmonics. Each GMF can be subject to amplitude and phase modulation by any multiple of a rotating component frequency. These modulation functions are periodic with the considered gear rotational frequency.

3 Signal Processing Background

3.1 Time Synchronous Averaging 'TSA'

The TSA is regarded as a powerful tool for the detection and diagnosis of epicyclic gear faults. It is known as the first order cyclostationary analysis. TSA is intended to extract a deterministic signal and enhance the synchronous vibration components related to the rotational frequency of the gear of interest while reducing non-synchronous components and noise [1, 2]. Generally, rotating machines are characterized by cyclic phenomena directly related to the speed, which explains why a cyclostationary analysis is very well suited to diagnose these machines. However, in a practical environment, the variation of the shaft speed will result in smearing of amplitude energy in the frequency domain. To reduce this effect and to reinforce cyclostationarity, it is very important to resample the time domain signal into the angular domain. For healthy gear, the TSA signal is normally composed of gear meshing component's waveforms modulated by some low shaft order functions. The introduction of some CIs to the resultant signal following the application of TSA would help in detecting the availability and severity of the fault.

3.2 AR Model Based Diagnosis

The idea behind mathematical modelling [6, 8] is to indicate changes in the vibration signal due to damage. It is done by filtering the vibration data collected from the operational transmission using a model that is based on data collected from a healthy machine. An AR model seeks to represent a time series by a linear regression of the series on itself plus a residual error. It is applied, for example, when demanding high spectral re solution estimation. The purpose of the residual error is to isolate the effect of the fault impulses coming from the faulty signal in order to detect and monitor the health condition of the gear. The principle of autoregressive model identification of undamaged gear is presented in Fig. 1. Then, the AR filter created through this model is used as a linear prediction filter to pass the future state averaged signal as shown in Fig. 2.

An AR model seeks to represent a time series x(n) by a linear regression of x on itself plus a residual error e(n) (Eq. 1). It's applied, for example, when demanding high spectral resolution estimation [6, 7].

$$x(n) = \sum_{k=1}^{p} a(k)x(n-k) + e(n) \tag{1}$$

where:
x(n) is the nth data sequence point,
a(k) is the kth coefficient in the AR model,

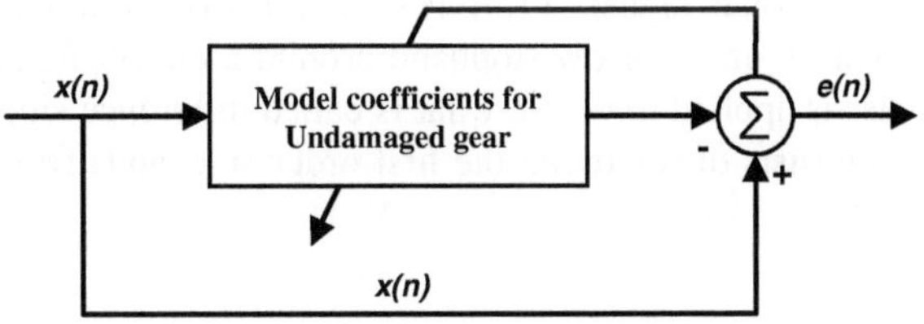

Fig. 1 Principle of AR modeling: model identification from undamaged gear

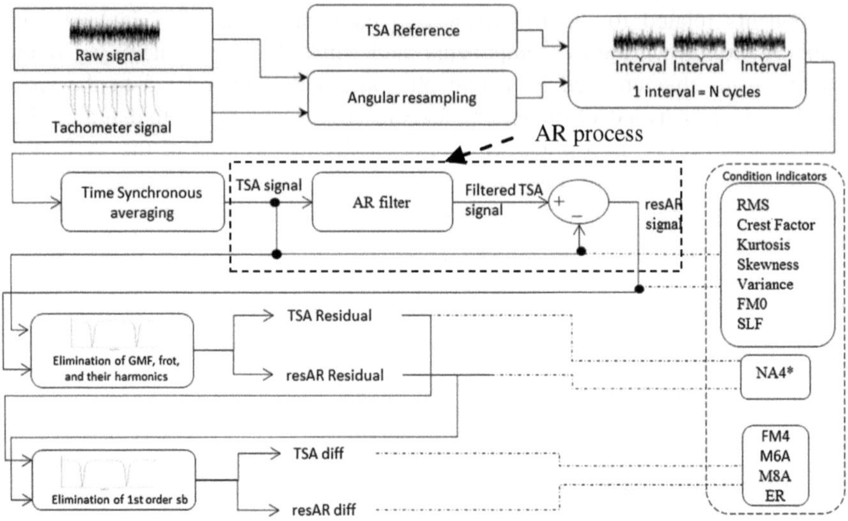

Fig. 2 Processing flow chart: feature extraction methods for vibration analysis

p is the order of the AR model,
e is the residual error

3.3 Features Extraction

Several indicators can be used to detect the mechanical abnormalities of a gear by taking into account the kinematic analysis of the machine and the signal processing (SP) techniques [9, 10]. For the success of any type of CM, effective features should be extracted. The literature includes some statistical or power CIs which track the amplitude of some frequencies or order components. The effectiveness of these CIs is generally related to the signal from which the computations are made. Figure 2 presents some CI used in this paper, associated with the different signal processing methods. It is worth noting that after applying the TSA and AR process, the resultant signals pass through a notch filter which eliminates the primary meshing and shaft components along with their harmonics to generate the residual signals (*TSA residual, resAR residual*). This filter attenuates frequencies in a narrow stopband around a cut-off frequency. Another filtering process is applied to obtain what is called difference signals (*TSA dif and resAR diff*). It consists of removing the first order sidebands from the residual signals.

4 Experimental Setup

This section presents the winch test bench, the Accelerated Life Time test (ALT), the use of different signal processing methodologies in order to extract appropriate CIs and finally to present their historical trends.

4.1 Test Bench

Figure 3 shows a global view of the winch used to accomplish the ALT and an overhead photo of its planetary gear. A tri-axe accelerometer is installed on the bearing housing at the gear side to measure the vibration induced by the transmission gearbox. In addition, a key phasor is used to get a phase reference of the rotor rotational speed. The load levels applied to the winch during the test are relatively high. The ALT began in September 2008 and a gear oil analysis carried out in July 2012 revealed excessive metal debris. In July 2012, inspection was conducted and the gear box was dismantled. Pitting has been found on the sun gear of the 1st stage and severe degradation on the sun of the 2nd stage were found, see Fig. 4.

4.2 Signal Processing

Before the extraction of the condition indicators, it is important to evaluate the performance of the TSA and the AR models and to compare the evolution of the vibration signals between the beginning and the end of the *ALT*.

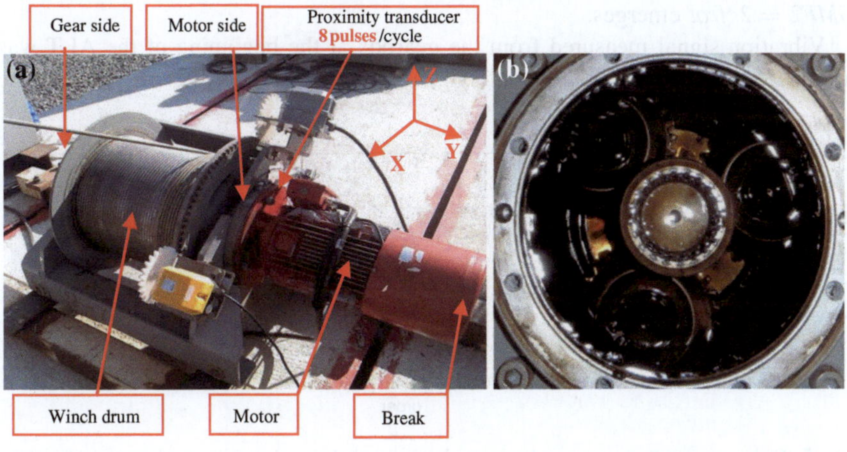

Fig. 3 Benchmark of the lifting crane. **a** General view. **b** Overhead photo of the gear

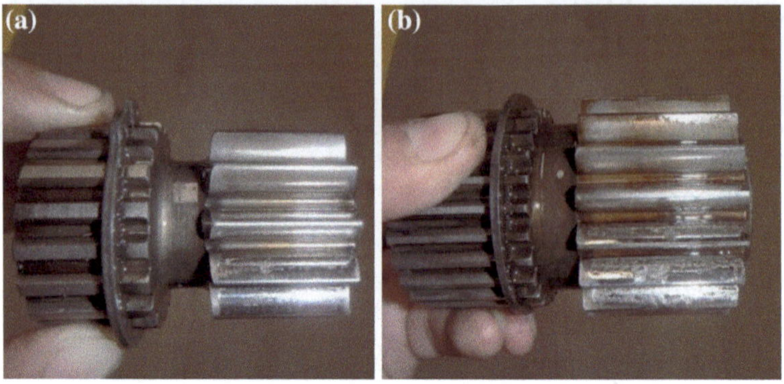

Fig. 4 Sun gears of epicyclic gearbox. **a** First stage. **b** Second stage

It is substantial to define synchronous components (frequencies that are multiples of a reference speed) since they are expected to clearly appear in the spectrum. Generally, the TSA can be applied to many epicyclic gearbox components taking an appropriate reference. However, several factors are to be respected which may influence the effectiveness of this approach, i.e. length of the signal, number of samples per revolution, number of averages, etc. Thus, for the winch monitoring, three main references are used: input shaft frequency of the 1st stage, that of the 2nd stage (respectively denoted 'frot' and 'fs2'), and Planet Pass Frequency of the first stage (PPF1). It should be kept in mind that 'fs2' is equal to the planet carrier frequency of the 1st stage *(fps1 = fs2)*. Foremost, TSA is conducted on the angular signal over 75 averages while taking *'frot'* as reference. The effectiveness of this technique to synchronize the signal with the considered frequency and to reduce asynchronous components and noise is shown in Fig. 5. It can be observed that *GMF* of the first stage '*GMF*1' = *10.64*frot* as well as synchronous frequencies are attenuated while the *GMF* of the 2nd stage *GMF*2 = *2*frot* emerges.

Vibration signal measured from the gearbox at the beginning of the ALT was used as the referential healthy-state signal in order to build the AR model. The

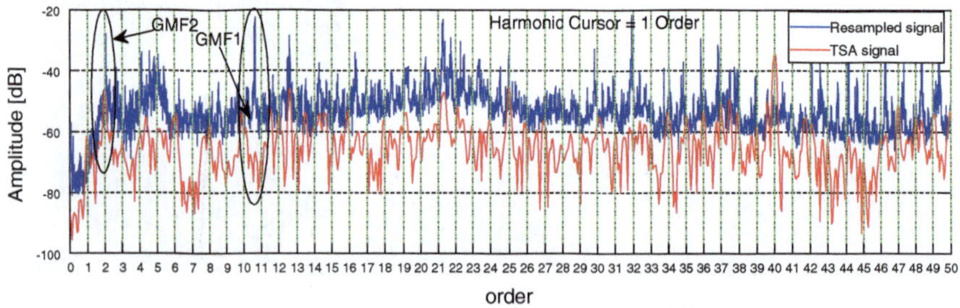

Fig. 5 Order spectrum comparison between the resampled signal and its synchronous part at the beginning of the ALT (Reference cycle)

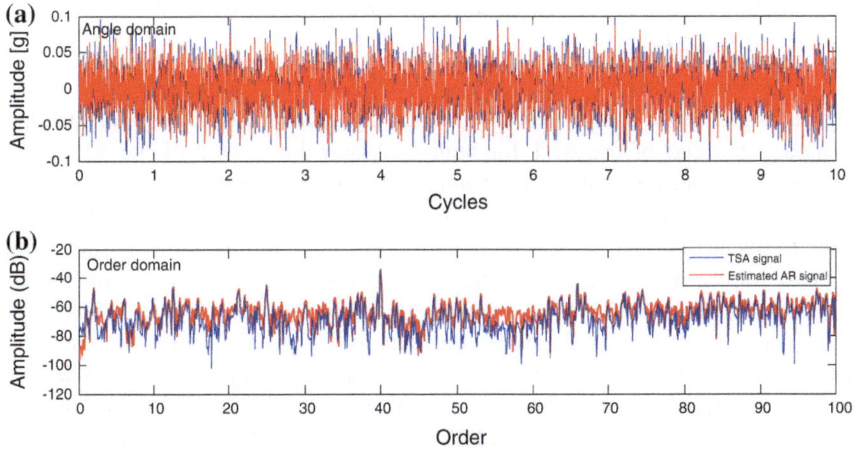

Fig. 6 Comparison between the TSA signal and estimated AR signal in the (**a**) angle domain and (**b**) frequency domain

optimum length of the AR filter was determined based on the minimal value of the Akaike Information Criterion (AIC) (corresponding to filter's order 700). Figure 6 illustrates a comparison between the estimated AR signal and the TSA signal in the time and frequency domain. This comparison shows that the filtered signal perfectly fits to its original version. The error between these two signals is considered as the residual AR signal (resAR) at the reference cycle 5,500; this error signal is depicted in Fig. 7b.

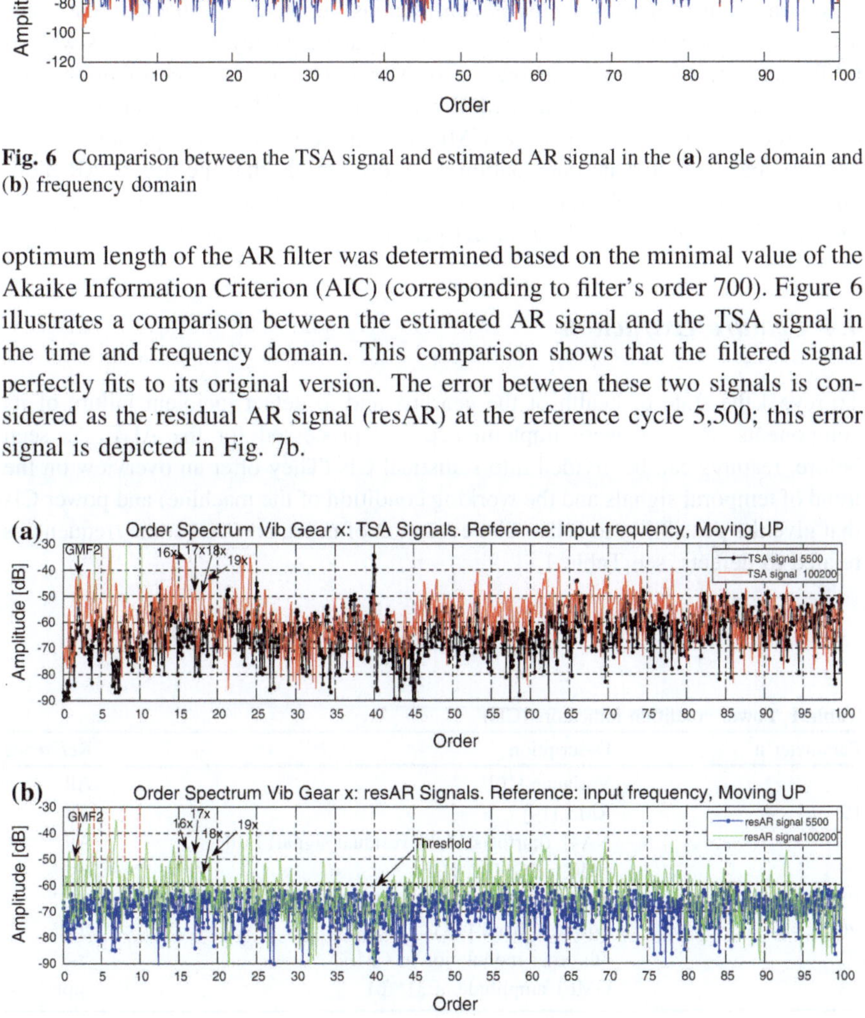

Fig. 7 Spectrum comparison of vibration signal at the beginning (no defects) and at the end of the ALT (with gear degradation). **a** TSA signals. **b** resAR signals

4.3 Signal Comparison at the Beginning and the End of the ALT

Figure 7a presents a comparison of vibration signals at the beginning (undamaged gear) and the end of the ALT for both TSA and the residue of the AR model 'resAR'. Order spectrum is used to compare the angular vibration signal between 0 and 100 orders. The spectral parameters are: 500 samples per revolution as sampling frequency and 6 cycles as the length of the hamming window. By analysing the two graphs, we notice an increase in the overall vibration levels. This feature reveals the degradation and the wear of the gearbox. It can also be deduced that power in some frequency bands is conspicuous i.e. [*0–25x*]. Specific gear characteristic frequencies such as harmonics of input frequency (*16x–19x*) as well as GMFs and their harmonics increase too. Moreover, the presence of pulse train frequencies with a period of the GMF2 can be noticed. This last phenomenon is a clear proof that there are impacts at GMF2 coming from the 2nd stage of the winch gearbox and revealing its degradation. It is discernible that the use of AR modelling has better highlighted the signature of the gear defects. According to Fig. 7b the use of a simple threshold to detect gear faults may appears to be effective.

4.4 Feature Extraction

To reveal the state of health of the gearbox and to detect incipient failure of its components, features were implemented and processed for the ALT. As seen before, features can be divided into statistical CIs (They offer an overview on the trend of temporal signals and the working condition of the machine) and power CIs that give the possibility to follow the evolution of energy around some frequencies related to defects, see Table 1.

Table 1 Power condition indicators (CIs)

Parameter n°	Description	Reference
P1	Variance [10]	All
P2	RMS [10]	All
P3	NA4* (kurtosis of the residual signal) [10]	All
P4	FM0 for GMF2. [10]	'fs1'
P5	4th order modulation at GMF2	'fs1'
P6	Power in band [48.fs2; 60.fs2]	'fs2'
P7	4th order modulation at GMF2	'fs2'
P8	GMF1 amplitude at 31*fp1	'ppf1'

5 Diagnosis, Results, and Discussion

The historical trends of all the defined CIs are depicted in Fig. 8. It should be clarified that to normalize the values of the CIs, they are divided by the value obtained from the reference cycle. Clearness and fluctuation level of these trends are considered as criteria for selecting relevant features from the generated parameters used for damage detection. The experimental result indicates that P1 and P2 have shown a remarkable increase over the ALT. They have clearly described and tracked the condition of the tested gearbox condition. Between January and March 2011, these two CIs increased rapidly, which may reflect the beginning of the degradation of the planetary gear. The start point of the wear is also confirmed by P3 (NA4*). Since P3 is calculated based on the residual signal, it indicates that a distributed fault could emerge on the winch.

CIs extracted from difference signals did not reflect any explicit trends. In general, these indicators detect heavy wears that give rise to high order sidebands. We ascribe their fluctuation to the fact that some heavy defects such as teeth breakages or gears cracks didn't affect the winch's gearbox.

Values of FM0 'P4' (usually sensitive to impulsive type defects) increase in the 1st half part of the ALT and then decrease. Contrarily, P5 that describes a modulation phenomenon is not sensitive to these types of defects and does not evolve like P4. Nevertheless, it increases in a linear pattern. P5 also reveals the presence of modulation activity and distributed degradation of the gearbox second stage. P6 and P7 are specialized in detecting defects on the second gear stage according to

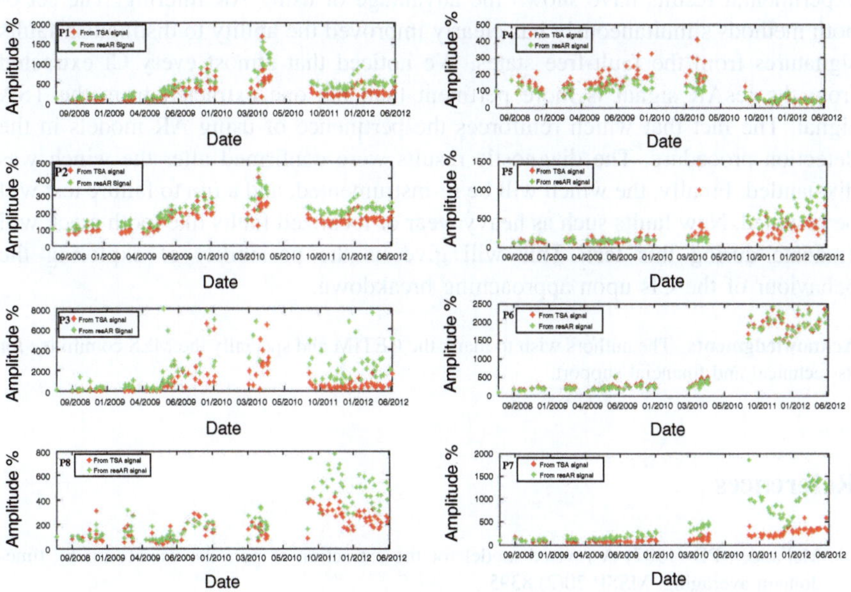

Fig. 8 Evolution of power CIs listed in Table 1

the adopted TSA reference (fs2). These parameters increase monotonically during the ALT and highlight the existence of faults related to the sun gear of the gearbox second stage. The CI which is specialized in detecting defects on the first stage with PPF1 taken as reference frequency is P8. It points out the amplitude of the GMF1. P8, the last CI, evolves in a linear manner along the ALT and is considered as pertinent, nevertheless its value is less significant than P6-P7 in relation to the GMF2. The value of P8 reflects that degradation is more important in the second stage than in the first one. This fact has been confirmed by the winch inspection, see Fig. 4. As shown below, the trend of CIs extracted from signals filtered using the AR process is sensitive to the type of the CI. Two cases are distinguished: No improvement (Values of CIs of AR process following the values of CIs extracted from TSA signal such as P4 or P6), or showing a clear improvement in the detectability of the faulty states like RMS, Variance, NA4* or those who represent the modulations phenomena (P5;P8).

6 Conclusion

One of the most important insights to be drawn from this work is that choosing a suitable reference for the TSA and special CIs could lead to a better diagnosis of wear in an epicyclic gearbox. This paper has shown that statistical CIs provide global information about the gearbox condition. However, power CIs have provided appropriate information about the condition of the gearbox components. The experimental results have shown the advantage of using AR filtering. The set of both methods simultaneously has clearly improved the ability to distinguish faults signatures from the fault-free states. We noticed that almost every CI extracted from the resAR signal is more pertinent than the one extracted from the TSA signal. The fact that which reinforces the pertinence of using AR models in the detection procedure. The diagnostic results were confirmed after the winch was dismantled. Finally, the winch will be re-instrumented, and a run to failure test will be resumed. New faults such as heavy wear or localized faults like tooth crack will emanate during this test. This will give us the possibility of exploring the behaviour of the CIs upon approaching breakdown.

Acknowledgments The authors wish to thank the CETIM and specially the MLS committee for its technical and financial support.

References

1. McFadden PD (1987) A revised model for the extraction of periodic waveforms by time-domain averaging. MSSP 20(2):8395
2. Forrester BD (1996) Method and apparatus for performing selective signal averaging. Australian Patent Number 672166

3. Wang WJ, McFadden PD (1996) Application of wavelets to gearbox vibration signals for fault detection. J Sound Vib 192(5):927–939
4. Yu J. (2011) Early fault detection for gear shaft and planetary gear based on wavelet and hidden markov modeling. Phd Thesis, University of Toronto
5. Antoni J, Randall RB (2006) The spectral kurtosis: application to the vibratory surveillance and diagnostics of rotating machines. MSSP 20(2):308–331
6. Wang W, Wong A (2000) A model-based gear diagnostic technique. Technical Report 1079, DSTO (Defense Science and Technology Organization)
7. Yip L (2011) Analysis and modeling of planetary gearbox vibration data for early fault detection. Phd thesis, University of Toronto
8. Wang W, Wong AK (2002) Autoregressive model-based gear fault diagnosis. Trans ASME J Vib Acoust 124:172–179
9. Harry D (2001) Gear diagnostic parameters NASA Glenn Research Center
10. Večeř P, Kreidl M, Smid R (2005) Condition indicators for gearbox condition monitoring systems. Acta Polytechnica 45(6):35–43

Gear Parameter Identification in Wind Turbines Using Diagnostic Analysis of Gearbox Vibration Signals

Nader Sawalhi and Robert B. Randall

Abstract The correct diagnosis of faulty components in rotating machines requires a pre-knowledge of the characteristics of the system being monitored and the identification of the frequencies of interest. In gearboxes, the number of gear stages and the number of teeth for each gear are required to calculate the gear mesh frequencies and monitor these frequencies and their sidebands. It is not always possible to have this information available, especially in old equipment. In this chapter a fresh approach is presented to deduce such crucial information from the measured vibration signal. The approach focuses on fine tuning harmonic/sideband cursors to capture different gear mesh families. The approach is illustrated on a signal taken from a wind turbine gearbox, which poses the extra challenge of the variable speed within the measurement record. Results show the possibility of identifying the number of teeth for the first two stages with much more confidence than the planetary stage, where a trial and error approach was used to decide on the most likely combination for the ring, sun and planetary gears. This chapter sets a good practice example for understanding the system characteristics by detailed analysis of the vibration signal using finely tuned harmonic and sideband cursors.

Keywords Harmonics · Sidebands · Gear mesh frequency · Order tracking · Spectrogram · Wind turbines

N. Sawalhi (✉)
Mechanical Engineering Department, Prince Mohammad Bin Fahd University (PMU), Al Khobar 31952, Dhahran, Saudi Arabia
e-mail: nsawalhi@pmu.edu.sa

R. B. Randall
School of Mechanical and Manufacturing Engineering, University of New South Wales, Sydney 2052, Australia
e-mail: b.randall@unsw.edu.au

G. Dalpiaz et al. (eds.), *Advances in Condition Monitoring of Machinery in Non-Stationary Operations*, Lecture Notes in Mechanical Engineering,
DOI: 10.1007/978-3-642-39348-8_23,

1 Introduction

Monitoring the health of gears in a gearbox, using vibration signals, requires a knowledge of the input or the output rotational speed, the number of stages as well as their arrangement, and the number of teeth for each gear. Often, this information is available to the vibration analyst by means of a speed reference signal and the manufacturer's details of the gearbox. Typically, a tachometer/encoder is installed at the high speed shaft to enable estimation of the shaft speed and order tracking of the signal [1] to remove speed fluctuations.

In cases where more limited information is available to the analyst, a harmonic cursor approach has been recommended for the blind determination of the numbers of teeth on a gear pair [2]. The use of a harmonic cursor in such occasions is possible if the gear pair represents a 'hunting tooth' design [2]. The 'hunting tooth design', which is considered as good practice, means that there is no common factor between the pinion and the gear and consequently their harmonics are completely separable except at the gear mesh frequencies (the closest the other harmonics can reach is $1/(m \times n)$, where m and n are the numbers of teeth on the two gears. In order to use a harmonic cursor, the machine speed is required to be stable to about 1:20000 [2]. If this is not the case, order tracking can be used to increase the degree of stability. The harmonic cursor approach also requires the presence of a reasonable number of harmonics of both shafts in the spectrum (logarithmic/decibel scale) including sidebands around the gear mesh frequency. The harmonic cursor approach as described in [2] starts by setting up a harmonic cursor on each shaft in succession, first on the low orders and then progressively adjusting it by zooming in higher frequency bands. This will eventually determine the fundamental frequency to the required accuracy. If lists of the two harmonic series are then compared, the gear mesh frequency corresponds to where they match to better than 1:10000 [2]. Even if the design is not hunting tooth, the approach can often still be used. The most likely common factor is 2 or 3, in which case the first correspondence will be at 1/2 or 1/3 of the actual mesh frequency. For the pinion, this will often lead to an unlikely minimum number of teeth, and in any case if there is an inspection port, the tooth pitch can be measured sufficiently accurately with a tape measure to exclude incorrect possibilities.

The situation presented in this chapter discusses the case where knowledge about the characteristics of the gearbox, of a wind turbine, was very limited and there was no speed signal. The challenge imposed in such a situation requires an accurate estimate of the different shaft speeds (or at least their ratios) and the numbers of teeth for each of the gears to calculate the gear mesh frequencies. The case is complicated by the fact that the wind turbine does not run at constant speed and in fact the measurements analyzed are actual signals from a full size wind turbine running under normal conditions and being subject to varying wind conditions and with no tachometer/speed reference signal available.

The chapter is organized as follows: after this introduction, the experimental data collected is described in Sect. 2. This is followed by an illustration of the data handling and processing in Sect. 3. Results are presented and discussed in Sect. 4. Finally the chapter is concluded in Sect. 5.

2 Experimental Data

The case presented in this chapter was taken from an actual wind turbine gearbox. The gearbox has three stages (two helical parallel stages and one planetary). The only information available about this gearbox was that it had three stages and an overall gear ratio close to 78.472. There was no speed reference signal. The generator was of the doubly fed induction type, for which the speed can vary by up to ± 30 %.

Ten minutes of vibration data, sampled at 51,200 Hz were measured by the company Mechanical Solutions Inc. (MSI) from a number of accelerometers mounted on the machinery of a wind turbine. The four accelerometers were placed as follows: one on the gearbox first stage (planetary), a second on the gearbox third stage (close to the high speed shaft) and the other two on the inboard and outboard bearings of the generator. No other information or data were made available. A detailed analysis of one of the signals measured on the gearbox was made in order to attempt to discover the numbers of teeth on the various gears. The signal measured at the Gearbox 3rd Stage was chosen for this.

3 Data Processing Stages

The different stages of processing the signal are described in Fig. 1. In the first stage, the signal from the third gearbox stage was decimated (down-sampled) by a factor of 8, partly to allow analysis of longer records (in terms of time) and partly because the information about the various gear mesh frequencies was contained in the valid frequency range to 2,500 Hz after down-sampling. The second stage utilized a Short Time Fourier Transform (Spectrogram) to observe the speed variation within the measured record and decide on a part where speed variation is small. The Spectrogram also gives very useful information about the different sets of gear mesh frequencies and shaft speeds.

Order tracking was used in the third stage to remove speed fluctuations from the signal and enable the tuning of harmonic/sideband cursors to a very high accuracy for a confident analysis. As no tachometer signal was provided, the signal itself was used to perform the order tracking. In cases where a tachometer signal is not available, a reference signal can be obtained from the vibration signal itself and used for order tracking [3, 4], Combet and Zimroz, [5, 6]. The approach used in this chapter was developed by [5] and is based on phase demodulation, with the

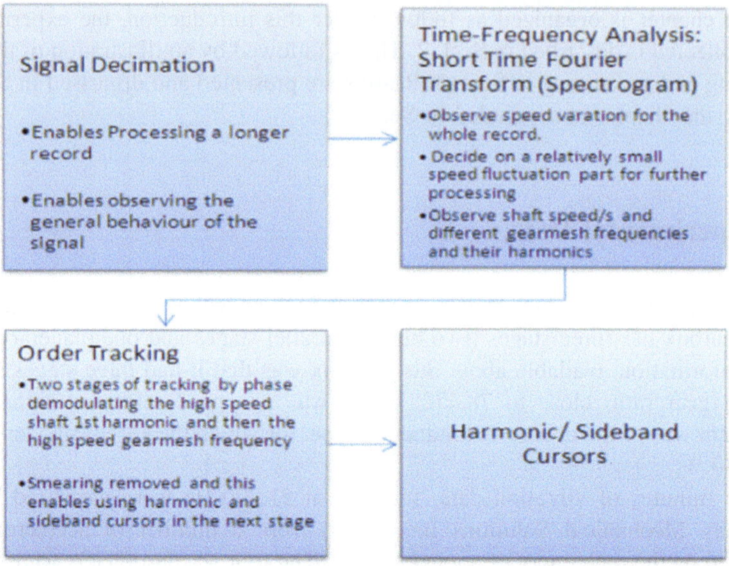

Fig. 1 Data processing stages

possibility to improve accuracy by successive iteration, perhaps starting with a very low harmonic for which a much larger speed variation can be handled. In the current application, phase demodulation around the first harmonic of the output shaft speed was initially attempted and was then refined by demodulating around the high speed stage gear mesh frequency. After the order tracking process, which removes the speed fluctuations from the signal, harmonics of different shaft families and gear mesh families become recognizable, and extremely localised. A harmonic/sideband cursor [2] could then be used to identify the different harmonic and sideband families, and thus determine the numbers of teeth on each gear.

4 Results

The Spectrogram (Short Time Fourier Transform) was initially used to examine the speed fluctuation within the gearbox 3rd stage signal as shown in Fig. 2a. A more detailed zoomed-in view (0–700 Hz) is shown in Fig. 2b. Figure 2a, b reveal a number of pieces of useful information for the analyst. It gives an indication about the speed fluctuations and the speed profile during the 600 s of measurement. Note that a reasonably small variation section of the measured data can be selected between 190 and 290 s.

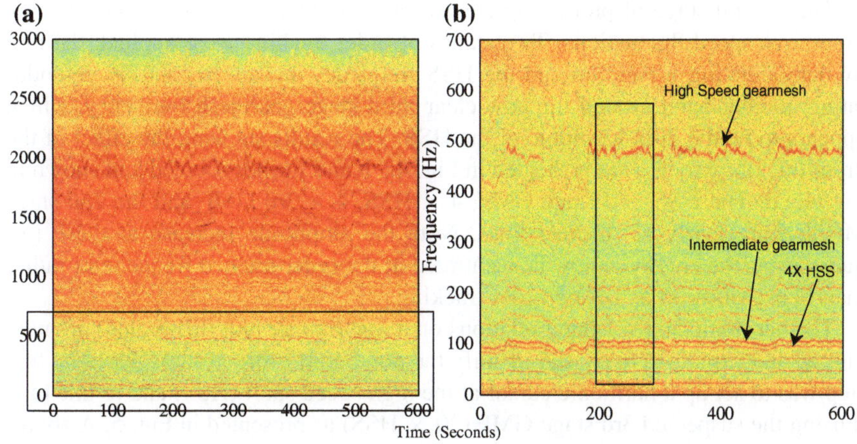

Fig. 2 **a** Spectrogram for gearbox 3rd stage signal. **b** Zoom up to 700 Hz

1. Presumed gear mesh frequencies and shaft speeds can be roughly identified. The clearest is the supposed third stage gear mesh frequency (GMF) appearing just shy of 500 Hz and the second stage gear mesh frequency around 100 Hz.
2. A further zoom in the low frequency region below 150 Hz, as presented in Fig. 3, shows harmonics of the high speed shaft (HSS). In particular, the 4th harmonic of the HSS is very clear and just slightly above it there appears to be the intermediate stage GMF.
3. A section for further processing is selected between the time 190 and 290 s, where speed fluctuation is minimum.

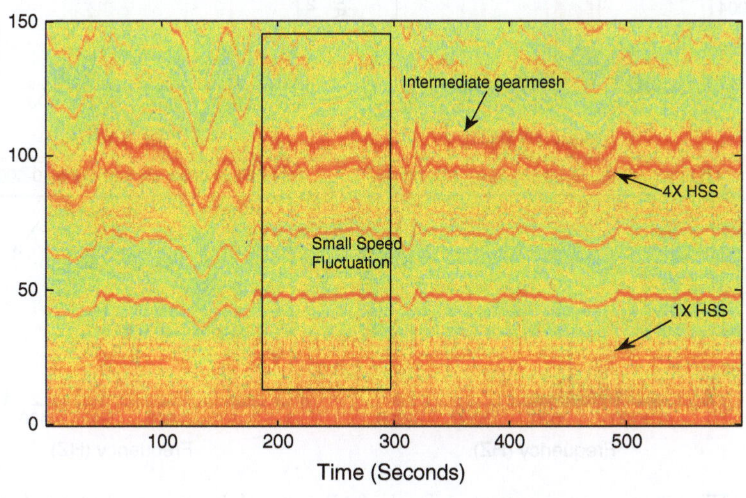

Fig. 3 Zoomed spectrogram in the lower frequency range (0–150 Hz)

The second stage of processing concentrated on this section for order tracking to remove speed fluctuations. Two stages of order tracking were applied: the first by demodulating a band around the HSS frequency and the second by demodulating another band around the now clear third stage gearmesh frequency, which appears to be the 20th harmonic of the HSS. The results showing the effect of the order tracking process are presented in Fig. 4a, b and further zoomed and shown in Fig. 4c, d. The HSS, 3rd stage GMF, 4 × HSS, 4 × 3rd GMF and the 2nd stage MF appear clearly as discrete lines. Note the amplitude increase at these frequencies as the energy is now concentrated at the order tracked frequencies rather than being smeared as before order tracking.

The harmonic and sideband relations on a logarithmic plot (dB scaled) are next plotted and explained to try and identify the number of teeth on each gear. The first step was to set up a harmonic cursor at the high speed shaft (HSS) and to fine tune it using the suspected 3rd stage GMF (20 × HSS) as presented in Fig. 5. A zoom-in around this 3rd stage GMF was used for an exact tuning of the frequency. A zoom-out was then used to ensure that all the HSS harmonics are well aligned with the harmonic cursors. Figure 5a shows the harmonic cursor of the HSS, while Fig. 5b shows the sidebands (also harmonics) spaced at the HSS around the 3rd stage GMF. A zoom-in around the 20th harmonic (3rd stage GMF) is shown in Fig. 5c to demonstrate the accuracy of the harmonic cursor. Note that an order scale is usually recommended after the order tracking process, however a

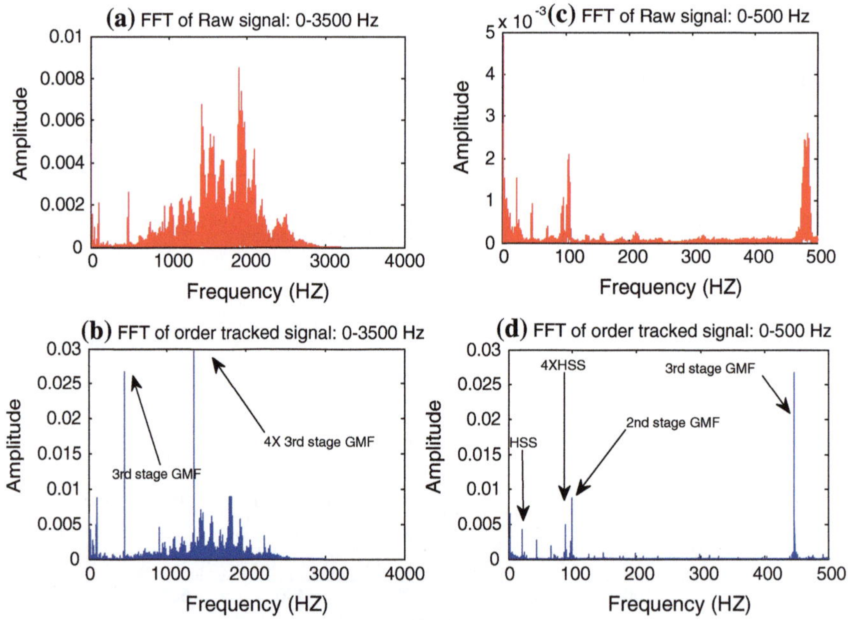

Fig. 4 FFT of the raw and order tracked signals: (**a**) Raw signal (**b**) Order tracked signal (**c**) Raw (0–500 Hz) (**d**) Order tracked (0500 Hz)

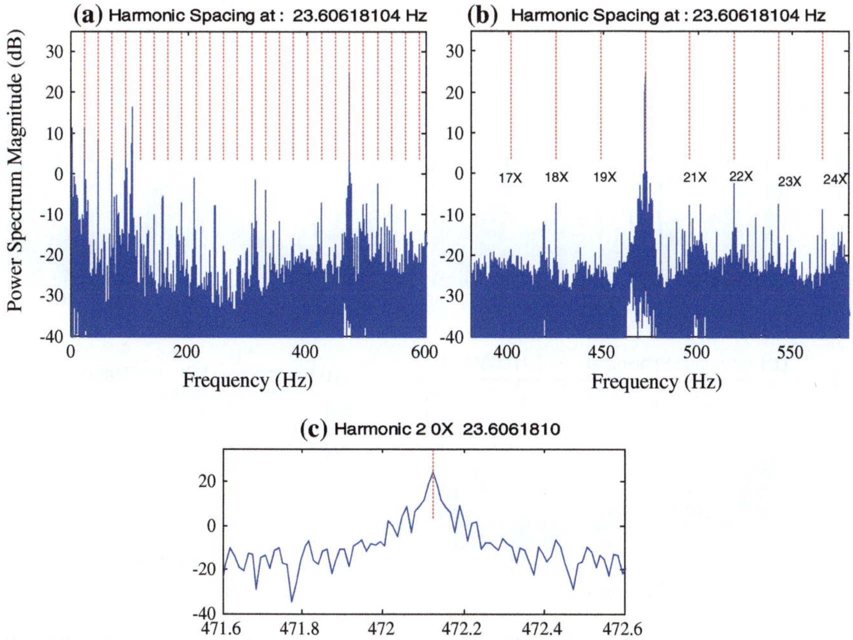

Fig. 5 HSS harmonics (**a**) the first 25th harmonics (0–600 Hz) (**b**) A zoom in showing harmonics 17–24 (**c**) A zoom in around the 20 × HSS (3rd stage GMF)

frequency scale has been used here and the mean output frequency is always 23.60618104 every time (order 1), even after resampling for each iteration.

In Fig. 6, it is shown how the 3rd stage GMF (20 × HSS) is in fact the 113th harmonic of a speed around 4.18 Hz. Figure 6b shows a family of sidebands around the 3rd stage GMF spaced at around the 4.18 Hz [believed to be that of the intermediate shaft speed (ISS)]. It can also be seen from Fig. 6c that the 25th harmonic of this ISS aligns with the suspected 2nd stage GMF. A zoom-in around the 3rd stage GMF shows how 113 × ISS aligns perfectly with the 3rd stage GMF. Note that Figs. 5 and 6 are exactly the same spectrum with the same adjustment of the cursor. Zooming-in the spectrum to adjust the harmonic cursor was performed to give the necessary alignment and fine tuning. Tiny adjustments of the cursor cause some harmonics to lose alignment.

In Table 1, the harmonics based on the two identified fundamental frequencies of the HSS and the ISS obtained from Figs. 5 and 6 respectively are shown to 11 figure accuracy. The 20 × HSS and the 113 × ISS are seen to be the same up to the seventh decimal figure indicating an accuracy of $1:10^{10}$. The number of figures shows how accurate the harmonic cursor can be after order tracking and gives the confidence in the number of teeth for the pinion and gear of the 3rd stage. The ratio 20 × 113 is hunting tooth. The closest other harmonics can come is $1/(20 \times 113) = 1:2260$ (compared with $1:10^{10}$).

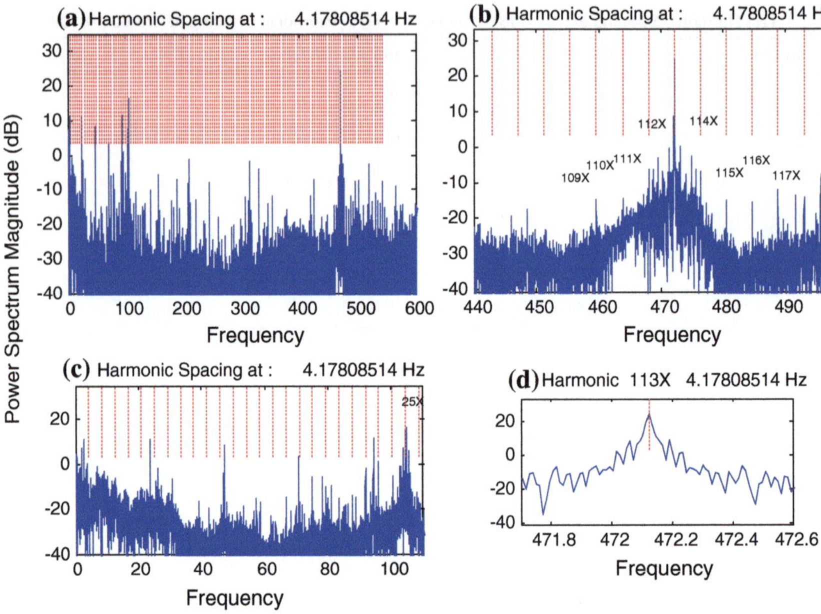

Fig. 6 ISS harmonics (**a**) the first 120 harmonics (**b**) A zoom in showing harmonics 106–119 (**c**) The 25 × ISS locks into the suspected 2nd GMF (**d**) A zoom in around the 113X ISS (3rd stage GMF)

Table 1 HSS harmonics (17–23) and ISS harmonics (105–120)

	Harmonic number	Frequency (Hz)		Harmonic	Frequency (Hz)
	17	401.30507768		105	438.69893970
	18	424.91125872		106	442.87702484
HSS fundamental	19	448.51743976		107	447.05510998
23.60618104	**20**	**472.12362080**		108	451.23319512
	21	495.72980184		109	455.41128026
	22	519.33598288		110	459.58936540
	23	542.94216392		111	463.76745054
			ISS		
fundamental	112	467.94553568			
			4.17808514	**113**	**472.12362082**
				114	476.30170596
				115	480.47979110
				116	484.65787624
				117	488.83596138
				118	493.01404652
				119	497.19213166
				120	501.37021680

The numbers of teeth for the second stage pinion (thought to be 25: see Fig. 6c) and matching sun shaft wheel (71) as well as the planetary stage are shown in Table 2. The bases for filling in the rest of Table 2 are explained using Figs. 7, 8, 9 and 10. This parallel stage is also hunting tooth, and 1/(m × n) is 1:1750.

Another series of harmonics was sought for the sun gear shaft, which should also have the 2nd GMF (104.45 Hz) as a member. In order to do this, the signal was order tracked again, but this time by demodulating a band around the 2nd GMF [a band taken between 104.53 Hz and 104.78 Hz, with the centre frequency being 104.67]. Figure 7 shows the 2nd GMF before and after the order tracking. Order tracking gives a more definite peak at the expected 104.45 Hz peak rather than at around 104.67 before the order tracking.

Figure 8a, c–e show the series of harmonics at 1.4711567 Hz with the 2nd gear mesh frequency as harmonic No. 71, and including a number of sidebands around the gear mesh frequency. Note also that there is a pair of strong sidebands spaced at about 0.902 Hz (Fig. 8b), with others at 1/3 of this spacing. These sideband families are however not sub- harmonics of the gear mesh frequency, and are suspected to be the input shaft speed (0.30 Hz) and its third harmonic the planet

Table 2 Estimated shaft and gear mesh frequencies (for generator mean speed 23.606 Hz)

Estimated Frequencies in the gearbox			
High shaft speed (Hz)	23.60618104		
HSS			3rd Stage GMF
pinion		20	472.12362080
gear		113	
Intermediate shaft speed	4.17808514		
ISS			2nd Stage GMF
pinion		25	104.45212850
gear		71	
Low shaft speed (sun)	1.47115674		
Ring gear teeth		105	Epicyclic mesh
Planet gear		39	31.59643452
Sun gear		27	
Output speed (carrier)	0.30091842		
Planet pass frequency	0.90275527		
Spin frequency	-0.81016499		
	overall planet stage	4.89	
	Total ratio		78.447

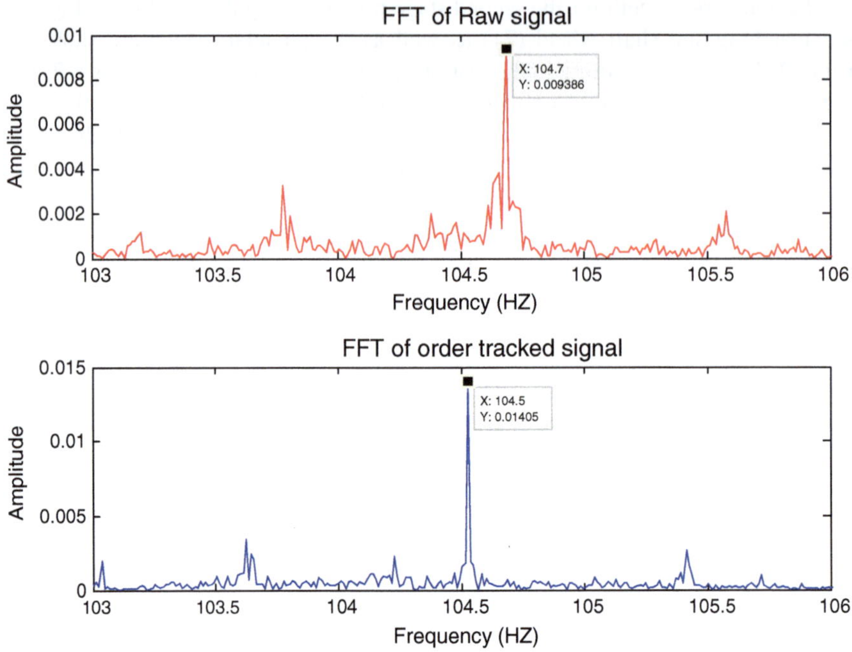

Fig. 7 *Top* 2nd GMF before being tracked. *Bottom* 2nd GMF after order tracking

pass frequency (0.90 Hz). It is thus thought that the sun gear speed is 1.4711 Hz. Note that the HSS is roughly around, but not exactly 16 × Input Shaft Speed.

Figure 9 lends weight to the theory that the input speed is 0.30 Hz, by showing some of the harmonics of a series spaced at 0.90 Hz (Nos. 28, 29, 30 and 57, 58, 59, respectively). Harmonics 29 and 58 have sub-harmonics at 1/3 of the spacing as well.

It remains to find the details of the gears in the planetary section. These are subject to considerable restraint, as follows. If S, P, and R are the numbers of teeth on the sun, planet and ring gears, respectively, then in general the following equations should be satisfied:

$$R = S + 2P \tag{1}$$

and either, R and S must be both divisible by 3 (simultaneous meshing with all planets) or the sum of R and S must be divisible by 3 (offset meshing of the planets). By trial and error, the set of gears indicated in Table 2 were arrived at; viz. R = 105, P = 39 and S = 27. These satisfy the rules given above, and also give approximately the right gear ratio. The overall ratio of the gearbox with the parameters in Table 2 is 78.447, whereas the actual is thought to be 78.472. It must be admitted that the hypothetical planetary section is the least likely to be correct, as little evidence could be found to confirm the details. For example, the planetary gear mesh frequency (31.59 Hz in Table 2) could not be located with certainty.

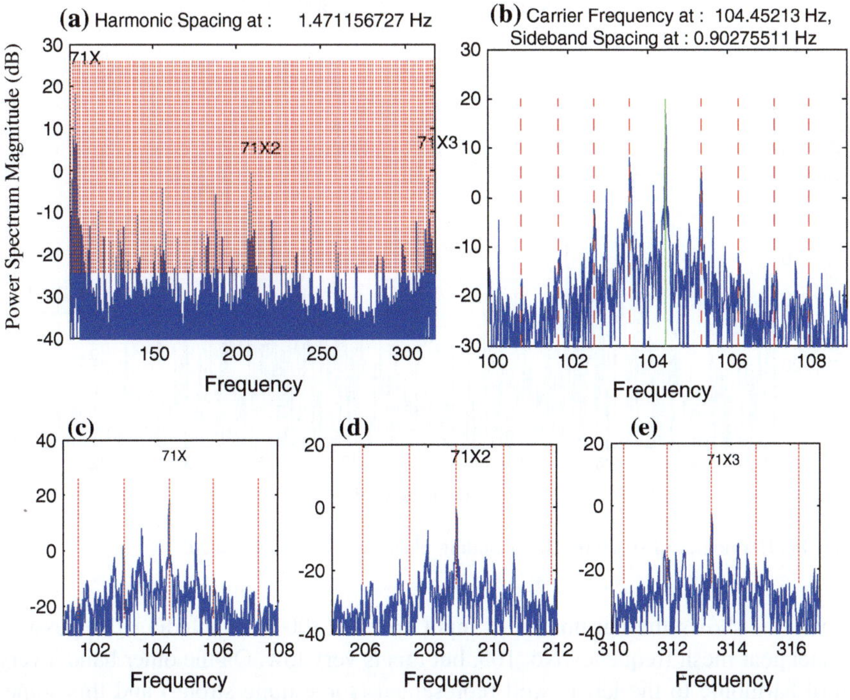

Fig. 8 **a** A series of harmonics spaced at 1.4711 Hz, which include the second gear mesh frequency (104.45 Hz) as harmonic No. 71. **b** Sidebands at 0.9 Hz (planet pass frequency). **c** A zoom in around the 2nd GMF. **d** A zoom in around the 2 × 2nd GMF. **e** A zoom in around the 3 × 2nd GMF

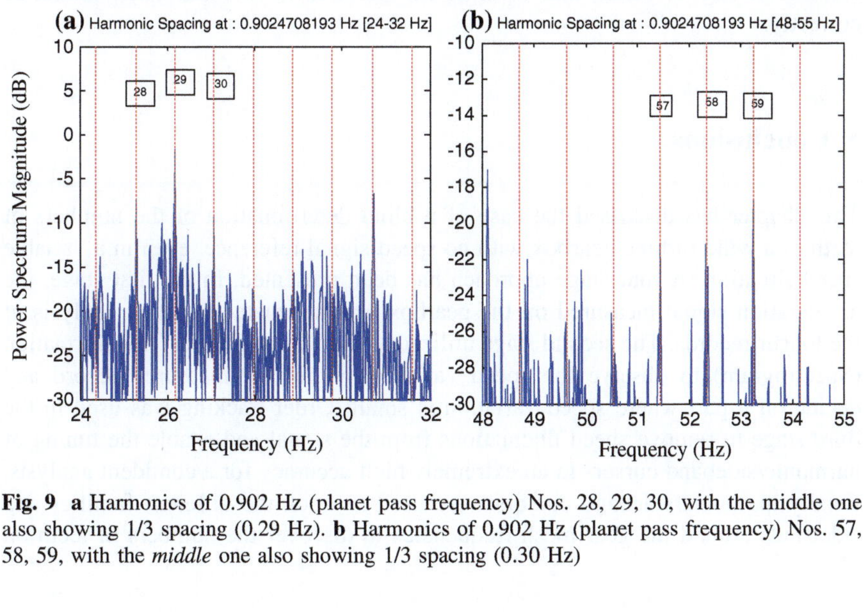

Fig. 9 **a** Harmonics of 0.902 Hz (planet pass frequency) Nos. 28, 29, 30, with the middle one also showing 1/3 spacing (0.29 Hz). **b** Harmonics of 0.902 Hz (planet pass frequency) Nos. 57, 58, 59, with the *middle* one also showing 1/3 spacing (0.30 Hz)

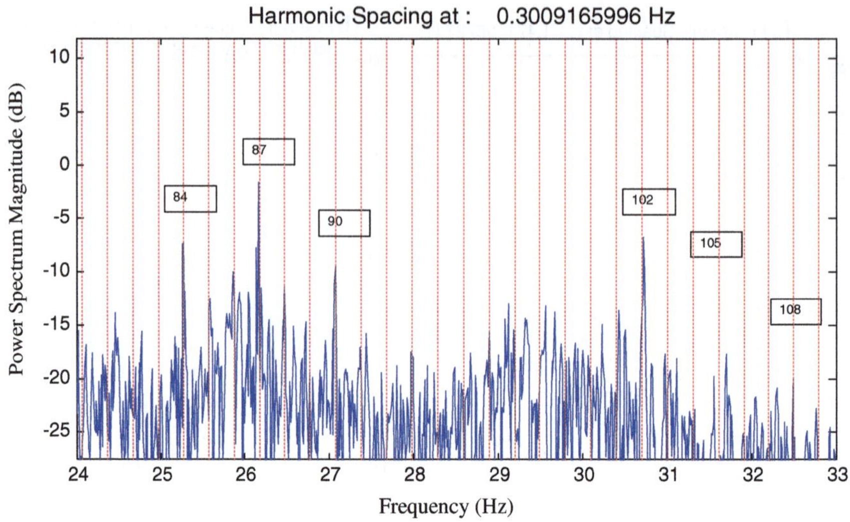

Fig. 10 Harmonics of 0.0.3009 Hz, including planetary gear mesh No.105

Figure 10 shows the harmonics of input speed 0.300 Hz, including the possible planet gear mesh frequency No. 105, but this is very low. On the other hand, every third harmonic to the left is well represented (some quite strong) and this sometimes occurs with planetary gears (i.e. the sidebands adjacent to the gear mesh frequency are stronger than the gear mesh itself) [McFadden]. Another point is that not all planetary gear sets obey the rules of Eq. (1), as other possibilities can be accommodated by addendum/dedendum modification. Even so, it is believed that there is strong evidence that at least the parallel section has been identified correctly.

5 Conclusions

This chapter has discussed the case of a blind determination of the numbers of teeth of a wind turbine gearbox with no speed signal reference, even in a variable speed situation. A four stage approach has been presented. In the first stage, the acceleration signal measured on the gearbox was decimated to allow analysis of the longer records. The second stage utilized a Short Time Fourier Transformation (Spectrogram) to observe the speed variation within the measured record and decide on a part where speed variation is small. Order tracking was used in the third stage to remove speed fluctuations from the signal and enable the tuning of harmonic/sideband cursors to an extremely high accuracy for a confident analysis. Finally harmonic/sideband cursors were used to track the different families and sidebands around the gear mesh frequencies to discover the numbers of teeth on each gear. Results indicated the ability to identify the numbers of teeth for the two parallel stages with much more confidence than for the planetary stage, where a trial and error approach was used to decide on the most likely combination for the ring, sun and planet gears.

Acknowledgments This work is partially supported by Mechanical Solutions Inc. and by Australia's Defence Science and Technology Organisation (DSTO). Signals used for demonstration purposes were provided by Mechanical Solutions Inc., NJ, USA, and were obtained under a research project funded by NYSERDA. NYSERDA has not reviewed the information contained herein and the opinions expressed in this chapter do not necessarily reflect those of NYSERDA or the State of New York.

References

1. Forrester BD (1991) Time-frequency domain analysis of helicopter transmission vibration, department of defence, Aeronautical Research laboratory, Propulsion Report 180
2. Randall RB (2011) Vibration-based condition monitoring: industrial, aerospace and automotive applications. Wiley, NY
3. Bonnardot F, El Badaoui M, Randall RB, Danière J, Guillet F (2005) Use of the acceleration signal of a gearbox in order to perform angular resampling (with limited speed fluctuation). Mech Sys Signal Process 19:766–785
4. Combet F, Gelman L (2007) An automated methodology for performing time synchronous averaging of a gearbox signal without speed sensor. Mech Sys Signal Process 21(6):2590–2606
5. Coats MD, Sawalhi N, Randall RB (2009) Extraction of Tacho information from a vibration signal for improved synchronous averaging, Paper presented at ACOUSTICS 2009, Australian Acoustical Soc, 23–25 Nov, Adelaide
6. Combet F, Zimroz R (2009) A new method for the estimation of the instantaneous speed relative fluctuation in a vibration signal based on the short time scale transform. Mech Sys Signal Process 23(4):1382–1397
7. McFadden PD, Smith JD (1985) An explanation for the asymmetry of the modulation sidebands about the tooth meshing frequency in epicyclic gear vibration., Proc. I.Mech. E. Part CJ. Mech. Eng. Sci. 199(1):65–70

Phase Monitoring by ESPRIT with Sliding Window and Hilbert Transform for Early Detection of Gear Cracks

Thameur Kidar, Marc Thomas, Mohamed Elbadaoui and Raynald Guilbault

Abstract The detection of cracks in gears may be considered as among the most complicated operations in the diagnosis of this type of machines. This paper presents the crack signature in the vibration signal through a numerical model. Then, a comparison of phase analysis is conducted between the phase estimated by the Hilbert method and the proposed technique Estimation of Signal Parameters via Rotational Invariant Technique (ESPRIT) by using a sliding window. This comparison was made on signals coming from both a numerical model of a cracked tooth and a multiplicative signal modulated in frequency. The proposed method gives very interesting results despite the existence of the amplitude modulation generated by the transmission error of the gear model.

Keywords Crack · Gear · Phase · Frequency and amplitude modulation · ESPRIT · Sliding window

T. Kidar · M. Thomas (✉) · R. Guilbault
Department of Mechanical Engineering, École de Technologie Supérieure,
1100, Notre-Dame street West, Montreal, QC H3C 1K3, Canada
e-mail: marc.thomas@etsmtl.ca

T. Kidar
e-mail: thameur.kidar@univ-st-etienne.fr

R. Guilbault
e-mail: raynald.guilbault@etsmtl.ca

T. Kidar · M. Elbadaoui
University of Lyon, University of Saint Etienne, LASPI EA-3059,
20 Av de Paris, 42334 Roanne Cedex, France
e-mail: mohamed.elbadaoui@univ-st-etienne.fr

G. Dalpiaz et al. (eds.), *Advances in Condition Monitoring of Machinery in Non-Stationary Operations*, Lecture Notes in Mechanical Engineering,
DOI: 10.1007/978-3-642-39348-8_24,

1 Introduction

The transmission of power by using gearboxes is present in almost all rotating machinery. We find them in most industrial sectors such as gearboxes of vehicles, aircraft engines and wind turbines. A lot of researchers are still interested in detecting gear degradation [1]. About the existing methods for the diagnosis of gears damage, the vibration monitoring is considered the most commonly used. Several algorithms have been developed to process the data acquired from sensors to detect vibration damages occurring at the level of teeth. Vibration analysis is often based on Fast Fourier Transform (FFT) to detect common faults (unbalance, misalignment, blade pass, shocks etc.) [2]. However, this classic approach is somewhat limited because it is not reliable to describe non-linearity introduced by wear or defects and may be asked for a high frequency resolution. Other techniques have been proposed to solve some of these problems, such as we find the time-frequency analysis (spectrogram, wavelets…) [3, 4] or high-resolution techniques (MUSIC, ESPRIT, WSF…) [5–7]. When cracks appear, they generate phase variations, non stationary vibrations, amplitude and frequency modulations [8, 9] that affect the modal properties [10] and that are not easy to detect. In order to understand the behavior of a cracked tooth and its vibratory signature, a gear tooth has been modeled to define its gear meshing stiffness [11]. For processing the signals resulting from these models, the technique of high-resolution ESPRIT [12] which allows for estimating the frequency components and their energies from vibration signals is used in this paper. In order to monitor the phase variation in time due to the cracks, a sliding window is introduced in this paper. Then, the results are compared to those obtained from the instantaneous phase calculated by the Hilbert technique. Since amplitude and frequency modulations appear, a multiplicative signal with frequency-modulation is first used to analyze the phenomenon of amplitude modulation. In the next section, we present the numerical model of a cracked tooth and the resultant vibratory signals. Then, in Sect. 3, we present the formulation of the data model. Next, we discuss the high-resolution ESPRIT technique with the sliding window in Sect. 4. In Sect. 5, we present the estimation of phases by the Hilbert method. Then, we performed a comparison between the ESPRIT and the Hilbert technique for estimating the phases of the two models in Sect. 6. Finally, we discuss the results and we present the conclusions in Sect. 7.

2 Tooth Crack Modeling

To understand and develop a model for the stiffness variation due to the crack, the tooth is considered as a beam with a breathing crack [8]. The tooth sections are 2.88×20 mm with a length of 6 mm. The application of a harmonic force on the tooth (Fig. 1) generates oscillations with compressive and traction stresses.

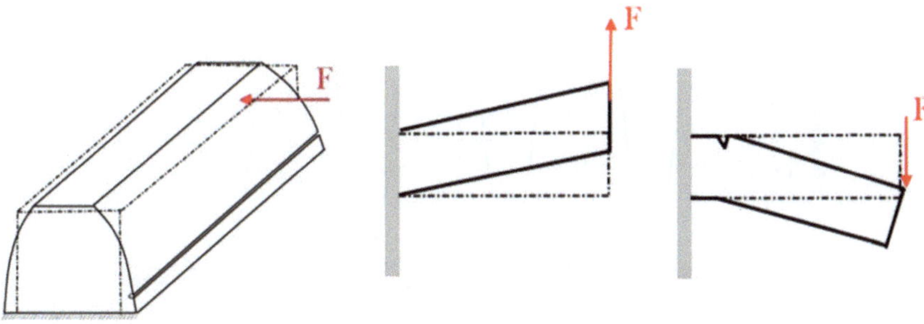

Fig. 1 Principle of the breathing crack

In compression, the crack is closed and the structure behaves like a healthy tooth. In traction the crack is opened. Consequently, the rigidity decreases as a function of time [8]. From Eq. (1), the variation in the stiffness may then be expressed as:

$$k(t) = \begin{cases} k_0 & if \quad t<T/2 \\ k_0 + \delta k \sin\left(\frac{2\pi t}{T}\right) & if \quad t \geq T/2 \end{cases} \tag{1}$$

where

- T is the period of gear meshing (7.5E–3 s).
- k_0 represents the mean stiffness for an intact tooth (1.1*E8 N/m).
- δk(1.25E7 N/m) is the amplitude of the stiffness variation due to the crack (11.4 %).

This model was validated by measurements in previous studies [9]. The values of the cracked tooth stiffness were extracted from finite element simulations [11]. By assuming a one degree of freedom system, the equation of motion gives a non stationary behavior after introducing the non-linear variation of stiffness, as follows:

$$0.396\ddot{x}(t) + 1744.1\dot{x}(t) + (1.1E8 + 1.25E7\sin(8325.22t))x(t) = F(t) \tag{2}$$

The variation of stiffness leads to a variation of the natural frequency and the critical damping, and thus the damping rate. Figure 2 shows the variation of the stiffness, the natural frequency and the damping rate in function of time.

The variations due to the presence of a breathing crack have thus an influence on the modal properties of the tooth, and their effect on the vibratory response. The non linearity generates harmonics of the excitation frequency. When exciting harmonically a structure at a frequency equal to half its natural frequency, amplitude of the second harmonic is thus amplified by the coincidence with the resonance (Fig. 3).

This results in amplitude and frequency modulations [8]. The crack generates a variation in the instantaneous phase and thus a variation of the natural frequency of the tooth. Figure 4 shows the phase shift of the acceleration signal between the healthy tooth and a defective one.

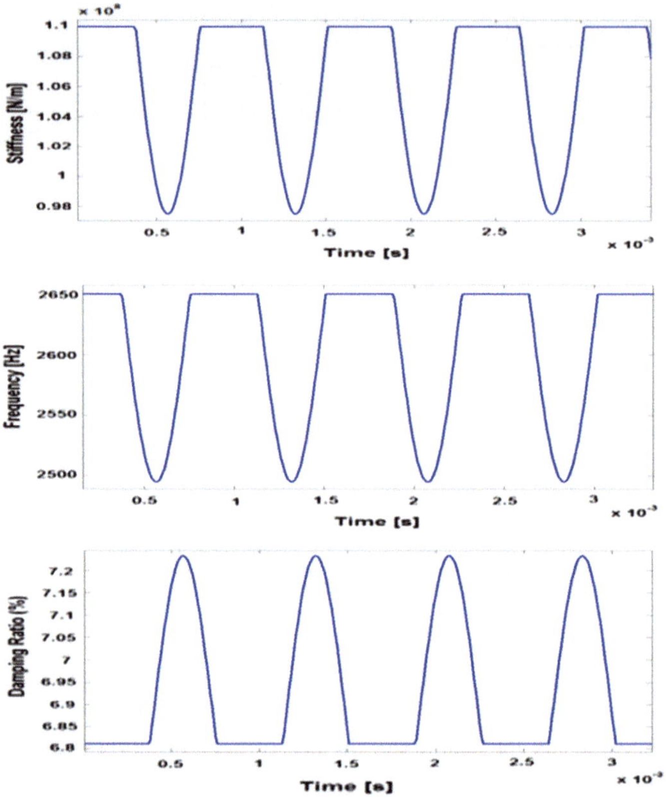

Fig. 2 Variations of the dynamic properties of a breathing cracked tooth

Fig. 3 Amplitude and frequency modulation due to a breathing crack

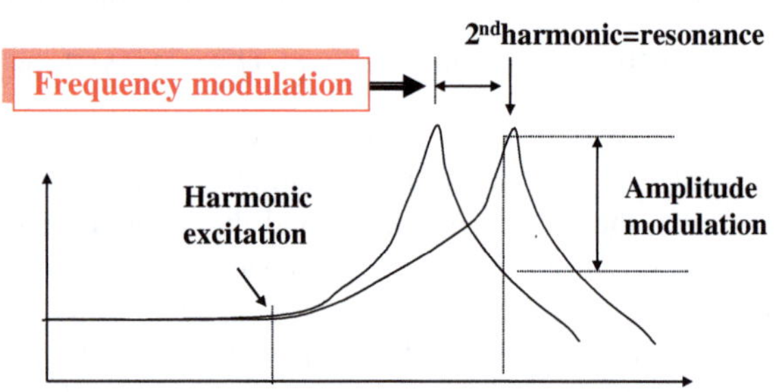

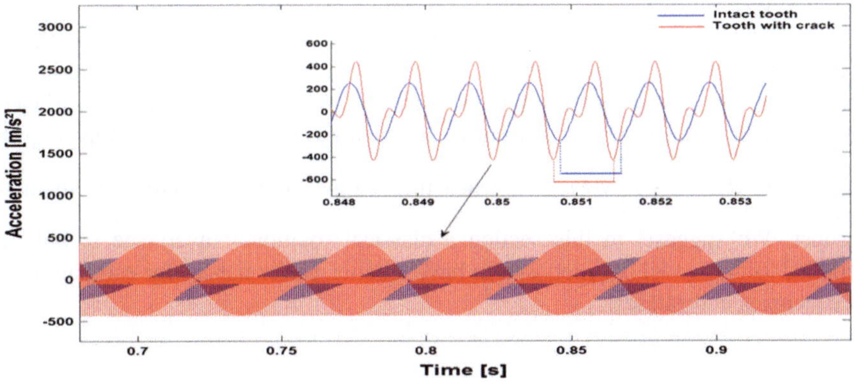

Fig. 4 Responses of the healthy and defective teeth

3 Data Model Formulation

This part is aimed to define the problem for estimating the frequencies of d complex sinusoids $S_k(t)$ of the form: $S_k(t) = \alpha_k e^{j(w_k t+\varphi_k)}$, where α_k is the real amplitude of the sin wave, ω_k is the searched frequency, and φ_k is the initial phase. We assume that N samples are available from a noisy measurement $z(t)$ defined as:

$$z(t) = \sum\nolimits_{k=1}^{d} S_k(t) + b(t), \qquad t = 1, \ldots, N \tag{3}$$

The vector $b(t)$ presents a complex noise, zero mean and Gaussian random vector with $\mathbb{E}[b(t)b^H(t)] = \sigma^2 I_m$, where I_m is a $(m \times m)$ identity matrix. As it intends to apply a subspace approach [6, 7] in the vibration domain whereas only one sensor is used, we define a data vector *y(t)* viewed as a windowed partition of the whole data set acquired, where $y(t) = (z(t)z(t+1)\cdots z(t+m-1))^T$, with m is the window length. The signal may hence be represented accordingly with the so-called matrix form:

$$y(t) = AS\omega(t) + b(t) \tag{4}$$

where

$$A = \begin{bmatrix} 1 & \cdots & 1 \\ e^{j\omega_1} & \cdots & e^{j\omega_d} \\ \vdots & \vdots & \vdots \\ e^{j\omega_1(m-1)} & \cdots & e^{j\omega_d(m-1)} \end{bmatrix},$$

$$S = diag\left(\alpha_1 e^{j\varphi_1}, \cdots, \alpha_d e^{j\varphi_d}\right), \text{ and } \omega(t) = \left[e^{j\omega_1 t}, \cdots, e^{j\omega_d t}\right]^T.$$

For subsequent use and simplifying notation, let us define $L = N - m + 1$. Finally, the L data windowed vectors $y(t)$ are stacked such as

$Y(t) = [y(t), \cdots, y(t+L)]$. Each elements of one column of $B(t) = [b(t), \cdots, b(t+L)]$ are independent whereas the columns are mutually correlated (with respect to the window length). We also point that the matrix A could be re-written in a centro-symmetric manner without loss of generality. The covariance matrix of the observed data is usually estimated by means of temporal averaging leading to $\hat{R}(t) = \frac{1}{L}\sum_{k=0}^{L-1} y(k)y^H(k)$ when the complex random vectors are reputed proper. We suggests in this section a classical rotational invariance technique applied to the model (4) supposed to have a better performances for estimation of frequencies and phases.

4 ESPRIT Approach

The ESPRIT approach is based on the exploitation of the subspace signal of the autocorrelation matrix $\hat{R}(t)$. This could be obtained by an Eigen- or singular value decomposition of $\hat{R}(t)$. When using this technique, we have:

$$A_{\downarrow}SD = A_{\uparrow}S \quad (5)$$

with $D = diag(e^{j\omega_1}, \cdots, e^{j\omega_d})$ and for any matrix B, the subscript $._{\downarrow}$ means the elimination of the last row of B and the subscript $._{\uparrow}$ means the elimination of the first row of B.

To improve the calculation time of this algorithm, [13] suggests a basic selection matrix based on Kronecker product for eliminating the appropriate rows.

Let:

$$\hat{R} = \hat{E}_s\hat{\Sigma}_s\hat{E}_s^H + \hat{E}_b\hat{\Sigma}_b\hat{E}_b^H \quad (6)$$

be an Eigen-decomposition of $\hat{R}$ where $\hat{\Sigma}_s$ is a diagonal matrix which carries the d largest eigenvalues, and columns $2m \times d$ of $\hat{E}_s$ correspond to their eigenvectors.

In the same way, $\hat{\Sigma}_b$ contains the $(2m-d)$ remaining eigenvalues and the columns of $\hat{E}_b$ are the associated eigenvectors. A rough estimation of the noise variance is given by $\hat{\sigma}^2 = \frac{1}{2m-d} Trace\left[\hat{\Sigma}_b\right]$. In this position, we can now rewrite the system (5) as follows:

$$\hat{E}_{s\downarrow}\hat{\Phi} = \hat{E}_{s\uparrow} \quad (7)$$

The $\hat{\Phi}$ matrix can be estimated by the Least Squares approach and the d frequencies are next achieved by taking the argument of the eigenvalues of $\hat{\Phi}$. On other hand, the matrix $\hat{S}$ can be estimated as follows:

$$\hat{S} = \left(\left(A^H A\right)^{-1} A^H\right).y \quad (8)$$

To well estimate the instantaneous phases by this technique, we propose to use a sliding window with a step of Δt along the signal (Fig. 5). For each window, we firstly estimate the desired frequency and then its phase by:

$$\varphi_k = angle(\delta_k) \tag{9}$$

where $\delta_k = \alpha_k e^{j\varphi_k}$ represent the elements of the first column of the matrix $\hat{S}$ given on (8). This will allow for controlling the slightest variation in the phase.

5 The Phase of Hilbert

Hilbert transforms are essential in understanding many modern modulation methods. Let $x(t)$ denote a real signal and $\breve{x}(t)$ its Hilbert transform. The analytic signal of $x(t)$ as proposed by [13] is given by:

$$x_{as}(t) = x(t) + j\breve{x}(t), \quad \text{where} \quad j = \sqrt{-1}. \tag{9}$$

The amplitude modulation (AM), phase modulation (PM) and frequency modulation (FM) are obtained by:

$$A(t) = |x_{as}(t)| = \sqrt{x^2(t) + j\breve{x}^2(t)} \tag{10}$$

$$\theta(t) = \tan^{-1}\left(\frac{\breve{x}(t)}{x(t)}\right) \tag{11}$$

$$\omega(t) = \frac{\partial(\theta(t))}{\partial t} \tag{12}$$

where $\tan^{-1}(.)$is the inverse of the tangent function, which gives the phase values in the range $[-\pi, +\pi]$.

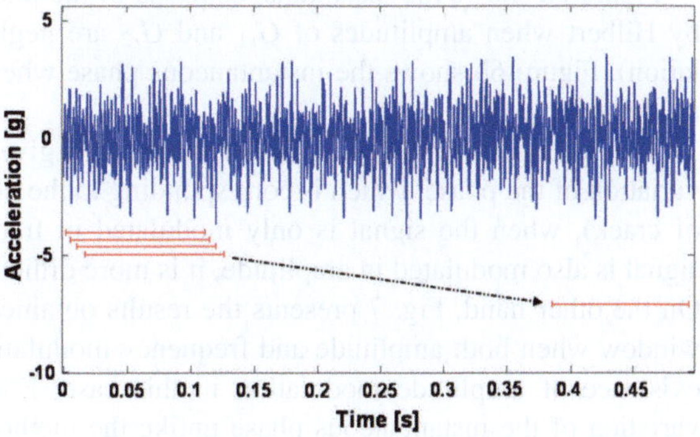

Fig. 5 Sliding window to estimate the instantaneous phase of the signal

In our study, we are only interested in the formulas (11) and (12) for the estimation of the instantaneous phases and frequencies.

6 Numerical Analysis

In this section, we present some theoretical numerical simulations to illustrate the performance of the technique ESPRIT with a sliding window for the estimation of phases and detecting cracks.

6.1 Phase Analysis of a Multiplicative Signal with Amplitude and Frequency Modulation

In order to investigate the efficiency of both methods (ESPRIT and Hilbert) when amplitude and frequency modulations are both presents, a multiplicative signal which has the same spectral shape than gear signals, has been generated. Consider a multiplicative signal as follows:

$$G(t) = \left(\sum\nolimits_{n=-\infty}^{+\infty} G_e((t - n\tau_e) + \emptyset_1(t))\right)\left(1 + \sum\nolimits_{m=-\infty}^{+\infty} G_{r1}((t - m\tau_{r1}) + \emptyset_1(t)) + \sum\nolimits_{p=-\infty}^{+\infty} G_{r2}((t - p\tau_{r2}) + \emptyset_1(t))\right) \tag{13}$$

where τ_e, τ_{r1} and τ_{r2} represent the meshing period and the rotational periods of the two wheels, respectively. $G_e(t), G_{r1}(t)\ and\ G_{r2}(t)$ represent the meshing signal and the modulation caused by the two wheels, respectively. $\emptyset_1(t)$ is a square wave which has the same pulsation than the defective pinion $G_{r1}(t)$.

The variation of the phase depends on $\beta = \frac{T_0}{T_{r1}}$, where T_0 is the opening time of the crack and T_{r1} presents the period of the pinion.$\emptyset_1(t)$ will affect the meshing phase and the phase of the second wheel.

Figure 6a shows the instantaneous phase of the multiplicative signal, estimated by Hilbert when amplitudes of G_{r1} and G_{r2} are negligible (no amplitude modulation). Figure 6b shows the instantaneous phase when there only is an amplitude modulation.

The results show that the Hilbert technique gives a good estimation of the variation of the phase, which is corresponding to the period of the pinion (presence of crack), when the signal is only modulated in frequency. However, when the signal is also modulated in amplitude, it is more difficult to identify this frequency. On the other hand, Fig. 7 presents the results obtained by ESPRIT with a sliding window when both amplitude and frequency modulations are present. Despite the existence of amplitude modulation in this case, ESPRIT clearly illustrates the variation of the instantaneous phase unlike the method of Hilbert.

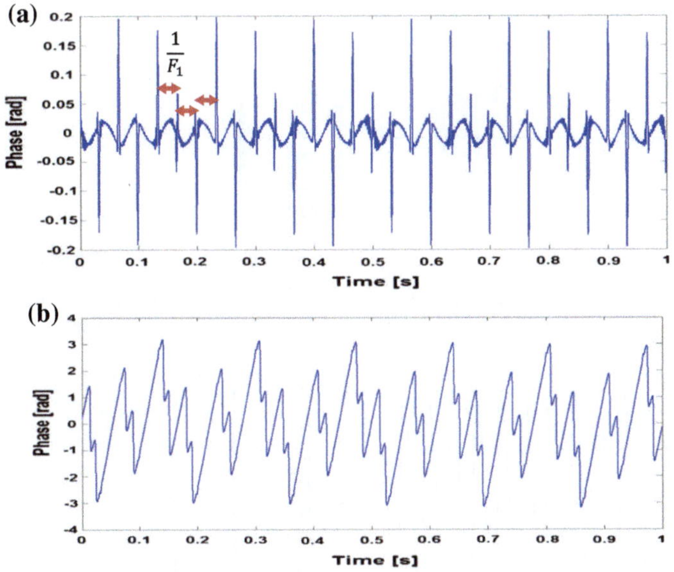

Fig. 6 Phase estimation by Hilbert for signal: **a** without amplitude modulation **b** with amplitude modulation

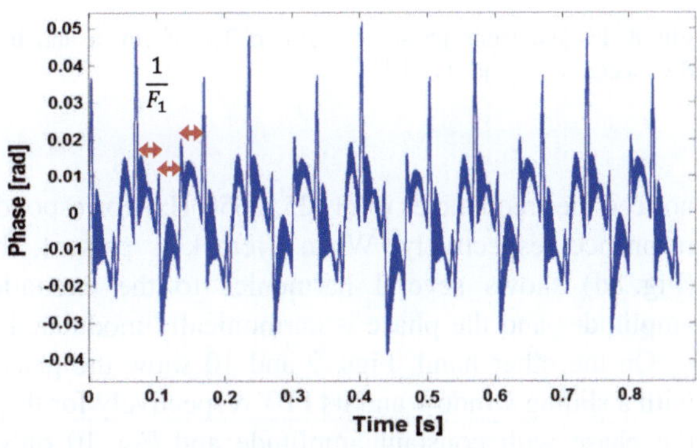

Fig. 7 Phase estimation by ESPRIT with sliding window, of a signal modulated in amplitude and frequency

6.2 Phase Analysis of a Breathing Cracked Tooth Model

By applying the principle to the cracked tooth model as described in Fig. 3, the tooth has been harmonically excited to half of its natural frequency (1325 Hz). The second harmonic which corresponds to the natural frequency is subjected to amplitude modulation and the variation of natural frequencies between 2350–3120 Hz with a mean value at 2650 Hz, produce frequency modulations.

Figure 8 presents the behavior of phases obtained through Hilbert transform (Eq. 11) for the case of intact tooth (Fig. 8a) and its FFT (Fig. 8b). It can be

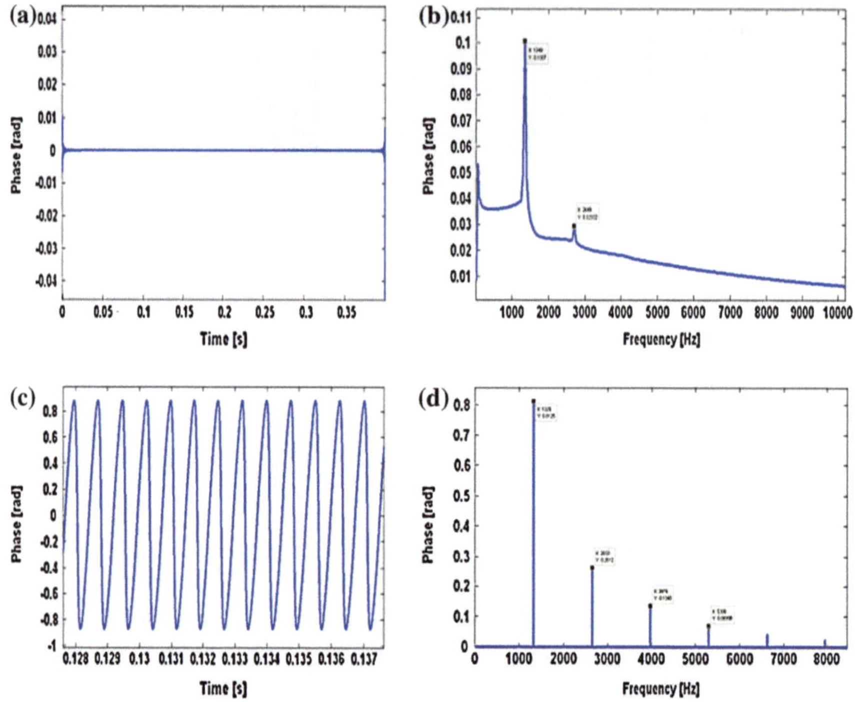

Fig. 8 Instantaneous phases by Hilbert Transform: **a** and **b** Intact tooth and its FFT. **c** and **d** Cracked tooth and its FFT

noticed the frequencies of (1325, 2650 Hz) corresponding to the excitation and the resonance respectively. When a crack is present, the FFT of the defect tooth (Fig. 8d) shows several harmonics to the excitation frequency, with higher amplitudes and the phase is harmonically modulated (Fig. 8c).

On the other hand, Figs. 9 and 10 show the phases as computed by ESPRIT with a sliding window and its FFT respectively for the intact tooth. Figure 9 shows the phase with constant amplitude and Fig. 10 only exhibits the mean natural frequency (2650 Hz). For the case of cracked tooth, Fig. 11 shows the phase with a large variation in amplitude and Fig. 12 shows the amplitude modulation around the resonance frequency due to the crack. Figure 10 exhibits the natural frequency when there is no crack while the natural frequency is modulated by the excitation frequency (Fig. 12) when a crack is present. In this last case, the amplitude at the natural frequency decreases with the crack depth while it increases at its modulation frequency as it is generally the case when investigation a phase modulation.

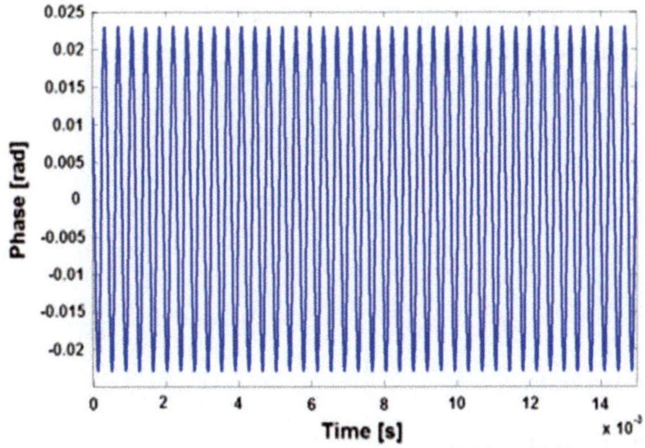

Fig. 9 Phase estimation by ESPRIT with sliding window (intact tooth)

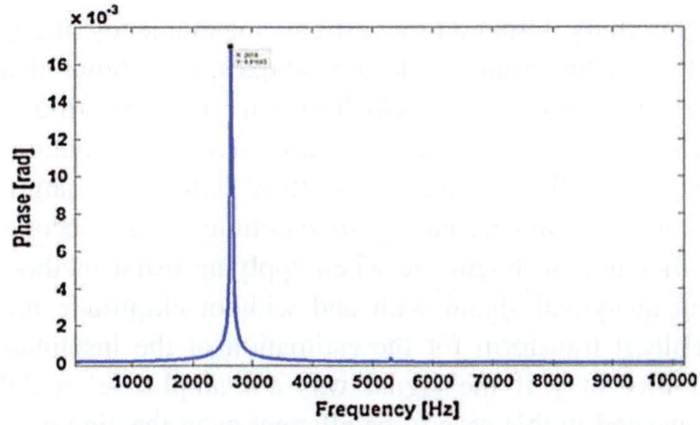

Fig. 10 FFT of the phase estimated by ESPRIT (intact tooth)

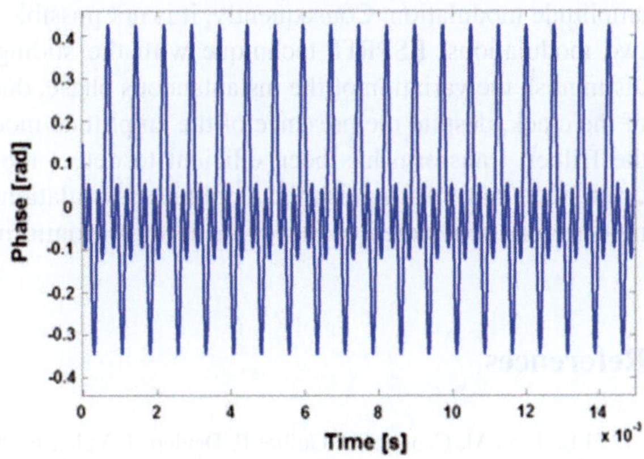

Fig. 11 Phase estimation by ESPRIT with sliding window (cracked tooth)

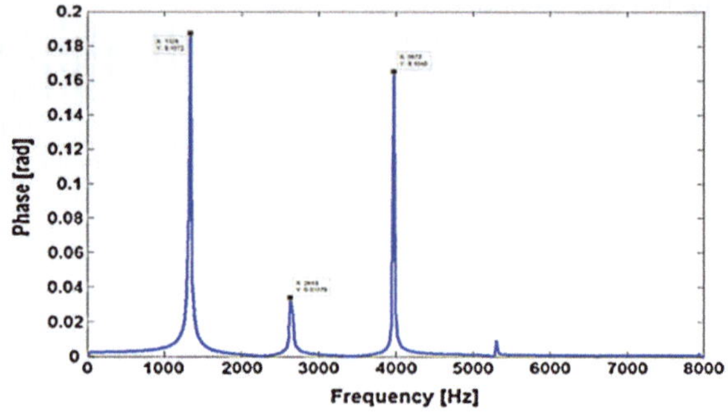

Fig. 12 FFT of phase estimated by ESPRIT (cracked tooth)

7 Conclusion

This study is aimed to investigate the efficiency of Hilbert and ESPRIT methods to detect a breathing crack in gearboxes. It is shown that a breathing crack affect the stiffness of the tooth, which produce non linearities. Since the non-linearity generates harmonics of the excitation frequency, a natural amplification is obtained by harmonically exciting the tooth at half of its natural frequency. This results in amplitude and frequency modulations at the second harmonic which affect the efficiency of diagnostic when applying usual methods (FFT, etc.). By generating an analytical signal with and without amplitude modulation, the application of Hilbert transform for the estimation of the instantaneous phase has proven reliability only if the signal was not amplitude modulated. The ESPRIT method revealed in this case to be efficient even the signal is modulated in amplitude and frequency. However, it is well known that the transmission error in gears produce amplitude modulation. Consequently, it is not possible to practically separate these two modulations. ESPRIT technique with the sliding window has been able to distinguish the variation of the instantaneous phase, due to the opening and closing of the crack, despite the presence of the amplitude modulation. On the other hand, the Hilbert transform has been efficient to detect the phase variation due to the crack but it can't distinguish the amplitude modulation. These two methods should be therefore considered in Structural Health Monitoring (SHM) of gears.

References

1. El Badaoui M, Cahouet V, Guillet F, Danière J, Velex P (2001) Modelling and detection of localized tooth defects in geared systems. ASME Mech Des 123:422–430
2. Dalpiaz G, Rivola A, Rubini R (1998) Gear fault monitoring: comparison of vibration analysis techniques. 3rd conference on acoustical and vibratory, surveillance, vol 2. Senlis, Fr., pp 623–637
3. Wang WJ, McFadden PD (1993) Early detection of gear failure by vibration analysis: calculation of the time-frequency distribution. MSSP 7(3):193–203

4. Safizadeh MS, Lakis AA, Thomas M (2002) Gear fault diagnosis using time frequency methods. In: Proceeding of the 20th seminar on machinery vibration, Québec, pp 7.18–7.27
5. Kay SM (1988) Modern spectral estimation. Prentice Hall, signal processing series, Englewood cliffs, New Jersey 07632
6. Skullestad A, Hallingstad O (1997) Identification of vibration parameters in a spacecraft using subspace methods. Control Eng Pract 5(4):507–516
7. Bengtsson M, Ottersten B (2001) A generalization of weighted subspace fitting to full-rank models. IEEE Trans Signal Process 49(5):1002–1012
8. Thomas M, Lakis AA (2002) Detection of breathing crack by time-frequency analysis, 30th computer and industrial engineering (CIE) conference, vol 2. Tinos (Greece), pp 905–910
9. Ouahabi A, Thomas M, Lakis AA (2006) Detection of damage in beams and composite plates by harmonic excitation and time-frequency analysis. 3rd SHM. Granada, Spain, pp 775–782
10. Thomas M, Lakis AA, Hamidi L, Massoud M (2002) Rotor health monitoring by modal analysis. In: Proceeding of the 20th seminar on machinery vibration (CMVA). Québec, pp 4.20–4.29
11. Palaisi D, Guilbault R, Thomas M, Lakis A, Mureithi N (2009) Numerical simulation of vibratory behavior of damaged gearbox. 27th machinery vibration, Vancouver, CB, p 16
12. Haardt M, Nossek JA (1995) Unitary ESPRIT: how to obtain increased estimation accuracy with a reduced computational burden. IEEE Signal Process 43(5):1232–1242
13. Gabor D (1946) Theory of communication. J IEE Part III 93:429–457

Part IV
Signal Processing for Machine Condition Monitoring

Performance of Time Domain Indicators for Gear Tooth Root Crack Detection and Their Noise-Sensitivity

Omar D. Mohammed and Matti Rantatalo

Abstract There are different statistical fault detection indicators applied in the time domain to detect crack propagation in the gear tooth root. 'TALAF' and 'THIKAT' are two newly presented indicators which have been designed and recommended to improve the performance of ball bearing fault detection after a certain stage of degradation. This chapter studies the performance of these two new indicators, together with the RMS, kurtosis and crest factor indicators, in the context of detecting faults in the gear tooth root. The chapter also presents an investigation of the performance of these indicators in the presence of three levels of random background noise. Gear mesh stiffness calculations and dynamic simulation have been performed using Matlab^{TM} to obtain the residual gear centre point displacement signals for different crack propagation cases. The simulations indicate that the RMS and kurtosis perform well for crack depths up to approximately 50 % of the tooth root thickness. Kurtosis and THIKAT show the most sensitive performance with an increasing noise level.

Keywords Crack propagation · Gear mesh stiffness · Vibration-based diagnosis · Fault detection indicators · Background noise

O. D. Mohammed (✉) · M. Rantatalo
Division of Operation, Maintenance and Acoustics, Lulea University of Technology, Luleå, Sweden
e-mail: omar.mohammed@ltu.se; enggomar@hotmail.com

M. Rantatalo
e-mail: matti.rantatalo@ltu.se

O. D. Mohammed
Mechanical Engineering Department, College of Engineering, University of Mosul, Mosul, Iraq

G. Dalpiaz et al. (eds.), *Advances in Condition Monitoring of Machinery in Non-Stationary Operations*, Lecture Notes in Mechanical Engineering,
DOI: 10.1007/978-3-642-39348-8_25, © Springer-Verlag Berlin Heidelberg 2014

1 Introduction

Applying fault detection indicators on the obtained vibration signal is considered as an efficient tool for detecting the faults or damages existing in gears, and then giving a warning as early as possible [1].

Background noise due to random impulses has an influence on the values of fault detection indicators [2]. There are different methods for de-noising, but there is still a possibility of some amount of background noise remaining. Actually, this depends on the accuracy of the de-noising method used.

Wu et al. [3] concluded that the residual signal is more sensitive for fault detection. The RMS is the best indicator when using the residual signal and kurtosis is the most robust indicator for all kinds of signals used. Chen and Shao [4] presented an analytical approach to modelling crack propagation with a non-uniform parabolic distribution. They studied the effect of the size of a crack propagating along the tooth width and the crack depth on the statistical indicators (the RMS and kurtosis). Many research studies have been conducted to study fault detection in ball bearings. The sensitivity of the scalar indicators (the crest factor and kurtosis) was investigated by Dron et al. [2], who studied the performance improvement of these indicators by applying a method of de-noising to reduce the background noise in the measured vibration signal. A comparison of three de-noising methods was presented in [5] for improvement of the sensitivity of the scalar indicators (the crest factor and kurtosis) which are used in detecting bearing defects. There are differences in the effectiveness of the de-noising methods implemented, and the existence of a certain amount of remaining background noise due to random impulses has an effect on the values of the scalar indicators.

The sensitivity of fault scalar indicators for bearing defects was analysed by Sassi et al. [6], who showed that ordinary time indicators can be used for the early prediction of a fault up to a certain stage of degradation. Thus these indicators become less sensitive as the damage increases and becomes very severe. To improve bearing fault diagnosis up to the level of probable catastrophic failure, the authors developed two new indicators called TALAF and THIKAT, which combine the ordinary indicators. However the performance of these new indicators in detecting gear tooth root cracks has not been investigated.

The current chapter studies the performance of these two new indicators, together with the RMS, kurtosis and crest factor indicators, in detecting gear tooth root cracks with and without the presence of random noise.

2 Gear Mesh Stiffness Modelling with a Crack in the Tooth Root

Different crack propagation scenarios were studied in (Mohammed et al. Submitted), where it was found that the crack model of a constant crack depth along the whole tooth width showed the most significantly increased performance for the RMS and kurtosis. Accordingly, to study how the indicator performance decreases with an increasing noise level, it is more meaningful to consider the crack propagation scenario where these indicators have shown the most significant performance. Therefore, in the current study the calculation model presented in [4] was applied under the assumption that the crack extended through the whole tooth width with a constant crack depth, as explained in Fig. 1.

3 Analytical Crack Propagation Model

The propagation case data of the crack propagation scenario which was considered in our analysis are explained in Table 1. The crack level CL represents the crack depth as a percentage to the root tooth thickness. A program was developed using Matlab™ to investigate the time-varying mesh stiffness for two gears in contact. The stiffness of a cracked tooth with a uniform crack distribution could be investigated as illustrated in Fig. 2.

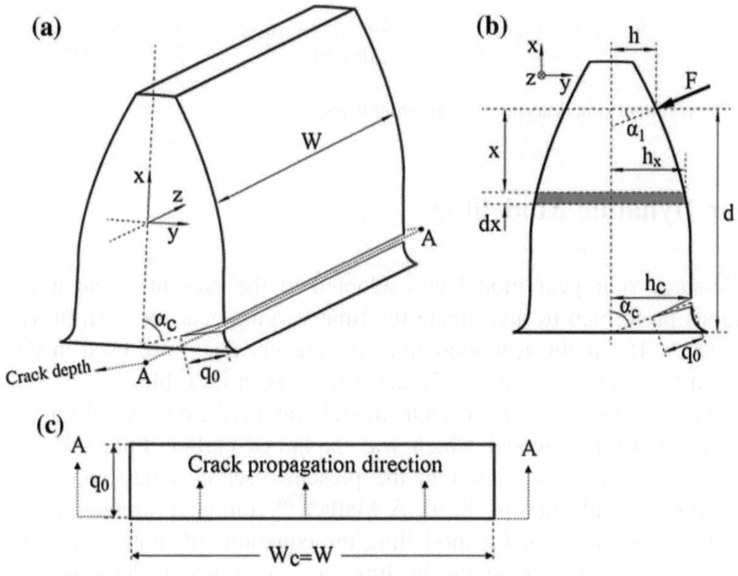

Fig. 1 Modelling of gear tooth crack. **a** Modelling of cracked tooth. **b** Tooth notation. **c** Uniform crack distribution

Table 1 Case data of the crack propagation scenario applied

Case	qo = q2 (mm)	CL %	Case	qo = q2 (mm)	CL %	
1	0	0	11	0.9	24.19	Wc = W
2	0.05	1.34	12	1.0	26.88	$\alpha_c = 70°$
3	0.1	2.68	13	1.1	29.56	
4	0.2	5.37	14	1.2	32.25	
5	0.3	8.06	15	1.3	34.94	
6	0.4	10.75	16	1.4	37.63	
7	0.5	13.44	17	1.5	40.32	
8	0.6	16.12	18	1.6	43.01	
9	0.7	18.81	19	1.7	45.69	
10	0.8	21.50	20	1.8	48.38	

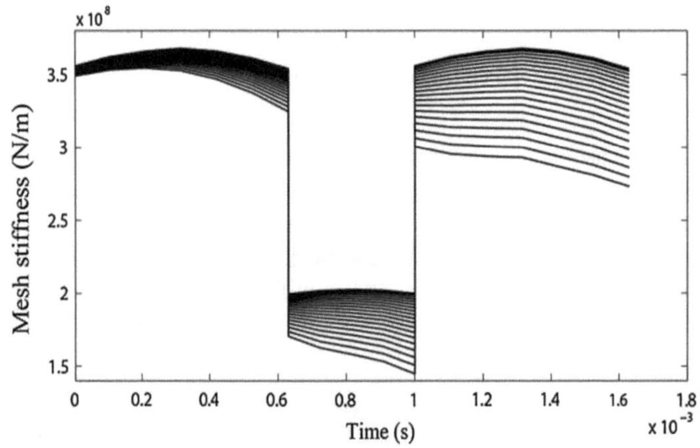

Fig. 2 The resultant time-varying gear mesh stiffness

4 Gear Dynamic Modelling

A single-stage spur gear model was adopted in the present research work for simulations performed to investigate the time-varying gear mesh stiffness of two mating gears. The main gear modelling parameters that were used in the study were taken from Chaari et al. [4, 7], and can be seen in Table.

A dynamic simulation of a 6 DOF model was performed based on the time-varying mesh stiffness model which was explained earlier. Figure 3 shows the dynamic model which was used in the present research study and which was adopted in Chen and Shao [4, 8, 9]. A Matlab™ computer simulation using the ODE45 function was used for modelling the equations of motion. The dynamic simulation was performed for the healthy case, after which the simulation was repeated to extract the dynamic behaviour for the faulty cases corresponding to the crack sizes (Table 2).

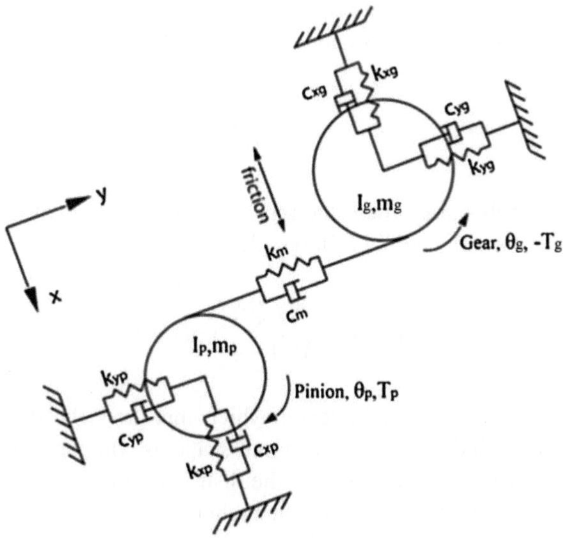

Fig. 3 Dynamic model of a reduction gear system with 6 DOF

Table 2 Parameters of gear-pinion set [4, 7]

Parameter	Gear	Pinion	Parameter	Gear	Pinion
Number of teeth	30	25	Mass (kg)	0.3083	0.4439
Module (mm)	2	2	Mass moment of inertia (kgm^2)	0.96×10^{-4}	2×10^{-4}
Teeth width (mm)	20	20	Radial stiffness of the bearing in x, y direction (N/m)	6.56×10^{8}	6.56×10^{8}
Contact ratio	1.63	1.63	Radial damping of the bearing in x, y direction (N/m)	1.8×10^{3}	1.8×10^{3}
Rotational speed (rpm)	2,000	2,400	Coefficient of friction	0.06	0.06
Pressure angle (degree)	20	20	Total damping between meshing teeth (Ns/m)	67	67
Young's modulus, E (N/mm^2)	2×10^{5}	2×10^{5}	Poisson's ratio	0.3	0.3

5 Statistical Fault Detection Indicators

Statistical indicators which are commonly applied to the obtained signal permit efficient fault detection, which is necessary to prevent a sudden breakage in the machinery system. In the current study, the indicators applied to the obtained signal are the RMS, kurtosis, the crest factor, TALAF and THIKAT.

The RMS is considered as one of the basic statistical indicators that measure the energy level of a signal. The RMS can be defined as follows [1]:

$$\mathrm{RMS} = \sqrt{\frac{1}{N}\sum_{n=1}^{N}(x(n))^2} \tag{1}$$

In the case where the mean value of the signal is not zero, the RMS indicator can be obtained as follows [3]:

$$\mathrm{RMS} = \sqrt{\frac{1}{N}\sum_{n=1}^{N}(x(n)-\bar{x})^2}, \text{ where } x = \frac{1}{N}\sum_{n=1}^{N}x(n). \tag{2}$$

Kurtosis is an indicator which measures the degree of peakiness of a distribution and describes the signal shape as compared to the normal distribution. The kurtosis value depends on the distribution tail length, so that the kurtosis value of the residual signal is much higher than that of the original signal. The kurtosis indicator can be defined as follows [1, 3]:

$$\mathrm{Kur} = \frac{\frac{1}{N}\sum_{n=1}^{N}(x(n)-\bar{x})^4}{\left[\frac{1}{N}\sum_{n=1}^{N}(x(n)-\bar{x})^2\right]^2} \tag{3}$$

The crest factor is the ratio between the maximum absolute value reached by the signal and the RMS of the signal. This indicator gives an idea as to whether any impacting can exist in the signal [1].

$$\mathrm{CF} = \frac{\max|x(n)|}{\mathrm{RMS}} \tag{4}$$

TALAF is a new indicator developed by Sassi et al. [6] and combining the RMS and kurtosis.

$$\mathrm{TALAF} = \log\left[\mathrm{Kur} + \frac{\mathrm{RMS}}{\mathrm{RMS_h}}\right], \text{ where } \mathrm{RMS_h} \text{ is for healthy case.} \tag{5}$$

THIKAT is a new indicator developed by Sassi et al. [6] and combining the RMS, kurtosis, the peak, and the crest factor.

$$\mathrm{THIKAT} = \log\left[(\mathrm{Kur})^{\mathrm{CF}} + \left(\frac{\mathrm{RMS}}{\mathrm{RMS_h}}\right)^{\mathrm{peak}}\right] \tag{6}$$

6 Results and Discussion

To study the influence of the remaining background noise on the five indicators' performance, we introduced three levels of noise with the SNR values 21.54, 13.11 and 8.85 dB, with respect to the original signal without any crack. These assumed noise levels were added to the simulated signal, and then the statistical indicators were calculated every time to investigate how the indicator sensitivity was affected by the noise level. In reality the background and measurement noise could differ from the assumed Gaussian noise used in this study. This could affect the fault detection abilities of the indicators if the noise has a time varying property. In this study the effect of time varying noise is not considered.

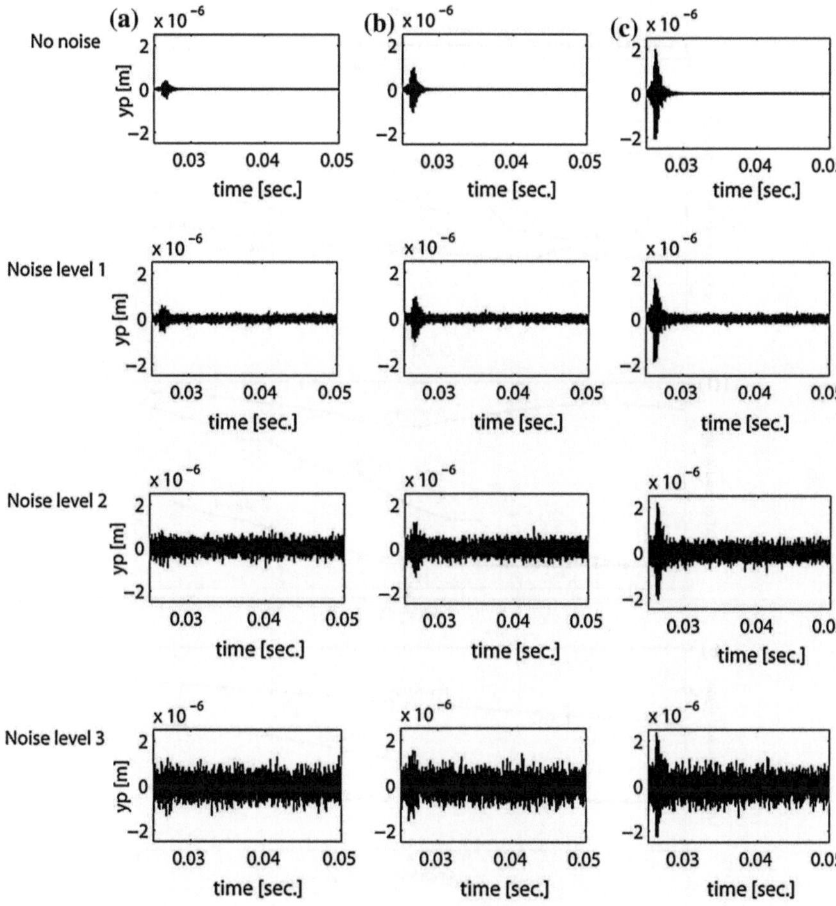

Fig. 4 One revolution of the residual signal for different crack seizes and noise levels. **a** Crack depth 0.6 mm. **b** Crack depth 1.2 mm. **c** Crack depth 1.8 mm

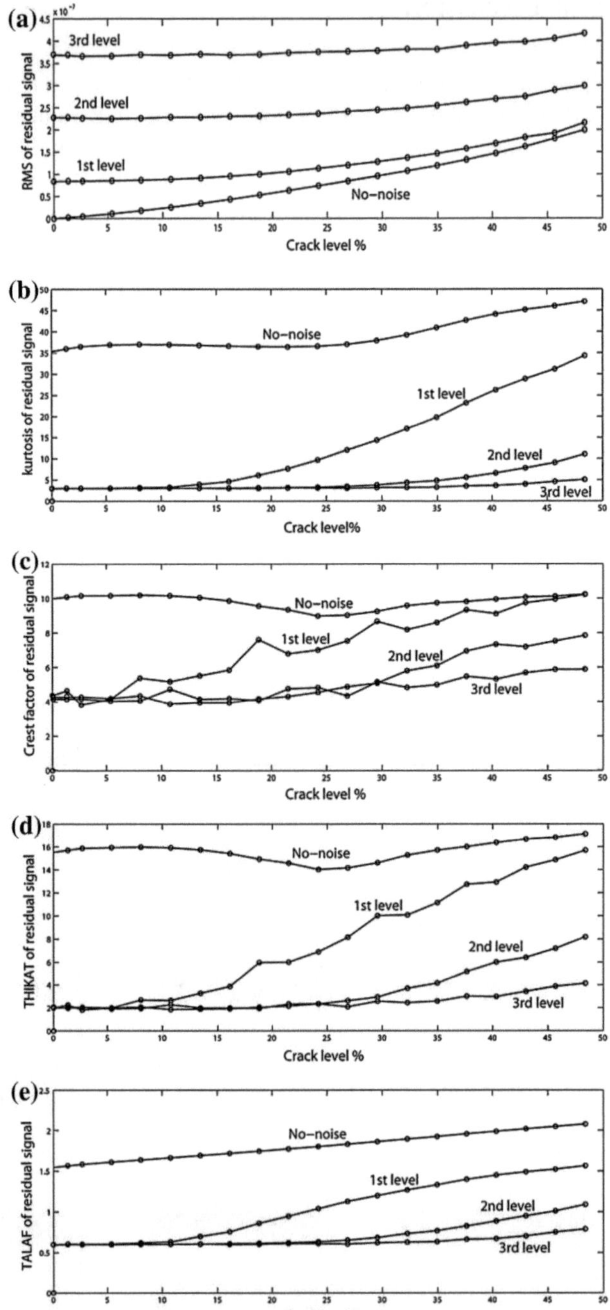

Fig. 5 Changes in the performance of the fault detection indicators with the crack propagation for the studied noise cases

Table 3 Percentage increases of the studied indicators obtained for the larger studied crack size

	RMS (%)	Kurtosis (%)	Crest factor (%)	THIKAT (%)	TALAF (%)
No-noise	7,211.56	31.03	1.44	9.15	32.66
1st noise level	147.93	1,032.7	120.19	603.49	60.52
2nd noise level	30.91	264.32	81.48	292.91	80.44
3rd noise level	13.17	67.33	42.02	108.33	31.04

Figure 4 shows the residual signals for three crack sizes, and for each crack size the effect of the four noise cases studied is given. Figure 5 shows the performance of the indicators for the no-noise case and the other three cases of added noise. For the no-noise case, the RMS performed very well and increased significantly, furthermore in the 1st noise level it could give an indication at a crack level of around 20 %. On the other hand kurtosis and the other indicators which depend on kurtosis showed a better and more significant performance in the 1st case of noise than in the no-noise case. The reason for this is that, for the residual signal in the case where an amount of random noise is added, there is a random signal with a normal distribution lying on the entire resultant residual signal. Thus the kurtosis starts with the value 3 and then starts increasing significantly as the spikes become stronger with the fault propagation. However, for the no-noise case there is definitely no random signal, and then the kurtosis value will rapidly be changed from zero for the healthy case to a higher level when a crack is started.

Kurtosis performed well and gave a significant indication especially in the 1st noise level, where the fault could be detected at a crack level of around 15 %. With an increase in the background noise being added to the signal, the indicator performance was affected correspondingly. In the 3rd noise level it was hard to detect the fault within the early stage, and it is obvious that the performances of all indicators deteriorated as can be seen in Table 3.

The plots show that the applied indicators kept performing normally and there was no particular stage in which their performances start decreasing. THIKAT performed well and better than the crest factor according to Table 3.

7 Conclusions

The two new indicators, TALAF and THIKAT, are applied together with the RMS, kurtosis and crest factor indicators for gear tooth fault detection up to approximately 50 % crack level. The plots resulting from the simulations performed in the present study show that kurtosis, the crest factor and the RMS kept performing normally and there was no particular stage in which their performances start decreasing like in the bearing case reported in [6]. Therefore, there is no limitation for using these indicators for gear fault detection. Kurtosis and THIKAT, as seen in Figure 5, show the most sensitive performance with an increasing noise level. However the Crest factor seems to be able to react on cracks in an early stage in the case of noise level 1.

Practically there is a possibility of some amount of background noise remaining. With an increase in the background noise being added to the signal, the indicator performance was affected correspondingly. In the 3rd noise-level case it was hard to detect the fault within the early stage, and it is obvious that the performances of all the indicators deteriorated.

References

1. Parey A, Tandon N (2010) Fault detection of spur gears using vibration monitoring, lambert. LAP LAMBERT Academic Publishing, Saarbrucken, Germany
2. Dron JP, Bolaers F, Rasolofondraibe I (2004) Improvement of the sensitivity of the scalar indicators (crest factor, kurtosis) using a de-noising method by spectral subtraction: application to the detection of defects in ball bearings. J Sound Vib 270:61–73
3. Wu S, Zuo M, Parey A (2008) Simulation of spur gear dynamics and estimation of fault growth. J Sound Vib 317(3–5):608–624
4. Chen Z, Shao Y (2011) Dynamic simulation of spur gear with tooth root crack propagating along tooth width and crack depth. Eng Fail Anal 18(8):2149–2164
5. Bolaers F, Cousinard O, Estocq P, Chiementin X, Dron JP (2011) Comparison of denoising methods for the early detection of fatigue bearing defects by vibratory analysis. J Vib Control 17:1983–1993
6. Sassi S, Badri B, Thomas M (2008) Tracking surface degradation of ball bearings by means of new time domain scalar indicators. Int J COMADEM 11:36–45
7. Chaari F, Fakhfakh T, Haddar M (2009) Analytical modelling of spur gear tooth crack and influence on gearmesh stiffness. Eur J Mech A Solid 28(3):461–468
8. He S, Cho S, Singh R (2008) Prediction of dynamic friction forces in spur gears using alternate sliding friction formulations. J Sound Vib 309(3–5):843–851
9. He S, Gunda R, Singh R (2007) Effect of sliding friction on the dynamics of spur gear pair with realistic time-varying stiffness. J Sound Vib 301(3–5):927–949
10. Mohammed OD, Rantatalo M, Aidanpaa J, Kumar U Submitted, Vibration signal analysis for gear fault diagnosis with various crack propagation scenarios. Accepted in Mech Syst Sig Proces

Cepstral Removal of Periodic Spectral Components from Time Signals

Robert B. Randall and Nader Sawalhi

Abstract The use of the cepstrum for removing components from a signal which manifest themselves as periodic spectral components has previously been described. These include discrete frequency components with uniform spacing such as families of harmonics and modulation sidebands, but also narrow band noise peaks coming from slight random modulation of almost periodic signals, such as higher harmonics of blade pass frequencies. The removal is effected by applying a notch "lifter" to the real cepstrum of the signal, thus removing the targeted components from the log amplitude spectrum, and then combining the modified amplitude spectrum with the original phase spectrum. Not much attention was previously paid to the type of notch lifter, but two different situations occurring in conjunction with analysis of signals from wind turbines showed that different lifters have advantages in different situations. This chapter describes two different approaches, illustrating them with the two examples of application.

Keywords Cepstrum · Harmonic removal · Sideband removal · Notch lifter · Wind turbines

R. B. Randall (✉)
School of Mechanical and Manufacturing Engineering, University of New South Wales, Sydney 2052, Australia
e-mail: b.randall@unsw.edu.au

N. Sawalhi
School of Mechanical Engineering, Prince Mohammad Bin Fahd University, Al Khober 31952, Kingdom of Saudi Arabia
e-mail: nsawalhi@pmu.edu.sa

G. Dalpiaz et al. (eds.), *Advances in Condition Monitoring of Machinery in Non-Stationary Operations*, Lecture Notes in Mechanical Engineering,
DOI: 10.1007/978-3-642-39348-8_26, © Springer-Verlag Berlin Heidelberg 2014

1 Introduction

The cepstrum is the inverse Fourier transform of a log spectrum, and thus concentrates periodic spectrum components, such as families of equally spaced harmonics and sidebands into a small number of components called "rahmonics" in the cepstrum. There are a number of types of cepstrum, but the most common are the complex cepstrum, formed from the complex spectrum, with both log amplitude and phase, and the real cepstrum (or power cepstrum) formed from the log amplitude only [2]. Because the complex cepstrum includes the phase of the spectrum, it is possible to reverse the whole procedure back to the time domain after editing (or "liftering") in the cepstrum. However, to calculate the complex cepstrum requires that the phase be "unwrapped" to a continuous function of frequency, and this is only possible for well-behaved functions such as frequency response functions, for which the phase is continuous and then only if the log amplitude stays within the dynamic range of the measurement.

Phase unwrapping is not possible for stationary signals, either forcing functions or responses, since by definition these consist of (mixtures of) discrete frequency components, for which the phase is undefined between the components, and stationary random components for which the phase varies randomly from one frequency to the next.

However, it was recently shown [5] that the log amplitude spectrum of stationary signals can be edited using the real cepstrum, and then the edited spectrum amplitudes combined with the original phase spectrum to return to the time domain. The procedure is illustrated in Fig. 1. This can be done for example to remove a family of discrete frequency components, since these occupy only a small proportion of the total spectrum. The phase will be incorrect at the frequencies where the amplitudes are edited, but the latter will typically be greatly reduced to the level of adjacent noise components, and are much fewer in number, so the effects of such disturbance are often negligible.

Figure 2 shows a typical result [5] using the described method to remove all shaft harmonics (including gear mesh components) from the signals from a gearbox with a faulty bearing, where the result is almost as good as that obtained by removing the synchronously averaged signal, synchronized with the shaft rotational speed (only one average was required because the gear ratio was 1:1). Even though the residual signal containing the effects of the bearing fault was slightly noisier for the cepstral method, the envelope analysis for diagnosis of the bearing fault was equally good.

In order to remove the harmonic/sideband family from the log amplitude spectrum, a notch lifter must be applied to the cepstrum, and this chapter discusses the different ways in which this can be done.

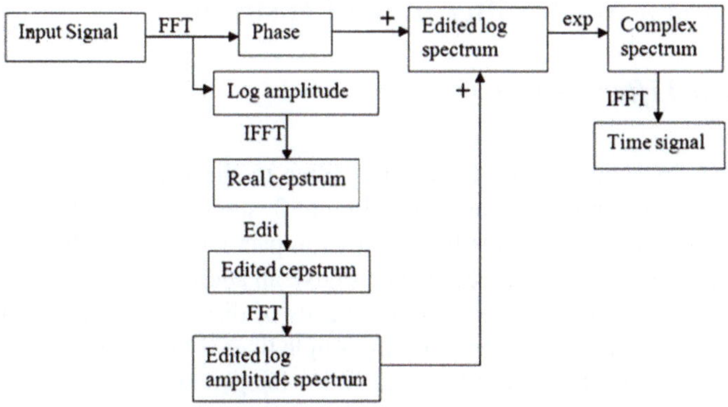

Fig. 1 Schematic diagram of the cepstral method for removing selected families of harmonics and/or sidebands from time signals

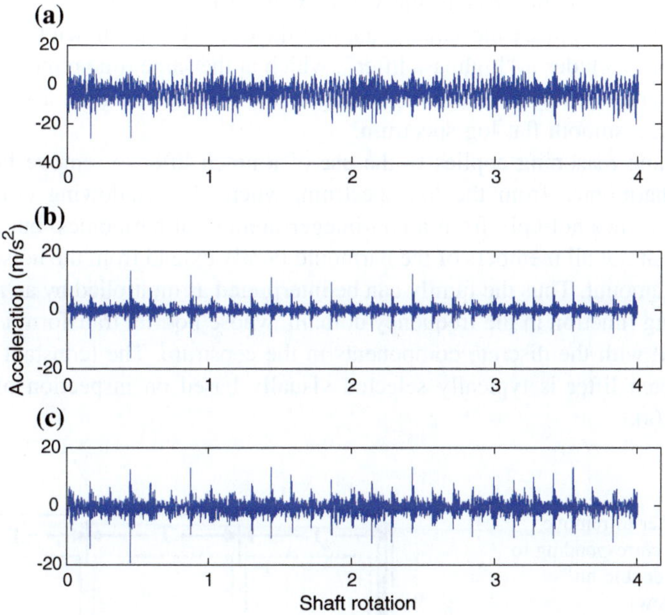

Fig. 2 Time domain signals for gearbox test rig. **a** Raw signal. **b** Residual signal (after removing the synchronous average). **c** Residual signal after editing the Cepstrum to remove the shaft rahmonics

2 Applying a Notch Lifter in the Cepstrum

2.1 Notch Lifter Type 1

One of the first applications of a notch lifter to the cepstrum was to remove rahmonics corresponding to an echo. Figure 3 shows a typical lifter used to remove the effects of a "double hit" with an impact hammer [3]. In principle, the effects of an echo in the log spectrum are to give an additive periodic component to the log spectrum, which translates into a series of discrete rahmonics in the cepstrum. However, these would only occupy single lines if the frequency range of the whole (2-sided) spectrum corresponded to an integer number of periods of this frequency spacing (the reciprocal of the echo delay time T), and in general it doesn't. Consequently, the cepstrum is convolved with the Fourier transform of a rectangular window, (a sinc function) and to remove all discrete rahmonics a notch window with a width corresponding to the effective width of this sinc function must be used. For this reason, the width Δ of all notches in the "lifter type 1" shown in Fig. 3 is made the same. Note that the overall lifter shown there (for that application) includes a "highpass lifter", which at the same time removes the very low quefrency components corresponding to the forcing function, a short impact force with a smooth flat log spectrum.

The same reasoning applies to the use of a notch lifter to remove families of discrete harmonics from the log spectrum, where the windowing effect in the spectrum comes not only from a non-integer number of harmonics, but also from the fact that not all members of the harmonic family extend from the noise level by the same amount. Thus the family can be interpreted as multiplied by an amplitude modulating function in the frequency domain, whose Fourier transform is likewise convolved with the discrete components in the cepstrum. The (constant) width Δ of the notch lifter is typically selected visually based on inspection of the first rahmonic(s).

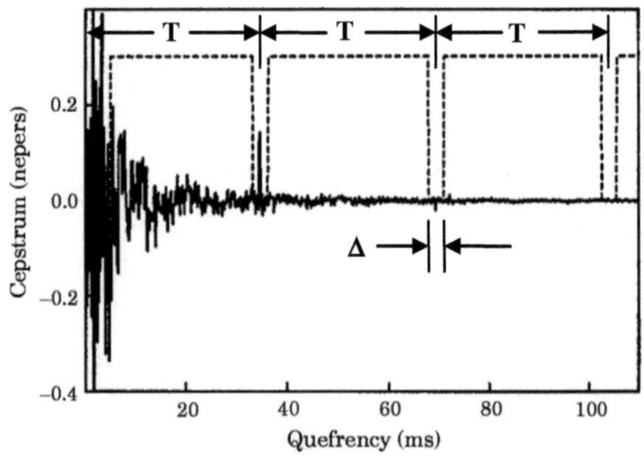

Fig. 3 Lifter to remove rahmonics corresponding to an echo ("double hit" hammer blow)

2.2 Notch Lifter Type 2

However, not all periodic spectrum components have this property of constant width in the spectrum, for example those resulting from a pseudo-cyclostationary signal such as that from local faults in a rolling element bearing. Because of random slip, the series of impulse responses resulting from impacts of the rolling elements on the fault are not uniformly spaced, and the intervals $\Delta T_i (= T_{i+1} - T_i)$ are random and result from a stationary Markov process $\{T_i\}$ [1]. This has been studied in detail in [1], where it was shown that the uncertainty of the arrival time of future impulses increases with time (measured from the one defined as the first), meaning that the probability density (of arrival time) increases in width, but decreases correspondingly in height, with increasing time from the reference point. The ΔT_i only vary by a small amount of the order of 1 or 2 %, and in this case the probability density can be modelled as Gaussian around the mean arrival time of each impulse, with mean spacing T. With this assumption the spectrum of a series of unit impulses was shown [4] to be given by:

$$F_1(\alpha) = [1 - \Phi * (\alpha)]^{-1} \quad \text{where } \Phi(\alpha) = \exp\left(-\sigma^2\alpha^2 - j\alpha T\right) \tag{1}$$

α is the radian frequency, and σ is a measure of the random variation (σ/T is 1 % for a standard deviation of 1 %).

Figure 4a shows the log amplitude spectrum for a 1 % variation, and it is seen that the width of the "harmonics" also increases with increasing frequency at the same time as the height decreases. Figure 4b shows the real cepstrum corresponding to Fig. 4a, and it is seen that the "rahmonics" also increase with increasing quefrency, though not proportionally to the latter.

Another case where this type of spectrum structure occurs, with uniformly spaced spectral components but of gradually increasing width, is for the blade

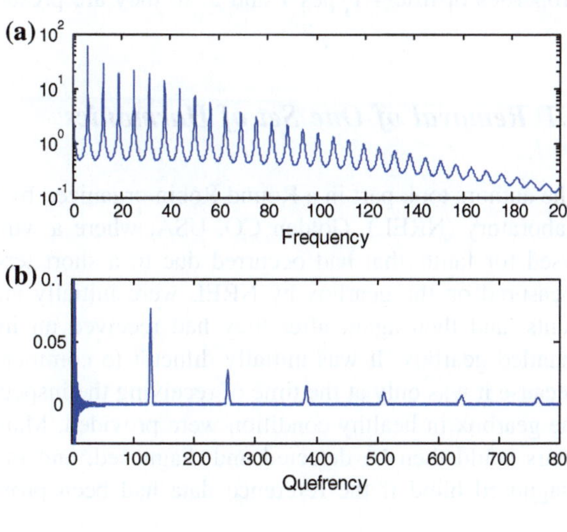

Fig. 4 Log spectrum and cepstrum for a pseudo-cyclostationary series of impulses with 1 % random variation in pulse spacing **a** log spectrum **b** cepstrum

passing harmonics of bladed machines such as turbines and fans, where there is a small amount of random frequency modulation of the perceived blade passage intervals, because the only connection between the blades and fixed points on the casing is via a turbulent fluid. Although coming from a much smaller percentage variation, the higher harmonics still gradually become broader.

This led to the definition of a notch lifter Type 2, where the notch width Δ was made proportional to quefrency. Even though not strictly necessary for the type of signal just discussed, it had the practical advantage that the notches of increasing width eventually merge into continuous zeros, giving a superimposed "lowpass lifter", which tends to reduce noise in the high quefrency part and smooth the corresponding log spectrum to some extent. Until recently, this notch lifter Type 2 was used for all our applications with the new cepstral method, and gave excellent results in a number of different applications. These include the removal of fan blade passage harmonics as well as other discrete shaft harmonics to reveal a bearing fault [5], where it was found that time synchronous averaging (TSA) removed the low harmonics of shaft speed but not all of the blade pass harmonics of a fan, because only the centre frequency of each "harmonic" peak was removed by TSA. It was also very successful in removing harmonics of the rotor blade pass frequency from response signals measured on a helicopter in flight, to facilitate performing operational modal analysis on the residual broadband signals [4]. In that case, a slightly improved result was achieved by combining the notch lifter with an additional exponential lowpass lifter in any case, to give even more smoothing of the spectrum.

3 Application to Wind Turbines

Two separate applications to wind turbines gave an appreciation of the different properties of lifters Types 1 and 2, so they are presented here.

3.1 Removal of One Set of Harmonics

The authors took part in a Round Robin organized by National Renewable Energy Laboratory (NREL), Golden CO, USA, where a wind turbine gearbox was analysed for faults that had occurred due to a short term loss of lubricant. Signals measured on the gearbox by NREL were initially analysed blind by the participants, and then again after they had received an inspection report on the dismantled gearbox. It was initially difficult to comment on the condition of gears, because it was only at the time of receiving the inspection report that spectra from the gearbox in healthy condition were provided. Many of the faults on the failed gears could then be detected and diagnosed, and most could in fact have been diagnosed blind if the reference data had been provided earlier. The results of analysis by all Round Robin teams have been published in an NREL Report [7]. The analysis was made more difficult than usual by the fact that the design was far from "hunting tooth", considered good design practice, where the numbers of teeth in mating pairs have no common factors, so that each tooth on one gear contacts every tooth on the other gear. In this case the final drive had a ratio of exactly 4:1 (88:22), meaning that the 22 teeth on the pinion always had contact with the same four groups of 22 teeth on the wheel. The drive from the sun gear to the intermediate shaft did however have a hunting tooth design (82:23).

In our contribution to the NREL report, we performed a time synchronous average (TSA) over the rotation period of the intermediate shaft (IS), which turned out to be dominated by the high speed (HS) gear mesh (harmonics 88, 176...) and strong sidebands around them at the HS shaft speed of approx. 30 Hz (4 $\times$ IS speed).

When compared with the spectra in original condition, the HS tooth mesh harmonics had increased indicating general wear, and the strong sidebands which had appeared were compatible with the severe scuffing that was reported around both pinion and wheel, but dominated by the location on the pinion. The four sections of 22 teeth were then averaged together to get the best estimate of the contribution of the pinion, this being subtracted from the total signal to get a residual dominated by effects other than from the HS pinion. This result is shown in Fig. 5, and it is seen that the spectrum of the residual is dominated by the harmonics of the IS pinion gear mesh (23 teeth), but without modulation sidebands, although the noise level is quite high because of the small number of averages. These mesh harmonics were also much higher compared with the original condition, and the lack of sidebands is compatible with the fact that the wear was distributed uniformly by the hunting tooth design, a point remarked on in the inspection report.

After the NREL report was completed it was decided to try the cepstral method to separate the effects of the two sets of gears, using an appropriate notch lifter. Normally, even without a hunting tooth design, the tooth ratio is not an exact

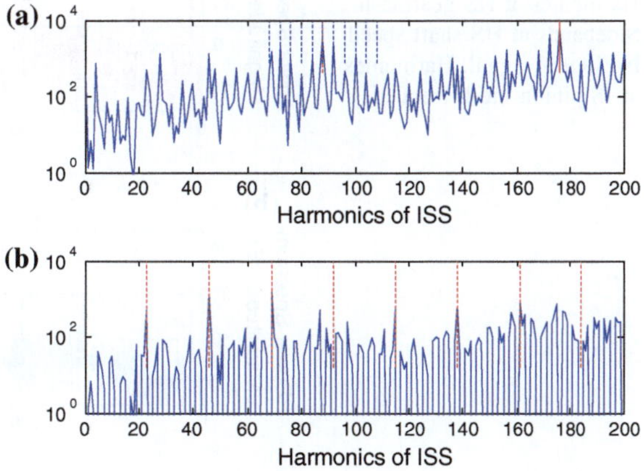

Fig. 5 Spectra of IS TSA average. **a** Original including four rotations of the HSS. Harmonics at 88 (HS mesh), sidebands at 4 (HS shaft). **b** Residual after removal of the HSS average. Harmonics at 23 (IS pinion mesh). From [7]

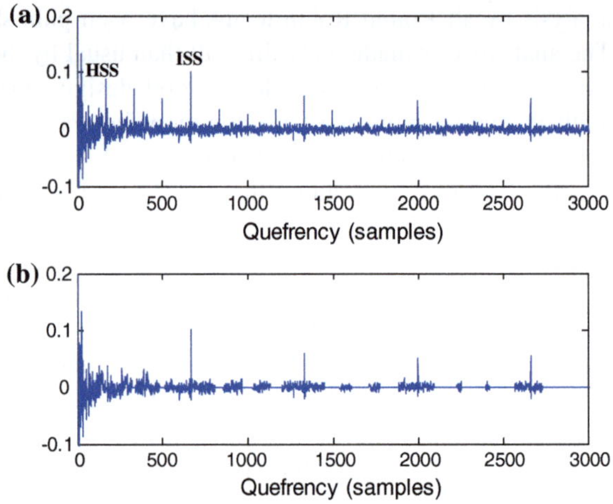

Fig. 6 Cepstra of the IS shaft signal before (**a**) and after (**b**) the application of a notch lifter Type 2 to remove HSS rahmonics

integer, so the two shaft speeds will have separate sets of rahmonics in the cepstrum. In this case, because of the exact 4:1 ratio, every fourth rahmonic of the HS shaft corresponded with a rahmonic of the IS shaft, so the code had to be modified to leave every fourth rahmonic. Complete separation cannot be achieved as it can be appreciated that remnants of the HS rahmonics are left in the IS rahmonics, but the same applied to the TSA method of Fig. 5. The cepstra before and after the application of the notch lifter are shown in Fig. 6, where it is seen that the Lifter Type 2 has been used, with increasing notch width. Note that the cepstra were obtained from the raw signals, without the requirement for order tracking or TSA, because the speed was reasonably stable.

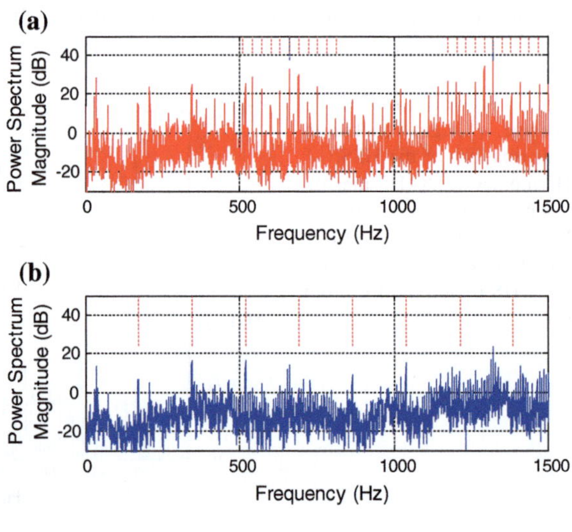

Fig. 7 Log spectra corresponding to the cepstra of Fig. 6. **a** Original signal. Harmonics at HS gearmesh. Sidebands at HS shaft speed. **b** Liftered signal. Harmonics at IS pinion mesh frequency

Figure 7 shows the log spectra corresponding to the cepstra of Fig. 6, and can be compared with Fig. 5. It is scaled in Hz rather than harmonic order, but the HS shaft speed is about 30 Hz (and IS shaft speed 7.5 Hz). It is seen that the results are very similar, but the cepstral method gives better resolution and better definition of the base noise level.

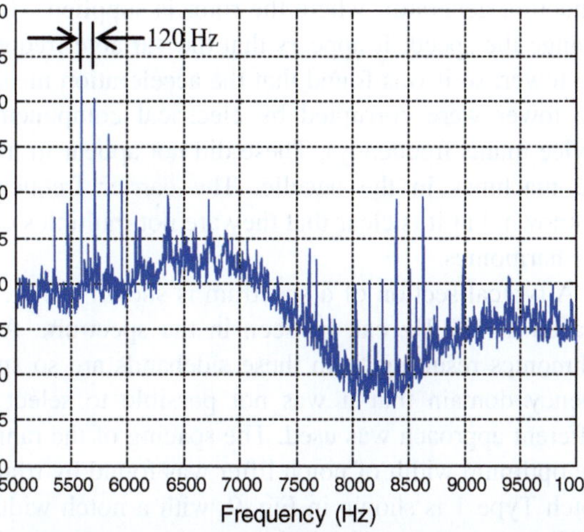

Fig. 8 Zoomed spectrum showing groups of sidebands spaced at 120 Hz

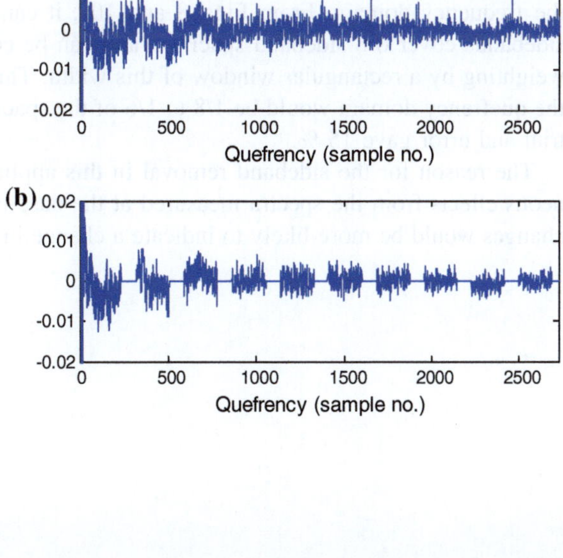

Fig. 9 Cepstrum of whole spectrum. **a** Original. **b** After liftering by a Type 1 lifter

3.2 Removal of Sidebands

One of the advantages of the cepstral method is its ability to remove families of uniformly spaced sidebands, something that cannot be done by TSA (unless the sidebands are also harmonics).

A study has been made by the company Mechanical Solutions Inc. NJ, USA, of the possibilities for making diagnosis of wind turbine machinery from measurements at the base of the tower. The initial measurements on a turbine with a squirrel cage induction generator were reported in [6] and were very promising. Later measurements were made on a variable speed machine with doubly fed induction generator, where the rotor is supplied with a varying frequency field to change the speed. It appears that the variable frequency supply is transmitted via the tower, so it was found that the acceleration measurements made at the base of the tower were corrupted by electrical components with a spacing of 120 Hz (twice mains frequency). These did not appear in measurements made directly on the machines in the nacelle. The carrier frequencies for the modulation are unknown, but it is clear that they are not multiples of 120 Hz, so the sidebands are not harmonics.

A typical section of a spectrum is shown in Fig. 8, where groups of sidebands spaced at 120 Hz can be seen in the spectrum. As can be seen in Fig. 9a, the rahmonics resulting from these sidebands are so spread by weighting in the frequency domain that it was not possible to select the first manually, and so a different approach was used. The spacing of the rahmonics was known exactly, so the optimum width of notch lifter was found by trial and error. The result, using a notch Type 1 is shown in Fig. 9, with a notch width of ± 15 % of the spacing.

The original and liftered spectra are compared in Fig. 10, and it can be seen that the remaining spectrum is little affected by the operation.

Initially a Type 2 notch lifter was used, as this had become standard practice, but this had such a strong lowpass effect that the resulting liftered spectrum was more distorted. It has since been realized that a good first estimate for the notch width can be obtained by considering the windowing of the relevant components in the frequency domain. From Figs. 8 and 10a, it can be seen that the groups of sidebands cover 6-8 sideband spacings, and can be considered as equivalent to a weighting by a rectangular window of this width. Thus the equivalent window in the quefrency domain would be 1/8 to 1/6 of the spacing, i.e. 12.5–16.7 %, where trial and error gave 15 %.

The reason for the sideband removal in this application was to remove extraneous effects from the spectra measured at the base of the tower, so that spectral changes would be more likely to indicate a change in condition of the machinery.

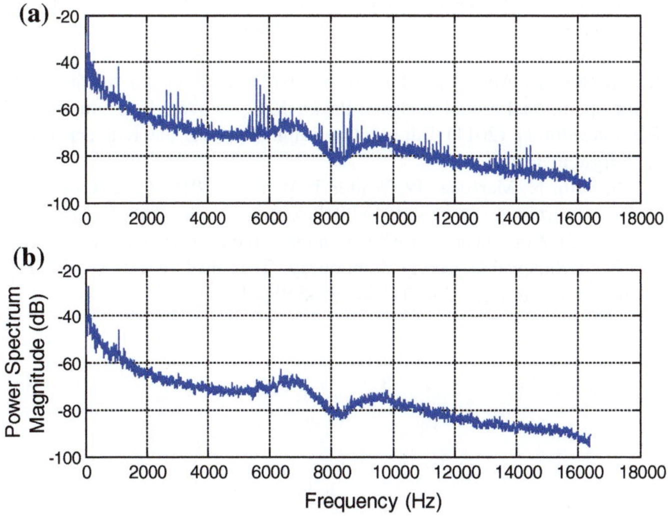

Fig. 10 Log spectra corresponding to the cepstra of Fig. 9. **a** Original signal. **b** After liftering

4 Conclusion

A new method has become available to edit time signals to remove components that have a periodic structure in the frequency domain, such as families of harmonics and sidebands. This requires a notch lifter to be applied in the real cepstrum to edit the spectrum amplitudes, which are then combined with the original phase to obtain edited time signals. Two different types of notch lifters have been found to have different advantages in different situations and these are illustrated by two applications to wind turbine machinery.

Acknowledgments This work is partially supported by Australia's Defence Science and Technology Organisation (DSTO). Signals used for demonstration purposes were provided by National Renewable Energy Laboratory, CO, USA, and Mechanical Solutions Inc., NJ, USA, the latter obtained under a research project funded by NYSERDA. NYSERDA has not reviewed the information contained herein and the opinions expressed in this chapter do not necessarily reflect those of NYSERDA or the State of New York.

References

1. Antoni J, Randall RB (2003) A stochastic model for simulation and diagnostics of rolling element bearings with localised faults. ASME J Vib Acoust 125:282–289
2. Childers DG, Skinner DP, Kemerait RC (1977) The cepstrum: a guide to processing. Proc IEEE 65(10):1428–1443

3. Gao Y, Randall RB (1996) Determination of frequency response functions from response measurements. Part I: extraction of poles and zeros from response cepstra. Mech Syst Sig Process 10(3):293–317
4. Randall RB, Peeters B, Antoni J, Manzato S (2012) New cepstral methods of signal pre-processing for operational modal analysis. ISMA 2012, Leuven, Belgium, pp 755–764
5. Randall RB, Sawalhi N (2011) Editing time signals using the real cepstrum. In: MFPT conference, Virginia Beach
6. Randall RB, Sawalhi N, Marscher W, Walter T, Weiss J (2011) Decomposition of structural vibration for wind turbine diagnosis. In: MFPT conference, Virginia Beach
7. Sheng S (ed) (2012) Wind turbine gearbox condition monitoring round robin study: vibration analysis. National Renewable Energy Laboratory. Technical report NREL/TP-5000-54530, July 2012. http://www.nrel.gov/docs/fy12osti/54530.pdf

The Local Maxima Method for Enhancement of Time-Frequency Map

Jakub Obuchowski, Agnieszka Wyłomańska and Radoslaw Zimroz

Abstract In this paper a new method of failure detection in rotating machinery is presented. It is based on a vibration time series analysis. A pure vibration signal is decomposed via the short-time Fourier transform (STFT) and new time series for each frequency bin are processed using novel approach called local maxima method. We search for local maxima because they appear in the signal if local damage in bearings or gearbox exists. Due to random character of obtained time series, each maximum occurrence must be checked for its significance. If there are time points for which the average number of local maxima is significantly higher than for the others, then the machine is suspected of being damaged. For healthy condition machinery, the vector of average number of maxima for each time point should not have outliers. The main attention is concentrated on the proper choice of required local maxima significance. The method is illustrated by analysis of very noisy both real and simulated signals. Also possible generalizations of this method are presented.

Keywords Time-frequency analysis · Enhancement · Feature extraction · Local maxima

J. Obuchowski (✉) · R. Zimroz
Diagnostics and Vibro-Acoustics Science Laboratory, Wroclaw University of Technology, Na Grobli 15, 50-421 Wroclaw, Poland
e-mail: jakub.obuchowski@pwr.wroc.pl

R. Zimroz
e-mail: radoslaw.zimroz@pwr.wroc.pl

A. Wyłomańska
Hugo Steinhaus Center, Institute of Mathematics and Computer Science, Wroclaw University of Technology, 50-370 Wroclaw, Poland
e-mail: agnieszka.wylomanska@pwr.wroc.pl

G. Dalpiaz et al. (eds.), *Advances in Condition Monitoring of Machinery in Non-Stationary Operations*, Lecture Notes in Mechanical Engineering,
DOI: 10.1007/978-3-642-39348-8_27,

1 Introduction

Time-frequency representation is very reasonable method of non-stationary signal analysis. Local damage in rotating machinery is a form of change of machine condition that causes cyclic, impulsive and non-stationary contribution in machine vibration response. Expected impulsive and cyclic disturbances in time domain in many industrial cases are masked by other vibration sources generated by machine [1–4]. Time-frequency representation of complex signal is much easier to interpret (impulsive excitation in time domain is represented by wideband disturbance of the spectrum) so this method is very frequently used technique for vibration based damage detection [5–12]. In fact, even time-frequency representation of complex signal may require some extra activities, i.e. enhancement of time-frequency plane readability before feature extraction and decision making. In this paper a novel method of spectrogram enhancement for local damage detection in rotating machinery is proposed. Spectrogram is used to present performance of the procedure because it is the simplest, intuitive and very often used in signal analysis [6]. In the method we analyze/process energy flow in time domain but for single frequency bin. It is assumed that complex raw vibration signal is decomposed into set of narrowband sub-signals with much smaller complexity. The paper is organized as follows: in Sect. 2 a proposal of new time-frequency map enhancement procedure is presented and next, in Sect. 3 and 4, validation to simulation and real data is provided.

2 Methodology

Before the further analysis we transform the original signal into time-frequency map. As a result we obtain the spectrogram which is denoted as $\{STFT(t,f)\}_{t\in[0,T],\,f\in F}$. Physically it is a two-dimensional array of real numbers calculated using a standard procedure. Parameters of spectrogram are very important. This issue will be discussed in real case study part.

The proposed procedure is based on the local maxima finding. More precisely, for each frequency band (i.e. time series adequate to selected frequency) we check the local maximum occurrence. We assume that local maximum occurs in given time point when value in the analyzed point is higher than the other values in its neighborhood of a length not less than a certain value (in the further analysis we assume the length of neighborhood equal to six). Discussion of the selection of the minimal neighborhood length is included in the data analysis sections. Then, for each frequency band we create a new binary time series which is a transformation of the original data into zero-one series. More precisely, we put one for this time point when the local maximum occurs and zero otherwise. Let us point that the binary values obtained in this way minimize influence of insignificant signals for local damage detection as well as maximize impact of characteristic signals for locally damaged machinery. Then, in our methodology for each time point we suggest to use the vector of weights (VoW), which is a vector of averaged maxima occurrence, i.e. VoW time series is defined as follows:

$$W(t_i) = \frac{1}{\#F} \sum_{f \in F} M(t_i, f), \tag{1}$$

where $M(t_i,f)$ represents binary valued time series of the local maxima occurrence for a time point t_i and frequency f. After multiplying each previously computed binary value by the value of VoW at the corresponding time point we obtain enhanced spectrogram. Therefore the enhanced spectrogram at point (t_i,f_j) is defined as follows:

$$ENH(t_i,f_j) = W(t_i)M(t_i,f_j). \tag{2}$$

In Fig. 1 we present the scheme of creating the enhanced spectrogram. One of the theoretical advantages of this method is its invariance under strictly increasing transformations. The maxima remain maxima after strictly increasing transformations such as logarithms, scaling, the square root or square (for positive numbers). Thus we do not need to compute logarithm of squared magnitude of $\{STFT(t,f)\}_{t\in[0,T], f\in F}$ as it is widely performed for spectrogram—only the magnitude value is needed. Moreover no normalization of each sub-signal is needed. Another advantage of the presented method is its possibility of generalization to non-stationary conditions, e.g. changes at the speed of potentially damaged rotating machinery parts. If the speed at point t increases, it will decrease time interval between expected wide-range excitations. Then the minimum neighborhood length must be reduced to avoid significant maxima omitting. The local maxima method can be also applied to another time-frequency representation where cyclic impulses are represented by broadband disturbances.

3 Simulated Data Analysis

In order to show the results of our method we have simulated complex raw vibration signal of a machine without and with defect. The model describes real

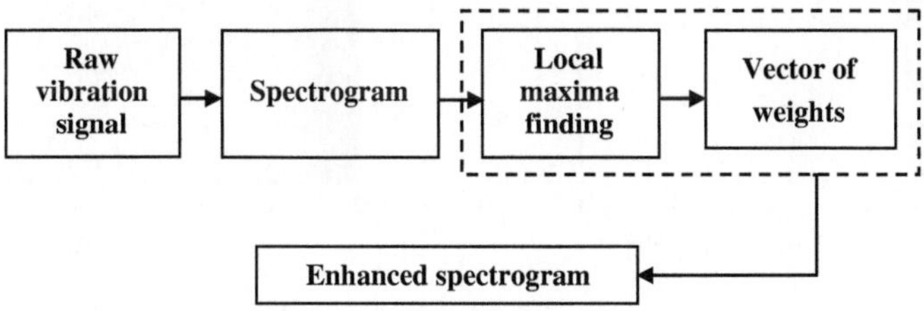

Fig. 1 Diagram of time-frequency map enhancement procedure

situations met in case of industrial signals analysis. Raw vibration signal is modeled as an additive mixture of: (1) deterministic contribution related to mesh frequency components, shaft components and other sources that generate narrowband deterministic signal with discrete spectrum, (2) Signal of Interest (SOI)—amplitude modulated wideband random process related to localized damage and (3) Gaussian noise.

The SOI is obtained by simple amplitude modulation with depth of modulation as a parameter stimulating damage growth, random wideband carrier with center frequency corresponding to structural resonance, cycle of impulsive disturbance related to fault frequency. Such a model was widely used in previous researches [1, 3]. Figure 2 shows time-frequency representation of simulations for "undamaged" and "damaged" signals. From the Fig. 2 (right) one may notice set of parallel horizontal lines in frequency range 2–6 [kHz]. They correspond to impulsive disturbance in time domain. Unfortunately, they are not so clear, and completely invisible in time domain due to low-frequency high-energy contribution related to first part of the signal model (note vertical concentration of energy in frequency range <1 [kHz]). Figure 3 presents result of time-frequency map enhancement obtained after application of the proposed procedure to simulated data.

One of the most important questions for this method is how to choose the minimal neighborhood length. We propose to make the choice dependent on expected time intervals between broadband excitations caused by local damage. The expected frequency of failure signals for our simulated data is 13 [Hz]. The distance between time points on the spectrogram is

$$\frac{STFT\ window\ length}{frequency\ sampling} = \frac{200}{20000\ [Hz]} = 0.01\ [s]. \tag{3}$$

Because of the expected fault frequency (or more precisely cycle related to failure) we choose minimal neighborhood length of 12 adjacent time intervals, i.e. 0.12 [s]. It means that a point of sub-signal is called a local maximum when its

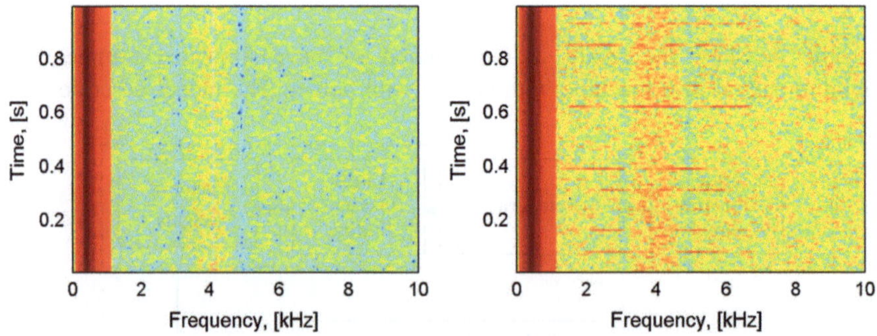

Fig. 2 Time-frequency representation of simulations for "undamaged" and "damaged" signals (*left* and *right* respectively)

magnitude is higher than past 6 and future 6 magnitudes. This choice is the maximum symmetrical neighborhood which does not exclude maxima at time points corresponding to the expected broadband excitations. Lower neighborhood length results in more noise, i.e. even minor maxima are observed.

Figures 2 and 3 show the standard spectrograms compared to the enhanced ones for both healthy and locally damaged machines. The spectrogram for signal of locally damaged machine contains horizontal lines specific for a failure signal. Because of magnitude values, these lines are colored similarly to the sub-signals that contain more energy than others (see at frequency band of 3–5 [kHz]). Hence, this band may be rejected as non-informative in further processing after standard spectrogram-based analysis. One can see that the enhanced spectrogram for a healthy signal contains no information of energy in each frequency band. Its behavior on low frequencies (<1 [kHz]) is very similar to the highest frequencies (>7 [kHz]). Information about energy is unnecessary in the wide-range excitation detection, so this feature, called energy-insensitivity, causes that these horizontal lines are better visible on the enhanced spectrogram. In particular, one can see that the frequency band of 3–5 [kHz], however, contains significant information about damage and should not be rejected. Better visibility can be noticed not only in the broader spectrum of excitations, but also in noise reduction between them. The second advantage can be seen especially when vertical lines on the enhanced spectrogram are almost continuous (see $t \sim 0.6$ [s] in Fig. 3, right panel).

Another output of the presented method is the vector of weights. VoW for healthy and unhealthy signal (Fig. 4) show the averaged local maxima occurrence for each time point. Local maxima of sub-signals are required to be higher than 6 past and future magnitudes, so 6 first and last values of VoW are zeros. Small value of the modulation depth causes that local maxima are visible only at about 30–40 % of frequency band spectrum. Such a narrow informative band causes that in the enhanced spectrogram one can observe many random maxima (especially in the band of >7 [kHz]) which do not follow any wide-band excitation.

Average number of local maxima between excitations for the unhealthy signal is significantly lower than the average number of local maxima at the corresponding

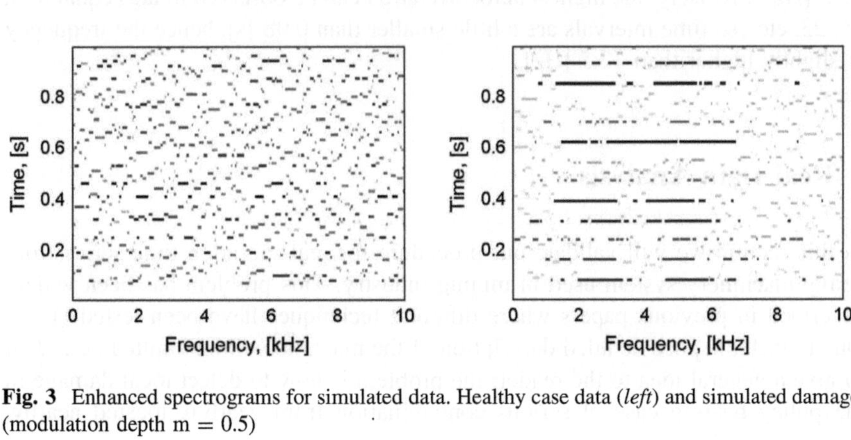

Fig. 3 Enhanced spectrograms for simulated data. Healthy case data (*left*) and simulated damage (modulation depth m = 0.5)

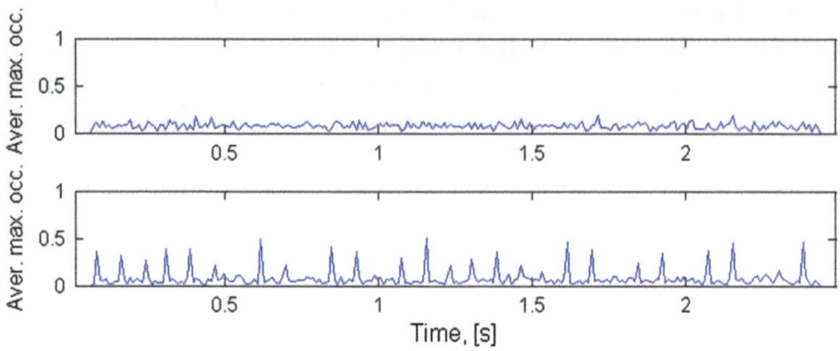

Fig. 4 Vectors of weights for simulated data—average numbers of maxima occurrence for each time point. Unhealthy data with modulation depth m = 0.5 (*bottom panel*) and healthy data (*top panel*)

time points for the healthy one. This is caused by noise reduction—local maximum at one point excludes local maximum at some time points before and after. After statistical analysis, this feature can be a base in noise reduction procedure. Values of the VoW significantly lower than those expected for a healthy signal certainly do not represent broadband excitation hence can be filtered out.

Vector of weights is a great instrument to compare healthy and unhealthy signal, but if we do not have comparative samples, we need a tool to say whether the analyzed VoW represents machinery in a good condition or not. We propose to use a famous tool widely-used in the stochastic analysis—the sample autocorrelation function (ACF). In this case we propose to measure correlation inside time series from the VoW. In Fig. 5, ACF for a signal which comes from damaged rotating machinery the VoW time series possesses significantly higher values than the confidence interval, so it seems to be correlated. For the healthy signal the VoW time series appears to be uncorrelated. One can see that VoW representing the damaged machinery is autocorrelated with lags close to a multiple of 8, which corresponds to time intervals close to a multiple of 0.08 [s] and frequency of 12.5 [Hz]. Precisely, the highest autocorrelation can be observed in lags equal to 8, 15, 23, etc., so time intervals are a little smaller than 0.08 [s], hence the frequency is slightly higher than 12.5 [Hz].

4 Real Data Analysis

In this section we will validate our procedure for real vibration time series from heavy machinery system used in mining industry. This problem has been widely described in previous papers where different techniques have been tested [1, 2]. Due to lack of space detailed description of the machine will be omitted here. Just to give a general idea to the reader: the problem is how to detect local damage in the pulley bearing case of serious contamination from gearbox located nearby.

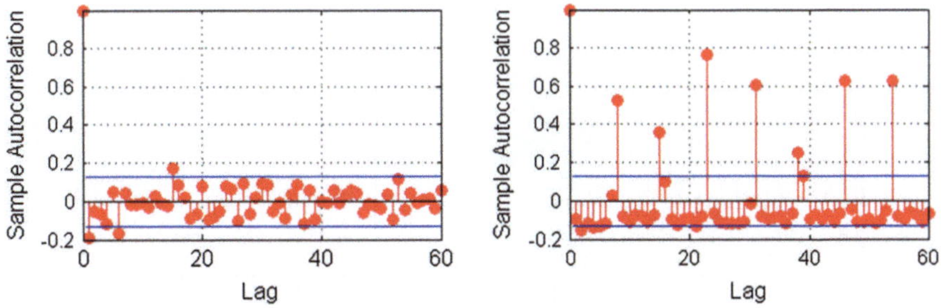

Fig. 5 Sample autocorrelation function for vector of weights for healthy simulated signal (*left panel*) and unhealthy data with modulation depth m = 0.5 (*right panel*)

Parameters of the signal acquisition are: sampling frequency 19.2 [kHz], duration 2.5 [s]. Time-frequency map has been obtained using Matlab spectrogram function with the following parameters: no overlapping, FFT length $NFFT = 512$ points and Kaiser's window of length 200.

Also the minimal neighborhood length used for enhancement procedure is fixed at 6 adjacent points. This is caused by the expected frequency of disturbance connected with damage which should be close to 12.69 [Hz]. Two cases were considered: bearings in good and in bad condition. Time-frequency map for both of them are presented in Fig. 6 (left panels). In these figures only a part of the signal is presented (first second only). As for simulations, some horizontal lines are visible for signal with damaged bearings (up to 6 [kHz]). Again, for frequencies less than 1 [kHz] a non-informative contribution of high energy may be found. It is related to gearbox vibration transmitted to the sensor through the shaft. Top and bottom plots on the right side of Fig. 6 present results of time-frequency map enhancement.

Looking at the spectrogram, one can divide whole spectrum into three energy-similar parts: high-energy (<1 [kHz]), middle-energy (1–7 [kHz]) and low-energy band (>7 [kHz]). Wide-band excitations do not always have enough energy to be easily visible at the edges of these bands. This means that spectrogram analysis fails in the band of variable energy (1 and 7 [kHz]). Analysis of local maxima in the spectrogram sub-signals deal with this problem. Enhanced time-frequency map for damaged bearing is characterized by many almost-continuous horizontal lines in the band of 7 [kHz] that are not visible in the standard spectrogram. Healthy signal represented on the top-right panel of Fig. 6 is totally random. No particular band with specific behavior can be distinguished. This feature confirms that the local maxima method is insensitive to energy, thus it is a better tool to search for excitations than the standard spectrogram.

Comparing VoW time series for healthy and unhealthy signals (Fig. 7) one can see that, again, the average maxima occurrence is lower than for healthy data between two damage signals. As in the simulated vibration signal case study, statistical analysis of healthy VoW distribution can provide rules to filter VoW and receive clearer information about damage.

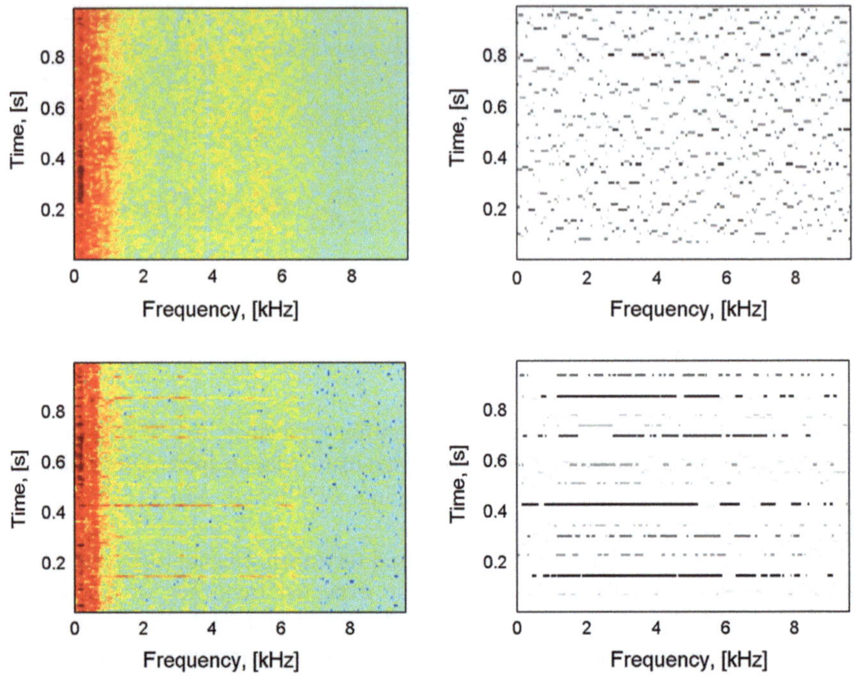

Fig. 6 Spectrograms (left panels) and their enhancements (*right panels*) for real unhealthy data (*bottom panels*) and healthy data (*top panels*)

What is worth mentioning, some of significant horizontal lines are barely visible on the standard spectrogram (compare bottom panels on Fig. 6, $t_1 = 0.3$ [s] and $t_2 = 0.6$ [s]). Vector of weights confirms that these lines are of enough wide-band character to be treated as a signal connected with damage (Fig. 7).

Figure 8 show that even in the real case study the sample autocorrelation function can be useful to recognize the occurrence of damage as well as its characteristic frequency. ACF for time series representing unhealthy bearing

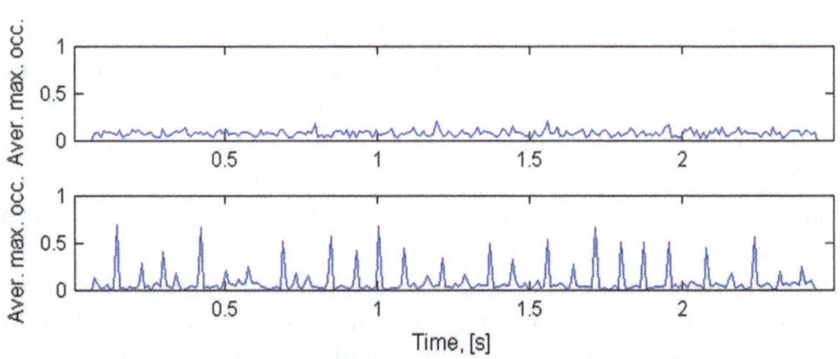

Fig. 7 Vectors of weights for real data—average numbers of maxima occurrence for each time point. Unhealthy data (*bottom panel*) and healthy data (*top panel*)

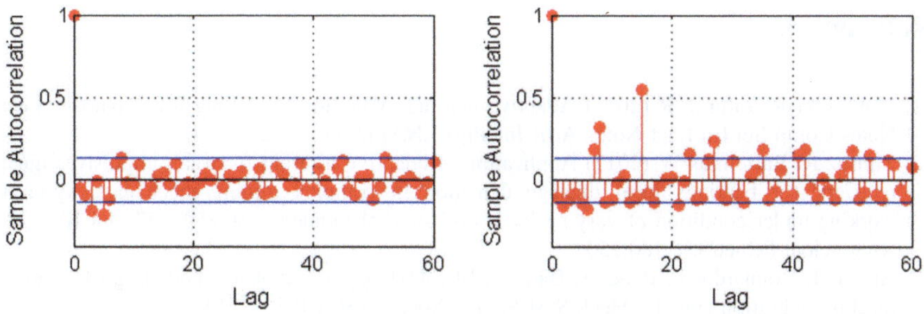

Fig. 8 Sample autocorrelation function for vector of weights for healthy bearing (*left panel*) and damaged bearing (*right panel*)

exceeds confidence intervals in many lags. Correlation is highest for time series shifted by 15, but ACF in lags of 7 and 8 is also well above confidence intervals. In the case analyzed in this Sect. 1 lag represents $\sim$ 0.0104 [s], i.e. time intervals between broadband excitations are close to 0.083 [s] (12 [Hz]). Recall that the characteristic damage frequency is 12.69 [Hz] which corresponds with $\sim$ 7.8 time intervals equal to 1 lag.

5 Conclusions

In the paper a novel method of time-frequency map processing for feature extraction for local damage detection is proposed. It should be mentioned that the procedure has been applied to spectrogram; however, it can be applied to other time-frequency representation. After processing, new time-frequency map contains clear information that can be also easily aggregated to one-dimensional time series with cyclic impulses. Original raw vibration signal contains strong non-informative contribution so detection of cyclic impulses related to damage is not possible. The advantage of the method is that after decomposition of raw vibration signal to two-dimensional time-frequency map, further processing is applied to simple sub-signal for each frequency bin. Non-informative sub-signals are in some sense neglected; sub-signals with cyclic disturbances are enhanced. It provides new time-frequency map with highlighted informative part.

Acknowledgments This work is partially supported by the statutory grant No. S20096 (J. Obuchowski and R. Zimroz).

References

1. Makowski R, Zimroz R (2011) Adaptive bearings vibration modelling for diagnosis. Lect Notes Comp Sci (in Lect Notes Artif Intellig) LNAI 6943:248–259
2. Zimroz R, Bartelmus W (2012) Application of adaptive filtering for weak impulsive signal recovery for bearings local damage detection in complex mining mechanical systems working under condition of varying load solid state phenomena 180:250–257. doi:10.4028/www.scientific.net/SSP.180.250
3. Antoni J, Bonnardot F, Raad A, Badaoui M (2004) Cyclostationary modelling of rotating machine vibration signals. Mech Syst Signal Process 18(6):1285–1314
4. Antoni J (2009) Cyclostationarity by examples. Mech Syst Signal Process 23(4):987–1036
5. Urbanek J, Barszcz T, Zimroz R, Antoni J (2012) Application of averaged instantaneous power spectrum for diagnostics of machinery operating under non-stationary operational conditions. Meas: J Int Meas Confederation 45(7):1782–1791. doi:10.1016/j.measurement.2012.04.006
6. Safizadeh MS, Lakis AA, Thomas M (2000) Using short-time Fourier transforms in machinery fault diagnosis. Int J COMADEM 3(1):5–16
7. Baydar N, Ball A (2001) A comparative study of acoustic and vibration signals in detection of gear failures using Wigner-Ville distribution. Mech Syst Signal Process 15(6):1091–1107
8. Lee SK, White P (2001) The L-Wigner distribution and its use as a tool for condition monitoring. Int J COMADEM 4(1):5–11
9. Dalpiaz G, Rivola A, Rubini R (2000) Effectiveness and sensitivity of vibration processing techniques for local fault detection in gears. Mech Syst Signal Process 14(3):387–412
10. Lee SK, White PR (1997) Higher-order time-frequency analysis and its application to fault detection in rotating machinery. Mech Syst Signal Process 11(4):637–650
11. Lin J, Zuo M (2003) Gearbox fault diagnosis using adaptive wavelet filter. Mech Syst Signal Process 17(6):1259–1269
12. Peng ZK, Chu FL (2004) Application of the wavelet transform in machine condition monitoring and fault diagnostics: a review with applications. Mech Syst Signal Process 18(2):199–221

Reconstruction of the Instantaneous Angular Speed Variations Caused by a Spall Defect on a Rolling Bearing Outer Ring Correlated with the Length of the Defect

Adeline Bourdon, Didier Rémond, Simon Chesné and Hugo André

Abstract In the framework of monitoring of rotating machinery, this paper proposes a simple signal processing tool to reconstruct the Instantaneous Angular Speed (IAS) variations caused by the presence of spalled bearing. This tool is applied to signals obtained on a specific test bench. Associated with an angular sampling, the analysis of these variations can identify the length of the defect whatever the mode of operation, particularly in non-stationary operating conditions in rotation speed.

Keywords Instantaneous Angular Speed (IAS) · Bearing · Fault · Monitoring · Fault length

1 Introduction

The works carried out for several years in the framework of monitoring and maintaining rotating machinery have shown the advantage of the angular sampling to overcome stationary operating conditions limitations [1, 2]. Angular sampled signals can be analyzed using conventional signal processing tools such as Fourier Transform that leads to spectrums of cyclic frequencies that are expressed as number of events per revolution (ev.rev^{-1}). These sampling techniques and tools have been applied to the Instantaneous Angular Speed (IAS) signal and shown that the presence of defects leads to the emergence of peaks in the associated spectrum [3–6]. For a spall defect on the outer ring of a bearing the frequency localization of these peaks corresponds to the harmonic of the rolling elements pass frequency on

A. Bourdon (✉) · D. Rémond · S. Chesné
Université de Lyon, CNRS, UMR5259 INSA-Lyon, LaMCoS, F-69621 Lyon, France
e-mail: Adeline.Bourdon@insa-lyon.fr

H. André
Maïa Eolis, Tour de LILLE, Bd de Turin 59777 LILLE, France

G. Dalpiaz et al. (eds.), *Advances in Condition Monitoring of Machinery in Non-Stationary Operations*, Lecture Notes in Mechanical Engineering,
DOI: 10.1007/978-3-642-39348-8_28,

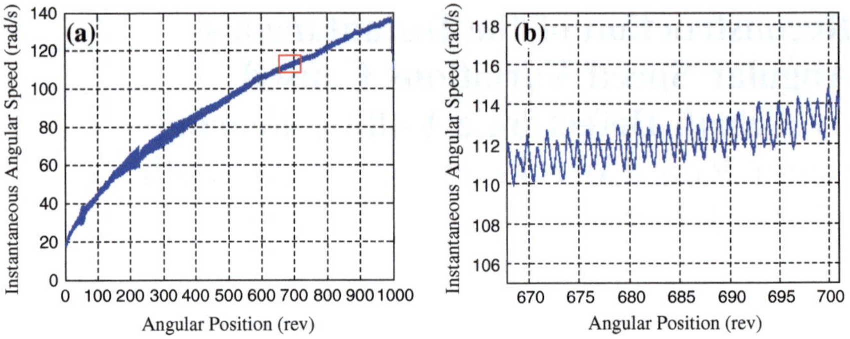

Fig. 1 Example IAS signal in non stationary operating conditions

the outer raceway (BPFO). This means that the presence of a fault on the outer ring of a bearing leads to variations of the Instantaneous Angular Speed [7, 8]. As an example Fig. 1 shows IAS measured on a shaft with a defective bearing in non-stationary operating condition. Zoom of Fig. 1a between 670 and 700 rev. is presented in Fig. 1b.

Figure 2a presents the associated cyclic frequencies spectrum. Harmonics associated with bearing defect (theoretical BPFO 7.26 $ev.rev^{-1}$) are reported in Fig. 2b.

Following these findings, studies have focused on the development of specific tools and indicators [9, 10]. Although useful for industrial use of IAS signal in rotating machinery maintenance, [11] these approaches are not sufficient to estimate the magnitude and shape of the IAS variations due to bearing defect. Indeed, as shown in Fig. 1b, even zooming does not enable to identify variations due to the presence of the defect in the IAS signal. In their current developments, these tools can not easily estimate the size of the defect. This last point is essential in order to follow the evolution of a default and, in the long term, to estimate the remaining life time of the bearing.

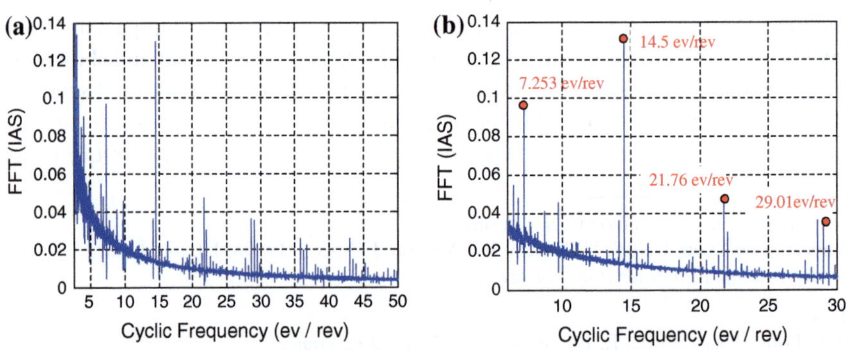

Fig. 2 Cyclic frequencies spectrum and harmonics associated with bearing defect

Works presented in this paper offer an innovative tool based on the reconstruction of a filtered signal to isolate and to identify the variations in Instantaneous Angular Speed caused by a defect on a bearing ring. The filters are rectangular windows applied on frequency domain and the signal is reconstructed using the usual inverse Fourier transform. In a second part of this paper, this tool is used to analyze experimental signals obtained for different levels of bearing defect and different operating conditions. These experimental applications show that it is possible to easily determine the length of the defect.

2 Reminders on Angular Sampling, Cyclic Frequencies and IAS

As the originality of this work is based on the angular sampling and IAS, few reminders seem necessary. For rotating machines, the angular sampling involves selecting one of the rotations of the machine as a reference. Signals are then sampled as a function of the angular position of this reference. Different methodologies are widely reported in the literature [1, 2]. One advantage of angular sampling is that the cyclic characteristic frequencies of the rotating elements are independent of the macroscopic speed. For bearings, invariance of these frequencies is only theoretical, in practice they may vary from 1 to 2 %. In the following we consider that despite these small changes we are able to localize and to identify these bearing frequencies in the spectrum.

In the present paper, Instantaneous Angular Speed signal is obtained by the elapsed-time method. This method has also been the subject of numerous publications [2–6]. It is based on the simultaneous use of an optical encoder with a resolution of R impulse per revolution and of a high-frequency clock (f_c). It consists in counting the number n_c of clock ticks between two successive pulses of the encoder signal. This allows estimating the average speed $\bar{\omega}_i$ for each angular step $\Delta\theta_i$ by Eq. 1. By construction this signal is angular sampled

$$\bar{\omega}_i = \frac{\Delta\theta}{\Delta t} = \frac{2\pi}{R}.\frac{1}{n_c}f_c. \qquad (1)$$

3 Reconstitution of the Speed Disturbance Using Inverse Fourier Transform

The proposed method is based on simple signal processing tools: Fourier and inverse Fourier transforms. The originality of this work lies on the reconstruction of IAS variations caused by the presence of a defect in bearing to estimate its magnitude and get information on the length of the defect. For that, IAS is seen as a combination of an undisturbed component and of a component resulting from the disruption caused by the defect. This second component appears with a constant angular period, which is reflected in the cyclic frequency domain by the emergence of peaks located at harmonics of the bearing characteristic frequency. The idea implemented in this paper is to extract in the frequency spectrum the harmonics associated with the disturbance and then to come back into the physical space using inverse Fourier transform.

The following methodology is proposed:

1. Measuring IAS signal by the help of the elapse time method,
2. Computing its cyclic frequency spectrum,
3. Identifying the cyclic frequency $f_{\theta C}$ associated to bearing default,
4. Extracting, in the cyclic frequency spectrum, m rectangular windows of length $2L_w$, centered around the first m harmonics of characteristic frequency $f_{\theta C}$.
5. Using inverse Fourier transform on this modified spectrum to obtain a filtered signal which represents IAS variations linked to the damage/default.

This methodology is based on the assumption that the cyclic frequency selected is associated only with the monitored bearing. For bearings, the characteristic frequencies are non-integer variables, which make the previous assumption realistic.

4 Application to Experimental Signals

The previous method is applied to the IAS signals measured on a test bench described below for various defects and different operating conditions. All measurements described below were performed using 6 million points. The Fourier and inverse Fourier Transforms are computed using Matlab FFT and IFFT functions. Calculations are performed on the first 2^{21} points that leads to cyclic frequency precision of 0.0024 $ev.rev^{-1}$. IAS measurement is done using an encoder with a resolution of 5000 impulses per revolution, that leads to a Nyquist's Frequency of 2,500 $ev.rev^{-1}$.

IAS variations are built using 15 rectangular windows of length $L_w = 0.0477$ $ev.rev^{-1}$. These characteristic quantities were defined empirically. We can note that the size of the passband windows is very small regarding the length of the spectrum, we suppose, but future work can refute this assumption, that the artifacts produced by the IFFT after this windowing are negligible.

4.1 Description of Test Bench

A specific test bench has been developed (Fig. 3a) to understand the relationship between the presence of a defect in a bearing and the Instantaneous Angular Speed variations. It consists in a shaft supported by two tapered roller bearings that may

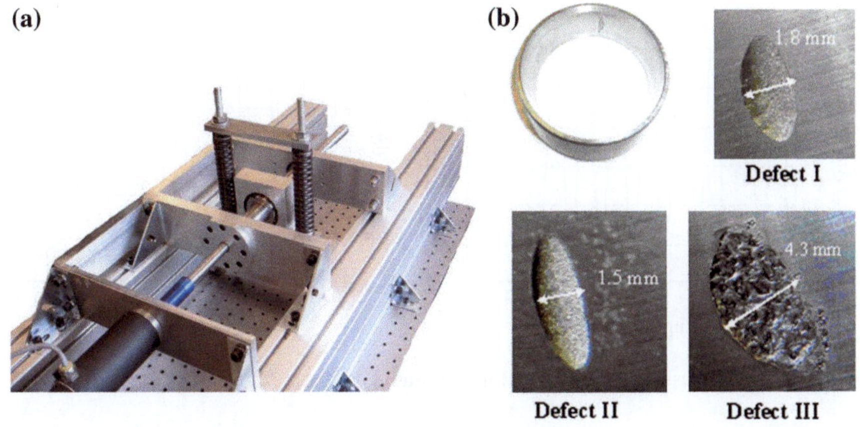

Fig. 3 **a** Specific test bench **b** Examples of bearing defects

have defects in their outer ring (Fig. 3b). These defects correspond to a theoretical characteristic angular frequency (BPFO) equals to 7.26 ev.rev^{-1}. This frequency can change a little bit with the load and operating conditions.

An original device composed of a spherical bearing and two springs allows applying adjustable radial loads independently of rotation speed. An axial preload can also be applied in the form of an imposed displacement of one bearing outer ring. A high resolution optical encoder (R = 5,000 pulses per revolution), not shown in Fig. 3a, coupled to an acquisition device with a clock frequency f_c of 80 MHz ensures the angular sampling and measurement of the Instantaneous Angular Speed as described before.

4.2 IAS Variations in Case of a Steady State Operation

In this section the instantaneous velocity is measured for a stationary operating regime with a bearing having defect I (Fig. 3b) on its outer ring. Despite a constant torque, the speed varies slightly and has a periodic phenomenon with a cyclic frequency of 1 ev.rev^{-1} as seen in Fig. 4.

4.2.1 First Analysis

Cyclic frequency of the defect $f_{\theta o}$ in these operating conditions is identified using the frequency spectrum (Fig. 5), which is the peak closer to the theoretical BPFO (7.26 ev.rev^{-1}). In that case it is located at 7.348 ev.rev^{-1} 1.2 % higher than the theoretical frequency.

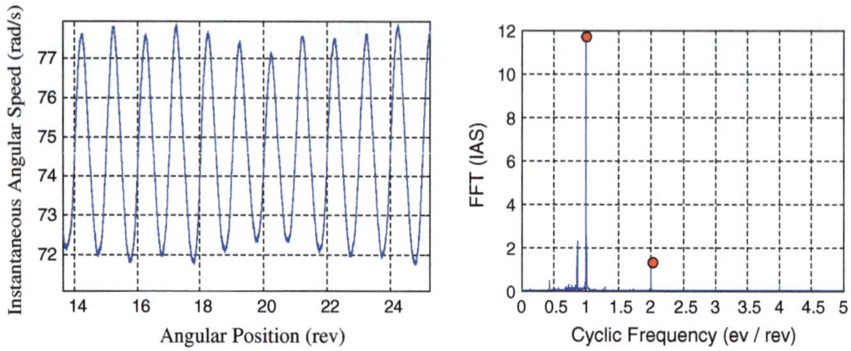

Fig. 4 Instantaneous angular speed and its spectrum

Fig. 5 Identification of effective BPFO in cyclic frequency spectrum

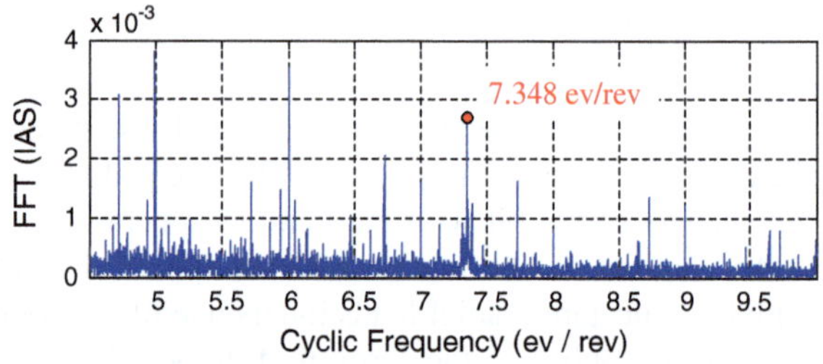

IAS variations reconstructed is shown in Fig. 6b for 2.5 periods of the fault. Figure 6a shows the initial IAS signal for the same angular duration. Influence of defect can not be detected in Fig. 6a.

IAS variation shows a localized phenomenon corresponding to a speed magnitude increase of about 0.045 $rad.s^{-1}$ or 0.06 % of the macroscopic speed (74 $rad.s^{-1}$) and with an angular duration about 0.03 rev.

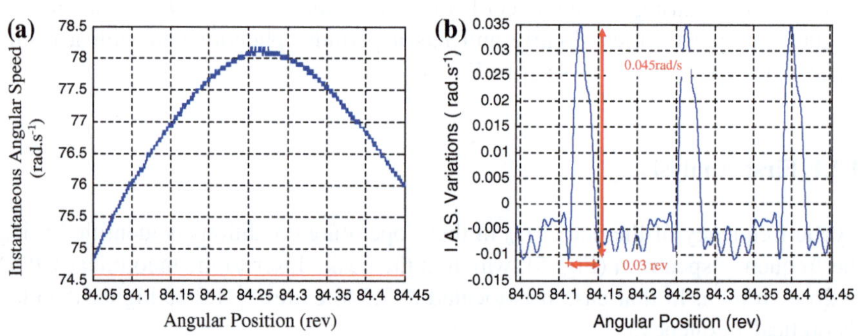

Fig. 6 Initial IAS and IAS variations for bearing with defect I

4.2.2 Correlation with the Size of the Defect

To correlate the angular duration of this phenomenon and the length of the defect present in the outer ring, this length must be translated into an angular duration of the reference shaft. A localized defect located on the outer ring with a length L_d at a diameter d has an angular length of θ_d. For a bearing with Z rolling elements, angular length between 2 rolling elements is θ_o. By definition of the cyclic frequency f_θ in reference to the rotation of the shaft, the angular duration between two consecutive rolling elements passing on a given point on the outer ring is α_o. The angular duration for a rolling element passing on the default will be α_d. These geometrical parameters are shown in Fig. 7 and their values given by:

$$\theta_d = \frac{2.L_d}{d};\ \theta_o = \frac{2.\pi}{Z};\ \alpha_o = \frac{2\pi}{f_{\theta o}};\ \alpha_d = \frac{L_d}{\theta_o}.\alpha_o. \tag{2}$$

In the studied case, bearing is a tapered roller bearing with $Z = 17$ rolling elements, with a spalled defect of length $L_d \approx 1.8$ mm located on the outer ring at a diameter $d = 40$ mm. Operating conditions lead to a default frequency of $7.348\ \mathrm{ev.rev}^{-1}$. In reference with the angular position of the shaft, the angular duration for a rolling element passing on the default is $\alpha_d = 0.033$ rev which is perfectly consistent with the width of the reconstructed velocity perturbation shown in Fig. 6b.

4.2.3 Repeatability of Analysis

This correlation between the length of the fault and IAS variations is found for other defects. Figure 8 shows IAS variations obtained for two other defects (II and III) under stationary operating conditions. The red horizontal line is used to display the angular duration for a rolling element passing on the default.

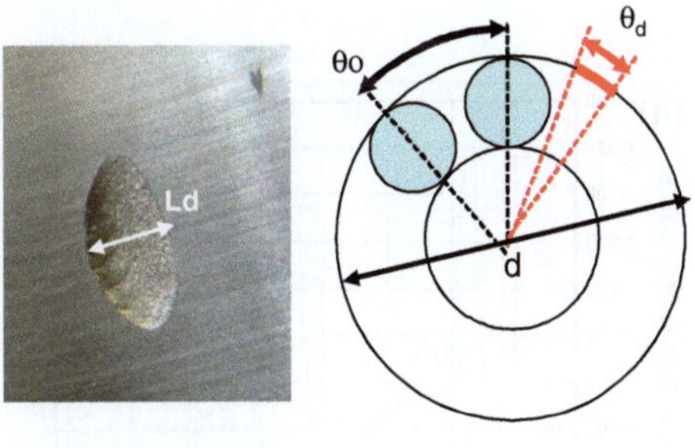

Fig. 7 Geometrical parameters linked with the defect

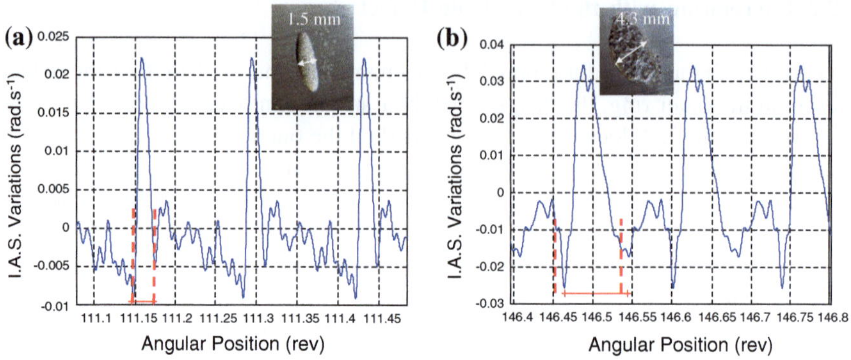

Fig. 8 Example of correlation between the length of the defect and IAS variations

4.2.4 Influence of Radial Load

For a given defect (II) and identical macroscopic speeds (60 rad.s^{-1}), analysis performs for various loads shows the influence of the radial load on the IAS variation (Fig. 9). Amplitude of variations increases with the applied radial load, which is 0.04 rad.s^{-1} for a radial load of 2,220 N and becomes 0.08 rad.s^{-1} for a radial of 3,210 N.

4.3 Measurements in Non-Stationary Operating Regime

The proposed approach is now applied to 2 non-stationary operating regimes for the default II. Results are presented in Figs. 10 and 11. Again a localized phenomenon, whose length corresponds to the passage of a body rolling on the

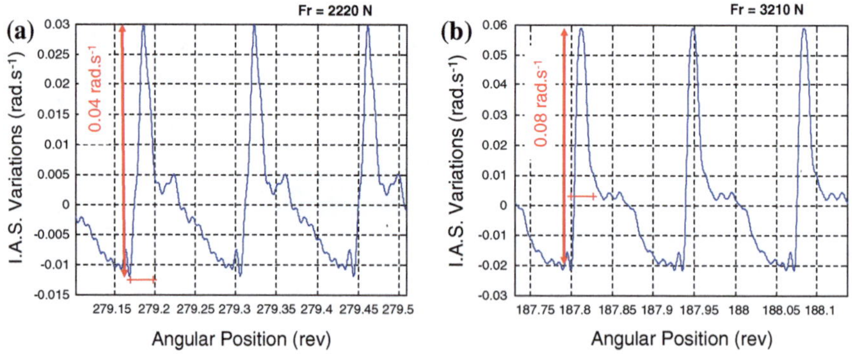

Fig. 9 Influence of radial load on IAS variations

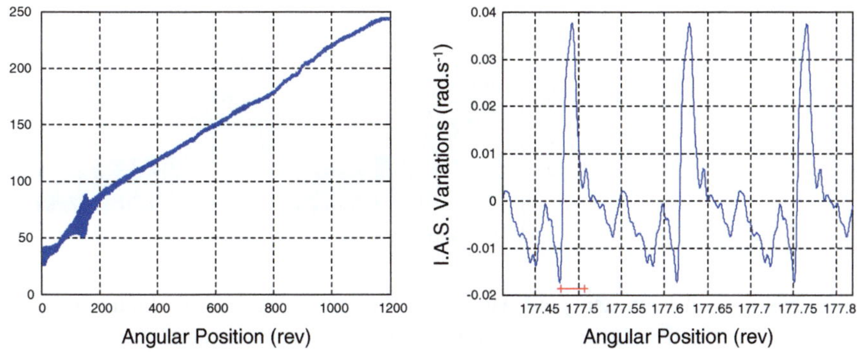

Fig. 10 IAS and IAS variation in case of machine run up

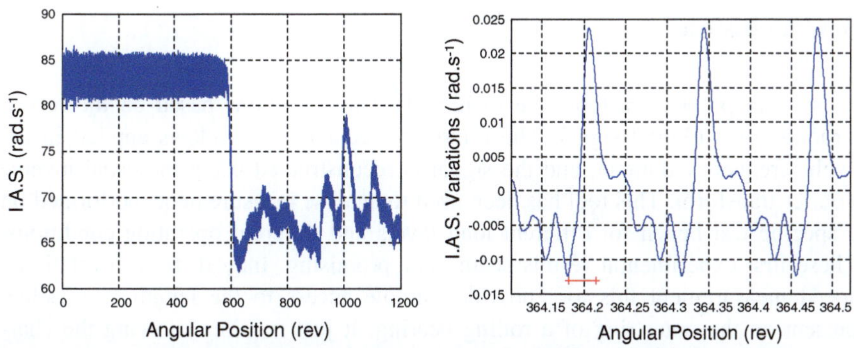

Fig. 11 IAS and IAS variations in case of large change in speed history

default, appears clearly on the reconstructed signal. It must be noticed that the non-stationary conditions can be very smooth like for machine run-up (Fig. 10) or very sharp with large changes in the speed history (Fig. 11).

4.4 Results Analysis

In presence of spalled bearing, IAS variations seem to be divided into two zones as shown schematically in Fig. 12. This remark must be related to the work of R.B. Randall and N. Sawalhi [12, 13].

A first area (in green) corresponds to the passage of a rolling body on the fault. This zone is associated with the largest amplitude of speed variation. For a fault and a given load condition, the general shape of this area does not seem very sensitive to operating conditions. These results are likely to be very constructive to understand the way of transfer between the defect and IAS. The second zone (in

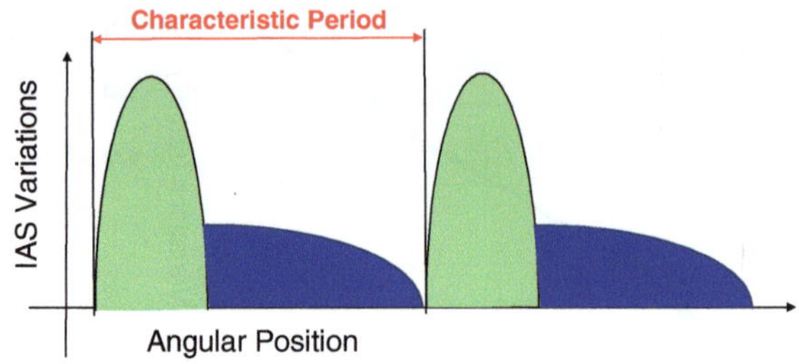

Fig. 12 Schematic representation of the IAS variations

blue) has not yet been analyzed. However, the amplitudes of the fluctuations are lower than in the first zone and the general shape seems to be more sensitive to the operating conditions.

5 Conclusion

This paper proposes a tool able to identify IAS variations based on the reconstruction of a filtered signal. The filters are rectangular windows applied in the cyclic frequency domain, and the signal is reconstructed using the usual inverse Fourier transform. This tool has been used to analyse measurements performed on a specific test bench for different load levels and different operating conditions. These first experimental results seem very promising. Indeed, from a rotational speed measurement this tool provides simple access to the length of a defect present on the outer ring of a rolling bearing. It just requires knowing the characteristic frequency associated with this defect. This tool also allows estimating the amplitude of IAS variations caused by the presence of the defect. Other works in progress show that this amplitude can be correlated with bearing loading conditions. Therefore it could be a first step towards the development of tools to analyze the residual life time of the bearing. This tool must still be refined and tested on other measurements and other devices. The influence of the characteristics of filtering windows (number and frequency bandwidth) on the results should be estimated. In parallel the experimental results can be compared to those obtained with phenomenological models in order to understand the pathway between the fault and its manifestation. More generally, these studies confirm the richness of the IAS signal associated to a very sensitive measurement method that both characterize speed variations less than 0.06 % of the macroscopic velocity.

References

1. Bonnardot F, Badaoui MEL, Randall R, Daniere J, Guillet F (2005) Use of the acceleration of a gearbox in order to perform angular resampling (with limited speed fluctuation). Mech Syst Signal Process 19:766–785

2. Rémond D, Mahfoudh J (2005) From transmission error measurements to angular sampling in rotating machines with discrete geometry. Shock Vib 12(2):149–161
3. Jianguo Y, Lijun P, Zhihua W, Yichen Z, Xinping Y (2001) Fault detection in a diesel engine by analysing the instantaneous angular speed. Mech Syst Signal Process 15(3):549–564
4. Gu F, Yesilyurt I, Yuhua L, Harris G, Ball A (2006) An investigation of the effects of measurement noise in the use of instantaneous angular speed for machine diagnosis. Mech Syst Signal Process 20(6):1444–1460
5. Yuhua L, Gu F, Harris G, Ball A, Benett N, Travis K (2005) The measurement of instantaneous angular speed. Mech Syst Signal Process 19(4):786–805
6. André H, Bourdon A, Rémond D (2011) On the use of the instantaneous angular speed measurement in non-stationary mechanism monitoring. In: Proceedings of the ASME 2011 international design engineering technical conferences', paper DETC2011/MECH-47470, Washington DC, USA, 28–31 Aug 2011
7. Renaudin L, Bonnardot F, Musy O, Doray JB, Rémond D (2010) Natural roller bearing fault detection by angular measurement of true instantaneous angular speed. Mech Syst Signal Process 24(7):1998–2011
8. Bourdon A, André H, Rémond D (2012) Numerical investigations on the magnitude of disturbance that could be detected through IAS. In: Proceedings of the international conference on condition monitoring of machinery in non-stationary operations
9. André H, Bourdon A, Antoni J, Rémond D (2012) Precision of the IAS monitoring system based on the elapsed time method in the spectral domain. In: Proceedings of the international conference on noise and vibration engineering ISMA 2012–ISAM 2012, Belgique
10. André H, Bourdon A, Rémond D (2012) Instantaneous angular speed monitoring of a 2 MW wind turbine using a parametrization process. In: Proceedings of the international conference on condition monitoring of machinery in non-stationary operations
11. Yu S, Zhang X (2010) A data processing method for determining instantaneous angular speed and acceleration of crankshaft in an aircraft engine-propeller system using a magnetic encoder. Mech Syst Signal Process 24(4):1032–1048
12. Sawalhi N, Randall RB (2011) Vibration response of spalled rolling element bearings: observations, simulations and signal processing techniques to track the spall size. Mech Syst Signal Process 25:846–870
13. Sawalhi N, Randall RB (2010) Simulation of the vibration generated by entry and exit to/from a spall in a rolling element bearing. In: Proceedings of 20th international congress on acoustics, ICA 2010, Sydney, Australia, 23–27 Aug 2010

2. Rémond D, Mahfoudh J (2005) From transmission error measurements to angular sampling in rotating machines with discrete geometry. Shock Vib 12:149–161
3. Jianguo Y, Lijun P, Zhihua W, Yichen Z, Xinping Y (2003) Fault detection in a diesel engine by analysing the instantaneous angular speed. Mech Syst Signal Process [illegible]
4. [illegible]

Instantaneous Angular Speed: Encoder-Counter Estimation Compared with Vibration Data

M. Spagnol and L. Bregant

Abstract In rotating machinery, actions of the moving parts take place at specific angular positions rather than at specific times. For this reason, having a geometrical reference, such as the one provided by an encoder, and studying the Instantaneous Angular Speed (IAS) variations can provide a large amount of information about the health status of the machine. In fact, from the variation of the IAS during the machine loads' cycle it is possible to identify defects and faults. The current work focuses on the estimation of the IAS through the Elapsed Time (ET) method, using a counter in order to measure the time elapsed between the pulses of an encoder. Both IAS and vibration measurement are conducted on an asynchronous four poles electrical motor driven by 50 Hz line current, without load. The study compares the order analysis of both signals. The bearing's Fundamental Train Frequency is detected using IAS estimation.

Keywords Instantaneous angular speed · Encoder · Elapsed time method · AC DC electrical machine · Slip effect

1 Encoder-Counter System

Among the different processing strategies to obtain IAS [1], in this study, the Elapsed Time (ET) method is used. In this case, the counter measures the time elapsed between two successive pulses of the encoder. With this approach there are as many measurement values as there are pulses/revolution of the encoder. The frequency of the counter and the number of pulses determine the resolution of

M. Spagnol (✉) · L. Bregant
Università degli Studi di Trieste, via A. Valerio 10, 34127 Trieste, TS, Italy
e-mail: mspagnol@units.it

L. Bregant
e-mail: bregant@units.it

G. Dalpiaz et al. (eds.), *Advances in Condition Monitoring of Machinery in Non-Stationary Operations*, Lecture Notes in Mechanical Engineering,
DOI: 10.1007/978-3-642-39348-8_29,

the IAS estimates. The method is strictly correlated to the real rotational angle of the shaft, except from the encoder tolerances [2–4]. A National Instruments [5] 80 MHz counter is used for the measurements described in the following paragraphs. The selected board allows a choice between three counting methods: the Elapsed Time (ET) method, using one counter, and the High-Frequency (HF) and Large-Range (LR) methods, using two counters. A comparison between these methods is reported in [6]. In this paper ET method is used: the rising edge of the input signal of the encoder triggers the counting of the timebase ticks (Fig. 1). Since the timebase is of a known frequency, the frequency of the input signal can be obtained as (3).

$$f_{count} = 80\,\text{MHz} \tag{1}$$

$$\Delta t_{count} = f_{count}^{-1} = 1.25 \times 10^{-8}\,\text{s} \tag{2}$$

$$f_{input} = \frac{1}{n_{count}\Delta t_{count}} \tag{3}$$

IAS measurement errors come from different sources. Generally speaking, the absolute error value, increases linearly with the speed and the resolution of the encoder, considering that the upper measured speed limit is the ratio between the encoder's resolution and the clock frequency of the counter.

The ideal encoder assumes exactly equal geometric segments and any variation causes the ET to be sampled on a non-uniform angular basis. Since the spacing pattern repeats itself after each revolution, the error manifests itself as high-level content at integer multiples of the shaft running speed. It is possible to use the synchronous averaged encoder passage times to correct for the uneven encoder spacing [7]. These errors are unavoidable, but the production standards are very high and great precision can be obtained.

The ET measurement depends on the achievable time resolution, governed by the clock rate and the zero crossing detection circuit. These lead to two main problems: the counting method and the clock stability. Different authors [2, 4], have analysed the problems and have suggested appropriate remedies.

Further errors can be experienced if the sensor undergoes lateral movement, if it is installed with eccentricity or misalignment, or if any light-path transmission variations are present, as shown from the test setup (Fig. 2).

Fig. 1 Digital signal frequency measured with ET method

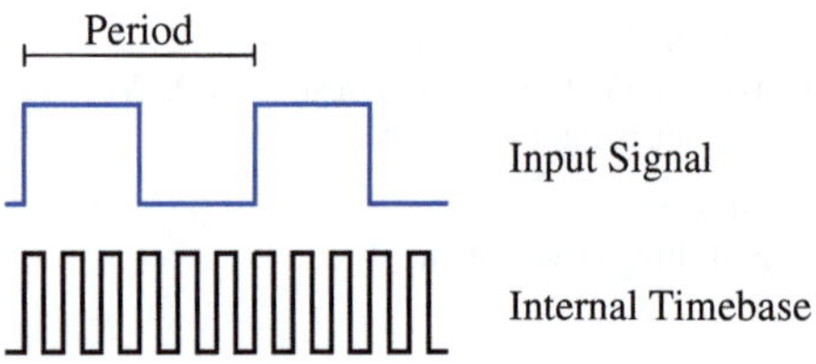

Fig. 2 Test setup: encoder 1,000 ppr, accelerometer PCB, 1 kW 4 poles induction motor

2 Experimental Results

In this paper, the IAS is compared to the vibration measurement obtained from accelerometer data. The motor used for the test is a 1 kW 4 poles electric induction motor. It is driven by 50 Hz line current 380 V. The test was conducted without load in order to emphasize speed variations. The encoder has 1,000 pulse per revolution (ppr) and its shaft is rigidly connected to the motor's one. The encoder's body is fixed at the baseplate through a spring. The accelerometer is a PCB 356A16 with the X axis in tangential, the Y in the axial and the Z in the radial direction. The acquisition system is based on National Instruments hardware. Analog signals are collected at 51.2 kHz, while the encoder's signal uses a 80 MHz counter. The MATLAB software does the necessary signal processing. Due to the motor type, the slip effect is present. It is possible to view this phenomenon in Fig. 3, where signals from an AC and a DC motor are compared. The main component of the signal is the electro-magnetic force seen by the rotor and its frequency is given by (9), where p is the number of poles, f_m is the mechanical rotational frequency and f_e is the electrical rotational frequency. In time/angular-domain, it appears as a shifting waveform due to the fact that there is a difference between the rotating speed to the magnetic field and the rotor. This effect doesn't appear with synchronous or DC machines because mechanical and electrical speeds are synchronized.

$$f_e = \frac{f_m}{1-s}\frac{p}{2} \Rightarrow o_e = \frac{1}{1-s}\frac{p}{2} \quad (4)$$

The motor under test has a working history. The measurements present electric phase imbalance since the fourth order is dominant. The characteristic frequencies of the installed bearing (type 6204-2RS1) are listed in Table 1.

In order to compare the measurements from the encoder and the accelerometer, an order analysis is performed. To visualize the results the order domain is chosen. The encoder signal is already in the proper domain, while the accelerometer output needs an angular resampling. The comparison (Figs. 4, 5) compares the IAS orders

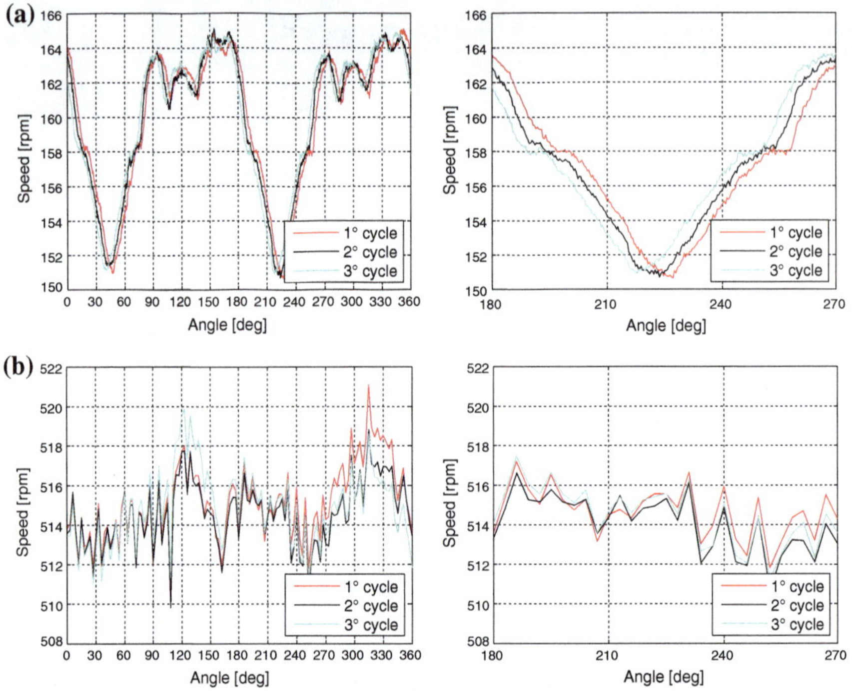

Fig. 3 Three cycles at 0-360° and zoom at 180-270°: **a** AC motor, 150rpm, encoder 1000ppr, with slip effect; **b** DC motor, 515rpm, encoder 120ppr, without slip effect

Table 1 Characteristic orders, bearing 6204-2RS1

Defect		Order
Bass pass frequency—Inner	BPFI	4.948
Bass pass frequency—Outer	BPFO	3.052
Fundamental train frequency (cage)	FTF	0.382
Ball spin frequency (rolling element)	BSF	1.992

obtained with the ET of the 1,000 ppr encoder signal and the accelerometer signals resampled at 1,000 ppr. The encoder signal shows that at low order, the signal is good, while at higher order there is some quantization error, [2]. The higher the speed, the bigger the quantization effect.

The order analysis from IAS and accelerometer data presents similar sidebands in the low order region. f_m is the mechanical frequency, f_e is the electrical supply frequency, f_s is the synchronous frequency and depends on the number of poles. The motor actual speed is 1,497 rpm while the theoretical is 1,500 rpm. The slip (6) between the two is 0.002 that appears in the orders plot.

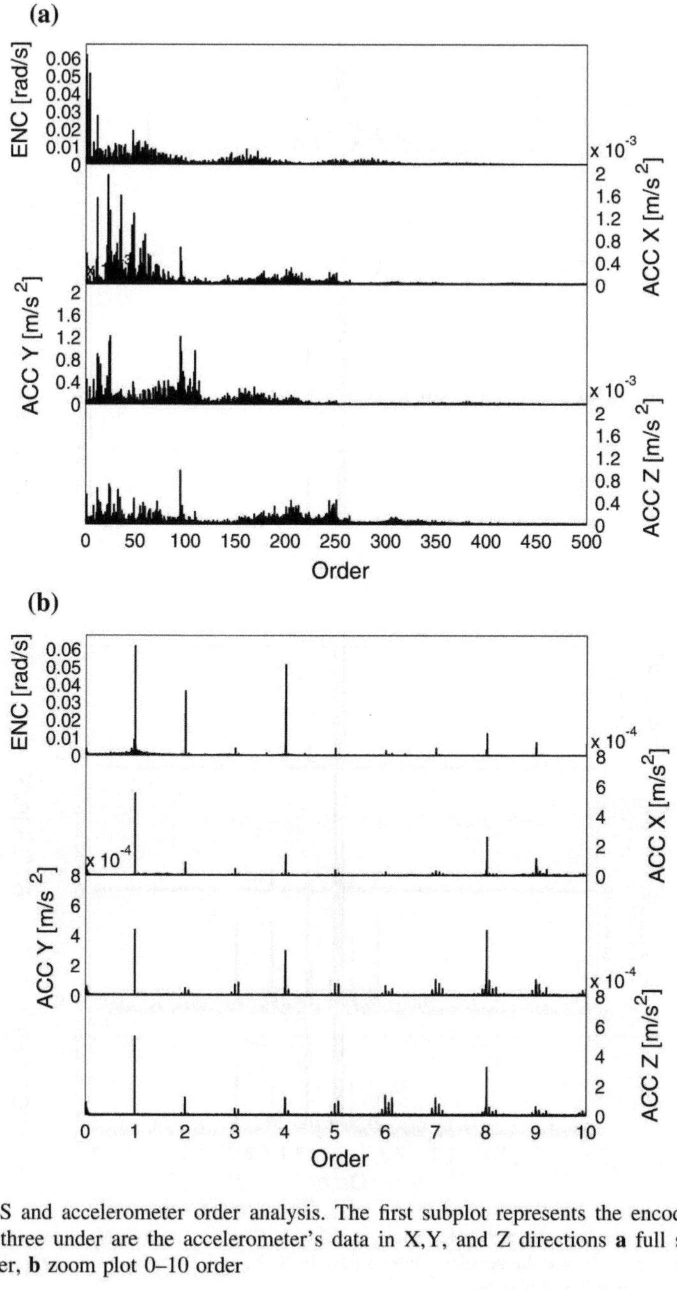

Fig. 4 IAS and accelerometer order analysis. The first subplot represents the encoder's data, while the three under are the accelerometer's data in X,Y, and Z directions **a** full scale plot, 0–500 order, **b** zoom plot 0–10 order

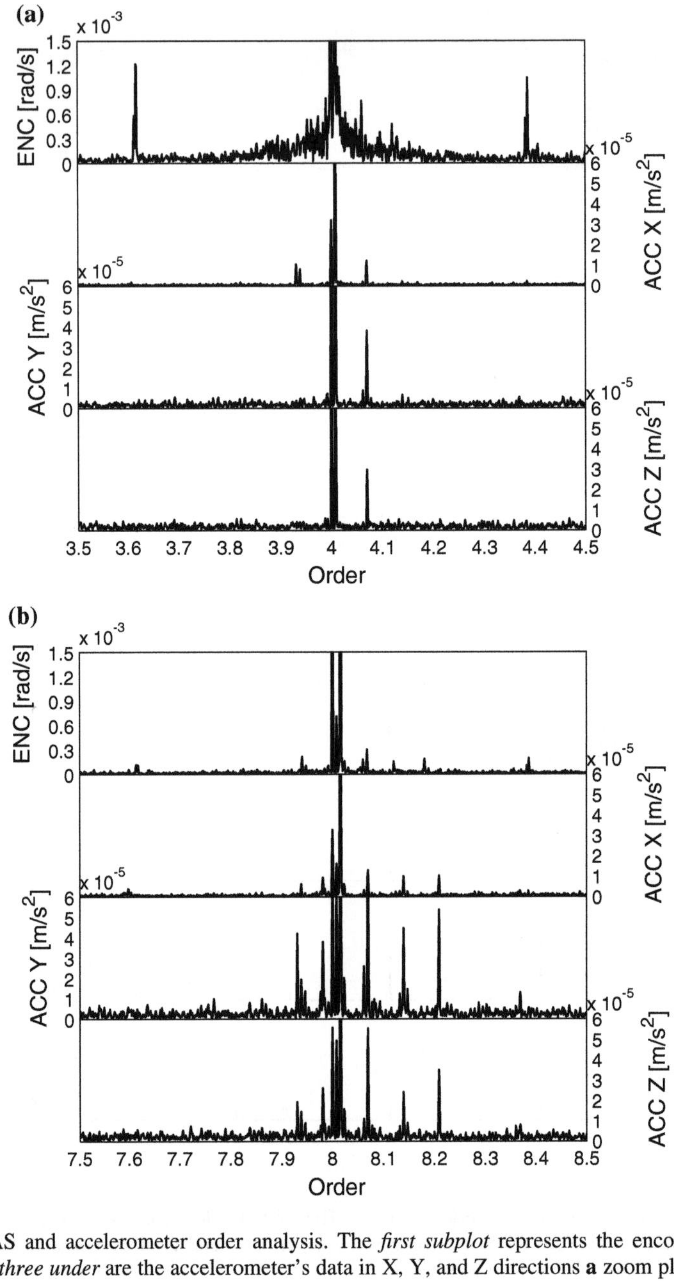

Fig. 5 IAS and accelerometer order analysis. The *first subplot* represents the encoder's data, while the *three under* are the accelerometer's data in X, Y, and Z directions **a** zoom plot, 3.5–4.5 order, **b** zoom plot 7.5–8.5 order

$$f_s = \frac{2f_e}{p} = \frac{2}{4} 50\,\text{Hz} = 25\,\text{Hz} \tag{5}$$

$$s = \frac{f_s - f_m}{f_s} = \frac{25 - 24.95}{25} = 0.002 \tag{6}$$

It is present an order 0.008 due to the slip and pole pass. In this case there are four poles, so *pps* is obtained in (7).

$$pps = p \cdot s = 4 \cdot 0.002 = 0.008 \tag{7}$$

Due to (4), the phase imbalance is present at $2o_e$, obtaining order 4.008, (9).

$$o_e = \frac{1}{1-s}\frac{p}{2} = \frac{1}{1-0.002}\frac{4}{2} = 2.004 \tag{8}$$

$$2o_e = 2 \cdot 2.004 = 4.008 \tag{9}$$

Figure 5 zooms around the fourth order where the 4.008 order appears in all signals. All the peaks are surrounded by sidebands at 0.008. These are more evident at higher orders, especially in accelerometer data. IAS also shows sidebands at FTF order 0.382. The same sidebands are present in accelerometer data, but at higher orders and the peaks are smeared. Other sidebands are present at order 0.070. These are related to the number of rotor bars, 35.

3 Conclusions

This research shows the capability of IAS to identify FTF bearing defect. The electrical behaviour is the principal element of IAS order analysis, so the method can be used for diagnosis for electrical and mechanical defects. Further analysis has to be done to improve the signal processing technique but the comparison between IAS and acceleration measurements showed interesting differences.

Acknowledgments This research is supported by Università degli Studi di Trieste, Nidec ASI S.p.A. and "Programma Operativo del Fondo Sociale Europeo 2007/2013 della Regione Autonoma Friuli Venezia Giulia".

References

1. Li Y, Gu F, Harris G, Ball AD, Bennett N, Travis K (2005) The measurement of instantaneous angular speed. Mech Syst Signal Process 19(4):786–805
2. André H, Bourdon A, Antoni J, Remond D (2012) Precision of the IAS monitoring system based on the elapsed time method in the spectral domain. In: Proceedings of the international conference on noise and vibration engineering ISMA 2012, Belgium, pp 1126–1142

3. Renaudin L, Bonnardot F, Musy O, Doray JB, Remond D (2010) Natural roller bearing fault detection by angular measurement of true instantaneous angular speed. Mech Syst Signal Process 24(7):1998–2011
4. Youssef W, Guillet F, Elbadaoui M (2011) Characterization of counter technique for instantaneous angular speed measurement: Application on gear box signal. In: Proceedings of the ICS6, Université Technologique de Compiègne, France
5. National Instruments (2013) Frequency measurements: how-to guide. Tutorial, 7111
6. Spagnol M, Bregant L, Boscarol A (2013) Instantaneous angular speed: comparisons between torsional laser vibrometer and encoder-counter estimations. In: Proceedings of COMADEM 2013
7. Resor BR, Trethewey MMW, Maynard. KPK (2005) Compensation for encoder geometry and shaft speed variation in time interval torsional vibration measurement. J Sound Vib 286(4–5):897–920

Non-linear Geometric Approach to Friction Estimation and Compensation

Marcello Bonfè, Paolo Castaldi, Nicola Preda and Silvio Simani

Abstract This contribution describes the application of differential geometry and nonlinear systems analysis to the estimation of friction effects in a class of mechanical systems. The proposed methodology, that has been developed for the more general problem of fault detection and diagnosis, relies on adaptive filters designed with a nonlinear geometric approach to obtain the disturbance de-coupling property. The classical model of an inverted pendulum on a cart is considered as an application example, in order to highlight the complete design procedure, including the mathematical aspects of the disturbance de-coupling method as well as the feasibility and the efficiency of the approach. Thanks to accurate estimation, friction effects can also be compensated by means of a controller designed to inject the on-line estimate of friction force to the control action calculated by classical linear state feedback. This strategy, which belongs to the class of so-called Active Fault-Tolerant Control Schemes, allows to maintain existing controllers and enhance their performance by introducing an adaptive estimator of unmodeled friction forces.

Keywords Fault-tolerant control · Friction compensation · Nonlinear dynamic systems · Adaptive filters

---

M. Bonfè (✉) · N. Preda · S. Simani
Department of Engineering, University of Ferrara, Via Saragat 1, 44122 Ferrara, Italy
e-mail: marcello.bonfe@unife.it

N. Preda
e-mail: nicola.preda@unife.it

S. Simani
e-mail: silvio.simani@unife.it

P. Castaldi
University of Bologna, Via Fontanelle 40, 47100 Forlì, Italy
e-mail: paolo.castaldi@unibo.it

G. Dalpiaz et al. (eds.), *Advances in Condition Monitoring of Machinery in Non-Stationary Operations*, Lecture Notes in Mechanical Engineering,
DOI: 10.1007/978-3-642-39348-8_30, © Springer-Verlag Berlin Heidelberg 2014

1 Introduction

Feedback control systems for mechatronics engineering applications are most of the times affected by challenges involved by friction in mechanical components. Despite the fact that detailed modeling of friction effects has been the subject of many research activities [7], robust on-line estimation of such effects is still an open problem, especially because the parameters characterizing the friction phenomena are commonly varying in time, as the effect of components ageing and insufficient maintenance. Therefore, there is a growing demand for reliable and adaptive algorithms for on-line estimation of friction forces in mechatronic systems.

Friction estimation and compensation has been addressed in literature also as a fault-tolerant control problem [8], namely the problem of designing a controller that has the ability to accommodate component faults automatically. A fault-tolerant control system can include a Fault Detection and Diagnosis (FDD) module, which is mainly used to fulfill the requirement of fault estimation. Over the last three decades, the growing demand for safety, reliability, maintainability, and survivability in technical systems has drawn significant research in FDD. Such efforts have led to the development of many FDD techniques, see for example the survey works [5, 9].

In general, fault tolerant control methods are classified into two types, i.e. Passive Fault Tolerant Control Scheme (PFTCS) and Active Fault Tolerant Control Scheme (AFTCS) [3, 6]. In PFTCS, controllers are fixed and are designed to be robust against a class of presumed faults, without explicit use of FDD schemes or controller reconfiguration. In contrast to PFTCS, AFTCS react to the system component failures actively by reconfiguring control actions so that the stability and acceptable performance of the entire system can be maintained, by means of FDD schemes providing on–line information about the true status of the system.

This chapter is focused on the application of an AFTCS to address the problem of friction compensation. The proposed AFCTS integrates a reliable and robust friction estimation module, implemented according to an FDD procedure relying on adaptive filters. The controller reconfiguration exploits a second control loop depending on the on-line estimate of the friction force. The advantages of this strategy are that a structure of logic-based switching controller is not required and, instead, an existing controller can be preserved and enhanced.

The FDD method is based on the NonLinear Geometric Approach (NLGA) developed by De Persis and Isidori [2]. By means of this framework, a disturbance de-coupled adaptive nonlinear filter providing the fault identification is developed. It is worth observing that the original NLGA FDD scheme based on residual signals cannot provide, in general, fault size estimation.

Both the NLGA Adaptive Filters (NLGA–AF) and the AFTCS strategy are applied to the well-known model of an inverted pendulum on a cart (also called cart-pole system), an underactuated mechanical structure that is commonly used as a benchmark system for control design and mechatronics prototyping. A simulation model for the complete AFTCS loop has been implemented in the Matlab® and Simulink® environments, and tested in the presence of nonlinear friction, disturbances, measurement noise and modelling errors.

The work is organised as follows. Section 2 provides the description of the cart-pole nonlinear benchmark system. Section 3 describes the implementation of the FDD scheme and the structure of the AFTCS strategy. Stability and robustness of the AFTCS method with are investigated in simulations reported in Sect. 4. Finally, Sect. 5 summarises contributions and achievements of the chapter, providing some suggestions for possible further research topics.

2 The Cart-Pole Nonlinear Model

The dynamic model of a pendulum (or pole) on a cart shown in Fig. 1 is a classical benchmark in Systems and Control Theory.

The interest in this mechanical system is motivated by the similarity between its dynamic properties and those of several real-world engineering applications like, for example, aerospace vehicles during vertical take-off, cranes, and many others.

Assuming that the cart has mass M and the pendulum mass m is concentrated at the tip of a pole, with neglectable inertia, of length L, the dynamic model obtained using Hamilton's principle is the following:

$$\begin{cases} (M+m)\ddot{x}_p + mL\ddot{\theta}\cos\theta - mL\dot{\theta}^2\sin\theta = F_a - F_{fric} \\ m\ddot{x}_p\cos\theta + mL\ddot{\theta} - mg\sin\theta = \tau \end{cases} \tag{1}$$

in which g is the gravity constant, F_a is the controllable actuator force, F_{fric} is the friction involved by the contact between cart and ground, and t is a torque acting directly at the base of the pole. Enumerating the state variables as $x = [x_1\, x_2\, x_3\, x_4]^T = \left[x_p\ x_p\ \theta\ \dot{\theta}\right]^T$ and considering $u = F_a$, as the control input, and $d = \tau$, as a disturbance, the model can be rewritten in the following state-space input affine form:

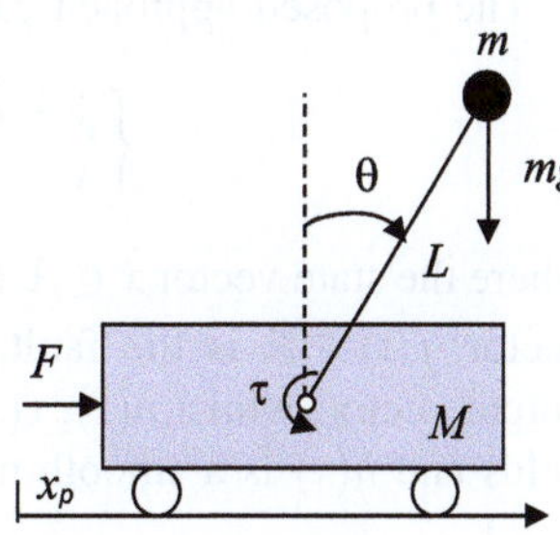

Fig. 1 Scheme of the inverted pendulum on a cart

$$
\begin{cases}
\dot{x}_1 = x_2 \\
\dot{x}_2 = \dfrac{mLx_4^2 \sin x_3 - mg \sin x_3 \cos x_3}{M + m \sin^2 x_3} + \dfrac{u - F_{fric}}{M + m \sin^2 x_3} - \dfrac{d}{(M + m \sin^2 x_3)L} \\
\dot{x}_3 = x_4 \\
\dot{x}_4 = \dfrac{(M+m)g \sin x_3 - mLx_4^2 \sin x_3 \cos x_3}{(M + m \sin^2 x_3)L} - \dfrac{(u - F_{fric}) \cos x_3}{(M + m \sin^2 x_3)L} \\
\qquad + \dfrac{(M+m)d}{m(M + m \sin^2 x_3)L^2}
\end{cases}
\tag{2}
$$

The dynamic model of the cart-pole system fulfils the structural requirements described in [2], which considers the following class of nonlinear systems:

$$
\dot{x} = n(x) + g(x)u + p(x)d \tag{3}
$$

where $n(x)$, $g(x)$ and $p(x)$ are smooth vector fields. In the following section, the proposed solutions to FDD and AFTC problems, based on the nonlinear geometric approach, will be developed.

It is worth noting that other types of disturbance terms, different from the τ signal, can be also considered in Eq. 2 (e.g. measurement noise on the linear and angular velocity signals). However, a disturbance described in terms of a torque acting at the base on the pole represents a realistic situation, since it may be related to the effect of an impact between the pole and some kind of obstacle.

3 FDD and AFTCS Design

The presented FDD scheme belongs to the NLGA framework, that allows to de-couple disturbances by means of a nonlinear coordinate transformation. Such transformation is then the starting point to design a set of adaptive filters that are able to both detect additive fault acting on a single actuator and estimate the magnitude of the fault. It is worth observing that in this chapter, we can consider the effect of friction as a fault affecting the actuator, so that thanks to the NLGA approach the friction estimate is de-coupled from disturbance d.

The proposed approach can be applied to nonlinear systems in the form:

$$
\begin{cases}
\dot{x} = n(x) + g(x)u + \ell(x)f + p_d(x)d \\
y = h(x)
\end{cases}
\tag{4}
$$

where the state vector $x \in \mathcal{X}$ (an open subset of $\mathbb{R}^{\ell_n}$), $u(t) \in \mathbb{R}^{\ell_c}$ is the control input vector, $f(t) \in \mathbb{R}$ is the fault, $d(t) \in \mathbb{R}^{\ell_d}$ the disturbance vector and $y \in \mathbb{R}^{\ell_m}$ the output vector, whilst $n(x)$, $\ell(x)$, the columns of $g(x)$, and $p_d(x)$ are smooth vector fields and $h(x)$ is a smooth map.

The model of Eq. 2 can be related to the form of Eq. 4 setting:

$$n(x) = \begin{bmatrix} x_2 \\ \frac{mLx_4^2 \sin x_3 - mg \sin x_3 \cos x_3}{M+m\sin^2 x_3} \\ x_4 \\ \frac{(M+m)g \sin x_3 - mLx_4^2 \sin x_3 cosx_3}{(M+m\sin^2 x_3)L} \end{bmatrix} \quad g(x) \equiv \ell(x) = \begin{bmatrix} 0 \\ \frac{1}{M+m\sin^2 x_3} \\ 0 \\ -\frac{\cos x_3}{(M+\sin^2 x_3)L} \end{bmatrix} \tag{5}$$

and

$$p_d(x) = \begin{bmatrix} 0 \\ -\frac{\cos x_3}{(M+m\sin^2 x_3)L} \\ 0 \\ \frac{M+m}{m(M+\sin^2 x_3)L^2} \end{bmatrix} \tag{6}$$

Assuming in addition that the full state vector is measurable (i.e. $h(x) = I_4 x$), the design of the strategy for the diagnosis of the fault f with disturbance decoupling is organised as follows:

- computation of Σ_*^P, i.e. the minimal conditioned invariant distribution containing P (where P is the distribution spanned by the columns of $p_d(x)$);
- computation of Ω^*, i.e. the maximal observability codistribution contained in $(\Sigma_*^P)^\perp$;
- if $\ell(x) \not\subset (\Omega^*)^\perp$, fault detectability condition, the fault is detectable and a suitable change of coordinate can be determined.

The computation Σ_*^P can be obtained by means of the following recursive algorithm:

$$\begin{cases} S_0 = \bar{P} \\ S_{k+1} = \bar{S} + \sum_{i=0}^{m} [g_i, \bar{S}_k \cap \ker\{dh\}] \end{cases} \tag{7}$$

where m is the number of inputs, $\bar{S}$ represents the involutive closure of S, $[g, \Delta]$ is the distribution spanned by all vector fields $[g, \tau]$, with $\tau \in \Delta$, and $[g, \tau]$ the Lie bracket of g, τ. It can be shown that if there exists a $k \geq 0$ such that $S_{k+1} = S_k$, the algorithm 7 stops and $\Sigma_*^P = S_k$ [2].

Once Σ_*^P has been determined, Ω^* can be obtained with the following algorithm:

$$\begin{cases} Q_0 = (\Sigma_*^P)^\perp \cap \text{span}\{dh\} \\ Q_{k+1} = (\Sigma_*^P)^\perp \cap \sum_{i=0}^{m} [L_{g_i} Q_k + \text{span}\{dh\}] \end{cases} \tag{8}$$

where $L_g \Gamma$ denotes the co-distribution spanned by all covector fields $L_g \omega$, with $\omega \in \Gamma$, and $L_g \omega$ the derivative of ω along g. If there exists an integer k^* such that $Q_{k^*} = Q_{k^*+1}$, Q_{k^*} is indicated as o.c.a$\left((\Sigma_*^P)^\perp\right)$, where the acronym o.c.a. stands for *observability co-distribution algorithm*.

It can be shown that $Q_{k^*} = \text{o.c.a.}\left((\Sigma_*^P)^\perp\right)$ represents the maximal observability co-distribution contained in $P^\perp$, i.e. Ω^* [2]. Therefore, with reference to the model 4, when $\ell(x) \notin (\Omega^*)^\perp$, the disturbance d can be de-coupled and the fault f is detectable.

As mentioned above, the considered NLGA to the fault diagnosis problem, described in [2], is based on a coordinate change in the state space and in the output space, $\Phi(x)$ and $\Psi(y)$, respectively. They consist in a surjection Ψ_1 and a function Φ_1 such that $\Omega^* \cap \text{span}\{dh\} = \text{span}\{d(\Psi_1 \circ h)\}$ and $\Omega^* = \text{span}\{d\Phi_1\}$, where:

$$\begin{cases} \Phi(x) = \begin{pmatrix} \bar{x}_1 \\ \bar{x}_2 \\ \bar{x}_3 \end{pmatrix} = \begin{pmatrix} \Phi_1(x) \\ H_2 h(x) \\ \Phi_3(x) \end{pmatrix} \\ \Psi(y) = \begin{pmatrix} \bar{y}_1 \\ \bar{y}_2 \end{pmatrix} = \begin{pmatrix} \Psi_1(y) \\ H_2 y \end{pmatrix} \end{cases} \tag{9}$$

are (local) diffeomorphisms, whilst H_2 is a selection matrix, i.e. its rows are a subset of the rows of the identity matrix. By using the new (local) state and output coordinates $(\bar{x}, \bar{y})$, the system 4 is transformed so that it exhibits an observable subsystem that is affected by the fault and not affected by the disturbance, as described in [2] :

$$\begin{cases} \dot{\bar{x}}_1 = n_1(\bar{x}_1, \bar{x}_2) + g_1(\bar{x}_1, \bar{x}_2)c + \ell_1(\bar{x}_1, \bar{x}_2, \bar{x}_3)f \\ \bar{y}_1 = h(\bar{x}_1) \end{cases} \tag{10}$$

In the case of the cart-pole system, the following result is obtained:

$$S_0 = \bar{P} = \text{cl}(p_d(x)) = \text{cl}\left(\begin{bmatrix} 0 \\ -\frac{\cos x_3}{(M+m\sin^2 x_3)L} \\ 0 \\ \frac{M+m}{m(M+\sin^2 x_3)L^2} \end{bmatrix}\right) \equiv p_d(x) \tag{11}$$

By recalling that $\ker\{dh\} = \emptyset$, it follows that $\Sigma_*^P = \bar{P}$ as $\bar{S}_0 \cap \ker\{dh\} = \emptyset$. Thus, the algorithm 7 stops with $S_1 = S_0 = \Sigma_*^P$.

Proceeding with algorithm 8, it can be observed that:

$$(\bar{P})^\perp = \left(\begin{bmatrix} 0 \\ -\frac{\cos x_3}{(M+m\sin^2 x_3)L} \\ 0 \\ \frac{M+m}{m(M+m\sin^2 x_3)L^2} \end{bmatrix}\right)^\perp = \begin{bmatrix} 1 & 0 & 0 & 0 \\ 0 & 0 & 1 & 0 \\ 0 & 1 & -\frac{mLx_4 \sin x_3}{M+m} & \frac{mL\cos x_3}{M+m} \end{bmatrix} \tag{12}$$

and $\text{span}\{dh\} = I_4$. From algorithm 8 it follows that $\Omega^* = \left(\Sigma_*^P\right)^\perp = (\bar{P})^\perp$, and $(\Omega^*)^\perp = \Sigma_*^P = \bar{P}$.. Therefore the fault f (i.e. the friction force) is detectable, since $\ell(x) \notin (\Omega^*)^\perp = \Sigma_*^P = \bar{P}$.

As $\dim\{\Omega^*\} = 3$, and $\dim\{\Omega^* \cap \text{span}\{dh\}\} = 3$, it follows that $\Phi_1(y) : \mathbb{R}^4 \to \mathbb{R}^3$. Moreover, as $\Omega^* \cap \text{span}\{dh\} = \text{span}\{d(\Psi_1 \circ h)\}$, $H_2 y : \mathbb{R}^4 \to \mathbb{R}^1$. Thus, as $h(x) = I_4 x$, the surjection $\Psi(y(x))$ is given by:

$$\Psi(y(x)) = \begin{pmatrix} \Psi_1(x) \\ H_2 x \end{pmatrix} = \left(\begin{bmatrix} x_2 + \frac{mLx_4 \cos x_3}{M+m} \\ x_1 \\ x_3 \end{bmatrix} \atop \begin{bmatrix} x_4 \end{bmatrix} \right) \tag{13}$$

where $H_2 = \begin{bmatrix} 0 & 0 & 0 & 1 \end{bmatrix}$, the diffeomorphism $\Phi_1(x)$ is given by $\Phi_1(x) = \Psi_1(x)$ and the state variable of the observable subsystem decoupled from disturbance is $\bar{x}_1 = \Phi_1(x)$.

Given this result, an NLGA–AF can be designed introducing the following new constraints [1] :

- the observable subsystem is independent from the $\bar{x}_3$ state components;
- there exists a proper scalar component $\bar{x}_{1s}$ of the state vector $\bar{x}_1$ such that the corresponding scalar component of the output vector is $\bar{y}_{1s} = \bar{x}_{1s}$ and the following relation holds:

$$\dot{\bar{y}}_{1s}(t) = M_1(t) \cdot f + M_2(t) \tag{14}$$

where $M_1(t) \neq 0, \forall t \geq 0$. Moreover $M_1(t)$ and $M_2(t)$ can be computed for each time instant, since they are functions just of input and output measurements. The proposed adaptive filter is based on the least-squares algorithm with forgetting factor [4], and it is described by the following adaptation law:

$$\begin{cases} \dot{P} = \beta P - \frac{1}{N^2} P^2 \breve{M}_1^2, & P(0) = P_0 > 0 \\ \dot{\hat{f}} = P \varepsilon \breve{M}_1, & \hat{f}(0) = 0 \end{cases} \tag{15}$$

with the following equations representing the output estimation, and the corresponding normalised estimation error:

$$\begin{cases} \hat{\breve{y}}_{1s} = \breve{M}_1 \hat{f} + \breve{M}_2 + \lambda \breve{\bar{y}}_{1s} \\ \varepsilon = \frac{1}{N^2} \left(\breve{\bar{y}}_{1s} - \hat{\breve{y}}_{1s} \right) \end{cases} \tag{16}$$

where all the involved variables of the adaptive filter are scalar. In particular, $\lambda > 0$ is a parameter related to the bandwidth of the filter, $\beta \geq 0$ is the forgetting factor and $N^2 = 1 + \breve{M}_1^2$ is the normalisation factor of the least-squares algorithm. Finally, the signals $\breve{M}_1$, $\breve{M}_2$, $\breve{\bar{y}}_{1s}$ are obtained by means of a low-pass filtering of the signals M_1, M_2, $\bar{y}_{1s}$.

In order to de-couple the effect of the disturbance d from the fault (i.e. friction) estimator, it is necessary to select from the observable subsystem the following state:

$$\bar{x}_{1s} = \bar{x}_{11} = x_2 + \frac{mLx_4 \cos x_3}{M+m} \tag{17}$$

whose differentiation allows to compute the full expression of M_1 and M_2.

4 Simulation Results

To compute with Matlab/Simulink the simulation results described in this section, the AFTCS design has been completed by means of an optimal state feedback control law, on the basis of the linear approximation of a frictionless version of Eq. 2, in a neighbourhood of $x_o = [x_{1d}\ 0\ 0\ 0]^T$, in which x_{1d} can be any value. In fact, the linear approximation is independent from x_1, so that the input vector of optimal controller can be calculated as $\tilde{x} = [(x_1 - x_{1d})\ x_2\ x_3\ x_4]$ and the cart-pole system will be stabilised in the upright position at any linear position reference.

The following values of the system parameters have been assumed: $M = 1$ kg; $m = 0.1$ kg; $L = 0.3$ m; $g = 9.81$ m/s^2. The optimal controller has been designed using the LQR approach in order to minimize the well-known quadratic cost function with $Q = 10I_4$ and $R = 1$.

The simulation of the mechanical system has been completed by a nonlinear model of friction affecting the linear motion of the cart, including viscous friction, with coefficient $F_x = 0.6$ N/m/s, and Stribeck friction with Coulomb part coefficient $F_c = 0.25$ N, static part coefficient $F_s = 0.4$ N and Stribeck velocity $v_s = 0.2$ m/s, using the same mathematical description of [8]. The shape of the nonlinear friction model is shown in Fig. 2.

The NLGA-AF used as a friction estimator has been designed assuming the nominal model of the cart-pole, but the simulated mechanical system included a mismatch of 10% in the values of M and m, a random disturbance torque d (which

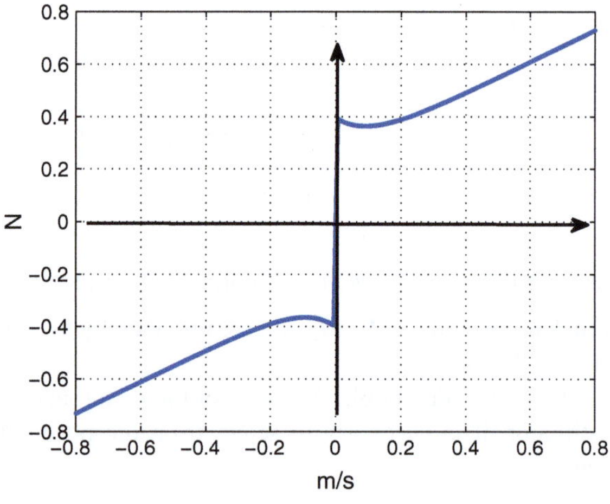

Fig. 2 Nonlinear model of Stribeck and viscous friction effects affecting motion of the cart

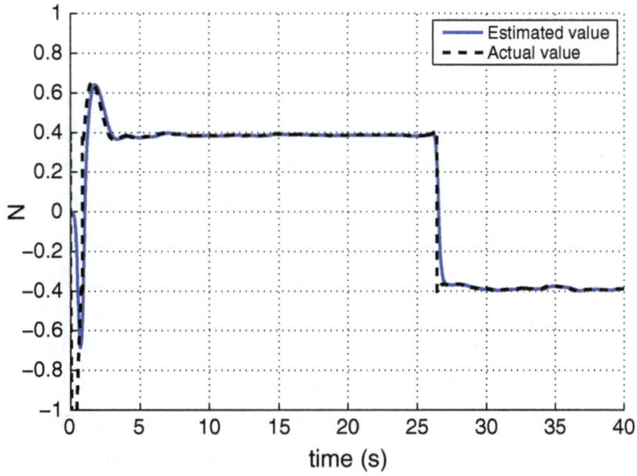

Fig. 3 Friction estimation obtained by means of the proposed adaptive filter

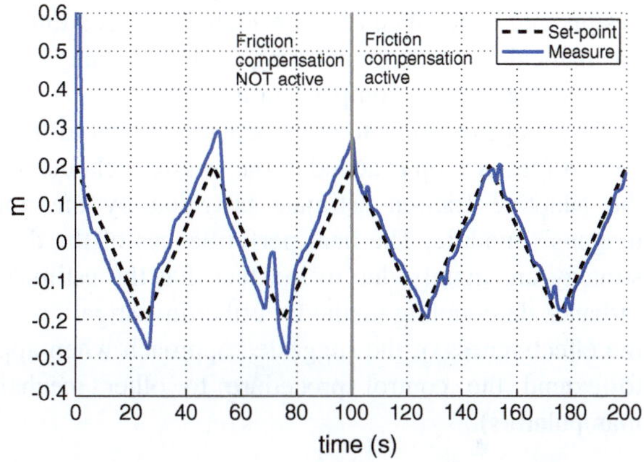

Fig. 4 Tracking of a sawtooth linear position reference, with (*right*) and without (*left*) friction compensation

is nevertheless de-coupled by design) and measurement noise on the state feedback signals. As can be seen from Fig. 3, the proposed NLGA-based estimator provides an accurate and robust measure of the actual friction force, even if the mathematical model of friction effects is not explicitly included in the design of the estimator. Such effective performance is achieved by means of a proper tuning of the adaptation mechanism of the filter, in particular $\beta = 12$ and $\lambda = 8.8$ have been fixed for the simulated case.

Thanks to this accurate reconstruction, friction can also be compensated by simply adding its estimated value to the output of the optimal controller. As a result, the tracking of a time-varying linear cart position reference is dramatically improved, as shown in Fig. 4. The figure shows that in the first half of the plot the optimal controller by itself is not robust with respect to the nonlinear friction disturbing action, while in the second part of the simulation, friction compensation on the basis of the NLGA–AF estimation is introduced and tracking error is reduced from a mean square value of 0.0128 m to a mean square value of 0.0007115 m.

5 Conclusion

This chapter described the development of an active fault tolerant control scheme for the purpose of friction compensation in mechanical systems, which integrates a robust fault diagnosis method providing accurate estimation of friction effects. The methodology relies on disturbance de-coupled adaptive filters designed via the nonlinear geometric approach. The fault tolerant strategy has been applied to a classical control design benchmark, namely the inverted pendulum on a cart, which was simulated in presence of nonlinear friction, disturbing torque acting on the pole pivot, measurement noise and modelling errors.

It is worth observing that the suggested active fault tolerant control was already developed in works by the same authors, but applied to aerospace examples. Thus, the contribution of this chapter consists of the application of the active fault tolerant control scheme to the well-known benchmark, in order to highlight the computational and mathematical aspects of the nonlinear disturbance de-coupling design, and hence it can be considered also as a tutorial for researchers working in mechanical systems monitoring and fault diagnosis, as well as fault tolerant control and friction compensation.

The proposed fault tolerant scheme allows to maintain the existing controller, since a further loop is added to the original scheme, thus providing the feedback of the adaptive friction estimate provided by the nonlinear geometric approach diagnosis module. The final performances of the developed fault tolerant control strategy are mainly due to the fact that the estimate is unbiased, thanks to disturbance de-coupling method. Further investigations will be carried out to evaluate the effectiveness of the suggested approach when applied to real case studies and to extend the control procedure to other mechanical systems (e.g. robotic manipulators).

References

1. Castaldi P, Geri W, Bonfè M, Simani S, Benini M (2010) Design of residual generators and adaptive filters for the FDI of aircraft model sensors. Control Eng Pract 18(5):449–459
2. De Persis C, Isidori A (2001) A geometric approach to non-linear fault detection and isolation. IEEE Trans Autom Control 45(6):853–865
3. Edwards C, Lombaerts T, Smaili H (eds) (2010) Fault tolerant flight control: a benchmark challenge. Lecture notes in control and information sciences. Springer, Berlin
4. Ioannou P, Sun J (1996) Robust adaptive control. PTR Prentice–Hall, Upper Saddle River, NJ, USA
5. Isermann R (2005) Fault–diagnosis systems: an introduction from fault detection to fault tolerance, 1st edn. Springer, Berlin. ISBN: 3540241124
6. Noura H, Theilliol D, Ponsart JC, Chamseddine A (2009) Fault-tolerant control systems: design and practical applications. Advances in Industrial Control, 1st edn. Springer, London
7. Olsson H, Astrom K, de Wit C, Gafvert M, Lischinsky P (1998) Friction models and friction compensation. Eur J Control 4(3):176–195

8. Patton R, Putra D, Klinkhieo S (2008) Friction compensation as a fault-tolerant control problem. In: European workshop on advanced control and diagnosis. Plenary Presentation
9. Simani S, Fantuzzi C, Patton RJ (2003) Model-based fault diagnosis in dynamic systems using identification techniques. Advances in industrial control, 1st edn. Springer, London. ISBN: 1852336854

Empirical Mode Decomposition of Acoustic Emission for Early Detection of Bearing Defects

Mourad Kedadouche, Marc Thomas and Antoine Tahan

Abstract Empirical Mode Decomposition (EMD) is one of the techniques that proved its efficiency for an early detection of defects in many mechanical applications like bearings and gears. The EMD methodology decomposes the original times series data into intrinsic mode functions (IMF), by using the Hilbert-Huang transform. In this study, EMD is applied to acoustic emission signals. The acoustic emission signal is heterodynined around a central high frequency in order to obtain an audible signal. Scalar statistical parameters such as Kurtosis and THIKAT are then used in this study. These statistical descriptors are calculated for each IMF. The technique is validated by experiments on a test bench with a very small crack (40 μm) on the outer race of a ball bearing. It is found that the application of time descriptors to different IMF decomposition levels gives good results for early detection of defects in comparison with the original time signal.

Keywords Acoustic emission · Fault diagnosis · Bearing · Empirical mode decomposition · Intrinsic mode function · Scalar time descriptors

M. Kedadouche (✉) · M. Thomas · A. Tahan
Department of Mechanical Engineering, École de Technologie Supérieure 1100, Notre-Dame Street West, Montreal, QC H3C 1K3, Canada
e-mail: mourad.kedadouche.1@ens.etsmtl.ca

M. Thomas
e-mail: marc.thomas@etsmtl.ca

A. Tahan
e-mail: antoine.tahan@etsmtl.ca

G. Dalpiaz et al. (eds.), *Advances in Condition Monitoring of Machinery in Non-Stationary Operations*, Lecture Notes in Mechanical Engineering,
DOI: 10.1007/978-3-642-39348-8_31,

1 Introduction

Bearing defect diagnostic has been a topic of research interest for over 30 years [1]. The most widely used technique for monitoring and diagnosing bearings is vibration analysis which is usually conducted in the low and medium frequency range ($\leq$20 kHz) [2], and which has conducted to many developments in improved methods for signal processing, as well as improvements in algorithms for anomaly detection, classification and failure prediction [3]. Most researches on the diagnosis can be classified in time, frequency or time–frequency domain. In industry, scalar descriptors extracted from time signals are very often used to detect a defect. The most popular descriptors are the *RMS, Peak, Crest Factor, Skewness, K factor* and *Kurtosis* [1, 2]. Since these descriptors are not always effective, especially when the defect increases, new descriptors, such as *TALAF* and *THIKAT* have been developed [4]. In the frequency domain, the envelope analysis, also known as amplitude demodulation has been widely used and proved to be very effective in detecting bearing characteristic frequencies [2]. Recently, a new method, designated as Empirical Mode Decomposition (EMD), has been proposed by Huang et al. [5]. By using EMD, any complex signal can be decomposed into a collection of Intrinsic Mode Functions (IMFs) based on the local characteristic time scale of the signal. The IMFs represent the natural oscillatory modes embedded into the signal. EMD is a self-adaptive method because the IMFs are determined by the signal itself rather than by pre-defined criteria. Therefore, EMD is highly efficient in non-stationary data analysis [5, 6]. Several publications on the use of EMD for fault diagnosis have appeared in literature [7]. Some applications of EMD in mechanical fault diagnosis have been studied, for example, in structural health monitoring [8] and gear fault diagnosis [9, 10]. The EMD may also be used for monitoring bearing defects by vibration. Several improved EMD methods have been proposed aiming to enhance the performance of EMD in bearing fault diagnosis [11, 12]. Finally, many researchers have applied EMD combining with other techniques to bearing fault diagnosis in recent years and achieved better diagnosis results compared with the use of EMD alone [13, 14]. However, all the studies based on vibration measurements were deficient for very early detection of defects, most of them being studying defects greater than 100 microns. In fact, when a defect arises, it creates vibration at very high frequencies and thus acoustic emission may be a measurement method more efficient than vibration for an early detection [15]. Some research studies have been published on the detection and diagnosis of bearing defects by acoustic emission [16–18], but also based their studies on relatively large defects (more than 100 microns). In this study, we proposed to use the EMD method from acoustic emission measurements. The scalar statistical parameters such as Kurtosis and THIKAT [4] are used in this study. These statistical descriptors are calculated for each IMF [14], and the results are compared with the same descriptors extracted from the original signal. All the experiments are conducted on a test bench with a very small defect (40 microns) artificially created on the outer race of a bearing.

2 Empirical Mode Decomposition

The Empirical Mode Decomposition (EMD) method decomposes the time signal into a finite set of oscillatory functions called Intrinsic Mode Functions (IMF). An IMF is a function that meets the following conditions

1. The number of extrema and the number of zero crossings must either equal or differ at most by one;
2. The value of the moving average envelope defined by local maxima and the envelope defined by local minima is zero.

An intrinsic mode is the embedded time scale in the signal. It is defined as the time between two successive extrema (Fig. 1).

It is not necessarily a harmonic function. In fact, it may include non-stationary amplitudes and modulated frequencies. The decomposition method in IMF may be summarized as follows [19]:

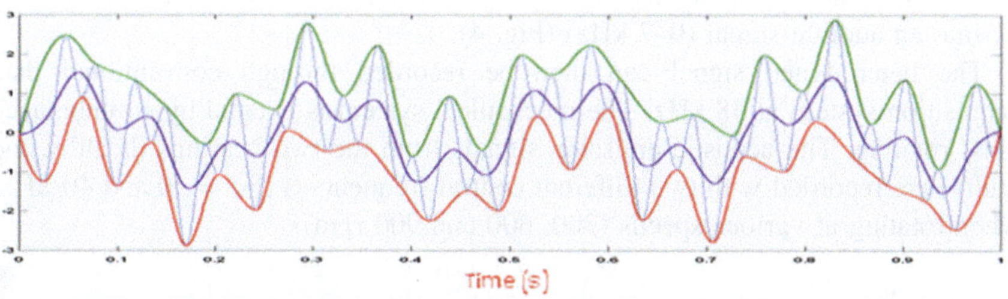

Fig. 1 Principle of EMD method

---

1. Initialize: $r_0 = x(t)$, and $i = 1$
2. Extract the ith IMF c_i
a. Initialize: $h_{i(k-1)} = r_{(i-1)}$, $k = 1$
b. Extract the local maxima and minima of $h_{i(k-1)}$
c. Interpolate the local maxima and the minima by cubic spline lines to form upper and lower envelops of $h_{i(k-1)}$
d. Calculate the mean $m_{i(k-1)}$ of the upper and lower envelops of $h_{i(k-1)}$
e. Let $h_{ik} = h_{i(k-1)} - m_{i(k-1)}$
f. If h_{ik} is an IMF then set $c_{ik} = h_{ik}$, else go to step(b) with $k = k + 1$
3. Define the remainder $r_{i+1} = r_i - c_i$
4. If r_{i+1} still has least 2 extrema then go to step (2) with $i = i + 1$, else, the decomposition process is finished and r_{i+1} is the residue of the signal

---

3 Experimental Investigation

3.1 Test Bench

The test bench used in this study is shown in Fig. 2a. A shaft is supported by two bearings and connected to a motor with a flanged coupling bolted rubber. Two system were investigated, one with a healthy bearing and the other with an artificially damaged bearing. The defect has been created as a long transversal strip and a small width of 40 microns through the ball passing way, as shown in Fig. 2b.

The equipment for acoustic emission data collection is shown in Fig. 3. It consists of an ultrasound detector (UE Systems UltraProb 10000). The sensor is mechanically connected to the bearing housing with an adaptor. The sensor is connected to an analogue digital converter (THOR Analyzer PRO: DT9837-13310) with a sampling frequency of 48 kHz.

The acoustic emission sensor used in this study operates in the lower ultrasonic spectrum from 20 to 100 kHz. A heterodyne circuit converts the high frequency acoustic emission signal as detected by the transducer around a central frequency Fc into an audible signal (0–7 kHz) (Fig. 4).

The heterodyned signal can then be recorded through conventional data acquisition system at 48 kHz. The mechanical system is excited by an unbalance mass rotating. The acoustic emission signals from the two bearings (healthy and faulty) are recorded with two different central frequencies (Fc = 30 and 40 kHz) when rotating at various speeds (300, 600 and 900 rpm).

Fig. 2 **a** Experimental test bed. **b** The artificial defect of outer ring

Fig. 3 Data acquisition system

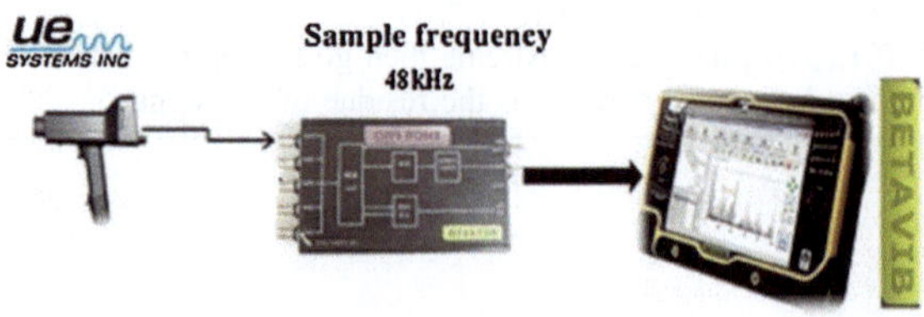

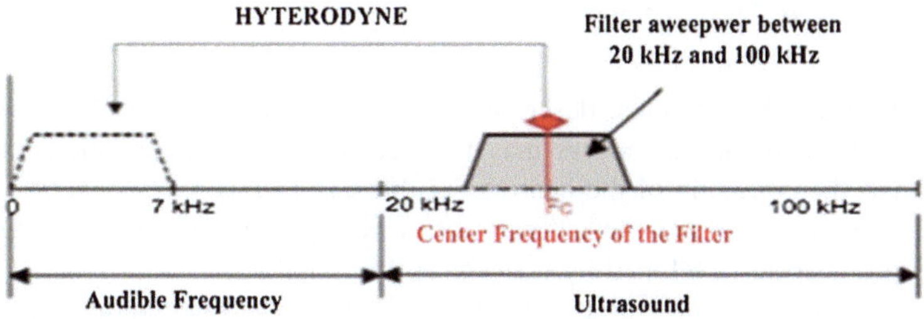

Fig. 4 Heterodyne principle

3.2 Application of EMD to Acoustic Emission Signals

Figure 5 show the EMD decomposition of the acoustic emission signal acquired at the central band of Fc = 30 kHz for the healthy bearing and for the defective one. The right curves show spectra of the damaged bearing at each IMF and from the original signal while the left curves show the spectra of the healthy one. We may notice that some IMF of the defective bearing present an increase in energy over a wide frequency band compared to the healthy bearing. These side bands are thus due to the defect, since at an early stage, the defect occurs around the resonances of the bearing at high frequencies. EMD decomposition showed this change.

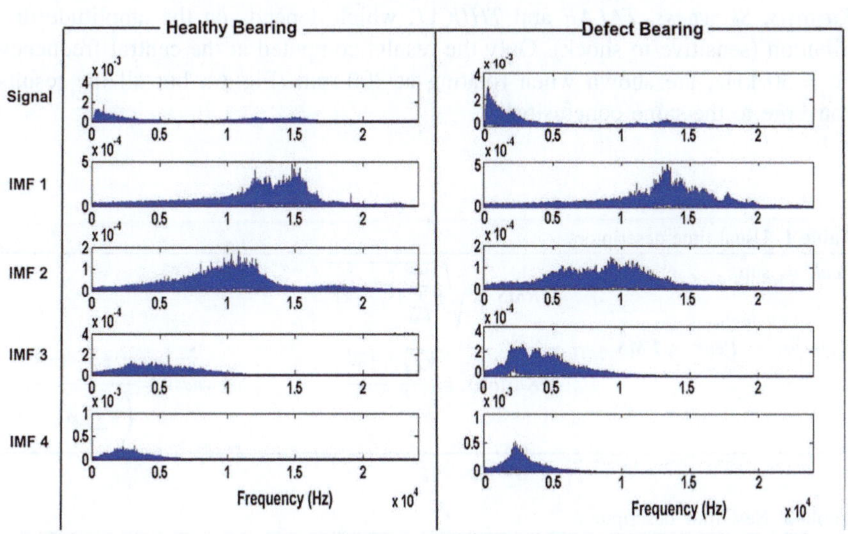

Fig. 5 EMD applied to the acoustic signal: comparison between the healthy bearing and the defective bearing (40 μm) for Fc = 30 kHz

3.3 Statistical Descriptors

In time domain, the statistical descriptors: *Peak* (Max), *RMS*, *Crest Factor* (*CF*), *K-Factor, Kurtosis* and *Skewness* may be used to compare their efficiencies for monitoring the condition of the bearings (Table 1) [20].

We have also used two new descriptors which proved their effectiveness for monitoring bearing developed by Sassi et al. [4] (Table 2).

In order to monitor these descriptors in time, we have used the ratio between the indicator calculated for the defect bearing and the same indicator for the healthy bearing.

$$Evolution\ of\ descriptor = \left[\frac{Indicator\ (defect\ bearing)}{Indicator\ (healthy\ bearing)}\right] \tag{1}$$

3.4 Results Analysis and Applications

3.4.1 Sensibility of Scalar Descriptors to Detect the Defect

In this part, the *Evolution of descriptor* as described by Eq. (1) has been calculated for each IMF (blue color), for the envelope of each IMF (red color) and also for the original signal (black color). The previous indicators may be divided into two families; the first group (group 1) is the *RMS*, *Crest factor*, *Peak* and *K factor*, which depends on energy content of the signal, while the second group (group 2) is *Kurtosis, Skewness, TALAF* and *THIKAT*, which depends on the amplitude distribution (sensitive to shock). Only the results computed at the central frequency Fc = 30 kHz, are shown when rotating at 300 rpm (Fig. 6), but all the results converge to the same conclusions.

Table 1 Usual time descriptors

$Peak = \sup_{1 \le i \le N} \lvert x_i \rvert$	$RMS = \sqrt{\frac{1}{N}\sum_{i=1}^{N}(x_i - \bar{x})^2}$	$CF = \frac{Peak}{RMS}$
$K\_factor = Peak \times RMS$	$Kurtosis = \frac{\frac{1}{N}\sum_{i=1}^{N}(x_i - \bar{x})^4}{RMS^4}$	$Skewness = \frac{\frac{1}{N}\sum_{i=1}^{N}(x_i - \bar{x})^3}{\left(\frac{1}{N}\sum_{i=1}^{N}(x_i - \bar{x})^2\right)^{3/2}}$

Table 2 New time descriptors

$TALAF = \log\left[kurtosis + \frac{RMS}{RMS_0}\right]$	$THIKAT = \log\left[(kurtosis)^{CF} + \left(\frac{RMS}{RMS_0}\right)^{Peak}\right]$

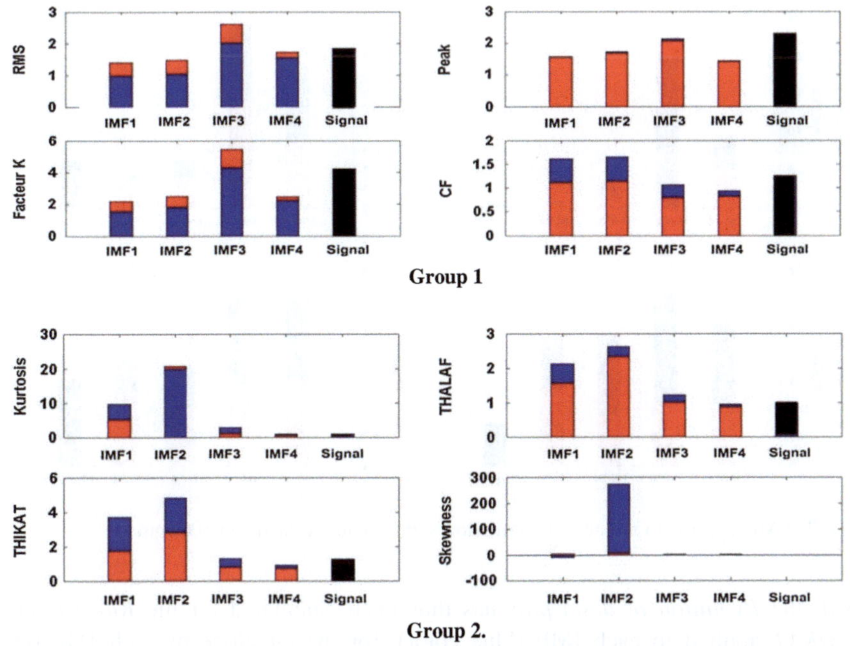

Fig. 6 EMD applied to the acoustic emission signal for Fc = 30 kHz (300 rpm)

- The results show that most of the indicators calculated on the IMF or on its envelope have a better sensitivity than indicators calculated on the raw signal, except the *Peak* that revealed to be less sensitive to the defect.
- The IMF was generally more efficient than the envelope IMF for most indicators but it is not true for all indicators.
- All the indicators were efficient to detect the default.
- The indicators of group 1 are more sensitive in the frequency content of IMF 3 (close to 4,000 Hz) while those of group 2 were more sensitive in the frequency content of IMF 1 (highest frequencies: close to 14,000 Hz) and IMF 2 (highest frequencies: close to 10,000 Hz).
- The indicators of IMF of group 2 are more effective than those of group 1 since they detect the damage at the first IMF 1.

3.4.2 Effect of Speed and Central Frequency on Descriptors to Detect the Defect

As the *Kurtosis* and *THIKAT* were found in the previous part more sensitive to discriminate the effect of the default at specific IMF than from the original signal, the following part describes the results obtained with these two descriptors. In this

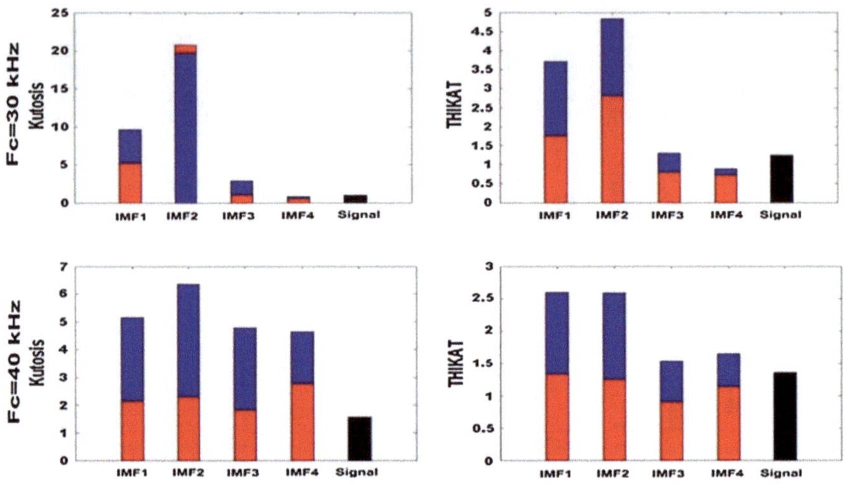

Fig. 7 EMD applied to the acoustic emission signal acquired at speed 300 rpm

part, the *Evolution of descriptor* has thus been calculated for the *Kurtosis* and *THIKAT* applied to each IMF (blue color), for the envelope of each IMF (red color) and to the original signal (black color) at two central frequencies (Fc = 30, 40 kHz) when rotating at three different speeds (300, 600 and 900 rpm). These results are shown in Figs. 7, 8 and 9 for the three rotating speeds.

For all speeds and central frequencies, both indicators revealed to be more sensitive than indicators calculated on the original signal (black). For all speeds, the results show that both descriptors computed at the 30 kHz central frequency are more sensitive at the IMF 2 (10,000 Hz), while those computed at the 40 kHz

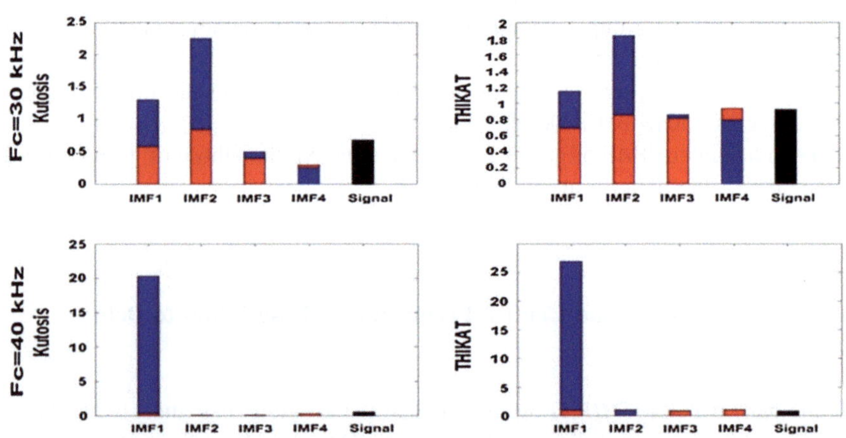

Fig. 8 EMD applied to the acoustic emission signal acquired at speed 600 rpm

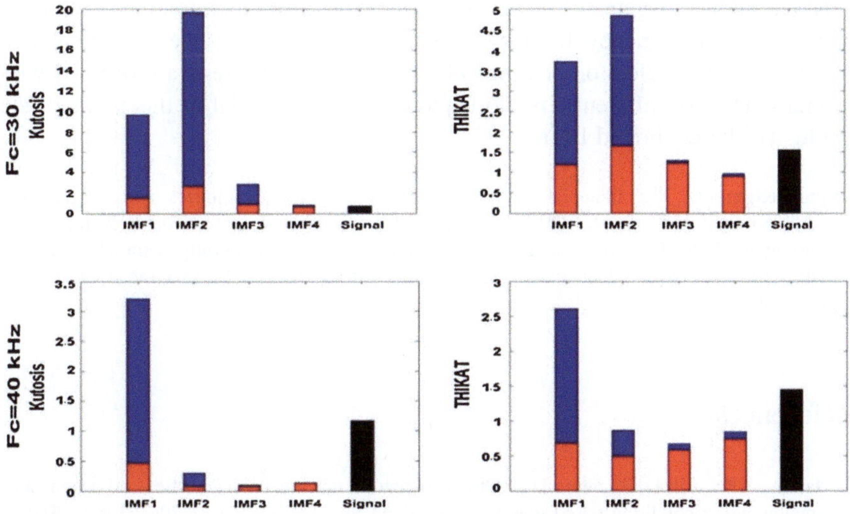

Fig. 9 EMD applied to the acoustic emission signal acquired at speed 900 rpm

central frequency are more sensitive to the defect at the IMF 1 (highest frequency: 14,000 Hz). The IMF 1 and IMF 2 revealed to reflect the frequency contents that are more excited by the defect. In fact, a defect when detected at its early stage generates shocks that excite first high frequency components due to resonances. As the EMD technique acts as frequency filters from the highest to the lowest frequency, the two first IMF are more sensitive to the defect while the original signal consider all the frequency spectra with noise. Since *Kurtosis* and *THIKAT* are more sensitive to shocks (Group 2), these two indicators revealed to be sensitive to the defect. There is no evidence that *THIKAT* is a better indicator than *Kurtosis* and vice versa.

4 Conclusions

The study is aimed to detect very small bearing defects (40 microns) at their very early stage of degradation. Since small defects produce vibratory signals in the high frequency domain at their early stage of degradation, acoustic emission has been used instead the vibration measurements. In this chapter, the EMD technique combined with statistical indicators has been applied to acoustic emission signals in order to improve the fault detection of ball bearing. This study showed the efficiency of the Empirical Mode Decomposition method for the early detection of mechanical defects. Using the EMD, the descriptors presents more evolution than the same descriptors computed on the original signal. The results show that the indicators *THIKAT* and *Kurtosis* descriptors applied to the IMF were the most sensitive at the first and 2 levels of IMF decomposition accordingly with the selected central frequency. It must be noticed that this study is the first one able to detect a so small defect on bearings (40 microns). Further tests are planned with several defects of different sizes to explore the full potential of this method; the results will be confirmed later.

Acknowledgments The financial support of NSERC (Natural Sciences and Engineering Research Council of Canada), FQRNT (Fonds Québecois de la Recherche sur la Nature et les Technologies), MITACS Canada and Pratt & Whitney Canada are gratefully acknowledged. The technical supports of R. Ginet and S. Koch from Pratt & Whitney Canada Inc. were greatly appreciated.

References

1. Jardine AKS, Lin D, Banjevic D (2006) A review on machinery diagnostics and prognostics implementing condition-based maintenance. Mech Syst Signal Process 20(7):1483–1510
2. Thomas M (2011) Reliability, predictive maintenance and machinery vibration (in french). *Presses de l'Université du Québec*, 633, D3357, ISBN 978-2-7605-3357-8
3. Batista L, Badri B, Sabourin R, Thomas M (2012) Detecting bearing defects under high noise levels: a classifier fusion approach. In: Proceedings of IECON, 38th annual conference on IEEE industrial electronics society, Montréal, pp 3880–3886
4. Sassi S, Badri B, Thomas M (2008) Tracking surface degradation of ball bearings by means of new time domain scalar descriptors. Int J COMADEM 11(3):36–45
5. Huang NE, Shen Z, Long SR (1998) The empirical mode decomposition and hilbert spectrum for nonlinear and non-stationary time series analysis. Proc. R. Soc. London, Ser A 454:903–995
6. Terrien J, Marque C, Karlsson B (2011) Automatic detection of mode mixing in empirical mode decomposition using non-stationarity detection: application to selecting IMFs of interest and denoising. EURASIP J Adv Signal Process 2011:1–8
7. Lei Y, Lin J, He Z, Zuo MJ (2013) A review on empirical mode decomposition in fault diagnosis of rotating machinery. Mech Syst Signal Process 35(1–2):108–126
8. Li H, Deng X, Dai H (2007) Structural damage detection using the combination method of EMD and wavelet analysis. Mech Syst Signal Process 21:298–306
9. Kidar T, Thomas M, Guilbault R, Badaoui MEl (2012) Comparison between the sensitivity of LMD and EMD algorithms for early detection of gear defects. In: Proceedings of the 3rd conference on experimental vibration (AVE), Blois (FR), p 8
10. Dorostghol A, Dorfeshan M (2012) Intelligent fault diagnosis via EMD method. J Appl Sci 12:1960–1965
11. Du QH, Yang SN (2007) Application of the EMD method in the vibration analysis of ball bearings. Mech Syst Signal Process 21:2634–2644
12. Rai VK, Mohanty AR (2007) Bearing fault diagnosis using FFT of intrinsic mode functions in Hilbert–Huang transform. Mech Syst Signal Pro 21:2607–2615
13. Li H, Zhang YP, Zheng HQ (2010) Bearing fault detection and diagnosis based on order tracking and Teager–Huang transform. J Mech Sci Technol 24:811–822
14. Yang Y, Yu DJ, Cheng JS (2007) A fault diagnosis approach for roller bearing based on IMF envelope spectrum and SVM. Measurement 40:943–950
15. Shiroishi J et al (1997) Bearing condition diagnosis via vibration and acoustic emission measurements. Mech Sys Signal Process 11(5):693–705

16. Choudhury A, Tandon N (2000) Application of acoustic emission technique for the detection of defects in rolling element bearings. Tribology Int 33(1):39–45
17. Dadouche A et al (2008) Sensitivity of air-coupled ultrasound and eddy current sensors to bearing fault detection. Tribol Trans 51(3):310–323
18. Kilundu B et al (2011) Cyclostationarity of acoustic emissions (AE) for monitoring bearing defects. Mech Syst Signal Process 25(6):2061–2072
19. Lei YG, Lin J, He ZJ, Zuo MJ (2012) A review on empirical mode decomposition in fault diagnosis of rotating machinery. Mech Syst Signal Process http://dx.doi.org/10.1016/j.ymssp.2012.09.015
20. Kedadouche M, Thomas M, Tahan A (2012) Monitoring bearings by acoustic emission: a comparative study with vibration techniques for early detection. In: Proceedings of the 30th seminar on machinery vibration, CMVA, Niagara Falls, Canada, p 17

Signal Processing Diagnostic Tool for Rolling Element Bearings Using EMD and MED

Steven Chatterton, Roberto Ricci, Paolo Pennacchi and Pietro Borghesani

Abstract The signal processing techniques developed for the diagnostics of mechanical components operating in stationary conditions are often not applicable or are affected by a loss of effectiveness when applied to signals measured in transient conditions. In this chapter, an original signal processing tool is developed exploiting some data-adaptive techniques such as Empirical Mode Decomposition, Minimum Entropy Deconvolution and the analytical approach of the Hilbert transform. The tool has been developed to detect localized faults on bearings of traction systems of high speed trains and it is more effective to detect a fault in non-stationary conditions than signal processing tools based on envelope analysis or spectral kurtosis, which represent until now the landmark for bearings diagnostics.

Keywords Bearing diagnostics · Empirical mode decomposition · EMD · Minimum entropy deconvolution · MED

1 Introduction

In the case of small size mechanical components, like rolling element bearings, signal processing of vibrating data is the most used approach to perform diagnostics procedures. Several signal processing techniques have been proposed during time in addition to the more traditional ones like the Fourier Transform. Time synchronous average was introduced to analyze data acquired on gearboxes whose functioning is typically periodic [1]. Envelope analysis [2], second order cyclostationary analysis [3, 4] and spectral kurtosis [5] are widely accepted and used for bearing monitoring and diagnostics. Others *smart* signal processing

S. Chatterton · R. Ricci · P. Pennacchi (✉) · P. Borghesani
Department of Mechanical Engineering, Politecnico di Milano,
Via G. La Masa 1, 20156 Milan, Italy
e-mail: paolo.pennacchi@polimi.it

G. Dalpiaz et al. (eds.), *Advances in Condition Monitoring of Machinery in Non-Stationary Operations*, Lecture Notes in Mechanical Engineering,
DOI: 10.1007/978-3-642-39348-8_32, © Springer-Verlag Berlin Heidelberg 2014

techniques proposed to highlight the non-stationary behaviour of a phenomenon are, for example, wavelet transform [6], which performs a scale-frequency analysis, and cepstrum analysis [7, 8], suitable to detect echoes in the signal.

The effectiveness of all the mentioned techniques for stationary signals has been proven in various papers by means of laboratory tests. Unfortunately, some difficulties arise when the considered signals are related to non-stationary phenomena or the systems work in unsteady operating conditions. Moreover, a loss of effectiveness can be observed in the case of complex systems, in which the measured signals can be a mixture of different sources, and in real applications in which the signals are affected by high level of noise. For these reasons, adaptive techniques became very attractive in the last years. These techniques, also called *data driven* techniques, are able to overcome the hypothesis of periodicity and stationarity of the signals. Empirical Mode Decomposition (EMD) has been probably the first technique belonging to this category. Proposed for the first time by Huang et al. in [9], EMD is used either as an independent tool [10] or as a preliminary method for the Hilbert-Huang Spectrum calculation with good results to signals measured on mechanical systems [11]. The Minimum Entropy Deconvolution (MED) [12] is another example of adaptive signal processing technique. Contrary to the EMD, which is based on the decomposition of the experimental signal in different functions corresponding to different signal *modes*, the MED algorithm enhances the peakedness of the signal by highlighting the transient components by means of a filter. The MED is particularly effective to separe subsequent bursts caused by shocks generated by localized defects in experimental signals [13].

Notwithstanding the EMD and MED signal analysis techniques have been proven in some laboratory applications, the diagnostics effectiveness of both of them must be still tested for non stationary real applications.

In this chapter, Empirical Envelope MED (EEMED) tool is proposed by the authors. The method has been developed and tuned for the detection of localized faults on bearings installed in traction equipments of high speed trains. The developed tool allows detecting the fault in transient conditions with better results than those given by envelope analysis or spectral kurtosis, which nowadays represent the landmark for bearings diagnostics.

2 EEMED Algorithm

As previously introduced, the proposed tool is a combination of the EMD and the MED algorithm. The former allows the extraction of the principal intrinsic mode function of the signal whereas the latter the enhancement of the signal peakedness. In the following, both of the two algorithms will be briefly described.

2.1 Empirical Mode Decomposition Algorithm

For a generic signal $x(t)$, the EMD is based on the research of its stationary points. By indicating with $s_{max}(t)$ and $s_{min}(t)$ respectively the spline functions interpolating the maxima and minima points of the signal $x(t)$, their mean value function $m(t)$ is removed from the original signal:

$$x_1(t) = x(t) - m(t) \tag{1}$$

The signal $x_1(t)$ is an intrinsic mode function (IMF) if the two particular conditions are satisfied, described in detail in [9]. If the resulting signal $x_1(t)$ is not an IMF, $s_{max}(t)$, $s_{min}(t)$ and $m(t)$ must be recomputed starting from $x_1(t)$ with an iterative procedure called *sifting* process. The sifting process stops when the first intrinsic mode function $C_1(t)$ is extracted. Therefore, $C_1(t)$ can be subtracted from the original signal:

$$r_1(t) = x(t) - C_1(t) \tag{2}$$

The residual signal $r_1(t)$ represents the input for the calculation of the second IMF by means of the sifting process. The EMD algorithm for the original signal $x(t)$ stops when the residual signal $r_N(t)$ is a constant or monotonic function, after the extraction of the N-th IMF. This stop condition of the decomposition algorithm can be expressed in terms of standard deviation threshold and number of extremes.

The EMD is a deterministic decomposition; therefore, the original data can be re-obtained by adding the extracted IMFs to the last residual signal.

2.2 Analytic Signal

The IMFs extracted by means of the EMD process can be considered a sort of sub-signals of the original one since they enclose only some components of it. Therefore, the IMFs represent favorable starting points for further analysis. With reference to roller bearing diagnostics, it is widely accepted that the *analytic signal* is particularly sensitive to the slippage occurring among the rolling elements and the rings of the bearings as effects of the presence of localized faults. The analytic signal $z(t)$ is defined as:

$$z(t) = x(t) + \mathrm{i}\,y(t) \tag{3}$$

where $x(t)$ is the experimental signal and $y(t)$ its Hilbert transform defined as:

$$y(t) = \frac{1}{\pi} \int_{-\infty}^{+\infty} \frac{x(\tau)}{t - \tau} d\tau \tag{4}$$

The analytic signal $z(t)$ could be expressed as:

$$z(t) = a(t)\mathrm{e}^{\mathrm{i}\theta(t)} \tag{5}$$

where $a(t)$ and $\theta(t)$ represent the envelope and the instantaneous phase of the analytic signal respectively. Usually, the signal analysis techniques using the Hilbert transform are devoted to the calculation of the instantaneous frequency (e.g. the Hilbert-Huang spectrum). The attention is here focused on the treatment of the analytic signal envelope $a(t)$ which can be obtained by root squaring the sum of the squares of the real and the imaginary components. The envelope of the analytic signal considers simultaneously the variation in the time domain of the original signal and the variations of the two components of $z(t)$ along the real and imaginary axis. In the proposed signal processing tool, the analytic signal is calculated on the IMFs. Since the bearings functioning is characterized by high frequency phenomena, the first IMF extracted by the EMD will be considered.

2.3 Minimum Entropy Deconvolution Algorithm

The aim of MED is to separate the components of a signal. The MED is based on the concept of signal entropy, namely the randomness degree of the signal. The MED algorithm reduces the randomness of a signal by minimizing its entropy. A general signal $v(t)$ could be considered as composed of two terms:

$$v(t) = h(t) * w(t) + \eta(t) \tag{6}$$

The first term is the convolution between the system behavior $h(t)$ and its excitation $w(t)$. The second term $\eta(t)$ considers the signal noise randomly distributed. If a filter $f(t)$ is applied to the signal $v(t)$, it is possible to obtain:

$$u(t) = v(t) * f(t) = h(t) * w(t) * f(t) + \eta(t) * f(t) \tag{7}$$

Expressing the previous equation as a function of sampled points, it follows:

$$u_n = v_n * f_n = \sum_{i=1}^{M} f_i \, v_{n+1-i} \tag{8}$$

where $n = 1, \ldots, N$ and $N = T + M - 1$.

The convolution between the signal, with a length of T, and the filter with length M, can be expressed as a sum of products. The crucial point of the approach is the selection of a proper filter length M.

Considering the output of Eq. (8), the final signal is a simple solution, characterized by the maximum possible order. A measure of the signal order is the Varimax norm V:

$$V(u) = \sum_{j=1}^{N} u_j^4 \Big/ \left(\sum_{j=1}^{N} u_j^2\right)^2 \tag{9}$$

By exploiting the Varimax norm definition, the signal entropy can be minimized by means of an iterative procedure as function of the filter $f(t)$. The norm maximization allows finding the output u_n which best fits the components of $h(t)$.

The MED algorithm is applied to the envelope of the first IMF analytic signal, obtaining the EEMED. In this is way, the MED is applied to simplified (due to EMD decomposition) and enhanced signals (due to the Hilbert envelope of the IMF), thus improving the effectiveness of MED itself.

3 Description of the Test-Rig

The EEMED algorithm described in the previous section has been applied to the vibration signals measured on the test-rig shown in Fig. 1a. The test rig is equipped with a complete traction system of a high speed train (HST): a 265 kW 4-poles asynchronous HST motor is connected to the input shaft of the HST gearbox by means of a toothed coupling. The braking torque is provided by a braking system composed of a braking motor and an industrial gearbox connected to the HST traction system through a double cardan shaft.

Different operating conditions, as train speeds and motor supplied torques, were tested. Stationary conditions, i.e. constant speed and torque due to cruise trip of the train travelling on a flat track, and non-stationary conditions with train speed and motor torque variations, i.e. train approaching or leaving the railway station, were considered. In the paper, only the signals measured during the tests performed in non-stationary conditions will be considered. Vibration signals were measured by means of several accelerometers placed on the traction motor and the gearbox and acquired with a sampling frequency of 20 kHz. The operating conditions are shown in Fig. 2a and only the vibration data acquired during an interval of 10 s of the test and indicated with [a] in Fig. 2b will be considered.

4 Experimental Results

Two damaged configurations will be compared with respect to a reference configuration with all brand new bearings. In each configuration all the bearings are new except the artificially damaged one:

- Configuration (1): localized damage on the inner ring of the four-contact point ball bearing (Fig. 3a) placed in the non-driven end (NDE) side of the high speed shaft of the gearbox (P5 in Fig. 2b);
- Configuration (2): localized damage on the outer ring of the roller bearing (Fig. 3b) placed in the driven-end (DE) side of the traction motor (P2 in Fig. 2b).

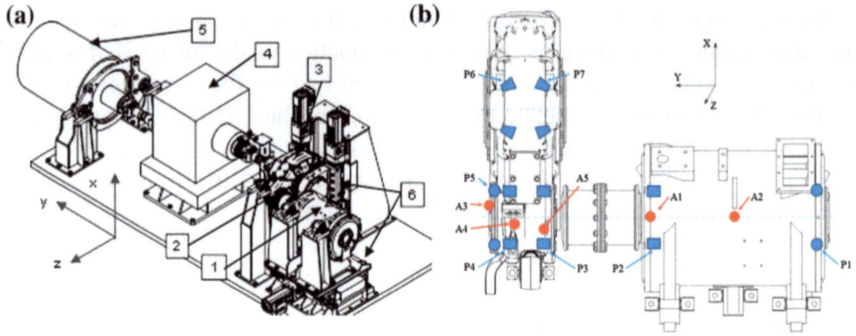

Fig. 1 **a** Test-rig layout: (*1*) HST motor, (*2*) HST toothed coupling, (*3*) HST gearbox, (*4*) Additional gearbox, (*5*) Braking motor; **b** Traction system arrangement with bearing position (from P1 to P7) and accelerometer position (from A1 to A5)

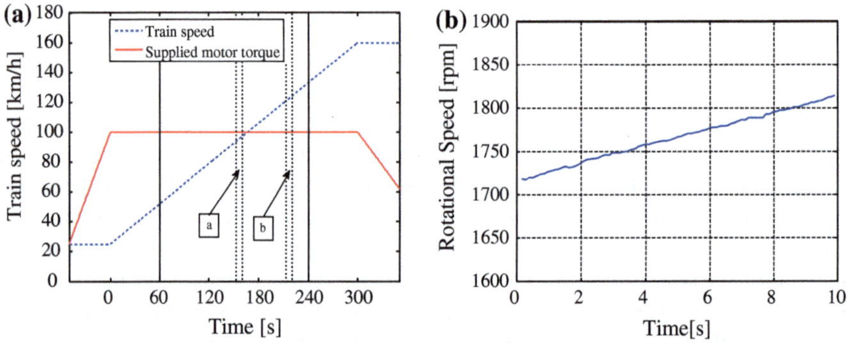

Fig. 2 **a** Operating conditions (train speed and supplied motor torque) for the considered test; **b** Rotational speed of the traction motor measured during the interval [a]

4.1 Configuration (1): Damaged Bearing on Gearbox NDE Side

The configuration (1) concerns a ball bearing installed on the high speed shaft of the gearbox (i.e. the shaft connected to the traction motor one) characterized by a localized fault on the inner ring. The artificial damage consists in a spall.

The vibration signals are measured by the accelerometer A3 (Fig. 2b) placed on the gearbox close to the considered bearing position and measuring the vibration along the *y*-direction that is the direction of the motor shaft axis.

By applying a widely accepted technique in bearing diagnostics as the envelope analysis to the measured signals, the results shown in Fig. 4 can be obtained. In Fig. 4a, for the healthy bearing, no significant frequency component is detectable

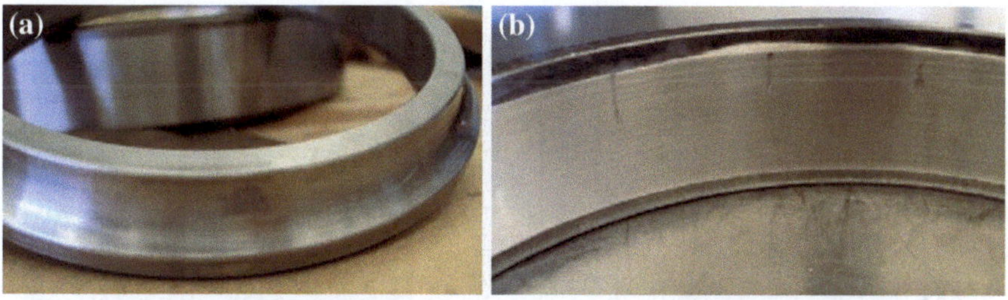

Fig. 3 **a** Bearing damage for the configuration (1); **b** Bearing damage for the configuration (2)

in the plot apart that corresponding to the average rotational speed of the motor (29.9 Hz). Frequency components with high amplitude can be detected on the contrary in Fig. 4b. All these components are harmonics of the motor rotational speed and again, no indication about the presence of the fault on the bearing inner ring can be obtained. It is worth noting that the same analysis performed in stationary operating conditions offers better results since the characteristic frequencies of the bearing fault can be traced in the envelope domain. This confirm that the envelope analysis, is ineffective when applied to data acquired in transient conditions and this remarks the necessity to develop a tool able to provide good results in this kind of conditions.

The final results provided by the EEMED algorithm in time domain are then treated with Fourier Transform as shown in Fig. 5, where the * symbol in the vertical labels indicates that the amplitude is modified by the MED filter.

In Fig. 5a the frequencies components with higher energy are those related to high order harmonics of the average supply frequency of the motor, whereas for the damaged bearing of configuration (1) (Fig. 5b), the highest frequency component is at the frequency of the passage of ball elements on the inner ring (with reference to the actual average rotational speed, the nominal BPFI is equal to 273 Hz).

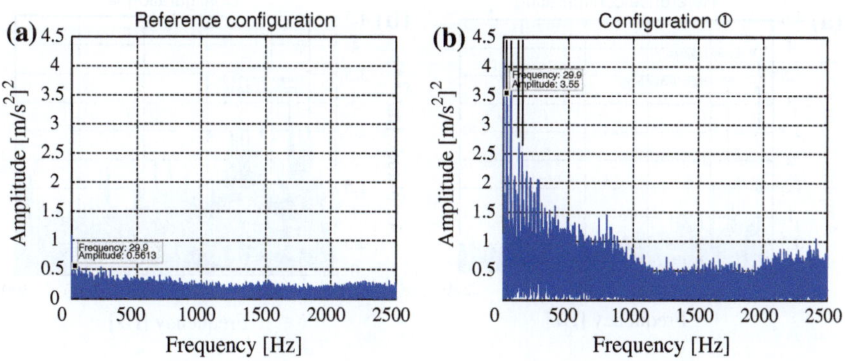

Fig. 4 Envelope analysis of the signal for the reference configuration and the configuration (1)

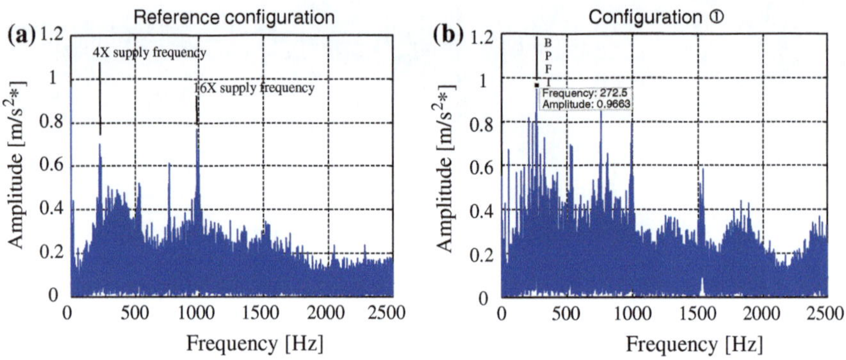

Fig. 5 Fourier transform of the EEMED signals for the reference configuration and the configuration (1)

4.2 Configuration (2): Damaged Bearing on Motor DE Side

In this case, a defect on the outer ring of the roller bearing of the traction motor is considered. The damages are the indentations shown in Fig. 3b. The vibration signals are acquired by the accelerometer A1 (in Fig. 2b) placed very close to the bearing and measuring the vibration along the z-direction.

For the sake of brevity, only the results provided by the EEMED algorithm are shown in Fig. 6. While for the reference configuration (Fig. 6a) the frequency components with high amplitude are those related to the high order harmonics of the average supply frequency of the motor (2X component at 120.1 Hz and 6X component at 359.7 Hz), for the configuration (2) with the damaged bearing (Fig. 6b), the frequency component with the maximum energy is that due to the passage of the rolling element on the bearing outer ring (BPFO). For the considered speed profile indeed, the nominal BPFO for that damaged bearing, evaluated with

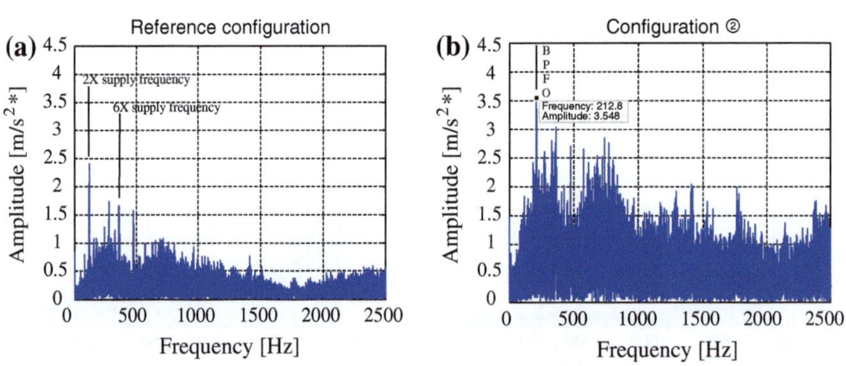

Fig. 6 Fourier transform of the EEMED signals for the reference configuration and the configuration (2)

the average value of the speed, is equal to 213 Hz. The identified frequency component located at 212.8 Hz in Fig. 6b is very close to the nominal BPFO. Therefore, also in this case, the EEMED tool is able to detect the bearing damage in non-stationary condition.

5 Conclusion

The application of the proposed EEMED tool for bearing diagnostics in non-stationary operating conditions has been discussed. The effectiveness of EEMED has been tested on vibration signals measured on a test-rig reproducing the functioning of high speed train traction equipments.

The application of EEMED to the vibration signals measured for different bearing damages and for non-stationary operating conditions has provided encouraging results. The comparison with the traditional and widely applied technique as the envelope analysis shows the suitability of the EEMED for the detection of bearing damages in non-stationary conditions.

References

1. McFadden PD (1987) Examination of a technique for the early detection of failure in gears by signal processing of the time domain average of the meshing vibration. Mech Syst Signal Process 1(2):173–183
2. Darlow MS, Badgley RH, Hogg GW (1974) Application of high frequency resonance techniques for bearings diagnostics in helicopter gearboxes. Technical Report, US Army Air Mobility Research and Development Laboratory, US, pp 74–77
3. Capdessus C, Sidahmed M, Lacoume JL (2000) Cyclostationary processes: application in gear faults early diagnosis. Mech Syst Signal Process 14(2000):371–385
4. Borghesani P, Pennacchi P, Ricci R, Chatterton S (2013) Testing second order cyclostationarity in the squared envelope spectrum of non-white vibration signals. Mech Syst Sig Proces http://dx.doi.org/10.1016/j.ymssp.2013.05.012
5. Antoni J, Randall RB (2009) The spectral kurtosis: application to the vibratory surveillance and diagnostics of rotating machines. Mech Syst Signal Process 23(2009):987–1036
6. Ohue Y, Yoshida A, Seki M (2004) Application of the wavelet transform to health monitoring and evaluation of dynamic characteristics in gear sets. Proc IMechE J Eng Tribology 218:1–11
7. Badaoui ME, Antoni J, Guillet F, Danière J, Velex P (2001) Use of the moving cepstrum integral to detect and localise tooth spalls in gears. Mech Syst Signal Process 15(5):873–885
8. Borghesani P, Pennacchi P, Randall RB, Sawalhi N, Ricci R (2013) Application of cepstrum pre-whitening for the diagnosis of bearing faults in variable speed conditions. Mech Syst Sig Proces 36(2):370–384
9. Huang NE, Shen Z, Long SR, Wu MC, Shih HH, Zheng Q, Yen NC, Tung CC, Liu HH (1998) The empirical mode decomposition and the Hilbert spectrum for nonlinear and non-stationary time series analysis. Proc Royal Soc London Ser A 454:903–995
10. Loutridis SJ (2004) Damage detection in gear systems using empirical model decomposition. Eng Struct 26:1833–1841

11. Ricci R, Pennacchi P (2011) Diagnostics of gear faults based on EMD and automatic selection of intrinsic mode functions. Mech Syst Signal Process 25:821–838
12. González G, Badra RE, Medina R, Regidor J (1995) Period estimation using minimum entropy deconvolution (MED). Signal Process 41:91–100
13. Pennacchi P, Ricci R, Chatterton S, Borghesani P (2011) Effectiveness of MED for fault diagnosis in roller bearings. Vibration Problems ICOVP, pp 637–342

Influence of Stopping Criterion for Sifting Process of Empirical Mode Decomposition (EMD) on Roller Bearing Fault Diagnosis

A. Tabrizi, L. Garibaldi, A. Fasana and S. Marchesiello

Abstract Empirical mode decomposition (EMD) is a self-adaptive data driven technique for analyzing nonlinear and non-stationary signals and decompose them into some elementary Intrinsic Mode Functions (IMFs). Although EMD method has been applied in various applications successfully, this method has some drawbacks, i.e. lack of a mathematical base, no robust stopping criterion for sifting process, mode mixing and border effect problem. Under the practical point of view, the most relevant is possibly the sifting stop criterion. Although sifting as many times as possible is needed to decompose the signal, too many sifting steps will reduce the physical meaning of IMFs. To preserve the natural amplitude variations of the oscillations, sifting must be limited to as few steps as possible. The proposed criteria so far are: Cauchy-type convergence, three-threshold, energy difference tracking, resolution factor, bandwidths, and orthogonality criterion. There is not a thorough study yet regarding the fault diagnosis application, to determine the effects of stopping criteria on the fault detection performance. In this chapter, the influence of different criteria to this purpose is investigated.

Keywords Empirical mode decomposition · Support vector machine · Roller bearing fault diagnosis · Sifting process

---

A. Tabrizi (✉) · L. Garibaldi · A. Fasana · S. Marchesiello
Dynamics and Identification Research Group, DIMEAS, Politecnico di Torino,
Corso Duca degli Abruzzi 24, 10129 Turin, Italy
e-mail: ali.tabrizi@polito.it

L. Garibaldi
e-mail: luigi.garibaldi@polito.it

G. Dalpiaz et al. (eds.), *Advances in Condition Monitoring of Machinery in Non-Stationary Operations*, Lecture Notes in Mechanical Engineering,
DOI: 10.1007/978-3-642-39348-8_33,

1 Introduction

Empirical mode decomposition (EMD), introduced by Huang et al. [1, 2] is a method for decomposing a multi-component signal into several elementary Intrinsic Mode Functions (IMFs). Some drawbacks of this method are lack of a mathematical base, stopping criterion for sifting process and border effect problem.

The sifting process actually serves two purposes: to eliminate riding waves and to make the wave profiles more symmetric with respect to zero. On the other hand, too many sifting steps will reduce the IMF to be a constant amplitude frequency-modulated function, which makes the results physically less meaningful. To preserve the natural amplitude variations of the oscillations, sifting must be limited to as few steps as are mathematically possible.

Different kind of criteria has been proposed so far. In the Cauchy-type convergence criterion, standard deviation of two consequent sifting results is used as the criterion (SD criterion) [1]. The main flaw of this approach named SD (because of using standard deviation) is that it is unrelated to the definition of IMFs. The mean fluctuations thresholds (MFT) utilize three thresholds to compare a specific defined fraction [3]. The shortcoming of MFT is that the thresholds do not adapt to the signal. The energy difference tracking (EDT) is based on the assumption that the residue and IMFs are orthogonal mutually [4]. The difference between total energy and energy of the original signal is tracked as the sifting stop criterion. The resolution factor (RF) is defined by the ratio between the energy of the signal at the beginning of the sifting and the energy of the envelopes means [5]. None of the before-mentioned criteria uses the frequency or phase information of the analyzed signal. Xuan and Xie [6] designed a new stop criterion-bandwidth criterion based on two types of bandwidth (BW): instantaneous bandwidth and frequency bandwidth which is caused only by frequency changes. Based on orthogonality definition, Lin and Hongbing defined orthogonality criterion (OC) [7].

In this study, we will investigate the influence of different criteria on fault diagnosis to obtain if using different kind of stopping criteria can affect the result of fault detections.

2 EMD Algorithm

EMD method decomposes a complex signal into a number of intrinsic mode functions (IMFs).Decomposition consists of following steps:

1. Identify all the local extrema, and then connect all the local maxima by an interpolation method. Repeat the procedure for the local minima to produce the lower envelope.
2. Determine the difference between the signal x (t) and the mean of upper and lower envelope value to obtain the first component. If it is an IMF, then it

would be the first component of x (t). Otherwise, it is treated as the original signal and step (1)–(2) are repeated. The sifting process can be stopped by any of the predetermined criteria which will be discussed in the next section.
3. Separate IMF from the original signal x (t) to obtain the residue and consider it as the new data and repeat the above described process.
4. Stop the decomposition process when the residue becomes a monotonic function from which no more IMF can be extracted.

3 IMF Sifting Stop Criteria

3.1 Cauchy-Type Convergence (SD)

Huang et al. [1] proposed a criterion where the size of the standard deviation of two consequent sifting results (h_n, h_{n-1}) should be limited and when it reaches a certain predefined value, sifting must stop:

$$SD = \sum_t \frac{[h_{n-1}(t) - h_n(t)]^2}{h_{n-1}^2(t)} < \in \tag{1}$$

3.2 Mean Fluctuations Thresholds

Rilling et al. [3] introduced a new criterion based on three thresholds $(\theta_1, \theta_2$ and $\alpha)$ aimed at guaranteeing globally small fluctuations in the mean while taking into account locally large excursions. For $(1 - \alpha)$ fraction of data, sifting will be continued when $\sigma(t) < \theta_1$ and for remaining fraction when $\sigma(t) < \theta_2$:

$$\sigma(t) = \left|\frac{m(t)}{a(t)}\right| \tag{2}$$

where: $a(t) = \frac{(e_{max}(t) - e_{min}(t))}{2}$, $m(t) = \frac{(e_{max}(t) + e_{min}(t))}{2}$ and e is envelope

3.3 Energy Difference Tracking

If h_1 (separated out from the signal) is an orthogonal component of $x(t)$, the sum of its energy and those of the residual signal (E_{tot}) is equal to the original signal energy E_x. Otherwise, there is a difference denoted as E_{err} [4]. When it reaches a certain minimum and the mean value of envelops becomes small enough, sifting process is completed:

$$E_{err} = E_{tot} - E_x = \int h_1^2(t)dt - \int x(t).h_1(t)dt \quad (3)$$

3.4 Resolution Factor

Rato et al. [5] applied the ratio between the energy of the signal at the beginning of the sifting (E_x) and the energy of average of envelopes (E_m). The stopping point is when the ratio ascends a predefined factor (resolution factor (R.F).

3.5 Bandwidth

Xie et al. designed a new stop criterion [6] which uses MFT criterion to get a result that almost satisfies the two conditions of IMF. Then the sifting process continues until to find the minimum of σ_f^2 (variance of instantaneous frequency) or until the difference of σ_f^2 between two consequent sifting results is very small:

$$\min \sigma_f^2 \quad \text{or} \quad \left|\left(\sigma_f^2\right)_{h_n} - \left(\sigma_f^2\right)_{h_{n-1}}\right| < \in \quad (4)$$

3.6 Orthogonality Criterion

Lin and Ji Hongbing [7] proposed the orthogonality criterion (OC) that when it reaches to a pre-defined value, the sifting will be terminated:

$$OC = \left|\sum_{t=1}^{N} \frac{m(t).x(t)}{m(t).(x(t) - m(t))}\right| \quad (5)$$

4 Support Vector Machine

Support vector machine (SVM) attempts to constructs a hyperplane that separates two different classes of samples and orients it in such a way that the margin is maximized. SVM could be applied in non-linear classification by mapping the data onto a high dimensional feature space where the linear classification is then possible. Various kernel functions could be used as the inner product of mapping functions, such as linear, polynomial or Gaussian RBF (Radial basis function). It is better for the smooth boundary to ignore few data points than be curved or go in loops, around the outliers. This is handled here by using slack variable (ξ) [8]:

$$\min \frac{1}{2}\|w^2\| + c\sum_{i=1}^{N} \xi_i \tag{6}$$

Subject to $y_i\langle w.X\rangle + b \geq 1 - \xi_i,\ \xi_i \geq 0$

5 Methodology

The goal of this study is to investigate which sifting stop criterion is more efficient for bearing damage identification. Normalized energy of IMFs is used as the feature vector [9]. The fault diagnosis method is given as the following:

1. Apply EMD with different sifting stop criteria to decompose the vibration signals into some IMFs. The first m IMFs which include the most dominant fault information are chosen to extract the feature.
2. Create a feature vector with the energy of the first m IMFs and normalize it:

$$FV_n = [E_1/E, E_2/E, \ldots, E_m/E] \tag{7}$$

where: $E_i = \int_{-\infty}^{+\infty} |c_i(t)|^2 dt$, $E = \left(\sum_{i=1}^{m} |E_i|^2\right)^{1/2}, i = 1, 2, \ldots, m$

3. Carry out the training procedure of SVM by utilizing the normalized feature vectors. The 60 % of data (40 samples) are used for training and the rest are taken as the test samples. After training the SVM successfully, it would be ready to test samples to identify the different work conditions and fault patterns.

6 Application

6.1 Experiments

The vibration signals of Roller Bearings were collected using the test rig (Fig. 1) developed and assembled by Dynamics & Identification Research Group (DIRG) at Department of Mechanical and Aerospace Engineering of Politecnico di Torino for three sizes of the artificial defect over the roller: A4(450 microns), A5(250 microns), A6(150 microns). The speed of the shaft and the sampling frequency were settled to 100 Hz and 102.4 kHz, respectively. This bearing test rig has been designed to perform accurate testing of bearings under different levels of damage in a controlled laboratory conditions, allowing speed, load and oil temperature control.

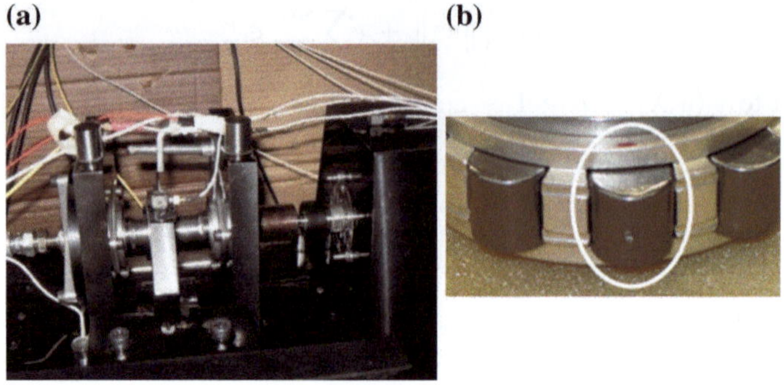

Fig. 1 DIRG test rig (**a**), damaged roller used in the tests with different defect sizes (**b**)

6.2 Analysis

Implementing the methodology described in Sect. 5, introduced feature vector for each sifting stop criterion and defect size is obtained. The 2D normalized energy of IMFs introduced as an efficient feature vector in fault diagnosis of roller bearing [9], has been adopted just using only first and second element of the feature vector. The results are shown in Fig. 2 for healthy (A0) and three different damage sizes (A4, A5 and A6). Obviously, for great damage sizes (A4 and A5), there is perfect distinction between the two classes of data and samples are easily recognized if they are healthy or defective. For very small defects such as A6, there is a confusion condition and classification will be with some errors in training or test. If first and third IMFs are used, the results will be less confused as it is shown in

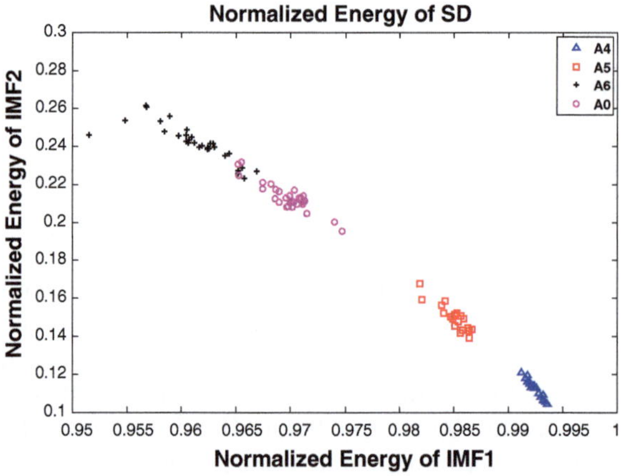

Fig. 2 Feature vector for 4 different bearing conditions

Fig. 3 for all criteria. The results of soft margin SVM automatic classification based on these two kind of feature vectors are proposed in Table 1(Gaussian kernel with gamma = 2 had better performance). As it was expected (from Fig. 3), for the all criteria, using the first and third IMFs has more accurate classification results than using the first and second IMFs. For example, in MFT method, prediction for both feature vectors are 100 %, whereas for training results are 91.7 and 95.8 % for first–second IMFs and first-third IMFs, respectively. Comparing these

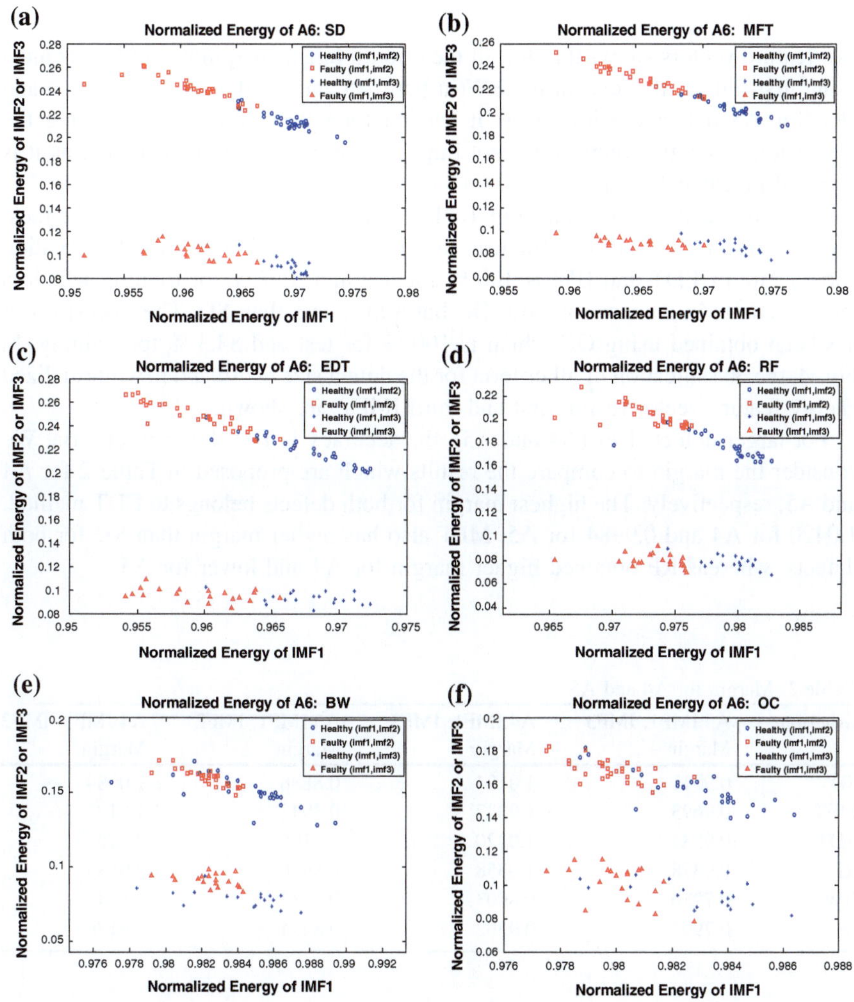

Fig. 3 Two different 2D feature vectors for all criteria (**a–f**)

Table 1 Classification accuracy A6

Criterion	2D:IMF1, IMF3		2D:IMF1, IMF2		3D	
	Train (%)	Test (%)	Train (%)	Test (%)	Train (%)	Test (%)
SD	93.3	100	91.7	100	95.8	100
MFT	95.8	100	91.7	100	95.8	100
EDT	91.7	100	86.1	100	91.7	100
RF	95.8	100	94.4	100	95.8	100
BW	91.7	100	75.0	95.7	100	93.3
OC	83.3	100	80.6	100	87.5	100.0

two feature vectors shows that for all the criteria larger margin would be obtained using first-third IMFs, except for BW(in both A5 and A4 damage conditions) and OC (for A5 defect condition), as it can be seen from Table 2. Increasing the dimension of feature vector does not improve the classification accuracy, as it is reported in Table 1.

Secondly, as it can be seen from Table 1, MFT and RF have obtained the most accurate results with 100 % for test and 95.8 % for training. The classification correct rate of EDT and BW is 100 % for test and 91.7 % for training which is more accurate in comparison with OC but a little less than SD. The lowest result has been obtained using OC which is 100 % for test and 83.3 % for training. In Fig. 4, the classification of all criteria for the defect size of A6 and new normalized energy feature vector (using first and third IMFs) are shown.

For larger defect sizes (A4 and A5), the accuracy is 100 % for all criteria. We consider the margin to compare the results which are proposed in Table 2 for A4 and A5, respectively. The highest margin for both defects belongs to EDT method, 1.0420 for A4 and 0.9984 for A5. MFT also has higher margin than SD for both defects, whereas RF obtained higher margin for A4 and lower for A5.

Table 2 Margin for A4 and A5

Criterion	A5:IMF1, IMF3 Margin	A4:IMF1, IMF3 Margin	A5:IMF1, IMF2 Margin	A4:IMF1, IMF2 Margin
SD	0.9011	1.0104	0.8888	1.0084
MFT	0.9693	1.0322	0.8941	1.0121
EDT	0.9984	1.0420	0.8105	0.9264
RF	0.9378	1.0358	0.8966	0.9683
BW	0.7286	0.8003	0.7887	0.8348
OC	0.7902	0.9302	0.8424	0.9176

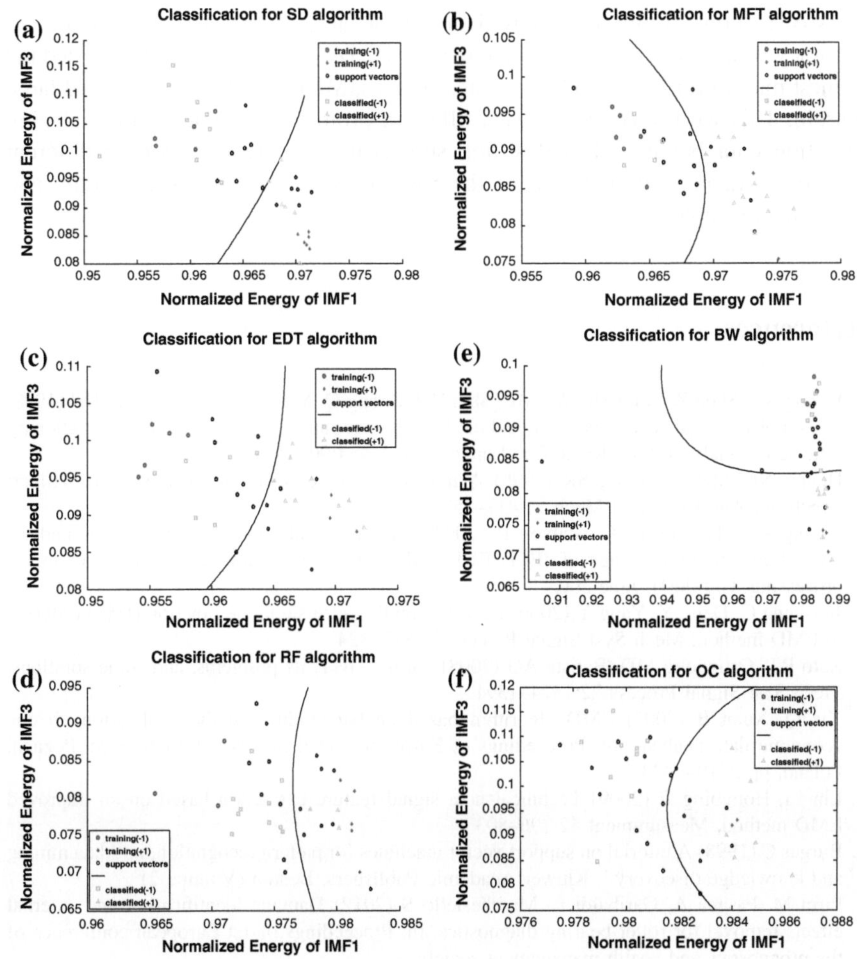

Fig. 4 Classification the feature vectors obtained by all criteria (**a–f**)

7 Conclusions

In this study, a comparison among different sifting stop criteria in EMD decomposition method is done. From damage identification point of view, it is important whether an algorithm is able to distinguish the healthy or faulty condition of a machine, as well as the severity of its damage. By using a normalized energy feature vector (first and third IMFs); it seems that samples are classified more properly with MFT and RF than others (high margin in larger defect sizes (A4, A5) and the most accurate classification rate for small damages (A6)). EDT is able to classify a sample, more accurately than SD for larger defect size. In addition, its accuracy is not far from SD for small defect size (A6). Although EDT obtained the highest margin for both two larger damage sizes, MFT and RF also achieved large margin. This work recognizes that EMD is a promising technique for analyzing vibration data but its selected sifting stop criterion may give rise to different solutions which, in turn, influence the SVM classification between healthy and faulty data samples.

References

1. Huang NE, Shen Z, Long SR, Wu ML, Shih HH, Zheng Q, Yen NC, Tung CC, Liu HH (1998) The empirical mode decomposition and the Hilbert spectrum for nonlinear and non- stationary time series analysis. Proc Royal Soc London Ser A 454:903–995
2. Huang NE, Shen Z, Long SR (1999) A new view of nonlinear water waves: the Hilbert Spectrum. Ann Rev Fluid Mech 31:417–457
3. Rilling G, Flandrin P, Goncalves P (2003) On empirical mode decomposition and its algorithms. In: Proceedings of IEEE EURASIP workshop on nonlinear signal and image processing, Gra do(I), June 2003
4. Junsheng C, Dejie Y, Yang Y (2006) Research on the intrinsic mode function (IMF) criterion in EMD method. Mech Syst Signal Process 20:817–824
5. Rato RT, Ortigueira MD, Batista AG (2008) On the HHT, its problems, and some solutions. Mech Syst Signal Process 22:1374–1394
6. Xie Q, Xuan B (2007) EMD algorithm based on bandwidth and the application on one economic data analysis. In: Proceedings of European signal processing conference, Poznán, Poland, pp 2419–2423
7. Lin Li, Hongbing Ji (2009) Technical note signal feature extraction based on an improved EMD method. Measurement 42:796–803
8. Burges C (1998) A tutorial on support vector machines for pattern recognition, in "data mining and knowledge discovery". Kluwer Academic Publishers, Boston (Volume 2)
9. Pirra M, Fasana A, Garibaldi L, Marchesiello S (2012) Damage identification and external effects removal for roller bearing diagnostics. In: Proceedings of 1st European conference of the prognostics and health management society

On the use of Vibration Signal Analysis for Industrial Quality Control: Part I

Gianluca D'Elia, Simone Delvecchio, Marco Malagò and Giorgio Dalpiaz

Abstract Vibration signals can be successfully captured and analyzed for quality control at the end of the production line. Various signal processing techniques and their applications are presented in this paper. These applications demonstrate the importance of selecting proper signal processing tools in order to extract the most reliable information from the signals. The presented applications regards tooth fault detection in helical gears and the detection of assembly faults in diesel engines by means of cold test technology.

Keywords Vibration · Condition monitoring · Quality control · Helical gears · Diesel engines · Cold test

1 Introduction

In industrial manufacturing rigorous testing is used to ensure that the delivered products meet their specifications. In the last few years, a great effort has been put into automating fault detection by using vibration measurements and processing techniques, due to its non-intrusive character and ability to detect a wide range of mechanical faults. In industrial environments there is an increasing demand for

G. D'Elia (✉) · S. Delvecchio · M. Malagò · G. Dalpiaz
Engineering Department in Ferrara (EnDIF), University of Ferrara, Ferrara, Italy
e-mail: gianluca.delia@unife.it

S. Delvecchio
e-mail: simone.delvecchio@unife.it

M. Malagò
e-mail: marco.malago@unife.it

G. Dalpiaz
e-mail: giorgio.dalpiaz@unife.it

G. Dalpiaz et al. (eds.), *Advances in Condition Monitoring of Machinery in Non-Stationary Operations*, Lecture Notes in Mechanical Engineering,
DOI: 10.1007/978-3-642-39348-8_34,

automatic on-line systems which are able to classify final products as pass or fail and/or to diagnose faults. Firstly, the monitoring procedure involves the acquisition of vibration signals by means of piezo-electric accelerometers. Since the selection of the acquisition parameters is critical, this data acquisition step is not of minor importance. Sometimes, several operations (i.e. correct selection of time histories, averaging and digital filtering) are needed in order to separate the most informative part of the signal from the environmental noise (electrical and mechanical). Secondly, signal processing techniques have to be implemented by taking into account the characteristics of the signal and the type of machine from which the signal is being measured (i.e. rotating or alternative machine with simple or complex mechanisms). Finally, several features have to be extracted in order to assess the physical state of the machine or to detect some incipient defects and to determine the causes of their presence.

Mechanical faults in machines often show their presence through abnormal vibration signals, thus techniques for machine condition monitoring based on the analysis of these signals are widely used [1, 2]. However, most studies have been carried out on simple mechanical parts, such as gears and rolling bearings, having well-determined dynamic characteristics. Therefore, gearbox condition monitoring and bearing defect analysis using vibration signatures are extensively reported ([3, 4] among a wide range of References). Moreover, some works related to the condition monitoring of machining processes are present in literature ([5, 6] refer to the drilling process as an example).

The aim of this paper is to present some quality control applications that are primarily based on vibration analysis. The use of processing techniques that can be considered well suited for implementation in on-line monitoring equipment at the end of the production line is proposed. The presented applications are: (a) tooth fault detection in helical gears; (b) the detection of assembly faults in diesel engines by means of cold test technology.

2 Tooth Fault Detection in Helical Gears

The first application regards the use of Time synchronous Average (TSA) vibration signal for the on-line vibration quality control of gear unites. The diagnostic capabilities of this simple technique have been tested on the basis of experimental results concerning two different tooth faults in helical gears: poor tooth surface quality and presence of tooth face bumps. The first fault condition concerns the presence of oxide residuals on the tooth surface due to the heat-treatment and the grinding process. The second one is caused by gear tooth impacts during gear conveyance before the heat treatment, see Fig. 1b. During the test campaign the faulty gears have been mounted on the first stage of a gear unit and the vibration signals have been acquired from the gearbox case.

Concerning fault detection, two statistical parameters, i.e. root-mean-square (RMS) and Kurtosis, extracted from faulty and sound TSA vibration signals have

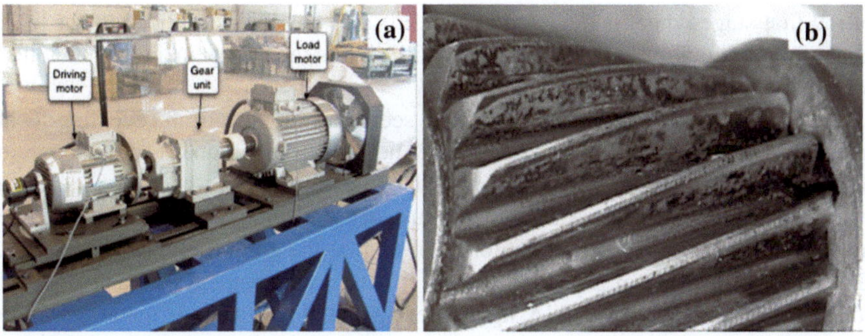

Fig. 1 **a** Test bench for gear units and **b** example of a tested gear

been evaluated in order to asses a reliable quality control strategy. The experimental apparatus (see Fig. 1a) consists of a base, including two induction motors controlled by inverters and a gear unit. The driving induction motor is controlled in feedback speed loop, whilst the loading motor is controlled in feedback torque loop. The gear unit contains two spur gear pairs, one having 18 and 71 teeth, the other one 12 and 55 teeth, giving a global speed reduction ratio of 18.1. More details concerning the test bench design can be found in [7].

Three kinds of test are performed:

- Test 1: all gears in sound condition;
- Test 2: the first stage gearbox pinion (18 teeth) exhibiting several oxide residuals on tooth faces, so poor tooth surface quality (distributed fault);
- Test 3: the first stage gearbox wheel (71 teeth) showing tooth face bumps (localized fault).

The results presented in this work have been obtained with a nominal driving motor speed of 3,000 rpm (50 Hz) and a output shaft nominal torque of 36.6 Nm. The vibration signal is captured from the gearbox case by means of a Brüel and Kjær piezoelectric accelerometer mounted close to the bearing support of the first stage pinion in the radial direction.

The sample frequency was 32,800 Hz, whilst the acquisition time was 60 s. In all the tests the TSA vibration signal is calculated over 80 revolutions for both first stage pinion and wheel, obtaining both pinion TSA and wheel TSA. Figure 2a and b plot the TSA of the first stage pinion and wheel for Test 1, which can be taken as a reference for the detection procedure. In particular, it is possible to notice that the main signal component is the meshing frequency and no signal alteration can be observed. Figure 2c shows the TSA of the first stage pinion for Test 2, concerning poor tooth surface quality. Comparing Fig. 2a and c it is possible to notice that this type of fault gives rise to an increase of the mean amplitude vibration level without any local alteration. Therefore by the visual inspection of the TSA this type of fault cannot be surely identified and so further analyses have to be performed, i.e. evaluation of statistical parameters of the TSA signal. Concerning Test 3, the TSA of the first stage wheel is depicted in Fig. 2d. It is possible to notice a clear local alteration of the vibration signal at about 230° due to the engagement of the faulted tooth with bumps.

Here it is clearly evident that the TSA seems a pivotal tool in revealing local alterations, i.e. localized fault in gear, nevertheless this techniques is not enough to reveal distributed faults. Moreover, in order to assess the presence of a defect a visual inspection of the TSA is needed; thus such a technique is not suitable to be implemented in an automatic monitoring system at the end of the production line. Ergo, for quality control purposes, we can link this achievement to the analysis of statistical parameters of the TSA signal, i.e. RMS and Kurtosis. As a matter of fact these parameters are of simple interpretation and they are linked to different physical properties of the analyzed signal. In particular, RMS takes into account the energy conveyed by signals and so it can be considered as an useful tool in order to detect an increase of the mean signal amplitude (distributed faults). On the other hand Kurtosis is exceedingly sensitive to local signal alteration (localized faults). Table 1 summarises the RMS and the Kurtosis values for both pinion and wheel TSA of the three different tests.

It is well know that these statistical parameters evaluated on the TSA signal, have to be compared on the basis of the same mechanical component, i.e. pinion or wheel. As a matter of fact TSA in practice extracts from the signal the genuine portion containing only the components which are synchronous with the revolution of the specific gear in question. Therefore, the statistical parameters evaluated on the pinion TSA signal and on the wheel TSA signal, differ because they are evaluated on the basis of different signal components.

As reported in Table 1, there is an increase of the RMS value of the pinion TSA for Test 2 with respect to Test 1 (i.e. 0.35 vs. 0.21 g), whilst the RMS of the wheel TSA does not show remarkable changes. Such an increase highlights the poor surface quality of the tooth pinion. Concerning Test 3 (wheel localized fault) the

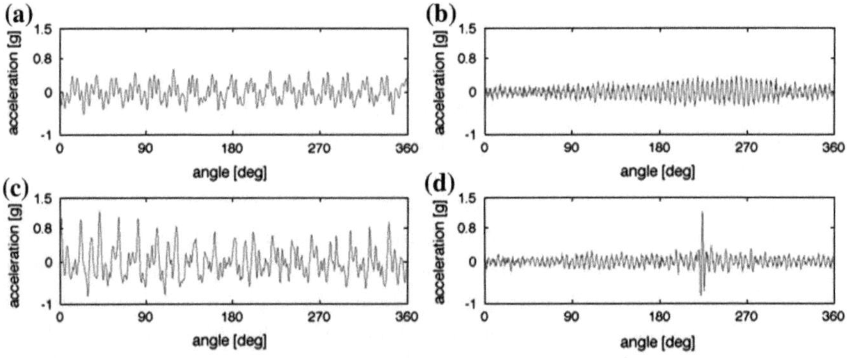

Fig. 2 Time-synchronous averages for the three tests: (**a**) Test 1: pinion TSA in sound condition, (**b**) Test 1: wheel TSA in sound condition, (**c**) Test 2: pinion TSA with pinion distributed fault and (**d**) Test 3: wheel TSA with wheel localized fault

Table 1 RMS and Kurtosis values of the TSA for the three tests

	Pinion TSA		Wheel TSA	
	RMS (g)	Kurtosis	RMS (g)	Kurtosis
Test 1 (pinion and wheel in sound condition)	**0.21**	**2.39**	**0.14**	**2.52**
Test 2 (pinion distributed fault)	**0.35**	3.15	0.19	2.60
Test 3 (wheel localized fault)	0.13	2.80	0.13	**17.34**

RMS of both pinion and wheel TSA does not reveal the fault presence, in fact the energy conveyed by signal remains roughly the same (see Fig. 2b and d for the wheel TSA). On the other hand, the Kurtosis of the wheel TSA for Test 3 shows a marked increase with respect to Test 1 (i.e. 17.34 vs. 2.52), highlighting the presence of a localized fault on the first stage wheel.

Concerning this application, we can consider TSA as a pivotal starting point for on-line gear quality control. However, this techniques has to be linked to proper statistical parameters, such as RMS value or Kurtosis coefficient, in order to detect poor tooth surface quality or localized gear tooth faults. In particular, because RMS value accounts for the energy conveyed by the signal, it is well suited for distributed fault localization, while Kurtosis coefficient, which is sensitive to local signal alterations, is a pivotal tool for the monitoring of localized faults.

3 Detection of Assembly Faults in Diesel Engine Cold Test

This application addresses the use of basic signal processing tools as means for the quality control of assembly faults in diesel engines through the cold test technology. Nowadays, the main part of engine manufacturers tests their engines by means of "hot tests", i.e. tests in which engine is firing. Hot tests mainly aim to determine engine performances. Recently some companies have introduced "cold tests" aiming at identify assembly anomalies by means of torque, pressure and vibration measurements: this method has to be further improved [8]. In hot tests, anomalies are detected through the deterioration of the engine performances. Moreover vibration analysis is sometimes used to detect faults affecting combustion efficiency. Cold tests are more oriented to identify the sources of anomalies since they are not affected by noise and vibration due to the firing.

The fault detection and diagnosis of i.e. engines can be carried out using different strategies. One strategy can consist in modelling the whole mechanical system using lumped [10–14] or finite element methods in order to simulate several faults and compare the results with what the experimental tests found. Another strategy can adopt signal processing techniques in order to obtain features or maps that can be used to detect the presence of the defect. Regarding this, a decision algorithm is needed for a visual or automatic detection procedure. Moreover the maps can be also analysed for diagnostic purposes. This method is the most commonly used and well suited for the judgment of expert technicians.

In this paper we try to investigate the problem of obtaining a strategy for the pass/fail decision at the end of the assembly line.

As presented in [9], the normalized time histories of acceleration signals concerning diesel engines can be represented as symmetrized dot polar (SDP) graphs. The formulation used for the transformation of a discrete signal to a polar coordinate graphs and the properties of this method are described in detail in [9]. Examples of the results are shown in Figs. 3 and 4. In order to implement this technique in the cold test procedure for fault detection, it is necessary to develop an image correlation system. The authors applied the algorithm of edge detection, illustrated in [9], which represents the most common approach for detecting meaningful discontinuities in intensity values. The basic idea behind edge detection is to find the points where the intensity rapidly changes. For each case, by applying this edge detection algorithm on the image of the visual symmetrized dot polar graph, the result will be a boolean matrix with entries equal to 1 (represented as white pixels in the grey scale) located at the edge points, detected on the image, and entries equal to 0, (black pixels) located elsewhere (Fig. 3c).

The goal is to identify a reference pattern that represents the normal condition and then compare the images obtained from all the test engines with this 'healthy pattern' by means of a similarity parameter. Among many possibilities, this parameter was chosen as the percentage of common white pixels with respect to

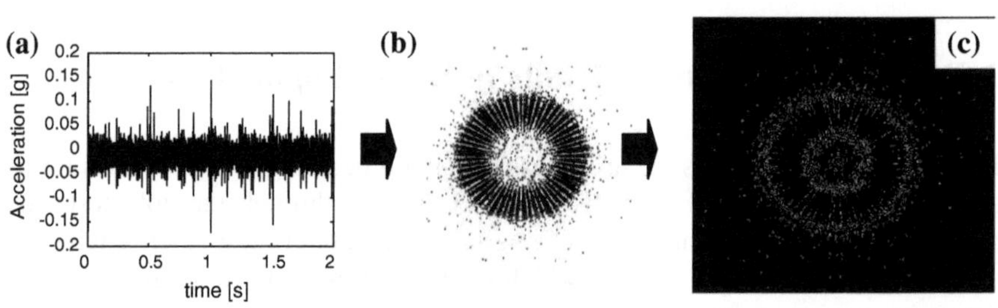

Fig. 3 SDP method: (**a**) time input waveform; (**b**) symmetrized dot polar graph and (**c**) image obtained after the application of edge detection algorithm

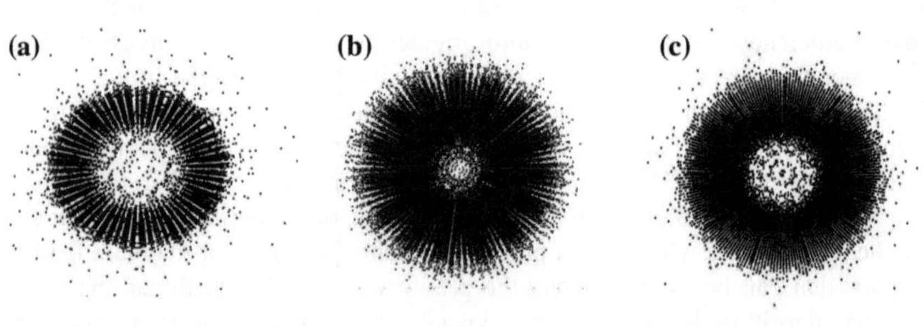

Fig. 4 Symmetrized dot polar graphs: (**a**) healthy engine (reference pattern); (**b**) and (**c**) faulty engines

Table 2 Symmetrized dot polar patterns: correlation values

Comparison between symmetrized dot polar patterns	Percentage of common white pixels (%)
Healthy engine—healthy engine	25.08 (CT)
Healthy engine—faulty engine (incorrectly tightened rod screws)	10.80

the total number of white pixels in the healthy engine pattern. Hereafter, this parameter will be called correlation.

In order to set the detection threshold for the correlation, firstly the correlation parameter is calculated for all the couples of engines from a sample of 21 healthy engines: the engine that presents the minimum correlation with each other is assumed as the reference pattern, unique for all investigations, and the corresponding minimum correlation of 25.08 % is fixed as the correlation threshold (CT). This pattern is compared with the images obtained from different faulty engines and so the correlations are calculated, verifying if they are lower than the CT in order to discriminate the faulty condition from the normal one.

Hereafter only the result concerning a specific fault is shown, as example. It deals with a connecting rod with incorrectly tightened screws: the rod screws are only tightened with the preload of 3 kgm, instead of the correct torque of 9 kgm.

Table 2 shows that this faulty engine gives a percentage of common white pixels lower than the CT. Similar results were found for other faulty conditions. It can be concluded that the percentage of common white pixels between each pattern and the reference one can be considered as a reliable parameter. Thus, it seems that this technique exhibits high sensitivity to faults, presenting some advantages, as fast computational implementation and easy output evaluation.

4 Concluding Remarks

This paper describes some applications of vibration analyses for the quality control of mechanical devices at the end of the production line. The following conclusions can be achieved based on the presented results.

- Vibration signal is useful for quality control of the above-described applications, but it should be properly processed in order to obtain reliable information.
- In case of rotating machines, Time Synchronous Average can be considered as a powerful tool in detecting the presence of faults. It also permits the characterization of the source of alteration.
- The quality control at the end of assembly line requires the assessment of simple parameters that can be easily used for implementing a strategy of pass/fail decisions.

Acknowledgments This work has been developed within the Advanced Mechanics Laboratory (MechLav) of Ferrara Technopole, realized through the contribution of Regione Emilia-Romagna—Assessorato Attività Produttive, Sviluppo Economico, Piano telematico—*POR-FESR* 2007–2013, Activity I.1.1.

References

1. Braun SJ (1986) Mechanical signature analysis. Academic Press, London
2. Collacot RA (1979) Vibration monitoring and diagnostic. Wiley, New York
3. McFadden PD (1986) Examination of a technique for the early detection of failure in gears by signal processing of the time domain average of the meshing vibration. Mech Syst Sig Process 1(2):173–183
4. Randall RB (1982) A new method of modelling gear faults. J Mech Des 104:259–267
5. El-Wardanay TI, Gao D, Elbestawi MA (1996) Tool condition monitoring in drilling using vibration signature analysis. Int J Mach Tools Manuf 36(6):687–711
6. Jantunen E (2002) A summary of methods applied to tool condition monitoring in drilling. Int J Mach Tools Manuf 42:997–1010
7. Dalpiaz G, Rivola A, Rubini R (2000) Effectiveness and sensitivity of vibration techniques for local detection in gears. Mech Syst Sig Process 14(3):387–412
8. Delvecchio S, D'Elia G, Mucchi E, Dalpiaz G (2010) Advanced signal processing tools for the vibratory surveillance of assembly faults in diesel engine cold tests. J Vibr Acoust 132(2):021008–021010. ISSN: 1048-9002. doi:10.1115/1.4000807
9. Gonzalez R, Woods R, Eddins S (2004) Digital image processing using matlab. Prentice Hall, New Jersey
10. Mucchi E, Dalpiaz G, Fernàndez del Rincòn A (2010) Elasto-dynamic analysis of a gear pump. Part I: pressure distribution and gear eccentricity. Mech Syst Sig Process 24:2160–2179. doi:10.1016/j.ymssp.2010.02.003
11. Mucchi E, Dalpiaz G, Rivola A (2010) Elasto-dynamic analysis of a gear pump. Part II: meshing phenomena and simulation results. Mech Syst Sig Process 24:2180–2197. doi:10.1016/j.ymssp.2010.02.004
12. Mucchi E, Dalpiaz G, Rivola A (2010) Dynamic behaviour of gear pumps: effect of variations in operational and design parameters. Meccanica 46(6):1191–1212
13. Mucchi E, D'Elia G, Dalpiaz G (2012) Simulation of the running in process in external gear pumps and experimental verification. Meccanica 47(3):621–637
14. Mucchi E, Di Gregorio R, Dalpiaz G (2013) Elastodynamic analysis of vibratory bowl feeders: modeling and experimental validation. Mech Mach Theory 60:60–72

On the use of Vibration Signal Analysis for Industrial Quality Control: Part II

Simone Delvecchio, Gianluca D'Elia, Marco Malagò and Giorgio Dalpiaz

Abstract Vibration signals can be successfully captured and analyzed for quality control at the end of the production line. Various signal processing techniques and their applications are presented in this paper. These applications demonstrate the importance of selecting proper signal processing tools in order to extract the most reliable information from the signals. The presented applications regards ball bearings and threshing process in harvesting machines.

Keywords Vibration · Condition monitoring · Diagnostics · Quality control · Ball bearings · Harvesting machines

1 Diagnostics of Distributed Faults in Ball Bearings

The study of localized failure detection in bearings started over two decades ago, embracing a large number of signal processing techniques that can be roughly subdivided with respect to their pertinence domain, i.e. time, frequency and time–frequency domain [1, 2]. The aim of this work is to apply cyclostationary metrics for the identification of both the appearance and the growth of distributed faults in ball bearings, in order to overcome the pitfall of the usual approaches. Non-

S. Delvecchio (✉) · G. D'Elia · M. Malagò · G. Dalpiaz
Engineering Department in Ferrara (EnDIF), University of Ferrara, Ferrara, Italy
e-mail: simone.delvecchio@unife.it

G. D'Elia
e-mail: gianluca.delia@unife.it

M. Malagò
e-mail: marco.malago@unife.it

G. Dalpiaz
e-mail: giorgio.dalpiaz@unife.it

G. Dalpiaz et al. (eds.), *Advances in Condition Monitoring of Machinery in Non-Stationary Operations*, Lecture Notes in Mechanical Engineering,
DOI: 10.1007/978-3-642-39348-8_35, © Springer-Verlag Berlin Heidelberg 2014

stationary signals can be defined as signals which satisfy a non-property, i.e. they do not satisfy the property of stationarity. It is not possible to define a general theory which treats non-stationary signals. The non-stationary behavior of each signal has to be individually evaluated [3–9]. An important cyclostationary tool is the Indicator of Second order cyclostationarity (ICS_{2x}) outlined in [10]. This indicator tries to quantify the distance of a second-order cyclostationary process from the closest stationary process having a similar power spectral density, giving an indication of the presence of second-order cyclostationary components within a signal. It is defined as (11):

$$ICS_{2x} = \sum_{\alpha \in A} \frac{\left|\left(\lim_{T\to\infty} \int_T \left|x^R(t)\right|^2 e^{-j2\pi\alpha t} dt\right) e^{2\pi\alpha t}\right|}{\lim_{T\to\infty} \int_T \left|x^R(t)\right|^2 dt} \quad (1)$$

where A is the set of all possible cyclic frequencies α and $x^R(t)$ is the residual signal. As reported by [10] this is a cumulant based estimator.

An experimental campaign was carried out on a ball bearing in order to obtain distributed faults on the outer race, by using a test-bed composed of an asynchronous 4-pole motor which moves a shaft by means of a driving belt. The shaft is supported by a couple of cone-shaped bearings, Fig. 1a. The bearing under test is a double-row self-aligning type SKF 1205; it is cantilever mounted on this shaft at the opposite to the pulley. A radial external load supplied by a leverage system acts on the test bearing. In the present test, the bearing was externally loaded with a force of 1962 N, while the shaft was rotating at 26.6 Hz. The bearing was degreased in advance in order to accelerate the wear process and then mounted on the test machine. Three accelerometers were used to measure the vibration signal. The vibration signals were acquired each 15 min with an acquisition time of 2 min, obtaining 21 acquisitions in total. The sampling frequency was 51.2 kHz. At the end of the test the bearing was unmounted to check the status of the

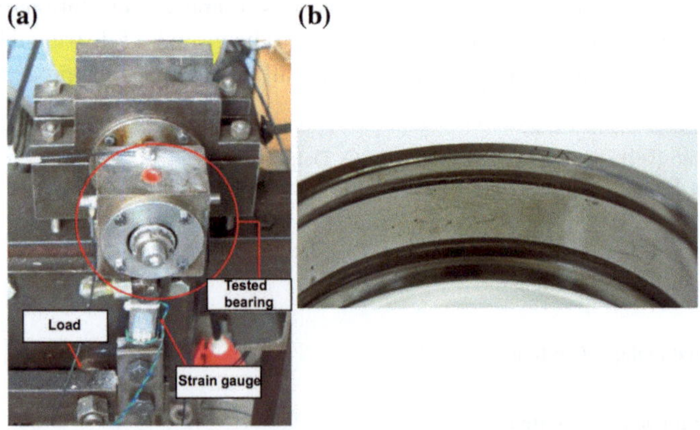

Fig. 1 **a** Test rig; **b** Bearing outer race at the end of the test

Table 1 Characteristic fault frequencies

Description	Symbol	[Hz]
Rotation frequency	*fr*	26.6
Outer race fault frequency	*fo*	127
Inner race fault frequency	*fi*	192
Cage fault frequency	*fc*	10.6
Ball fault frequency	*fb*	121

surfaces. The outer race of the bearing presented a groove corresponding to the passage of the balls, see Fig. 1b. The groove length took more or less half of the outer race circumference. The expected frequencies of the typical faults are computed from the usual formulae [11]. Their values for the bearing under test are collected in Table 1. Since the damage mainly involves the outer race in the actual case, the outer race fault frequency is briefly referred as "fault frequency" in the following.

In this paper it is considered only one of the three acceleration signals: it deals with the vibration signal that has be proven to be less affected by the transmission path. Figure 2 depicts the time signal captured during acquisitions n. 1 (first) and 21 (last) for a complete revolution of the shaft. As expected, the overall amplitude level strongly increases from the first to the last acquisition, but no impulsive content can be observed. Actually a distributed fault is not related to an impulsive content, but to an increase of the signal energy. Therefore, the evaluation of the RMS value can be an useful parameter for condition monitoring. As depicted in Fig. 3 the RMS value gives an alarm on acquisition 5 where probably the condition of the bearing is changed. However the RMS remains high until the end of the acquisitions giving no information about the evolution of the fault. In addition, due to its global nature, the RMS cannot give any information about the fault position. In order to better investigate the fault behavior, the classical envelope analysis is carried out by following these steps:

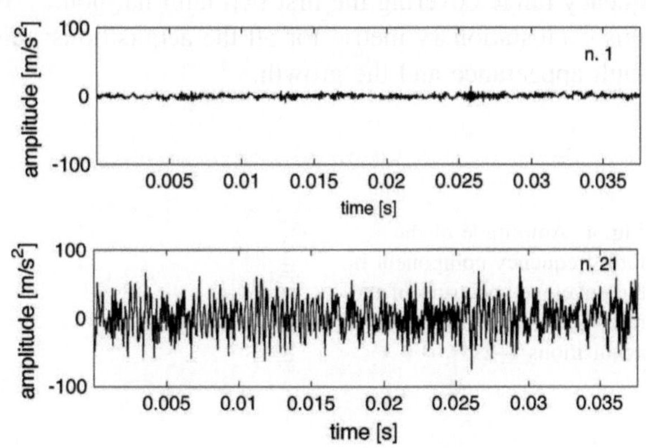

Fig. 2 Time signal for one shaft rotation: n. 1 first acquisition; n. 21 last acquisition

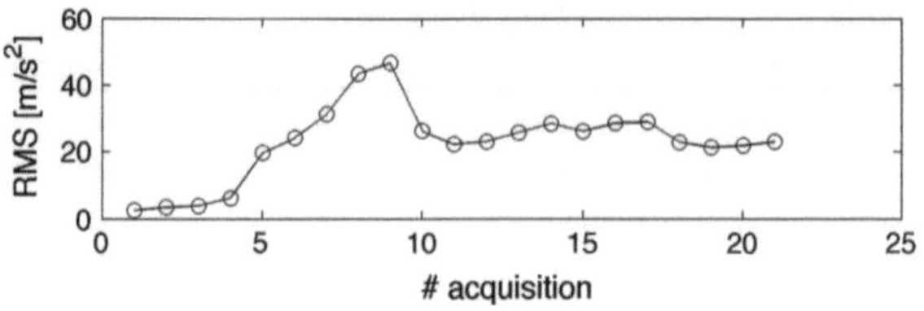

Fig. 3 RMS trend for acquisitions 1–21

- Band-pass filtering around a suitable frequency;
- Computation of the analytical signal, which is a complex quantity having the acquired signal as real part and its Hilbert Transform as imaginary part;
- Computation of the Spectrum of the absolute value of the analytical signal.

It can be noted that the envelope analysis (Fig. 4) confirms what found by the RMS value analysis trend. Moreover the envelope can identify the presence of a localized fault on the outer race around acquisition n. 5. In more details this localized fault gives origin to very slight but detectable impacts. The envelope is sensitive to these impacts showing the ball passing frequency on the incipient localized defect. Unfortunately this technique cannot supply further information concerning the increase of severity and extension of the wear. As a matter of fact, when the localized defect on the outer race grows, becoming a distributed fault, no impulses are generated; thus, this technique cannot highlight the fault evolution.

At this stage the cyclostationarity analysis [12] has been applied as a powerful tool in order to obtain information concerning the fault evolution. This type of analysis is well suited in describing the vibration response signal captured from a ball-bearing with a distributed fault. In fact, it is reasonable that this signal has non-stationary properties with a second-order cyclostationary content, due to the periodic variation of the bearing configuration. For example, the action of the balls passing on a distributed fault on the outer race produces a cyclostationary vibration with fundamental cyclic frequency corresponding to the ball passing frequency. The ICS_{2x} is evaluated in order to have a simple cyclostationary metric that can track the fault evolution. In particular, this metric is computed in the cyclic frequency range covering the first two fault harmonics. Figure 5 depicts the trend of this cyclostationary metric for all the acquisitions: this metric highlights both the fault appearance and the growth.

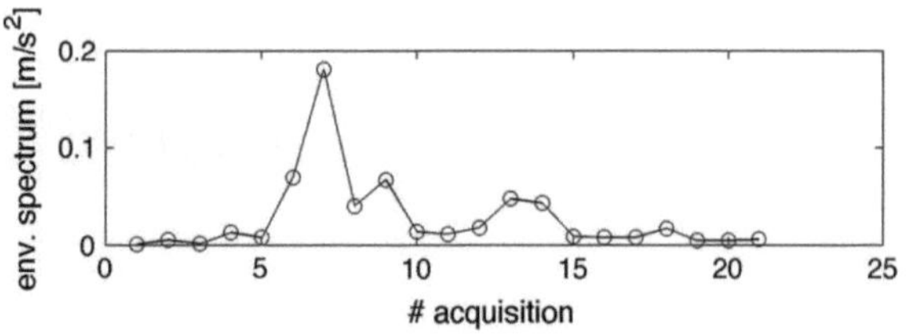

Fig. 4 Amplitude of the fault frequency component in the envelope spectrum of the raw signal: trend for acquisitions 1–21

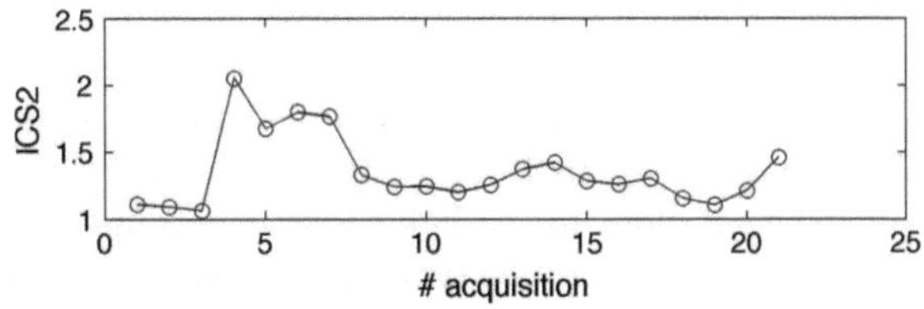

Fig. 5 *ICS2x* trend for acquisitions 1–21

2 Condition Monitoring of the Threshing Process in Harvesting Machines

This section of the paper concerns the analysis of the relationship between the threshing process in an axial flow harvesting machine and its vibro-acoustic behaviour. Several indoor and outdoor measurements are performed in various operational conditions in order to evaluate their influence on the vibration response of the threshing unit. The main goal was to identify a possible link between sound/vibration and crop distribution in threshing machines by using different signal processing tools.

The combine harvesting machine is a machine which "combines" the tasks of harvesting, threshing and cleaning grain plants. The desired result is the seed or grain (including corn, soybeans, flax, oats, wheat, or rye among others). In an axial combine harvester, as the crop spirals around the rotor, rasp bars mounted on the rotor rub the grains out. An helicoidally flux of the crop is favoured by splitting the incoming crop flow over two counter-rotating rotors. The threshing process is mainly given by two principal effects: (1) the "grain over grain" effect (i.e. the threshing of the grain kernels among themselves); (2) the threshing between kernels and concave (i.e. the rotor cage). The crop processing unit of an axial flow harvesting machine performs the following operations (Fig. 6): (1) after cutting the crop the threshing unit performs the threshing and separation activities making the kernels free from chaff and straw; (2) the cleaning unit cleans the crop separating the kernels from other small particles like chaff and short straw. The first operations are performed by the rotors that induce an helicoidal flux of the crop. Spirally arranged rasp bars and friction elements mounted on the rotors favor the friction of the crop against the concaves. There are four concaves, two for each rotor. The concave size, the number as well as the shape of the rasp bars changes as they go from the threshing zone to the separation zone.

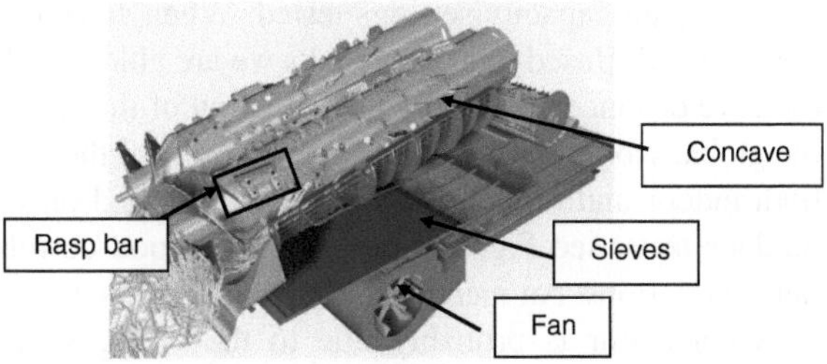

Fig. 6 Threshing unit: mechanical parts involved in the threshing process

The frequency analysis was firstly carried out for signals measured by 3 concave accelerometers in the radial direction. This direction was expected to be the most excited by the threshing flow. The middle signal presents higher amplitudes than the other signals from the front and rear concave accelerometers since they are mounted close to the bolt joints connecting the concave to the frame. The change in amount of crop processed between the concave front (beginning of the threshing zone) and rear positions (end of the threshing zone and start of the separation zone) does not affect the frequency behaviour, which is the same in the first 600 Hz for both positions. Thus, it is proven that the signal amplitudes are strongly dependent on the transducer location and not on the change of the amount of processed crop between the two positions. The relevant characteristic frequencies is supposed to be determined by the periodical rasp bar interactions of the rasp bars with the concave. Because of their helicoidally placement along the rotor, one rasp bar interacts with the concave three times during each rotor revolution. Therefore, it is useful to define a characteristic frequency as F_{bar} that can be expressed in Hz as $F_{bar} = 3n_{rotor}/60$, where n_{rot} is the nominal rotor speed in rpm.

On the basis of the first measurements the radial direction signal of the accelerometer positioned on the concave middle could be considered as the best correlated with the threshing process and it can be taken as the reference for the condition monitoring procedure. Based on the frequency behaviour, this signal can be decomposed into three different components: a component measured in idle condition due to the vibration of the mechanical parts in operation; a sinusoidal component given by the superposition of the F_{bar} and its harmonics; and a broadband random component. It is worth noting that the broadband spectrum component holds the response of the concave to two vibration sources: turbulence due to the threshing flow and hits of the kernels impacting the concave. For this reason the noise must not to be neglected during the analysis but has to be considered as "operational" noise generated by the process itself. In order to evaluate the influence of the operational parameters some vibration metrics are extracted from the raw vibro-acoustic signals: Root Mean Square value (RMS) reflecting the signal energy; Crest Factor (CF), Temporal Kurtosis (TK) and Impulse Factor (IF) reflecting the level of signal peakness. In particular the correlations among acceleration features and concave distance have been reported here. Looking at Fig. 7a, b and c it can be observed a very good correlation existing between the energy of all concave accelerometer signals and the concave distance: as the concave distance increases the Av. RMS decreases in a less than linear way. This is true for all capacity settings tested. When capacity is set at 10 % higher values are obtained. Based on these results we are able to link the concave vibration to the concave distance. Through the evaluation of the Av. RMS trends we get an idea of the global vibration of the concave. The TK of the concave middle radial signal for both indoor and outdoor tests (but here plotted only for signals measured during outdoor tests, see Fig. 7d) presents also good correlations with the concave distance variations. An increase of the concave distance causes an increase of the TK. This behaviour is probably due to the changes in crop distribution: at higher

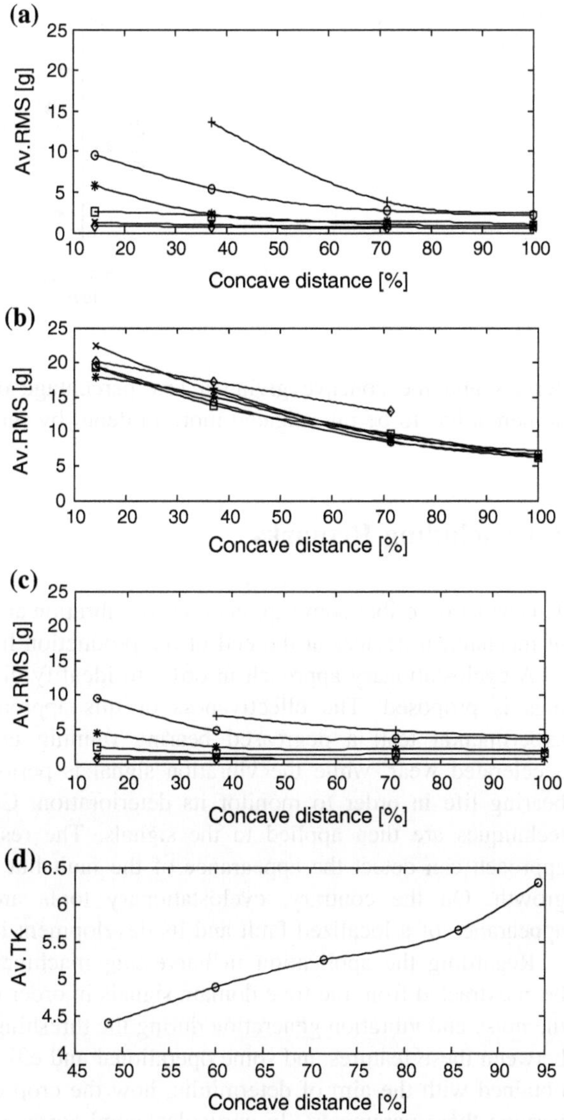

Fig. 7 Features versus concave distance: **a** Av. RMS at rear position; **b** Av. RMS at middle position; **c** Av. RMS at front position; **d** Outdoor tests: Av. TK at middle position

concave distances the crop is more spread less densely between the concave and the rotor. Therefore more space is available for the crop which interact more easily the concave. Hence we can refer to a sort of saturation effect occurring at low concave distances. Moreover, as confirmed by Fig. 8, it seems that the TK parameters can be correlated with the amount of broken grains. This correlation can be interpreted assuming that an increase of the concave distance, indicated by higher TK values, may cause a decrease in the impact force existing between the

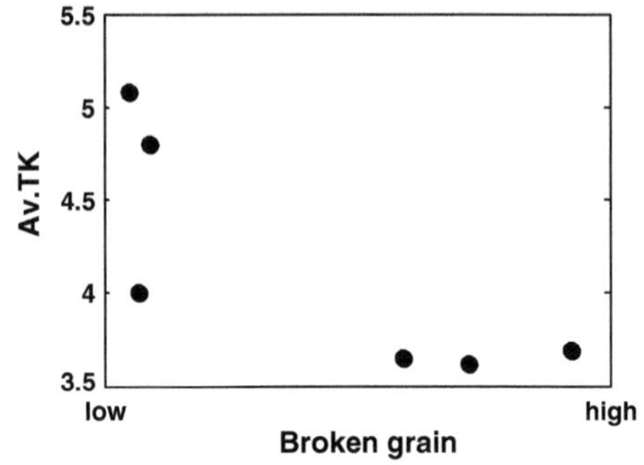

Fig. 8 Field tests: Av. TK versus broken grains

kernels and the concave giving a low percentage of broken grains. This phenomenon has to be investigated more in detail by further tests.

3 Concluding Remarks

This paper describes some applications of vibration analyses for the quality control of mechanical devices at the end of the production line.

A cyclostationary approach in order to identify distributed faults in ball bearings is proposed. The effectiveness of this approach is assessed through an experimental test: a degreased bearing running under radial load developed accelerated wear, while the vibration signal is periodically captured during the bearing life in order to monitor its deterioration. Classical and cyclostationary techniques are then applied to the signals. The results indicate that the usual approach can detect the appearance of the fault but cannot track the successive growth. On the contrary, cyclostationary tools are able to detect both the appearance of a localized fault and its development in a distributed fault.

Regarding the application in harvesting machines, appropriate metrics have been extracted from the time domain signals in order to explain the mechanism of the noise and vibration generation during the threshing process. Some correlations between these features and some operational and efficiency parameters have been obtained with the aim of determining how the crop distribution is influenced by varying these parameters. In particular, good correlations have been obtained for the concave middle radial signal, calculating the trends of RMS and TK. Some of these metrics can be assumed as good indexes that predict the efficiency of the process. Moreover, the concave vibration has been proved to be well correlated with the concave distance that is tuneable by the user during the field operations.

Acknowledgments This work has been developed within the Advanced Mechanics Laboratory (MechLav) of Ferrara Technopole, realized through the contribution of Regione Emilia-Romagna—Assessorato Attività Produttive, Sviluppo Economico, Piano telematico—*POR-FESR* 2007-2013, Activity I.1.1.

References

1. Delvecchio S, D'Elia G, Mucchi E, Dalpiaz G (2010) Advanced signal processing tools for the vibratory surveillance of assembly faults in diesel engine cold tests. J Vib Acoust 132(2):021008–021010, ISSN: 1048-9002, doi:10.1115/1.4000807
2. Mucchi E, Vecchio A (2010) Acoustical signature analysis of a helicopter cabin in steady-state and run up operational conditions. Measurement 43:283–293
3. Mucchi E, Dalpiaz G, Fernàndez del Rincòn A (2010) Elasto-dynamic analysis of a gear pump. Part I: pressure distribution and gear eccentricity. Mech Syst Signal Process 24:2160–2179. doi:10.1016/j.ymssp.2010.02.003
4. Mucchi E, Dalpiaz G, Rivola A (2010) Elasto-dynamic analysis of a gear pump Part II: meshing phenomena and simulation results. Mech Syst Signal Process 24:2180–2197. doi:10.1016/j.ymssp.2010.02.004
5. Mucchi E, Dalpiaz G, Rivola A (2010) Dynamic behaviour of gear pumps: effect of variations in operational and design parameters. Meccanica 46(6):1191–1212
6. Mucchi E, D'Elia G, Dalpiaz G (2012) Simulation of the running in process in external gear pumps and experimental verification. Meccanica 47(3):621–637
7. Mucchi E, Di Gregorio R, Dalpiaz G (2013) Elastodynamic analysis of vibratory bowl feeders: Modeling and experimental validation. Mech Mach Theory 60:60–72
8. Pierro E, Mucchi E, Soria L, Vecchio A (2009) On the vibro-acoustic operational modal analysis of a helicopter cabin. Mech Syst Signal Process 23:1205–1217
9. Mucchi E (2012) Experimental evaluation of modal damping of automotive components in different boundary conditions. Meccanica 47(4):1035–1041
10. Raad A, Antoni J, Sidahmed M (2008) Indicators of cyclostationarity: theory and application to gear fault monitoring. IEEE Trans Power Delivery 22:574–587
11. McFadden PD, Smith JD (1984) Vibration monitoring of rolling element bearings by the high-frequency resonance technique. a review. Tribol Int 17:3–10
12. Antoni J (2009) Cyclostationary by examples. Mech Syst Signal Process 23:987–1036

Part V
Experimental and Numerical Modeling of Machine Dynamics

Non-Clustering Method for Automatic Selection of Machine Operational States

Adam Jablonski, Tomasz Barszcz and Piotr Wiciak

Abstract A reliable evaluation of technical condition of machinery working under non-stationary conditions requires a rigorous tracking of operational parameters. Therefore, modern condition monitoring systems (CMS) enable reading and registering of process parameters (e.g. speed, load, pressure, etc.) in parallel with acquisition of vibroacoustic signals. Although few tries have been undertaken to develop state-free analysis of vibration signals, currently installed systems still do rely on state-preclassified data. The paper shows how the process, referential data might be automatically transformed into proposition of optimal machine operational states in terms of their number and their range. As indicated by the title, the paper shows common pitfalls coming from implementation of popular clustering approach. The proposed algorithm illustrates is verified on real data from a pitch-controlled wind turbine.

Keywords Machine operational states · Condition monitoring of machinery working under non-stationary parameters

1 Introduction

In the field of vibrodiagnostics, a machine operational state is understood as an accepted range of machine's operational points enabling referential analysis [1]. In practice, machine operating point is defined by values of available measurements

A. Jablonski (✉) · T. Barszcz · P. Wiciak
AGH University of Science and Technology, A. Mickiewicza 30 30-059 Krakow, Poland
e-mail: ajab@agh.edu.pl

T. Barszcz
e-mail: tbarszcz@agh.edu.pl

P. Wiciak
e-mail: wiciak@student.agh.edu.pl

G. Dalpiaz et al. (eds.), *Advances in Condition Monitoring of Machinery in Non-Stationary Operations*, Lecture Notes in Mechanical Engineering,
DOI: 10.1007/978-3-642-39348-8_36,

of physical quantities such as speed, load, pressure, temperature, etc., usually called “machine process parameters” [2]. Typically, from each vibration record, a set of diagnostic indicators is calculated known as a “trends”. Each trend point is a combination of representation of true machine technical condition and behavior, machine’s current operating point, measurement error and random factor. In a typical condition monitoring set up, each trend is tracked against a precalculated threshold value. In this case, operational states (hereafter called “states”) are used for data classification during the data acquisition and consequently in data analysis process. Based on these states, data is combined into sets, which are assumed to represent a particular machine.

As a result, the overall number of defined diagnostic indicators and estimators is equal to the number of indicators and estimators calculated from machine kinematic configuration multiplied by the number of states. Therefore, from operator’s point-of-view, it is desirable to have as little states as possible in order to minimize the diagnostic workload. On the other hand, from reliability point-of-view, in order to minimize the fluctuation of machine’s operating points, it is desirable to define ranges of states as narrow as possible. In the latter case, the state configuration would result either in (1) a single operational state with low permissible fluctuation of operational parameters, (2) a large number of operational states with low permissible fluctuations of operational parameters within a state. The first solution might cause a major part of data to be omitted, whereas the latter solution, as it was mentioned, constitutes a major burden to both, computational capabilities as well as to operators of condition monitoring systems.

In this paper, the authors aim in redefining the problem of configuration of machine operational states enabling automatic selection of optimum states’ ranges. The optimization is to be carried out with respect to the range of permissible fluctuation of significant, predefined process parameters.

2 State of the Art

As demonstrated in [3], in many industrial branches, efficient condition monitoring and fault detection of machinery working under non-stationary conditions is a crucial factor in their overall revenue. As an example, the authors give popular wind farms. In authors’ opinion, the subject of automated wind turbine data classification has been explored in literature with the inclination towards clustering methods. So far, it has been shown that clustering algorithms [4] and artificial neural networks might be used for the purpose of states’ selection. According to clustering algorithms, both probability-based methods [4] and so-called “*k-means*” algorithms [5] have already been applied to wind turbines data. Both algorithms are well known clustering methods, which might be easily applied to large data sets as well. However, within this paper, the authors claim that implementation of popular clustering approaches leads to common pitfalls due to mathematical assumptions which are not met in practice. Finally, the authors would like to mention that although few tries have been undertaken to develop state-free analysis of vibration signals [6], currently installed systems still do rely on state-preclassified data.

3 Redefinition of the Problem of Selection of Machine Operational States

3.1 Number of States

According to a most popular design, the number of machine operational states within a CMS is predefined. For instance, in case of wind turbines, it is common to implement two operational states based on theoretical power curve, as endorsed by norms [7]. On the other hand, as illustrated in [5], clustering-based methods calculate the number of states (clusters) on the run.

However, as authors' practices shows, it is very convenient to take advantage of practical aspects of CMS performance, namely to make the number of states flexible, yet to set the number of states to a reasonable number from a diagnostic engineer point-of-view. In this way, the number of operational states of an industrial object is desired to be from 1 to 3. Therefore, apart from scientific justification, it seems permissible to design the algorithm, which will illustrate three resultant propositions of operational states definitions, one for a single state, one for two states, and one for three states.

3.2 Shape of State-Defining Sets

The second aspect concerning practical implementation of machine operational states is the requirement of array-shaped sets, or regions. In this approach, it is desired that selected ranges of process parameters must always create arrays. Considering an exemplary two-variable state definition (commonly power and speed), a selected state might be presented as a particular area, as illustrated in Fig. 1. In Fig. 1, the operational state is defined for Speed variable from S1 to S3 and for Power variable from P1 to P3, but in case when Speed is from S2 to S3, the Power must not be greater than P2.

Although graphical representation of these dependencies is easy to grasp, it introduces unaccepted level of complexity into architecture beneath the logic of industrial CMS. Unfortunately, as shown in Chap. 5, such selections are practically inevitable for clustering-based methods.

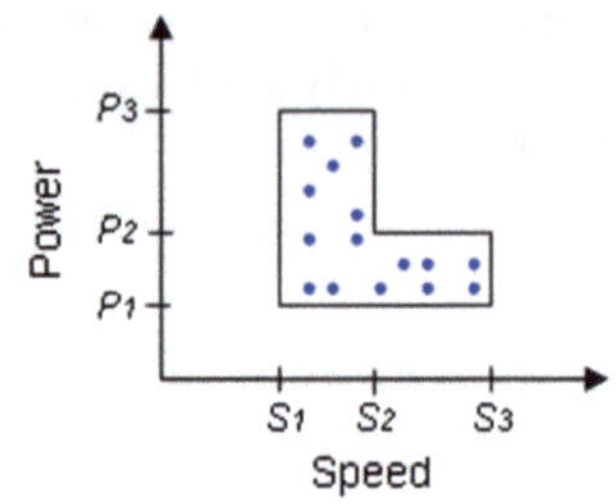

Fig. 1 Representation of non-array state definition

3.3 Intersection of States

The third aspect of states selection refers to intersection of ranges of process channels. From authors' research it is concluded that machine operational states should not have any common ranges. Considering an exemplary two-variable state definition (commonly power and speed), two selected state might be presented as areas with a common part, as illustrated in Fig. 2b.

In this case, as illustrated in Fig. 2a, the first process variable, e.g. "speed" varies from low to high, while the second process variable, e.g. "power" remains at a low level. Such situation might be encountered for instance in wind turbines, where the relation between wind speed, rotor speed, and generated power is modified by complex control systems [8]. At this point, the authors would like to draw a conclusion that from CMS point-of-view, for a given machine, only non-intersecting sets of process parameters should be taken into account because other variations are simply a transient state, which is to be avoided for data analysis. Referring to Fig. 2b, the state characterized by low speed and low power should be considered as valid, whereas the potential state characterized by high speed and low power should be considered as transient, and consequently invalid. Apart from scientific descriptions, in practice, conclusions concerning "normal" operation parameters are easily available from machine operators. Unfortunately, as shown in Chap. 5, such intersections are practically inevitable for clustering-based methods as well.

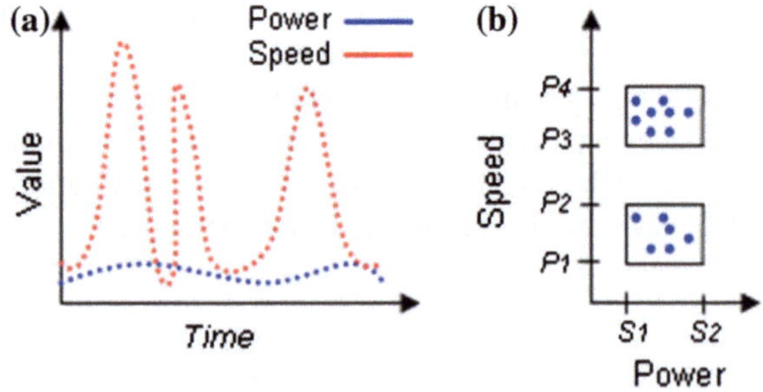

Fig. 2 Representation of intersecting states definition

3.4 Practical Parameterization of the Problem

Finally, the authors propose that states should cover as little multidimensional space as possible, yet enough to meet required number of signals to be registered. In this way, the input parameter to the method is the minimum required number of instances (final time signals), whereas the output includes up to three states defined by ranges of process parameters. In practice, considering process parameters data from a given time interval (for instance three months), it is convenient to specify the minimum (sufficient) number of samples as a percentage of this time. In this way, the CMS designer might build states so that at least a particular percentage of all possible data is to be state-defined.

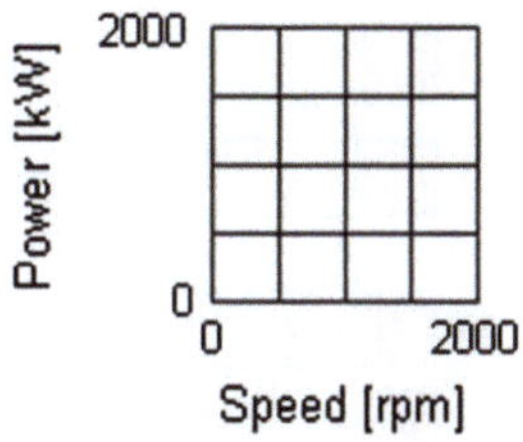

Fig. 3 A schematic representation of transforming 2-dimensional process data into a 2-dimensional grid

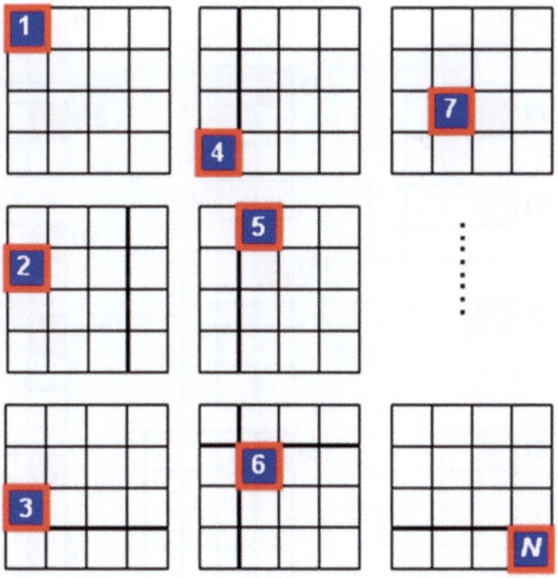

Fig. 4 Representation of scanning of the 2-D grid within the main, outer loop

4 Algorithm Description

The algorithm is illustrated on the example of a wind turbine process data, where operational states are to be calculated based on pairs of generator output power and generator speed assigned to a grid, as illustrated in Fig. 3.

Next, the grid is scanned in two loops. The first loop, main loop covers consecutives areas within the grid, as illustrated in Fig. 4. The second, inner loop performs scanning of all array-shaped combinations, as illustrated in Fig. 5. For each combination, a quality coefficient (QC) is calculated as the number of

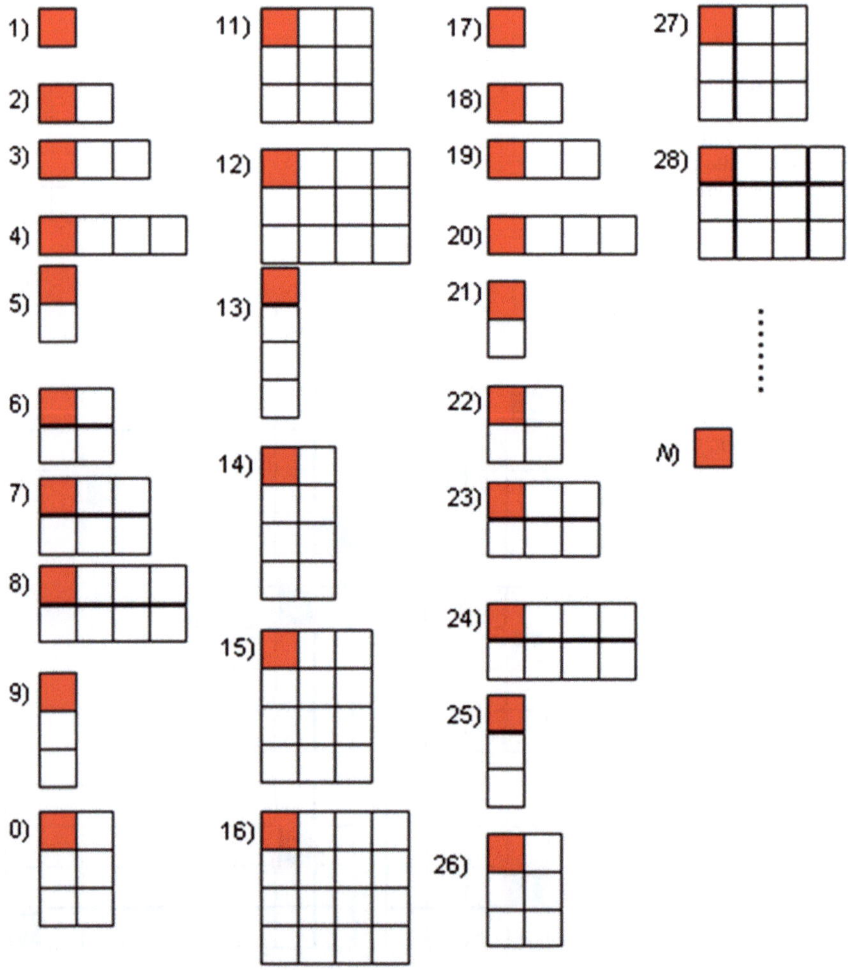

Fig. 5 Representation of scanning of all possible array-shaped combinations. Combinations 1–16 are calculated for cell "1" in Fig. 4, while combinations 17–28 are calculated for cell "2" in Fig. 4, etc

instances, in this examples pairs (power, speed), divided by the range (conveniently by the number of covered grids).

For a square configuration of a grid, the total number of array combinations to be checked (i.e. the length of the quality coefficient) is calculated by:

$$K(n) = \begin{array}{ll} 1 & for\, n = 1 \\ \sum_{i=2}^{n} i^3 + 1 & for\, n \geq 2 \end{array} \cdot \quad (1)$$

Although for a reasonable grid step for real WT data (e.g. 10 Hz, 10 kW), the number of configurations to be checked reaches millions (Table 1), it is easily reduced by implementation of maximum variability of each process parameter within a single array (for instance speed fluctuation less than 200 rpm).

5 Case Study: Wind Turbine Data

The exemplary data comes from a 1.5 MW pitch-controlled wind turbine. Figure 6 illustrates an XY (or scatter) plot of power with respect to generator speed.

Analyzing Fig. 6, it is extremely difficult to draw a conclusion about detailed data distribution. Figure 7 illustrated the same data platted as a color map (note that all cells with more than 30 points are plotted with the same dark-red color.

Table 1 Exemplary number of necessary scanning steps

Size of main grid (*n* by *n*)	Number of configurations to be checked
3	36
10	3,025
20	44,100

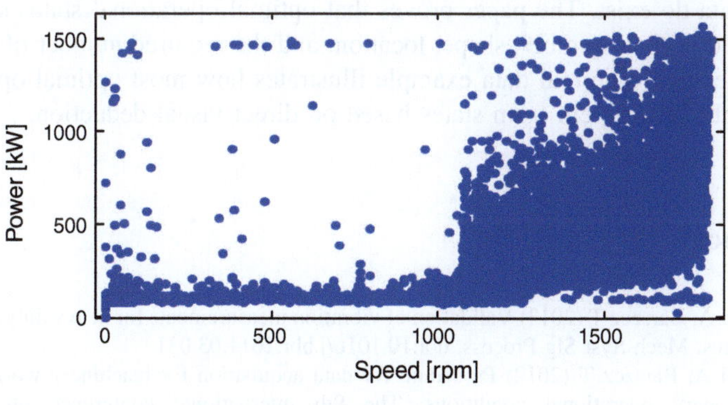

Fig. 6 A scatter plot of power versus generator speed

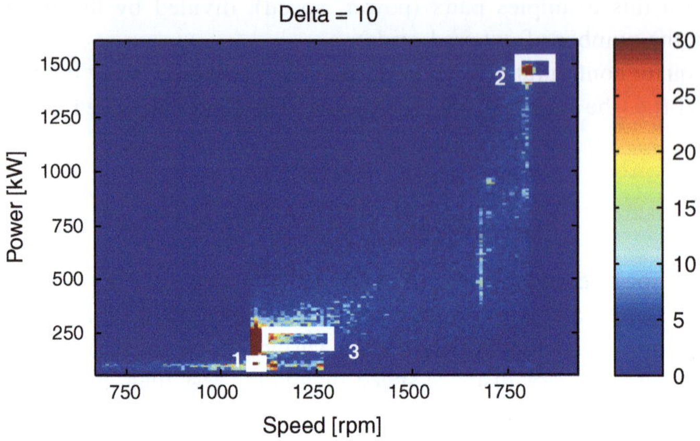

Fig. 7 Proposition of optimal operational states according to their "quality coefficient" (*zoomed figure*), the grid was constructed with 10 rpm and 10 kW resolution

In Fig. 7, the ranges of first three optimal states are marked with white rectangles. The minimum required number of points to be covered was based on daily acquisition need (1,600 out of 30,000 total). Moreover, the maximum allowable range of speed and power within state were set to 200 rpm and 200 kW. The first automatically calculated state (1) covers 4 cells and has a QC1 equal to 496. The second state (2) covers 54 cells with QC2 equal to 30. The third state (3) covers 112 cells with QC3 equal to 15.

6 Conclusion

The application of the algorithm to real data has shown that operational states of machinery should not be selected where we would like them to be but where the desired data do exist. The paper proves that optimal operational states should be constrained in their number, shape, location, and the required number of data in a given time span. The real data example illustrates how most optimal operational states might be different from states based on direct visual deduction.

References

1. Jablonski A, Barszcz T (2013) Validation of vibration measurements for heavy duty machinery diagnostics. Mech Syst Sig Process. doi:10.1016/j.bbr.2011.03.031
2. Jablonski A, Barszcz T (2012) Procedure for data acquisition for machinery working under non-stationary operational conditions. The 9th international conference on condition monitoring and machinery failure prevention technologies, London, 12–14 June 2012

3. Hameed Z et al (2009) Condition monitoring and fault detection of wind turbines and related algorithms: a review. Renew Sustain Energy Rev 13(1):1–39
4. Mykhaylyshyn V et al. (2010) Classification of wind turbine generator operation states for the purpose of advanced vibration monitoring' (Mita-Teknik, Denmard). European wind energy conference and exhibition (EWEC), Warsaw, Poland, 20–23 April
5. Broda D, Jablonski A, Barszcz T (2012) Optimisation of operational state definition for wind farms: part 1: development of algorithms. The 9th international conference on condition monitoring and machinery failure prevention technologies, London, 12–14 June
6. Zimroz R. et al. (2012) Statistical data processing for wind turbine generator bearing diagnostics. The 2nd international conference condition monitoring of machinery in non-stationary operations, Hammamet, Tunisia, 26–28 March
7. Gellermann T (2003) Requirements for condition monitoring systems for wind turbines, AZT Expertentage, Allianz, Nov. 10–11
8. Hau E (2006) Wind turbines: fundamentals, technologies, applications, economics, 2nd edn. Springer, Berlin

Simple Relations for Estimating the Unknown Functions of Incomplete Experimental Spectral and Correlation Response Matrices

Jose Antunes, Laurent Borsoi, Xavier Delaune and Philippe Piteau

Abstract In this paper we suggest two simple approximate methods to estimate the unknown terms of incomplete spectral or correlation matrices, when the cross-spectra or cross-correlations available from multiple measurements do not cover all pairs of transducer locations. The proposed techniques may be applied whenever the available data includes the auto-spectra at all measurement locations, as well as selected cross-spectra which implicates all measurement locations. The suggested formulae can also be used for checking the consistency between the spectral or correlation functions pertaining to measurement matrices, in cases of suspicious data. After presenting the proposed formulations, we discuss their merits and limitations. Then we illustrate their use on a realistic simulation of a multi-supported tube subjected to turbulence excitation from cross-flow.

Keywords System identification · Vibrations · Acoustics · Spectra · Correlation

---

J. Antunes (✉)
Applied Dynamics Laboratory, Campus Tecnologico e Nuclear, Instituto Superior Tecnico, Universidade Tecnica de Lisboa. ADL/CTN/IST/UTL, Estrada Nacional 10, 2695-066 Bobadela, Portugal
e-mail: jantunes@ctn.ist.utl.pt

L. Borsoi · X. Delaune · P. Piteau
Commissariat à l'Energie Atomique et aux Energies Alternatives, CEA, DEN, DM2S, SEMT, Laboratoire d'Etudes de Dynamique, F-91191 Gif-sur-Yvette, France
e-mail: laurent.borsoi@cea.fr

X. Delaune
e-mail: xavier.delaune@cea.fr

P. Piteau
e-mail: philippe.piteau@cea.fr

G. Dalpiaz et al. (eds.), *Advances in Condition Monitoring of Machinery in Non-Stationary Operations*, Lecture Notes in Mechanical Engineering,
DOI: 10.1007/978-3-642-39348-8_37,

1 Introduction

Given a set of R response measurement locations in vibratory or acoustical experiments, several system identification techniques make use of a complete set of $R \times R$ cross-spectral response estimates $S_{ij}(f)$, with $i,j = 1,2,\ldots,R$. Examples of such techniques include the SVD decomposition of cross-spectral measurement matrices $[S_{YY}(f)]$, in order to identify the system modal responses:

$$[S_{YY}(f)] = [U(f)][\Sigma(f)][U(f)]^H \tag{1}$$

where, at each frequency, the columns of matrix $[U(f)]$ contain the singular vectors of $[S_{YY}(f)]$ and the terms of the diagonal matrix $[\Sigma(f)]$ are the corresponding singular values. Recently, such technique became widely used for the purpose of the so-called operational modal identification, when dealing with large structures subjected to unmeasured random excitations, such as bridges and towers, under the excitation of wind or traffic [2–4, 7].

Obviously, because of reciprocity, one has $S_{ji}(f) = S^*_{ij}(f)$ (where the starred function refers to complex conjugate), meaning that only the $(0.5(R-1)+1)R$ cross-spectra of the upper triangular terms must be measured, instead of the full R^2 terms of the spectral matrix. Even so, this may represent a significant amount of measurements to be performed, involving either many transducers or many transduce location changes, and possibly the need for very long signal cables. All these features are certainly inconvenient and often non-feasible. Therefore, we believe that a procedure which would enable the "filling" of incompletely measured spectral matrices might be very convenient and useful in many instances. In the following we will show that, provided some essential assumptions apply to the analyzed system and data, simple approximate methods may be devised for such purpose. Two straightforward estimation methods are suggested and discussed in this paper, which can be applied in many situations of practical interest.

In the proposed approaches, the assumed minimum of available experimental data includes the R auto-spectra $S_{ii}(f)$ from all the measurement locations, as well as $R-1$ cross-spectra involving all transducer locations. A specific set of available cross-spectra, which will be of particular interest here because it fits the experimental constraints of our work, involves measurements between consecutive transducers: $S_{12}(f)$, $S_{23}(f), \ldots, S_{i(i+1)}(f), \ldots S_{(R-2)(R-1)}(f)$, $S_{(R-1)R}(f)$. The simple formulations detailed in the following apply to spectral matrices $[S_{YY}(f)]$, manipulations being easier in the frequency domain. Nevertheless results also apply to correlation matrices $[R_{YY}(\tau)]$, as the former are related to the later through the inverse Fourier Transform $R_{ij}(\tau) = F^{-1}(S_{ij}(f))$.

The suggested formulae may also be used for checking the consistency between the spectral or correlation functions pertaining to measurement matrices, in cases of suspicious experimental data. After presenting the proposed approximate formulations, we discuss their merits and limitations. Then we illustrate their use on a realistic simulation of a multi-supported tube subjected to turbulence excitation from cross-flow.

2 Estimation of Unknown Cross-Spectra

In order to estimate the unknown terms of spectral matrices, consider three generic measurement locations i, j and k, from which the experimental cross-spectra $S_{ij}(f)$ and $S_{jk}(f)$ are known, but not the cross-spectrum $S_{ik}(f)$. It will be discussed in the following that, provided that none of the modes present near-identical frequencies, the unknown cross-spectrum may be estimated through the approximate relation:

$$S_{ik}(f) \approx \frac{S_{ij}(f)S_{jk}(f)}{S_{jj}(f)} \tag{2}$$

In order to establish the validity of Eq. (2), let us express the spectral products of the physical responses $S_{ij}(f)S_{jk}(f)$ and $S_{ik}(f)S_{jj}(f)$ in terms of the system modal responses, assuming N significantly excited modes. Then, if the response signals are not contaminated by significant noise levels, we have:

$$S_{ij}(f) \equiv S_{YY}(x_i,x_j,f) = \sum_{n=1}^{N}\sum_{m=1}^{N} \phi_n(x_i)\,\phi_m(x_j)S_{q_nq_m}(f), \quad i,j = 1,2,\ldots,R \tag{3}$$

and, further assuming that the modal responses are nearly uncorrelated—which is most often true for structures subjected to distributed random loads—then the cross-terms can be neglected, $S_{q_nq_m}(f) \ll S_{q_nq_n}(f)\ ,\forall m,n$. We may thus approximate (3) through the single summation:

$$S_{ij}(f) \approx \sum_{n=1}^{N} \phi_n(x_i)\,\phi_n(x_j)S_{q_nq_n}(f), \quad i,j = 1,2,\ldots,R \tag{4}$$

In matrix terms, this means that the exact expression:

$$\begin{bmatrix} S_{11}(f) & S_{12}(f) & \cdots & S_{1R}(f) \\ S_{21}(f) & S_{22}(f) & \cdots & S_{2R}(f) \\ \vdots & \vdots & \ddots & \vdots \\ S_{R1}(f) & S_{R2}(f) & \cdots & S_{RR}(f) \end{bmatrix} = [\Phi] \begin{bmatrix} S_{q_1q_1}(f) & S_{q_1q_2}(f) & \cdots & S_{q_1q_N}(f) \\ S_{q_2q_1}(f) & S_{q_2q_2}(f) & \cdots & S_{q_2q_N}(f) \\ \vdots & \vdots & \ddots & \vdots \\ S_{q_Nq_1}(f) & S_{q_Nq_2}(f) & \cdots & S_{q_Nq_N}(f) \end{bmatrix} [\Phi]^T \tag{5}$$

may be approximated through:

$$\begin{bmatrix} S_{11}(f) & S_{12}(f) & \cdots & S_{1R}(f) \\ S_{21}(f) & S_{22}(f) & \cdots & S_{2R}(f) \\ \vdots & \vdots & \ddots & \vdots \\ S_{R1}(f) & S_{R2}(f) & \cdots & S_{RR}(f) \end{bmatrix} \approx [\Phi] \begin{bmatrix} S_{q_1q_1}(f) & 0 & \cdots & 0 \\ 0 & S_{q_2q_2}(f) & \cdots & 0 \\ \vdots & \vdots & \ddots & \vdots \\ 0 & 0 & \cdots & S_{q_Nq_N}(f) \end{bmatrix} [\Phi]^T \tag{6}$$

where the columns of matrix $[\Phi]$ are the system modeshapes $\phi_n(x_i)$, with $n = 1,2,\ldots,N$ and $i = 1,2,\ldots,R$.

In general, the cross-spectra $S_{ij}(f)$ obtained from (5) are complex quantities. However, if we assume that the modeshapes $\phi_n(x_i)$ are real, notice that the $S_{ij}(f)$ obtained from (6) must be real, as they are built from the real auto-spectra of the modal responses $S_{q_nq_n}(f)$. This may appear somewhat surprising, but comes as a natural consequence of (6) whenever the modal responses are nearly decoupled, which is actually the case for many situations of practical interest.

Accounting for the previous analysis, one can estimate the products $S_{ij}(f)S_{jk}(f)$ and $S_{ik}(f)S_{jj}(f)$ as:

$$\begin{aligned} S_{ij}(f)S_{jk}(f) &\approx \left[\sum_{n=1}^{N} \phi_n(x_i)\phi_n(x_j)S_{q_nq_n}(f)\right]\left[\sum_{m=1}^{N} \phi_m(x_j)\phi_m(x_k)S_{q_mq_m}(f)\right] \\ &= \sum_{n=1}^{N} \phi_n(x_i)\phi_n(x_j)\phi_n(x_j)\phi_n(x_k)\left[S_{q_nq_n}(f)\right]^2 + \\ &+ \sum_{n\neq m=1}^{N}\sum_{m\neq n=1}^{N} \phi_n(x_i)\phi_n(x_j)\phi_m(x_j)\phi_m(x_k)S_{q_nq_n}(f)S_{q_mq_m}(f) \end{aligned} \tag{7}$$

$$\begin{aligned} S_{ik}(f)S_{jj}(f) &\approx \left[\sum_{n=1}^{N} \phi_n(x_i)\phi_n(x_k)S_{q_nq_n}(f)\right]\left[\sum_{m=1}^{N} \phi_m(x_j)\phi_m(x_j)S_{q_mq_m}(f)\right] \\ &= \sum_{n=1}^{N} \phi_n(x_i)\phi_n(x_k)\phi_n(x_j)\phi_n(x_j)\left[S_{q_nq_n}(f)\right]^2 + \\ &+ \sum_{n\neq m=1}^{N}\sum_{m\neq n=1}^{N} \phi_n(x_i)\phi_n(x_k)\phi_m(x_j)\phi_m(x_j)S_{q_nq_n}(f)S_{q_mq_m}(f) \end{aligned} \tag{8}$$

and it follows from (7) and (8) that the “diagonal” terms of both spectral products are identical, but not the “cross” terms. However, following our assumption of decoupling of the modal responses, one may assume that, in the frequency region Δf_n encompassing each modal peak f_n, the corresponding modal term $S_{q_nq_n}(\Delta f_n)$ is dominant when compared with other neighbor modes with responses $S_{q_mq_m}(\Delta f_n)$, with $m \neq n$. Then, one can write:

$$S_{ij}(\Delta f_n)S_{jk}(\Delta f_n) \approx \phi_n(x_i)\phi_n(x_j)\phi_n(x_j)\phi_n(x_k)\left[S_{q_nq_n}(\Delta f_n)\right]^2 \tag{9}$$

$$S_{ik}(\Delta f_n)S_{jj}(\Delta f_n) \approx \phi_n(x_i)\phi_n(x_k)\phi_n(x_j)\phi_n(x_j)\left[S_{q_nq_n}(\Delta f_n)\right]^2 \tag{10}$$

which are identical. Hence, for sufficiently decoupled modes, one obtains:

$$S_{ij}(f)S_{jk}(f) \approx S_{ik}(f)S_{jj}(f) \tag{11}$$

from which follows the approximation (2).

In spite of its simplicity and potential practical usefulness, we could not find in the literature any mention to expression (2), other than a single work in the context of modal identification [6], where a similar expression $H_{jj}(f_n)H_{ik}(f_n) = H_{ij}(f_n)H_{jk}(f_n)$, based on the system transfer functions at modal frequencies, was suggested. This technique, designated as "triangular measurement", was proposed as a practical alternative to avoid the measurement of collocated transfer functions, considered by these authors prone to errors, or to avoid measurements which would imply the use of very long signal cables, which is in our view perhaps the most interesting application of such formulae.

Notice that, from the approximation (2), a more general expression can be deduced relating the cross-spectra at four locations i, j, k and l. Considering a relation similar to (2), but with respect to location l, we deduce:

$$S_{ik}(f) \approx \frac{S_{il}(f)S_{jk}(f)}{S_{jl}(f)} \tag{12}$$

which obviously reduces to (2) when $l \equiv j$. For the purposes of estimating the unmeasured terms of spectral matrices, we will further generalize (2) to approximate every other term of a spectral matrix, assuming that the auto-spectra and cross-spectra of the following tri-diagonal subsystem are known:

$$[S_{YY}(f)] = \begin{bmatrix} S_{11}(f) & S_{12}(f) & ? & ? & ? & ? & ? \\ S_{21}(f) & S_{22}(f) & S_{23}(f) & ? & ? & ? & ? \\ ? & S_{32}(f) & \ddots & S_{(i-1)i}(f) & ? & ? & ? \\ ? & ? & S_{i(i-1)}(f) & S_{ii}(f) & S_{i(i+1)}(f) & ? & ? \\ ? & ? & ? & S_{(i+1)i}(f) & \ddots & S_{(R-2)(R-1)}(f) & ? \\ ? & ? & ? & ? & S_{(R-1)(R-2)}(f) & S_{(R-1)(R-1)}(f) & S_{(R-1)R}(f) \\ ? & ? & ? & ? & ? & S_{R(R-1)}(f) & S_{RR}(f) \end{bmatrix} \tag{13}$$

Under the assumptions of non-noisy signals and of modal non-degeneracy, any given term $S_{ik}(f)$ may be approximated using (2), however this can be achieved in several manners. For instance, considering a 5×5 matrix:

$$[S_{YY}(f)] = \begin{bmatrix} S_{11}(f) & S_{12}(f) & S_{13}(f) & S_{14}(f) & S_{15}(f) \\ \vdots & S_{22}(f) & S_{23}(f) & S_{24}(f) & S_{25}(f) \\ \vdots & & S_{33}(f) & S_{34}(f) & S_{35}(f) \\ \vdots & & & S_{44}(f) & S_{45}(f) \\ S_{51}(f) & \cdots & \cdots & \cdots & S_{55}(f) \end{bmatrix} \tag{14}$$

the unknown term $S_{25}(f)$ may be alternatively estimated as:

$$S_{25}(f) \approx \frac{S_{23}(f)S_{35}(f)}{S_{33}(f)} \approx \frac{S_{23}(f)}{S_{33}(f)} \frac{S_{34}(f)S_{45}(f)}{S_{44}(f)} \tag{15}$$

$$S_{25}(f) \approx \frac{S_{24}(f)S_{45}(f)}{S_{44}(f)} \approx \frac{S_{23}(f)S_{34}(f)}{S_{33}(f)} \frac{S_{45}(f)}{S_{44}(f)} \tag{16}$$

both (15) and (16) leading to the same result:

$$S_{25}(f) \approx \frac{S_{23}(f)S_{34}(f)S_{45}(f)}{S_{33}(f)S_{44}(f)} \tag{17}$$

which can be generalized, for any term $S_{mn}(f)$:

$$S_{mn}(f) \approx \frac{S_{m(m+1)}(f)\; S_{(m+1)(m+2)}(f)\; \ldots\; S_{(n-2)(n-1)}(f)\; S_{(n-1)n}(f)}{S_{(m+1)(m+1)}(f)\; S_{(m+2)(m+2)}(f)\; \ldots\; S_{(n-2)(n-2)}(f)\; S_{(n-1)(n-1)}(f)} \tag{18}$$

This estimation formula will be tested in Sect. 5, based on a realistic simulation of a tube subjected to turbulence excitation. However, before proceeding, notice that (18) reveals a potential for numerical problems if the product of auto-spectra in the denominator leads to excessively low values. Such situation may arise at several frequencies, whenever the locations chosen for measuring the auto-spectra are near one or more nodal modeshapes. As the actual $\phi_n(x_i)$ are unknown at this stage, such unfortunate situation can easily arise, leading to ill-conditioning of (18) and possibly wrong estimates of $S_{mn}(f)$. Of course, one may attempt to mitigate this difficulty using some form of regularization, such as Tikhonov technique. However, we will develop in the following section another simple formulation alternative to (18), in order to bypass the problem.

3 An Alternative Estimation Technique

The starting point for this approach is that, if as assumed the response signals are not contaminated by significant noise and the modal responses are sufficiently decoupled, then cross-spectral coherency at response peaks will approach the maximum value. Then, we may write in the vicinity of the modal frequencies:

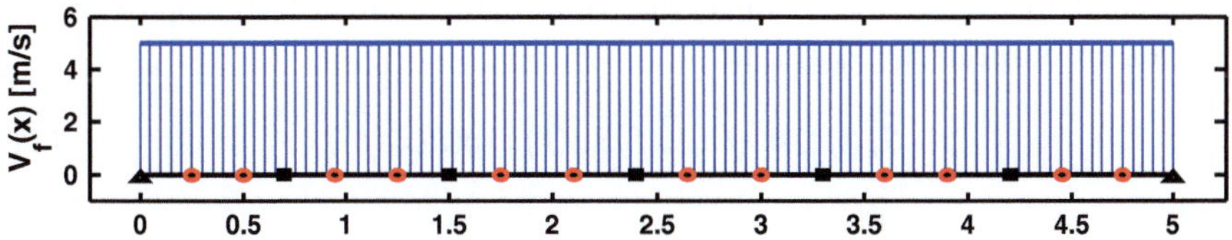

Fig. 1 Multi-supported tube subjected to flow turbulence excitation (uniform flow velocity profile)

$$\gamma_{ik}^2(\Delta f_n) = \frac{|S_{ik}(\Delta f_n)|^2}{S_{ii}(\Delta f_n)S_{kk}(\Delta f_n)} \approx 1 \Rightarrow |S_{ik}(\Delta f_n)| \approx \sqrt{S_{ii}(\Delta f_n)S_{kk}(\Delta f_n)} \tag{19}$$

therefore, if one adopts this assumption, the cross-spectra may be estimated as:

$$S_{ik}(f) = |S_{ik}(f)|\exp(i\theta_{ik}(f)) \approx \sqrt{S_{ii}(f)S_{kk}(f)}\exp(i\theta_{ik}(f)) \tag{20}$$

where the phase angle $\theta_{ik}(f)$ may be estimated based on (2):

$$S_{ik}(f) \approx \sqrt{S_{ii}(f)S_{kk}(f)}\exp(i\theta_{ik}(f)) \approx \frac{S_{ij}(f)S_{jk}(f)}{S_{jj}(f)} \tag{21}$$

or:

$$S_{jj}(f)\sqrt{S_{ii}(f)S_{kk}(f)}\exp(i\theta_{ik}(f)) \approx \left|S_{ij}(f)\right|\exp\left(i\theta_{ij}(f)\right)\cdot\left|S_{jk}(f)\right|\exp\left(i\theta_{jk}(f)\right) \tag{22}$$

so that, the phase of auto-spectra being always nil:

$$\exp(i\theta_{ik}(f)) \approx \exp\left(i\theta_{ij}(f)\right)\exp\left(i\theta_{jk}(f)\right) \Rightarrow \theta_{ik}(f) = \theta_{ij}(f) + \theta_{jk}(f) \tag{23}$$

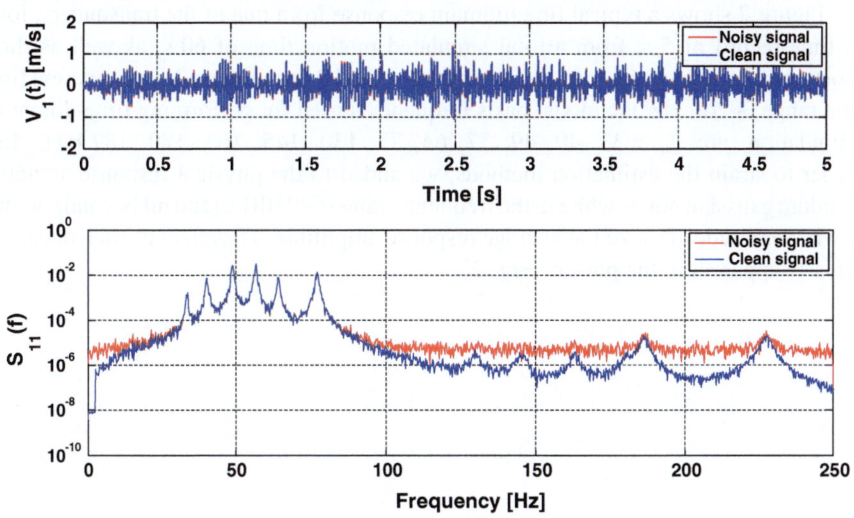

Fig. 2 Typical time-domain response sample and auto-spectrum (transducer 1)

Generalizing these results, an alternative formulation to (18) for approximating a given cross-spectrum $S_{mn}(f)$, is:

$$S_{mn}(f) \approx \sqrt{S_{mm}(f)S_{nn}(f)}\exp(i\theta_{mn}(f)) \tag{24}$$

with:

$$\theta_{mn}(f) = \theta_{m(m+1)}(f) + \theta_{(m+1)(m+2)}(f) + \cdots + \theta_{(n-2)(n-1)}(f) + \theta_{(n-1)n}(f) \tag{25}$$

which does not entail numerically treacherous computations, whatever the values of the modeshapes at the measurement locations.

4 Illustrative Simulated System

We will now illustrate how the previous formulations perform in a realistic situation. The following results are based on computations performed when simulating the vibratory response of a multi-supported tube subjected to the distributed random excitation by a turbulent cross-flow with constant velocity profile (velocity 5 m/s), as shown in Fig. 1. Effective techniques for modeling the turbulence excitation and the dynamics of such system have been thoroughly described elsewhere by the authors [1, 2].

As shown in Fig. 1, the tube with total length 5 m, has 6 spans, each one provided with a transducer (supports are shown using triangles and squares, while measurement locations are shown using circles). Time-domain simulation of the tube excitation encapsulates all effects connected with turbulence, namely the spectral properties and space correlation of the turbulence eddies—see the aforementioned references for modeling and implementation details.

Figure 2 shows a typical time-domain response from one of the transducers, for a time-record of 5 s, from a total simulated motion time of 60 s, as well as the corresponding auto-spectrum, where the many modal response peaks are apparent (in the range 0–200 Hz, the modal basis frequencies used for performing time-domain simulation are $f_n = 33$, 40, 49, 57, 64, 77, 130, 145, 163, 180, 187 Hz). In order to strain the estimation methods, we added to the physical response signals random gaussian noise, white in the frequency range 0–250Hz, (and nil beyond), with RMS amplitude 10 % of the average response amplitude. The effect of such noise is clearly apparent in the plots of Fig. 2.

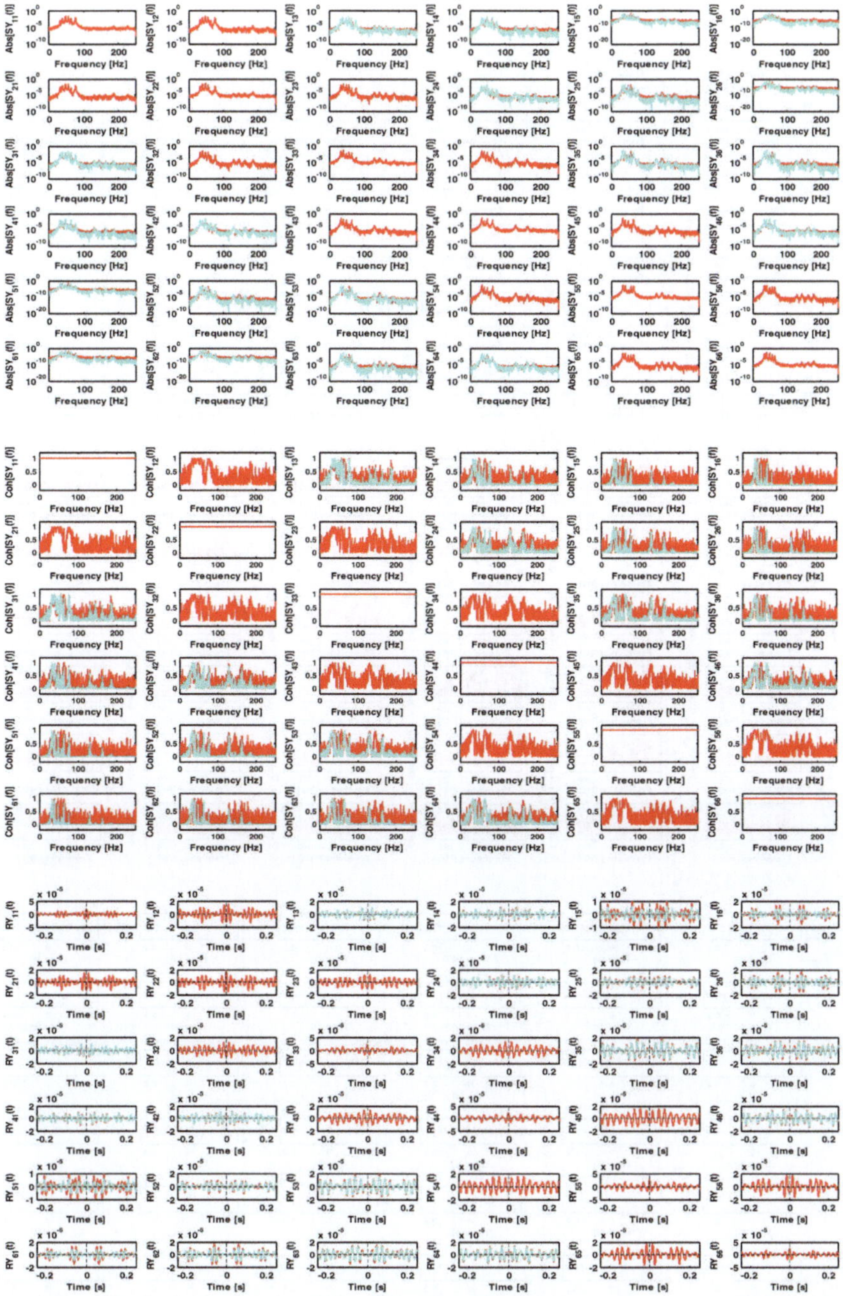

Fig. 3 Absolute value and coherency of the spectral matrix, as well as the corresponding correlation functions estimated from the response vibrations at six locations along the tube; *Red* = actual noisy data, *Cyan* = estimated data using Eq. (18)

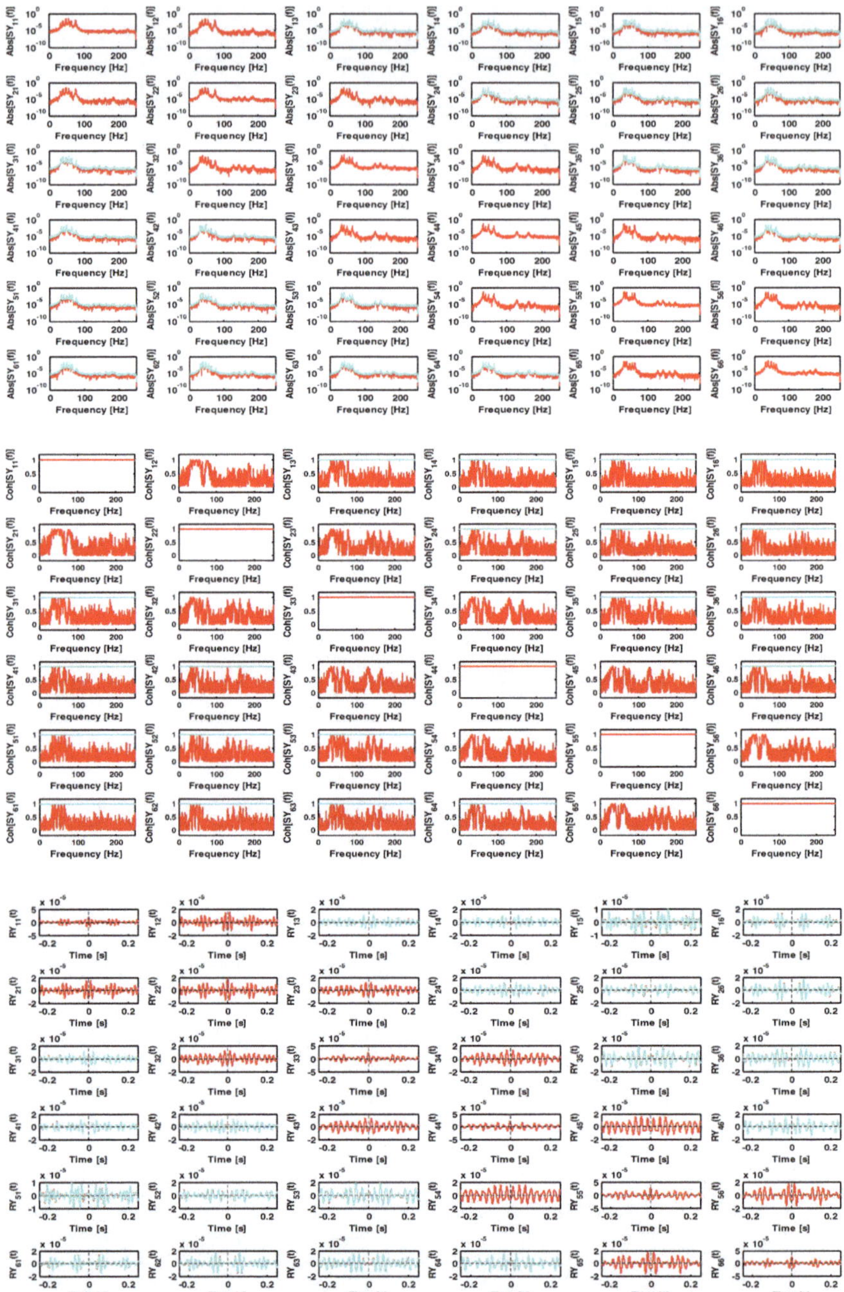

Fig. 4 Absolute value and coherency of the spectral matrix, as well as the corresponding correlation functions estimated from the response vibrations at six locations along the tube; *Red* = actual noisy data, *Cyan* = estimated data using Eqs. (24–25)

5 Evaluation of the Estimation Techniques

The results from the approximation formula (18) are shown in Fig. 3, while those obtained using approximation (24–25) are presented in Fig. 4. Shown are the real noisy response data (Red) and estimated data (Cyan) of the 6×6 response matrices, in the following manner: absolute value of the spectra (upper plots), coherency of the spectra (central plots) and a zoom of the corresponding correlation functions (lower plots).

Comparing the real cross-spectra and cross-correlation functions shown in these plots—which were obtained directly from the time-domain simulation responses—with the corresponding estimations, it appears that the proposed approximation procedures behave fairly well, even if a significant amount of noise is added to the system response signals. The most illuminating plots shown in Figs. 3 and 4 are perhaps those of the cross-correlations $R_{ij}(\tau)$, which best display differences between the real and estimated data. In particular, it is clear that the second proposed approximation method (24–25) is more robust than the basic technique (18), as expected. However, the coherency functions thus obtained are always unity, see (19), a feature with no further unpleasant consequences.

6 Conclusion

In this work we proposed two straightforward methods that, under mild restrictions, can be used either: (1) to estimate unknown cross-spectra in incomplete spectral or correlation matrices, or (2) to cross-check suspicious experimental data. The proposed methods were tested using realistic numerical simulations of a multi-supported tube subjected to the excitation by flow turbulence. The results obtained are quite convincing and the presented analysis stresses the idea that, for coherent vibratory or acoustical response measurements from linear systems, the full spectral matrix can be approximated using only the diagonal auto-spectra and a limited set of cross-spectra. However, if some of the modal frequencies are near-degenerate, the estimated spectra will be erroneous in the frequency ranges of the mode doublets. More sophisticated techniques are needed to address this problem.

References

1. Antunes J, Borsoi L, Piteau P, Delaune X (2012) The equivalent spectrum concept for turbulence conveying excita-tions. In: 10th international conference on flow-induced vibration & flow-induced noise (FIV 2012), Dublin, Ireland, 2–6 July 2012
2. Brincker R, Zhang L, Andersen P (2001) Modal identification of output-only systems using frequency domain decomposition. Smart Mater Struct 10:441–445

3. Gentile C, Saisi A (2007) Ambient vibration testing of historic masonry towers for structural identification and damage assessment. Constr Build Mater 21:1311–1321
4. Magalhães F, Cunha A (2011) Explaining operational modal analysis with data from an arch bridge. Mech Syst Signal Process 25:1431–1450
5. Piteau P, Delaune X, Antunes J, Borsoi L (2012) Experiments and computations of a loosely supported tube in a rigid bundle subjected to single-phase flow. J Fluids Struct 28:56–71
6. Schwartz B, Richardson M (2003) Scaling mode shapes obtained from operating data. Sound Vibr 37 :18–22
7. Zhang L, Wang T, Tamura Y (2010) A frequency-spatial domain decomposition method for operational modal analysis. Mech Syst Signal Process 24:1227–1239

Condition Monitoring Under Non-Stationary Operating Conditions using Time–Frequency Representation-Based Dynamic Features

O. Cardona-Morales, D. Alvarez-Marin and G. Castellanos-Dominguez

Abstract Condition monitoring is useful to describe the machine state under current operating regimes, especially, when non-stationary operating conditions appears. Nevertheless, in actual applications the faulty data are not always available. This paper proposes a novel methodology for condition monitoring using dynamic features and one-class classifiers. The dynamic features set comprises the spectral sub-band centroids and linear frequency cepstral coefficients computed from time–frequency representations. A one-class classification stage is carried out to validate the performance of the dynamic features and commonly used statistical features as descriptors of the machine state. Proposed methodology is evaluated by using a test rig, which is composed by outliers (unbalance and misalignment) and target objects (undamaged state). The data set is obtained under variable speed conditions including start-up and coast-down. The attained results outperform other state-of-the-art extracted parameters and the methodology is robust to large speed fluctuations in the machine.

Keywords Condition monitoring · Dynamic features · Non-stationary · Vibration signals · One-class classification

1 Introduction

In modern industries, fault detection in rotating machinery is a fundamental issue since it helps to reduce the unnecessary expenditures in repairs while improving machine performance. On this matter, the main challenge is to determine the current state of the machine from a set of measurements, so called Condition-

O. Cardona-Morales (✉) · D. Alvarez-Marin · G. Castellanos-Dominguez
Universidad Nacional de Colombia, Signal Processing and Recognition Group,
Campus La Nubia, 17001000 Manizales, Colombia
e-mail: ocardonam@una.edu.co

G. Dalpiaz et al. (eds.), *Advances in Condition Monitoring of Machinery in Non-Stationary Operations*, Lecture Notes in Mechanical Engineering,
DOI: 10.1007/978-3-642-39348-8_38,

Based Maintenance (CBM). Usually, the state of the machine is assessed using vibration analysis because it gives a high precision and it has a low cost. However, two problems arise: the first one is regarded to the time-varying machine conditions, either by changes in speed or load, yielding to non-stationary vibration signals; the second problem is associated to the amount of available data characterizing the different states of the machine, since in most of the cases the recordings of the damaged machine are the only ones available; this fact hinders the application of conventional classification techniques due to the strong imbalance of the faulty/normal classes (states of the machine).

With regard to the former problem, several authors have proposed methods to characterize the dynamic behavior of the machine under non-stationary operating conditions taking into account an insufficient amount of damaged condition signals. Therefore, describing the time-varying behavior supposes the use of time–frequency representations (TFR) in order to analyze frequency changes through the time domain and how to associate the spectral components to some kind of damage. For example, the full-spectrum and the continuous wavelet transform (CWT) are used in [1, 2], respectively, in order to identify the dynamic behavior of the machine. However, they visually show the difference between vibration signals acquired from a test rig for several machine states. In [3] summarizes different methods for estimating the energy concentration from several TFR according to a set of test rig faults. Recently, a promissory TFR has been developed aiming to reduce the problems associated to the interpretation of conventional TFR, which means both improved concentration and eliminated misleading interference terms [4]. Thereby, a wavelet-based TFR enhancement method is proposed in [5], known as synchrosqueezing, which reallocates the energy around of instantaneous frequency in the scale dimension. Nonetheless, those TFR comprise a lot of points, and their post-processing imply a high computational cost [6]. On the other hand, there are some statistical features, extracted from time and frequency domain, such as mean, standard deviation, skewness, kurtosis, root mean square (RMS), crest factor, shape factor and others, are commonly used because of their computation easiness [7–9].

With regard to the second problem, one-class classification (OCC) techniques have been used in order to determine when the machine state ceases to be normal and when the first symptoms of damage appear. For instance, several standard one-class classifiers such as the normal distribution classifier, the *k*-nearest neighbor (*k*-NN) classifier and support-vector data description (SVDD) are compared in [10]. Considered classifiers are trained and tested employing vibration signals at different constant speeds by using a set of statistical-based features. However, the data description performance is low. Therefore, several authors have proposed different methodologies for improving the classification performance by estimating statistical features from piecewise segmented non-stationary vibration signals. Some of the most commonly used methodologies are weighted SVDD [11], moving-average model [12], wavelet packet transform [13] and subspace reduction by principal component analysis (PCA) [14]. Those approaches reach a high classification performance but, in some practical cases, the signal segmentation entails a loss of information either in time or frequency [15].

In this paper, a novel methodology for mechanical systems description with non-stationary behavior is introduced. In particular, the proposed approach uses the spectral sub-band centroids and the linear frequency cepstral coefficients, all of them extracted from a TFR. Due to the large number of features obtained from the TFR, a feature selection process is carried out in order to determine the amount of most relevant dynamic characteristics. Finally, resulting dynamic features are validated by a OCC and the performance is compared against state-of-the-art statistical features. Proposed methodology is tested with a dataset collected in a test rig for normal, unbalanced and misaligned assemblies. Recordings are acquired for several intensity levels and under variable speed conditions including machine Start-up and Coast-down.

2 Background

2.1 Generation of Time-Varying Features

Let $\{x(l); l = 1,\ldots,L\}$ be a given vibration signal, defined along $L \in \mathbb{R}$ time instants and $S_x(l,k) \in \mathbb{R}^{L\times K}$ its respective time–frequency representation (TFR), where $\{k; k = 1,\ldots,K\}$ denotes the ordinal index in frequency. The basic goal of a TFR is to determine the energy concentration along the frequency axis, $\{f = k\Delta f; f \in (0, f_s/2)\}$, at a given time instant, where f_s denotes the sample frequency. Nonetheless, either TFR holds a huge amount of non-relevant information. Therefore, it is of primal importance the definition of methods allowing to extract salient and discriminant information from the vibration signal [6].

Taking into account that the time–frequency (*t–f*) analysis models the signal spectral density as a function of time under the assumption that the spectral content remains stationary within the small time intervals of computation, then short–time parameters extracted from TFR can be considered. Adequate candidates for feature extraction methods are the *Spectral Sub–band Centroids* (SSC) [16] and the *Linear Frequency Cepstral Coefficient* (LFCC) [17]. Both of TFR–based short–time parameters are extracted by using a filter–banked decomposition, when efficiently combining frequency and magnitude information from the short–term power spectrum input signals. Time–variant outputs of these filters, that might be chosen so as to cover the most relevant part of the frequency range, are regarded as the set of time–variant features $\mathbf{y}_n = \{y_n(l) : n = 1,\ldots,n_f\}$, with $y_n(l) \in \mathbb{R}$. Therefore, sampled vector over discrete time, l, of each narrow–band feature, $\mathbf{y}_n$, is attained by filter bank modeling. For instance, using the set of LFCC that is extracted by Discrete Cosine Transform of triangular log–filter banks, $\{h_m(k) : m = 1,\ldots,M\}$, linearly spaced in the frequency domain:

$$y_n(l) = \sum_{m=1}^{M} \log(s_m(l)) \cos\left(n\left(m - \frac{1}{2}\frac{\pi}{n_f}\right)\right) \quad (1)$$

where n_f is the number of desired LFCC features to be considered, and $s_m(l) \in \mathbb{R}$ is the weighted sum of each frequency filter response set, $s_m(l) = \sum_{k=1}^{K} S_x(l,k)h_m(k)$, being m, l, and k the indexes for filter ordinal, time, and frequency axes, respectively; K stands for the number of samples in the frequency domain.

Other effective way of generating t–f-based time–variant features can be achieved through computation of the histograms of the SSC that are estimated for each filter in the frequency domain, $h_n(k)$, by:

$$y_n(l) = \frac{\sum_{k=1}^{K} kh_n(k)S_x^{\gamma}(l,k)}{\sum_{k=1}^{K} h_n(k)S_x^{\gamma}(l,k)} \quad (2)$$

where $\gamma \in \mathbb{R}^+$ is a parameter that represents the dynamic range of the spectrum, used in the computation of the centroid, and the filters $h_n(k)$ are linearly distributed along the spectrum. In addition, the energy around each centroid can be also considered as time–variant feature that for a fixed bandwidth Δk is computed by means of:

$$y_n(l) = \sum_{k=\hat{y}_n(l)-\Delta k}^{\hat{y}_n(l)+\Delta k} S_x(l,k) \quad (3)$$

where $\hat{y}_n(l)$ is the actual value of the time–variant centroid estimated by 2.

2.2 One-Class Data Inference

Based on optimal signal detection inferring whether the signal is present, different approaches to distinguish one class from the rest of the data feature space, $\mathbf{Z} = [\mathbf{z}_1, \ldots, \mathbf{z}_i, \ldots, \mathbf{z}_q]; \mathbf{Z} \in \mathbb{R}^{r \times q}$, had been developed, (being r the data dimension and q the number of available objects). Particularly, the measured data space is related to just one of the classes (termed target) that can be properly characterized as well as compactly clustered, in such a way as to guarantee the discrimination of other possible objects (that is, non-target class from which no measurements are available) distributed outside of the target class. Therefore, to circumscribe the target class within concrete bounds two concepts are introduced: (*i*) the distance, $d(\mathbf{z}_i) \in \mathbb{R}^+$, that measures the closeness of a $\mathbf{z}_i$ object to the target class, and (*ii*) the threshold $\theta \in \mathbb{R}^+$ on this distance, that fixes the decision boundary of the target class, that is:

$$d(\mathbf{z}_i) < \theta, \mathbf{z}_i \rightarrow \textit{target class}$$

$$d(\mathbf{z}_i) > \theta, \mathbf{z}_i \rightarrow non\ target\ class \tag{4}$$

The definition of the adequate classification boundary around target class remains as the most important issue. Moreover, the threshold θ should accept as many objects as possible from the target class, while minimizing the chance of accepting non-target (or outlier) objects [18]. In practice, used distances range from the simplest Euclidean to more elaborated ones, by instance the statistical-based distances. As regards the former distance, it may face several restrictions describing a poor dense volume of the hyper-sphere, no mentioning that this algorithm requires the tuning of number of neighbors k The latter distances are more robust since they impose a model to the OCC that allows to provide a highly dense volume of the decision hyper-sphere. Specifically for implementing the OCC, the Gaussian distribution classifier (using Mahanalobis distance) [19] and the Support Vector Data Description (using kernel based square distance) [10], are employed.

3 Experimental Analysis

3.1 Collected database from experimental test rig

A set of experiments is performed with the supplied test rig shown in Fig. 1a, which includes a 2 HP Siemens electromotors with a maximum speed of 1,800 rpm. The motor is connected to the shaft by means of a rigid coupling. The shaft has two supports, each one with a SKF-6005 NR ball bearing and two drilling wheels designed for simulating either static or dynamic unbalance. The sensors are located in both horizontal and vertical directions. The vibration signals are acquired by ACC102 accelerometers, with a measurement range of 0 – 10 kHz

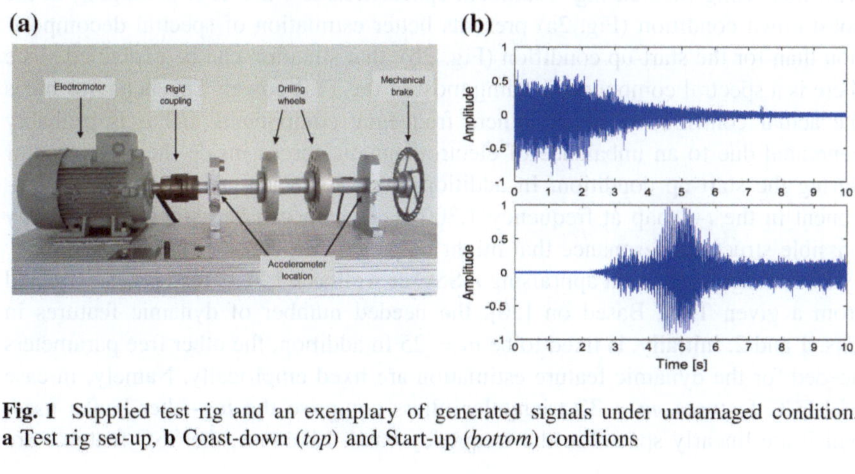

Fig. 1 Supplied test rig and an exemplary of generated signals under undamaged condition. **a** Test rig set-up, **b** Coast-down (*top*) and Start-up (*bottom*) conditions

and 100 mV/g of sensibility. A National Instruments USB-6009 data acquisition card is employed using a sampling frequency of 20 kHz.

The data set holds the following types of acquired outliers regarding the considered machine states: (1) two static and one dynamic unbalance, and (2) two angular and two parallel misalignments. The data collection also includes an undamaged condition type, which is taken as the only target class. The machine state is measured for start-up and coast-down conditions, where each recording under coast-down condition (Fig. 1b-top) contains three phases: (1) maximum speed (1,800 rpm), (2) turning motor off, and (3) steady–state regime. The start-up condition case (Fig. 1b-bottom) has the same phases, but in reverse order. Each recording is ten seconds long and half of the time the machine spends to change its phase. It is worth noting that considered working phases are not synchronized each to other, that is, the second phase may begin at different times within each recording.

As a result, 20 recordings were acquired at each horizontal measurement point for each one of the 8 considered types of machine states, i.e., in total 160 recordings were collected; nonetheless, each recording is downsampled to 4 kHz, that is, the length of the recording is $L = 40{,}000$.

3.2 Estimation of Dynamic Features

In this work, two different TFR, the Short-Time Fourier Transform (STFT) [3] and the Synchrosqueezing Wavelet Transform [5] are used. For the STFT, the TFR matrix has dimension $S^{40{,}000\times 256}$, where $K = 256$ is chosen as to exceed a reasonable resolution of 0.1 Hz. Given the sampling frequency of 4,000 Hz, in case of SWT, the number of scales is fixed according to [5], that is, $K = 375$, ranging from 0.0001 to 1.649 So, TFR matrix has dimension $40{,}000 \times 375$ Figure 2 shows the spectral decomposition of each signal using the TFR described above, i.e. STFT (top) and SWT (bottom). As seen, spectral components expand (contract) with increasing (decreasing) rotational speed. Besides, the TFR belonging to the coast-down condition (Fig. 2a) presents better estimation of spectral decomposition than for the start-up condition (Fig. 2b); that situation can be explained since there is a spectral component holding most of the TFR power. This behavior hides the actual contribution of the others frequency components and it is probably generated due to an unbalance of electromagnetic forces inside the electromotor during the start-up condition. In addition, there is an appreciable spectral component in the *t–f* map at frequency 1,300 Hz, which can be associated with any possible structural resonance that might be present on the test rig.

In the beginning, a set appraising n SSC as well as n LFCC features is extracted from a given TFR. Based on [20], the needed number of dynamic features in Eqs. 1 and 2, initially, is fixed to be $n_f = 25$ In addition, the other free parameters needed for the dynamic feature estimation are fixed empirically. Namely, in case of LFCC features, $m = 32$ triangular filters comprise the log–filter banks used, which are linearly spaced in the frequency band. Likewise, the SSC features are

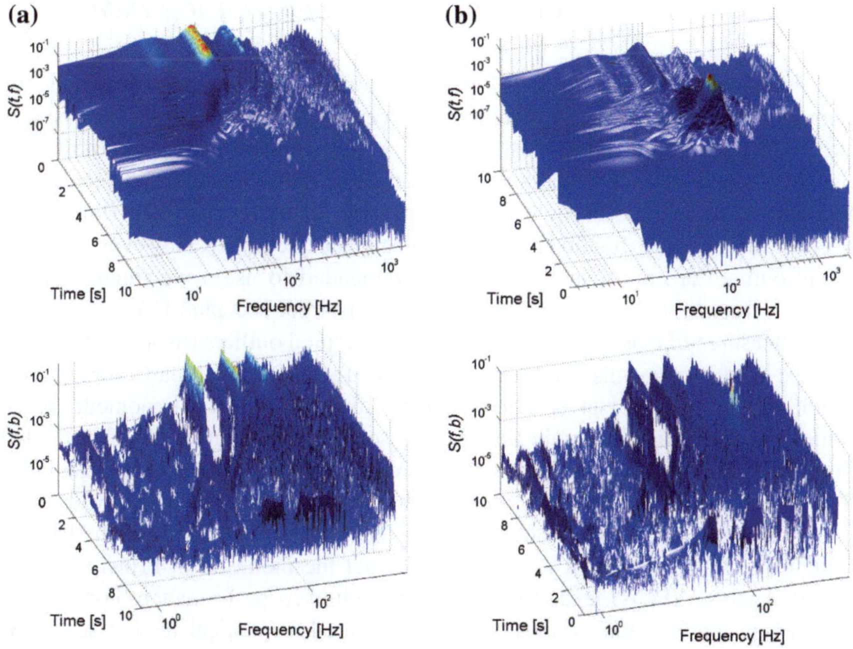

Fig. 2 Time-frequency representations from acquired signals on the test rig under two conditions, using STFT (*top*) and SWT (*bottom*). **a** Coast-down, **b** Start-up

estimated using a dynamic range $\gamma = 1$ to preserve the spectral power of the TFR. Consequently, 50 dynamic features are extracted from considered TFRs; each one having 40,000 samples of length. Therefore, extracted feature set holds a huge amount of information to be fed into the classifier.

From the above huge achieved dynamic feature set, it follows that there is a need for an adequate dimension reduction, which in this cases, is provided in two stages: Firstly, each dynamic feature is adequately represented by the first statistical moment, i.e., $\mathcal{E}\{\mathbf{z}_n(l) : \forall l \in L\}$, that is, $\mathbb{R}^{L\times 1} \mapsto \mathbb{R}^1$, being $\mathcal{E}\{\cdot\}$ the expectation operator. Secondly, the obtained feature space matrix has dimension 25×160 for each considered TFR. Therefore, a multivariate latent approach is used to select the data set to be fed into the one-class classifier, for either TFR. Lastly, we determine the optimal number of features, n_f that are required to properly characterize the vibration signal, inasmuch as this parameter controls the computational cost and the performance within the classification framework. To select an adequate number of features, the following criterion of multivariate reconstruction error is carried out: $\min_{\forall n}\{\mathcal{E}\{\| x(t) - \hat{x}(t) \|_2\} - \eta\} = 0$, where $\hat{x}(t)$ is the reconstructed signal, h is a small enough real positive value, $\| \cdot \|_2$ is the norm squared value. As result, the minimum number of LFCC features to achieve an explained variance value of 97.6 % is $n_f = 16$, while $n_f = 20$ for SSC parameters are needed for to reach a variance of 96.2 %. Roughly speaking, the explained variance can be fixed as 0.976 Therefore, the value $\eta \approx 0.024$.

3.3 Target Class Classification Validated on Test Rig Data

For classifier validation, the OCC error performance is computed by using the commonly used 10-fold cross-validation procedure, where the target data is split into 70 % for training and the rest 30 % is merged with the outlier data (nontarget class). When outliers are available, an acceptable discrimination between target and outlier objects means both a small fraction of outlier accepted (false-positive—f_p) and a large fraction of target objects accepted [10]. However, in cases where outliers are not available, it is recommended to use a weighted measure known as *recall*, which means the proportion of target accepted [15].

As measures of classifier performance, the accepted outliers (false positive—f_p) and the classifier recalls (*rec*) are shown in Table 1 (upper part) for start-up condition. The feature set is extracted using the first statistical moment (mean value) reducing each dynamic feature. As seen, the estimated LFCC extracted from STFT present the best performance regardless of the considered classifier. Nonetheless, it is worth noting that the higher performance is achieved using Gauss classifier since it does not misclassify outliers while preserving the highest value of achieved recalls. It is worth noting that the inclusion of Principal Component Analysis (PCA) leads to improve the achieved performance. For sake of comparison, performance values are also computed for statical feature set (SFS) (24 features statical extracted in time and frequency domains [8, 9]); values achieved in this case exhibit worse outcomes.

Table 1 Outlier accepted (f_p) and recall (*rec*) in (%), using the recordings from test rig under considered non-stationary operating conditions

Features	Start-up				Coast-down			
	Gauss		SVDD		Gauss		SVDD	
	f_p	*rec*	f_p	*rec*	f_p	*rec*	f_p	*rec*
STFT-LFCC	**0**	**100**	0	92	0	92	0	80
STFT-LFCC-PCA	**0**	**100**	0	88	0	94	1	82
STFT-SSC	0	92	0	98	**0**	**98**	0	86
STFT-SSC-PCA	0	98	0	94	0	90	0	86
SWT-LFCC	0	90	0	88	0	68	0.5	48
SWT-LFCC-PCA	0	90	0	88	0.1	70	1.1	56
SWT-SSC	0	74	0	64	1.5	56	3.5	64
SWT-SSC-PCA	0	82	0	78	0	68	0.7	44
SFS	2.3	94	0.5	44	0.8	92	0	52
SFS-PCA	4.2	88	3.1	80	20	86	0.9	28

The outcomes using the proposed dynamic features (upper part) and the state-of-the-art statistics features (lower part)

4 Discussion

After testing of discussed training methodology for rotating machine diagnosis, some assertions can be stated regarding the time–frequency representations from the dynamic features are extracted as follows:

- With respect to used TFRs, the best performance results are obtained when the STFT is employed, despite the SWT holding better energy concentration, at least, for both tested experimental databases. Although, the difference in term of achieved classifier performance remains close for both TFRs when the dynamics of input vibration signal becomes slow enough. The STFT preserves high rates of classification, having in most of the cases no outliers accepted; that performance should be considered as an acceptable result in CBM, since it reduces the possible expenses associated to any rotating machine damage.
- The static features (SFS) are useful when changes observed in the dynamics are of smooth nature, since this type of features are strongly dominated by the amplitude and the localized frequencies in the signal. However, it is possible to reach more precision in the condition monitoring if the estimated features are related with the energy. Also, it is worth noting that the generalization capability reached by using statistical features is low to distinguish normal samples from outliers. Therefore, this type of features may face serious restrictions to be considered for fault diagnosis, specially, in attempting to track for time-variant damages within the CBM program. Instead, the proposed dynamic features achieve better performance, particularly, if they are extracted from the STFT.

5 Conclusions

A novel methodology for CBM based on the estimation of dynamic features such as LFCC and SSC from TFR is proposed. The methodology improves the characterization of the dynamic data behavior allowing to get a high classification of target class. The proposed characterization process allows completely to analyze the vibration signal reducing the loss of information caused by the signal segmentation, and it estimates the amount of required dynamic features in order to describe the behavior of rotative machines offering a properly reconstruction. Using the dynamic features several statistical features are estimated in order to compare with the best statistical features proposed in the state-of-the-art. As a result, the methodology overcomes the obtained performance by means those statistical features. In general, the proposed dynamic features allow to classify different time-variant behaviors and it presents a high performance, either rejecting outliers or accepting targets. Finally, using this methodology, it is possible to reduce the classifier complexity improving the data description based on the presented dynamic features. In order to do further research, this methodology should include more wide class of dynamic features. Also, the methodology will be test with other types of mechanical systems.

The authors acknowledge to Colciencias and the Universidad Nacional de Colombia for the financial support of the research projects "*Sistema autónomo de monitoreo de vibraciones para diagnóstico de fallas no estacionarias*" (with code 1101-521-28792), and "*Sistema embebido para el análisis de señales de vibraciones mecánicas orientado al diagnóstico de fallas en máquinas rotativas*".

References

1. Fengqi W, Meng G (2006) Compound rub malfunctions feature extraction based on full-spectrum cascade analysis and SVM. Mech Syst Signal Process 20(8):2007
2. Wu JD, Chen JC (2006) Continuous wavelet transform technique for fault signal diagnosis of internal combustion engines. NDT E Int 39(4):304
3. Sejdic E, Djurovic I, Jiang J (2009) Time-frequency feature representation using energy concentration: An overview of recent advances. Digit Signal Process 19(1):153
4. Li C, Liang M (2012) Time-frequency signal analysis for gearbox fault diagnosis using a generalized synchrosqueezing transform. Mech Syst Signal Process 26(0):205. doi:10.1016/j.ymssp.2011.07.001. http://www.sciencedirect.com/science/article/pii/S0888327011002895
5. Daubechies I, Lu J, Wu HT (2011) Synchrosqueezed wavelet transforms: An empirical mode decomposition-like tool. Appl Comput Harmonic Anal 30(2):243. doi:10.1016/j.acha.2010.08.002. http://www.sciencedirect.com/science/article/pii/S1063520310001016
6. Cardona-Morales O, Angel-Orozco A, Castellanos-Domínguez G (2010). Damage detection in vibration signals using non parametric time-frequency representations. In: Proceedings of ISMA2010, international conferences on noise and vibration engineering
7. Honarvar F, Martin H (1997) New statistical moments for diagnostics of rolling element bearings. J Manuf Sci Eng 119(3):425
8. Lei Y, He Z, Zi Y (2008) A new approach to intelligent fault diagnosis of rotating machinery. Expert Syst Appl 35(4):1593
9. Villa LF, Reñones A, Perón JR, de Miguel LJ (2012) Statistical fault diagnosis based on vibration analysis for gear test-bench under non-stationary conditions of speed and load. Mech Syst Signal Process 29(0):436. doi:10.1016/j.ymssp.2011.12.013. http://www.sciencedirect.com/science/article/pii/S0888327011005292
10. Tax DMJ, Duin RPW (2004) Support Vector Data Description. Machine Learning 54(1):45
11. Zhang Y, Liu XD, Xie FD, Li KQ (2009) Fault classifier of rotating machinery based on weighted support vector data description. Expert Syst Appl 36(4):7928. doi:10.1016/j.eswa.2008.10.062. http://www.sciencedirect.com/science/article/pii/S0957417408007914
12. McBain J, Timusk M (2009) Fault detection in variable speed machinery: Statistical parameterization. J Sound Vibr 327(3–5):623. doi:10.1016/j.jsv.2009.07.025. http://www.sciencedirect.com/science/article/pii/S0022460X09005896
13. Pan Y, Chen J, Li X (2010) Bearing performance degradation assessment based on lifting wavelet packet decomposition and fuzzy c-means. Mech Syst Signal Process 24(2):559. doi:10.1016/j.ymssp.2009.07.012. http://www.sciencedirect.com/science/article/pii/S0888327009002490
14. McBain J, Timusk M (2011) Feature extraction for novelty detection as applied to fault detection in machinery. Pattern Recog Lett 32(7):1054. doi:10.1016/j.patrec.2011.01.019. http://www.sciencedirect.com/science/article/pii/S0167865511000341
15. Bartkowiak A, Zimroz R (2011) Outliers analysis and one class classification approach for planetary gearbox diagnosis. J Phys: Conf Ser 305(1):012031

16. Paliwal K (1997) Spectral subband centroids as features for speech recognition. In: Proceedings of IEEE workshop on Automatic speech recognition and understanding, pp 124–131
17. Rabiner L, Juang B (1993) Fundamentals of speech recognition. Prentice Hall, Englewood Cliffs, NJ
18. Khan S, Madden M (2010) in *Artificial Intelligence and Cognitive Science*, *Lecture Notes in Computer Science*, vol. 6206, ed. by L. Coyle, J. Freyne (Springer Berlin Heidelberg, 2010), pp. 188–197. doi:10.1007/978-3-642-17080-5_21. http://dx.doi.org/10.1007/978-3-642-17080-5_21
19. Bishop C (1995) Neural networks for pattern recognition, Oxford University Press, Walton Street, Oxford OX2 6DP
20. Nelwamondo FV, Marwala T, Mahola U (2006) Early classification of bearing faults using Hidden Markov Models, Gaussian Mixture Models, Mel-Frequency Cepstral Coefficients and Fractals. Inter J of Innov Comp Inf and Cntl 2(6):1281–1299

Comparison of Torsional Vibration Measurement Techniques

Karl Janssens and Laurent Britte

Abstract Noise and vibration performance plays an important role in the development of rotating components, such as engines, drivelines, transmission systems, compressors and pumps. The presence of torsional vibrations and other specific phenomena require the dynamic behaviour of systems and components to be designed accurately in order to avoid comfort and durability related problems. This paper provides an overview of the instrumentation and challenges related to torsional vibration testing. The accuracy and performance of various measurement techniques are investigated by measurements on a Fiat Punto 1.4 l engine. The potential sources of error are discussed for each technique.

Keywords Torsional vibration testing · Measurement techniques

1 Introduction

With eco-engineering comes a new range of NVH issues to solve. New powertrain designs like start-stop systems, downsized engines, advanced torque lock-up strategies and the generic trend for weight reduction of the powertrain raise the importance of an in-depth understanding of torsional vibrations as they negatively impact comfort and ultimately engine and driveline efficiency. Torsional vibrations are of importance whenever power needs to be transmitted using rotating shafts or couplings, like is the case for e.g. automotive, truck and bus drivelines, recreation vehicles, marine drivelines or power-generation turbines.

Torsional vibrations are angular vibrations of an object, typically a shaft along its axis of rotation. As mainly rotational speeds are measured, torsional vibrations are assessed as the variation of rotational speed within a rotation cycle. These

K. Janssens (✉) · L. Britte
RTD Test Division, LMS International, Interleuvenlaan 68 3001 Leuven, Belgium
e-mail: karl.janssens@lmsintl.com

G. Dalpiaz et al. (eds.), *Advances in Condition Monitoring of Machinery in Non-Stationary Operations*, Lecture Notes in Mechanical Engineering,
DOI: 10.1007/978-3-642-39348-8_39, © Springer-Verlag Berlin Heidelberg 2014

RPM variations are typically induced by a non-smooth driving torque or a varying load. Structural sensitive frequencies along a driveline may then amplify and transfer these phenomena, leading to comfort, durability or efficiency problems.

Despite tremendous progress in modeling accuracy, overall system complexity still necessitates accurate qualification and quantification of these torsional vibrations, under controlled or real-life operating conditions, in order to better understand and refine counter measures.

This paper provides an overview of the instrumentation and challenges related to torsional vibration testing. The accuracy and performance of various measurement techniques are investigated by measurements on a engine test bench.

2 Torsional Vibration Measurement Sensors

Various measurement techniques are available for torsional vibration testing. The best sensor can be selected for each individual case based on the physical quantity to be measured, the type of analysis, the accessibility of the shaft, the ease of instrumentation and the required accuracy.

2.1 Direct Measurements

2.1.1 Linear Accelerometers

Two linear accelerometers are fixed in a face-to-face configuration on the rotating shaft. The two accelerometers will measure the tangential acceleration. As they have opposite direction in the fixed system of the rotation axis, any translational acceleration of the shaft is cancelled by taking the average of both accelerometer signals.

Advantages:

- High dynamic range directly determined by the dynamic range of the accelerometers
- Low sensitivity to shaft translational vibrations when the accelerometers are well aligned.

Disadvantages:

- Expensive telemetry system or sensitive slip rings are needed to transfer the acceleration signals from the rotating shaft to the measurement hardware
- Mass loading for relatively small shafts influencing the structural behaviour of the shaft, e.g. causing torsional resonances to shift in frequency or shaft unbalance
- Bigger shafts at relatively high RPM cause centrifugal forces that may lead to dangerous loss of accelerometers and measurement equipment when not sufficiently well glued

- Since acceleration is measured, and angle and speed can only be derived by integration, no absolute angular position is available. Angle domain processing will not be possible.

2.1.2 Dual-Beam Laser Interferometers

Laser interferometers can be used as well to measure torsional vibrations. Laser manufacturers typically propose specific systems for rotating measurement based on dual beam techniques to cancel the effect of translational movement of the shaft. The angular velocity is computed from the velocity measured in the direction of the laser beams on the two pointed areas through the Doppler shift.

Advantages:

- Contactless measurement
- Low sensitivity to shaft translational vibration
- Low sensitivity to the shape of the shaft
- Easy instrumentation.

Disadvantages:

- Expensive device. Since it is often required to measure torsional vibrations at different shaft locations simultaneously, this is often a large drawback
- Exact angular speed and position are not known. Since velocity is measured, angle can only be derived by integration, no absolute position is available. Angle domain processing for example will not be possible
- The size of the device does not allow using it in a confined environment. Its use in real-life mobile conditions is very difficult or nearly impossible.

2.2 Coder-Based Techniques

Coder-based techniques make use of equidistantly spaced markers on the shaft or rotating component. The system measures every time a marker passes in front of a sensor and the time difference between two markers is used to estimate the angular velocity. The coder-based techniques have the advantage to deliver RPM and discrete angle position. The data resolution is determined by the number of markers: the more markers, the more accurate information.

Different types of coders are used, for example stripes drawn or glued on the shaft or the teeth of gears. Also different sensors are available to detect the markers, such as electro-magnetic pick-ups or optical sensors. Incremental encoders are devices combining the coder and sensor in one single hardware.

2.2.1 Magnetic Pick-ups

Magnetic pick-ups detect changes in the magnetic field or magnetic flux, typically resulting from metallic teeth passing the sensor. They are often used in industrial applications because of their robustness and low sensitivity to ambient dust. Set-ups for this are often very practical as well, since existing gear sets on the machine can be used as coder, e.g. the gear teeth on flywheels of transmissions. Resulting from that, magnetic pick-ups are very popular for measuring torsional vibrations, as they are easy to set up, as they work very well with existing gear teeth and as they are very robust. Most combustion engines today are equipped, by default, with these sensors to transfer the different shaft positions to the engine or gearbox controllers.

Advantages:

- Price. Mass production of magnetic pick-ups for automotive and industrial applications has a very positive influence on their end-user price
- Simplicity of instrumentation. The sensor is typically fixed on non-rotating components, which avoids the need for e.g. telemetry. Coders are mostly generated by existing gear sets
- Robust sensors with low sensitivity to ambient dust.

Disadvantages:

- The number of gear teeth sets limits to the number of pulses per revolution which could be insufficient to capture all torsional content
- The accuracy of the measurement is very much dependent on the machining accuracy and deformation of the gear teeth
- The sensor must be fixed very close to the rotating shaft which is sometimes difficult, e.g. when the shaft has an important translational movement
- Relative displacements between the magnetic pick-up and the shaft, due to shaft bending or due to displacement of the sensor attached on a too soft mounting, influence the quality of the measured pulses and generate a fictive torsional vibration.

2.2.2 Optical Sensors

Many types of optical sensors can be found on the market, however most of them are designed for object detection. To measure torsional vibrations, the sensor not only needs to be able to detect a high rate of events per second, also the timing accuracy of the detection is very important and this accuracy is often insufficient.

Optical sensors generate an electric signal proportional to the received light intensity. Optical fibers are used to conduct the light from the emitter to the sensor head and back from the sensor head to the receptor. They can be configured in reflection or transmission configuration.

Optical sensors can be used with many different types of coders as long as the visible contrast between the stripes is sufficient. Most optical sensors deliver a TTL output signal.

Advantages:

- The instrumentation is very simple as the sensor is typically fixed on non-rotating components. Only the coder needs to rotate.
- Optical sensors can be directly instrumented on gears as is the case with magnetic pick-ups, under condition that the reflection of the material gear surface is sufficient
- Coders can easily be implemented on shafts with contrasted paint or zebra tape
- The fast response and good phase accuracy of high-quality optical sensors allow the measurement of very high pulse rates.

Disadvantages:

- The sensitivity to ambient light and/or the quality of the material reflection complicate the direct instrumentation of the gears in gearboxes
- The sensor must be fixed very close to the rotating shaft which is sometimes difficult when the access is limited or when the shaft has some translational movement
- Relative displacements between the optical sensor and shaft influence the quality of the measured time stamps and lead to torsional vibration measurement errors.

Black and white tapes are more and more used to quickly implement a coder on a shaft. They can be used to create a coder when no gear wheel is available or when a higher number of pulses per rotation is required. There are two families of tape depending on whether it must be glued around the shaft (zebra tape) or on the extremity (zebra disc). Zebra tapes and discs exist in multiple stripe width to adapt the number of pulses per revolution in function of the shaft diameter.

Although zebra tape is very easy to instrument, an error will be introduced onto the measurement at the location where the two zebra tape endings come together. When this point passes the optical sensor it will introduce a discontinuity in the RPM signal. A butt joint correction should therefore be applied before analysing the measurement signal (Janssens et al. [1]).

Zebra discs do not suffer from this butt joint problem, however proper care needs to be taken to properly center the disc. Since perfect centering is never possible, torsional order 1 is typically not reliable when using this coder set-up.

2.2.3 Incremental Encoders

Incremental encoders are devices typically used in automat or robotic applications for accurate detection of shaft positions. Their high accuracy makes them very attractive for torsional vibration analysis applications as well. Often based on optical technology, incremental encoders combine the coder and the sensor in one single device. They consist of both a rotating (rotor) and static (stator) component and the full sensor needs to be mounted on the set-up. Incremental encoders come in many different shapes and sizes, to cover all required applications.

The incremental encoder makes use of three embedded coders: one detecting one single pulse per revolution, called index, as absolute angle reference and two more high resolution encoders called A and B. The A and B signals have exactly the same number of pulses but the B signal is phase shifted with a quarter of a pulse cycle (90°) compared to A. The combination of these two coder signals allows detecting the sense of rotation of the coder.

Advantages:

- The fully integrated approach allows developing accurate coders with potentially very high pulse rate. Incremental encoders can be delivered with the appropriate number of pulses, depending on the application and desired accuracy typically 50–500
- The sense of rotation can be a great advantage, e.g. for the investigation of the start/stop behaviour on engines
- The integrated index signal allows duty cycle related analysis with accurate TDC identification (e.g. engine combustion analysis).

Disadvantages:

- The relative complex instrumentation limits their usage for in-vehicle or mobile measurements. Incremental encoders are mainly used when working on test benches where the instrumentation makes part of the test bench equipment.

3 Test Campaign

3.1 Test Set-up

An experimental test campaign was carried out on a 4-cylinder Fiat Punto 1.4 l engine (Fig. 1 left). The engine was driven by an electric motor controlling the speed profile. The engine crankshaft and electric motor were connected by a pulley. Torsional vibration measurements were conducted on the camshaft (Fig. 1 right). The shaft was instrumented as follows:

- Dual beam laser (RLV 5500 Polytec system)
- High speed incremental encoder (Heidenhaim ROD 426, 1,024 pulses per rev)
- Zebra tape (142 stripe pairs, 1 mm stripe width, Optel Thevon optical sensor)
- Zebra disc (120 pulses per revolution, Optel Thevon optical sensor).

Various constant speed and run-up tests were carried out in the following two conditions: (1) 4 cylinders open (spark plugs removed): rather small torsional

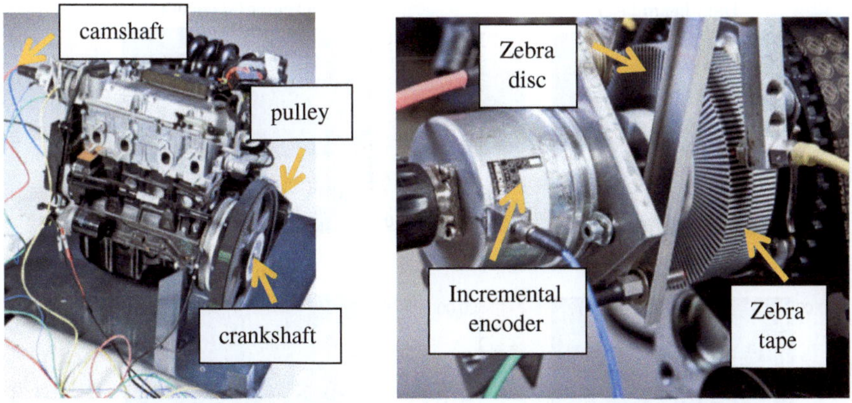

Fig. 1 Fiat Punto engine set-up (*left*), instrumentation of camshaft (*right*)

content related to the piston movements; and (2) 1 cylinder closed: large torsional content due to the high pressure changes (compression and expansion) in the closed cylinder, once per rotation of the camshaft.

3.2 Test Results

3.2.1 Four Open Cylinders

Figure 2 shows the RPM-frequency colormap of the torsional vibration measurements on the camshaft by the dual beam laser, incremental encoder, zebra disc and zebra tape in a run-up test with 4 open cylinders. There is obviously a good match between the analysis results within a 30 dB dynamic range. The dominant orders 4, 12 and 16 are well captured by all methods. Only slight differences are noticeable for the less significant orders. The zebra disc results are of lower quality in the 70–90 Hz frequency range, possibly due to the vibrations of the optical probe which is not idealy fixed. One can also notice that the zebra disc is not perfectly centered on the shaft, causing an order 1 misalignment error.

Figure 3 shows the amplitude and phase profile of torsional orders 4 and 16 as obtained with the 4 measurement techniques. There is a good match in amplitude and phase all along the RPM axis, even when the order becomes less important with amplitudes lower than 0.01°.

Figure 4 compares the RPM time variations obtained from the zebra tape measurements with those of the incremental encoder before (lower graph) and after (upper graph) application of a zebra tape butt joint correction. The spikes in the RPM data are clearly removed, illustrating the effectiveness of the correction method. The corrected zebra tape measurements and incremental encoder RPM data match very well.

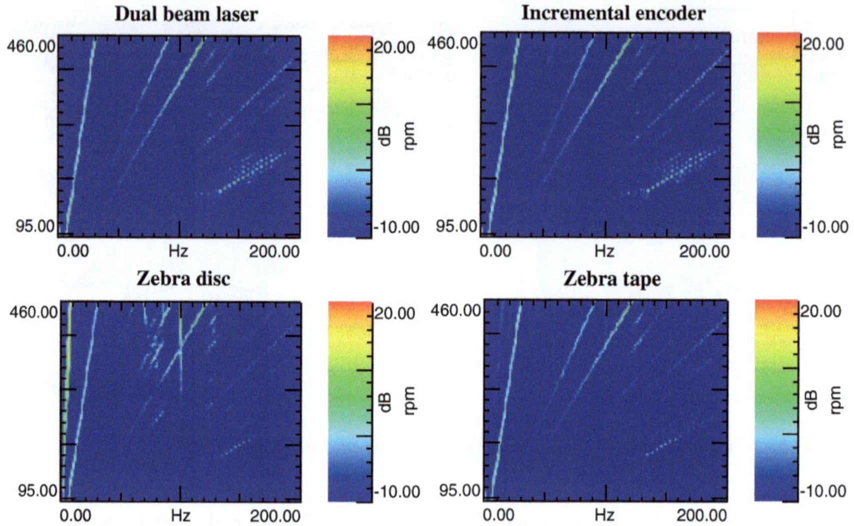

Fig. 2 RPM-frequency map of camshaft torsional vibration measurements in run-up test with 4 open cylinders

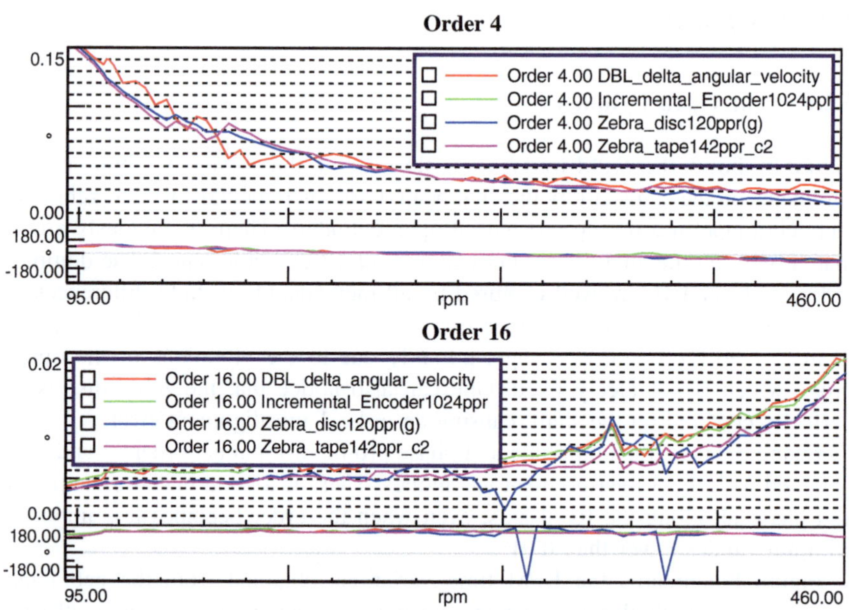

Fig. 3 Camshaft torsional order cuts in run-up test with 4 open cylinders

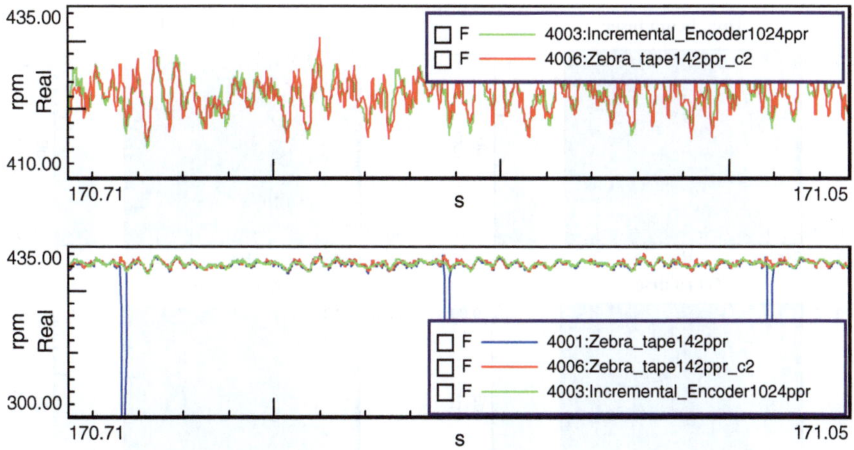

Fig. 4 Camshaft RPM fluctuations measured with zebra tape (with and without butt joint correction) and incremental encoder in a test with 4 open cylinders

3.2.2 One Cylinder Closed

The torsional vibration results for a run-up test with 1 cylinder closed are shown in Figs. 5, 6, and 7. The large pressure changes (once per camshaft revolution) in the closed cylinder cause strong RPM variations which are clearly visible in Fig. 5. The large RPM drop every rotation yields multiple torsional orders as shown in Fig. 6. Here again, the torsional vibration results of the 4 measurement methods are similar within the 30 dB dynamic range. This is also clear from the order sections in Fig. 7, showing a similar amplitude and phase profile.

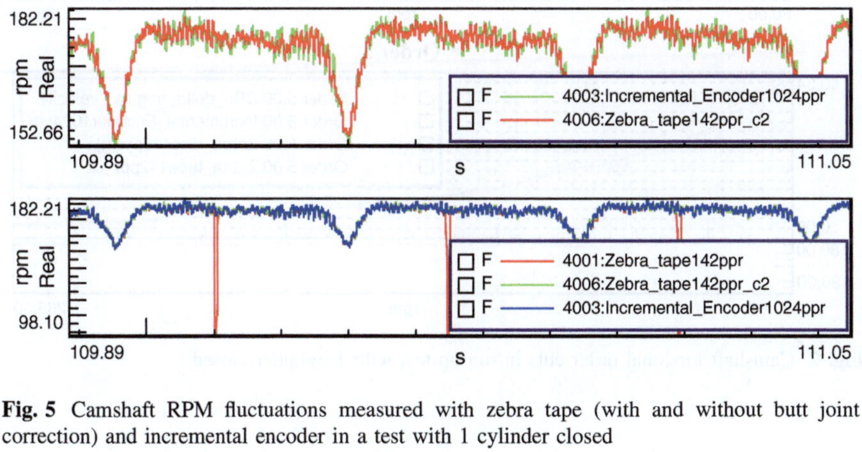

Fig. 5 Camshaft RPM fluctuations measured with zebra tape (with and without butt joint correction) and incremental encoder in a test with 1 cylinder closed

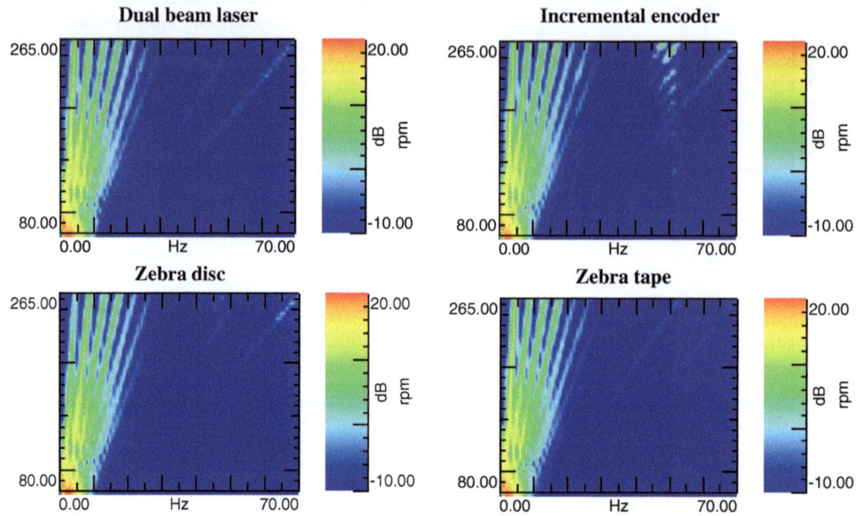

Fig. 6 RPM-frequency map of camshaft torsional vibration measurements in run-up test with 1 cylinder closed

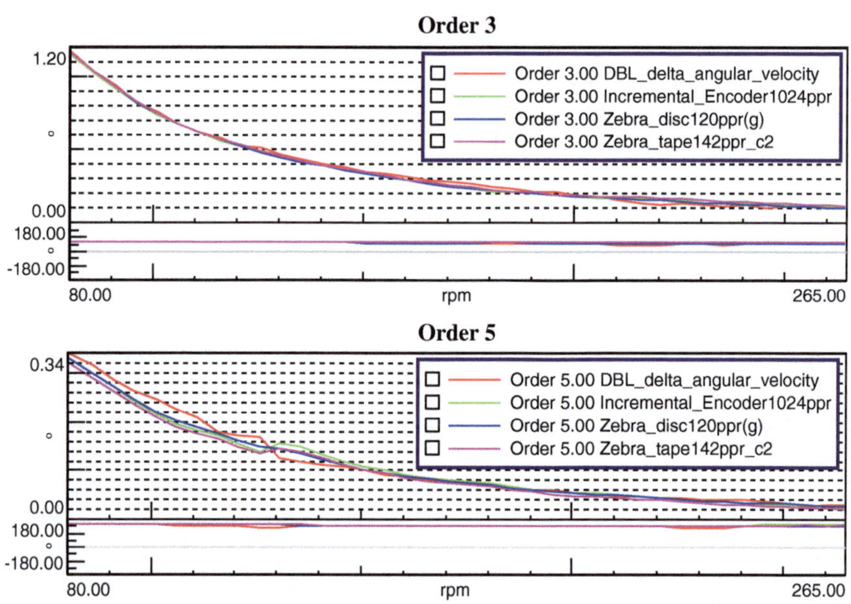

Fig. 7 Camshaft torsional order cuts in run-up test with 1 cylinder closed

4 Conclusions

It is known since a long time that the dual beam laser and incremental encoder are good and accurate measurement techniques for torsional vibration testing. However, the set-up space of the laser system and relative complex instrumentation of the incremental encoder limit their usage for in-vehicle and mobile measurements. Zebra tape and disc measurements do not suffer from these limitations which is obviously a benefit. Next to this, they also perform well in terms of accuracy. With various experimental tests on a Fiat Punto engine, we have demonstrated that the torsional vibration results obtained with these measurement techniques correspond very well to those of the dual beam laser and incremental encoder.

Acknowledgments This work has been carried out in the framework of the COST Action TU 1105 "NVH analysis techniques for design and optimization of HYBRID and ELECTRIC vehicles".

References

1. Janssens K, Van Vlierberghe P, Claes W, Peeters B, Martens T, D'Hondt PH (2010) Zebra tape butt joint correction algorithm for rotating shafts with torsional vibrations. ISMA Conf., Leuven, Belgium, 20–22 Sept 2010

Numerical Investigations on the Accuracy of an Automated Modal Identification Technique

Carlo Rainieri and Giovanni Fabbrocino

Abstract Systems and techniques for fast damage detection play a fundamental role in the development of effective Structural Health Monitoring systems. Modal-based damage detection algorithms are well-known techniques for structural health assessment but they need reliable and accurate automated modal identification and tracking procedures in order to be effective. In this chapter, the performance of a recently developed algorithm for automated output-only modal parameter estimation is assessed. An extensive validation of the algorithm for continuous monitoring applications is carried out based on simulated data. Different levels of damping are considered. The numerical study demonstrates that the algorithm provides fairly robust, accurate and precise estimates of the modal parameters, including damping ratios.

Keywords: Automated operational modal analysis · Damping measurement

1 Introduction

Several algorithms for automated identification [1] and tracking [2] of modal parameters based on Operational Modal Analysis (OMA) methods have been developed in recent years. As a consequence, modal based damage detection techniques [3] are again gaining in popularity, even if they suffer limitations in terms of damage localization and quantification, as well as drawbacks related to sensitivity to measurement quality and to environmental and operational factors.

C. Rainieri (✉) · G. Fabbrocino
StreGa Lab, DiBT Department, University of Molise, Via Duca degli Abruzzi 86039 Termoli, Italy
e-mail: carlo.rainieri@unimol.it

G. Fabbrocino
e-mail: giovanni.fabbrocino@unimol.it

G. Dalpiaz et al. (eds.), *Advances in Condition Monitoring of Machinery in Non-Stationary Operations*, Lecture Notes in Mechanical Engineering,
DOI: 10.1007/978-3-642-39348-8_40,

Nevertheless, continuous monitoring of modal parameters has a large potential in performance and health assessment of civil engineering structures [4]. Applications range from prompt detection of damage and degradation phenomena to post-earthquake health assessment and emergency management.

The available solutions for automated output-only modal identification show some drawbacks [1] which can affect their reliability and accuracy in the framework of continuous monitoring:

- threshold based peak and physical pole detection;
- need of a preliminary calibration phase at each new application;
- static settings of thresholds and parameters which may be unsuitable to track the natural changes in modal properties of structures due to damage or environmental effects;
- sensitivity to noise, problems of false or missed identification.

Moreover, a number of algorithms do not provide damping estimates; whenever they are able to estimate modal damping, the resulting values are usually very scattered. The fairly large scatter associated to damping estimates, compared with natural frequency and mode shape estimates, is well documented in the literature. Even if it can be partially addressed to inherent limitations of the estimators and the adoption of an equivalent viscous damping model [5], appropriate data processing procedures have to be adopted in order to minimize the estimation error. As a consequence, a careful design of the algorithm is fundamental to enhance reliability, robustness and accuracy of automated modal identification procedures.

A thorough performance assessment of automated modal identification algorithms is rarely reported in the literature but it is a fundamental step in view of proper post-processing of modal parameters for damage detection and performance evaluation purposes.

In the present chapter, the performance of a recently developed procedure for fully automated output-only modal identification is analyzed. The performance of the algorithm is assessed through the analysis of the modal identification results obtained from four case studies, each one consisting of 10,000 simulated datasets. The analysis of the obtained results and the comparisons with the reference values of the modal properties show that the algorithm is characterized by a very high success rate and that it is able to provide accurate estimates of the modal parameters, including damping. The performance assessment of the algorithm is still in progress. However, the herein illustrated results are definitely promising, so the algorithm has a potential in the continuous vibration based monitoring of civil structures.

2 The Automated OMA Algorithm

The proposed algorithm (Fig. 1) is based on the combination of different OMA techniques in order to faclitate the analysis and interpretation of the stabilization diagram. The method is based on Stochastic Subspace Identification (SSI) [6] and

Fig. 1 Flowchart of the automated OMA algorithm

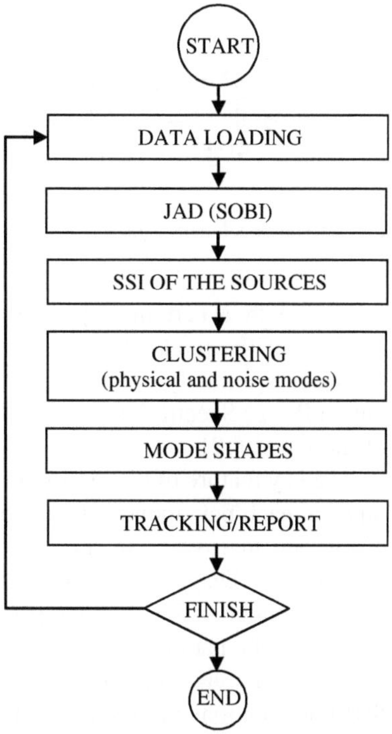

the selection of physical poles in the stabilization diagram by clustering techniques, but it takes advantage of the Blind Source Separation (BSS) [7], operated at a preliminary stage according to the Second Order Blind Identification (SOBI) [8, 9]) procedure, in order to simplify the interpretation of the stabilization diagram.

In fact, as a result of the BSS phase, the raw data associated to the measured structural response are transformed into sources which can be well-separated (they show the contribution of a single mode to the structural response), not well-separated (noise or minor contributions from other modes could be superimposed to the contribution of the main mode) or just noise sources [9]. Thus, the BSS simplifies the analysis of the data and the interpretation of the stabilization diagram, allowing the extraction of the modal information from the individual sources and not the multivariate time series of raw data. The interpretation of the stabilization diagram, therefore, becomes easier since it basically reports information about only one mode at the time. The sources are analyzed one-by-one according to the SSI method and the physical poles are separated from the spurious ones by means of clustering techniques [10] and mode validation criteria. The final values of the natural frequency and damping ratio for the identified modes are obtained by a sensitivity analysis with respect to the number of block rows in SSI, for a fixed value of the maximum model order in the stabilization diagram. The cluster characterized by the minimum variance of the estimates when i ranges in a certain

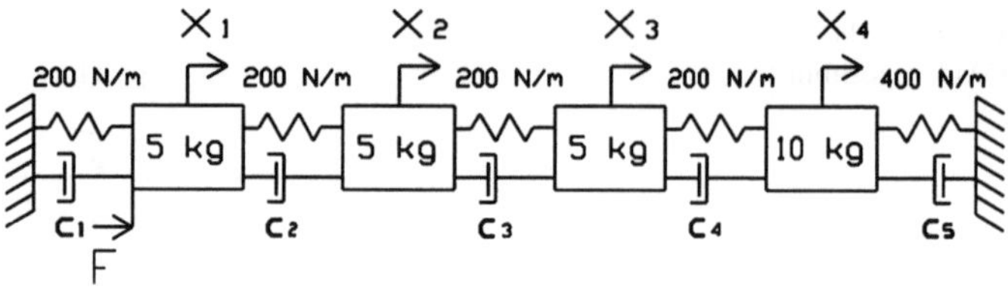

Fig. 2 The benchmark 4-DOF system

interval with a certain step Δi is then selected as the one providing the best estimate of the modal parameters for a given structural mode [11]. Mode shape estimates are finally obtained from Singular Value Decomposition (SVD) of the output Power Spectral Density (PSD) matrix at the previously estimated frequency of the mode [12].

The key feature of the algorithm is the absence of any analysis parameters to be tuned at each new monitoring application. In fact, it has been shown [13] that the results are insensitive to the setting of the two parameters governing the Joint Approximate Diagonalization (JAD) [14] for source extraction, while the influence of the number of block rows [6] is taken into account by sensitivity analyses. Moreover, the parameters governing the JAD play a primary role in the control of response time and computational burden [13]. This is relevant, in particular, for SHM applications in seismically prone areas [2].

3 Case Studies and Analysis of Results

The performance of the proposed algorithm in terms of accuracy and reliability of estimates has been investigated through a statistical analysis of the results obtained from simulated data continuously generated through the excitation of a 4-DOF system by a Gaussian white noise.

The mass and stiffness properties of the system are reported in Fig. 2. Rayleigh damping is adopted. The modal properties of the system are reported in Table 1. The four case studies differ for the assumed values of damping or the signal-to-noise ratio (SNR).

The system matrices and, therefore, the associated modal parameters are the same in all runs, since only the uncertainties associated to inherent limitations of the estimator are the key focus of the investigation. The performance of the method in the presence of uncertain system matrices, that is when modal parameters can slightly change at each run as an effect of the deviation of the system matrices from their nominal values, is out of the scope of the present paper. Specific numerical analyses are under development and will be discussed elsewhere.

The system response to Gaussian white noise N(0, 1) has been simulated 10,000 times. The input has been applied at DOF #1. Each simulated dataset consisted of

Table 1 Modal properties of the simulated 4-DOF system

Mode #	Natural frequency [Hz]	Damping ratio [%]			
		Case study #1 (SNR = 5 dB)	Case study #2 (SNR = 15 dB)	Case study #3 (SNR = 5 dB)	Case study #4 (SNR = 5 dB)
I	0.668	1.00	1.00	2.00	2.00
II	1.137	0.88	0.88	1.31	1.76
III	1.526	0.92	0.92	1.09	1.84
IV	1.879	1.00	1.00	1.00	2.00

Table 2 Success rate of automated modal identification over 10,000 runs

Mode #	Success rate [%]			
	Case study #1 (SNR = 5 dB)	Case study #2 (SNR = 15 dB)	Case study #3 (SNR = 5 dB)	Case study #4 (SNR = 5 dB)
I	99.79	100.0	99.71	99.87
II	99.96	100.0	99.97	100.0
III	99.95	100.0	99.98	100.0
IV	100.0	100.0	100.0	100.0

four measurement channels; the total record length was 3,600 s and the sampling frequency was 10 Hz. Gaussian white noise has been added to the system response in order to simulate the effect of measurement noise. The SNRs are reported in Table 1. Each dataset has been then processed using the described algorithm in order to automatically extract the modal parameters of the system. The analysis of the simulated datasets has been carried out considering a number of block rows i ranging between 20 and 80 with $\Delta i = 2$ and considering a maximum model order of 16 in the construction of the stabilization diagram for each analyzed source.

The analysis of the obtained results has pointed out that the algorithm carries out automated output-only modal identification in a very robust way. In fact, a success rate [15] larger than 99 % has been obtained for all modes (Table 2). Just in a few runs missed identification of the modal properties of a mode occurred. This was probably due to a combined effect of weak excitation and low SNR, which affected the quality of separation and stabilization.

The results in terms of natural frequency and damping estimates are summarized in Tables 3 and 4. Very accurate natural frequency estimates, associated also to a low standard deviation σ, have been obtained. The error in natural frequency estimates is much lower than 1 % in the 95 % of the runs for all case studies. The accuracy of estimates slightly improves when the SNR increases. The analysis of the coefficient of variation γ associated to the natural frequency estimates in the cluster selected by the sensitivity analysis with respect to the number of block rows in each run points out that the estimates are also very precise ($\gamma_{\mathrm{f,\ cluster}}$ much lower than 0.1 % in the 95 % of the runs). When damping increases, the error associated to the estimates also increases, but the results are still very accurate.

Table 3 Automated modal identification results: natural frequencies

Case study	Mode #	$f_{nominal}$ [Hz]	f_{av} [Hz]	σ_f [Hz]	Δf [%] 50th centile	Δf [%] 95th centile	$\gamma_{f,cluster}$ [%] 95th centile
1	I	0.668	0.668	0.000730	0.072	0.216	0.0254
1	II	1.137	1.137	0.000911	0.052	0.156	0.0285
1	III	1.526	1.526	0.001157	0.047	0.143	0.0269
1	IV	1.879	1.879	0.001568	0.048	0.151	0.0363
2	I	0.668	0.668	0.000665	0.067	0.196	0.0319
2	II	1.137	1.137	0.000879	0.050	0.152	0.0234
2	III	1.526	1.526	0.001090	0.047	0.140	0.0232
2	IV	1.879	1.879	0.001425	0.047	0.149	0.0296
3	I	0.668	0.668	0.001275	0.121	0.359	0.0598
3	II	1.137	1.137	0.001324	0.070	0.219	0.0509
3	III	1.526	1.526	0.001309	0.054	0.166	0.0345
3	IV	1.879	1.879	0.001469	0.048	0.148	0.0346
4	I	0.668	0.668	0.001368	0.117	0.358	0.0729
4	II	1.137	1.137	0.001636	0.086	0.278	0.0814
4	III	1.526	1.526	0.002054	0.083	0.260	0.0798
4	IV	1.879	1.879	0.002825	0.085	0.277	0.0991

Table 4 Automated modal identification results: damping ratios

Case study	Mode #	$\xi_{nominal}$ [%]	ξ_{av} and (mode, median) [%]	σ_ξ [%]	Δξ [%] 50th centile	Δξ [%] 95th centile	$\gamma_{\xi,cluster}$ [%] 95th centile
1	I	1.00	1.02 (1.0, 1.02)	0.108	7.2	21.6	7.0
1	II	0.88	0.89 (0.9, 0.89)	0.080	6.0	18.0	4.4
1	III	0.92	0.93 (0.9, 0.93)	0.072	5.1	15.3	2.5
1	IV	1.00	1.01 (1.0, 1.01)	0.075	4.9	14.9	3.1
2	I	1.00	1.01 (1.0, 1.01)	0.099	6.7	19.4	2.2
2	II	0.88	0.88 (0.9, 0.88)	0.075	5.7	16.6	1.6
2	III	0.92	0.92 (0.9, 0.92)	0.073	5.3	15.5	1.7
2	IV	1.00	1.00 (1.0, 1.00)	0.074	4.7	14.6	2.1
3	I	2.00	2.03 (2.0, 2.03)	0.179	5.9	17.9	7.7
3	II	1.31	1.32 (1.3, 1.32)	0.109	5.4	16.7	5.1
3	III	1.09	1.10 (1.1, 1.10)	0.081	4.9	14.7	2.4
3	IV	1.00	1.01 (1.0, 1.00)	0.075	4.9	14.9	2.8
4	I	2.00	2.03 (2.0, 2.02)	0.183	6.0	18.3	6.7
4	II	1.76	1.78 (1.8, 1.78)	0.138	5.2	15.6	4.6
4	III	1.84	1.85 (1.8, 1.85)	0.129	4.5	13.9	3.0
4	IV	2.00	2.02 (2.0, 2.02)	0.143	4.3	14.0	3.6

Damping estimates are fairly accurate and characterized by moderate uncertainty (σ lower than 0.2 %). In particular, the variability of estimates slightly increases when the nominal damping values increase. Larger errors are associated to damping estimates with respect to natural frequencies. However, the scatter with respect to the nominal values is lower than 10 and 20 % in the 50 and 95 % of the runs, respectively. The errors slightly decrease when the SNR increases. Damping estimates are also fairly precise ($\gamma_{\xi,\ \mathrm{cluster}}$ much lower than 10 % in the 95 % of the runs). The mean, mode and median of the identified damping ratios after 10,000 runs are very close each other and to the nominal values of modal damping ratios. Taking into account the uncertainty associated to damping estimates, the mode of damping values is given with one decimal place only.

Marginal refinements can be obtained by outlier removal, confirming the robustness and accuracy of the algorithm.

4 Conclusions

A hybrid automated OMA algorithm has been described in the present paper. Its performance has been assessed against simulated data generated by a 4-DOF for different damping levels and SNRs. The system was excited by a Gaussian white noise. The results obtained from 10,000 runs have been analyzed and robustness, accuracy and precision of the algorithm have been checked. Encouraging results have been obtained, in particular as the possibility to estimate damping ratios in an accurate and fully automated way is concerned. Further investigations are in progress to assess the performance of the algorithm in the case of uncertain system matrices in view of continuous, long term vibration based SHM applications.

Acknowledgments The present work is carried out in the framework of AT2—LR 2—Task 3 of the ReLUIS-DPC Project 2010-2013, whose support is gratefully acknowledged.

References

1. Rainieri C, Fabbrocino G (2010) Automated output-only dynamic identification of civil engineering structures. Mech Syst Signal Process 24:678–695
2. Rainieri C, Fabbrocino G, Cosenza E (2011) Near real-time tracking of dynamic properties for standalone structural health monitoring systems. Mech Syst Signal Process 25:3010–3026
3. Doebling SW, Farrar CR, Prime MB, Shevitz DW (1996) Damage identification and health monitoring of structural and mechanical systems from changes in their vibration characteristics: a literature review. Technical Report LA-13070-MS, UC-900, Los Alamos National Laboratory, New Mexico 87545
4. Rainieri C, Fabbrocino G, Cosenza E (2010a) Integrated seismic early warning and structural health monitoring of critical civil infrastructures in seismically prone areas. Struct health monit—Int J 10:291–308
5. Rainieri C, Fabbrocino G, Cosenza E (2010b) Some remarks on experimental estimation of damping for seismic design of civil constructions. Shock Vib 17:383–395
6. Van Overschee P, De Moor B (1996) Subspace identification for linear systems: theory—implementation—applications. Kluwer Academic Publishers, Dordrecht
7. Ans B, Hérault J, Jutten C (1985) Adaptive neural architectures: detection of primitives. *COGNITIVA'85*:593-597
8. Belouchrani A, Abed-Meraim K, Cardoso JF, Moulines E (1997) A blind source separation technique using second-order statistics. IEEE Trans Signal Process 45:434–444

9. Poncelet F, Kerschen G, Golinval JC, Verhelst D (2007) Output-only modal analysis using blind source separation techniques. Mech Syst Signal Process 21:2335–2358
10. Tan P-N, Steinbach M, Kumar V (2006) Introduction to data mining. Pearson Addison-Wesley, Reading, MA
11. Rainieri C, Fabbrocino G, Cosenza E (2010c) On damping experimental estimation. In: Proceedings of 10th international conference on computational structures technology, Valencia
12. Brincker R, Zhang L, Andersen P (2000) Modal identification from ambient responses using frequency domain decomposition. In: Proceedings of 18th SEM international modal analysis conference, San Antonio
13. Rainieri C, Fabbrocino G (2012) A hybrid automated modal identification algorithm for Structural Health Monitoring: a comparative assessment. In: Proceedings of international conference on noise and vibration engineering ISMA2012, Leuven
14. Cardoso JF, Souloumiac A (1996) Jacobi angles for simultaneous diagonalization. SIAM J Matrix Anal Appl 17:161–164
15. Magalhaes F, Cunha A, Caetano E (2012) Online automatic identification of the modal parameters of a long span arch bridge. Mech Syst Signal Process 23:316–329

The Principles of Operation and Equipment Design in Modelling of Separating-System Dynamics

Vladimir D. Anakhin and Timur V. Anakhin

Abstract This paper describes the principles of operation of equipment design which may be used in the field of application of vibration technology for efficient separation of free-flowing dry materials on the thickness differences of the solids present in the mixture. An initial prototype equipment is constructed to simulate actual operating conditions for monitoring the condition of machinery and for selecting a suitable design of equipment characterized by properly selected types of components. The inventive aspect of process design includes development of completely new separation method and the arrangement of the equipment design for application in operation processes depending essentially on characteristics of oscillation in a mechanical system. Through mathematical modelling, a better understanding of the processes motivated by the rectangular acceleration of translational vibrational motion and steps in exact solution of transcendental equations of particle movement are presented. Design methods using analytical procedures is profitable in general for theoretical prediction. Technical feasibility of some types of suitable designs are used for the purpose of studying the effects of rectangular acceleration of vibration on particle size separation which may be attractive from engineering perspective.

Keywords Condition of Machinery · Equipment Design · Vibration Technology

1 Introduction

In the field of the mechanical design and application of vibration technology for efficient separation of minerals, chemicals, pharmaceuticals, foods, plastics, grinding and other non-coherent dry materials and powder products in particle sizes from 0.02 to 15 mm, the pioneering works of authors in the late 1990s have

V. D. Anakhin · T. V. Anakhin
University of Ferrara, Ferrara, Italy
e-mail: anakhin@mail.ru

G. Dalpiaz et al. (eds.), *Advances in Condition Monitoring of Machinery in Non-Stationary Operations*, Lecture Notes in Mechanical Engineering,
DOI: 10.1007/978-3-642-39348-8_41,

developed completely new separation methods and a wide variety of possible choices of separation processes in some industries is then summarized. A vibratory separator (VS) is distinguished from a well-known screen and a sieve in that it is complete with a non-perforated vibrating deck (separating surface) which includes provisions for separating the particles directly on it. Conceptually, the time-history of deck vibration may be considered to be sinusoidal or simple harmonic in form. For a particular multicomponent separation, there might be several ways in which products could be separated. In general, separability is dependent on two tilts of the deck: [1] longitudinal and [2] lateral. Either tilt could be the controlling factor. Positioning is estimated by doing an optimal balance. The particle size and shape, and the friction coefficient are main factors with the separation processes. Many variations of parallel-decks constructions are now available: some have flat decks, some use concave decks. Most of the decks are placed on two tilt angles to allow the components of the mix to disengage as they pass through the deck section. Parallel decks added to VS can greatly improve separation quality and increase capacity.

Advantages of the new screenless methods used in dry process of VS and application of vibration technology for all types of products for efficient separation by particle size or shape are evident. This method can be effectively used, particularly, in the abrasive industry to produce materials in which more than 90 % of the grains is isometric. Grindings wheels made from such grains are twice as effective as those made from regular grains which are unclassified by shape. Lead, zinc or copper powders generally are separated without forming dust. In the powder metallurgy, steel powders can be effectively separated. Also, spherical particles can be collected to form the bronze powders used in production of metal-ceramic filters. In the diamond tool industry, VS are used for selecting isometric, plane and needle-shaped diamond grains. In the agriculture and food industry, these vibrating decks can utilized for removing harmful inclusions in grains.

The deck vibration may be a combination of two sinusoidal quantities, the second frequency and amplitude components having an integral multiple of the lowest frequency and the amplitude. Graphical methods have been applied in studies of these effects on the behavior of solids in the mix. Since vibration separation processes are based on the creation of velocity differences within and between solid particles, whether minerals or powder products, a collection of useful relations was obtained to monitor the condition of machinery such as its performance, the selection of the appropriate vibration parameters, particularly, for separating the mixture into desirable products, by products, and waste materials. The analysis of rectangular acceleration of vibration when the single degree-of-freedom system of VS is acted upon by the rectangular step excitation involves certain physical concepts that are different from those applied to the analysis of harmonic and biharmonic vibration. Available vibration exciters of biharmonic and rectangular acceleration of vibrations which may be involved in the operation of such equipment as conveying and separating machines are described in.

The thickness factor of the solids present in the mix is also essential with the screenless separation motivated by the rectangular acceleration of vibration. The design of equipment for this particular separation is less developed aspects of vibration technology. Selection of the separation technique or techniques to be used for a particular system can be also broken down into the task to use the translational vibration motion in the direction of the Y axis of the deck of VS (longitudinal vibration). The technical feasibility and engineering perspectives of a given method must be attractive.

2 Types of Dynamic Models and Principles of Operation

Mixtures of dry solids can be separated by the specific thickness differences of the components. The proper introduction to feed to a vibratory separation is one of the keys to its performance. The batching and removal of solids require a good control. For some critical designs, the performance cannot be predicted theoretically and such systems require experimental work to determine concentration of feed, sizing (spacing for equipment), material of construction, operating conditions and costs, quality required, etc. For consistence performance dynamic models are useful in evaluating optimal performance. The model shown in Fig. 1 demonstrates practicality of separating mixtures according to the particle thickness.

The principal parts of such a VS are bar (part1) which can be turned into three positions of 10, 20, and 30^0, and deck (part 3). The disturbance force from electromagnetic exciter (part 4) moves the deck and the bar in opposite directions. The deck and reactive frame (part 5) are supported by elastic damper (part 8) to prevent vibrations from being transferred from the upper part of the machine to the turning frame (part 6).The machine rests on base (part 7). The base usually comprises handwheel (part 9) and screw mechanism (part 10). The initial mixture is separated into fractions: finer grains (thin) pass a clearance under the bar (part 1); larger grains (thick) with dimensions exceeding the gap are caught by the bar and directed upwards.

Characteristics of vibratory separator (see Fig. 1)
The frequency range of the simple harmonic motion 3,000 min^{-1}
The amplitude of the displacement 1 mm
The angle of vibration β 25°
Longitudinal tilt of the deck 10–20°
Lateral tilt of the deck 0–8°

Simultaneous oscillatory movements of the bar and the deck in opposite directions prevent the narrow gaps from being clogged and promote passing fine particles through. This mechanism is provided by the device shown in Fig. 2. A schematic diagram of a typical motion is shown in Fig. 3. The gap may be calculated by the expressions:

$$\delta_{y.max} = \delta + (A_1 + A_2)\,\text{Sin}\,\beta$$
$$\delta_{y.min} = \delta - (A_1 + A_2)\,\text{Sin}\,\beta$$

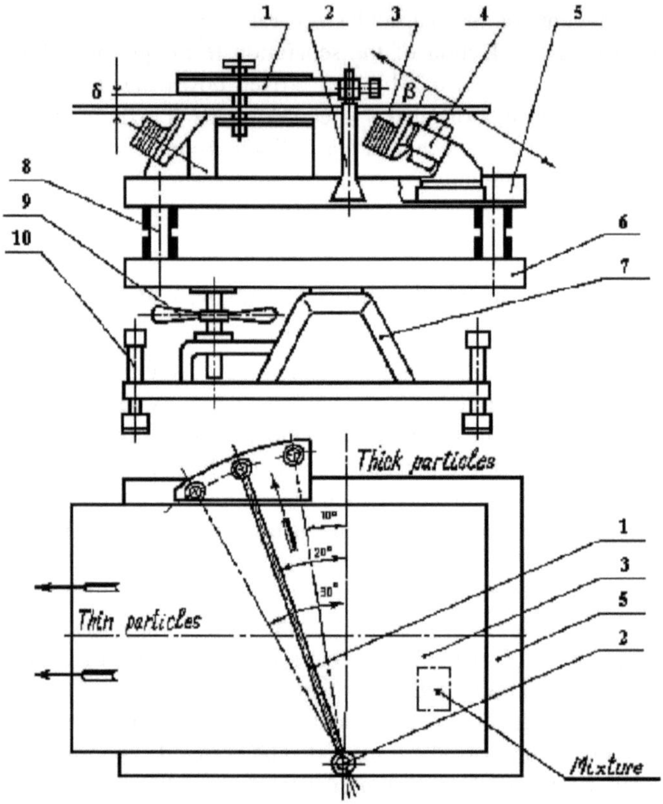

Fig. 1 Vibratory device for sorting particles according to their thickness *1* = Separating bar; *2* = Lever; *3* = Deck; *4* = Electromagnetic exciter; *5* = Reactive frame; *6* = Turning frame; *7* = Base; *8* = Damper; *9* = Hand wheel; *10* = Screw mechanism

where $\delta_{y.max}$ is the maximum linear shift in the direction of Y- axis;$\delta_{y.min}$ is the minimum linear shift in the direction of Y- axis;δ is a gap between the bar (part 1) and the deck (part 2) as shown in Fig. 3.;

- A_1 is a shift amplitude of the deck;
- A_2 is a shift amplitude of the bar, and
- β is the angle of vibration.

The schematic representation of the displacement amplitudes of the mechanical parts of VS used in the design and operation of separation process arising from differences in particle thickness illustrates possibility of passage of thin particles through opening δ between the bar and deck plane. The device has been tried experimentally but are not in use commercially. This horizontal device can be used to estimate the performance of VS of various sizes and it can be used to predict the effect of specific thickness difference of solids, and allows prediction of capacities

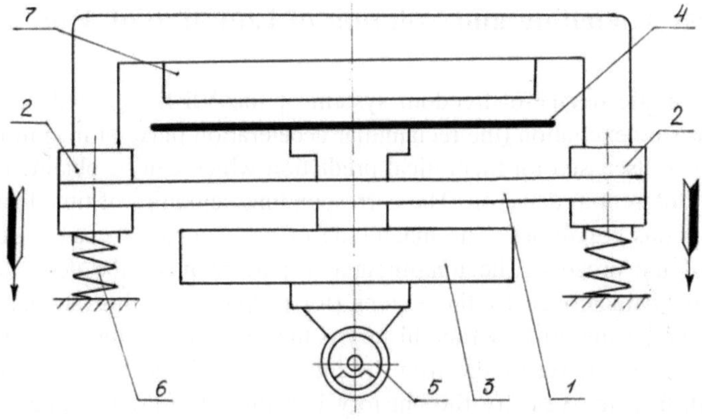

Fig. 2 Electromotive device: *1*-Leaf spring; *2*-Reactive frame; *3*-Active frame; *4*-Deck plane; *5*-Vibration exciter; *6*-Isolators; *7*-Separating bar

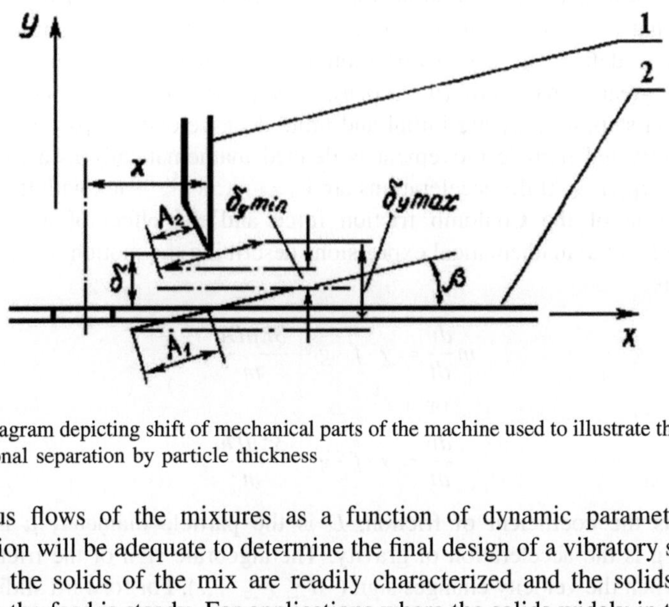

Fig. 3 Diagram depicting shift of mechanical parts of the machine used to illustrate the principle of vibrational separation by particle thickness

at various flows of the mixtures as a function of dynamic parameters. This information will be adequate to determine the final design of a vibratory separator, provided the solids of the mix are readily characterized and the solids concentration in the feed is steady. For applications where the solids widely in thickness, a test program should be undertaken. The VS of this type is applicable where separated solids are expected to be very dry. In horizontal vibratory separators mathematical modeling is useful in evaluating their performance.

Theoretical calculations are recommended for potential design of VS with the deck generally placed horizontally and acted upon by the rectangular step excitation. Alternate forms of the excitation and methods of solution in terms of one of them are referred to [1-5]. The mathematical basis for separation motivated by the rectangular acceleration of translational vibration motion and steps in exact solution of transcendental equations of particle movement are presented below.

3 Process Modelling and Solution of Equation of Motion

Motion of single degree-of-freedom system of the VS deck acted upon by the rectangular step excitation (the rectangular acceleration pulse of magnitude w and duration t) is discussed for theoretical prediction which can be obtained from the governing differential equation. The corresponding velocities of time histories are also for various conditions. The magnitude of the velocity u change defines the intensity of the process. The longitudinal displacement of the deck during the vibration is characterized by three steps (for purposes of illustration in the following examples the primary time history is that of acceleration, time-histories of velocities may be derived there from by integration). If the velocity u is zero at time $t = 0$, then the velocity time-history is a line of constant slope, the corresponding acceleration time-history is the acceleration step of constant value as was shown in [3]. The first step is defined as a forward motion which has value u of zero and a value of w slightly greater than zero $(w_1 > 0, u_{1H} = 0)$; the second step describes forth and back motion for the conditions $(w_2 < 0, u_{2H} = u_{1H} > 0, u_{2K} < 0)$; that is the acceleration step has a value less than zero significantly; the third step is defined as backwards motion which has a value of acceleration somewhat greater than zero $(w_3 > 0, u_{3H} = u_{2K} < 0, u_{3K} = 0)$, where w_i is the acceleration step, u_{iH}, u_{iK} are initial and finite deck velocity steps.

The motivated particle movement is defined mathematically as a function of $w_1, w_2, w_3, u_{1K}, u_{2K}$. If the accelerations are $w_1 = w_3$, $w_1 < -w_2$, with the simplest representation of the Coulomb friction force and the effect of air resistance $(F_c = 3\pi\mu D\upsilon)$ the mathematical expressions describing the motion of a particle of mass m are

$$m\frac{d\upsilon}{dt} = \chi \cdot f \cdot g - \frac{3\pi\mu D\upsilon}{m},$$
$$\text{or} \tag{1}$$
$$\frac{d\upsilon}{dt} = \chi \cdot f \cdot g - \frac{3\pi\mu D\upsilon}{m},$$

where f is the coefficient of friction, D is the particle diameter, μ is the air viscosity, g is the acceleration of gravity. The algebraic sign of the friction term changes when the velocity changes sign $(-I \leq \chi \leq +I)$. For forward sliding when $\upsilon < u$ it must have a positive sign $\chi = +I$; for backwards sliding : $\upsilon > u, \chi = -I$; at rest $\upsilon = u$ it is $\chi \neq \pm I$.

By performing transformation of the latter equation the following differential equation of the particle motion is obtained

$$\frac{d\upsilon}{dt} = \chi \cdot f \cdot g - n\upsilon; \tag{2}$$

where n = 3πμD/m.

Rewriting,

$$d\upsilon = \chi \cdot f \cdot g \cdot dt - n\upsilon \cdot dt.$$

The solution for the latter equation is of the form

$$\upsilon = \chi \frac{fg}{n} + \left(\chi \frac{fg}{n} - \upsilon_{iH} \right) \exp[-n(t - t_H)]; \tag{3}$$

$$l = \chi \frac{fg}{n} (t - t_H) - \frac{1}{n} \left(\chi \frac{fg}{n} - \upsilon_{iH} \right) \{1 - \exp[-n(t - t_H)]\}; \tag{4}$$

where the required values of particle displacement l and its velocity υ are found.

Consider the following operating conditions:

(1) $w_1 < -w_2 << fg$ is a stationary rate.
(2) $w_1 < fg < -w_2$. A brief review of the complete solution evaluated from a knowledge of these starting conditions is given as follows: the particle size is a somewhat factor with the moving and separation motivated by the rectangular vibration. They occur at values of velocities not greatly different from each other; hence, attention is devoted to the next case.
(3) $fg < w_1 < w_2$. For the forward sliding mode ($\upsilon < u$) the highest possible value of the velocity is described by the following equation:

$$\upsilon_{1K} = \frac{fg}{n} - \left(\frac{fg}{n} - \upsilon_{2K} \right) \exp\left[-n\left(T - T'\right)\right], \tag{5}$$

and the expression for the response particle displacement is

$$l_1 = \frac{fg}{n} \left(T - T_1'\right) - \frac{1}{n} \left(\frac{fg}{n} - \upsilon_{2K} \right) \left\{1 - \exp\left[-n\left(T - T'\right)\right]\right\}. \tag{6}$$

For the backwards sliding mode ($\upsilon > u$) the terminal velocity and displacement function are defined by

$$\upsilon_{2K} = -\frac{fg}{n} + \left(\frac{fg}{n} - \upsilon_{1K} \right) \exp\left(-nT'\right); \tag{7}$$

$$l_2 = -\frac{fg}{n} T' + \frac{1}{n} \left(\frac{fg}{n} - \upsilon_{1K} \right) \left[1 - \exp\left(-nT'\right)\right], \tag{8}$$

where $T' = t_2 - t_1$.

The foregoing equations are alike, mathematically, and a solution may be applied to any of the others by making simple substitutions. Therefore, the "Eqs. (5, 6, 7 and 8)" may be expressed in the general form of Eq. (9):

$$\frac{\left(T' - \frac{T}{2}\right)[1 - \exp(-nT)]}{\frac{fg}{n}} = \frac{1 + \left(\exp(-nT) - 2\exp\left[-n(T - T')\right]\right)}{|w_2|}$$

$$-\frac{1 + \exp(-nT) - 2\exp\left(-nT'\right)}{};$$

$$\upsilon_{1K} = \frac{fg}{n[1 - \exp(-nT)]} \cdot \left\{1 + \exp(-nT) - 2\exp\left[-n\left(T \frac{w_1}{-} T'\right)\right]\right\} = w_1\tau - |w_2| \cdot (t_1 - \tau);$$

$$\upsilon_{2K} = \frac{fg}{n[1 - \exp(-nT)]} \cdot \left[-1 - \exp(-nT) + 2\exp\left(-nT'\right)\right] = w_1(T_2 - t_2);$$

$$l = l_1 + l_2 = \frac{fg}{n}\left(T - 2T'\right)$$

$$\upsilon_y = \frac{l}{T}$$

If the air resistance is negligible ($n = 0$), the Eq. (9) reduces to the form of Eq. (10):

$$\upsilon_{1K} = fg\tau;$$

$$\upsilon_{2K} = fg\left[\tau - \frac{T}{2}\right];$$

$$\upsilon_y = fg\left[\tau - \frac{T}{4}\right]$$

Within each case there are variations and differences of effects. For the third case 3, separation depends essentially on the size differences of the particles present in the mix. The maximum value of the velocities will hold in this case: $w_1 \geq fg$. $w_1 \geq fg$ $(n \to 0)$ w_2.

The particle velocity forward the deck plane directly related to the friction coefficient which is a function of the particle shape and size [4]. Rectangular pulse excitation: the excitation function given by τ and T includes the natural period of the responding system and a significant period of the excitation. The excitation may be defined in terms of various physical quantities, and the response factor may depict various characteristics of the response. The purpose is to compare vibration motions, to design equipment and to obtain useful information. Care must be taken to assure that the same VS and its performance can be predicted theoretically but such vibratory separators require experimental work. The mechanical design, however, may be used commercially by the application of vibration considered to be sinusoidal or simple harmonic in form. Alternate form of the excitation may be applied after making simple substitutions of vibration exciters.

4 A Suitable Design of Equipment and its Operation

An improved design of VS and for its application for efficient separation of mixtures on the thickness differences of the solids is shown in Fig. 4. The VS consist of active frame (part 1) and two top and bottom plate decks (part 2) and

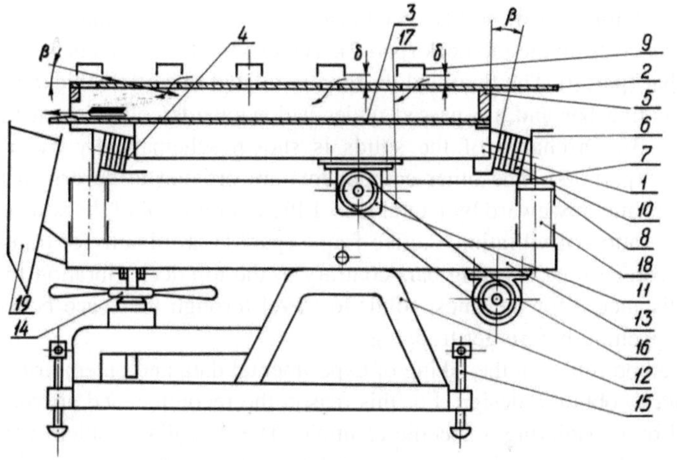

Fig. 4 The mechanical design for vibratory separation on the thickness differences of the solids: *1* = active frame; *2* = top deck; bottom deck; *4* = boss; *5* = distant strip; *6* = reactive frame; *7* = boss; *8* = barrel; *9* = dividing bar; *10* = springs; *11* = vibration exciter; *12* = bed; *13* = turning frame; *14*, *15* = screws; *16* = electric motor; *17* = drive belt; *18* = isolator; *19* = receptacle

(part 3), respectively; reactive frame (part 6) with bosses (part 7) for attaching springs, barrels (part 8) for shock-absorbers, and dividing bars (part 9); stack of springs (part 10); the direct-drive vibration exciter (part 11) consists of a rotating unbalanced mass driving a positive linkage connection between the base and decks of the machine; the bed (part 12) with turning frame (part 13) and screw mechanisms (parts 14 and 15) for regulation of longitudinal tilt of the decks and lateral tilt of the bed; electric motor (part 16) with step belt pulley (part 17); rubber shock-absorbers (part 18) isolating the upper part of the machine from the base; receptacles (part 19) for removing separated components from a mixture. A constant-displacement vibration machine of this type attempts to maintain constant-displacement amplitude while the frequency is varied. Similarly, a constant-acceleration vibration machine attempts to maintain a constant-acceleration amplitude as the frequency is changed. The primary independent parameters that influence the ability of VS of this type to make separation are the thickness differences of the solids and mixtures loading.

The VS is applicable where the amount of the material to be processed is reduced to the thin layer and where the separated solids are expected to be dry. Separations using horizontal VS are very similar in principle to those for tilted VS. In this situation, the force driving the separation could also be several orders of magnitude greater than that of gravity. The proper introduction to feed to a separation is provided by a special vibrating feeder. The feed material should be introduced uniformly across the active cross-section of the deck and done in a way of metering its amount.

An initial mixture of solids that must be separated into individual products inflows or arrives on the top deck (part 2). The driving force directs the mix toward dividing bar (part 9). The finest solids directly related to particle thickness selected by the dividing bar and are passed under it downwards collection at the bottom deck (part 3). Discharge of the solids is shown schematically by pointers to receptacles (part 19). The other components are ordered to be driven along the driving bars and downward by lateral side. Efficient removal of the solids increases with acceleration of vibration. Active frame (part 1) is suitable for transmitting a vibratory force to the dividing bars contrary to the top deck vibration to correlate vertical distance δ for the finest solids to travel through the space between them and deck opening before get trapping.

The method relies on the taking of experimental data and on empirical analysis of the data to obtain a design. For this reason the recommended approach is best presented by considering a specific example. The potential problem with separation of clear bulb glass from open-circuit primary crushing of electric lamps is probably best supported by this type of VS. For very efficient operations, two-stage combinations of separation are used. The VS is applicable for efficient removal of needed thin bulb glass (in a way shown in Fig. 4 by arrow-pointers). The second stage may be secondary crushing with screening or vibratory processing to obtain metal wires. Caps, their fillings and glass stems are non-utilizable waste. The VS, however, may be used commercially in some other industries.

5 Conclusions

1. In the design and operation of separation processes depended essentially on the thickness differences of the solids present in the mix a problem is approached logically by first preparing an initial design. A subsequent analysis point to desirable modifications.
2. An initial prototype equipment has been designed and is then constructed in which actual operating conditions preferably are determined and considered from practical point of view.
3. From the analytical point of view and to provide some perspective for later improvement of existing processes the single degree-of-freedom system of the horizontal VS model acted upon by rectangular step excitation is considered with mathematical method of analysis to obtain useful information. The technical feasibility of a given separation method might be essentially attractive.
4. The optimum design must arise from careful consideration of all feasible alternatives and represents the further inventive aspect of process design.

References

1. Anakhin V, Pliss D, Monakhov V (1991) Vibrational separators. Nedra, Moscow, p 157
2. Anakhin V (1993) Vibrational separation of powdered materials. Irkutsk State University, East Siberia, p 208
3. Anakhin V, Pliss A (1999) Procedures for the design of vibratory separating equipment. In: Andrew G. Starr, Andrew Y. Leung, Jan R. Wright, David J. Sandoz (eds) DYMAC'99: University of Manchester. Proceedings of the 1st international conference on the integration of dynamics, monitoring and control, Manchester, United Kingdom, September 1999. Integrating dynamics, condition monitoring and control for the 21st Century. A.A. Balkema, Rotterdam, Brookfield, 1999, pp 451–457
4. Anakhin V (2001) Graphical methods in the analysis of biharmonic vibration. In: International conference on condition monitoring, Oxford, June 2001. Proceedings of the international conference on condition monitoring. Coxmoor Publish. Co., Oxford, UK, pp 220–228
5. Anakhin V, Anakhin T (2005) Process consideration and concepts in data analysis. In: International conference on condition monitoring, July 2005. Proceedings of the International conference on condition monitoring. Coxmoor Publish. Co., Cambridge, pp 85–90

Test Bench for the Analysis of Dynamic Behavior of Planetary Gear Transmissions

A. Fernández del Rincón, R. Cerdá, M. Iglesias, A. de-Juan, P. García and F. Viadero

Abstract In this paper a back-to-back test bench for experimental characterization of planetary gear transmissions is described. Some considerations related to the design of this kind of devices as well as the compromises and general features of the proposed design are presented. Particular attention is given to the instrumentation layout and the alternatives foreseen for measurement. In the last chapter, some results are presented in order to show the capabilities of the bench configuration.

Keywords Planetary Testing · Gear Dynamics · Vibration

1 Introduction

Power transmission systems based on planetary gears are nowadays under a great demand to enhance their performance in terms of power level, speed, efficiency and compactness, with the subsequent increase in the load value they must withstand. The proportional growth in the computing cost of these requirements, as well as higher and higher demanding expectations in terms of durability and reliability has resulted in an increasing interest in the development of statistical models of prediction that would allow the prognosis of the systems in progress [1]. Besides, different physical models have been developed trying to simulate the dynamic behaviour of the gear transmissions [2]. However, both types of model, either those ones based on previous statistical data or the behavioural physical models; both require a detailed experimental study which can provide this field of

A. Fernández del Rincón (✉) · R. Cerdá · M. Iglesias ·
A.de-Juan · P. García · F. Viadero
Department of Structural and Mechanical Engineering, University of Cantabria,
Cantabria, Spain
e-mail: fernandra@unican.es

G. Dalpiaz et al. (eds.), *Advances in Condition Monitoring of Machinery in Non-Stationary Operations*, Lecture Notes in Mechanical Engineering,
DOI: 10.1007/978-3-642-39348-8_42,

knowledge with valuable and reproducible data about the dynamic behavior of the transmissions. Along the same lines, the current state of the art denotes the increasing interest on the different designs of test benches [3, 4], that will allow the adequate experimentation of planetary gear trains for the study of its dynamic behavior.

The first decision to be made when designing a test bench for gear trains is the type of layout, referring to the power flow through the bench. According to this criterion, the bench can adopt two different configurations: the direct and the power circulating one. The direct configuration is the simplest layout to conceive, with the Gears Under Test (GUT) being placed between a motor and a resistant load, the latter allowing the establishment of the transmitted torque. In the direct configuration, the power loop is open and the energy consumed in the tests is lost by heat dissipation in the load, which generally is a frictional brake, hydraulic or magnetic, depending on the scale of test power.

According to its installation, this direct configuration is the easiest and the least expensive layout. Furthermore, it is highly versatile as the system can easily adapt to different test specimens just by changing the distance between shafts. However, test benches with direct configuration are not appropriate for high power tests (e.g. wind turbine gearboxes), as the whole transmitted power is to be dissipated in the load, leading to a system with high energy consumption. This fundamental drawback motivates the appearance and usage of power circulating test benches.

In circulating power rigs, the load receiving the power form the gear train is also an energy recovery system, so that this energy is transformed (if necessary) and reintroduced at an initial point of the power loop, closing it. This feedback point where the power is re-injected can be a point somewhere before the actuator typical for hydraulic [5] or electric circulating systems, or after the actuator (dashed line), a type of loop common in systems with mechanical power circulation.

Test benches with electric power circulating systems are versatile and efficient, but their cost is a problem. Both the generator that acts as a recovery device and the whole electrical system that conditions and transforms the signal before its re-entry to the motor make cost to rise heavily. Moreover, in the test case of planetary gear trains, it will always be necessary to install another auxiliary planetary gearbox in a back-to-back configuration with the first one. This is caused by the high gear ratio that planetary gearings present, which would force the motor and generator to work out of their nominal speed range.

From the energy viewpoint, the most efficient circulating solution is the one carried out by purely mechanic means [6]. In this type of circulating systems (four-square gear tester [7]) the power follows a closed loop inside the bench, with the only external energy to supply being the one due to the internal losses.

The GUT is installed parallel to an auxiliary train (mechanically oversized to avoid it from getting damaged during the test), with identical gear ratio as the first one. The movement and the loss power are introduced by the motor while, in order to introduce the torque, the torsion of one of the shafts is needed as it is indicated in Fig. 1, thus preloading the whole group.

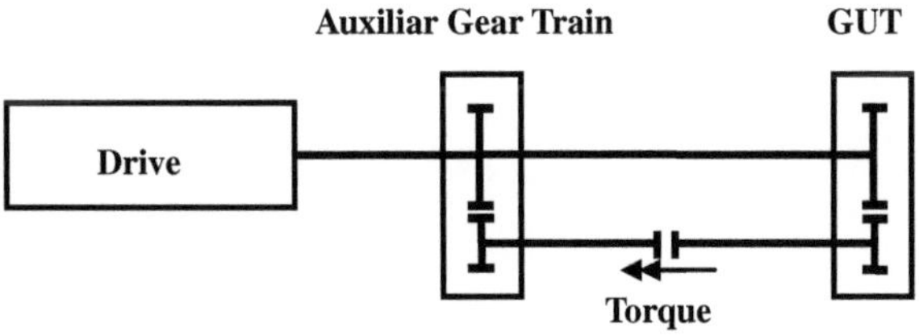

Fig. 1 Test bench with mechanic power circulating (Four-square)

The number of non-pure mechanical components in this type of configuration is reduced to just one: the motor. This motor has to be chosen only to supply the losses of the system, thus being significantly smaller than if the direct configuration is implemented. As a main drawback, the versatility of these test benches with mechanical circulation is very limited, as the whole recirculating system is determined by the test train itself. Moreover, the presented method of bringing in the torque is impractical and prevents the existence of a variable torque during the test and its introduction with the train in motion in most types of train gearings (those with a single degree of freedom). In order to avoid this drawback some devices [8] have been proposed to allow torque variations.

2 Test Bench Description

Based on the different design alternatives that have been discussed previously, and taking into account the particularities of planetary gear trains, the adopted layout for the test bench is presented below. Firstly, the characteristics of the GUT must be considered. These are reduction gearboxes for agricultural machinery, with the number of teeth presented in Table 1. As the test campaign is centered on this gearbox model in particular, versatility of the bench is not a priority. On the other hand, the gear size (module greater than 4) focuses part of the design decision on the needed test power, which has to be enough to work for load levels capable of causing deterioration in the tested train. Both for economic and energy efficiency criteria, the adopted layout will be the one with mechanical power circulation. In this particular case, this configuration is also more compact, consuming less space than any of the other alternatives.

Considering a four-square layout with mechanical recirculation, the first step of the installation consists in two planetary transmissions with the same gear ratio, the auxiliary gearbox and the GUT. Unlike the relatively easy assembly for trains with external gears, the internal gear of the ring of a planetary transmission does

Table 1 External torque required in order to achieve the failure

	Ring	Sun	Planet
Number of teeth	65	16	24
Ring torque (Nm) bending	–	8,534	5,548
Ring torque (Nm) pitting	–	7,982	8,535

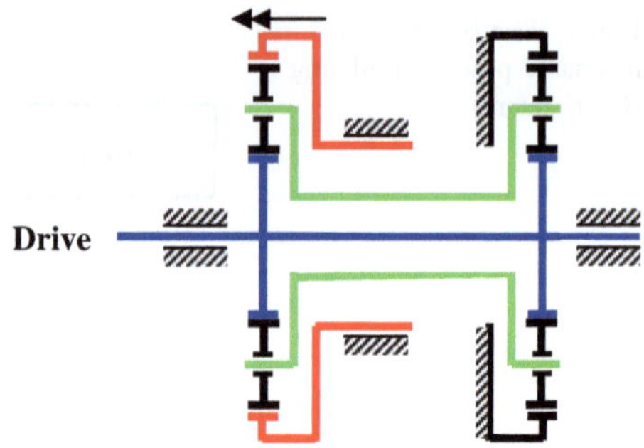

Fig. 2 Back-to-back layout as assembled in the bench

not allow for the same coaxial configuration shown in Fig. 1, so the introduction of two more auxiliary trains is needed.

Making these two additional auxiliary ordinary trains to have the same 1:1 gear ratio, it is easy to prove that both the carriers and the suns are joined by shafts rotating at the same speed. Thus, a simplifying option to avoid the usage of ordinary gear trains could be intuited. The two planetary transmissions could be placed facing each other and sharing the suns and carriers as it is shown in Fig 2 [9].

Moreover, this layout allows an alternative way to introduce the torque. Instead of the torsional pre-stress of the shaft that has been presented in the previous section, as the planetary transmission has two degrees of freedom compared to the single one of ordinary transmissions, the degree of freedom of the auxiliary planetary box ring can be left unconstrained in order to introduce the torque using that ring (Fig. 3).

3 Motor-Drive Selection: Power Losses

As the goal of the presented test bench is to allow the study of the dynamic behavior of planetary gears, both in normal working conditions and in limit situations, it is vital to determine the power and torque values that the bench must provide to the testing specimen. The pair power-torque must be sufficient not only to propagate and increase the typical defects of the gears, but even to create them. This study has focused its attention in the mechanical failure of the teeth caused by bending and pitting. Thus, an estimation of the minimum tangential forces that will produce the mechanical failure in case of teeth bending and pitting for the sun, planets and ring gears is necessary.

Following the ISO 6336-2 and 3 [10, 11], and considering a safety factor of one the tangential force required to produce the bending or pitting failure in the teeth of each gear is determined. Once the minimum tangential force values required to reach each type of failure in all the gears are determined, these data are converted into exterior torque that need to be supplied through the ring, arriving to the values presented in Table 1.

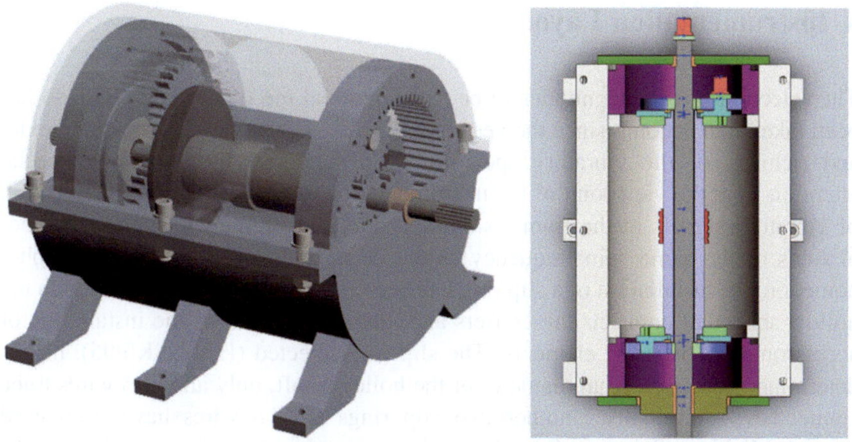

Fig. 3 General view and top section of the test bench assemble

Aiming a torque value able to cause both type of defects of our study, the value 7,982 Nm was chosen, as it causes pitting failure both in the sun and the planet and bending failure in the planet. This will be the maximum torque value that the bench must be able to supply and, therefore, the power of the required motor will be determined based on it (only the associated losses). Two devices were defined for applying the external torque. The first one was a lever arm with a dead weight at the end. This approach provides a good control of the input torque but don't allow a big load and moreover modifies the rotational mass of the free ring and therefore the resonance frequencies. As second option an automotive jack was disposed, pushing the arm in the opposite direction. This device provides higher loads although it does not provide the same control on the applied torque level.

In order to determine the motor power, it is necessary to take into account that the power to supply corresponds just to the losses in the bench. Therefore, the knowledge of the system efficiency is a primary and essential condition before selecting the motor. The efficiency of the system is obtained following the method described by Molian [12] that produces an efficiency value of 0.9584. Knowing that the input speed to the system (to the sun of the gear train to test) is 1,500 rpm and that the maximum external torque is 7,982 Nm, the power is calculated as:

$$P_{Sun} = T_{Sun} \cdot \omega_{Sun} = 317\text{kW} \rightarrow P_{MOTOR} = P_{Sun} \cdot (1 - \eta_{SIST}) = 16\text{kW} \quad (1)$$

At this point of the selection process, it is remarkable the difference in the required driving powers, either with the bench adopting a circulating one or in the layout without recirculation, with a power system of even 20 times in this new case. The impact that the selection of the bench layout has in the cost of the components and energy used turns to be even more evident now.

4 Instrumentation Layout

The selection of the magnitudes to be measured and the transducers to do so has been taken into account since the very initial phases of the design. The complexity and richness of the vibratory spectrum of planetary gears demand a special attention from the viewpoint of the instrumentation to be installed. The acquisition of vibratory signals in the frame is strongly affected by the appearance of modulations of the same spin frequency as the carriers. In order to avoid this phenomenon, the installation of a slip ring has been conceived. It spins jointly with the hollow shaft that connects the carriers and, therefore, it allows the installation of accelerometers on these elements. The slip ring selected (HBM SK5/95), highly determined by the external diameter of the hollow shaft, only allows 5 wires to be connected. Additionally, another two slip rings with 8 wires have been used (model: Michigan Scientific S8). One of them spins jointly with one of the planets, while the other one is placed at the free end of the suns' shaft. The adopted layout is presented in a schematic way in Fig. 4. In this figure, an empty circle represents a strain gauge channel whilst an empty square represents an acceleration measure point. The dashed lines (coloured in red) represent the connections with the slip rings whereas solid lines are direct connections to the acquisition system. The combination of the slip ring on the hollow shaft connecting the carriers and the one that spins jointly with one of the planet provides a great versatility regarding the measurement options and multiple scenarios can be observed:

- Register of the torque by the installation of a complete strain gauge bridge (of 4 wires) in the hollow shaft.

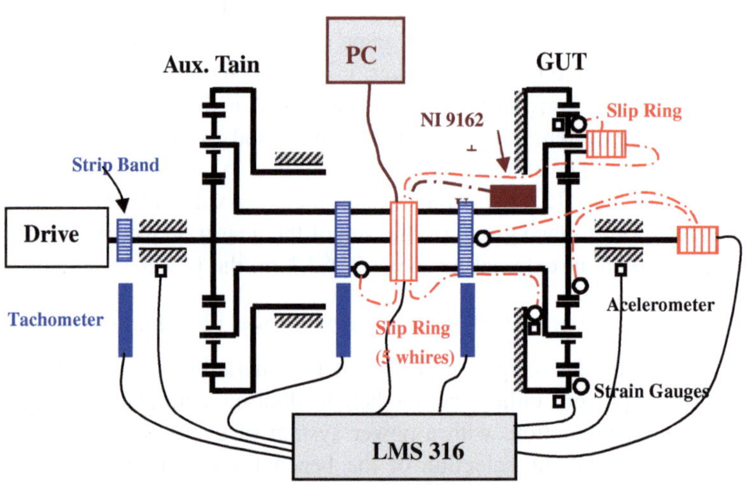

Fig. 4 Schematic representation of the instrumentation layout

- Register of the acceleration on the suns' bearings in the carrier. Two accelerometers (4 wires each) can be installed with different orientations: radial or tangential.
- Register of the angular acceleration of the carriers. Using two accelerometers place at the same radial distance (4 wires), orientated in the tangential direction averaging out both signals.
- Measure of the acceleration in the planet up to two directions combining both slip rings in such way that four wires in each one are used.
- Measure of strains in the base of the planet's teeth by two quarter—bridge (4 + 4 wires).

The slip ring at the end of the shaft that connects the two suns will be used to measure the actuating torque in the gap between both by a full bridge (4 wires) as well as to measure the strains at the base of the teeth. The signals coming from the slip rings located on the hollow shaft and on the shaft between the suns will be acquired by a LMS SCADAS 316 system. The use of this system permits the acquisition of up to 12 dynamic channels simultaneously, 8 coming from ICP sensors and 4 from strain gauges, and to combine them with two tachometric channels. Furthermore, in this way, apart from the signals from the slip rings, accelerometers could be placed on the frame and measure the strain at the base of the teeth of the annulus gear.

The main disadvantage of the slip rings is the possible pollution of the electric signal with noise due to the contact friction. This will be specially critic in the case of the signals obtained by means of strain gauges. Moreover, the limited number of available tracks on the slip ring located on the hollow shaft restricts the signals that can be sent out to its acquisition. To avoid these drawbacks, and taking into account the dimensions of the hollow shaft and its rated speed, the possibility of installing a low-cost acquisition module which could spin jointly with it has been considered. In particular, a NI USB-9162 device that allows the control of different acquisition modules with a USB connection (5 wires) with a PC has been selected. This device can be combined with a NI9233 module for the acquisition of signals coming from ICP transducers or with a NI9237 module for the acquisition of signals coming from strain gauges. Thus, the acquisition and conditioning of the dynamic signal would be done within the system, by sending the resulting information in a digital form via the USB connector using the slip ring for its transfer to the PC. This way, the noise inherent to this kind of elements would not affect the resultant measurement. This possibility has been represented in Fig. 5 with a bold dotted line from the acquisition device to the slip ring and with a continuous line in order to represent the final connection with the PC. Additionally, three optic tachometers (Compact VLS7) were used. Two of them were placed on two distant positions along the hollow shaft while the third will be placed in the motor axis. These tachometers will be combined with pulse tapes along the axes in order to measure their instantaneous angular velocity.

The actuating torque can be obtained by determining the length and the suspended mass of the arm but also from the electrical power consumed by the drive.

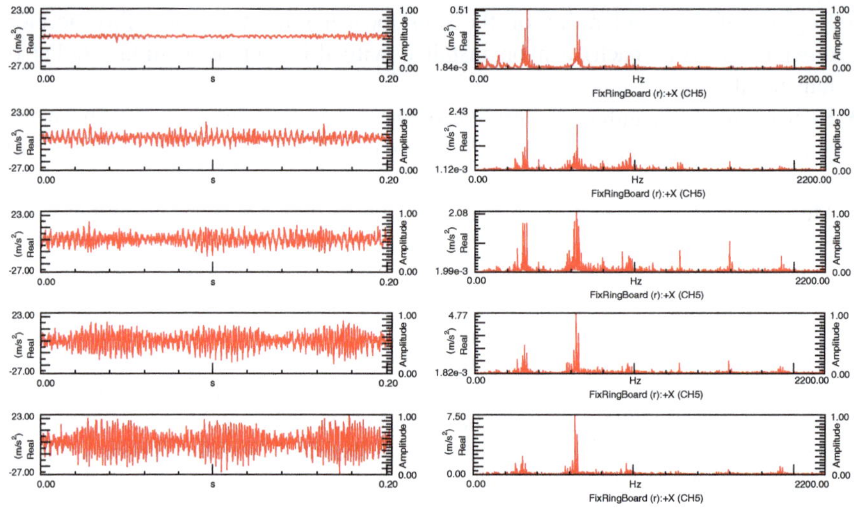

Fig. 5 Recorded data on time and frequency domain in radial direction for several torque levels (upper to lower 0; 900; 1,900; 2,900 and 3,900 Nm)

Regarding the velocity control, this is achieved by a variable-frequency drive whereas the power consumption can be estimated from the drive parameters.

5 Preliminary Results

The test bench described previously has been used in a first experimental campaign in order to identify the most important features of the vibratory signal. For these preliminary tests, only acceleration signals were recorded, in radial and tangential direction, with triaxial accelerometers located on the fix and free rings. Moreover, the lever arm and dead weights were used to calibrate the relationship between the applied torque measured by the motor controller and the torque actually applied to the system. Thus, the same level of torque was introduced by using the automotive jack, although this time the torque was applied in the opposite direction. These preliminary tests allowed to discern the impact of the loading system on the vibration at the auxiliary gear box and particularly at the GUT arriving at the initial conclusion that the jack constraints the movement of the free ring reducing the vibration amplitude with minor consequences on the GUT.

Then, the first task was to analyze the resulting vibration for the system under several torque levels. With this aim, five torque levels were applied (0; 900; 1,900; 2,900; and 3,900 Nm) through the jack. Accelerometer signals were recorded using a sampling frequency of 6,400 Hz jointly with a tachometric signal from the hollow shaft which rotates with an average speed (equal to the carrier speed ω_c) of 300 rpm. The left column of Fig 5 shows the radial acceleration recorded on the

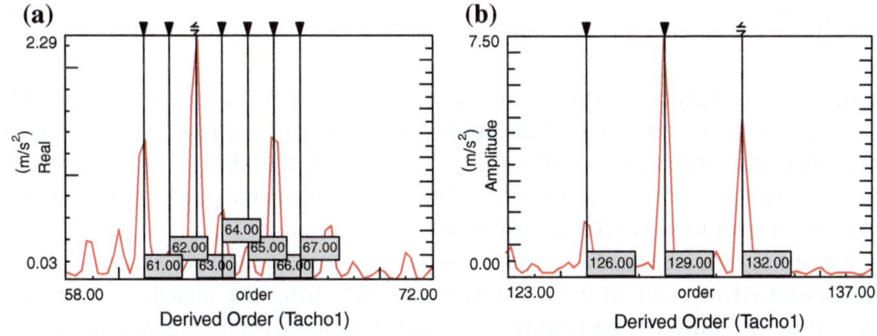

Fig. 6 Order spectra at radial direction in the vicinity of the (**a**) first GMF order and (**b**) second GMF order (Torque 3,900 Nm and 300 rpm)

GUT's fix ring corresponding to a rotation of the carrier for each applied torque. As expected, the time signal becomes higher as the torque is increased but also it stands out the signal modulation due to the carrier rotation. Regarding the corresponding spectra on the right column, it is clear the change on the relative amplitudes of the different harmonics of the Gear Mesh Frequency (GMF) which is equal to $Z_r\ \omega_C$. In particular, the increment of the torque exchanges the maximum peak between the first two GMF harmonics. A more detailed view of the order spectra for the maximum torque is presented in Fig. 6 to discern better where are located the higher peaks for the first and second GMF harmonics. For the first GMF (see Fig. 6a) which is located at the order 65, the highest peak appears at order 63 followed by the 66 while the amplitude for the 65 order is reduced. This fact agrees with the theoretical predictions provided in [13]. According to these authors, planetary gear sets can be classified in five different types based on the sideband activity. The configuration of the GUT belongs to the category of equally-spaced planets and sequentially phased gear meshes as it verifies.

$$\psi_i = \frac{2\pi(i-1)}{N}; \frac{Z_r\psi_i}{2\pi} \neq n; \text{and} \sum_{i=1}^{N} Z_r\psi_i = m\pi(m{,}n = \text{integers}) \quad (2)$$

where ψ_i is the angular position of planet *i*, *N* is the number of planets and Z_r is the ring's number of teeth. Following this classification the higher amplitudes correspond to orders which are multiple integers of N in the vicinity of the GMF harmonics. This fact is more remarkable for the second GMF (see Fig. 6b) where the higher peaks correspond to orders 42, 43 and 44 N. Nevertheless, more peaks can be appreciated for the first GMF distributed in orders 59–69, probably related with planet position errors and eccentricities which cannot be explained with the theoretical approach described in [13]. Moreover, the higher amplitude should be obtained for the multiple integer of N closer to GMF which in Fig. 6a corresponds to the order 66. Nevertheless, the largest peak is obtained at order 63 although this situation changes when the applied torque is lower.

6 Conclusions

In this work a planetary test rig with mechanical power recirculation is presented. It allows for the modification of the applied torque during the test, with the aim to reproduce non-stationary conditions. Details about the design are described, highlighting the procedure followed for determining the drive power, which in this type of configuration is related only to the power losses.

Regarding the measurement capabilities of the test rig, several slip rings have been disposed in order to extract dynamic signals from the planets and carrier. Moreover, two different possibilities of acquisition have been considered. On one hand, the vibratory signals can be extracted through the slip rings and sent to the acquisition device. On the other hand, the acquisition can be done on-board, and only digital signals are extracted through the slip rings to the computer.

Tests have been carried out and preliminary results have been discussed. The acceleration spectra under several torques have been assessed and the results have been positively correlated with the behavior found in the literature.

For the future it is foreseen the use of the testing bench for validation of numerical models, but also to assess condition monitoring techniques particularly those which involve non-stationary conditions as load susceptibility [14].

Acknowledgments This work has been carried out in the framework of the COST Action TU 1105 "NVH analysis techniques for design and optimization of HYBRID and ELECTRIC vehicles".

References

1. Heng A, Zhang S, Tan ACC, Mathew J (2009) Rotating machinery prognostics: state of the art, challenges and opportunities. Mech Syst Signal Process, 724–739
2. Fernández del Rincón A, Viadero FA (2009) Dynamic model for the study of gear transmissions. WIT transactions on modelling and simulation
3. Dalpiaz G, D'Elia G, Delvecchio S (2007) Design of a test bench for the vibro-acoustical analysis and diagnostics of rotating machines. In: Proceedings of the second world congress on engineering asset management and the fourth international conference on condition monitoring 2007, 11–14
4. Haigh J, Fawcett JN (2000) The UK 8 MW facility for gear noise and vibration research. In: Proceedings of the institution of mechanical engineers, part K: journal of multi-body dynamics, 132–131
5. Sweeney PJ (1994) Transmission error measurement and analysis. Thesis doctoral. The University of New South Wales, New South Wales
6. Adams ML (2009) Rotating machinery vibration:from analysis to troubleshooting. CRC Press,US
7. Limmer JD (1997) Improved methods of vibration measurement, gear fault detection and bearing fault detection for gearbox diagnostics. Thesis doctoral. Rensselaer Polytechnic Institute, New York
8. Mihailidis A, Nerantzis I (2009) A new system for testing gears under variable torque and speed. Recent patents on mechanical engineering, 2, 179–192, 2009. Bentham Science Publishers Ltd, Sharjah

9. Ligata H (2007) Impact of system-level factors on planetary gear set behaviour. Thesis doctoral. The Ohio State University, Columbus
10. ISO 6336-3 (2007) Calculation of load capacity of spur and helical gear, part 3: calculation of tooth bendind strength. International Standard
11. ISO 6336-2 (2006) Calculation of load capacity of spur and helical gear, part 2: calculation of surface durability (pitting). International Standard
12. Molian S (1997) Mechanism design: the practical kinematics and dynamics of machinery. Elsevier Science, Oxford
13. Inalpolat M, Kahraman A (2009) A theoretical and experimental investigation of modulation sidebands on planetary gear sets. J Sound Vib 323:3–5
14. Bartelmus W, Zimroz R (2009) A new feature for monitoring the condition of gear-boxes in non-stationary operating conditions, Mech Syst Signal Process 23:1528–1534

A Novel Gear Test Rig with Adjustable Shaft Compliance and Misalignments Part I: Design

A. Palermo, J. Anthonis, D. Mundo and W. Desmet

Abstract This paper describes the design aspects for a new gear test rig aimed at adjusting the influence of shaft compliance on gear meshing stiffness, while allowing the operator to impose gear misalignments. Static and dynamic testing is possible for the most important gear-related physical quantities: Transmission Error, relative displacements, tooth root strain, transmitted torque. The discussed test rig has a mechanical power circulation arrangement, where two sides can be identified. One side is dedicated to testing a cylindrical gear pair; the other side is needed for retaining a torque preload in the system by means of a second gear pair. Relative misalignment can be imposed between the test gears in the five possible degrees of freedom (three parallel misalignments, two angular misalignments). Shaft compliance can be adjusted by setting the axial position of the gears before fastening them to the shafts.

Keywords Angular Misalignment · Gear Test Rig · Lead Modifications · Parallel Misalignment · Profile Modifications · Shaft Stiffness · Vibration Isolation

1 Introduction

Experimental dynamic characterisation of mechanical transmissions requires measurement of vibrations related to motion or forces in response to excitations. Gear test rigs allow overcoming the limitations related to typically poor

A. Palermo (✉) · D. Mundo
Department of Mechanical Engineering, University of Calabria, Ponte Pietro Bucci 87036 Rende, Italy
e-mail: palermo.antonio@gmail.com

J. Anthonis
LMS International, Interleuvenlaan 68 B-3001 Leuven, Belgium

A. Palermo · W. Desmet
Department Mechanical Engineering, K.U. Leuven, Celestijnenlaan 300 B B-3001 Heverlee, Belgium

G. Dalpiaz et al. (eds.), *Advances in Condition Monitoring of Machinery in Non-Stationary Operations*, Lecture Notes in Mechanical Engineering,
DOI: 10.1007/978-3-642-39348-8_43,

accessibility of gears in service and offer the chance of heavily instrumenting the gears to seek validation of numerical gear meshing models.

Considering mechanical power, gear test rigs described in literature can be divided in two main categories: power absorption and power circulation. In both cases, electric machines are typically used to power the system thanks to their flexibility in speed and torque ranges and thanks to their suitability for control units. For a cylindrical gear system with one input shaft and one output shaft, power absorption test rigs rely on a motor which provides input power and a generator which absorbs the output power. Power circulation test rigs are self-balanced systems and only one motor is required to overcome losses to spin the system.

Design choices are discussed in the next paragraphs for a power circulation test rig for parallel cylindrical gear pairs.

2 Objectives and Global Design

Objectives for the discussed gear test rig are the static and dynamic measurement of typical gear-related physical quantities, under imposed conditions of misalignment and with a given degree of influence of shaft compliance on gear meshing stiffness. One of the final goals is to experimentally validate the gear multibody element proposed in [1]. Focus for the experimental campaign is on a gear pair belonging to families of spur or helical cylindrical gears. Both gear families have different amounts of profile and lead microgeometry modifications for the tooth surfaces (Sect. 4). Parallel or angular misalignments, can be imposed along the five degrees of freedom for relative orientation of the gears (Sect. 5). Shaft torsional and bending stiffness are by nature placed in series with gear mesh stiffness and affect the total stiffness of the gearing. The ratio between the shaft stiffnesses and the gear pair stiffness can be adjusted to impose which of the two elements will dominate the dynamic response (Sect. 6). All the tests can be performed in a range of speed, for a

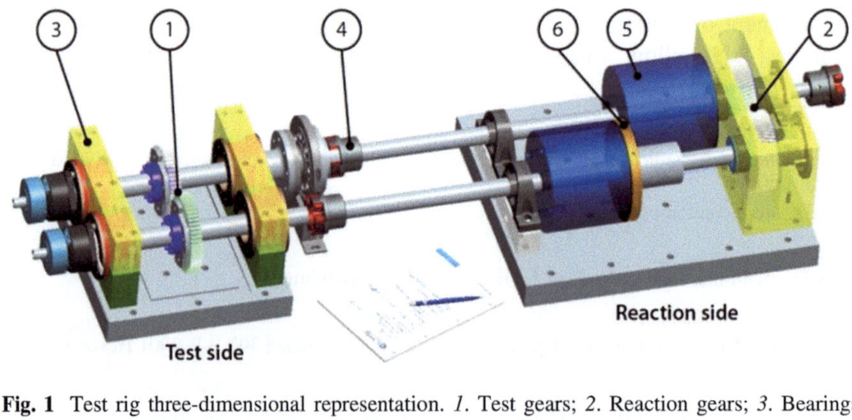

Fig. 1 Test rig three-dimensional representation. *1*. Test gears; *2*. Reaction gears; *3*. Bearings support plates; *4*. Flexible couplings; *5*. Flywheels; *6*. Clutch flange for preload

Table 1 Test rig specifications

Parameter	Range	Uncertainty
Speed	0–4,500 rpm (0–75 Hz)	Measured
Torque	0–500 Nm	± 0.05 %
Angular misalignments	0–2 mrad	0.1 mrad
Parallel misalignments	0–0.3 mm	0.020 mm

set of transmitted torque values. The CAD model of the test rig is represented in Fig. 1, while its specifications are reported in Table 1.

The test rig is mounted on a concrete base suspended on air springs which fulfils three functions. First, it provides reference pin holes to align the test side and the reaction side. Second, it provides rigid anchoring islands for the two sides and isolates the overall system. Third, it separates the two sides to avoid propagation of stray vibrations from the reaction side to the test side; such vibrations are damped through the concrete. A temperature-controlled pressurized lubrication system feeds oil to the test and reaction side bearings and gears. Bearings are lubricated through channels leading to the lubrication groove in their external race. Gears are lubricated by oil jet. A Plexiglas enclosure collects lubrication oil for the test gears which is returned to the lubrication circuit. Safety grids, acting also as safety retention elements for the flywheels, cover the rotating components between the test and the reaction side.

The quantities to be measured are: gear Transmission Error, speed, torque, tooth root strain along the axial direction, relative displacements of the gear bodies, shaft displacements and imposed misalignments. Detailed discussion of the measurements and the related instrumentation is addressed in the second part of this paper.

Speed can be set by means of a 9 kW asynchronous motor controlled by a frequency inverter. Each torque value is imposed as a preload in the system by using a torque wrench and a set of weights, executing the following sequence: the flywheel connected to the clutch is fixed to the base by using a pin; the clutch is released untightening the fastening screws; the torque wrench is fitted on the clutch disk and the required weights are hung; the fastening screws are tightened at the prescribed torque; the weights and the torque wrench are removed. Shafts are verified at the maximum gear torque of 500 Nm and provide a static safety factor of at least 4 to yield point, using the Von Mises criterion, considering combined bending moment and torsion at the worst locations. Materials are a ductile medium-carbon C45 steel (360 MPa yield stress) for the massive parts and a low-alloy 1.2312 steel (820 MPa yield stress) for shafts.

Bearings (Fig. 2) are chosen to maximise stiffness and their rating provides infinite service life under maximum loading conditions. In particular, high-precision spherical roller bearings are chosen to support test gears; more details on such bearings are provided in *Par. 5* and *6*. For improving turning accuracy, a conical adapter sleeve is used to adjust inner race interference and bearing clearance. Light interference fit is used for the external race to improve turning accuracy. Both bearings are in a locating configuration and require careful tuning of their relative

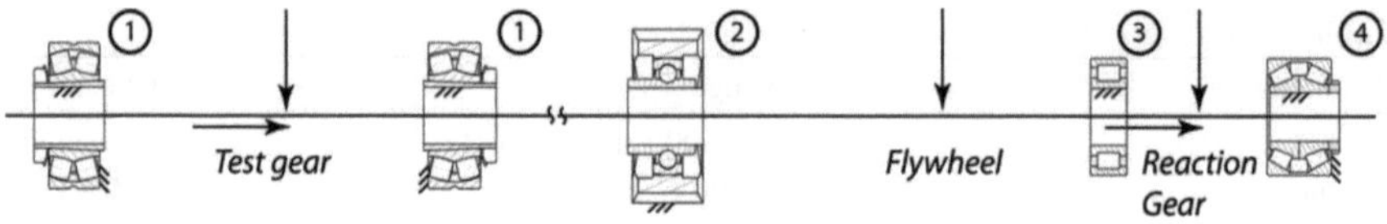

Fig. 2 Bearings arrangement for one shaft branch of the test rig and average bearing stiffness in the loading range. *1*. High-precision spherical roller bearing (1×10^9 N/m); *2*. Y-bearing unit; *3*. Wide-face single-row cylindrical roller bearing (6×10^8 N/m); *4*. Double-row tapered roller bearing (1×10^9 N/m radial, 3×10^8 N/m axial)

distance. Bearings for the reaction side are chosen to support flywheels and reaction gears. Radial loads from the flywheels are supported by Y-bearing units and by wide-face single-row cylindrical roller bearings. Y-bearings are chosen because they do not require precise alignment of their supporting base; cylindrical roller bearings are chosen with flanges on both races and are placed in a non-locating configuration allowing light clearance fit on the external race (which is stationary with respect to load). The same cylindrical bearings support radial loads originating from reaction gears, together with double-row tapered roller bearings in a locating configuration. The locating configuration is achieved using a housing shoulder on one side and a bearing cap with laminated shim on the opposite side. Tapered rollers allow absorbing thrust originated by the reaction gears.

3 Vibration Isolation Between Test and Reaction Sides

Vibration isolation between the two sides is achieved by decreasing torsional stiffness and increasing rotary inertia for the reaction side. The overall effect is to obtain the lowest possible rotational natural frequencies for the reaction side, so that the reaction gears meshing excitation can be filtered out as soon as possible. Torsional stiffness is decreased by using flexible couplings and by increasing shaft lengths. When preloading the system, low torsional stiffness implies high twist angle (1.5° at 500 Nm). This condition ensures that rotational vibrations generated by the test gears (hundredths of degree) do not influence the torque preload, since the latter comes from an elastic deformation. Furthermore, flexible couplings prevent non-torsional loading between the two sides, allowing eccentricity up to 0.28 mm and angular misalignment up to 35 mrad between the two sides.

Analytical equations for a rotational model (Fig. 3) were solved to obtain the rotational natural frequencies along with their related mode shapes. Four degrees of freedom were used in total, allowing rotation for each gear.

The term k_{12} represents the equivalent torsional stiffness for the series between the test gear pair mesh stiffness (equal to $1.6 \cdot 10^6 Nm/rad$), the two shafts torsional stiffness and the two (adjustable, *Par. 6*) shafts bending stiffness. The term k_{34} represents the torsional stiffness for the reaction gear pair (equal to $1.8 \cdot 10^6 Nm/rad$), since supporting shafts there are very short. The terms k_{13} and

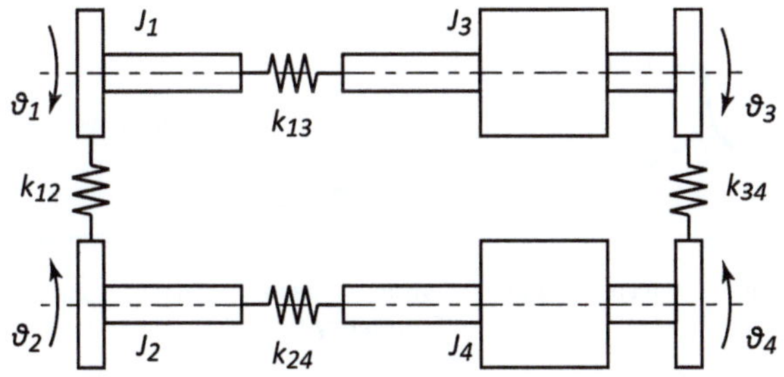

Fig. 3 Illustration of the 4-DOF rotational model of the test rig

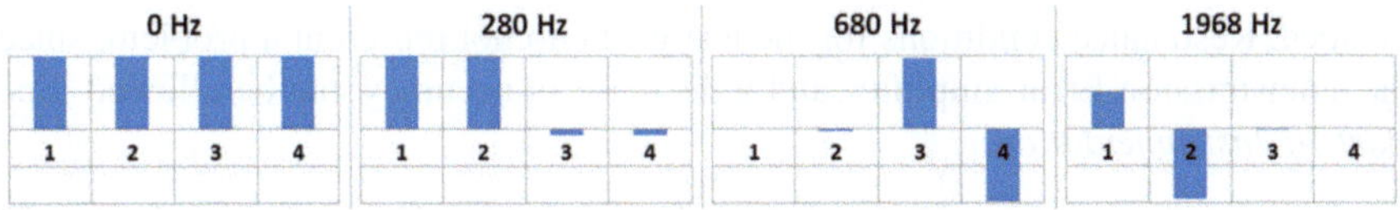

Fig. 4 Histogram representation of mode shapes for the 4-DOF rotational model of the test rig

k_{24} represent the torsional stiffness of the flexible couplings. The terms J_i represent the moments of inertia around the rotation axis for each gear plus the related shafts. Moments of inertia for the test side ($J_1 = 0.0218\,Kg \cdot m^2$, $J_2 = 0.0118\,Kg \cdot m^2$) are two orders of magnitude lower than the ones for the reaction side ($J_3 = J_4 = 0.2\,Kg \cdot m^2$). The obtained rotational mode shapes and their natural frequencies are illustrated in Fig. 4.

The first 0 Hz natural frequency is due to the allowed rotational degree of freedom. The second mode at 280 Hz involves in-phase rotation for both the test and the reaction gear pairs with opposite rotations for the two sides. Therefore, this mode involves minimal mesh deflections in proportion to the deformation of the flexible couplings. The third mode at 680 Hz involves reaction gears with out-of-phase rotation (namely mesh deflection) and very limited motion of the test gears (3 % of the reaction gears modal amplitude). The fourth mode at 1,968 Hz involves test gears similarly with very limited motion of reaction gears (0.2 % of the test gears modal amplitude). The third and fourth represent modes for the reaction gear pair and the test gear pair respectively. Limited motion of the test gears at a resonance for the reaction gears, and vice versa, is a first evidence of decoupling between the two sides.

The same model was used to obtain the forced-response Transmission Error for the test gear pair, applying twice the maximum variable contact force expected at the reaction side. This quantity allows estimating how much of the reaction gears excitation is able to reach the test gears. A typical value of 5 % was chosen for gears modal damping ratio [2]. Results in Fig. 5 show an induced Transmission Error higher than the encoder measurement resolution only in correspondence of gear resonances. Resonance conditions for the reaction gears must therefore be

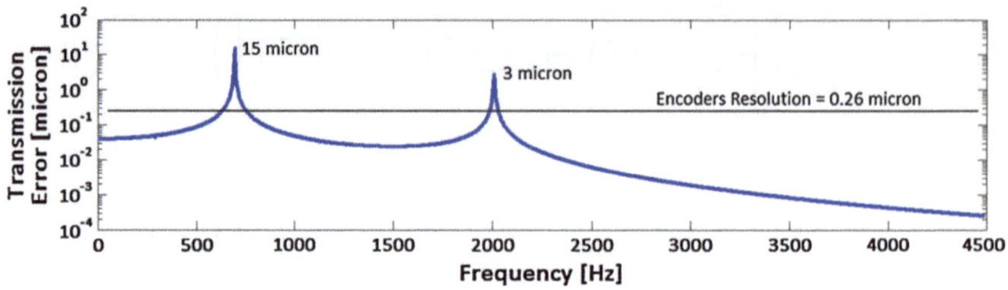

Fig. 5 Forced-response transmission error for the test gear pair when exciting the reaction gears with twice the maximum expected dynamic contact force

avoided. Resonance conditions for the test gears do not represent a problem, since the Transmission Error amplifies and a decrease in accuracy can be allowed (See *Part II: Instrumentation*).

4 Gear Design

Gear design is performed following different purposes for the test side and the reaction side, which have specifications reported in Table 2. Different tooth numbers are chosen for test and reaction gears to be able to distinguish excitation belonging to each gear pair.

Main requirements for test gears are identified in:

1. Providing very repeatable meshing excitation;
2. Possessing a gear mesh stiffness calculated to determine a desired;
3. gear pair resonance frequency;
4. Allowing room for mounting root strain gauges;
5. Not failing under load.

Table 2 Gear specifications

	Test gears (Spur)	Reaction gsears
Number of teeth	57	64
Normal module	2.60 mm	2.25 mm
Normal pressure angle	20 deg	14 deg
Helix angle at theoretical pitch circle	0 deg	13 deg
Tip diameter	154.50 mm	155.00 mm
Root diameter	141.70 mm	142.00 mm
Face width	23 mm	41 mm
Normal circular tooth thickness at theoretical pitch circle	3.780 mm	3.074 mm
Total contact ratio	1.456	3.0
Average gear mesh stiffness along the line of action	3.3×10^8 N/m	3.5×10^8 N/m

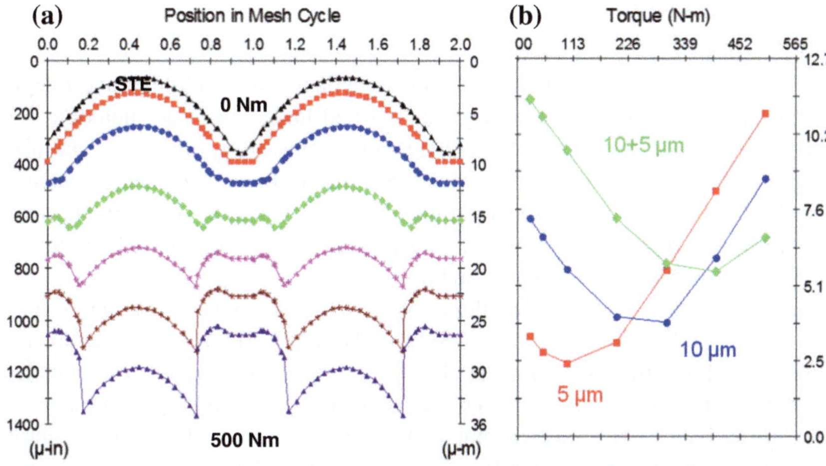

Fig. 6 **a** Static transmission error curves from 0 to 500 Nm for a test spur gear pair having 10 μm profile crowning modification. **b** Example of sub-optimal (5 μm), optimal (10 μm) and above-optimal (10 + 5 μm) profile modifications for minimizing peak to peak static transmission error, with respect to the median applied load. Simulations performed using LDP [6]

Repeatability is substantially enhanced by two choices: first, the gear ratio is equal to one; second, gears are precision ground to ISO 1328 Quality 3. Unity gear ratio implies that one tooth will always mesh with the same tooth of the mating test gear. This excludes deviations introduced by manufacturing variability. Tolerances for quality ISO 3 were outperformed in the manufacturing execution, with typical values within ±1 micron for both lead and profile. All the teeth for all the gears were measured on both flanks along the mid-plane profile and the operating pitch circle lead to ensure tolerance quality. To achieve a stable and durable tooth surface, before grinding, gears were made of 16 MnCr5 alloy special steel and case-hardened to 60 HRC until 0.8 mm depth. The value reached for test gear mesh stiffness, combined with shaft stiffness, provides a gear pair resonance frequency which can be adjusted in the range described in *Par. 6*. Mesh stiffness is tuned mainly by selecting appropriate tip and root circles diameters, operating centre distance and tooth thickness. Tip and root diameters and centre distance are also chosen to leave sufficient room for the installation of the root strain gauges.

Both families of spur and helical test gears are composed of two unmodified gears, two with parabolic profile modifications and two with parabolic lead modifications, which can be combined to achieve extensive validation of microgeometry effects. One modification value is designed to be sub-optimal for the median applied load, the other value to be optimal and the combination to be above-optimal (Fig. 6).

Main requirements for reaction gears are identified in:

1. Generating the lowest possible meshing excitation;
2. Being insensitive to small misalignments;

3. Providing the lowest possible mesh stiffness;
4. Having long service life.

Meshing excitation is limited first by using helical gears with high and integer total contact ratio (the latter property was verified under load and including microgeometry effects). Subsequently, profile modifications are optimized to minimize gear transmission error peak to peak value. Sensitivity to angular misalignment is reduced by applying end relief and parabolic lead crowning to the teeth surfaces. Mesh stiffness is reduced (to lower the gear pair resonance frequency) by reducing root circle diameter, tooth thickness and normal pressure angle.

Gear strength was verified using three software tools: Ansol Helical 3D [3] for a Finite Element full-field verification, OSU Gearlab LDP [2] for a confirmation and KISSsoft [4] for rating according to the norm ISO 6336. Ratings are provided hereafter for the already manufactured family of spur gears and for reaction gears. Helical gears have been designed, but will be manufactured after experimentally verifying the design for the spur family. According to the ISO 6336 norm, spur gears have high static safety factor (3.7 against yield at tooth root), but short service life (8.5 h at maximum speed and load) due to sharp handover between tooth pairs. This compromise is reached for test gears to verify the above-mentioned requirements, since test gears are appointed to operate for a limited time and then to be changed. Reaction gears have high static safety factor (5.3 against yield at tooth root) and infinite service life at maximum speed and load. Stresses were lowered for reaction gears by increasing the facewidth value. Higher facewidth value provides the additional advantage of increasing the total contact ratio, but also the disadvantage of increasing the mesh stiffness. The trade-off is balanced by tuning the helix angle.

5 Setting Misalignments

Angular misalignment in an arbitrary plane and parallel misalignment in the transverse plane can be set by a technique proposed by [5]. Bearings are mounted in eccentric caps; these caps can be turned to orient the eccentricity at a given angle. Same angles for the two caps on the same shaft result in imposing a parallel misalignment; opposite angles result in imposing an angular misalignment. Angles in-between result in a combination of angular and parallel misalignment. Angular misalignments motivate the compulsory need of spherical roller bearings. Axial parallel misalignment can be imposed thanks to expansion locking units to fasten the gears on the shafts. Each unit is composed of two self-centring tapered conical elements with cylindrical bore for the shaft and cylindrical outer surface for the gear. Gears can first be axially displaced of a known quantity by using a calibrated thickness and then fastened to the shafts.

Given the significant sensitivity of gears to small amounts of misalignments, manufacturing and assembly tolerances acquire paramount importance. Geometric tolerance constraints were calculated quantitatively and will be discussed here only qualitatively for the sake of conciseness. Holes in bearings support plates require tight positioning tolerance to avoid initial misalignments and cylindricity to centre bearing caps; cylindrical surfaces for bearing caps require tight cylindricity to centre bearings and accurate eccentricity to set misalignment; bearings and shafts require tight turning accuracy with very limited radial runout. Alignment and cylindricity for bearing support plates holes are achieved by assembling the plates and finishing rough holes with a single operation of wire electrical discharged machining (EDM). Wire EDM is also used to manufacture bearing caps. High precision bearings are selected to restrict to 30 % the normal amount of runout for the inner race (K_{ia} parameter). High precision pre-ground and hardened shafts are selected for minimized radial runout and improved cylindricity.

6 Adjusting Shaft Compliance

Expansion locking units are used to allow axial positioning of the gears and thus to adjust shaft stiffness. In particular, bending and torsional stiffness are maximum when the loaded shaft segment is shortest (i.e. the test gears are closest to the bearings within the loaded path). Two configurations will be used: gears at half shaft length and gears closest to the bearings. In the maximum shaft stiffness configuration, bending and torsional shaft stiffness values are substantially higher than the mesh stiffness. In the minimum shaft stiffness configuration, bending and torsional shaft stiffness values are substantially lower than the mesh stiffness. Using the beam theory of elasticity, bending and torsional stiffness assume values reported in Fig. 7. Corresponding resonance frequencies for the test gear pair are expected to be close to 1,968 Hz for maximum shaft stiffness and close to 800 Hz for minimum shaft stiffness.

Two additional aspects need to be clarified: the effects of bearing stiffness on test gears resonance frequency and on load-dependent misalignment. Average bearing stiffness in the loading range is three times the mesh stiffness for the test

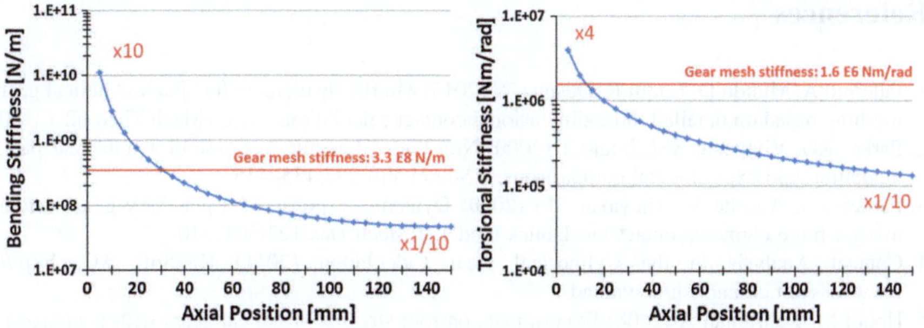

Fig. 7 Comparison of gear mesh stiffness with shaft bending and torsional stiffness with respect to the axial position of the gears

gears. Load dependent misalignment can be induced only in the minimum shaft stiffness configuration. In the maximum shaft stiffness configuration gears are equidistant from bearings and rotation induced by bending is zero. In the minimum shaft stiffness configuration, the closest bearing absorbs almost the complete meshing loads, while the opposite bearing is virtually unloaded. Bearing deflections at maximum load are below 10 microns, leading to a maximum misalignment of 0.03 mrad (three times below uncertainty on imposed misalignment).

7 Conclusions

The main steps of the design process for a cylindrical gear pair test rig have been described. Solutions already available in literature have been used to impose angular misalignment in an arbitrary plane and parallel misalignment in a transverse plane. A new solution has been proposed to impose axial misalignment and to select the influence of shaft compliance on gear mesh stiffness. The use of flexible couplings and flywheels has been discussed to isolate the test side from stray excitation belonging to the reaction side. The main criteria identified for gear design have been reported for the test and the reaction side. A 4-degrees-of-freedom rotational model has been used to estimate rotational resonance frequencies of the test rig. Contributions on mesh stiffness coming from shafts have been considered for test gears. Special care has been paid to tolerances, for ensuring repeatability of measurements and accurate values of imposed misalignments.

Acknowledgments The authors gratefully acknowledge the IWT (Agency for Innovation by Science and Technology in Flanders) for the financial support through the project HEV-NVH (Agr. 110360).

The authors are also grateful to Dr. Giorgio Bonori, Italian distributor for Ansol LLC, for providing free academic license and support for the software Transmission 3D. Similar acknowledgement goes to Mr. Ivan Saltini, Italian distributor for KiSSsoft AG, for providing free academic license for the software KISSsoft.

References

1. Palermo A, Mundo D, Hadjit R, Desmet W (2013) Multibody element for spur and helical gear meshing based on detailed three-dimensional contact calculations. Mech Mach Theo 62:13–30
2. Parker RG, Vijayakar SM, Imajo T (2000) Non-linear dynamic response of a spur gear pair: modelling and experimental comparisons. J Sound Vib 237:435–455
3. Parker RG, Agashe V, Vijayakar SM (2000) Dynamic response of a planetary gear system using a finite element/contact mechanics model. J Mech Des 122:304–310
4. Contact Analysis in the Cylindrical Gear Calculation (2011) KissSoft AG, http://www.kisssoft.ch/english/downloads/
5. Hotait M, Kahraman A (2008) Experiments on root stresses of helical gears with lead crown and misalignments. J Mech Des 130:1–5
6. Load Distribution Program (2002) Users Manual, Ohio State University, GearLab

A Novel Gear Test Rig with Adjustable Shaft Compliance and Misalignments. Part II: Instrumentation

A. Palermo, J. Anthonis, D. Mundo and W. Desmet

Abstract This paper describes the instrumentation aspects for a new parallel cylindrical gear test rig aimed at adjusting the influence of shaft compliance on gear meshing stiffness, while allowing the operator to impose gear misalignments. Static and dynamic testing is possible for the most important gear-related physical quantities. Transmission Error can be measured in a wide frequency range by high-resolution analogue encoders, low-cost digital encoders and accelerometers attached to the test gears. The latter are also used to obtain dynamic relative displacements between the gears in 6° of freedom. Tooth root strain is measured by a linear pattern of strain gauges, either wired in a quarter-bridge configuration or with direct measurement, to capture axial strain distribution. Signals from rotating sensors are fed out to the stationary acquisition system by means of slip rings. Transmitted torque is measured using a rotating torque flange with contactless signal transmission to its stator.

Keywords Accelerometers · Dynamic TE Measurement · Incremental Encoders · Misalignment · Strain Measurement at Tooth Root · Transmission Error

---

A. Palermo (✉) · D. Mundo
Department of Mechanical Engineering, University of Calabria,
Ponte Pietro Bucci, 87036 Rende, Italy
e-mail: palermo.antonio@gmail.com

J. Anthonis
LMS International, Interleuvenlaan 68, B-3001 Leuven, Belgium

A. Palermo · W. Desmet
Department Mechanical Engineering, K.U. Leuven, Celestijnenlaan 300 B,
B-3001 Heverlee, Belgium

G. Dalpiaz et al. (eds.), *Advances in Condition Monitoring of Machinery in Non-Stationary Operations*, Lecture Notes in Mechanical Engineering,
DOI: 10.1007/978-3-642-39348-8_44, © Springer-Verlag Berlin Heidelberg 2014

1 Introduction

The evaluation of mechanical transmission dynamics is a prerequisite for ensuring good noise and vibration (N&V) and durability behaviour. Issues in both areas are typically related to gear Transmission Error (TE), a quantity describing the deviation between the actual and the theoretical kinematic transmission of motion as gears rotate. Gears allow transmission of mechanical power with very limited mesh deflections (10^{-3}–10^{-2} mm), or, equivalently stated, with very high mesh stiffness (10^5–10^6 N/mm) [1]. This is the reason why oscillating mesh deflections, even of a few microns, due to variable TE are relevant for N&V and durability evaluations. Measuring such small displacements requires high measurement accuracy and represents a significant challenge, especially at high speed. Further complications arise from gear sensitivity to manufacturing errors [2] and variations of transmitted load [3], angular misalignment [4] and centre distance [5]. This holds not only from a measurement point of view, but also from a design and simulation perspective. A major design focus is on selecting values for deliberate microgeometry modifications of tooth surfaces to compensate for variations of the above-mentioned factors. Modifications along the tooth lead are typically designed to minimise stress peaks assuming one value of misalignment and an expected excursion on top of this value (arising from deflections or assembly/manufacturing errors) [6]. Modifications along the profile of the teeth are designed to minimize static and dynamic TE [7]. The underlying design considerations, when not obtained by expensive trial and error approach on prototypes, are based on simulations which first estimate the variations of transmitted load, angular misalignment and centre distance and then their effects on key gear-related quantities, combined with microgeometry modifications and manufacturing errors. In a nutshell, issues in design and simulation stand in the validation of meshing static and dynamic behaviour, which involve non-linearity due to contact; issues in measurements stand in closely reproducing the conditions of transmitted load, misalignment and centre distance to be tested and to set up a measuring system which is insensitive to their variations.

The authors aim at solving these issues in the proposed gear test rig, which will be used, among other research purposes, also to validate experimentally the proposed multibody gear dynamics element proposed in [8]. A more detailed discussion on the motivations for a dedicated test rig are reported in Part I: Design of this paper; however, given the complexity of the problem, the need of tailor-made design to accurately set operating conditions and to isolate a simple subsystem appears evident. The focus of this paper is on the definition of the most important gear-related quantities to be investigated, on the measuring systems and methods to be used, and on the implications that measurements determine on the test rig design.

TE has been already mentioned as being the most relevant quantity for N&V and durability evaluations; this quantity has been measured in literature using a wide variety of techniques [9, 10]. The two most established methodologies currently use encoders and tangentially-mounted accelerometers [11]. TE provides information on gear dynamics along the line of action, however gears are also coupled to the supporting structure and are typically affected by its vibration. A method to measure dynamic gear translations and rotations in 6° of freedom was recently proposed using accelerometers [12]. Together with quantities related to motion, it is relevant to analyse quantities related to load. Measurement of transmitted torque and tooth strain allow direct assessment of the loading conditions of the gears. Torque measurement was performed by Raja Hamzah and Mba [13] using shaft-mounted strain gauges in a half-bridge configuration. Strain measurement was performed by Haigh and Fawcett [6], Hotait and Kahraman [14]. In [6] separate strain gauges were placed along the tooth root, while in [14] pre-arranged linear patterns of strain gauges were used.

Details will be provided in the present paper also about the acquisition system used to perform the measurements and the slip rings used to feed signals from rotating components to the stationary equipment.

2 Transmission Error and Motion Measurement

Target for this section is the measurement of gear displacements, considering 6° of freedom per gear, together with shaft and bearing orbits. TE can be measured by using incremental encoders and accelerometers over the angular speed range spanned by the test rig. Measurement redundancy is foreseen to increase trust in the measurement (overlapping frequency bands in Fig. 1). Two types of encoders are used; one to have a reference high-resolution measurement in quasi-static conditions, performed by established techniques [1], and one to investigate low-cost encoders suitability for both quasi-static and dynamic TE measurement. Accelerometers can be used only for dynamic TE measurement.

High-resolution encoders (Heidenhain RON 285C) have 18,000 divisions and output a corresponding number of sinusoidal voltage waves. Each sinusoid repeats

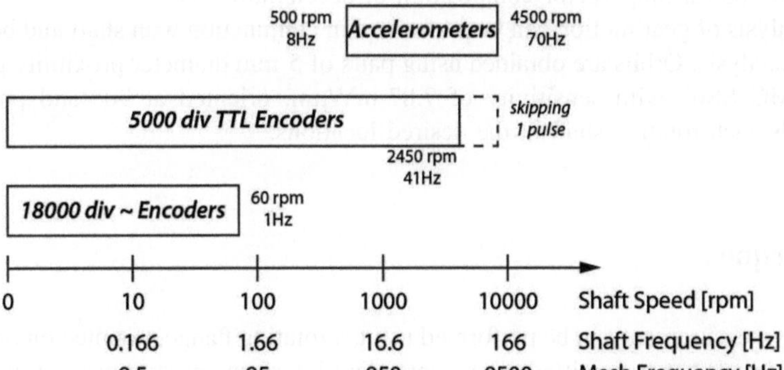

Fig. 1 Frequency bands of TE measurement by different measurement instruments

accurately within ±1 % of the grating period, so that interpolation on this analogue signal can be performed to read an angle increment of 0.0001° (0.36 arcsec) on each encoder. This yields a combined resolution of 0.24 μm on the TE along the line of action. Low-cost encoders (Heidenhain ERN 120) have 5,000 division and output a corresponding number of digital TTL voltage square-waves. The pulse timing method is used on each square wave rising edge to measure TE up to a speed of 2,450 rpm. At this speed the maximum acquisition frequency is reached. However, using the pulse-skip functionality of the acquisition system, one rising edge will be skipped (halving therefore the pulse rate) and the same encoders can be used for TE measurement up to the full speed range of 4,500 rpm.

High-resolution and low-cost encoders are mounted together in pairs on the same shaft. This allows a direct comparison of the angle measurements for accuracy evaluations. Particular care has been taken for encoder mounting tolerances. For each shaft and the supporting bearings, the measured combined radial run-out tolerance lies within 10 μm. This value is circa half of the one suggested for the high-resolution encoders by the manufacturer. Stator and rotor of the low-cost encoders are aligned on each shaft by means of high-precision bearings embedded in the encoders themselves. Stator and rotor of the high-resolution encoders are instead floating with respect to each other; since the rotor is mounted on the shaft it is intrinsically aligned. To align the stator, measured positioning references are taken on the low cost encoders to form a single assembly.

Together with low-cost encoders, two or four uniaxial tangentially-mounted accelerometers can be used to measure the alternating TE components in dynamic conditions [15] by double integration of angular acceleration. Four seats are prepared on an accelerometers carrier (Fig. 2) to be fastened on each of the two gears being tested. Particular care has been taken to ensure symmetry of the accelerometers around the gear body, since the acceleration signals from the different accelerometers need to be combined. Measured distances from the axis of the carrier bore are accurate up to ±5 μm, perpendicularity of seat surfaces lies within 0.1°, coaxiality between the gear collar and the gear axis lies within 0–5 μm. Triaxial accelerometers are used to obtain alternating motion for each gear in 6° of freedom, performing vector composition of accelerations [12].

Analysis of gear motion can be performed in conjunction with shaft and bearing orbits analysis. Orbits are obtained using pairs of 5 mm diameter proximity probes SKF MC SS65 with sensitivity of 7.87 mV/μm, oriented at 90° and pointing towards each rotating shaft at the desired locations.

3 Torque

Torque measurement can be performed using a rotating flange mounted on one test shaft. Signals are transmitted via a contactless interface across a gap between the rotating flange and the stator (Fig. 3), allowing misalignments to be accommodated.

The torque sensor covers three functions:

Fig. 2 Accelerometers carrier and its centring on the gear collar

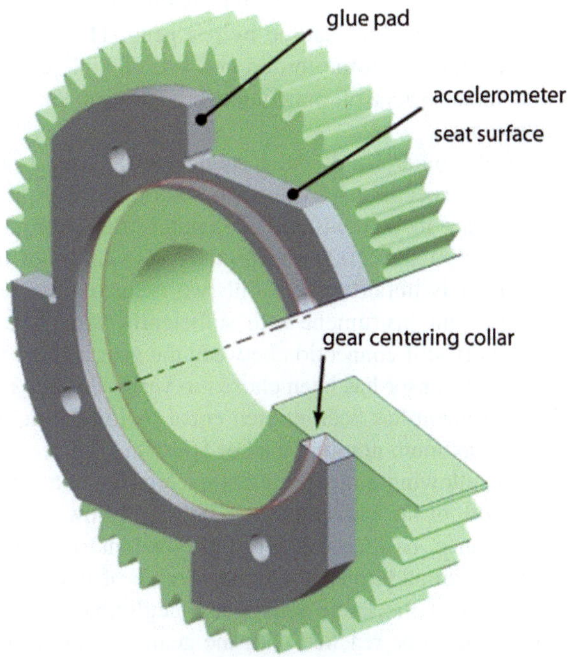

1. Measure and verify torque preload retention in the recirculation loop;
2. Measure instantaneous torque transmitted by the test gears;
3. Verify tooth root strain measurements.

Verification that the torque preload maintains constant average value is required to ensure that no (micro) slip is happening in the clutch flange and in the expansion bushings used to fasten the gears. It is worthwhile to point out how, besides

Fig. 3 Torque-measuring rotating flange and its stator for contactless signal transmission

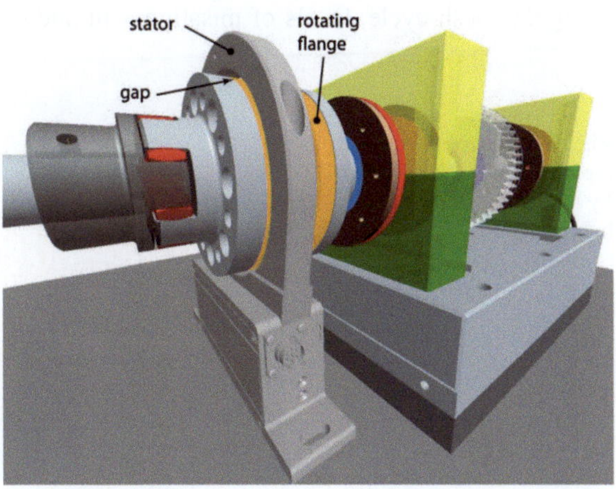

isolating the test and the reaction sides of the test rig, flexible couplings ensure that the twist fluctuations caused by test gears TE excitation do not determine significant fluctuation of the torque preload. Flexible couplings in fact store an elastic twist three orders of magnitude bigger than the one caused by variable mesh deflections.

On the other hand, the instantaneous torque transmitted by test gears does fluctuate, especially in resonance conditions. Torque overloads in the latter case are relevant to be measured. To capture these torque fluctuations at a frequency between 800 and 2,000 Hz (primary resonances of test gears reported in Part I: Design) it is important to fulfil two main requirements: firstly, use an accurate measurement instrument with sufficiently wide bandwidth; secondly, achieve a sufficiently stiff connection between the gear and the measurement instrument. An HBM T40 sensor has been chosen to verify the first requirement. Rated torque for the instrument has been chosen equal to 1,000 Nm, allowing room for overloads on the maximum nominal load of 500 Nm. The accuracy class of the instrument is 0.05 %, allowing a measurement uncertainty in a similar order of magnitude. The bandwidth of the instrument is 6,000 Hz, allowing to measure torque fluctuations at resonance. Evaluations using the torsional model illustrated in Part I: Design, including an additional rotational degree of freedom for the torque sensor, show however that the latter adds a non-negligible rotary inertia and prevents a stiff connection to be reached with the gear. In particular, the torque sensor torsional stiffness is 25 % lower than the equivalent torsional stiffness of the test gear pair. This yields a new resonance frequency due to the torque sensor below the original test gear pair resonance frequency. Moreover, modes of vibration related to the torque sensor induce significant rotation on the test gear pair and vice versa; therefore the measured torque is in both cases affected by resonance of the measurement instrument. To cope with this issue, the torque sensor can first be used to calibrate a torque-sensing rosette of strain gauges mounted on the test shaft and can then be replaced by a flexible coupling as shown in Fig. 4.

Different levels of measured transmitted torque are required for correlating tooth root strain measurements against static simulations for different positions along the mesh cycle, levels of misalignment and centre distance.

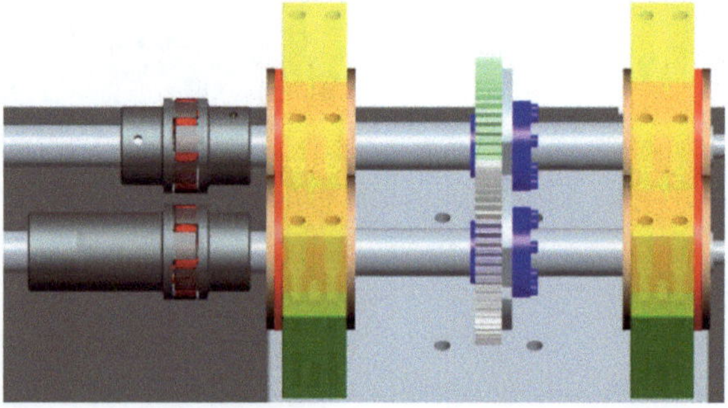

Fig. 4 Torque sensor replaced (on the *bottom left*) by a long flexible-coupling hub

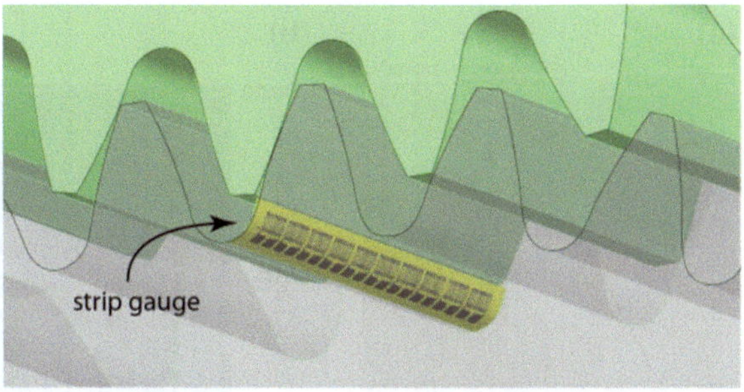

Fig. 5 Strip gauge positioning at the root of a test gear tooth

4 Strain and Slip Rings

Strain measurement at tooth root provides a direct evaluation of instantaneous tooth loading conditions. Since the total contact ratio of the test gear pairs is between 1 and 2, two teeth of one test gear will be instrumented to measure strain (access and recess phases) over simultaneously meshing tooth pairs. Moreover, to assess the effects of lead microgeometry modifications and imposed misalignment, multiple strain gauges will be placed along the face width. Strip gauges of type Vishay Micro-Measurement EA-06-031PJ-120 will be used. Each strip gauge has 22.7 $\times$ 4.8 mm dimensions and consists of a linear pattern of 10 strain gauges (Fig. 5) each having base length of 0.79 mm and width of 1.78 mm.

A short base length is chosen to have a local measurement of deformation. Accurate local measurement however requires careful gauge positioning and alignment [16] because of the steep strain gradients at the tooth root. Two measurement methods will be used to obtain dynamic strain: a conventional quarter-bridge configuration and a direct measurement of resistance (described in the next section). In both cases limitations may be caused by the data acquisition system and by the slip rings used to transfer the signals from the rotating components. The acquisition system needs to provide sufficient sensitivity to gauge resistance fluctuations, dynamic range and bandwidth. These characteristics are discussed in the next section. Slip rings need to cause the least possible resistance variation at their interface between races and brushes, since this directly results in an apparent strain. Two models of slip rings were chosen: a 12-channels HBM SK-12 and a 20-channels Michigan Scientific SR-20. Specifications from the manufacturer for the two models of slip rings report resistance fluctuations below respectively 0.002 and 0.1 Ω. The number of channels differs because the SK-12 will be used only to connect accelerometers, while the SR-20 will be used both for accelerometers and strain gauges. Slip rings performance was however preliminarily assessed in both cases connecting an unloaded strain gauge (Fig. 6a) and measuring apparent strain at different angular speeds up to 1,500 rpm. Considering a true strain to be measured in the order of 500 μm/m, the test provided encouraging results for both the slip rings. Highest amplitude of 6.3 μm/m was recorded at the first order of rotation (Fig. 6b) and noise vanishes from the fourth order onwards.

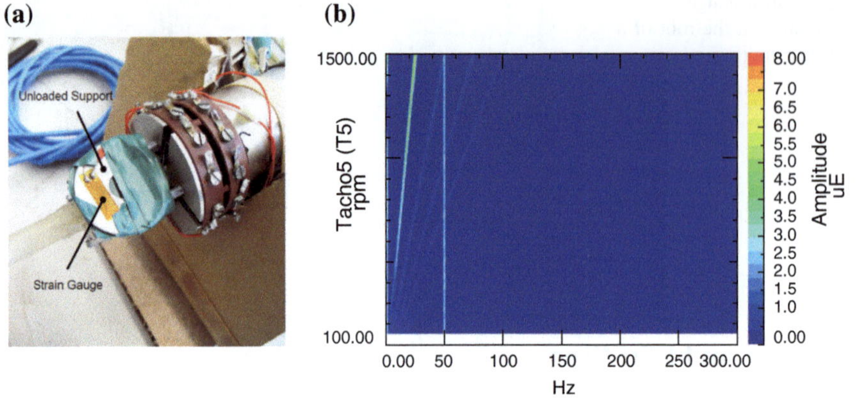

Fig. 6 (**a**) Unloaded strain gauge, measuring apparent strain, mounted on the rotating part of the SK-12 slip ring; (**b**) Waterfall diagram of the apparent strain caused by slip ring resistance variation. The *vertical line* is related to electromagnetic noise induced by the electric grid

5 Data Acquisition System

All the signals are acquired with a single LMS SCADAS acquisition system and managed using the LMS Test.Lab software platform. Channels are synchronized and can be sampled up to 204.8 kHz, each with a 24-bit analog-to-digital converter having 150 dB dynamic range. Direct strain measurement is performed using the BDS-4 module: instantaneous value of resistance, with a bandwidth of 80 kHz, is obtained via the Ohm's law injecting a constant current in the strain gauge and measuring voltage. The RV4 module is used to acquire incremental encoder signals: a clock running at 820 MHz yields a time resolution equal to 1.2 ns on encoder pulse events, with a maximum input pulse rate of 204.8 kHz or 1 MHz in combination with the pulse skip functionality.

6 Conclusions

Instrumentation aspects for a new parallel cylindrical gear test rig have been discussed. Relevant gear-related physical quantities were identified and measurement methods were outlined with focus on the main factors affecting the accuracy of each measuring chain. In particular, misalignments can be accommodated by the arrangement foreseen for the measurement instruments. The available measurements will allow, on a first stage, achieving validation of gear contact models and obtaining insights on gear dynamics over a broad frequency range and for various operating conditions for the gears.

Acknowledgments The authors gratefully acknowledge the IWT (Agency for Innovation by Science and Technology in Flanders) for the financial support through the project HEV-NVH (Agr. 110360).

References

1. Smith JD (2003) Gear noise and vibration. Marcel Dekker, Cambridge
2. Li S (2007) Effects of machining errors, assembly errors and tooth modifications on loading capacity, load-sharing ratio and transmission error of a pair of spur gears. Mech Mach Theo 42:698–726
3. Harris SL (1957) Dynamic loads on the teeth of spur gears. Proc. IMechE 172:87–112
4. Koide T, Oda S, Matsuura S, Kubo A (2003) Equivalent misalignment of gears due to deformation of shafts, bearings and gears. JSME Int J 46:1563–1571
5. Kim W, Lee JY, Chung J (2012) Dynamic analysis for a planetary gear with time-varying pressure angles and contact ratios. J Sound Vib 331:883–901
6. Haigh J, Fawcett JN (2003) Effects of misalignment on load distribution in large face-width helical gears. Proc. IMechE Part K 217:93–98
7. Lin HH, Oswald FB, Townsend DP (1994) Dynamic loading of spur gears with linear or parabolic tooth profile modifications. Mech Mach Theo 29:1115–1129
8. Palermo A, Mundo D, Hadjit R, Desmet W (2013) Multibody element for spur and helical gear meshing based on detailed three-dimensional contact calculations. Mech Mach Theo 62:13–30
9. Munro RG (1990) A review of the theory and measurement of gear transmission error. In: Proceedings of IMechE international conference gearbox noise and vibration, pp 3–10
10. White R, Palan V (2007) Measurement of Transmission Error Using Rotational Laser Vibrometers. In: Proceedings of ASME IDETC/CIE 2007, paper DETC2007-34430
11. Houser DR, Blankenship GW (1989) Methods for measuring gear transmission error under load and at operating speeds, SAE Technical Paper 891869
12. Kang MR, Kahraman A (2012) Measurement of vibratory motions of gears supported by compliant shafts. Mech Syst Signal Pr 29:391–403
13. Raja Hamzah RI, Mba D (2009) The influence of operating condition on acoustic mission (AE) generation during meshing of helical and spur gear. Trib Int 24:3–14
14. Hotait MA, Kahraman A (2011) An investigation of root stresses of hypoid gears with misalignments. J Mech Des 133:1–9
15. Blankenship GW, Kahraman A (1995) Steady state forced response of a mechanical oscillator with combined parametric excitation and clearance type non-linearity. J Sound Vib 185:743–765
16. Vishay Micro-Measurements (2010) Errors due to misalignment of strain gages, Document number 11061

A Distributed Control System for a Field of Spin-Elevation Heliostats

Alessandro Carandina, Mirko Morini, Claudio Pavan and Michele Pinelli

Abstract The aim of this paper is to describe the architecture of a distributed control system designed to handle several heliostats in a solar concentration field. The system foresees the use of solar power to achieve the necessary temperature to activate a dry reforming process in order to convert landfill gasses into hydrogen that will be exploited for energy production. Between the manifold solar concentration technologies the Central Tower Heliostat Field was adopted. This paper includes a first section regarding the most common geometry of solar receivers and concentrators and a set of equations for the main sun tracking techniques. The chosen hardware components and the software solutions will be illustrated as well.

Keywords Spin-elevation solar tracking · Central tower heliostat field · Distributed control system · LabView · Arduino

Nomenclature

α	elevation angle between the heliostat frame and the Zenith-North-East frame
ζ	azimuth angle (clockwise) between the heliostat frame and the Zenith-North-East frame
λ	elevation angle between the target and the North-East plane in the Zenith-North-East frame
ϕ	azimuth angle (clockwise) from North on the North-East plane in the Zenith-North-East frame
Φ	local latitude
δ	declination angle of the sun
ω	solar hour angle of the sun

A. Carandina (✉) · M. Morini · C. Pavan
MechLav Università di Ferrara, Ferrara, Italy
e-mail: alessandro.carandina@unife.it

M. Pinelli
Dipartimento di Ingegneria, Università di Ferrara, Ferrara, Italy

G. Dalpiaz et al. (eds.), *Advances in Condition Monitoring of Machinery in Non-Stationary Operations*, Lecture Notes in Mechanical Engineering,
DOI: 10.1007/978-3-642-39348-8_45,

θ_i incidence angle of sunrays on the reflector of heliostat
ρ_{H} rotation along the normal to the heliostat surface (Azimuth angle in the case study)
θ_H rotation along the direction normal to the previous one and tangent to the heliostat surface (Elevation angle in the case study).
ΔH_{298} Enthalpy of reaction at 298 K, [kJ·mol^{-1}].

1 Introduction

Nowadays, more than ever, the global energy demand, along with the awareness of the depletion of non-renewable energy sources and a pronounced sensitivity to the environment, has led to the development of several techniques for the exploitation of solar energy with growing efficiency. Several geometries have, indeed, been developed over the years, and each one provides a different solution for specific end-users with the best efficiency available. This led to the fulfillment of parabolic trough concentrators, compact linear Fresnel receivers, parabolic dishes, central tower heliostat fields and so on. Each one of these systems must track the sun during the day rotating around one or more axes as a function of the geometry of the receiver and of the mounting solution adopted. The project consists in the development of a sun powered reformer to crack the CH_4 and CO_2 produced by a garbage dump. These gases, if released in the atmosphere, would have a severe greenhouse effect. The synthetized H_2 gas produced, can then be exploited for both electric an thermal energy production in specific distributed energy conversion systems, such as fuel cells. The CO2 reforming of methane, (or dry reforming) is an endothermic reaction, like steam reforming, but it yields a syngas with a lower ratio of H_2/CO_2, for a complete conversion. The stoichiometric balance can be written as:

$$CH_4 + CO_2 = 2H_2 + 2CO$$
$$\Delta H_{298} = +247 \text{ kJ} \cdot \text{mol}^{-1}$$

Dry reforming is favored by high temperatures and low pressures. Thermodynamically, it can occur at temperatures higher than 640 °C but, technically and assuming a fixed stoichiometric ratio of CH_4/CO_2 in the feedstock, temperatures higher than 800 °C are needed to achieve acceptable conversions. The choice of a central tower heliostat field is due to the necessity of achieving these high temperatures. Many heliostats can be installed on the same field to increase the solar concentration onto the reformer, and each one will be positioned at a specific distance from the target. Hence, the tracking angles will be different for each heliostat as a function of the relative position of the heliostat compared to the tower.

2 Main Solar Receivers and Concentrators

In the described system, a Central Tower Heliostat Field is adopted. However, a number of different technologies can used. The most common are reviewed and described in the following.

Parabolic Trough Concentrator. This technology uses reflectors curved around one axis using a linear parabolic shape, which has the property of collecting parallel rays along a single line focus and almost-parallel rays from the solar beam in a line image [1]. A pipe receiver is placed on the focus of the parabolic trough and is usually made of two concentric components: a metallic pipe in which the heat transfer fluid flows, and an external glass pipe to provide greenhouse effect inside the receiver. To avoid both convection and conduction thermal loss, these receivers are usually evacuated. Internal pipes can have black chrome selective coatings to increase the energy absorption, or even metal-ceramic coatings which allow the achievement of higher temperatures, such as 391° C [1]. This is the most developed and commercially tested solar concentration-based technology. Since parabolic trough concentrators are used to produce steam to generate electricity with a conventional Rankine cycle, these systems can be readily hybridized with fossil fuel (typically natural gas) Rankine cycles, so the plant can produce electricity even when the sun is not shining [2].

Compact Linear Fresnel. The basic idea is to exploit the compact shape of a Fresnel reflector, that is to split up a parabolic receiver in a series of segments positioned at the same height. The solar beams are reflected in the same focus as in the case of the parabolic trough receiver, so the absorber is fixed in space above the mirror field and the reflector is composed of many long row segments which focus collectively on an elevated long tower receiver running parallel to the reflector rotational axis [1]. The use of flat mirrors in a Fresnel configuration is much less expensive than building a parabolic shaped solar mirror. Unlike the parabolic trough concentrator the tubular absorber is not coupled with the receiver, but is fixed at a defined height; this avoids the issues linked to a flexible connection of the receiver with the installation pipes. A particular kind of Fresnel receiver is the CLFR (compact linear Fresnel), that employs a strategy based upon multiple parallel receivers which allows area coverage to be significantly reduced by partially intermeshing two adjacent single tower arrays [1].

Parabolic Dish. Several types of this kind of receiver have been developed over the years, but the shape is basically a paraboloid reflecting the solar beams onto the focusing point. Here a direct end user can be positioned, such as concentration photovoltaic cells or a Stirling engine, able to directly convert thermal power to mechanical power. This kind of concentrator can only exploit direct solar radiation, that is the portion of the incoming solar radiation that travels from the sun to the Earth's surface in an essentially straight line, without being reflected, deflected or absorbed and transmitted by particles or gasses in the atmosphere. Thus a parabolic dish must have a two-axis tracking system moving continuously to face the sun throughout the day. The end user converts the thermal power into electricity without producing steam as an intermediate step. Individual parabolic dishes coupled with Stirling engines typically produce 3–25 kW each. The high fluid temperature attainable by the two axis tracking solar parabolic dish leads to high conversion efficiency of solar power to electricity. The association of a parabolic dish with a Stirling engine can be used as a relatively small distributed power source, and with an array of many units, a MW plant of electricity from solar power can be set up. In addition, this system has a very minimal water requirement since the Stirling engine is air cooled. Otherwise, due to the distributed nature of the parabolic dish, this type of system does not lend itself well to thermal energy storage, to allow electricity generation when the sun is not shining [3].

Double Parabolic Reflection Dishes. The configuration is basically the same as in a normal parabolic dish, but in the focus of the first parabolic receiver, instead of the end user, a second parabolic reflector is positioned. In this way the solar beam is reflected twice, striking the target positioned on a clamp. The high concentration level allows the achievement of very high temperatures in a precise spot and the second receiver can be calibrated by micrometric movements to compensate eventual optical aberration errors due to physical imperfections of the primary reflector. Otherwise the double reflection involves a lower optical efficiency and requires the use of a secondary reflector operating at elevated temperatures. The shading effect is another undesired issue due to the projection of the structures holding the secondary reflector, but the effect is proportional to the size ratio between the two reflectors (Fig. 1).

Double Parabolic Trough Concentrator: the trough design is suitable for applications requiring concentrations up to about 200 suns, such as steam generated power or concentrated photovoltaic operations, while the bowl design is suited for high temperature applications, such as Stirling engine power generation, high temperature water splitting, and thermochemical fuel production [4]. The point focusing double parabolic trough concentrator is made from two reflective parabolic troughs, a primary and a secondary, with longitudinal axes in perpendicular directions [4]. This configuration offers great flexibility because of the high temperature and high concentration achievable, together with a relative ease of realization, compared to a paraboloidal bowl. Unlike a single parabolic trough concentrator, this system requires two axes for tracking since it is basically a point focusing device. One of the peculiarities of this concentrator is the oval shape of

Fig. 1 Double parabolic reflection dishes

the spot due to the different path length of rays from the primary to the secondary reflector, because of the elongated shape of the trough collectors in the direction perpendicular to the parabolic section and parallel to the focus line.

Central Tower Heliostat Field: this kind of concentrator uses a field of two axes, tracking solar mirrors called heliostats. Each heliostat reflects the incoming solar radiation and focuses it on a receiver at the top of a tall tower. These systems allows the achievement of very high temperatures and the spots of the heliostats are all concentrated on a heat exchanger on the top of the tower. The steam produced in the heat exchanger can be exploited to feed a classic Renkine cycle. This system leads to the creation of sizable power plants, and because of the very high temperatures achieved, diathermic oil or eutectic salts are employed as a thermal vector to convey the thermal energy from the heat exchanger to the evaporator of the Renkine cycle. Moreover this system can be easily integrated with a salt heat storage and with a fossil fuel power plant (typically natural gas) in order not to interrupt the electricity production even in the case of a cloudy sky. Although the heliostat solar tower approach to solar power production is not as commercially developed as the solar parabolic trough system, it is more commercially developed than either the parabolic dish-stirling engine or linear fresnel systems [3].

Distributed Multiple Towers Heliostat Field: multi-tower solar array (MTSA) goes one step further toward an urban tower technology. The MTSA is based on a unique, optical concentration technology which allows extremely closely spaced reflectors (>90 % of ground area) and high delivered output from an area of roof or ground [1]. The reflector field uses extremely closely spaced silver-plated glass reflectors of a special shape to allow extremely close spacing. There are multiple receivers of radiation that advanced thermal and photovoltaic absorber technology placed above the reflector field between 8 and 12 m high. For the receiver of radiation, MTSA can simultaneously use high-temperature and photovoltaic absorbers in parallel by splitting the incoming solar beam spectrally. The high exploitation of the ground allows the achievement of high concentration levels and the possibility of splitting the solar radiation on different receivers, as a function of the beam specter which leads to an electricity production optimization but the visual impact on the territory is considerable and the high density of the heliostats may cause possible maintenance issues.

Solar Furnace: the idea is to split the solar concentration into two levels. One or more heliostats track the sun and reflect the solar beams on a fixed secondary parabolic receiver that focalizes them onto a target positioned in the focal point of the parabola. The aim is to ensure that the rays come in parallel on the parabolic receiver. This allows a lower precision for the primary reflectors, since the fine focusing can be carried out by micrometric screw actuators in the focal point of the parabolic receiver. The advantage of this solution lies in the fact that the fine focusing occurs in the static portion of the system avoiding optical aberration due to the sun tracking movements of the heliostats. By using a large number of heliostats the temperatures achieved can be rather high and this can be exploited for several usages:

- about 1,000 °C for metallic receivers producing hot air for next generation solar towers as will be tested at the Themis plant with the Pegase project [5]
- about 1,400 °C to produce hydrogen by cracking methane molecules [6]
- up to 2,500 °C to test materials for extreme environments such as nuclear reactors or space vehicle atmospheric reentry
- up to 3,500 °C to produce nanomaterials by solar induced sublimation and controlled cooling, such as carbon nanotubes [7]or zinc nanoparticles [8].

3 Main Tracking Methods

A key aspect for the correct design of a solar concentration field is the tracking method to be used. Basically the solar tracking methods can be divided into two categories: the first one is based on the feedback response of some kind of sensor (PV cells, pireliometers, piranometers..) positioned on the heliostat or on the receiver, and capable of measuring the solar radiation and moving the heliostat move to the desired position. This is called *closed loop.* The other category, based on tables of astronomic data, the ephemeris, can predict the apparent trajectory of the sun day by day without any feedback. This second one is called *open loop.* The most advanced systems are based on an integration of these two methods to provide a fine tracking. Another way of classifying the tracking methods is by the number of axes: so we have *one axis tracking methods* (e.g., for polar or latitude mounted heliostats) and *two axis tracking methods* (e.g. for alt-azimuth or spin-elevation heliostats). Referring to the [9] a complete set of solar tracking formulae for an arbitrarily oriented heliostat toward an arbitrarily located target is presented. The paper of Chen explains how to locate an object in the sky by two angles, $\boldsymbol{\rho_H}$, the rotation along the normal to the heliostat surface and $\boldsymbol{\theta_H}$, the rotation along the direction normal to the previous one and tangent to the heliostat surface.

The orientation of the normal vector to the heliostat's reflecting surface (N_H) is computed by using as inputs:

- the relative angles between *Heliostat Frame* (R,H,U) and *the Earth Surface Frame* (North, East, Zenith)
- the local hour angle and the declination of the sun.

If we consider an absolute frame whose axes are oriented to North, East and Zenith and the heliostat frame in the Fig. 2, we can write the conversion matrixes α and ζ to transpose a generic vector from one frame to another:

$$\alpha = \begin{bmatrix} cos\alpha & 0 & sin\alpha \\ 0 & 1 & 0 \\ -sin\alpha & 0 & cos\alpha \end{bmatrix}$$

$$\zeta = \begin{bmatrix} 1 & 0 & 0 \\ 0 & cos\zeta & -sin\zeta \\ 0 & sin\zeta & cos\zeta \end{bmatrix}$$

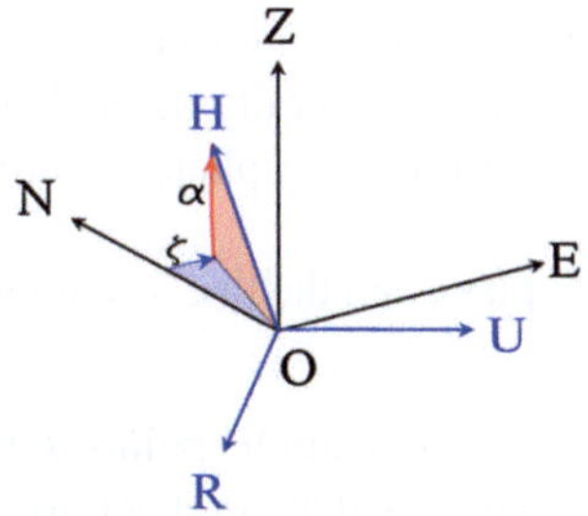

Referring to [9], we can write $\boldsymbol{\rho_H}$ and $\boldsymbol{\theta_H}$ for the general case as:

$$\theta_H = \frac{\pi}{2} - arc\sin\left[\frac{\begin{array}{c} -\sin\alpha(\cos\Phi\cos\delta\cos\omega + \sin\Phi\sin\delta) - \cos\alpha\sin\zeta\cos\delta\sin\omega \\ +\cos\alpha\cos\zeta(-\sin\Phi\cos\delta\cos\omega + \cos\Phi\sin\delta) + \\ +\sin\alpha\sin\lambda + \cos\alpha\sin\zeta\cos\lambda\sin\phi + \cos\alpha\cos\zeta\cos\lambda\cos\phi \end{array}}{2\cos(\theta_i)}\right]$$

$$\rho_H = arc\sin\left[\frac{-\cos\zeta\cos\delta\sin\omega + \sin\zeta(\sin\Phi\cos\delta\cos\omega - \cos\Phi\sin\delta) + \cos\zeta\cos\lambda\sin\phi - \sin\zeta\cos\lambda\cos\phi}{2\cos(\theta_i)\cos\beta_H}\right]$$

with:

$$\theta_i = 0.5\arccos[-\sin\lambda(\cos\ \Phi\cos\ \delta\cos\ \omega + \sin\ \Phi\sin\ \delta) \\ -\cos\lambda\sin\phi\cos\delta\sin\omega + \cos\lambda\cos\phi(cos\Phi\sin\delta - -\sin\Phi\cos\delta\cos\omega)]$$

Fig. 2 Heliostat frame

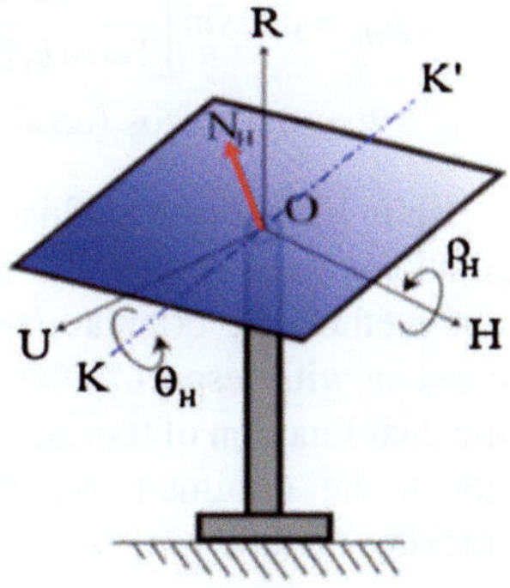

The following cases derive from these general equations, which are simplified by substituting some specific angles values. Always referring to [9] we have:

Alt-Azimuth: the observer is positioned in the center of a hemisphere identified by a plane, whose normal is the vector directed from the Earth's centre to the observer's head, and the portion of the sky above him. In this hemisphere it is possible to draw an ideal arch in the sky from South to North, called *Local Meridian*. The observer in the centre of this system, upright on this plane and looking northbound, can localize an object in the sky by two angles:

- *The Azimuth angle*, defined as a horizontal angle measured clockwise from a north base line or *meridian*. *Azimuth* has also been more generally defined as a horizontal angle measured clockwise from any fixed reference plane or easily established base direction line.
- *The Elevation angle*,which is included between the plane and the line joining the object to the centre of the system.

Referring to the [9] the heliostat is oriented in such a way that OH points to the zenith and the azimuth angle is denoted as ρ_H while the elevation is related to the angle θ_H. Thus the angle α is set at $\alpha = -90°$ and the equations become, respectively:

$$\theta_H = \frac{\pi}{2} - arc\sin\left[\frac{-\sin\lambda + \cos\Phi\cos\delta\cos\omega + \sin\Phi\sin\delta}{2\cos(\theta_i)}\right]$$

$$\rho_H = arc\sin\left[\frac{\cos\lambda\sin\phi - \cos\delta\sin\omega}{2\cos(\theta_i)\cos\beta_H}\right]$$

$$\begin{aligned}\theta_i = 0.5\arccos[&(\sin\Phi\cos\delta\cos\omega - \cos\Phi\sin\delta)(-\cos\lambda\cos\phi)\\ &- \cos\delta\sin\omega\cos\lambda\sin\phi - \sin\lambda(\cos\Phi\cos\delta\cos\omega + \sin\Phi\sin\delta)]\end{aligned}$$

Polar Mounted: in this type of heliostat, one of the axes is always parallel to earth rotation axis. Mathematically it means that $\alpha = \Phi$ and $\zeta = 180\,°$. Referring to a case with $\phi = 180\,°$, the angles of the two tracking axes can be written as:

$$\begin{aligned}\theta_H = \ & 0.5\arccos[(\sin\Phi\cos\delta\cos\omega - \cos\Phi\sin\delta)(-\cos\lambda\cos\phi)\\ & - \cos\delta\sin\omega\cos\lambda\sin\phi - \sin\lambda(\cos\Phi\cos\delta\cos\omega + \sin\Phi\sin\delta)]\end{aligned}$$

$$\rho_H = arc\sin\left[\frac{\cos\delta\sin\omega}{2\cos(\theta_i)\cos\beta_H}\right]$$

$$\theta_i = 0.5\arccos[(\cos\delta\cos\omega\sin(\Phi - \lambda) - \cos(\Phi - \lambda)\sin\delta]$$

Spin Elevation: in this case the target is set to be aligned with one of the rotational axes of the heliostat and this rotational axis is called *spinning axis.* In this method the orientation of the tangential and sagittal planes are always kept constant with respect to the reflector surface during tracking [9]. The heliostat has the dual function of focusing and tracking, and in this configuration it is possible to obtain the minimum optical aberration [10]. The constraints of this kind of tracking method are : $\lambda = \alpha$ and $\phi = \zeta$, so the equations become:

$$\theta_H = \frac{\pi}{2} + \\ - arcsin\left[\frac{1 - \sin\lambda(\cos\Phi\cos\delta\cos\omega + \sin\Phi\sin\delta) - \cos\lambda\sin\phi\cos\delta\sin\omega + \cos\lambda\cos\phi(-\sin\Phi\cos\delta\cos\omega + \cos\Phi\sin\delta)}{2\cos(\theta_i)}\right]$$

$$\rho_H = arc\sin\left[\frac{-\cos\phi\cos\delta\sin\omega + \sin\phi(\sin\Phi\cos\delta\cos\omega - \cos\Phi\sin\delta)}{2\cos(\theta_i)\cos\beta_H}\right]$$

$$\theta_i = 0.5\arccos[-\sin\lambda(\cos\Phi\cos\delta\cos\omega + \sin\Phi\sin\delta) - \cos\lambda\sin\phi\cos\delta\sin\omega + \cos\lambda\cos\phi(\cos\Phi\sin\delta - \sin\Phi\cos\delta\cos\omega)]$$

Latitude Oriented: this is a special case of spinning elevation in which the target is defined such that $\lambda = \alpha = \Phi - 90\,^\circ$ and $\phi = \zeta = 180\,^\circ$. With these special conditions we obtain:

$$\theta_H = \frac{\pi}{2} - arcsin\left[\frac{1 - \sin\lambda(cos\Phi\cos\delta\cos\omega + \sin\Phi\sin\delta) - \cos\lambda\sin\phi\cos\delta\sin\omega + \cos\lambda\cos\phi(-\sin\Phi\cos\delta cos\omega + \cos\Phi\sin\delta)}{2\cos(\theta_i)}\right]$$

$$\rho_H = arcsin\left[\frac{-\cos\phi\cos\delta\sin\omega + \sin\phi(\sin\Phi\cos\delta\cos\omega - \cos\Phi sin\delta)}{2\cos(\theta_i)cos\beta_H}\right]$$

$$\theta_i = 0.5\arccos[\cos\delta\sin\omega]n$$

Polar Oriented:in this method one of the rotation axes is always parallel to the Earth's rotation axis. This type of heliostat fulfills the requirement of constant-velocity single-axis tracking since the rotation speed of spinning axis is clock rate, while the movement of for elevation is very small throughout a tracking day. This configuration can be obtained by substituting $\alpha = \lambda = \Phi$and $\phi = \zeta = 180°$.

$$\theta_H = \frac{\pi}{4} + \frac{\delta}{2}$$

$$\rho_H = -\omega$$

4 Distributed Control System

The idea at the basis of the tracking system is to achieve the union between a centralized system for the heliostats angles calculation and a distributed system for movements of the actuators. The centralized system consists of a server on which runs a program in LabVIEW for the calculation of the position of the sun according to the Solar Position Algorithm (SPA) and the calculation of the relative positions of all heliostats. The distributed system consists instead of *n* units (DCU), each one placed on a given heliostat. Each unit is responsible for receiving data, sent from the server (the angles of Spin and Elevation), and read the actual angular position of the heliostat by two absolute encoders. An electronic board (Arduino) computes the deviation between these angles and controls the linear actuators to align the heliostat in the correct position. The communication between the server and the DCU is on a LAN using a TCP-IP protocol. The "brain" of each DCU is a control board ARDUINO MEGA 2560, appropriately programmed, with a different IP address assigned. Thus, the program running on the server can control each unit by sending a data packet to a specific IP address. In order to connect the server to all 42 units is therefore necessary to add a network switch.

Thanks to this centralized system, the operator can choose whether move automatically the heliostats according to the SPA calculations or by defining the angles manually.

On each heliostat a distributed control unit is mounted and it consists of:

- A 220 Vac to 5 Vdc transformer from for powering ARDUINO boards and the encoders
- Two absolute encoders for measuring the actual angles of SPIN and ELEVATION
- An ARDUINO MEGA 2560 board with Ethernet Shield
- 4 Solid State Relay (SSR) for the power contacts of the actuators (two for the Spin actuator extension and contraction and two for the Elevation actuator extension and contraction)
- 4 limit-switches to cut the power to the actuators in case of exceeding the limit angles
- A terminal block to connect the whole cables

5 Packet Data

There are two types of data packets that the server can exchange with the distributed control unit:

- **Handling Packet: t**he packet handling is constituted by a string of 5 fields of 4 bits of scalars (integer 32) with a value between 0 and 1024, as follows: the angles of SPIN and ELEVATION are translated into two numbers between 0 and 1023 and sent to the controller as such. The third and the fourth field together represent the duty cycle of the square wave for handling the heliostat and are expressed in milliseconds. The last field concerns the tolerance angle within which the movement ceases. The conversion angle is according to the expression:

  [(Maximum range in degrees)/1024] * target angle in degrees.

  For example, if the maximum excursions of Spin both of which are Elevation of 60° and the packet data is:

  0560 0455 0500 0100 0020

it means that the heliostat of must lead to an angle of spin equal to (60/1024) * 560 = 32.81°, and to an elevation angle equal to (60/1024) * 455 = 26.66° with jerking movement that provides a feed for 100 ms every 500 ms. The movement will end when the angles will be in a range of (60/1024) * 20 = 1.17°. The system encoders mounted on the Spin and Elevation axes measure the actual angle position of the heliostat and, which is compared by the ARDUINO microcontroller with the target provided by the server by means of the packet data. Thus the whole motion control is carried out by the DCU, while the server has the only function of computing the S-E angles for each heliostat and monitoring the proper functioning of the whole system.

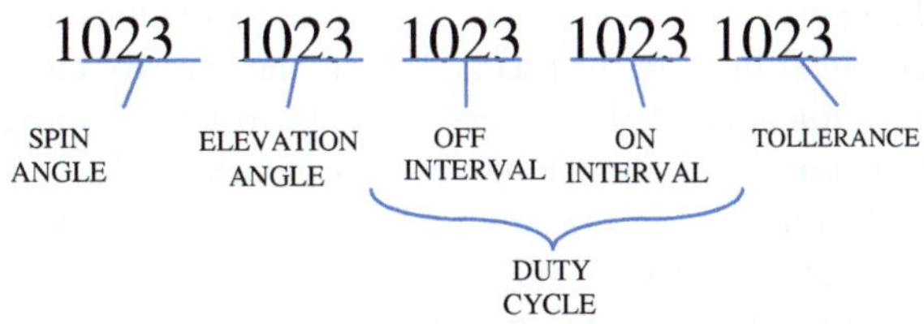

- **Read-only Packet**: in case of operator doesn't want an heliostat to move but just wants to know its angular position, another packet data can be sent: when all the values in the fields equal to zero:

  0000 0000 0000 0000 0000

the control unit sends back a packet to the server with the angles measured by the encoders. This same package can also be sent to stop at once a the motion of an heliostat. Whether no packets are sent to the DCU, the microcontroller just keeps the heliostat in the last position and doesn't feed the actuators. The position can be kept thanks to the irreversibility of the motion due to the pressure angle and to the transmission ratio of the gear worm inside the actuator itself.

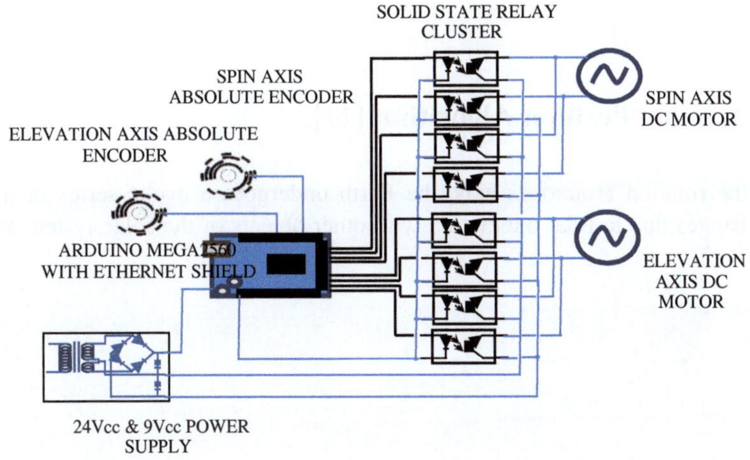

6 The Server Control Program

The chosen software to carry out the solar position computation was LabVIEW. The choice was mainly due to the program implementation simplicity and to the communication libraries availability for the ARDUINO device interface. The tracking program that runs on the server relies on Solar Position Algorithm (SPA): according to the location of the heliostat field (latitude, longitude and altitude), the date and time of the day, it can determine the position of the sun in the sky. Then, considering the relative position of the heliostats compared to the target, the Spin and Elevation angles are computed for each heliostat. In the first screen of the control software allows the operator to monitor the real-time calculations of the SPA and to set the summer/winter time. In the second screen, the operator can see the corners of Spin and Elevation computed for every single heliostat and activate (by means of Boolean buttons) or deactivate the movements of each of them. The program creates and sends data packets to one heliostat after the other cyclically with a frequency that con be set via software (Fig. 3).

In the last screen, eventually, it is possible to set the movement parameters including:

- The delay between two subsequent movements of a heliostat
- The duty cycle
- The angular tolerance for stopping the movement

The control software also allows to monitor the system status, the solar radiation, the number of working heliostat and the temperatures in several points of the plant.

Each angle is computed though the equations (10) to (12) and using as inputs the data of the solar position given by the NREL Solar Position Algorithm (SPA) (Fig. 4).

7 NREL Solar Position Algorithm [11]

Beyond the rotation around its axis, the Earth undergoes a cyclic series of trajectory changes due to tidal interaction with other objects of the solar system and

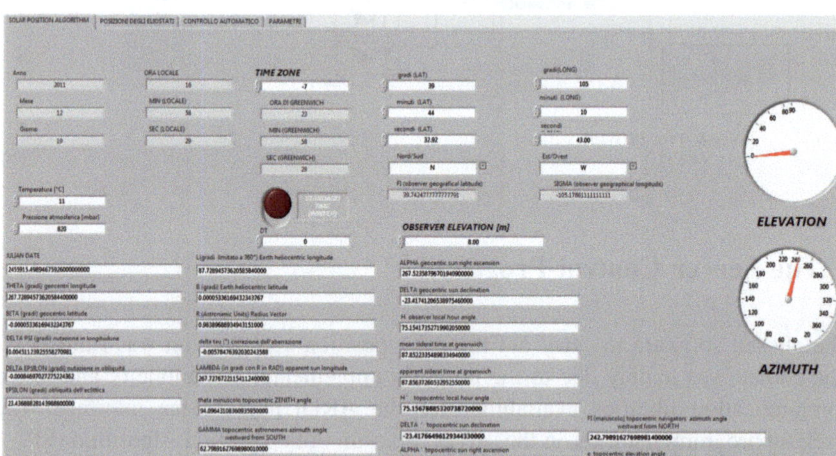

Fig. 3 Main screen of the lab view implemented solar position algorithm program

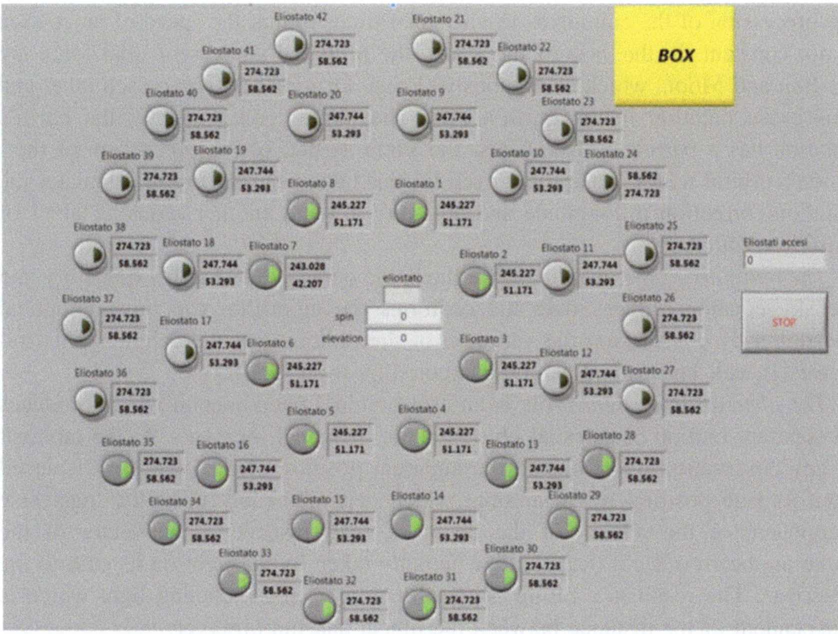

Fig. 4 Heliostat field control panel

to the shape of the ecliptic around the Sun. The data collected in the Ephemeris are the basis of several sun-tracking algorithms developed in the years, but one of the most widely used is the Solar Positioning Algorithm [11] drafted by Ibrahim Reda and Afshin Andreas and edited by the Nationa Renewable Energy Laboratory. The algorithm considers a high number of variables to locate the Sun during the day. First of all, the relative position of the Earth to the Sun is used. To translate it into an angle we must consider the ephemeris and refer to an absolute time: the first step is hence to calculate the Julian date (JD). Julian day is a system of time measurement for scientific use by the astronomy community, presenting the interval of time in days and fractions of a day since January 1, 4713 BC Greenwich noon. The use of this timing system is usually considered to be incorrect although it is widely used anyway. The Julian date (JD) is the interval of time in days and fractions of a day. Thus each moment is translated into a different progressive number (considering the year, the month, the day, hours, minutes and seconds). Several tables of empiric coefficients are used to calculate the Earth's heliocentric longitude, latitude and radius vector with high accuracy, so that the algorithm can adjust the angles by knowing the relative position of the Earth to the Sun.

Once the Earth's position in the heliocentric frame is known, the algorithm considers a series of disturbing factors of the Earth's trajectory:

The nutation, that is a rocking and swaying motion in the axis of rotation of a largely axially symmetric object, and occurs because of tidal forces that cause the precession of the equinoxes to vary over time, so that the speed of precession is not constant. In the case of the Earth, the principal sources of tidal force are the Sun and Moon, which continuously change location relative to each other and thus cause nutation in the Earth's axis. The largest component of the Earth's nutation has a period of 18.6 years, the same as that of the precession of the - Moon's orbital nodes. To account even for this factor, the algorithm calculates the nutation correction in longitude and obliquity through another series of tables of empiric coefficients.

The obliquity of the ecliptic. Since the angle between the Earth's trajectory and the sky equator changes over the centuries the algorithm can prevent optical aberration by considering the variation of the obliquity of the ecliptic. This trajectory is well known and can be computed by using the JD.

The aberration of light. This is an astronomical phenomenon which produces an apparent motion of celestial objects about their real locations. At the moment of any observation of an object, the apparent position of the object is displaced from its true position by an amount which depends solely upon the transverse component of the velocity of the observer, with respect to the vector of the incoming beam of light (i.e., the line actually taken by the light on its path to the observer). The result is a tilting of the direction of the incoming light which is independent of the distance between the object and the observer.

The apparent sidereal time correction. Since a mean sidereal day is about 23 h, 56 min, 4.091 s (23.93447 h or 0.99726957 mean solar days), and not 24 h like the solar day.

The altitude of the observer and the climatic conditions. The refraction angle depends on the air density, which, in turn is a function of temperature and pressure.

The Solar Positioning Algorithm returns the position of the Sun in the sky in the topocentric frame of the observer, in terms of the topocentric sun right ascension and the observer local hour angle or even in the geocentric frame in terms of the geocentric sun right ascension and the geocentric sun declination. These angles can be converted as a function of the chosen tracking method to allow the concentration of the solar beams upon the specific target.

8 Conclusions

The control system can be improved by refining the parameters of the motion control via software. A smoother movement of the heliostats can be achieved varying the delay between two subsequent movements of a heliostat, the duty cycle of the square wave that controls the steps of the actuators and the angular tolerance for stopping the movements. These whole parameters must be set as a result of a series of field tests: the dynamic behavior of the heliostats can't, indeed, be foreseen since a great verity of parameters rules the actual way of moving (wind, weight, misalignment of axle loads).Moreover each actuator could have a different response even if fed with the same voltage. Once carried out the final enhancement, the system could concentrate a very high energy over the target, reaching several hundred degrees, contributing to the reduction of greenhouse gases from the landfill.

References

1. Mills D (2004) Advances in solar thermal electricity technology. Sol Energy 76, 19–31
2. NREL (National Renewable Energy Laboratory) concentrating solar power projects. www.nrel.gov/cvp/solarpaces/
3. Harlan B, PhD in chemical engineering. www.brighthub.com/environment/renewable-energy/articles
4. Quincy CM (2001) Point focusing double parabolic trough concentrator. Sol Energy, 70(2):85–94, Kentucky Mountain Bible College, P.O. Box 2, Vancleve, KY 41385-0002, USA—Received 24 Feb 2000; revised version accepted 5 Oct 2000
5. PEGASE project home page
6. Solhycarb EU funded project, ETHZ official page
7. Flamant G, Luxembourg D, Robert JF, Laplaze D (2004) Optimizing fullerene synthesis in a 50 kW solar reactor. Sol Energy 77(1):73–80
8. Ait Ahcene T, Monty C, Kouam J, Thorel A, Petot-Ervas G, Djemel A (2007) Preparation by solar physical vapor deposition (SPVD) and nanostructural study of pure and Bi doped ZnO nano powders. J Eur Ceram Soc 27(12):3413–3424
9. Chen YT, Lim BH, Lim CS (2006) General sun tracking formula for heliostats with arbitrarily oriented axes. J Sol Energy Eng Copyright © 2006 by ASME MAY 2006, 128/245
10. Chen YT, Kribus A, Lim BH, Lim CS, Chong KK, Karni J, Buck R, Pfahl A, Bligh TP (2004) Comparison of two sun tracking methods in the application of a Heliostat field, 638 126, Feb 2004—Copyright © 2004 by ASME
11. Solar Positioning Algorithm for Solar Radiation Applications – NREL/TP-560-34302 – Revised January 2008 - Ibrahim Reda, Afshin Andreas – National Renewable Energy Laboratory, 1617 cole Boulevard, Golden – Colorado – 80401-3393. NREL is a U.S. Department of Energy Laboratory operated by Midwest Research Institute – Battelle – Betchel
12. Delgado-Torres AM, García-Rodríguez L (2010) Analysis and optimization of the low-temperature solar organic Rankine cycle (ORC). Energy Convers Manage. Elsevier Ltd. All rights reserved
13. Kongtragool B, Wongwises S (2003) A review of solar-powered stirling engines and low temperature differential stirling engines. Renew Sustain Energ Rev 7, 131–154
14. Kelly A, Gibson TL (2009) Improved photovoltaic energy output for cloudy conditions with a solar tracking system. Nelson Sol Energy 83, 2092–2102
15. Royne A, Dey CJ, Mills DR (2005) Solar hydrogen from photovoltaic-electrolyzer systems. Sol Energy Mat Sol Cells 86, 451–483
16. Duff WS, Hodgson DA (2005) A simple high efficiency solar water purification system. Sol Energy 79:25–32
17. García-Rodríguez L, Gómez-Camacho C, Hazim Mohameed Q (2001) Perspectives of solar-assisted seawater distillation technologies. Desalination, 136(1):213–218
18. Indarto A, J. Palgunadi J (eds) (2011) Syngas: production, applications and environmental impact. Nova Science Publishers, Inc

Part VI
Mechanical Systems Diagnostics

Preliminary Investigations on Automatic Detection of Leaks in Water Distribution Networks by Means of Vibration Monitoring

Alberto Martini, Marco Troncossi, Alessandro Rivola and Davide Nascetti

Abstract The efficiency of water supply networks is an important issue. In order to reduce water losses, policies of leak reduction are essential. The paper deals with a preliminary study on the use of vibration monitoring tools for the detection of leaks in water service pipelines. The long-term project is the development of a system for automatically detecting burst leaks occurring in service pipes. Preliminary experimental tests were performed on both a test rig and an actual service pipe of the water distribution system. Three main objectives were achieved: firstly, the effectiveness of vibration monitoring for leak detection purposes was assessed providing a positive response; then, a prototypal detection procedure was studied, implemented and tested on the preliminary experimental data; finally, the specifications for a prototypal acquisition equipment were also determined. This paper illustrates the experimental campaign and its main results.

Keywords Water Supply Networks · Monitoring · Leaks · Vibration

1 Introduction

All the utilities concerned with drinking water supply are making efforts to increase the efficiency of their distribution systems. Official statistics about the North America show that water losses typically reach 30 % of the production [1].

A. Martini (✉)
CIRI—Interdepartmental Centre for Industrial Research: Advanced Applications in Mechanical Engineering and Materials Technology, V.le del Risorgimento 2, 40136 Bologna, Italy
e-mail: alberto.martini6@unibo.it

M. Troncossi · A. Rivola
DIN—Department of Engineering for Industry, V.le del Risorgimento 2, 40136 Bologna, Italy

D. Nascetti
Hera S.p.A., V.le C. Berti Pichat 2/4, 40127 Bologna, Italy

G. Dalpiaz et al. (eds.), *Advances in Condition Monitoring of Machinery in Non-Stationary Operations*, Lecture Notes in Mechanical Engineering,
DOI: 10.1007/978-3-642-39348-8_46,

In some Italian areas water losses exceed 60 %, the national average being about 40 % [2].

Water losses include both background leaks (very small leaks at pipe joints and fittings, generally undetectable) and burst leaks (resulting from pipe holes and damages). While the former are mostly assumed as irremovable (i.e. associated with the normal system functioning), the latter should be detected and repaired. Indeed non revenue water represents a very high cost. Moreover water losses represent a danger for the public health since the leaking flow may damage the foundations of buildings and roads, and contamination by pollutants may occur. Last, but not least, water is a limited resource that should not be wasted at all.

In order to reduce and prevent water losses, a policy of leak detection and reduction is essential.

Conducting periodic water audits helps monitoring water losses of the whole distributing system. District Metered Areas (DMA) management [3] generally offers better leak detection performance and allows focusing the efforts to locate leaks on restricted portions of the network. Optimized distribution of noise loggers over the water supply network may be also adopted [4], although their economic viability and leak detection effectiveness are not guaranteed [5].

Several techniques and equipments for pinpointing leaks are available [5–8]. The most common equipments are listening devices (such as listening rods and geophones, whose efficiency largely depends on the operator skills) and noise correlators (which automatically pinpoint leaks by means of signal correlation techniques), both generally relying on vibration or pressure transducers.

This study deals with the detection of burst leaks occurring in service pipes, by using vibration measurements. The long-term project, planned by the R&D department of the multi-utility Hera S.p.A. (Bologna, Italy), aims at developing a device for the automatic early detection of bursts located in the customer connection branches running from the mains to the user's water meter [9].

The research (still at an early stage) was started by performing the preliminary activities that are described in this paper: evaluation of the effectiveness of vibration monitoring for leak detection purpose; study and development of a prototypal algorithm for leak detection; definition of the hardware specifications for the development of a prototypal monitoring device to be adopted in a massive experimental campaign in the actual water distribution network. In order to accomplish these tasks, experimental tests were carried out on both a test rig and an actual service pipe.

2 Materials and Methods

The desired system for leak detection should measure vibrations, analyze the recorded signals and automatically send a warning to the control centre in case of leak detection. This implies the detection algorithm to be executed on board. The system is not precisely intended for locating bursts. Indeed it is conceived for operating on water service pipes, which are relatively short pipes (up to 10 m): operations of leak pinpointing by maintenance teams (through standard listening devices) are reasonably expected to be straightforward, once leak detection has been achieved. Low cost represents an essential requirement of the hardware components, due to the planned widespread installation. In order to meet cost requirements the system should operate with the simplest hardware as possible, thus requiring a very careful definition of all the technical specifications to optimize costs and to guarantee adequate performance. Moreover the algorithm should involve only rather basic analysis techniques in order to require limited computational resources, thus further limiting costs the of hardware components.

A preliminary experimental campaign was performed for pursuing the following goals:

- verifying the leak detection capabilities of different approaches based on the analysis of vibration signals, and assessing how the variation of boundary and functioning conditions affects the detection performance;
- identifying the most suitable metrics for revealing the presence of water leaks and implementing a trial detection procedure based on it;
- defining the specifications of a prototypal acquisition device for further executing extensive measurements on real burst leaks occurring in the water supply network managed by the utility.

The preliminary experiments took into account the following variable factors:

- distance of the burst from the transducer (from 0 to 10 m);
- water pressure of the network (from 3 to 7 bar);
- presence of a water flow in the service pipe due to consumer water usage;
- value of the flow rate related to consumer water usage (from 0 to 54 l/min);
- non-zero flow rate in the main pipe induced from water consumption in service connections different from the monitored one (from 0 to 100 l/min);
- material of mains and service pipes (PE and PVC);
- type of soil where the pipes are buried (wet soil, dry soil, sand);
- transducer sensitivity (100, 1000 mV/g);
- transducer mounting (petroleum wax and cyanoacrylate adhesives);
- A/D bit resolution (from 8 to 24 bit).

Two different kinds of tests were carried out: experiments on a test rig, specifically developed and arranged for this study; measurements on an actual service connection of the public water supply network.

2.1 Test Rig Experiments

The experiments carried out on the test rig were designed for investigating all the above mentioned factors under controlled conditions.

An experimental facility was arranged in a buried testing site of the Hera R&D department. Two PVC pipes with diameter of 90 mm (DN 90) were used as water mains. Several small diameter pipes (DN 32) were connected to the distribution pipes for simulating typical customer connections running from the mains to the metering points. Each connection was about 10 m long (this value being derived from the statistics concerning the entire company water distribution network) and laid 0.5 m under the ground. Two valves located at the proximal and terminal extremities of each branch allowed flow regulation.

A pressure tank fed the system, thus permitting modifications of the input pressure parameter. The transducers were mounted on the pipe inside the terminal valve chamber.

First tests were carried out with no leaks, considering different combinations of functioning parameters and boundary conditions. Damages were then artificially caused, one per service pipe, at different distances from the junctions to the main pipe. The crack, a longitudinal cut 20 mm long, reproduced the most common pipe damage which typically characterizes burst leaks.

2.2 *Actual Service Pipe Experiments*

Experimental tests on a real customer service pipe were meant to monitor vibration signals in real working conditions. In order to have a good control of the experiment parameters, a leak was artificially introduced in the connection pipe. The buried service pipe of a residential customer was intentionally damaged by making a 20 mm long longitudinal cut (Fig. 1a). A special device for controlling the water flow through the crack was built and installed (Fig. 1b). The device, made of an adjustable sleeve and an operating lever, was arranged in a manhole, thus permitting to regulate the water loss (till completely stop the leak) even after that soil

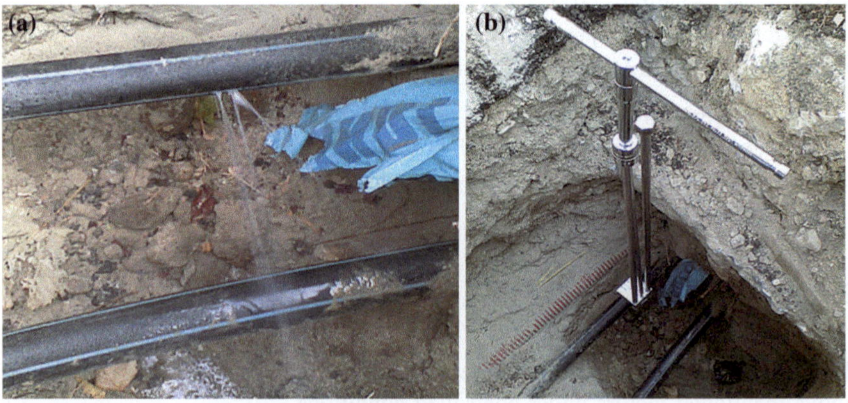

Fig. 1 Tests performed on the real customer service pipe: **a** leak close up, **b** control device

and asphalt had been restored, i.e. permitting to keep unchanged boundary conditions when a transition from a non-leaking to a leaking state was induced.

All the other parameters of interest were fixed for the experiments, with the exception of the water flow inside the monitored pipe. Indeed the consumer water consumption could induce a non-zero flow rate.

The transducers (adhesively mounted on the service pipe) and the acquisition system were located close to the customer meter. Measurements were performed during 8 consecutive nights, recording 90 acquisitions per night. For the first 7 nights the water leak was kept closed. On the last night the leaking condition was set by operating the regulation device.

2.3 Data Acquisition and Analysis

Two PCB piezoelectric accelerometers with sensitivity of 100 and 1,000 mV/g were used. A National Instruments 24 bit A/D card (NI USB-9233) was adopted for acquiring and conditioning the signals. As for the acquisition parameters, a sampling frequency of 5 kHz was adopted. The single acquisition duration was set to 2 and 10 s for the test rig campaign and the real service pipe tests respectively. Tens of acquisitions were performed for each test condition.

The experimental data were investigated by using different approaches in order to possibly identify the most suitable analysis for an effective detection of leaks. Complex analysis techniques were disregarded. Indeed the final leak detection algorithm will require a relatively simple analysis procedure to be implemented.

The raw signals were processed in the time domain where the most common statistics, such as mean value, standard deviation, RMS value, crest factor, skewness, and kurtosis were computed. A second analysis was performed in the frequency domain, where the signal power spectra (PSD) were calculated.

3 Results and Discussion

A huge database was collected from the experiments. Only the most relevant results are presented and discussed hereinafter.

Figure 2a reports the time histories of two signals recorded from the test rig with and without water leak, respectively, and with all the terminal valves being closed. The corresponding PSD in the frequency range 0–2.5 kHz are shown in Fig. 2b. Time domain analysis permits to clearly detect the presence of the burst due to a significant increase of the vibration level induced by the leaking water. Apart from the amplitude, the PSD comparison shows that the signal in leaking condition exhibits characteristic frequency content clearly distinguishable from the background noise, thus allowing leak detection in the frequency domain as well. These results are also supported by the analysis of experimental data obtained from tests on the actual service connection (Fig. 4), and are therefore promising with respect to the overall research goals.

Two factors appeared to mainly affect the recorded signals, i.e. the presence of a non-zero flow rate inside the service pipe, due to water usage, and the distance of the leak from the vibration transducers.

The effects on test rig signals of a non-zero water flow inside the monitored pipe (obtained by opening its terminal valve) are reported in Fig. 3. Figure 3a displays the comparison in the time domain between vibration measurements carried out with and without water leak, for a constant flow rate. The signals exhibit comparable amplitudes and trends. Vibration levels appear to be two orders of magnitude higher than those measured in case of leak and closed terminal valve (Fig. 2a).

Figure 3b reports the corresponding PSD. Frequency domain analysis shows that the presence of water flow causes higher vibration levels which cover the effects of the leak. In particular the frequency content characterizing the leak can be no more identified.

Hence the water leak is no longer detected. It is worth noticing that even very small values of the flow rate prevent the burst to be revealed.

Experimental data from the tests performed on the actual service connection revealed equivalent behaviour. The complete time histories relative to the fifth and the eighth days (obtained by aggregating all the daily 90 records) are shown in Fig. 4a and b respectively, as examples.

The condition of non-zero flow rate induced by customer water usage causes the abrupt and temporary increment of the measured vibrations, which, as already seen for the previous tests, makes the detection of leak impossible. The effects generated by water consumption may be then considered as the result of generic perturbations of the system superimposed to the "normal" steady-state signal. Hence such effects will be referred to (and treated) as "perturbations" hereinafter.

Distance between leak and transducers considerably influences measurements as well. Its increment causes the perceived vibration levels to significantly lower.

The PSD of the signals obtained by varying the distance between transducers and water leak are reported in Fig. 5. As the distance increases a consistent

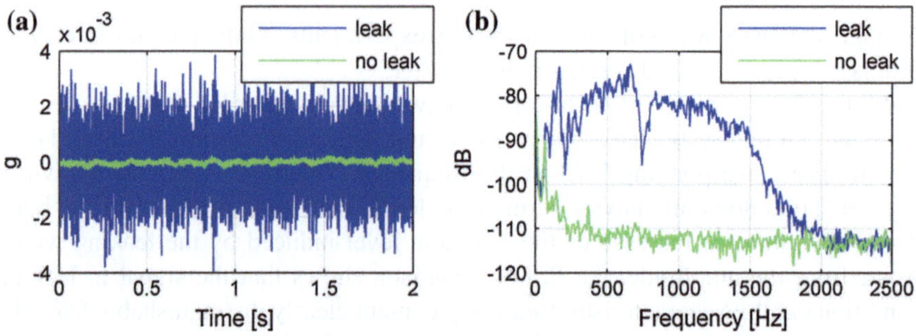

Fig. 2 Leaking and non-leaking conditions in case of zero flow rate: **a** time history, **b** PSD

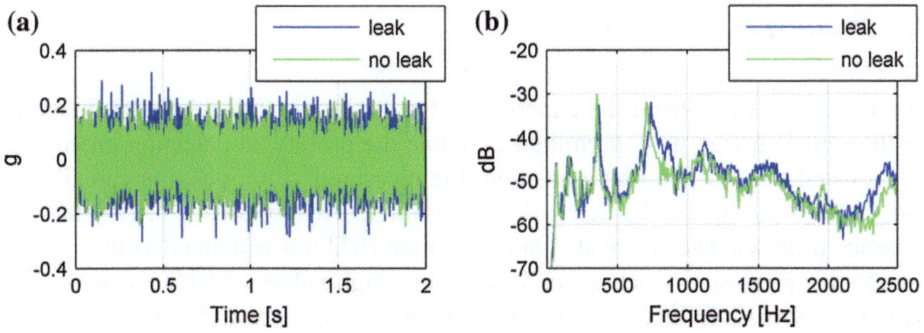

Fig. 3 Leaking and non-leaking conditions, with maximum flow rate: **a** time history, **b** PSD

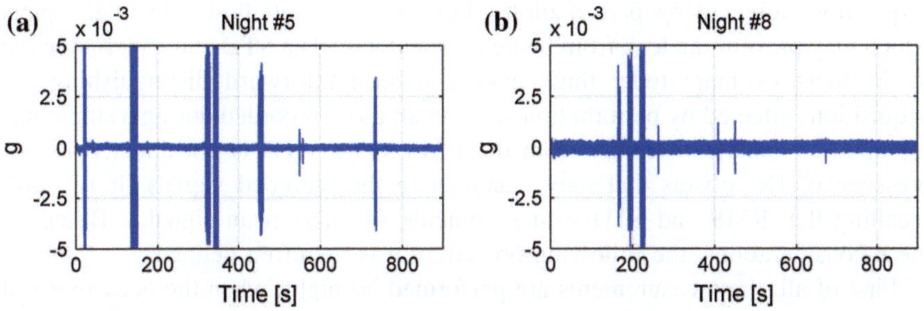

Fig. 4 Real service pipe measurements, time histories: **a** non-leaking and **b** leaking conditions

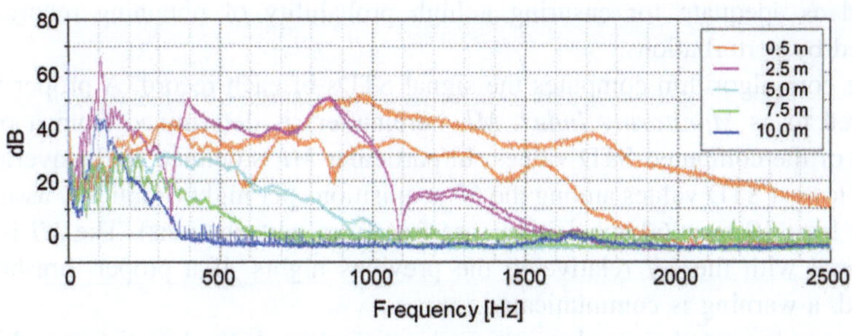

Fig. 5 PSD of signals recorded (by two sensors) for different leak-transducers distances

reduction of the vibration levels is observed. In particular the damping of higher frequency spectral components appears to rapidly increase.

All the other investigated factors did not exhibit appreciable influence on vibration measurements.

The use of vibration monitoring thus appeared to be a viable approach for detecting burst leaks.

3.1 Prototypal Algorithm

The results obtained from the experimental tests, in particular the measurements on the actual service pipe, permitted to define the prototypal detection procedure.

Provided that no perturbations affect the measured vibrations, the signal level resulted as the simplest characteristic that could be used to satisfactorily detect the presence of leaks. The signal standard deviation (STD) was defined as the metrics used to this purpose: in fact, as the RMS value, high values of STD are associated with high vibration levels. A strategy based on the signal STD computation appeared to be potentially profitable due to some essential advantages. Firstly, only basic signal processing is required, since STD can be easily computed from the raw signal. In addition STD allows detecting (and consequently discarding) all acquisitions affected by perturbations. Indeed a non-null flow rate in the pipe, which may prevent the leak from being revealed, entails a STD value increment up to 2 orders of magnitude, thus making it straightforward distinguishing the acquisitions affected by perturbations among all those recorded during a single day (or night). Finally, differently from the RMS value, STD is not affected by the presence of DC offsets that may characterize the recorded signals (it is worth recalling that RMS and STD values coincide for zero-mean signals). Based on these considerations, the following procedure was therefore defined.

First of all, the measurements are performed by night, when the occurrence of external perturbations (above all due to water usage by the customer) is expected to be less frequent.

An amount of 60 acquisitions per night is executed. Such quantity was considered as adequate for ensuring a high probability of obtaining records not affected by perturbations.

The core algorithm computes the signal STDs of each record. A proper index (referred to as *Monitoring Index*, *MI*, hereinafter) is determined from a proper subset of the computed STD values. In particular *MI* is defined as the average of the 10 lowest STD values among the 60 acquisitions per night (implicitly assuming that at least 10 over 60 acquisitions are free from perturbations). The *MI* is then compared with the *MI* relative to the previous nights. If a proper threshold is crossed, a warning is communicated.

The implemented procedure showed satisfactory leak detection capabilities considering the available experimental data. Figure 6a reports the *MI* (normalized to the highest value) computed by running the algorithm on the records from the real service pipe experiment: the leak can be easily identified, being the correspondingly computed *MI* almost twice the values assumed in no-leak conditions. Also the data from the test rig experiments led to the same conclusions. Further tests in other real facilities will be performed in order to verify and improve the algorithm effectiveness.

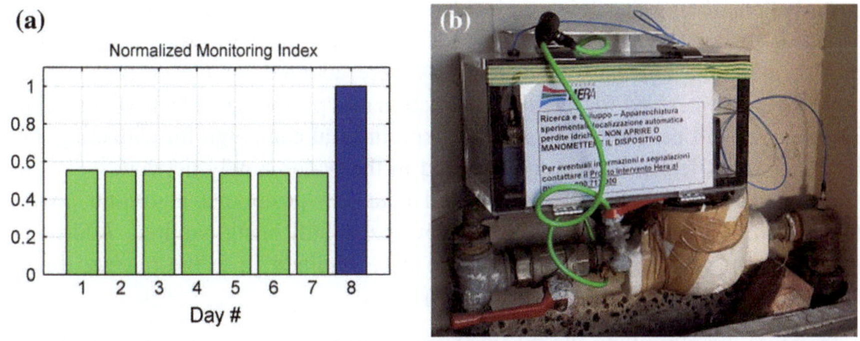

Fig. 6 **a** detection algorithm results, **b** prototypal measuring device

3.2 Prototypal Measuring Device

The results obtained from both the test rig and the real service pipe contributed to the definition of the hardware specifications of the prototypal measuring device devised for the next experimental campaign.

The sample frequency was defined at 4 kHz by considering that no relevant frequency content was observed over 1.5 kHz.

The analysis of experimental data showed that vibrations may exhibit very low levels, in particular when the leak is distant from the measuring point. On this point, the experimental data were further post-processed for simulating bit resolutions lower than the 24 bit of the NI board. The conclusion of that investigation was that a minimum resolution of 16 bit is required to the A/D card in order to limit the quantization distortion. Moreover accelerometers with high sensitivity are required in order to increase the signal to noise ratio. A signal amplifier was considered advisable as well: its actual adoption will depend on the corresponding incidence on the final system cost.

Some prototypal devices were then manufactured by a third party. They will be adopted in future extensive tests for performing acquisitions in case of real bursts occurring in the water distribution network. The collected data will be used both for further testing the detection algorithm and for refining the specifications of the definitive leak detection system.

4 Conclusions

The preliminary investigations concerning a project for leak detection in water distribution pipes by using vibrations were presented. An experimental campaign was carried out on both a test rig and an actual service pipe by artificially inducing burst leaks. The performed tests showed that if the system is free from external perturbations (being the flow rate inside the service pipe the most important), the detection of burst leaks may be achieved by using proper statistics computed on raw acceleration signals. The first results were adopted for the design of some acquisition devices and for the implementation of a prototypal detection algorithm. The devices will be adopted in a larger experimental campaign for collecting data from real burst leaks located and repaired in the water supply network.

Limited to the available experimental data, the prototypal algorithm proved effective for leak detection, further improvements still being required. Additional investigations will be thus performed on data obtained from the planned future tests in order to achieve the following goals:

- to assess the algorithm effectiveness on a statistically representative database;
- to increase the algorithm robustness in case of signals affected by noise or numerous perturbations, thus maximizing its detection effectiveness.

Different analysis techniques will be also tested in order to examine alternative approaches that may enable an effective detection of water leaks.

The leak detection device should be integrated, in the future, with the system for automatic meter reading (AMR), i.e. they should share the same power supply and data transmission network. The leak detection system is expected to provide early warnings in case of occurrence of bursts. Hence the leak runtime should be significantly reduced by decreasing awareness time. In addition the device is also expected to enable the detection of several unreported bursts, due to the foreseen widespread installation. Hence a global reduction of costs related to water leaks is expected thanks to the relevant reduction of water losses, integration with the AMR, and the possibility of remote data handling.

References

1. United States Environmental Protection Agency (2010) Control and mitigation of drinking water losses in distribution systems. EPA 816-R-10-019
2. Portolano D (2008) Il controllo delle perdite nei sistemi acquedottistici: criteri innovativi di gestione. PhD thesis, University of Napoli
3. International Water Association (2007) District metered areas guidance notes
4. Li W et al (2011) Development of systems for detection, early warning, and control of pipeline leakage in drinking water distribution: a case study. J Environ Sci 23(11):1816–1822
5. Hunaidi O, Wang A (2006) A new system for locating leaks in urban water distribution pipes. Manage Environ Qual Int J 17(4):450–466
6. Hunaidi O et al (2000) Detecting leaks in plastic pipes. J AWWA 92(2):82–94
7. Gao Y et al (2005) On the selection of acoustic/vibration sensors for leak detection in plastic water pipes. J Sound Vib 283(3–5):927–941
8. Khulief Y et al (2012) Acoustic detection of leaks in water pipelines using measurements inside pipe. J Pipeline Syst Eng Pract 3(2):47–54
9. Leoni G et al (2009) Method for detecting the presence of leaks in a water distribution network and kit for applying the method. Patent EP 2107357

An Application of Statistical Tools in the Identification of the Transient Vibrations of Bucket-Wheel Excavators Under Random Loads

Weronika Huss

Abstract Due to fatigue cracks appearing in atypical places at construction of bucket-wheel excavator's body, it was supposed that they are a consequence of transient vibrations associated with impulse loadings. This work presents a procedure for identification of such phenomenon. Operating loadings as well as vibration of bucket-wheel excavators' structures are strongly random and non-stationary, what makes any analysis difficult. Moreover they reveal continuous changes in the structure of the frequency spectrum. For that reason the procedure was based on statistical measures and relativity of power spectral densities of vibration in succeeding time periods. As a result it was found that indeed transient vibration accompany impulse loading, but it isn't full correlation, and it also happens for quite low impulses.

Keywords Excavators · Transient Vibrations · Statistical Tools

1 Description of Excavator and Excitations

Bucket-wheel excavators (Fig. 1), with their mass often exceeding 2,000 tons, form a group of the largest engineering machines. In Poland such machines are commonly used in lignite mining.

An operation of bucket-wheel excavators is based on four working motions. Each of them resulting from rotation of appropriate subassemblies. Hence one could expect the operational vibrations of an excavator to be highly cyclical. However, this is not the case since besides the bucket-wheel excavator the considered system includes the slope being mined (particularly the overburden).

W. Huss (✉)
Institute of Machines Design and Operation, Wroclaw University of Technology,
ul. Lukasiewicza 7/9, 50-371 Wroclaw, Poland
e-mail: Weronika.huss@pwr.wroc.pl

G. Dalpiaz et al. (eds.), *Advances in Condition Monitoring of Machinery in Non-Stationary Operations*, Lecture Notes in Mechanical Engineering,
DOI: 10.1007/978-3-642-39348-8_47,

Fig. 1 Excavator SRs-1200

The overburden in Poland is highly heterogeneous as regards its grading. The largest of the fractions—boulders (more than 200 mm in diameter) are it's permanent component. Also excavators often encounter boulders much larger than a single bucket. The presence of such inclusions is the cause of the high randomness, nonstationarity and even transients in the vibrations of the whole excavator structure. This has been known for years and attempts have been made to take it into account in the modelling of the vibrations of bucket-wheel excavators and in operational recommendations [1, 2]. But breakdowns and disasters, especially under pulse loads, have continued to occur.

Therefore a hypothesis was proposed that during operation some impulse loadings are accompanied by transients, which are responsible for fatigue cracks in unpredictable places and for accelerated propagation of existing cracks.

It should be noted that as regards the observation of the vibrations of the whole bucket-wheel excavator body the most representative measuring section is located in the bucket-wheel boom. It is "a direct recipient of the external load" [3]. Through this assembly the load is transferred to the other parts of the body. It's vibrations (in frequency domain) are the most complicated. For this reason the vibrations of the bucket-wheel boom in the orthogonal directions (vertical bending, horizontal bending and compression) are most often used to model the dynamics of such constructions. The vibrations are recorded by means of strain gauges. Due to the character of work of bucket-wheel excavators no active experiment can be carried out. This means only passive observations of many-hour continuous recording can be employed in this situation. The procedure presented further below is illustrated with a case of vertical vibrations of an SRs-1200 excavator (Fig. 1).

2 Existing Model

The history of the modelling of bucket-wheel excavator load-bearing structure vibrations dates back to the 1950s. Beginning with deterministic linear models [4], through statistical models [1], attempts have been made to estimate service loads in order to predict the residual reliability of the existing machines and to use the obtained results in the design of new machines. In particular, attention has been drawn to incidentally occurring extreme loads. For the first time this was mentioned in D. Dudek's monograph [2], but without looking into the nature and effects of such loads on the durability of the constructions. The data were used to built statistical models of maximum loads, which were found to be coherent and reproducible for each of the investigated bucket-wheel excavators. The research conducted since that time resulted in another comprehensive work by the same author [3], in which the term "stroke load", understood as a load rapidly increasing to abnormal magnitudes, was introduced. It was determined on the basis of numerous measurements that stroke loads produce stresses in the load-bearing structures of the excavators, which may exceed the maximum peaceful operation loads as much as five times. At the same time it was found that the stroke load waveforms were highly random and so—unrepeatable. The study culminated in a probability distribution of the stresses generated by stroke loads, constructed for an exemplary machine.

The building of the above model started with the statistical distributions of maximum stresses, obtained from three orthogonal load waveforms. After the distributions were superimposed on the Gumbel grid, they could be approximated with straight lines. Than an assumption that maximum stresses in the boom's cross section constitute sums of the moduli of bending stresses (horizontal and vertical vibrations) and compressive stresses, was made. Next a distribution of the probability of occurrence of that sum was also superimposed on the Gumbel grid (Fig. 2). The values obtained in this way were approximated with three straight lines with different slope coefficients. The points of contact between the successive line segments were adopted as the boundaries defining the difficulty of mining conditions. Similar probability distributions, differing in only their linear regression coefficients, were obtained then for each of the machines. In each of the cases, the last so defined interval was referred to as "exceptional loadings".

Fig. 2 Probability distribution of extreme stresses for different difficulty of operating conditions

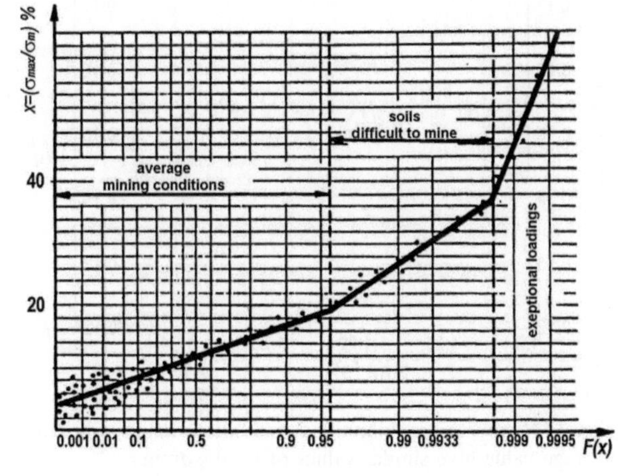

The above model is valid for absolute load values and constitutes a good measure enabling one to quickly determine the current mining conditions for a given machine in terms of its capacity. However, the model does not take into account the individual character of the extreme loads and so it does not provide a full picture of the actual dynamics of the construction.

Yet, the vibrations of the bucket-wheel excavator structure are nonergodic relative to the average [6]. It is not difficult to notice that the model does not take this into account and treats vibrations having a narrow range, but high instantaneous average values, in the same way as large deviations whose instantaneous values are lower (Fig. 3).

It often happens that changes in the instantaneous average value are at the level of dynamic amplitudes. Hence the above model gives information about the absolute values of strains (or stresses), but not about the latter's character. Moreover, it is obvious that the particular frequencies are bound with certain forms of the construction's vibration and the shape of the forms indicates the way in which the construction is loaded. Thus the neglect of this aspect significantly weakens the analysis of the strain of the excavator structure. Also the impact of low-amplitude vibrations and that of the high-amplitude vibrations are different. Therefore no conclusions about the dynamics of the strains can be drawn from the values of the local extrema alone.

3 Transient Vibrations in Load-Bearing Structures

The frequency spectra of the operational vibrations of the bucket-wheel excavator structures are characterized by high values in the neighbourhood of the constant component (wide fluctuations of the average). In addition, because of the high

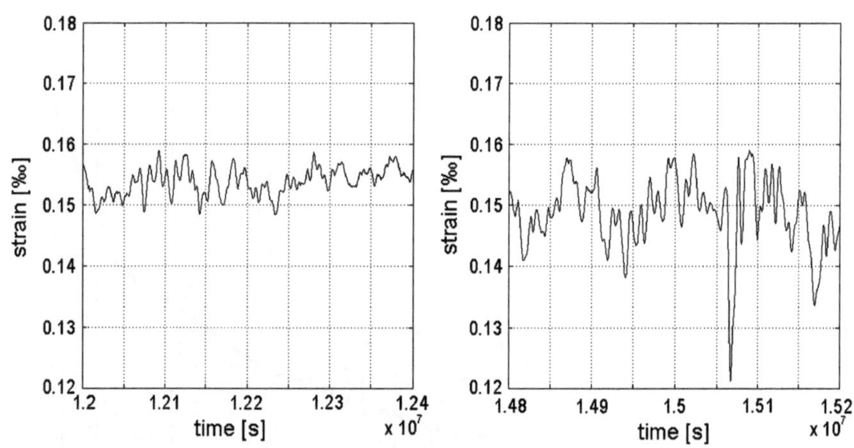

Fig. 3 Narrow-range vibrations and wider-range vibrations but at different instantaneous average value give similar values of local extrema

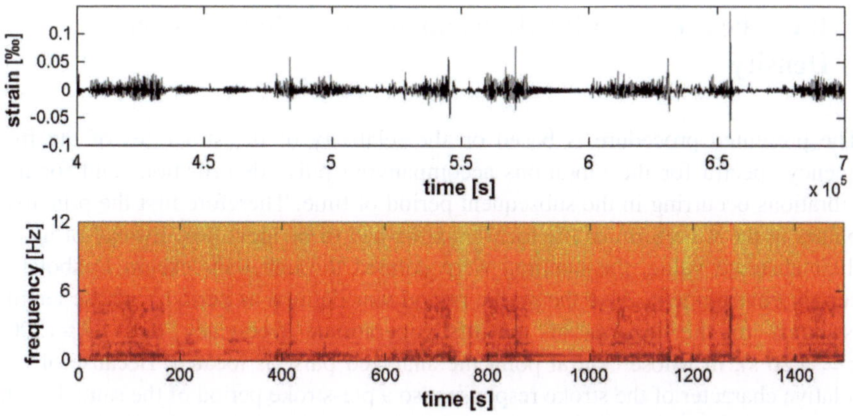

Fig. 4 Sample waveform of vertical vibration (0.19–12 Hz) of excavator SRs-1200 bucket-wheel boom and its spectrogram

inertia of machines, no vibrations with frequencies higher than 20 Hz arise in such structures. For this reason, particularly for the SRs-1200 excavator power spectral density in a range from the first operational frequency $f_O = 0.19$ to 12 Hz were used.

On the basis of the definition of transient vibrations given in [5] it can be said that they correspond to the system's free vibrations while steady vibrations correspond to forced vibrations from external excitation. If the system is (asymptotically) stable, after sufficiently long time the free vibrations are damped and then one can observe only the forced vibrations having the same frequency as the external excitation. Therefore when one examines this phenomenon in a spectrogram one should expect instantaneous disturbances in its structure. Several such spectrograms were generated for the vertical vibrations of the bucket-wheel boom of the SRs-1200 excavator (ex. in Fig. 4). It is characteristic that the structure of the spectrum undergoes practically continuous transformations. Particularly significant changes can be observed in the places corresponding to the distinct local maxima. The dark brown colour represents the maximum height of the fringes and according to the spectrogram, it appears in the places of the supposed strokes. Against the whole spectrogram, such events greatly disturb the earlier image of the spectrum. Even though a wide-range signal clearly results in high fringes in the frequency spectrum, it does not have to lead to a change in it's structure. This happens, however, when some pulse deformations occur in the construction.

4 Identification of Vibrations on Basis of Power Spectral Density

The presented procedure is based on the relativity of the structures of the frequency spectra for the vibrations accompanying pulse deformations and for the vibrations occurring in the subsequent period of time. Therefore first the principal pulses in the waveform of the local maxima had to be identified. Instead of using their absolute value, the maxima were defined through their excess D above a certain environment. Also the stroke period was defined as period τ whose length is equal to the inverse of the first operational frequency (for SRs-1200: $\tau = 5.26$ s), in whose central point the analyzed pulse is located. Because of the relative character of the stroke response also a pre-stroke period of the same length had to be defined (Fig. 5). By adopting the length of period τ to be the inverse of the first operational frequency (for SRs-1200: $f_O = 0.19$ Hz) it became possible to capture the most significant changes in the continuously changing spectrum and at the same time to correctly read all the frequencies above f_O.

In order to identify the pulse strains dominant among the local maxima in the analyzed waveforms, the following conditions which a potential stroke strain must meet were specified:

(1) the strain has the highest value from among the maxima which have not been analyzed yet in the course of the procedure,
(2) there exists such a neighbourhood of the analyzed maximum that at instant t_O when the latter arises there exists a waveform in the interval $[t_O - 1.5\tau, t_O + 0.5\tau]$; in other words, there exists a vibration waveform in the stroke period and in the preceding (pre-stroke) period;
(3) no local maximum higher than the considered one occurs in the interval $[t_O - 1{,}5\tau, t_O + 0{,}5\tau]$.

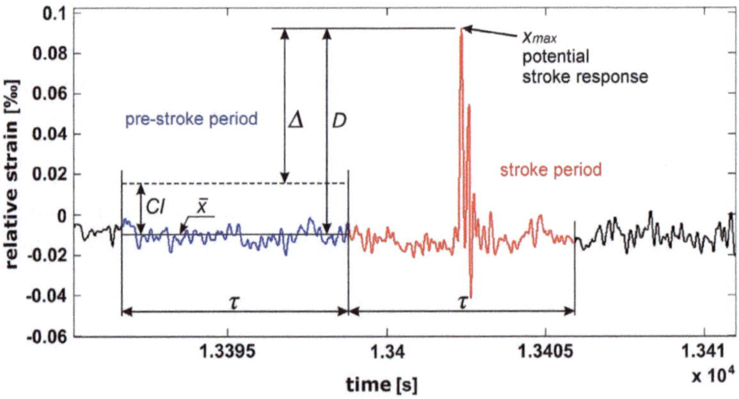

Fig. 5 Strain pulse as potential stroke response along with pre-stroke and stroke period

Local maxima satisfying the above conditions will be called *potential stroke strains* and will be denoted as x_{max}, and the accompanying vibrations will be referred to as *potentially significantly transient vibrations*. In the course of 427 h of observation 115 140 x_{max} values were found for the SRs-1200 excavator, which amounts to 0.7 % of all the local maxima.

Let us express *excess D* of local maximum x_{max} through a difference between the average ($\bar{x}$) from the local maxima in the pre-stroke period and the value of the considered extremum x_{max}:

$$D = x_{\max} - \bar{x} \tag{1}$$

The power spectral densities in both the stroke periods and the pre-stroke periods were reduced to a cumulative and normalized form in order to make comparison of their structures without absolute values possible. These are dimensionless quantities (functionals) and they were denoted as respectively: $G_u(f)$ and $G_p(f)$. One should bear in mind that for the SRs-1200 excavator, argument $f \in [0{,}19,\ 12]$ Hz.

The total signal power (H_u) contained in the stroke interval was adopted as the measure of the force of the impact of the vibrations accompanying pulse loads. Because of the adopted relative strain unit (per mille) the power is expressed in the latter's square (‰^{2}).

It should be noted that after it is presented in the logarithmic coordinate system (Fig. 6a) the dependence of power H_u on excess D becomes almost linear (Pearson's $r = 0.96$). Whereas the dependence of H_u on x_{max} does not assume an explicit form regardless of the adopted coordinate system (Fig. 6b and c). This calls for the use of excess D instead of the x_{max} value to identify the analyzed strain pulses.

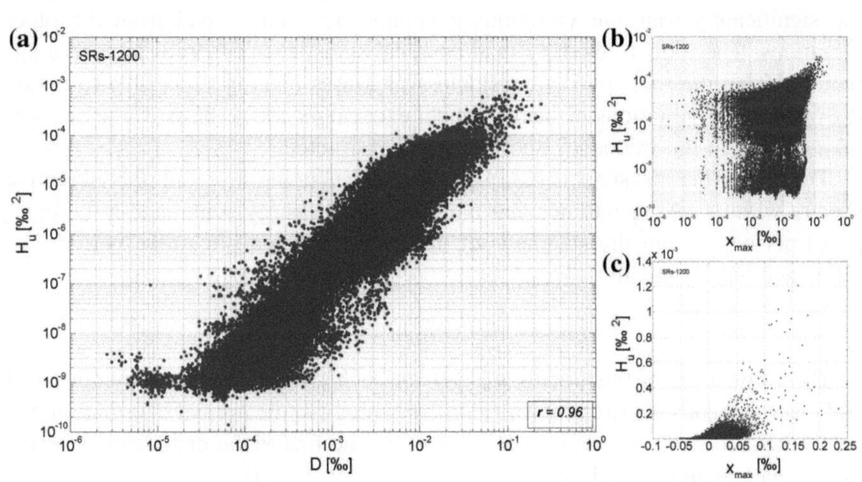

Fig. 6 Correlograms of excess D and value x_{max} relative to total power H_u in stroke period

The procedure is based on the assumption that the occurrence of a transient state in the vibration of the construction is correlated with local maxima x_{max}.

According to conditions i. ÷ iii, the values of local maxima x_{max} are the highest among the local maxima situated in the respective pre-stroke and stroke periods. Exploiting the relativeness between the periods one can say that if value x_{max} is within the confidence level for the average value, built on the local maxima in the pre-stroke period, then the character of the vibrations in the two periods is similar and the probability that significantly transient vibrations will occur is slight.

Thus for each value x_{max} one can calculate the excess:

$$\Delta = x_{\max} - ci \tag{2}$$

where *ci* is the upper limit of the confidence interval for the average of the maxima from the pre-stroke period for the considered x_{max}. Because of the considerable variation in the vibrations of the bucket-wheel excavator structure and their inherently transient character, when constructing limits *ci* the relatively high confidence level of 99 % was adopted.

Then the pre-stroke and stroke periods accompanying the x_{max} values for which the following condition is satisfied:

$$\Delta \leq 0 \tag{3}$$

can be referred to as *relatively peaceful operation* and they form a *standard set*. For the SRs-1200 excavator the set contains 864 observations (0.75 % of all the x_{max} in the set).

By characterizing in the frequency domain the dependence between the periods for which the above condition is satisfied, one could prove the hypothesis through, as it were, negation. However, because of the transient character of the operational vibrations of the bucket-wheel excavators it is impossible to sharply distinguish the significantly transient vibrations generated by a pulse load from the other vibrations. The tools offered by mathematical statistics are helpful in such a situation. Using the tools two dimensionless measures making it possible to divide the operational vibrations of bucket-wheel excavators on the basis of the changes taking place in the frequency spectra were constructed.

The first of the measures is based on the changes in the structure of consecutive spectra without taking into account the total signal energy contained in the compared periods. Using the standard set, for the stroke and pre-stroke periods the difference was calculated:

$$d_{\Delta \leq 0}(f) = G_{u,\Delta \leq 0}(f) - G_{p,\Delta \leq 0}(f) \tag{4}$$

The formed set of functionals $d_{\Delta \leq 0}(f)$ for each frequency *f* has a symmetrical and heterogeneous distribution [6]. Thus by analogy to the normal distribution, the triple standard deviation value was adopted as the first value demarcating steady vibrations and significantly transient vibrations for $d_{\Delta \leq 0}(f)$:

$$c_1(f) = 3 \cdot s_{d,\Delta\leq 0}(f) \tag{5}$$

where:

$c_1(f)$—a threshold of the permissible variation in the structure of the power spectral density over the length of the considered transform, $s_{d,\Delta\leq 0}(f)$—the standard deviation calculated from differences $d_{\Delta\leq 0}(f)$.

Similar, but magnitude differences were calculated for all the normalized cumulative power spectral density in the pre-stroke and stroke periods—$d(f)$ and were compared with threshold $c_1(f)$:

$$d(f) \leq c_1(f) \tag{6}$$

If for the given x_{max} the inequality [6] is satisfied, this means that the change in the shape of the spectrum, produced by this impulse, is wholly within the limits of the vibrations defined as *relatively steady*. The rate of change in the shape of the spectrum is within the acceptable limits for this state. Hence such a maximum can be removed from further analysis. Condition [6] for the data used is illustrated in Fig. 7. In this way for the SRs-1200 excavator, 95,082 observations (82.6 % of the initial set) from all the local maxima x_{max} were rejected as being relatively steady.

The changes taking place in the spectra can be expressed not only through their shape, but also through the transformations of the total power contained in them. Considering that pre-stroke periods and stroke periods, the increase of amplitudes translates into the rate of changes in the power of the signal. This rate can be expressed in a simple way through a ratio of the total signal power in the stroke period (H_u) to the total signal power in the pre-stroke period (H_p). In order to determine the permissible value of this ratio, again the standard set was used:

$$b_{\Delta\leq 0} = \frac{H_{u,\Delta\leq 0}}{H_{p,\Delta\leq 0}} \tag{7}$$

The value of the 9th decile (from the set of $b_{\Delta\leq 0}$ ratios) specifies the maximum permissible increase in signal power (c_2). For the SRs-1200 machine: $c_2 = 1.36$.

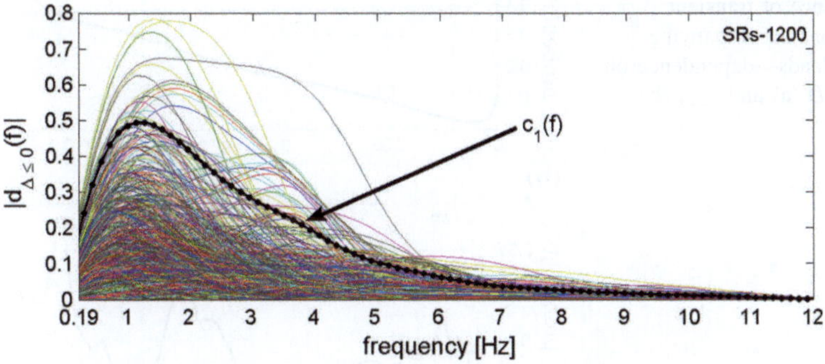

Fig. 7 Differences *d(f)* relative to threshold of permissible variation in structure of power spectral density $c_1(f)$ (*dotted line*)

A similar ratio was calculated for the periods remaining in the set of potentially significant stroke strains—b and compared to the permissible value of c_2:

$$b \leq c_2 \tag{8}$$

the stroke periods for which condition (8) is satisfied were classified as *transient vibrations with a relatively constant power* while the local maxima x_{max} accompanying them were classified as *stroke strains not accompanied by a significant increase in vibration amplitude*. The number of both stroke periods and stroke strains is 11,924 (10.4 % of the initial size of the set).

The other 8116 observations (7 % of the initial size of the set) can be called *significantly transient vibrations* and the local maxima x_{max} accompanying them can be referred to as *stroke strains*. The vibrations are accompanied by changes in the spectrum structure, and the significance of the changes is expressed through the threshold of the permissible variation in power spectral density ($c_1(f)$) and through the maximum permissible increase in signal power (c_2).

Figure 8a shows the proportions between the so defined *stroke strains* and the values of D from the initial set of potential transient strains. This graph is at the same time a graph of the conditional probability that the given excess D is accompanied by insignificantly transient vibrations. It should be noted that this probability increases linearly with excess D.

A similar graph (Fig. 8b) of the dependence on x_{max} as in Fig. 8a does not show a distinct correlation, which is also confirmed by PCC: $r = 0.74$ for x_{max} and $r = 0.98$ for D. Again this is evidence of the greater importance of the relativity of load values, in comparison with absolute load values, in the characterization of the vibrations of bucket-wheel excavator structures.

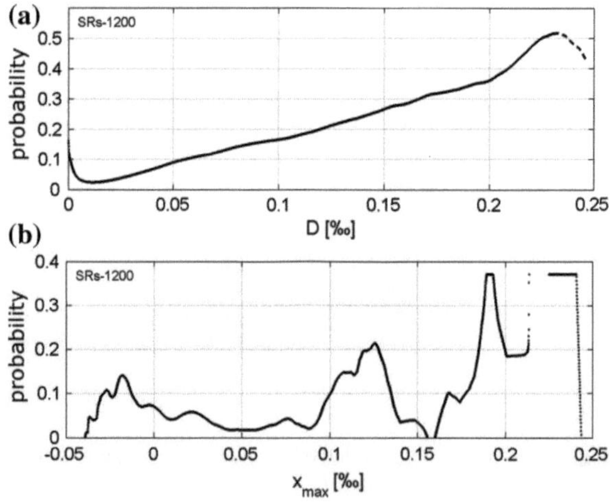

Fig. 8 Conditional probability distributions of occurrence of transient vibrations accompanying stroke loads—dependence on excess D (**a**) and x_{max} (**b**)

5 Conclusion

The triple standard deviation and the 9th decile, used to construct measures $c_1(f)$ and c_2 constitute the significance of transients in bucket-wheel excavator's vibrations. If other values are adopted, the hypothesis still will be true. Only a change in proportions between amounts of elements in the sets of *relatively steady vibrations*, *transient vibrations with a relatively constant power* and *significantly transient vibrations* will happen. Depending on the purpose of further study, the measures can be made more adjusted to the context and needs of the analysis.

If the presented algorithm is used in the diagnosis of bucket-wheel structures, the determination of the conditional probability distribution, for each excavator individually, and the identification of all the excavator's modal parameters are the prerequisites for evaluating the strain of the construction and estimating the unknown loading history.

References

1. Alenowicz J, Onichimiuk M, Wygoda M (2009) Extreme loads in design and operation of bucket-wheel excavators working in difficult to mine soils in open pit lignite mines (in Polish), Górnictwo i geoinżynieria, Year 33 (No 2), pp 31–41
2. Dudek D (1987) Statistical models of operational loads in mining with bucket-wheel excavators (in Polish). OWPWr, Wrocław
3. Dudek D (1994) Fundamentals of dynamics of surface mining machines. Signal acquisition, system analysis (in Polish), Wrocław University of Technology Publishing House, Wrocław
4. Hawrylak H, Sobolski R (1967) Principle surface mining machines (in Polish). Śląsk Publishing House, Katowice
5. Hayashi CH (1968) Nonlinear vibrations in physical systems (in Polish). WNT, Warsaw
6. Huss W (2012) Method of identifying nonstationary states of bucket-wheel excavator structure under random loads (in Polish), PhD thesis, Wrocław

Effectiveness of Advanced Vibration Processing Techniques for Fault Detection in Heavy Duty Wheels

Marco Malagó, Emiliano Mucchi and Giorgio Dalpiaz

Abstract This paper assesses the application of different processing techniques on acceleration signals extracted from faulty heavy duty wheels. Heavy duty wheels are used in applications as automatic vehicles and are mainly composed of a polyurethane tread glued to a cast iron hub. The adhesive application between tread and hub is the most critical assembly phase, since it is completely made by an operator and a contamination of the link area may happen. Furthermore the presence of rust on the hub surface can contribute to worsen the adherence interface, reducing the operating life. Several wheels with different types of faults have been manufactured 'ad hoc' with anomalies similar to the ones that can really be originated. Synchronous average is calculated over the wheel rotation in order to highlight the phenomena that have the wheel rotation as periodicity (e.g. the contact between defect and test bench drum). Successively, cyclostationary theory is applied to extract information from the frequency/order domain of the processed signals. Eventually, well-suited indicators/coefficients are applied to the processed signals, objectifying the anomaly presence and defining pass-fail reference values based on the non-statistical Tukey's method.

Keywords Condition Monitoring · Heavy Duty Wheel · Synchronous Average · Cyclostationary

M. Malagó (✉) · E. Mucchi · G. Dalpiaz
Universitá degli studi di Ferrara, Ferrara, Italy
e-mail: marco.malago@unife.it

E. Mucchi
e-mail: emiliano.mucchi@unife.it

G. Dalpiaz
e-mail: giorgio.dalpiaz@unife.it

G. Dalpiaz et al. (eds.), *Advances in Condition Monitoring of Machinery in Non-Stationary Operations*, Lecture Notes in Mechanical Engineering,
DOI: 10.1007/978-3-642-39348-8_48,

1 Introduction

This article, presented as a continuation and development of a previous work reported in [1], addresses a methodology and a procedure for the condition monitoring and diagnostics of heavy duty wheels based on vibration measurements. In particular, Synchronous Average (SA) and Cyclostationary analysis have been applied.

The Synchronous Average of the vibration signal allows to attenuate the periodic events not synchronous with the rotating component of interest and to reduce background noise [2–4]. The resulting signal average is the ensemble average of the angle domain signal, synchronously sampled with respect to the component rotation. The main advantage of SA is to extract from a complex vibration signal, its deterministic part, i.e. the events that are repeated periodically with the rotation of interest.

A further advanced tool adopted for investigating the relationships between spectral components is the cyclostationary theory [5, 6]. More specifically, a signal is cyclostationary if some of its statistics present periodicities. As previously described, averaging can make it possible to extract the deterministic part of the signal. If the signal obtained after subtracting this deterministic part from the synchronized signal does not exhibit cyclostationarity, the signal is said to be cyclostationary at an order 1. In general terms, a signal is cyclostationary at an order n if its statistical properties at order n are periodic.

The pass/fail thresholds are generally defined starting from the analysis of a large group of healthy components applying statistical techniques with the a priori knowledge of the real data distribution. However, in case of a small number of samples, it is advisable to use non statistical techniques. This is the case adopted in this research activity in which Tukey's non statistical method has been applied to a group of 15 healthy wheels. The Tukey's method [7] is a simple but effective procedure for the identification of anomalies in a distribution of data. Unlike common statistical procedures, Tukey's method is a non-parametric technique that does not consider any distributional assumptions about the statistical behavior of the data.

In Sect. 2, a preliminary description of the analyzed heavy duty wheels and the test set up used in the experimental analysis are given. Section 3 presents the application of different processing techniques: synchronous average is calculated over the wheel rotation in order to highlight the phenomena that have wheel rotation as periodicity. Moreover, Kurtosis and Root Mean Square parameters are utilized as statistical coefficients in order to define the state of health of a wheel and to obtain upper thresholds for the pass/fail decision. Successively, ciclostationary theory is applied to extract information from the frequency/order domain of the processed signal.

The originality of this activity consists of the attempt to study the presence of anomalies in heavy-duty wheels and to compare the detection capability of synchronous average and ciclostationary using well-suited indicators.

2 Test Set up

The wheels being studied are composed of a polyurethane tread and a cast iron hub, with 100 mm radius, 50 mm of width and 1,200 kg of maximum load. The adhesive application between tread and hub is the most critical assembly phase, since it is completely made by an operator and a contamination of the link area may happen. The operator has to be particularly careful to not touch the surface of the adhesive in order to prevent contamination with impurities that would lead a failure of the polyurethane-hub adherence. If this processing phase is not done properly some defects can arise as non-uniform adhesion of the adhesive on the metal surface or not complete wettability of the metal surface.

In this context, wheels with different types of faults have been manufactured ad hocwith anomalies similar to real ones. Such anomalies consist of:

- incorrect adherence zones between tread and hub (namely MA) of about 0.3 cm dimensions in the axial and tangential directions;
- localized and distributed rust presence in the hub surface (namely LR) of about 5 cm dimension in axial and 2 cm in tangential direction and distributed rust on the entire hub (namely DR).

Eventually, a set of 15 wheels without any defect has been analyzed for the estimation of a reliable reference pattern. The vibration parameters relative to these 15 wheels are referred as healthy wheels (HW).

The test bench consists of a bottom support, including a drum driven by an electric motor controlled by an inverter and an upper part composed by two hydraulic pistons that apply the load to the wheel under test, as depicted in Fig. 1. During tests, the vibration signal is acquired by means of a piezoelectric tri-axial

Fig. 1 Test bench

accelerometer (PCB 356A01, frequency range 1–10,000 Hz). The signals are acquired with a sample frequency of 12,800 Hz for a duration of about 23 s by using LMS hardware and software instrumentation [8]. Simultaneously with the acquisition at constant sample frequency, an on-line order tracking analysis has also been performed through the use of two optical tachometer sensors (KEYENCE-LVS series) both positioned near the wheel-drum contact.

3 Application of Advanced Signal Processing Techniques

3.1 Synchronous Average Analysis

The Synchronous Average technique has been applied on the measured signals in order to highlight the phenomena that are linked to the presence of faulty wheels. The Synchronous Average $m_x(\vartheta)$ of a measured signal $x(\vartheta)$, synchronized with the rotational element in the angle domain ϑ, is evaluated as the ensemble average over a number of rotations M, each corresponding to one angular period $\Theta = 360$ deg, as follows:

$$m_x(\vartheta) = \frac{1}{M}\sum_{l=0}^{M-1} x(\vartheta + l\Theta), \quad \text{with } 0 \le \vartheta < \Theta \tag{1}$$

where $M\Theta$ is the whole length of the signal. Synchronous Averaging (SA) is a signal processing technique which enables periodic waveforms to be extracted from noisy signals.

Two different Synchronous Averages have been computed. Firstly, the Synchronous Average of the acceleration signal over the drum revolution (called *SAd*) has been performed starting from the synchronized acceleration signal, by using the tachometer signal of the drum as a reference. Then, the purified signal has been calculated as the difference between the synchronized signal and the *SAd*, obtaining a new signal with reduced periodicities related to the driving drum and increased information concerning the manufactured faults linked to the wheel periodicities. Consequently, the purified signal has been synchronized (by using the tachometer of the wheel) and averaged over the wheel rotation obtaining the Synchronous Average of the acceleration signal over the wheel revolution (namely SAw). Eventually, the residual signal was determined as the difference between the above purified signal and *SAw*.

As an example of results of such a procedure, Fig. 2 depicts the raw acceleration signal measured at the operational condition of 4 km/h (speed) and 10,00 kg (load) in the case of a missing adherence localized defect and the further processing phases. The purification process gives a strong reduction of the components related to the driving drum rotation and the possibility of analyzing the residual signal depurated from all the known periodicities of the system.

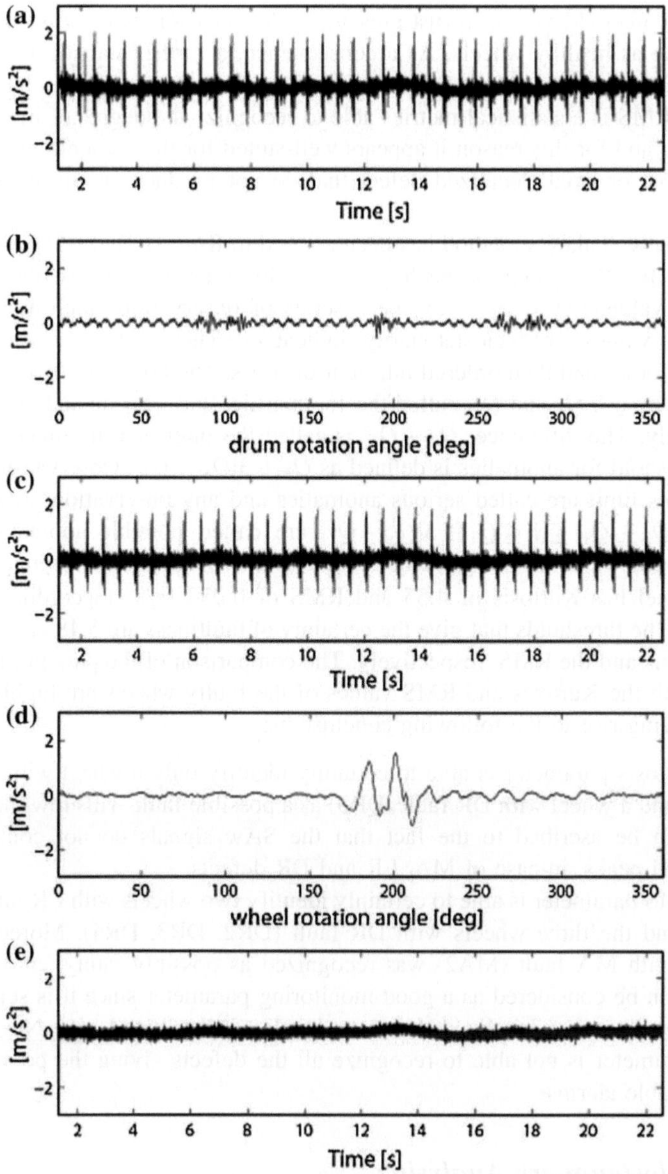

Fig. 2 Processing procedure: (**a**) raw acceleration signal, (**b**) SAd, (**c**) purified signal, (**d**) SAw and (**e**) residual signal at the operational condition of 4 km/h and 1,000 kg

The Kurtosis parameter can be considered as a monitoring feature for faults producing impulsive excitations and it can be used to obtain a reliable upper threshold [1]. Although this parameter is well suited for the recognition of extended missing adherence localized faults, it is not really sensitive to small localized defect (MA) and to rust presence (LR and DR), as it shows almost the same value as healthy wheels. As a consequence, a further statistical parameter, the root mean square (RMS) has been considered and compared to the previous one. The RMS is a statistical metrics able to recognize the degree of irregularities of a signal and for this reason it appears well-suited for the recognition of defects such as rust or small localized defect that do not produce significant localized peaks.

Eventually, Tukey's method has been applied to Kurtosis and RMS parameters evaluated for the group of healthy wheels in order to identify the pass/fail threshold value. Let $x_1, x_2, \ldots, x_n$ be a series of observations such as statistical parameter values or cyclostatonarity indicators. These data are arranged in ascending order and then ordered into four quarters. The boundary of each quarter is defined by Q_1, Q_2 and Q_3, called the 1st quartile, 2nd quartile and 3rd quartile, respectively. The difference $|Q_3 - Q_1|$ is called the inter-quartile range. The Tukey's threshold for anomalies is defined as $Q_3 + 3|Q_3 - Q_1|$. Observations falling beyond this limit are called serious anomalies and any observations x_i such that $Q_3 + 1.5|Q_3 - Q_1| \leq x_i \leq Q_3 + 3|Q_3 - Q_1|$ are called possible anomalies. With this approach, the threshold value that discriminates healthy wheel from a possible faulty wheel has Kurtosis of 4.55 and RMS of 0.033 m/s^2 (accordingly to the method). The thresholds that give the certainty of faultiness are 5.48 and 0.039 for the Kurtosis and the RMS, respectively. The comparison of the pass/fail threshold values with the Kurtosis and RMS values of the faulty wheels are highlighted in Fig. 3, giving rise to the following conclusions:

- the Kurtosis parameter is able to certainly identify only a wheel with LR fault (LR1) and a wheel with DR fault (DR3) as a possible fault. This low monitoring skill can be ascribed to the fact that the SAw signals do not contain high localized peaks, in case of MA, LR and DR defects.
- The RMS parameter is able to certainly identify two wheels with LR fault (LR1, LR2) and the three wheels with DR fault (DR2, DR3, DR4). Moreover, one wheel with MA fault (MA2) was recognized as possibly faulty. As result the RMS can be considered as a good monitoring parameter since it is sensitive to missing adherence localized defect or localized-distributed rust. Nevertheless this parameter is not able to recognize all the defects giving the possibility of undesirable alarms.

3.2 Cyclostationary Analysis

A further investigation has been carried out based on the study of cyclostationarity properties of the signal.

The first order cyclostationarity content has been computed as the Discrete Fourier Transform (DFT) of the sampled purified signal synchronized with the wheel rotation $x[n]$, called first order cyclic cumulant ($\hat{C}^{\alpha}_{1x}$) [9–11]:

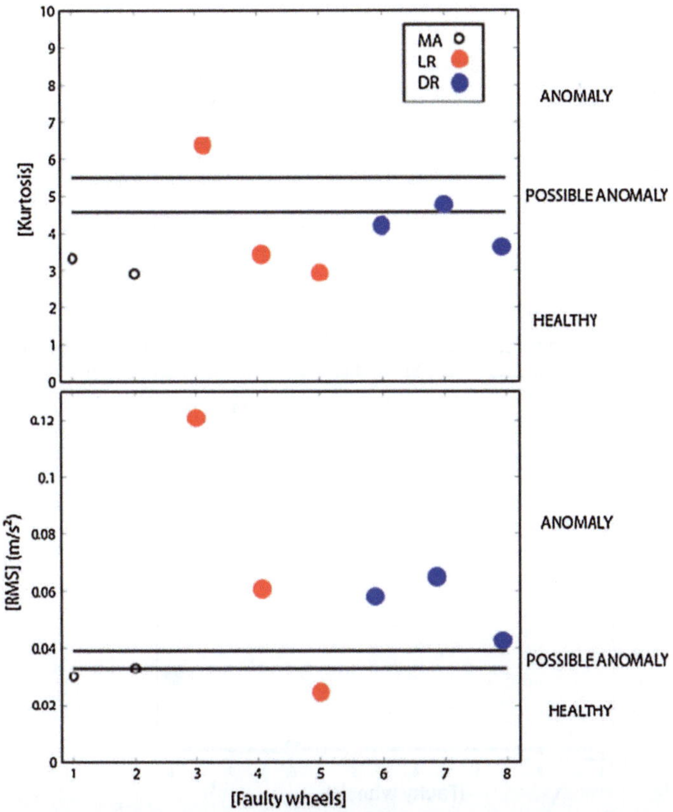

Fig. 3 Kurtosis and RMS pass-fail distributions in case of MA, LR and DR defects

$$\hat{C}_{1x}^{\alpha} = N^{-1}DFT\{x[n]\}(\alpha) \tag{2}$$

The second order cyclostationarity content of the signal has been evaluated through the second order cyclic cumulant ($\hat{C}_{2x}^{\alpha}$) estimated as the Discrete Fourier Transform of the squared residual signal $r^2[n]$:

$$\hat{C}_{2x}^{\alpha}(\varphi = 0) = N^{-1}DFT\{r^2[n]\}(\alpha) \tag{3}$$

It has to be noted that Eqs. 2 and 3 are consistent estimators of the cyclic cumulants at the zero angle lag ($\varphi = 0$) for a sampled signal. Moreover, the first and second order cyclic cumulants can be conveniently used to summarize the information related to first- and second-order cyclostationary contents by defining the following indicators of cyclostationarity [6]:

$$ICS_{1x} = \frac{\sum\limits_{\alpha \neq 0} \left|\hat{C}_{1x}^{\alpha}\right|^2}{\left|\hat{C}_{2x}^{0}(0)\right|} \tag{4}$$

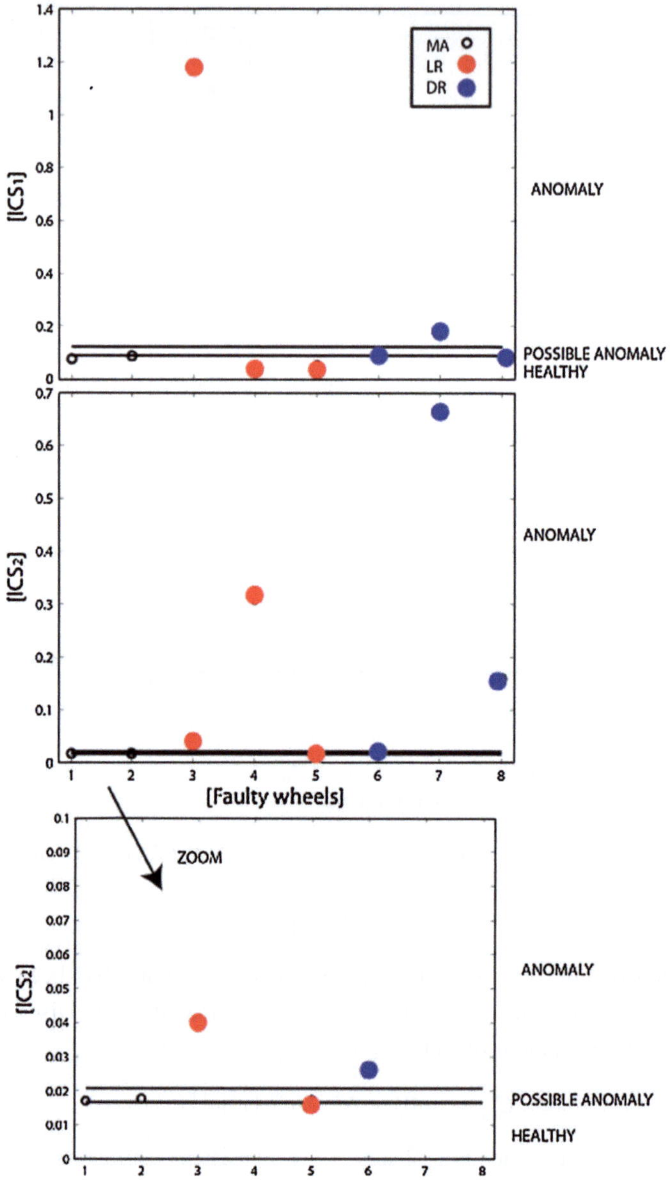

Fig. 4 $\hat{C}^{\alpha}_{1x}$ and $\hat{C}^{\alpha}_{2x}$ pass-fail distributions in case of MA, LR and DR defects

$$ICS_{2x} = \frac{\sum_{\alpha \neq 0} \left| \hat{C}^{\alpha}_{2x}(0) \right|^2}{\left| \hat{C}^{0}_{2x}(0) \right|^2} \tag{5}$$

where $\alpha \in \bar{A}$ and $\bar{A}$ is the set of wheel orders α presenting non-zero Fourier series coefficients. It is worth noting that indicators ICS_{1x} and ICS_{2x} are dimensionless, as they are normalized by the energy of residual signal $\hat{C}^0_{2x}(0)$; they quantify the presence of first and second-order cyclostationary components within the signal, respectively. In the particular case, α has been considered in the range 1st–100th order since the higher spectrum amplitudes lie in this order range.

As done for the statistical parameters, the Tukey's method has been applied to the *ICS*1*x* and *ICS*2*x* metrics evaluated for the group of 15 healthy wheels in order to identify the threshold pass/fail values. Possible faults can occur for values between 0.0893 and 0.1226 in the case of *ICS*1*x* and for values between 0.0167 and 0.208 for *ICS*2*x* metrics. Moreover assured faults can occur for values exceeding to 0.1226 and 0.0208 for *ICS*1*x* and *ICS*2*x*, respectively. The comparison between the threshold values and the results of the faulty wheels, highlighted in Fig. 4, gives rise to the following conclusions:

- the *ICS*1*x* parameter is able to detect only two certain faults (*LR*1 and *DR*2). The main reason of this apparent discrepancy between the monitoring capability of the first cyclic order cumulant and its indicator is due to the large value of dispersion characterizing the healthy wheel, which causes a large increase of the threshold level.
- the *ICS*2*x* is able to recognize the presence of anomalies or of possible anomalies for all the known faulty wheels, in the case of both missing adherence and rust defect. In case of healthy components, the *ICS*2*x* values do not present this dispersion effect so, as a result, the *ICS*2*x* monitoring skill is particularly sensitive to fault recognition.

4 Concluding Remarks

This research addresses a methodology and a procedure for the condition monitoring and diagnostics of heavy-duty wheels based on vibration measurements, passing through a simplified explanation of the physical phenomena that cause the faulty signal signature. Although this method and relative results are referred in this paper to wheels, they can be applied to a large variety of mechanical systems and give useful guidelines for similar applications.

A number of different processing techniques are developed and applied in order to recognize faults in heavy-duty wheels. Defects of different dimensions reproducing missing adherence between the polyurethane tread and the hub are artificially created. In these defects, the adhesive was not correctly applied due to the local absence of adhesive or due to the presence of rust in the hub, as can happen during the manufacturing process. These defects cause incorrect wheel rotations and fast failure. The synchronous average (SAw) has been evaluated and statistical parameters for fault detection (Kurtosis and RMS) have been used. Furthermore, the cyclostationary nature of the signal has been investigated through the first and second order cumulants and relative indicators. A non statistical approach (Tukey's method) has been used in order to calculate threshold values for the sound/fault discrimination (Figs. 3 and 4). This activity has enabled the conclusion that the ICS_{2x} of the residual signal represents a very useful indicators of tread/hub connection anomalies both for missing glue and rust (Fig. 4). Thus, ICS_{2x} can be considered as the key parameter to be adopted in a monitoring test station at the end of the production line.

Considering the presented research activity, the main original contributions concern the application of advanced vibration processing techniques to monitoring and diagnostics of heavy-duty wheels and the assessment of their effectiveness.

References

1. Malago' M, Mucchi E, Dalpiaz G (2009) Condition monitoring and diagnostics in heavy-duty wheels: a first experimental approach. In: Proceedings of the ASME, 2009 international design engineering technical conferences and computers and information in engineering conference, IDETC/CIE 2009, 30 Aug–2 Sep 2009, San Diego, California
2. Combet F, Gelman L (2007) An automated methodology for performing time synchronous averaging of a gearbox without speed sensor. Mech Syst Signal Process 21:2590–2606
3. McFadden PD, Toozhy MM (2000) Application of synchronous averaging to vibration monitoring of rolling element bearings. Mech Syst Signal Process 14(6):891–906
4. Delvecchio S, D'Elia G, Mucchi E, Dalpiaz G (2010) Advanced signal processing tools for the vibratory surveillance of assembly faults in diesel engine cold tests. J Vibr Acoust 132(2):021008–10, ISSN: 1048-9002, doi:10.1115/1.4000807
5. Antoni J (2009) Cyclostationary by examples. Mech Syst Signal Process 23:987–1036
6. Raad A, Antoni J, Sidahmed M (2008) Indicators of cyclostationarity: theory and application to gear fault monitoring. Mech Syst Signal Process 22:574–587
7. Tukey J (1977) Exploratory data analysis. Addison Wesley, Boston
8. LMS International (2007) LMS Test Lab, Rev 7B
9. Gardner WA (1994) The cumulant theory of cyclostationary time-series, Part I: Foundation. IEEE Trans Signal Process 42(12):3387–3409
10. Gardner WA, The cumulant theory of cyclostationary time-series, Part II: development and Applications. IEEE Trans Signal Process 42(12):3409–3429
11. Capdessus C, Sidahmed M, Lacoume JL (2000) Cyclostationary processes: application in gear faults early diagnosis. Mech Syst Signal Process 14(3):371–385
12. Mucchi E, Dalpiaz G, Fernàndez del Rincòn A (2010) Elasto-dynamic analysis of a gear pump Part I: pressure distribution and gear eccentricity. Mech Syst Signal Process 24:2160–2179. doi:10.1016/j.ymssp.2010.02.003
13. Mucchi E, Dalpiaz G, Rivola A (2010) Elasto-dynamic analysis of a gear pump Part II: meshing phenomena and simulation results. Mech Syst Signal Process 24:2180–2197. doi:10.1016/j.ymssp.2010.02.004
14. Mucchi E, Dalpiaz G, Rivola A (2010) Dynamic behaviour of gear pumps: effect of variations in operational and design parameters. Meccanica 46(6):1191–1212
15. Mucchi E, D'Elia G, Dalpiaz G (2012) Simulation of the running in process in external gear pumps and experimental verification. Meccanica 47(3):621–637
16. Mucchi E, Di Gregorio R, Dalpiaz G (2013) Elastodynamic analysis of vibratory bowl feeders: modeling and experimental validation. Mech Machine Theory 60:60–72

Chatter Marks and Vibration Analysis in a S6-High Cold Rolling Mill

Maria Cristina Valigi, Sergio Cervo and Alessandro Petrucci

Abstract S6-high rolling mill is an advanced mode to work the steel: it allows the use of very small work rolls laterally guided by individually adjustable side support rolls, which are supported by two rows of roller bearings mounted in cassettes. In this paper the vibrations generated in a S6-high cold rolling mill are analyzed with the aim to investigate the problem of skid marks generation. Such marks are regular, parallel marking across the width of strip metal that not only significantly affects the mill performance, but also reduces surface quality of the strip steel. The defects of the strip are the consequence of insurgence of vibrations, generically denominated 'chatter'. The analyzed rolling mill has six rolls able to roll steel strip coming directly from hot rolling mill train. The purpose of the present work is to identify the reason of the excitation in order to limit the problem. A solution based on empirical observations, vibration analysis and considerations of a model is proposed.

Keywords Chatter vibration · Cluster rolling mill · Vibration analysis · Cold rolling mill · Lumped parameter model

S. Cervo · A. Petrucci
Acciai Speciali Terni-Viale Benedetto Brin, 171-05100 Terni, Italy
e-mail: Sergio.cervo@thyssenkrupp.com

A. Petrucci
e-mail: alessandrozzu@gmail.com

M. C. Valigi (✉)
Dipartimento di Ingegneria Industriale, Università degli Studi di Perugia-Via G. Duranti, 93-06125 Perugia, Italy
e-mail: mc.valigi@unipg.it

G. Dalpiaz et al. (eds.), *Advances in Condition Monitoring of Machinery in Non-Stationary Operations*, Lecture Notes in Mechanical Engineering,
DOI: 10.1007/978-3-642-39348-8_49,

1 Introduction

In this work the authors will be discussed some problems relative to the vibrations generically denominated ‘chatter’, occurring in a S6-high cold rolling mills. Chatter can result in not good surface finish for some applications and, rare cases, in gauge variations in the rolled strip. It manifestation is the classical regular regular, parallel marking across the width of strip metal called “chatter marks” [1–5]. Chatter in rolling is considered to be the result of interactions between the mill structure and the rolling process. The dynamic forces which are generated in the rolling process deflect the structure of the mill, leading to variations in the roll gap, rolls speed, tension,etc. These, in turn, result in further variations in the rolling forces [6–8]. Chatter is a particular case of self-excited vibrations. Self-excited systems begin to vibrate of their own accord spontaneously, the amplitude increasing until some non-linear effect limits any further increase. The alternating force that sustains the motion is created by the motion itself and stops when the motion stops. Three basic types of rolling chatter have been observed in rolling mills which causes significant chatter bands across the strip and small thickness fluctuations [9]. Torsional chatter, occurs in the 5–25 Hz range. Third-octave-mode chatter, which produces large thickness variations and strip rupture, lies in the 125–240 Hz range. Fifth-octave-mode chatter occurs in the 500–800 Hz range. The third-octave-mode chatter is considered the most critical because it generates large gauge variations in the rolled materials. It therefore has the most detrimental effects in terms of loss productivity due to the lower rolling speeds required to avoid the phenomenon. To understand the conditions which lead to the dynamic instability of the rolling process, the interaction between the structural dynamics of the mill and the dynamics of the rolling process must be investigated. This investigation is often carried out by modelling the rolling mill and the rolling process and their interaction [9–12]. Lumped parameter models have been widely used to represent the mill dynamic [11]. In this paper a new rolling mill, S6-high rolling mill, is studied; it is able to realize in one pass, high reduction of strip (until 1:12). The analyzed plant presents the problem of chatter marks on the strips that cause a short life of grinded side support rolls. The objective of this paper is to investigate the reason of chatter-marks using a vibration analysis, in order to understand which of parameters is involved in the self-exiting behaviour and how a rolling mill can be adjusted to ensure maximum productivity and highest quality.

2 S6-High Rolling Mill

A new two-stands S6-High rolling mill, designed for highest quality standards, has been installed at the entry section of the Hot Annealing&Pickling line LAC10 of Acciai Speciali Terni S.p.A., in order to reduce the thickness of the incoming hot rolled strips to produce new products and to serve new range dimensions for the following processes.With the integrated tandem S6-High rolling mill positioned at the beginning of the Annealing&Pickling stainless steel line, the hot strip transported to the line is cold rolled, heat treated, shot blasted and chemically pickled in one step. With this process it is possible obtain a finished product (ZI finishing) or a material able to be re-workable on a Sendzimir cold rolling mill for new thickness reduction and new surface quality (2D-2B finishing). The main advantage for the ZI finishing is the availability of strips with geometrical tolerances comparable to the standard cold rolled strips but with lower production costs due to a shorter cycle, while an increasing of Sendzimir productivity is the benefit for the thickness reduction on strips to be re-rolled for 2B–2D surface finishing.

The S6-high rolling mill (Fig. 1) has six rolls with different diameters arranged horizontally one above the other, symmetrically to the neutral rolling plane. In addition to these six rolls (two work rolls WR, two intermediate rolls IMR and two back-up rolls BUR) there are four cassettes fixed with an additional cylinder called "Side Support Roll" (SSR). The aim of the SSR is to support the horizontal load created on the work roll during the process. The vertical rolling force is transmitted through an hydraulic system that acts on back-up rolls which then transmit the force up to the work rolls.

The motion is transmitted from a motor to the intermediate rolls through the gearboxes linked by mean of spindles. While the others ones are moved by rolling motion, the IMR are the only motorized and transmit the motion to the stand.

They are tapered on one extremity in order to reduce the high thickness reduction on the strip edges, and they can shift perpendicularly and horizontally with respect to the rolling direction to assume proper positions according to the thickness and the width to roll. The real rolling process is realized by WRs that are

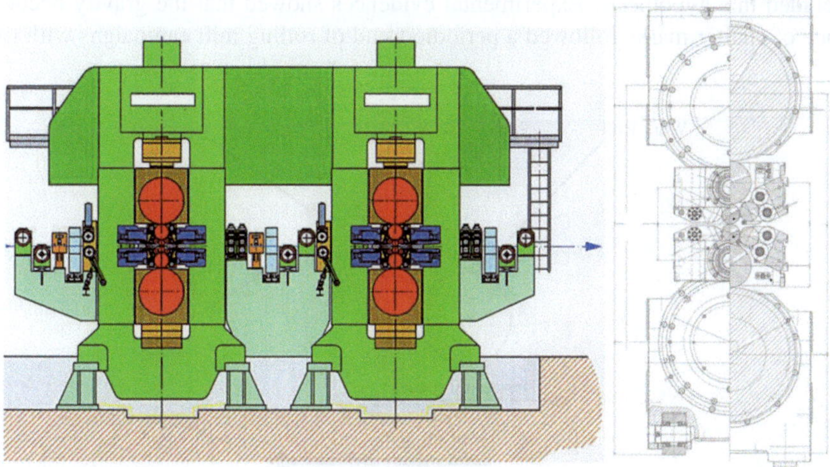

Fig. 1 High rolling mill: transversal section (on the *left*) and stand section (on the *right*)

idle and maintained in vertical axis with the BUR and IMR by the adjustable SSR of the cassettes and held sideways by two rows of roller bearings (Fig. 2).

The cassettes have the function to get a rolls packing condition helpful to provide a sufficient compression inside the rolling stand. The two rows of roll bearings have the axis parallel to the side support roll and they have the function to reduce the strokes originated by work rolls during the process. During the rolling process the WR is pushed on SSR because of the horizontal force and transmitted to the roll bearings with the aim to restrain the force. Little fluctuation on the process parameters take to little fluctuations on the forces values so that WR is pushed towards SSR with a vibration mode. When the WR pushes SSR towards its roll bearings with little force, SSR must react to follow the WR contact in order to damp the stroke effects of the following load increasing. This is the important role played by the springs located on the extremities of the SSRs necks into the cassette. The SSRs rolls have a limited grinded life of about some hundreds of rolled strip because they have to resist to the horizontal loads and because they are made with a softer material than the WRs.

3 Chatter Marks and Vibration Analysis

Transversal marks have been note on the rolled strip surface since the start-up of the plant: a series of chatter marks, perpendicular to the rolling direction, compromising the aesthetically quality of the product. The frequency of the defect was about two marks per centimeter that for a rolling speed of 40 m/min gives a characteristic frequency of about 130 Hz. Usually the main cause of chatter marks is the inaccurate grinding of involved rolls in the process, but a promptly analysis of the grinding process parameters, measurements and inspection reports checks excluded this hypothesis. Experimental evidences showed that the gravity occurrence of chatter marks followed a periodic trend of rolling mill campaigns with to

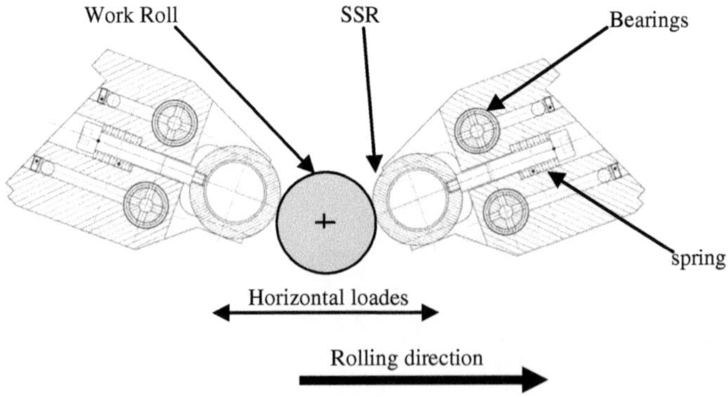

Fig. 2 Side support cassettes

the replacement of SSR cassettes. In particular an early deterioration of the SSRs was noted (chatter marks were on their surfaces, the so called facets) so that after just 150 km of rolled strip the cassettes had to be replaced. Disappearing the problem with the replacement of the cassettes, the immediate relationship between the age of the SSR rolls and the skid marks on the strip was deduced and in order to investigate the influences of individual parts of the plant, a vibration monitoring of the motor gear-box systems and the rolling mill stands have been evaluated for the rolling speeds 40 m/min. Since the vibration behavior of the gear-box systems presents a prevalent and admissible component of 108 Hz attributable to tooth mesh frequency the gear-box system was left out the possible origin of self-excited vibration.

The vibration measurements made directly on the rolling stand before the change of the cassettes, showed a critical value of 124.5 Hz at 40 m/min rolling speed (Fig. 3). This result show the origin of the chatter inside the stand and specifically in the cassettes.

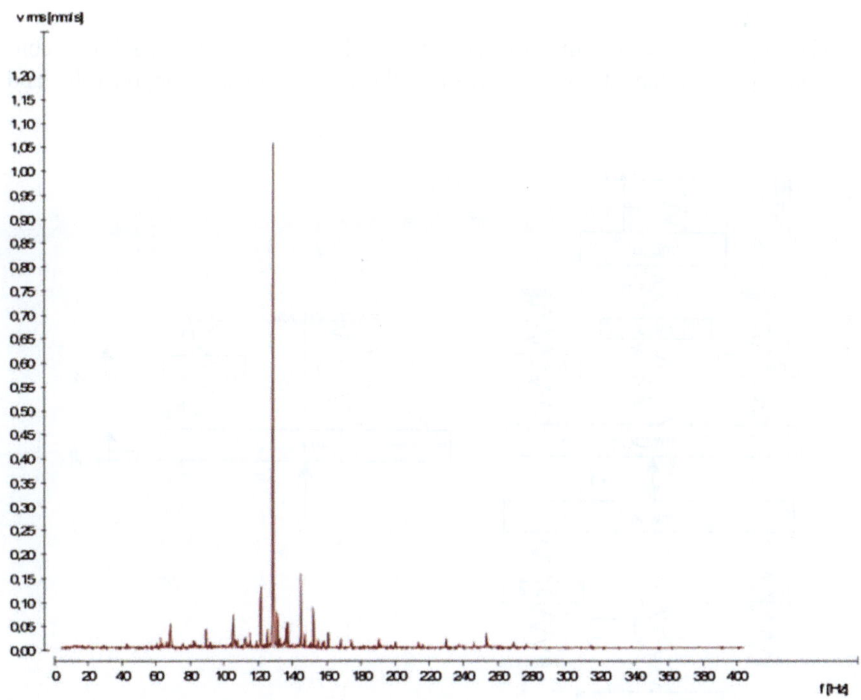

Fig. 3 Vibration in the stand at 40 m/min before the sides support rolls grinding

4 Model and Proposed Solution

In order to investigate the influence of individual parameters on the dynamic instability, a lumped parameter model of the rolling mill have been proposed, as the conventional, linear-mass–damping-spring system [13]. Firstly a model having ten degree of freedoms (Fig. 4a) where the masses are reduced to the ten rolls involved in the process can be considered but, to simplify the problem, the stand was assumed symmetrical in relation to the rolled strip and symmetrical to the vertical axis of the stand so that the ten degrees of freedoms model was reduced into the simplest system with two degree of freedoms (Fig. 4b). Equations for the two degree of freedoms model are:

$$\begin{bmatrix} m_1 & 0 \\ 0 & m_2 \end{bmatrix} \begin{Bmatrix} \ddot{y}_1 \\ \ddot{y}_2 \end{Bmatrix} + \begin{bmatrix} c_1 & 0 \\ 0 & 0 \end{bmatrix} \begin{Bmatrix} \dot{y}_1 \\ \dot{y}_2 \end{Bmatrix} + \begin{bmatrix} k_1 + k_2 & -k_2 \\ -k_2 & k_2 + k_3 \end{bmatrix} \begin{Bmatrix} y_1 \\ y_2 \end{Bmatrix} = \begin{Bmatrix} F \\ 0 \end{Bmatrix} \tag{1}$$

This model takes into account the equivalent mass m_1 of the working rolls, the intermediate rolls and the back-up rolls and the equivalent mass m_2 of the side support rolls [13].

Then k_1 e c_1 represent the stiffness and the damping of the mass frame connection, k_2 the contact stiffness between working rolls and side support rolls, and

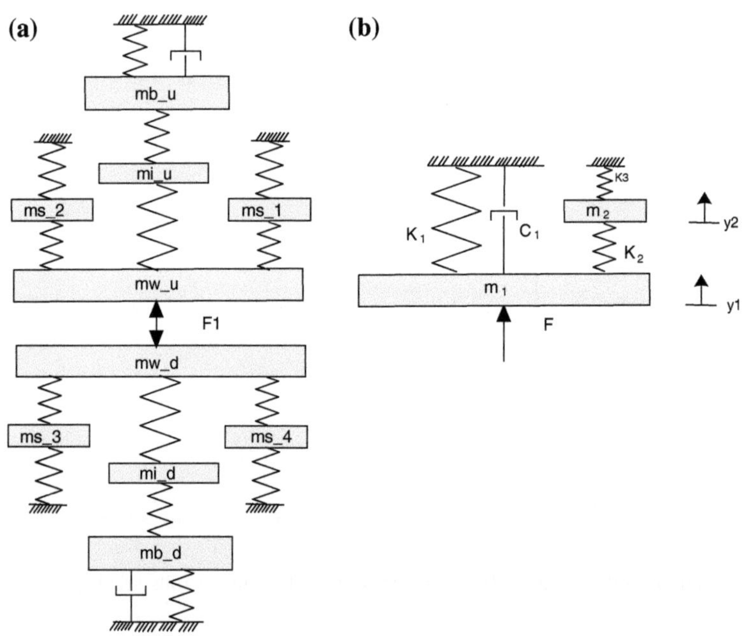

Fig. 4 Lumped parameter model with 10 DOF (**a**) reduced to a model with 2 DOF (**b**)

k_3 is the stiffness of spring in the cassettes. The vertical component F of the rolling force acting between strip and working rolls can be evaluated by a wide used model: the slab theory (Fig. 5).

$$dp = 2\frac{k}{y}dy \pm \frac{\mu p}{y} + 2dk \tag{2}$$

It supposes infinitesimal segments in deformation delimited by two surfaces that remain flat during the process where p is the rolling pressure,μ is the coefficient of friction and k is the mean yield stress in plane strain [1, 12]. The model determines the rolling force that, assuming that the roll radius is constant, may be implicitly written as a function of several variables:

$$F = F(y_{e,}y_d, \sigma_{xe}, \sigma_{xd}, \mu, \lambda) \tag{3}$$

where y_e, y_d, are the half thickness of rolled strip at the entry and at the exit of the stand, σ_{xe} σ_{xd}, are the horizontal tensile stress at entry and at exit of the stand and λ is the resistance to deformation dependent on strain hardening characteristics [14].

The disturbances and the variations of the strip thickness due to roll vibration generate the dynamic component of the rolling force. This dynamic component deflects the structure of the mill leading to variations in the roll gap, y_1, which in turn result in further variations in the rolling force. Under certain conditions, however, this interaction between the structure and process leads to dynamic instability. By applying the Laplace transform to the above Eq. (1), the following relationships are obtained:

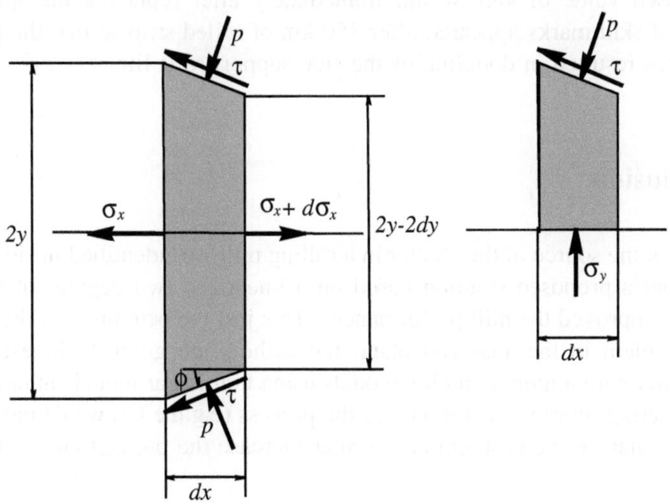

Fig. 5 Slab method principle

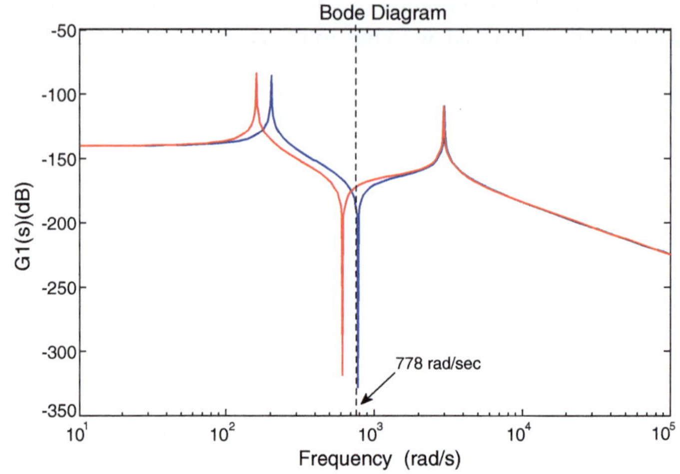

Fig. 6 Frequency response G1(s):the *blue line* with the proposed value of k_3 the *red line* with the current value of k_3

$$\begin{cases} \frac{L[y_1]}{L[F]} = G1(s) = \frac{(m_2 s^2+k_2+k_3)}{(m_1 s^2+c_1 s+k_1+k_2)(m_2 s^2+k_2+k_3)-k_2} \\ \frac{L[y_2]}{L[F]} = G2(s) = \frac{k_2}{(m_2 s^2+k_2+k_3)} \frac{(m_2 s^2+k_2+k_3)}{(m_1 s^2+c_1 s+k_1+k_2)(m_2 s^2+k_2+k_3)-k_2} \end{cases} \tag{4}$$

As partial solution of the problem, a change of spring stiffness in the cassettes was realized in order to have an anti-resonance at 124 Hz.

The Fig. 6 shows the diagrams of G1(s) with two different values of k_3: the red line regards the system with the current value of 260 N/mm and the blue one with the proposed value of 485 N/mm. Immediately after replacing the springs the presence of skid marks appeared after 350 km of rolled strip so that the proposed solution has resulted in doubling of the side support rolls life.

5 Conclusion

In this work the source of the chatter in a rolling mill was identified in side support rolls so that a proposed solution based on a linearized two degrees of freedoms model has improved the mill performances. This just the beginning of the study of chatter problem in the analyzed plant; the authors are going to investigate the phenomenon with a more complex models using non-linear models in closed loop and considering more refined model of the process (e.g.the Orowan's model) with the aim to analyze the problem and further increase the productivity of the mill.

References

1. Yun IS, Wilson WRD, Ehmann KF (1998) Review of chatter studies in cold rolling Int J Mach Tools Manuf 38:1499–1530
2. Tlusty J, Chandra G, Critchley S, Paton D (1982) Chatter in cold rolling. Ann CIRP 31:195
3. Tamiya T, Furui K, Hida H (1980) Analysis of chattering phenomenon in cold rolling. In: Proceedings of international conference on steel rolling, vol 2, pp 1191–1207
4. Keller NL (1992) Spindle involvement in rolling mill vibration. Iron steel Eng 79–84
5. Nessler GL, Cory JF (1993) Identification of chatter sources in cold rolling mills. Iron steel Eng 70:40–45
6. Evans PR, Hill DE, Vaughan ND (1996) Dynamic characteristics of a Rrolling mill. Proc Instn Mech Engrs 210:259–271
7. Rinchi M, Rindi A (1999) Fenomeni vibratori autoeccitati nei laminatoi finitori. In: XIV Congresso Nazionale AIMETA, Como, 6–9 ottobre 1999
8. Malvezzi M, Rinchi M (2000) Control systems of strip tension in hot steel rolling mills trains. In: Proceedings of 19th IASTED international conference—MIC 2000—modeling, identification and control, Innsbruck, Austria, 14–17 Feb 2000
9. Pei-Hua Hu1 Huyue Zhao1K F Ehmann (2006) Third-octave-mode chatter in rolling. Part 1: Chatter model. Proc Inst Mech Eng Part B J Eng Manuf 220:1267–1277
10. Chen Y, Liu S, Shi T, Yang S, Liao G (2002) Stability analysis of the rolling process and regenerative chatter on 2030 tandem mills. Imeche Part C J Mech Eng Sci 216:1225–1235
11. Bland DR, Ford H (1948)The calculation of roll force and torque in cold strip rolling with tensions. Proc Mech Eng 159:144–163
12. Misonoh K (1980) Analysis of chattering in cold rolling of steel strip. J JSTP 21(238):1006–1010
13. Swiatonoswki A, Bar A (2003) Parametrical excitement vibration in tandem mills-mathematical model and its analysis, J Mater Process Technol 134:214–224
14. Malvezzi M, Valigi MC (2008) Cold rolling mill process: a numerical procedure for industrial applications. Meccanica 43(1):1–9

Advanced Testing of Heavy Duty Gearboxes in Non-Stationary Operational Conditions

Paweł Kępski, Bartłomiej Greń and Tomasz Barszcz

Abstract Paper presents the approach of testing of heavy duty gearboxes dedicated to mining machinery. Main focus is on non-stationary operational conditions and advanced monitoring of vibration, process parameters and other signals connected with gearbox operation such as temperatures, pressure and flow in cooling system. Famur's Group advanced test rig is presented with description of its technical parameters and advanced control and monitoring system. The system features unique capabilities in scope of system dynamics enabling to simulate overloads of impacts often present in real life mining machine load characteristics. The case study shows exemplary test performed on heavy duty gearbox of 250 kW power both with recorded parameters and analysis results. Observed dependencies between vibration and time-varying process parameters were presented and discussed. Currently Famur Group develops and implements several signal analysis algorithms, taking into account recent research in this field. Further development of the test rig will allow to verify suitability of these methods in industrial systems for monitoring machines working in non-stationary operational conditions. This verification is possible thanks to ability of simulation of conditions comparable with real-life machine operation.

Keywords High power test rig · Gearbox testing · Mining machines · Vibration analysis · Non stationary operational conditions

P. Kępski (✉) · B. Greń (✉)
FAMUR Institute, Sp. z o.o. ul. Armii Krajowej 51 40-698 Katowice, Poland
e-mail: pkepski@famur.com.pl

B. Greń
e-mail: bgren@famur.com.pl

T. Barszcz
Department of Robotics and Mechatronics, AGH University of Science and Technology, Al. Mickiewicza 30 30-059 Cracow, Poland
e-mail: tbarszcz@agh.edu.pl

G. Dalpiaz et al. (eds.), *Advances in Condition Monitoring of Machinery in Non-Stationary Operations*, Lecture Notes in Mechanical Engineering,
DOI: 10.1007/978-3-642-39348-8_50,

1 Introduction

Gearboxes are part of nearly every drive in industry and its performance is critical especially in mining applications. Heavy duty drives used in conveyors and shearer loaders are subject to varying operational conditions (Kępski and Barszcz 2012), [1–3]. It is important to test gearboxes in near to real life conditions before deploying to ensure their reliable performance and provide data to prognostics and health management (PHM) systems [4]. Above mentioned facts lead to a conclusion that such testing should be a part of gearbox quality control in the production stage.

Famur Group as a leading mining machines producer developed a high power test rig for testing heavy duty gearboxes with complex monitoring of its parameters like temperature and vibration. Advanced control system enables to simulate load and speed profiles simulating the real life conditions of tested machines.

2 Test Rig Features

Famur's Group test rig consists of two Siemens 1,8 MW motors having 11,569 Nm of nominal torque. In order to increase the accuracy of torque control in lower power range test rig is additionally equipped in smaller Siemens 340 kW motor with nominal torque of 1865 Nm. All motors operate at 1,485 RPM nominal speed. Control accuracy parameters are presented in Table 1.

During operation one of the drives works in speed control mode. Second drive is in torque control mode as a breaking motor. Energy generated in braking motor is used in loop cycle and its excess can be returned to the power grid (Fig. 1).

Test rig design enables to perform tests with different kinematic configurations. Motors are moveable on dedicated base plate (13.2 × 9.6 m) and by using additional instrumentation it is possible to test all the gearboxes manufactured by Famur. Exemplary configuration is presented on Fig. 2.

Every motor has dual encoder −1,024 and 4,096 pulses per revolution and temperature sensors mounted on bearings and winding. All motors have dedicated cooling system. Implementation of chilled water plant ensures stable temperature in cooling system entry.

Standard procedure of gearbox testing includes break-in mode. Break-in procedure is divided into steps defined by speed and torque and its duration. This

Table 1 Control system parameters

Parameters	
Torque control accuracy for high power drive	±1.5 % of nominal torque
Torque control accuracy for lower power drive	±3.6 % of nominal torque
Speed control accuracy	±0.001 % nominal speed
Time of torque stabilization on drive shaft	2.6 ms

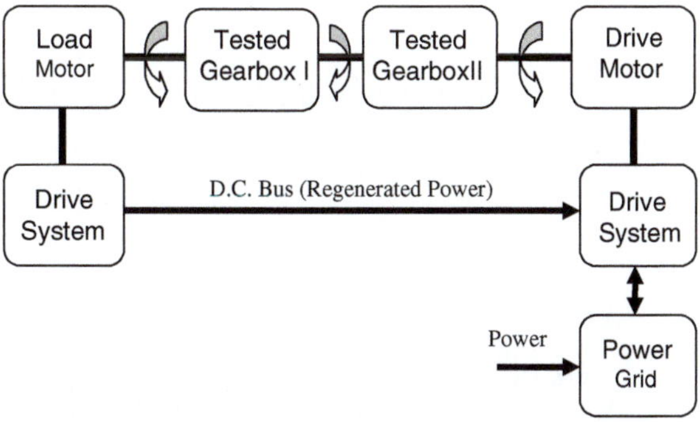

Fig. 1 Electrically closed loop drive configuration scheme

Fig. 2 Exemplary test rig setup

mode enables protection against rapid growth of speed or moment by control characteristics which are responsible for smooth transitions between speed and torque levels.

Special tests involve brake, rapid start under load and impact load mode. During braking mode one of the drives can generate 150 % of nominal torque which enables to stop gearbox having moment of inertia of 8 kg*m^2 from speed of 1,500 RPM in 0,6 s. Rapid start under load enables to accelerate to 1500 RPM in the same time.

Impact load mode enables to operate with load profiles consisting of pulses with amplitude up to 30 % of defined nominal value with frequency up to 10 Hz.

Additionally it is possible to operate in expert and LabVIEW modes which enable to generate nearly arbitrary load and speed profiles.

3 Dynamic Capabilities of Test Rig

Test rig has unique capabilities in terms of its dynamic behavior. This enables to simulate load profiles with characteristic often observed on mining machinery. Figure 3 shows exemplary load profile recorded on underground machine during

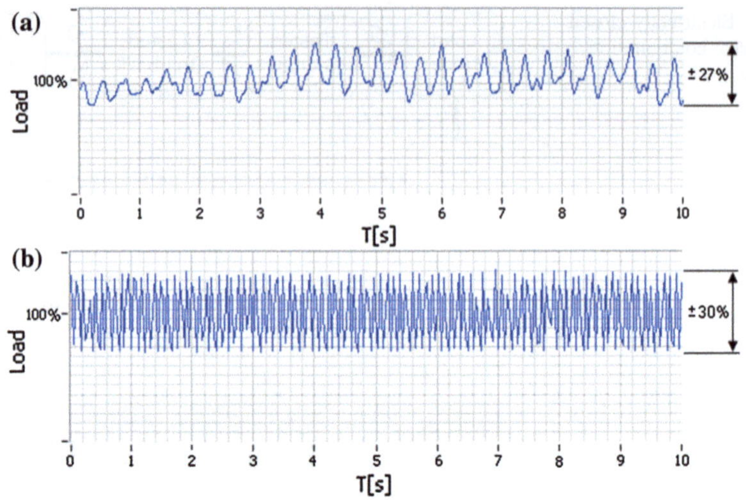

Fig. 3 Comparison of load profiles obtained form real-life operation on underground machine (**a**) and exemplary load profile generated on test rig (**b**)

its normal operation and for comparison load recorded during exemplary run on test rig. It can be seen that test rig enables to generate load profile with higher frequency and comparable amplitude dynamics. This feature is critical for testing for example shearer loader arms.

4 Case Study

As a case study a test of 250 kW heavy duty gearbox is discussed. Gearbox was subject to break-in procedure with different load levels. Certain parameters were monitored including temperatures, flow in the cooling system, pressures and vibration signal with tacho channel. Gearbox included one bevel stage, two spur stages and planetary stage at output.

Break-in procedure was a part of production plan and was the last stage of quality control before deploying the gearbox in target machine. Main goal of the test was to examine its dynamic and thermal behavior in nominal and overload conditions and collect vibration data with process parameters to examine correlation between vibration estimates and those parameters. This allows to select a method suitable for vibration based diagnostics of such machines possible and reasonable to implement in online monitoring systems.

First start of the gearbox setup is presented on Fig. 4 with load profile and oil temperatures in gearbox chambers.

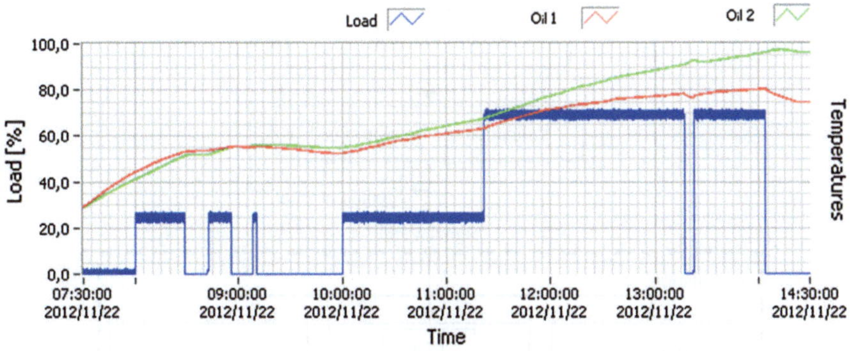

Fig. 4 First start of the gearbox with loads less than 100 % of the nominal

5 Vibration Analysis

Collected vibration data was analyzed, and according to methodology from [5–8], certain CI's (Condition Indicators) were computed. Calculation was performed only on signals valid due to validation algorithms [9–11], Kępski and Barszcz 2012) and (Jabłoński). Besides standard vibration estimates like acceleration peak to peak, RMS and velocity RMS according to ISO, residual and differential parameters where obtained.

Residual and difference signal where calculated according to definitions provided by [12], where residual signal is the TSA data with the shaft, mesh and their harmonic orders removed. The difference signal is obtained from residual signal by removing the first order sidebands. To trace bearing condition and potential tooth cracks during tests spectral kurtosis methodology according to [13] was adapted. Example of the results of this analysis is presented on Fig. 5. Scheme of

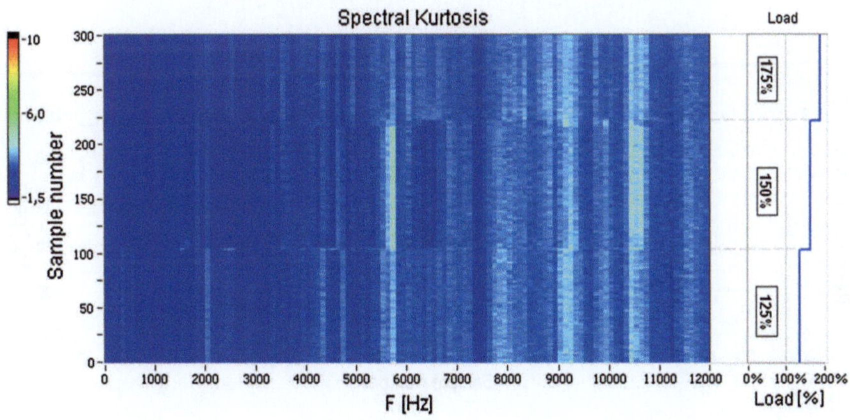

Fig. 5 Example of spectral kurtosis analysis results over time for overload steps for planetary stage

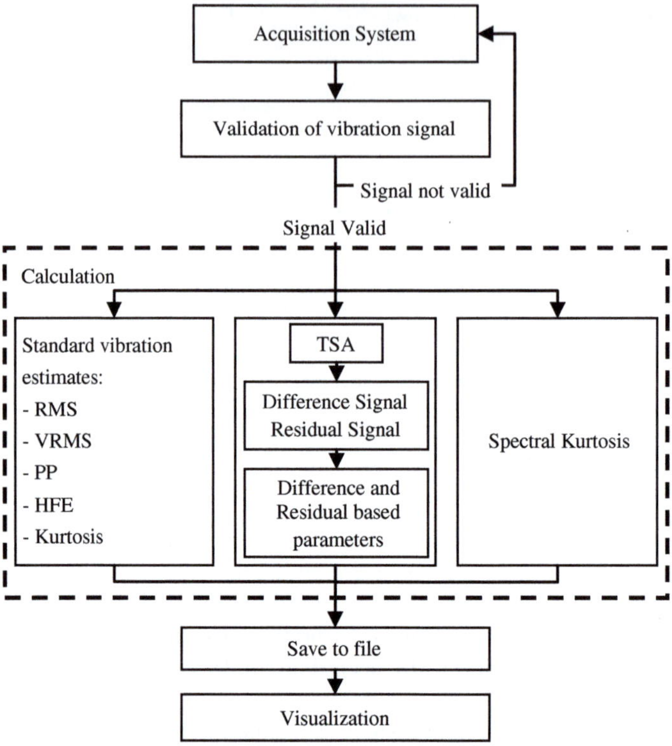

Fig. 6 Scheme of the analysis algorithm

analysis algorithm flow is presented on Fig. 6. Behavior of CIs was investigated with correlation to test conditions.

None of the CI's showed damage symptoms. Visual inspection of the gearbox after testing confirmed that gearbox is in good condition.

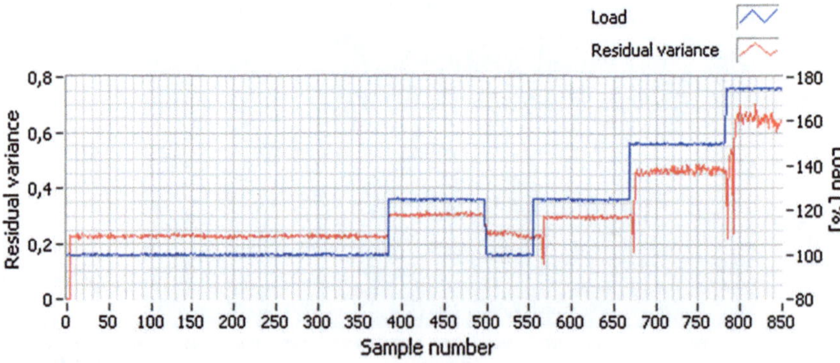

Fig. 7 Correlation between load and residual variance of signal synchronized with rotation of one of the spur gear

Influence of load on residual and differential parameters was tested. Also spur gear stages where observed closely as the ones that produced highest vibration.

Analysis of the residual and differential signal parameters showed that its variance is strongly correlated with load level Fig. 7. This conforms to the observations presented in [14]. Further findings on correlation between vibration signal parameters and load are presented of Fig. 8. It can be seen, that parameters like simple vibration signal peak to peak, residual signal variance and difference signal variance are load level dependent.

For all measurement points of the gearbox in question analysis of spectral kurtosis was performed. It showed no warning signs connected with any damages of the gears or bearing which was also confirmed by visual inspection after test.

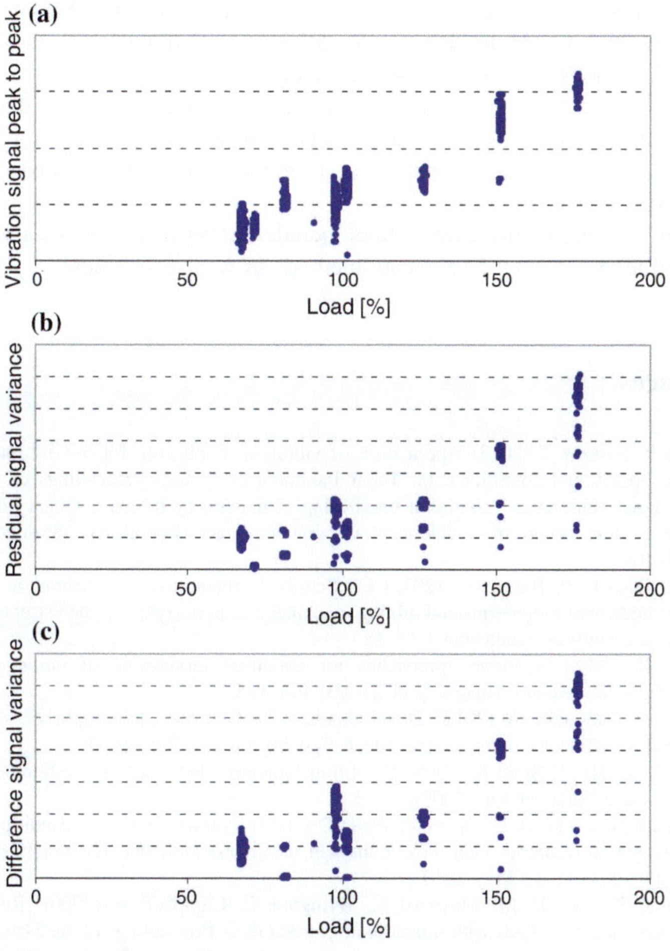

Fig. 8 Dependency between load and vibration parameters **a** Peak to Peak, **b** residual signal variance, **c** difference signal variance

Figure 5 present spectral kurtosis over time calculated for signals recorded during the overload stages of test.

By registering a large number of vibration signals (with sample rate of 25,600 Hz) it is possible to perform statistical analysis, and examine correlation between vibration data and process data, which could be used to determine machine state based only on vibration signals. This issue is the subject of ongoing work of authors.

6 Conclusion

Gearbox testing is important part of production quality control. Simulation of near to real-life operating conditions enables to evaluate gearbox behavior on target machine. Collecting of vibration data in different operating conditions is essential in optimal methods selection for online monitoring.

In discussed case of 250 kW heavy duty gearbox test it was shown that residual and differential signal are load dependent. Further study is recommended to investigate the opportunity of load level information reconstruction basing on the vibration data.

Further research planned by authors include testing of gearboxes after service before and after overhaul to evaluate methods of wear assessment.

References

1. Kępski P, Barszcz T (2012) Application of vibration monitoring for mining machinery in varying operational conditions. In: Tahar Fakhfakh et al. (ed) Proceedings of the second international conference "condition monitoring of machinery in non-stationary operations" condition monitoring of machinery in non-stationary operations. Springer, Berlin, pp 461–469
2. Greń B, Kępski P, Barszcz T (2012) Challenges in maintenance of vibration monitoring systems dedicated to underground mining machinery. Diagnostyka: applied structural health, usage and condition monitoring 1:13–18
3. Zimroz R (2008) Adaptive approaches for condition monitoring of mining machines. Gospodarka Surowcami Mineralnymi 24(4/3):103–115
4. Rezaei A, Dadouche A (2012) Development of a turbojet engine gearbox test rig for prognostics and health management. Mech Syst Sig Process 33:299–311
5. Vecer P, Kreidl M, Smid R (2005) Condition Indicators for gearbox condition monitoring systems. Acta Polytech 45(6/2005):35–43
6. Antolick L, Branning J, Dempsey P, Wade D (2010) Evaluation of gear condition indicator performance on rotorcraft fleet. Proceedings of the American helicopter society 66th annual forum, Phoenix, 11–13 May 2010
7. McClintic K, Lebold M, Maynard K, Byington C, Campbell R (2000). Residual and difference feature analysis with transitional gearbox data. Proceedings of the 54th meeting of the society for machinery failure prevention technology, Virginia Beach, p 635–645

8. Bechhoefer E, Kingsely M (2009) A review of time synchronous average algorithms. Annual conference of the prognostics and health management society 2009
9. Jabłoński A, Barszcz T, Bielecka M (2011) Automatic validation of vibration signals in wind farm distributed monitoring systems. Measur J Int Measur Confederation 44(10):1954–1967
10. Kępski P, Barszcz T (2012) Validation of vibration signals for diagnostics of mining machinery. Diagnostyka: applied structural health, usage and condition monitoring 4:25–30
11. Jabłoński A, Barszcz T (2011) Validation of vibration signals for heavy duty machinery diagnostics. Mechanical Systems and Signal Processing Special Issue, submission No. MSSP11-665
12. Zakrajsek JJ (1995) Detecting gear tooth fracture in a high contact ratio face gear mesh 49th meeting of the society for machinery failure prevention technology. Virginia Beach, Virginia, 18–20 April 1995
13. Barszcz T, Randall RB (2009) Application of spectral kurtosis for detection of a tooth crack in the planetary gear of a wind turbine. Mech Syst Sig Process 23(4):1352–1365
14. Dempsey PJ, Zakrajsek JJ (2001) Minimizing load effects on NA4 gear vibration diagnostic parameter. NASA TM-2001-210671, Glenn Research Center, Cleveland, Feb. 2001

Spatial Acceleration Modulus for Rolling Elements Bearing Diagnostics

Michele Cotogno, Marco Cocconcelli and Riccardo Rubini

Abstract Rolling Elements Bearing (REB) condition monitoring is mainly based on the analysis of acceleration (vibration) signal in the load direction. This is one of the three components of the acceleration vector in 3D space: the main idea of this paper is the recovery of additional fault information from all the three acceleration vector components by combining them to obtain the modulus of the spatial acceleration (SAM) vector. The REB diagnostic performances of the SAM are investigated and compared to the load direction vibration by means of two rough estimators of the "Signal-to-Noise" ratio (SNR) and the Spectral Kurtosis. The SAM provides a higher SNR than the single load direction. Finally, Spectral Kurtosis driven Envelope analysis is performed for further comparison of the two signals: its results highlight that demodulation of the SAM isn't strictly necessary to extract the fault features.

Keywords Rolling element bearing · Diagnostic · Spatial acceleration modulus · SAM · SNR · Spectral Kurtosis

M. Cotogno (✉) · M. Cocconcelli · R. Rubini
Department of Science and Engineering Methods, University of Modena and Reggio Emilia, Via Amendola, 2—Morselli Building 42122 Reggio Emilia, Italy
e-mail: michele.cotogno@gmail.com

M. Cocconcelli
e-mail: marco.cocconcelli@unimore.it

R. Rubini
e-mail: riccardo.rubini@unimore.it

G. Dalpiaz et al. (eds.), *Advances in Condition Monitoring of Machinery in Non-Stationary Operations*, Lecture Notes in Mechanical Engineering,
DOI: 10.1007/978-3-642-39348-8_51, © Springer-Verlag Berlin Heidelberg 2014

1 Introduction

Rolling Elements Bearing (REB) condition monitoring is mostly based on the analysis of the vibration signal [1, 2]. This is typically obtained from an accelerometer which measures the intensity of vibration (i.e.: acceleration) along the load direction. Many signal processing techniques are available for the REB fault features extraction. These techniques mainly deal with the fact that the measured vibration signal is the sum of different components (mechanical/electrical noise, the effect of vibration path from the REB to the sensor, vibrations coming from other moving parts in the machinery, etc.) which mask the fault signature. The main idea of this paper is the recovery of additional fault information from all the three acceleration components by analyzing the modulus of the spatial acceleration vector (SAM throughout the rest of this paper). In particular, it is supposed that one effect of the vibration path could be the spread of the signal of interest along other axes of acceleration rather than preserving it in the load direction only. Indeed, a faulty bearing produces a series of vibration pulses that excite all the vibration modes of the system, including the transversal modes (with respect to the load direction): thus traces of this pulse series could also be embedded in vibration data along the non-load directions. The SAM should embody this extra information, and is obtained for every instant t by the following Eq. 1:

$$SAM(t) = \sqrt{(x(t) - \mu_x)^2 + (y(t) - \mu_y)^2 + (z(t) - \mu_z)^2} \quad (1)$$

where $x(t)$, $y(t)$ and $z(t)$ are the acceleration vector projections acquired by the triaxial accelerometer at time t, and μ_x, μ_y and μ_z are the temporal averages of the respective axis. In this paper the SAM is compared with the acceleration signal along the load axis (i.e.: the classically analyzed signal in REB condition monitoring) from the diagnostic ability point of view. Equation 1 highlights that the SAM will exhibit larger (and non-negative) values than any single axis of acceleration, and this consideration can be extended also to the respective amplitude spectra. Consequently, a direct quantitative comparison between the SAM and any single acceleration axis (including their spectra) can't be completely truthful. In order to overcome this issue, some attempts of estimation of the Signal to Noise ratio are performed on real REB data and reported in the next section.

2 SNR Estimation

The bearing used in this series of experiments is SKF1204ETN9. The data are recorded at 25 kHz sampling frequency by a PCB356A01 triaxial accelerometer, from healthy and faulty bearing at five different bearing rotation speed (f_r = 15, 25, 35, 45 and 60 Hz). The (artificial) fault is made on the outer race. The only load applied is weight, directed along the y axis of the accelerometer: this axis of

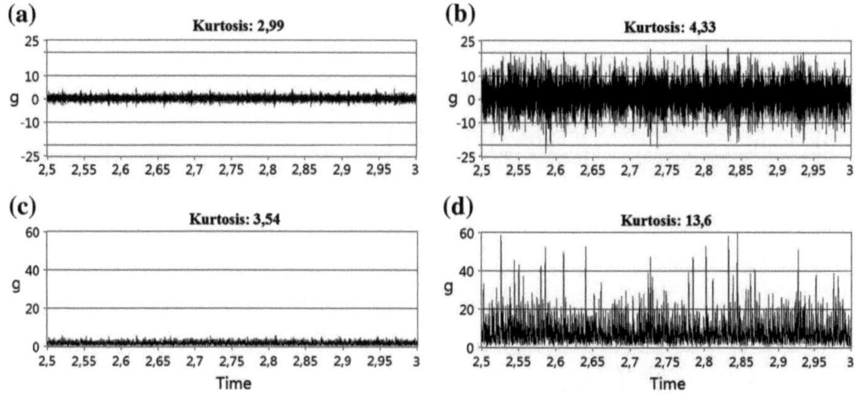

Fig. 1 Time sections of vibration data. **a** load axis, healthy bearing; **b** load axis, faulty bearing; **c** SAM, healthy bearing; **d** SAM, faulty bearing

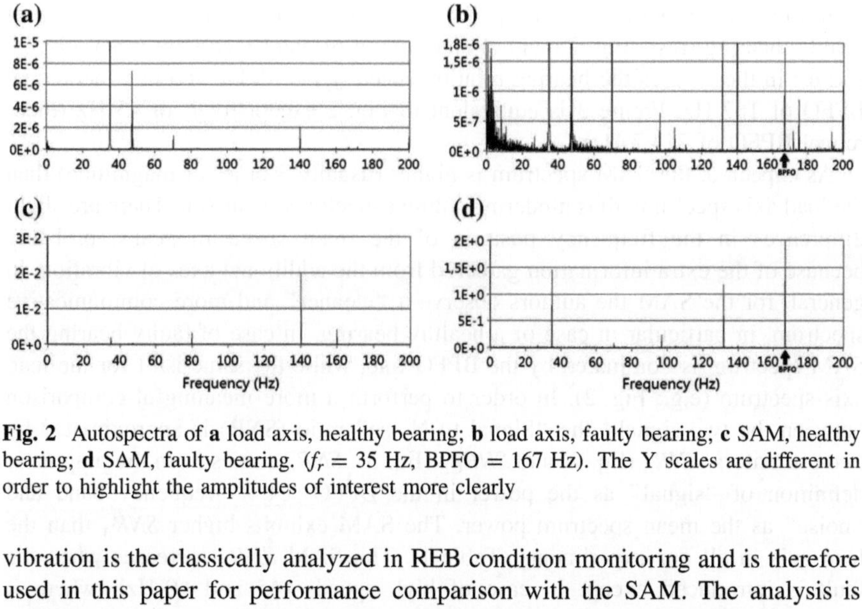

Fig. 2 Autospectra of **a** load axis, healthy bearing; **b** load axis, faulty bearing; **c** SAM, healthy bearing; **d** SAM, faulty bearing. (f_r = 35 Hz, BPFO = 167 Hz). The Y scales are different in order to highlight the amplitudes of interest more clearly

vibration is the classically analyzed in REB condition monitoring and is therefore used in this paper for performance comparison with the SAM. The analysis is performed on the raw data in order to evaluate the amount of information spontaneously embodied in the signals. In Fig. 1a (healthy bearing) and Fig. 1b (faulty bearing) extracts of the vibration data along the load axis (y axis) are reported and compared to the correspondent time sections of the SAM (Fig. 1c, d).

Figure 1 confirms that the amplitude of the signal is larger for the SAM. Moreover, it seems to have a more "spiky" nature than the load axis, particularly in presence of a fault (Fig. 1d): this observation is acknowledged by the global Kurtosis index (4,33 for the load axis and 13,6 for the SAM in case of faulty bearing); i.e., the SAM is representing the pulses of vibration in a clearer manner.

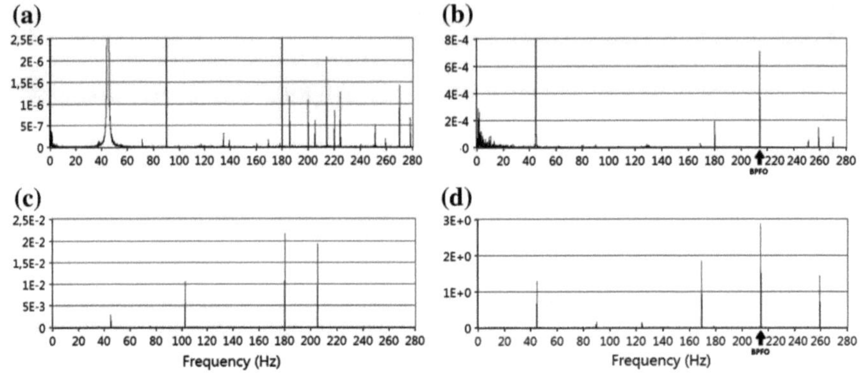

Fig. 3 Autospectra of **a** load axis, healthy bearing; **b** load axis, faulty bearing; **c** SAM, healthy bearing; **d** SAM, faulty bearing. (f_r = 45 Hz, BPFO = 214,7 Hz). The Y scales are different in order to highlight the amplitudes of interest more clearly

In Fig. 2 are reported the autospectra of the load axis and the SAM in case of healthy bearing (resp. Fig. 2a and c) and in case of outer race fault (resp. Fig. 2b and d): in these cases the bearing rotation speed (f_r) is 35 Hz giving a theoretical BPFO of 167 Hz. Figure 3 is equivalent to Fig. 2 except for f_r of 45 Hz (theoretical BPFO of 214,7 Hz).

As expected, the SAM spectrum is higher (usually 3 order of magnitude) than the load axis spectrum, thus hindering a direct spectra comparison. There are slight differences in the frequency position of the main spectrum peaks, probably because of the extra information gathered from the additional axes of vibration. In general, for the SAM the authors observe a "cleaner" and more communicative spectrum, in particular in case of a healthy bearing. In case of faulty bearing the SAM spectrum is dominated by the BPFO line, while the same isn't for the load axis spectrum (e.g.: Fig. 2). In order to perform a more meaningful comparison between the two signals, the "Signal to Noise" ratio (SNR) is approximated by two estimators, SNR_A (Eq. 2) and SNR_{CF} (Eq. 3). SNR_A comes from the pragmatic definition of "signal" as the power in the BPFO ± 1 % frequency band and "noise" as the mean spectrum power. The SAM exhibits higher SNR_A than the load axis in all our experiments (Table 1). The SAM SNR_A increases when the fault is introduced except in case of high speeds (45 and 60 Hz), where it decreases; this behavior is almost the exact opposite of the load axis SNR_A (which exhibits also larger variations). This could be explained by the larger energetic content of the vibration signals at higher speeds which increases the denominator in Eq. 2, thus lowering the SAM SNR_A indicator; in contrast, the load axis SNR_A increases probably because at higher speeds its BPFO line shows a larger amplitude (Fig. 3b with respect to Fig. 2b) thus increasing the numerator in Eq. 2. The sum of these two effects and the slightly different nature of the two signals could explain the SNR_A trends highlighted in Table 1. SNR_{CF} attempts to evaluate the SNR via the Crest Factor of the spectrum, which is obtained dividing the amplitude of the highest spectral peak by the RMS of the spectrum. Consequently, estimating the SNR with the Crest Factor is equivalent to defining "signal" as the spectral component with maximum amplitude and "noise" as the RMS of the spectrum. SNR_{CF} is again always higher for the SAM (Table 1) than for the load axis. Interestingly, the SNR_{CF} of the SAM is almost constant (approx. 30.8 dB) in every experiment while a wide range of variation (17,2–29 dB) is observed for the SNR_{CF} of the load axis. The uniformity of the SAM SNR_{CF} and the variations of the load axis SNR_{CF} can be explained by the fact that the SAM represent the total vibration energy flow at the measurement point while a single axis of vibration represent only a fraction of this flow. In the conducted experiments the load was only gravitational and this could cause a larger spread of the energy flow along the non-load directions, because a light load have (consequently) a lighter "polarizing" effect on the directions of vibration. This joined with the fact that different speeds produce different excitations should explain the behaviors in Table 1. SNR_A and SNR_{CF} values are interpreted by the authors as (light) symptoms of the SAM actually embodying more information or making it more readable than the load axis.

$$SNR_A = \frac{\text{power in the BPFO} \pm 1\ \% \text{ frequency band}}{\text{mean spectrum power}} \tag{2}$$

$$SNR_{CF} = \frac{\text{amplitude of the main peak in the spectrum}}{RMS \text{ of the spectrum}} \tag{3}$$

Table 1 Estimates of the signal-to-noise ratio

f_r (Hz)	Signal	SNR_A (dB)		SNR_{CF} (dB)	
		Healthy bearing	Faulty bearing	Healthy bearing	Faulty bearing
15	Load axis	8.9	0.4	26.2	17.2
	SAM	49.7	51.4	30.9	30.8
25	Load axis	−9.7	−12.2	27.8	29.0
	SAM	38.2	43.9	30.6	30.9
35	Load axis	−9.8	−18.5	24.6	18.8
	SAM	40.8	41.7	30.9	30.8
45	Load axis	1.4	−13.5	27.3	27.7
	SAM	43.9	40.8	30.8	30.8
60	Load axis	0.8	2.5	27.8	25.4
	SAM	43.7	34.1	30.9	30.8

A more refined way to estimate the SNR is represented by the Spectral Kurtosis (SK) [3, 4], a useful tool in condition monitoring [5] that can be used in REB diagnostics to find the optimum frequency band for the signal filtering prior to the Envelope analysis [2, 6, 7]. Antoni and Randall [4] reveals that the SK is proportional to the square of the Wiener filter *W(f)*, which is defined by the following Eq. 4:

$$W(f) = \frac{1}{1 + \rho(f)} \tag{4}$$

where $\rho(f)$ is the "Noise-to-Signal" ratio, i.e. the inverse of the SNR. "Signal" here is interpreted as "randomly occurring pulses of random amplitude" [4] which is a model that has proven its effectiveness in describing the vibration signal of a faulty REB. Thus, larger SK values are related to larger SNR of the data. In this paper, the SK is obtained by the following formula (Eq. 5), which is an unbiased STFT-based SK estimator proposed in [8]:

$$SK(f) = \frac{M}{M-1}\left[\frac{(M+1)\sum_{i=1}^{M}|X_i(f)|^4}{\left(\sum_{i=1}^{M}|X_i(f)|^2\right)^2} - 2\right] \quad (5)$$

where M is the number of non-overlapping blocks of the STFT and $X_i(f)$ is the DFT of the ith block of data. As highlighted in [4], the STFT block length (i.e.: the STFT window length, N_w in the rest of this paper) must be set to have a temporal duration lower than the expected mean time between impacts (i.e.: BPFO) in order to get consistent results: this yields $N_w = 128$ samples in case BPFO = 167 Hz and $N_w = 64$ samples in case BPFO = 214,7 Hz. In Fig. 4 are reported the SK for these two experiments.

Figure 4 shows that in case of healthy bearing (Fig. 4a and b) the SK of the SAM is lower (and closer to zero) than the SK of the load axis, and it increases more decisively when a fault is present (Fig. 4c and d, where the SAM SK graph is always over the load axis SK). In absence of the fault, the SAM SK overcomes the load axis SK in the 6,5–12,5 kHz (when $f_r = 35$ Hz, Fig. 4a) and in the 8–12,5 kHz (when $f_r = 45$ Hz, Fig. 4b) frequency bands. Zero-valued SK indicates that the signal is more similar to Gaussian noise rather than a series of

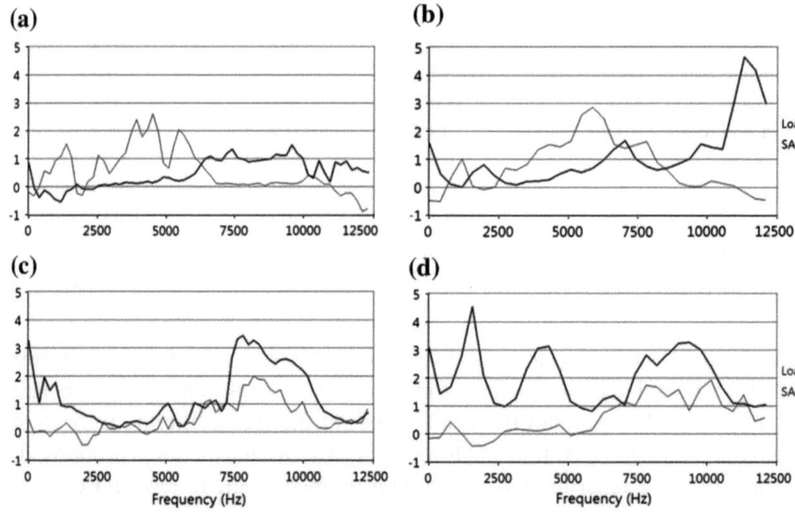

Fig. 4 Spectral Kurtosis of load axis and SAM; **a** healthy bearing, $f_r = 35$ Hz; **b** healthy bearing, $f_r = 45$ Hz; **c** faulty bearing, $f_r = 35$ Hz; **d** faulty bearing, $f_r = 45$ Hz

transients, while the presence of the latter is highlighted by higher SK values: thus, the SAM SK is communicating more explicitly the existence of a fault. The frequency bands with higher Kurtosis indicated by the two SK are similar when the fault is present, and correspond to a system resonance band (approximately 7,5–9,5 kHz). It's notable that in the majority of the experiments the SAM SK exhibits the highest values also in the low frequency zone (i.e.: near the DC component), thus communicating that demodulation isn't strictly necessary to extract the fault signature from the SAM. This is in accord with the raw signals spectra comparison (Figs. 2, 3 and Table 1). These considerations about the SAM SK seem to confirm the higher SNR of the SAM than the load axis. In the next section the REB diagnostic ability of the SAM is tested via the Envelope analysis.

3 Envelope Analysis

Envelope Analysis [1, 2, 7] is the benchmark signal processing tool for REB diagnostics. It performs a demodulation of the bandpass filtered signal. In this paper the filtering band is selected as the band in which the Spectral Kurtosis is maximized. The selected bands in case of $f_r = 35$ Hz are 7,7–8,8 kHz for the load axis and 7,5–8,5 kHz for the SAM; in case of $f_r = 45$ Hz they are 9,5–10,5 kHz for the load axis and 900–1900 Hz for the SAM. In Fig. 5 the Envelope spectra of the filtered load axis and SAM are reported for the case of faulty bearing rotating at $f_r = 35$ Hz.

It can be seen that the Envelope spectra of the SAM exhibit larger values, highlighting almost the same frequency of the load axis. The dominating frequency is the BPFO (167 Hz) for both signals, but other notable frequencies (e.g.: the $f_r = 35$ Hz frequency) exhibit higher magnitude in the SAM Envelope

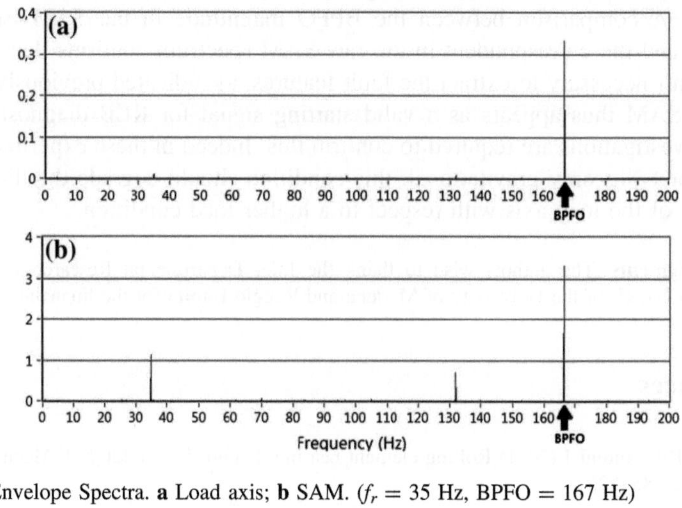

Fig. 5 Envelope Spectra. **a** Load axis; **b** SAM. ($f_r = 35$ Hz, BPFO = 167 Hz)

spectrum. This is about one order of magnitude higher than the load axis at those frequencies, while the average difference between the autospectra of the two signals is about 3 orders of magnitude (these considerations are valid for all the experiments). The BPFO magnitude in the SAM Envelope spectrum is about two times the correspondent magnitude in the raw SAM spectrum (respectively, 3,1 and 1,7); in contrast, the BPFO magnitude of the load axis Envelope spectra is about one million times the correspondent magnitude in the raw spectra (respectively, 0,3 and $1{,}75 \cdot 10^{-6}$). This observation confirms the indication of the Spectral Kurtosis of presence of high SNR in the low frequency zone of the SAM spectrum, signifying that demodulation isn't necessary for the SAM for the fault signature extraction.

4 Conclusion

In this paper the modulus of the spatial acceleration vector (SAM) is presented as a starting signal for rolling element bearings (REB) condition monitoring. Concerning the diagnostic capability, the SAM is compared to the acceleration component directed along the load axis, which is the classical analyzed signal in REB diagnostics. The comparison is performed on real raw REB data to evaluate the information spontaneously embedded in the signals. Considering the amplitude difference between the two signals and their spectra (about three orders of magnitude), two rough approximation of the Signal-to-Noise ratio (SNR) are used to compare them: these SNR estimators indicate always a higher SNR for the SAM. This is confirmed also by the Spectral Kurtosis (SK): higher SK is related to higher SNR and the SAM exhibit higher SK in case of faulty bearing. Finally Envelope analysis is performed and it confirms the diagnostic capability of the SAM; its Envelope spectrum is one order of magnitude higher than the load axis Envelope spectrum. A comparison between the BPFO magnitude in the SAM Envelope spectrum and the correspondent in the raw SAM spectrum confirms that demodulation isn't necessary to extract the fault features, as indicated previously by the SK. The SAM thus appears as a valid starting signal for REB diagnostics, but further investigations are required to confirm this. Indeed in these experiments the load applied was only gravitational: this condition should degrade the diagnostic capability of the load axis with respect to a higher load condition.

Acknowledgments The authors wish to thank the Inter Departmental Research Center INTERMECH MoRE of the University of Modena and Reggio Emilia for the financial support.

References

1. Randall RB, Antoni J (2011) Rolling element bearing diagnostics—a tutorial. Mech Syst Sig Proces 25:485–520

2. Randall RB (2011) Vibration-based condition monitoring: industrial, automotive and aerospace applications. Wiley, West Sussex
3. Antoni J (2006) The spectral Kurtosis: a useful tool for characterizing non-stationary signals. Mech Syst Sig Proces 20:282–307
4. Antoni J, Randall RB (2006) The spectral Kurtosis: application to the vibratory surveillance and diagnostics of rotating machines. Mech Syst Sig Proces 20:308–331
5. Barszcz T, Randall RB (2009) Application of spectral Kurtosis for detection of a tooth crack in the planetary gear of a wind turbine. Mech Syst Sig Proces 23(4):1352–1365
6. Cocconcelli M, Zimroz R, Rubini R, Bartelmus W (2012) Kurtosis over energy distribution approach for STFT enhancement in ball bearing diagnostics. In: Proceedings of the second international conference "condition monitoring of machinery in non-stationary operations" CMMNO'2012, pp 51–59
7. McFadden PD, Smith JD (1984) Vibration monitoring of rolling element bearings by the high-frequency resonance technique—a review. Tribol Int 17(1):3–10
8. Vrabie VD, Granjon P, Serviere C (2003) Spectral Kurtosis: from definition to application. IEEE-EURASIP workshop on nonlinear signal and image processing, Grado, Italy, 8–11 June 2003

Artificial Neural Networks-Based Decoupling Approach in the Vector Control Block of the Single-Phase Induction Machine

Kenza Bouhoune, Krim Yazid and Mohamed S. Boucherit

Abstract The vector control of single phase induction machine using conventional decoupling approaches has a remarkable decrease of rotor flux when the machine is powered by a real voltage source inverter. To solve this problem, we propose as an efficient solution, a decoupling approach based on artificial neural networks in the vector control block of single phase induction machine. The application of this approach to the single-phase machine increases its dynamic performance and constitutes a contribution to the study of this machine. Indeed, this type of machine has not yet taken its whole share from various works present until now, compared to the three-phase induction machine. For three different decoupling approaches with two types of supply: perfect voltage source and real voltage source inverter; a comparative study through numerical simulations is presented. The simulation results show the feasibility and good performance obtained by the proposed approach.

Keywords Single-phase induction machine · Vector control · Artificial neural networks · Neural decoupling approach · Neural controllers · Static converter

K. Bouhoune (✉) · K. Yazid
Power and Industrial Systems Laboratory, Faculty of Electronics and Informatics, University of the Sciences and Technology, Houari Boumediene, B.P 32, El Alia, Bab Ezzouar 16111 Algiers, Algeria
e-mail: k_bouhoune@yahoo.fr

K. Yazid
e-mail: k_yazid@yahoo.fr

M. S. Boucherit
Process Control Laboratory, Polytechnic National School, BP 182, 10 Hassan Badi, El-Harrach 16200 Algiers, Algeria
e-mail: ms_boucherit@yahoo.fr

G. Dalpiaz et al. (eds.), *Advances in Condition Monitoring of Machinery in Non-Stationary Operations*, Lecture Notes in Mechanical Engineering,
DOI: 10.1007/978-3-642-39348-8_52,

1 Introduction

Single-phase induction machines have been used for a long time because of their simple construction and because the single-phase power supply is available in almost every household. They are found in washers, air conditioners and many other applications. They are traditionally used in constant speed home appliances, usually in locations where only single-phase energy supply is available without any type of control strategy. Now, variable speed controls of electric motors are necessary in industrial applications. However, there is more research on the three-phase motor which is used in industrial frequency conversion control technology than single-phase motor control technology research relatively. Actually, the single-phase structure is simpler than three-phase motor, production costs are compared cheap, and more application, so single-phase motor control research is significant particularly important [1–3]. Moreover, the cost reduction and high efficiency of power electronic are motivating to implement single-phase motor drives in both industrial and domestic applications. These years, research laboratories focus on variable speed drives, especially for single phase induction machine, and major improvements are underway for achievement [4–11].

Control principles are considered as very important for the overall performances of electrical drives and power electronics. In the past two decades, the controls based on the tools of the artificial intelligence such as: artificial neural networks (ANN), fuzzy logic and genetic algorithm, expert systems, and particle swarm optimization; are distinguished from the others existed controls. They have made great progress in recent years to solve problems encountered in control system or just improve their operation, especially in electric motors because roughly half of all electricity generated is consumed in motors. Indeed, recent efforts to apply artificial intelligence to improving electric motors are receiving attention worldwide and are attracting much research; some applications can be founded in [3, 12–16] and even in the renewable energy field, some applications have already started to develop [17, 18]. Since most of modern applications are using AC electrical machinery, two key variables have to be feedback-controlled in motor: the flux and the torque. Invented in the early 70s, the field-oriented control is the most popular control technique since it allows AC electrical machines to have good transient performances. Nowadays, it is also known as vector control since it is more general. The vector control of single phase induction machine using conventional decoupling approaches has a remarkable decrease of rotor flux when the machine is powered by a real voltage source inverter (VSI) of PWM type with voltage control. These voltages contain variable terms that can deteriorate the desired functioning. They lead to lack of accuracy in current control; so, remarkable flux decreasing. To solve this problem, we propose by the means of this chapter, a new decoupling approach in the vector control block, based on artificial neural networks. A comparative study through numerical simulations between the proposed decoupling approach and the decoupling conventional approaches is presented, for two types of power: perfect voltage source and real voltage source inverter. The simulation results show the effectiveness and good performances obtained by the proposed approach.

The remainder of this chapter is organized as follows: a presentation of the single phase induction machine is presented in the Sect. 2 follows by its mathematical model at the Sect. 3. The proposed approach based on the artificial neural networks is developed in the Sects. 4, 5, 6, where, it also presented, as comparison, two conventional decoupling approaches. The Sect. 7 is devoted to illustrating by simulation the performances of the proposed approach. Conclusion and reference list end the paper.

Nomenclature

as	auxiliary winding index
d, q	synchronously rotating reference frame index
ms	main winding index
n	rated value index
r	rotor index
ref	reference value index
s	stator index
sp	speed index
AC	alternative current
AI	artificial intelligence
ANN	artificial neural networks
C_{sp}	speed controller
f	frequency
f	viscose friction coefficient, Nm.s/rad
L_M	mutual inductance, H
L_r	rotor inductance, H
L_s	stator inductance, H
R_r	rotor resistance, Ω Ω
R_s	stator resistance, Ω Ω
I	current, A
J	moment of inertia, $Kg.m^2$
NC	neural controller
NC_{cd}	neural current I_{ds} , l_{qs}
NC_{cq}	controller
PI	proportional integral
p	pole pair number
P_n	rated power, W
PWM	pulse width modulation
SPIM	Single-Phase Induction Machine
SPIM	WC Single-Phase Induction Machine Without Capacitor
T_{em}	electromagnetic and
T_L	load torque, N.m
$T_r = L_r/R_r$	rotor time constant, s
$T_s = L_s/R_s$	stator time constant, s

V	voltage, V
V_n	rated voltage
V_r	rectified voltage
VSI	voltage source inverter
σ	leakage coefficient
Ψ	flux, Wb
θ	angle, rad
ω_r	rotor angular frequency, rad/s
Ω_r	mechanical speed, rad/s
ω_s	stator angular frequency, rad/s.

2 Presentation of the Single-Phase Induction Machine

The single-phase induction machine (SPIM) consists of: Stator (fixed part) with a single-phase winding; supplied with a single-phase voltage; and rotor (revolving part); did not supplied by any kind of electrical energy. The current in the rotor is only induced by the synchronously field [19].

The stator winding is supplied by a single-phase voltage. The current of this winding creates a pulsatory magnetic field [20]. It can be divided into two fields of same amplitudes, equal to half of the amplitude of the pulsatory field, turning in opposite directions at the same speed. In a single-phase induction motor, the currents induced in rotor by the two fields, create equal and opposed electromagnetic torques. Thus, the resulting motoring torque is null and the motor can not start-up [21].

To have no null starting up torque which makes turning the motor; one solution is to place an auxiliary winding on the stator. That the field produced by the auxiliary winding will be perpendicular to the field produced by the main winding. The auxiliary winding is supplied through a capacitor which shifts the currents in the two windings of an angle lower than $\pi/2$, but sufficient to create a synchronous rotating field [19, 21].

So, the single-phase induction machine consists of a stator with two windings: the main winding and the auxiliary winding; displaced of 90 electrical and mechanical degrees. The rotor is squirrel cage that can be represented by two short-circuited windings, as shown in Fig. 1. If the two windings do not have the same parameters, we speak about an asymmetrical single-phase induction machine. In the contrary case, they have identical parameters; we speak about a symmetrical single-phase induction machine.

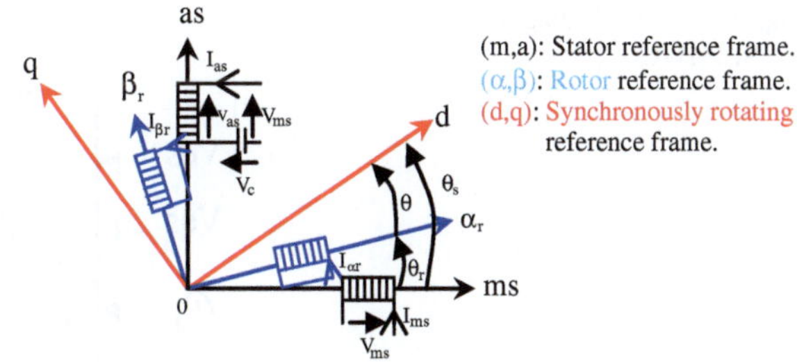

Fig. 1 Space representation of the single-phase induction machine

3 Mathematical Model of the Single-Phase Induction Machine

In the single-phase induction machine without capacitor (SPIMWC), we supply separately the two stator phases of the machine by voltages shifted of 90° (V_{ms}, V_{as}). We speak about two-phase induction machine. This operation can provide significant improvement to its performance [7]. The model of single-phase induction motor without capacitor in the stator reference frame is represented by the Eq. (1). Rotor flux, stator currents and rotor speed are considered as the state variable of the system.

$$\begin{cases} \dfrac{d\Psi_{mr}}{dt} = -\dfrac{R_r}{L_r}\Psi_{mr} - p\Omega_r\Psi_{ar} + \dfrac{R_r}{L_r}L_M I_{ms} \\ \dfrac{d\Psi_{ar}}{dt} = p\Omega_r\psi_{mr} - \dfrac{R_r}{L_r}\Psi_{ar} + \dfrac{R_r}{L_r}L_M I_{as} \\ \dfrac{dI_{ms}}{dt} = \dfrac{L_M R_r}{\sigma L_s L_r^2}\Psi_{mr} + \dfrac{pL_M\Omega_r}{\sigma L_S L_r}\Psi_{ar} - \left(\dfrac{L_M^2 R_r + L_r^2 R_s}{\sigma L_s L_r^2}\right) I_{ms} + \dfrac{V_{ms}}{\sigma L_s} \\ \dfrac{dI_{as}}{dt} = -\dfrac{pL_M R_r}{\sigma L_S L_r}\Psi_{mr} + \dfrac{L_M R_r}{\sigma L_S L_r^2}\Psi_{ar} - \left(\dfrac{L_M^2 R_r + L_r^2 R_s}{\sigma L_s L_r^2}\right) I_{as} + \dfrac{V_{as}}{\sigma L_s} \\ \dfrac{d\Omega_r}{dt} = \dfrac{PL_M}{JL_r}(\Psi_{mr} I_{as} - \Psi_{ar} I_{ms}) - \dfrac{f}{J}\Omega_r - \dfrac{T_L}{J} \end{cases} \tag{1}$$

By using the transformation matrix (2), we obtain the model of two-phase induction motor in the synchronously rotating reference frame (3).

$$\mathrm{A} = \begin{bmatrix} \cos\theta_s & \sin\theta_s \\ -\sin\theta_s & \cos\theta_s \end{bmatrix} \quad with: \quad \theta_s = \int \omega_s \, dt \tag{2}$$

Fig. 2 The block diagram of the SPIMWC

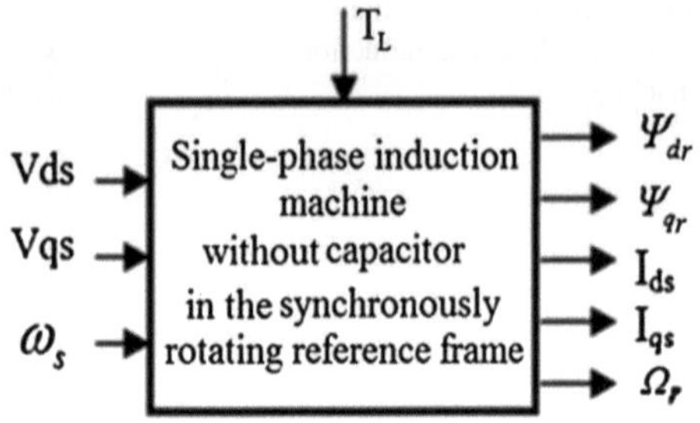

$$\begin{cases} \dfrac{d\Psi_{dr}}{dt} = -\dfrac{R_r}{L_r}\Psi_{dr} + (\omega_s - p\Omega_r)\Psi_{qr} + \dfrac{R_r L_M}{L_r} I_{ds} \\ \dfrac{d\Psi_{qr}}{dt} = (p\Omega_r - \omega_s)\Psi_{dr} - \dfrac{R_r}{L_r}\Psi_{qr} + \dfrac{R_r L_M}{L_r} I_{qs} \\ \dfrac{dI_{ds}}{dt} = \dfrac{L_M R_r}{\sigma L_S L_r^2}\Psi_{dr} + \dfrac{pL_M\Omega_r}{\sigma L_S L_r}\Psi_{qr} - \left(\dfrac{L_M^2 R_r + L_r^2 R_s}{\sigma L_s L_r^2}\right) I_{ds} + \omega_s I_{qs} + \dfrac{V_{ds}}{\sigma L_s} \\ \dfrac{dI_{qs}}{dt} = -\dfrac{pL_M\Omega_r}{\sigma L_S L_r}\Psi_{dr} + \dfrac{L_M R_r}{\sigma L_S L_r^2}\Psi_{qr} - \omega_s I_{ds} - \left(\dfrac{L_M^2 R_r + L_r^2 R_s}{\sigma L_s L_r^2}\right) I_{qs} + \dfrac{V_{qs}}{\sigma Ls} \end{cases} \tag{3}$$

Mechanical equation:

$$\frac{d\Omega_r}{dt} = \frac{pL_M}{JL_r}(\Psi_{dr} I_{qs} - \Psi_{qr} I_{ds}) - \frac{f\Omega_r}{J} - \frac{T_L}{J} \tag{4}$$

The block diagram of the single-phase induction machine without capacitor is shown in Fig. 2.

4 Vector Control of the Single-Phase Induction Machine

By considering the electromagnetic torque T_{em} and the rotor flux Ψ_r as reference variables, we invert the system, Eq. (3). After, having directed the flux along the d-axis [22], we obtain the Eq. (5a).

$$I_{ds} = \frac{L_r}{R_r L_M} \cdot \frac{d\Psi_{r\,ref}}{dt} + \frac{1}{L_M}\Psi_{r\,ref} \tag{5a}$$

$$I_{qs} = \frac{L_r}{L_M} \cdot \frac{Tem_{ref}}{\Psi_{r\,ref}} \tag{5b}$$

$$\omega_s = \omega_r + \frac{R_r L_M}{L_r \Psi_{r\,ref}} I_{qs} \tag{5c}$$

$$V_{ds} = \sigma Ls\frac{dI_{ds}}{dt} + \left(\frac{L_M^2 R_r + L_r^2 Rs}{L_r^2}\right)I_{ds} - \frac{L_M R_r}{L_r^2}\Psi_{r\,ref} - \omega_s \sigma L_s I_{qs} \tag{5d}$$

$$V_{qs} = \sigma L_s\frac{dI_{qs}}{dt} + \left(\frac{L_M^2 R_r + L_r^2 Rs}{L_r^2}\right)I_{qs} + \frac{L_M}{L_r}p\Omega_r\Psi_{r\,ref} + \omega_s \sigma L_s I_{ds} \tag{5e}$$

with: $\omega_s = \omega_r + \omega_{s1}$ and $\omega_{s1} = (R_r L_M / L_r \Psi_{r\,ref})I_{qs}$

- The reference electromagnetic torque $T_{em\ ref}$ is obtained through a speed controller C_{sp}, type PI.
- The reference rotor flux $\Psi_{r\,ref}$ is obtained through a field weakening block; operates as follows:

$$\Psi_{r\,ref} = \begin{cases} \Psi_n & |\Omega_r| \le \Omega_{rn} \\ \Psi_n \frac{\Omega_{rn}}{\Omega_r} & |\Omega_r| > \Omega_{rn} \end{cases} \tag{6}$$

- The reference control voltages $V_{ds\ ref}$, $V_{qs\ ref}$ are calculated using three different decoupling approaches.

5 Decoupling Approaches

First Conventional Approach

The control voltages are used directly from Eqs. (5d), (5e).

$$\begin{cases} V_{ds\,ref} = \sigma L_s\frac{dI_{ds}}{dt} + \left(\frac{L_M^2 R_r + L_r^2 R_s}{L_r^2}\right)I_{ds} - \frac{L_M R_r}{L_r^2}\Psi_{r\,ref} - \omega_s \sigma L_s I_{qs} \\ V_{qs\,ref} = \sigma L_s\frac{dI_{qs}}{dt} + \left(\frac{L_M^2 R_r + L_r^2 R_s}{L_r^2}\right)I_{qs} + \frac{L_M}{L_r}p\Omega_r\Psi_{r\,ref} + \omega_s \sigma L_s I_{ds} \\ \omega_s = \omega_{s1} + p\Omega_r \end{cases} \tag{7}$$

Second Conventional Approach

The variables V_{ds}, V_{qs} of the Eqs. (5d), (5e) are approximated. The reference control voltages become:

$$\begin{cases} V_{ds\,ref} = R_s I_{ds} - \omega_s \sigma L_s I_{qs} \\ V_{qs\,ref} = R_s I_{qs} + \omega_s \sigma L_s I_{ds} \\ \omega_s = \omega_{s1} + p\Omega_r \end{cases} \tag{8}$$

Proposed Approach

The calculation of this method is presented as follows:

- From Eq. (5a), we have:

$$\Psi_{r\,ref} = L_M I_{ds} - \frac{L_r}{R_r} \cdot \frac{d\Psi_{r\,ref}}{dt} \tag{9}$$

By replacing the Eq. (9) in the Eq. (5d), we obtain:

$$V_{ds} = R_s I_{ds} + \sigma L_s \frac{dI_{ds}}{dt} + \frac{L_M}{L_r} \frac{d\Psi_{r\,ref}}{dt} - \omega_s \sigma L_s I_{qs} \tag{10}$$

- From Eq. (5c), we have:

$$\omega_r = \omega_s - \omega_{s1} = \omega_s - (R_r L_M / L_r \Psi_{r\,ref}) I_{qs} \tag{11}$$

The replacement of the Eq. (11) in the Eq. (5e), p $\Omega_r = \omega_r$, gives:

$$V_{qs} = R_s I_{qs} + \sigma L_s \frac{dI_{qs}}{dt} + \omega_s \frac{L_M}{L_r} \Psi_{r\,ref} + \omega_s \sigma L_s I_{ds} \tag{12}$$

In order to decouple the Eqs. (10) and (12), they are putting in the form:

$$\begin{cases} V_{ds} - \dfrac{L_M}{L_r} \dfrac{d\Psi_{r\,ref}}{dt} + \omega_s \sigma L_s I_{qs} = R_s I_{ds} + \sigma L_s \dfrac{dI_{ds}}{dt} \\ V_{qs} - \omega_s \dfrac{L_M}{L_r} \Psi_{r\,ref} - \omega_s \sigma L_s I_{ds} = R_s I_{qs} + \sigma L_s \dfrac{dI_{qs}}{dt} \end{cases} \tag{13}$$

We introduce new variables V_{ds1}, V_{qs1}, such as:

$$\begin{cases} V_{ds1} = R_s I_{ds} + \sigma L_s \dfrac{dI_{ds}}{dt} \\ V_{qs1} = R_s I_{qs} + \sigma L_s \dfrac{dI_{qs}}{dt} \end{cases} \tag{14}$$

Finally, we obtain the reference control voltages $V_{ds\,ref}$ and $V_{qs\,ref}$ by correcting the error introduced during decoupling.

$$\begin{cases} V_{ds\,ref} = V_{ds1} + \dfrac{L_M}{L_r} \dfrac{d\Psi_{r\,ref}}{dt} - \omega_s \sigma L_s I_{qs} \\ V_{qs\,ref} = V_{qs1} + \omega_s \left(\dfrac{L_M}{L_r} \Psi_{r\,ref} + \sigma L_s I_{ds} \right) \end{cases} \tag{15}$$

The variables V_{ds1}, V_{qs1} are obtained through neural current controllers NC_{cd}, NC_{cq}. The overall diagram of the proposed system is shown in Fig. 3.

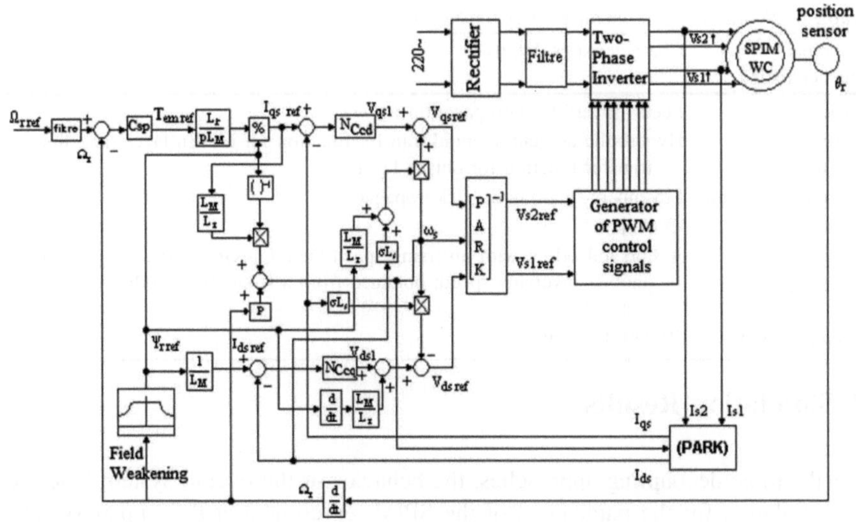

Fig. 3 The overall diagram of the proposed approach for control the SPIM

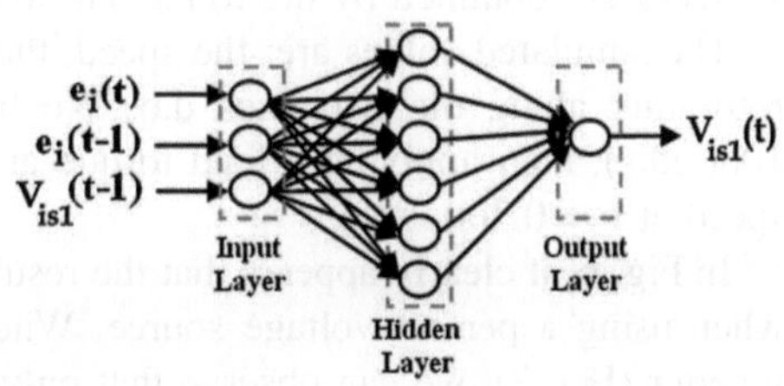

Fig. 4 Graphical representation of the neural currents controllers. with: $i = d, q$; ei = I_{is} - $I_{is\ ref}$: error; V_{is1} : control tension

6 Synthesis of the Neural Controllers

The adopted control law is based on the discrete approach:

$$O(t) = O(t - 1) + P_0 I(t) + P_1 I(t - 1)\ldots \tag{16}$$

I(t), I(t-1)	Inputs controller calculated respectively at the times t, t-1
P_0, P_1	Controller parameters
O(t), O(t-1)	Output controller calculated respectively at the times t, t-1

After several tests, concerning input number, hidden layers number, number of neurons in the hidden layers, training algorithm, squared error and type of learning examples. We obtain the parameters presented in Fig. 4 and Table 1.

Table 1 Neural network characteristics for NCcds, NCcqs

Principal characteristics	Characteristics details
Type	Feed-forward backpropagation network
Activation functions	Hyperbolic tangent sigmoid transfer function for hidden layers and linear transfer function for output layer.
Training algorithm	Levenberg–Marquardt backpropagation
Architecture	[3 6 1]
Nature of examples	A start unloading with inversing speed then applying a load torque with another inversing speed; obtained from a simulation on the fuzzy speed controller.
Examples number	30,000 iterations

7 Simulation Results

For the three decoupling approaches, the behavior of the overall system is tested by simulation for the parameters of the SPIM represented at the Appendix. The simulation results are presented in Figs. 6 and 7. We consider two types of supply: perfect voltage source and real voltage source by two-phase inverter with four legs controlled by the PWM technique, shown in Fig. 5. The control signals of the switches are obtained by the triangular-sinusoidal strategy.

The simulated values are: the speed, the rotor flux and the two components of rotor flux along the two axes d,q. We have simulated a start-up unloaded at 100(rad/s), then applying a load torque at t = 0.4(s) and we finish by inversing speed at t = 0.7(s).

In Fig. 6, it clearly appears that the results for the three approaches are similar, when using a perfect voltage source. When we use the real voltages source by inverter (Fig. 7), we can observe that only in the case of the third approach (the proposed), the rotor flux and the flux along the axis d, follow the value of the reference flux. However, in the other approaches a remarkable decrease of flux is observed. We can notice also at the figure of speed, a good performance obtained by the proposed control. Namely, improved response time and robustness of the system against the disruption caused by the application of a load torque or speed rungs on the dynamics of the global system.

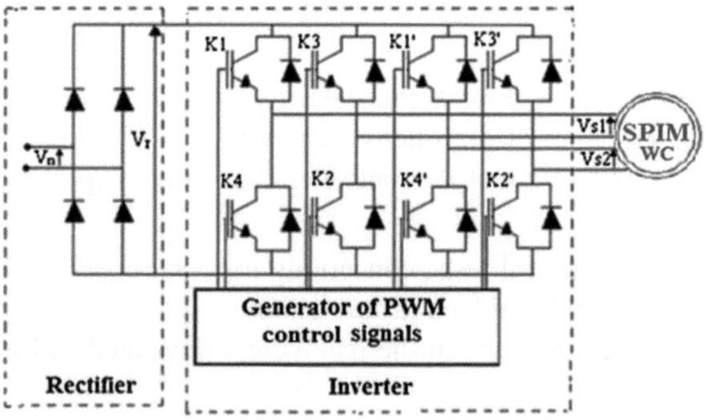

Fig. 5 Diagram of association SPIMWC - static converter

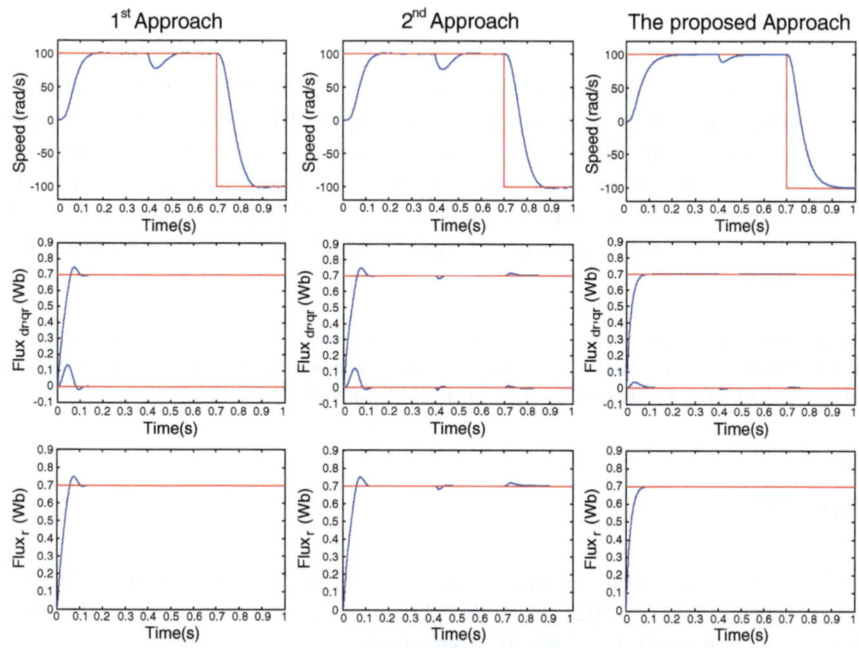

Fig. 6 Simulation results for three different decoupling approaches: a supply by perfect voltage source

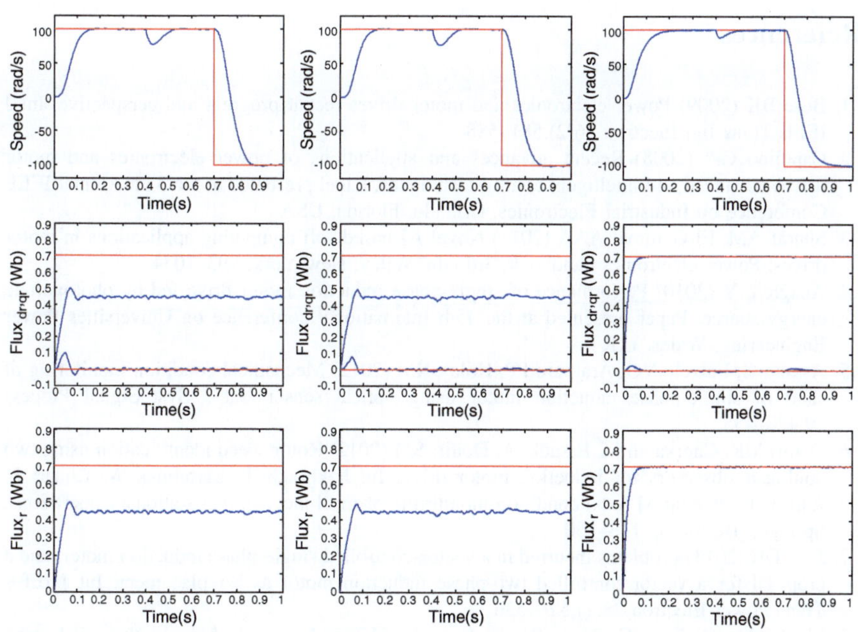

Fig. 7 Simulation results for three different decoupling approaches: a supply by a real voltage source inverter

8 Conclusion

This study compares the performance of the control of SPIM for three different decoupling approaches with two types of supply. The three approaches have similar results for an ideal voltage source. When, it is a real voltage source by inverter, the proposed approach based on ANN presents a clear improvement in the flux; that allowed us to validate this approach. The Analysis of these results shows that the proposed approach is simple, fast, robust and may be implemented in real time.

Acknowledgments This work is supported by the national research project under the J0200220100059 Code. The authors also gratefully acknowledge the helpful comments and suggestions of the reviewers, which have improved the presentation.

Appendix

Single-phase induction machine data: $P_n = 37(W)$, $f = 50(Hz)$, $P = 2$, $V_n = 220(V)$, $R_s = 115(\Omega)$, $R_r = 90.5(\Omega)$, $L_r = L_s = 1.71(H)$, $L_M = 1.41(H)$, $J = 1.2\ 10^{-4}(Kg.m^2)$, $f = 7.63\ 10^{-4}(Nm.s/rad)$.

References

1. Bose BK (2009) Power electronics and motor drives recent progress and perspective. Int J IEEE Trans Ind Electron 56(2):581–588
2. Capolino GA (2008) Recent advances and applications of power electronics and motor drives-advanced and intelligent control techniques. Panel presented at the 34th Annual IEEE Conference on Industrial Electronics, Orlando, Florida, USA
3. Sharaf AM, El-Gammal AAA (2011) Novel AI-based soft computing applications in motor drives. Power electronics handbook, 3rd edn. Wiley, New York, 993–1034
4. Anagreh, Y (2010) Performance of single-phase induction motor drive fed by photovoltaic energy source. Paper presented at the 45th International conference on Universities Power Engineering, Wales, U.K
5. Asmita G, Fulzele VG, Arajpure PP, Holay NM (2012) Mechanical condition monitoring of shaft of single-phase induction motor using optical sensor. Int J Syst Signal Process 29:428–435
6. Douiri MR, Cherkaoui M, Essadki A, Douiri SM (2012) Rotor speed identification using two nonlinear observers for induction motor drive. In: Fakhfakh T, Bartelmus W, Chaari F, Zimroz R, Haddar M (eds) condition monitoring of machinery in non-stationary operations. Springer, Berlin, pp 143–150
7. Jang DH (2013) Problems incurred in a vector-controlled single-phase induction motor, and a proposal for a vector-controlled two-phase induction motor as a replacement. Int J IEEE Trans Power Electron 28(1):526–536
8. Liu L, Ren D, Jiang C (2011) Research on the SVPWM control of single-phase induction motor. Paper presented on the IEEE International conference on Consumer Electronics, Communications and Netwoks, China

9. Pankaj Zope KSP, Prashant S (2011) Development of single phase Z-source inverter using ARM7 for speed control of induction motor. Int J Comput Intell Inf Technology, In: Das VV, Thnakachan N (eds) Communications in Computer and Information Science 250, Springer, New York, pp 440–446
10. Sharaf M, Elgamal AA (2010) Power efficient PID controller of wind driven induction generation single-phase induction motors for electric energy saving applications. Paper presented at the IEEE International conference on Electric Power and Energy, NS Canada
11. Sujitjorn S, Raweekul S, Kulworawanichpong T (2012) Control of paralleled single-phase motors for a crop chopping machine. Int J Control Eng Pract 20:663–673
12. Cabal-Yepez E, Valtierra-Rodriguez M, Romero-Troncoso RJ, Garcia-Perez A, Osornio-Rios RA, Miranda-Vidales H, Alvarez-Salas R (2012) FPGA-based entropy neural processor for online detection of multiple combined faults on induction motors. Int J Mech Syst Signal Process 30:123–130
13. Liu X, Ma L, Mathew J (2009) Machinery fault diagnosis based on fuzzy measure and fuzzy integral data fusion techniques. Int J Mech Syst Signal Process 23:690–700
14. Marichal GN, Mariano Artes M, Garcıea Prada JC, Casanova O (2011) Extraction of rules for faulty bearing classification by a neuro-fuzzy approach. Int JMech Syst Signal Process 25:2073–2082
15. Meneceur R, Metatla A, Meneceur N (2012) State space modeling theory of induction motors for sensorless control and motoring purposes. In: Fakhfakh T, Bartelmus W, Chaari F, Zimroz R, Haddar M (eds) condition monitoring of machinery in non-stationary operations, Springer, Berlin, pp 481–496
16. Xu H, Chen G (2013) An intelligent fault identification method of rolling bearings based on LSSVM optimized by improved PSO. Int J Mech Syst Signal Process 35:167–175
17. Hasanien HM, Muyeen SM (2012) Speed control of grid-connected switched reluctance generator driven by variable speed wind turbine using adaptive neural network controller. Int J Electr Power Syst Res 84:206–213
18. Kulaksız AA, Akkaya R (2012) A genetic algorithm optimized ANN-based MPPT algorithm for a stand-alone PV system with induction motor drive. Int J Solar Energy 86:2366–2375
19. Muljadi E, Zhao Y, Liu TH, Lipo TA (1993) Adjustable AC capacitor for single-phase induction motor. Int J IEEE Trans Ind Appl 29(3):479–485
20. Wildi T, Sybille G (2005) Electrotechnique. Laval University Press 4th edn. E.S.K.A, de Boek Belgium
21. Kostenko M, Piotrovski L (1969) Machines électriques II. MIR ed, Moscou
22. Blaschke F (1972) The principal of field orientation as applied to the transvector closed-loop system for rotating field machine. Siemens Rev 34:217–220

Fault Identification on Electrical Machines Based on Experimental Analysis

H. Balan, M. I Buzdugan and Karaisas P.

Abstract The paper reviews the main faults identification of electric machines based on amplitude versus time and amplitude versus frequency vibration analysis. The aim of the paper is to introduce a new method in determining rotor faults based on stator vibration. The analysis method uses a multi-channel vibration analyzer provided with three accelerometers, set in a plane perpendicular to the stator axis. The method is suitable for determining rotor faults generated by the stator vibrations, other than those due to current harmonic components or supply voltage unbalance. The SvanPC software allowed the determination of the characteristics regarding velocity vibration versus time and versus frequency. Their interpretation permitted to determine the fault points on electrical machines in two industrial applications.

Keywords Motor vibrations signature analysis (MVSA) · Motor current signature analysis (MCSA) · Experimental analysis (EVA)

1 Introduction

Monitoring and diagnosis of electrical machines significantly reduce maintenance costs and unexpected fault hazard, allowing an early detection of their improper operation.

H. Balan (✉) · M. I Buzdugan
Technical University of Cluj Napoca, 28, Memorandumului Street, Cluj Napoca, Romania
e-mail: horia.balan@eps.utcluj.ro

M. I Buzdugan
e-mail: mircea.buzdugan@insta.utcluj.ro

K. P.
T. E. I. of Piraeus, 250th Petrou Ralli & Thivon Street, 12244 Egaleo Athens, Greece
e-mail: karaisas@teipir.gr

G. Dalpiaz et al. (eds.), *Advances in Condition Monitoring of Machinery in Non-Stationary Operations*, Lecture Notes in Mechanical Engineering,
DOI: 10.1007/978-3-642-39348-8_53,

Incorrect operating conditions [1, 2], may lead to faults on the main parts of an electrical machine: stator, rotor, slide bearings and other types of faults.

Almost 38 % of the faults are located in the stator [3]. Stator windings are placed in slots and a lower insulation resistance leads to overloads or short-circuit currents. Faulty currents determine the stator overheating, leading to unbalanced magnetic fields in the air gap and unacceptable magnitudes of the vibration of the machine.

On the other hand, rotor faults [4, 5, 19], represent approximately 10 % of the total amount of faults (usually faulty rotor bars which lead to local overheating and vibration of the rotor [6]).

Most electrical machines are provided with slide bearings (ball bearings) [20], their faults representing 40 % of the total amount of faults, determining undesired rotational speed changes [23] and bearing eccentricities [21].

Other types of faults (12 % of the total amount of faults) are represented by rotor eccentricities with respect to the stator determining air gap non-uniformities. Changes in the air gap modify the magnetic field distribution and consequently a magnetic force on the minimum air gap direction, leading to mechanical vibrations.

Monitoring and diagnosis techniques of the electric machines operation are divided in passive and active techniques. The passive ones consist in measurements which do not affect the normal operation of the machines, while the active ones use control elements (e.g. inverters), modifying their operating conditions.

There are different monitoring techniques, depending on the signals involved:

- electrical signals monitoring (currents or voltages) or of the angular difference between the rotor and the armature, is suitable in detecting and evaluating electric faults [7].
- magnetic values monitoring, is based on measurements of the induced voltage in the rotating coils placed in the armature slots or outsides them, similar to Rogowski coils [8].
- vibrations monitoring, uses accelerometers measuring the velocity and the acceleration due to mechanical unbalances; this method is used in mechanical fault diagnosis, this type of faults being difficult to detect using electrical signals measurement.
- temperature monitoring can be performed using direct measurements or analysis performed on different models, because modifications in the temperature's field derive from electrical or mechanical faults [9, 19].
- acoustic emission monitoring, is based on the acoustic noise emitted by electrical machines in operation [22]. This type of monitoring is quite difficult to be used in industrial environment, due to the ambient noise which alters the measuring process.

The difficult part comes after monitoring the process, consisting in the analysis of the monitored data. The most known techniques on this issue are the time domain analysis [10, 11] and in the frequency domain analysis [12].

Lately several new techniques in locating the fault were developed:

- use of wavelet transform technique [13, 14].
- analysis of the dynamic time warping technique[15].
- vector techniques [16, 17].

2 Motor Vibrations Signature Analysis-MVSA

Analysis performed in amplitude versus time and amplitude versus frequency domains are the basis of electric machines vibration signature. The limit values in electric machines standards set vibration velocity as a compulsory value to be measured. The diagrams representing vibration velocity versus frequency permit to determine the RMS value of the vibration damping v_{eff}, according to the following procedure.

$$m \cdot \frac{d^2x}{dt^2} + c \cdot \frac{dx}{dt} + k \cdot x = F(t) \tag{1}$$

where:

m	mass of the system;
k	dynamic stiffness coefficient
c	dynamic viscosity coefficient
$x(t)$	vibration of vertical amplitude
$F(t)$	vibration excitation force

Neglecting the damping of the system, the instantaneous amplitude of vibration $x(t)$ becomes:

$$x(t) = \frac{F_0}{k(1 - r^2)} \cdot \sin \omega t \tag{2}$$

where:

$$r = \frac{\omega}{\omega_n} \text{ and } \omega_n = \sqrt{\frac{k}{m}} \tag{3}$$

If the system presents natural oscillations (oscillates at the resonance frequency), the vibration amplitude is high even for a low excitation force. The frequency value is:

$$f_n = \frac{\omega_n}{2 \cdot \pi} = \frac{2}{2 \cdot \pi} \sqrt{\frac{k}{m}} \tag{4}$$

and the corresponding speed of the critical frequency:

$$RPM_{cr} = 60 \cdot f_n = \frac{60}{2 \cdot \pi} \cdot \sqrt{\frac{k}{m}} \tag{5}$$

The force transmitted to the floor is:

$$F_T = k \cdot x \tag{6}$$

In the case of harmonic vibrations, with the instantaneous value of the vibration velocity:

$$v_i = \hat{v} \cdot \cos \omega_1 t \tag{7}$$

or of complex vibrations (i.e. a superposition of harmonic vibrations), vibration's intensity is defined as the root mean square value (RMS value) of the oscillation velocity.

The RMS value is calculated from the time diagrams of the vibration's velocity using Eq. (8):

$$v_{eff} = \sqrt{\frac{1}{T} \int_0^T v^2(t) \cdot dt} \tag{8}$$

Acceleration, speed and displacement amplitudes $(\hat{a}_j; \hat{v}_j; \hat{s}_j; \quad j = 1, 2\ldots, n)$ can be derived as functions of the rotating speed $(\omega_1, \omega_2, \ldots \omega_n)$. The RMS value was calculated according to ISO 10816 Standard, using Eq. (9):

$$\begin{aligned} v_{eff} &= \sqrt{\frac{1}{2} \cdot \left[\left(\frac{\hat{a}_1}{\omega_1} \right)^2 + \left(\frac{\hat{a}_2}{\omega_2} \right)^2 + \cdots + \left(\frac{\hat{a}_n}{\omega_n} \right)^2 \right]} \\ &= \sqrt{\frac{1}{2} \left(\hat{s}_1^2 \cdot \omega_1^2 + \hat{s}_2^2 \cdot \omega_2^2 + \cdots + \hat{s}_n^2 \cdot \omega_n^2 \right)} \\ &= \sqrt{\frac{1}{2} \cdot \left(\hat{v}_1^2 + \hat{v}_2^2 + \cdots + \hat{v}_n^2 \right)} \end{aligned} \tag{9}$$

If there are no vibration components with slightly close frequencies which could generate a beat phenomenon, the RMS value of the vibration velocity can be calculated as:

$$v_{eff} = \sqrt{\frac{1}{2} \cdot \left(\hat{v}_{\max}^2 + \hat{v}_{\min}^2 \right)} \tag{10}$$

where:

$\hat{v}_{\max}$ the peak value at the maximum value of the envelope curve

$\hat{v}_{\min}$ the peak value at the minimum value of the envelope curve

If the RMS value is measured using a measuring device, the RMS velocity will be approximately:

$$v_{eff} = \sqrt{\frac{1}{2} \cdot \left(R_{\max}^2 + R_{\min}^2\right)} \tag{11}$$

where:

$R_{\max}$ the maximum value indicated by the measuring device
$R_{\min}$ the minimum value indicated by the measuring device

As the RMS value calculated according to (11) is not an accurate one, it is advisable to use this method only for very low beat frequencies.

The above analysis allows setting vibration's velocity in the limits imposed in the standards. If the limits are exceeded, analysis of the vibration velocity versus frequency permits the fault identification on the electrical machine [26]:

- in the case of dynamic unbalance (Fig. 1a), a 1X harmonic component, different from the rotational speed occurs;
- rotor and stator misalignment issue (Fig. 1b) determines in the frequency spectrum three components, i.e. 1X, 2X and 3X;
- improper connections, insulation faults between laminates and phase unbalance (Fig. 1c) determine the frequency components 1X, 2X and 2FL, the latest having the same frequency as the supply voltage;
- rotor eccentricity (Fig. 1d), caused by faults on rotor bars, commutator or bearings, determine the frequency components 1X, 2X, 3X and –PPF and +PPF (due to adjacent poles);
- bearing faults (Fig. 1e) determine a continuous high frequency spectrum, beyond the frequency 1X.

3 Motor Current Signature Analysis-MCSA

The purpose of this approach is the fault detection on electrical machines using vibrations' signature analysis as a diagnosis and maintenance tool, so electrical values (supply voltage unbalance or current's harmonics) are not taken into account. The point is to locate non-electrical faults, i.e. mechanical faults, their location being established after measuring the current signature determined by vibrations.

The stator current amplitude-frequency signature highlights electrical and mechanical faults:

- Current unbalance (Fig. 2) is determined by stator infrastructure faults: insulation damage of the laminated sheets or high contact resistance and overloads or short-circuits.

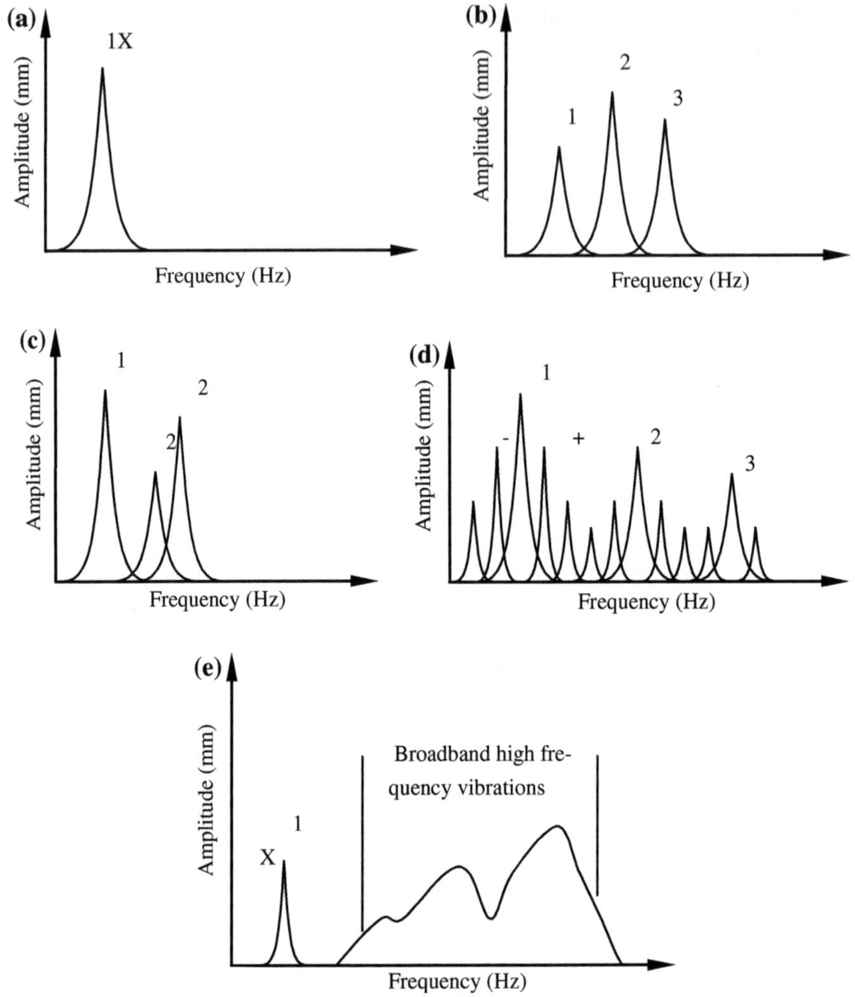

Fig. 1 The fault identification on the electrical machine. **a** Dynamic unbalance. **b** Rotor and stator misalignment c-stator eccentricity d-rotor eccentricity e-faulty bearings

- Vibrations due to asymmetries can be highlighted using an asymmetric supply system of the electric motor provided by a programmable source (15003iX-CTS, California Instruments). The RMS voltages in the supply lines are: line a- 220 V, line b- 165 V and line c- 110 V. The vibration amplitude versus time waveforms are depicted in Fig. 3, the vibration measurements being performed using three transducers coplanar mounted on the stator, displaced by 120°.
- Vibrations due to current harmonics can be generated using the same programmable source. Harmonic components determined by the motor or by the source (Fig. 3) cause vibrations. As the ratio between the harmonic components

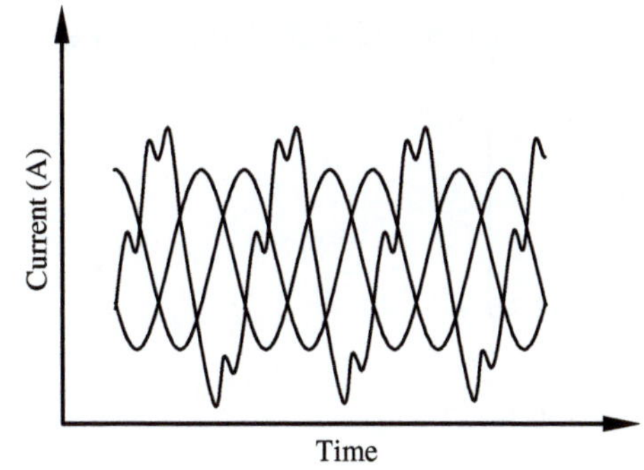

Fig. 2 Stator current asymmetry

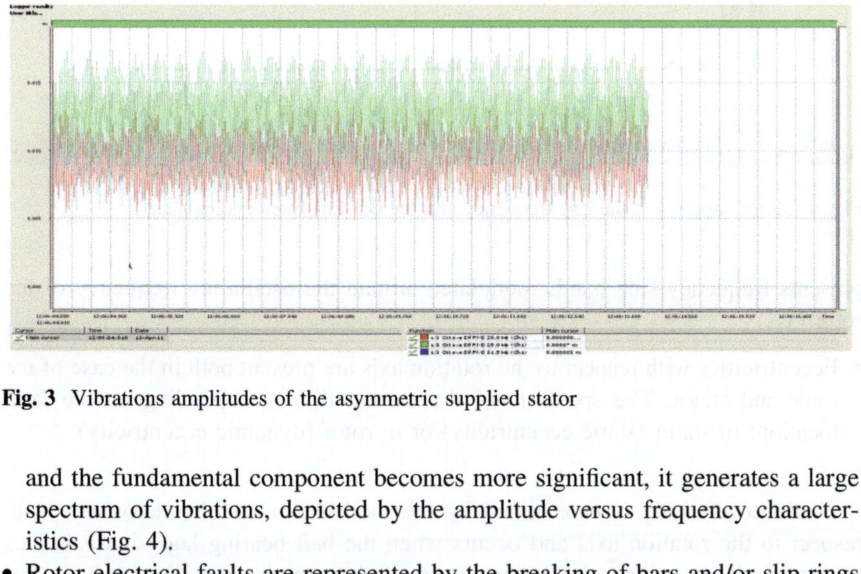

Fig. 3 Vibrations amplitudes of the asymmetric supplied stator

and the fundamental component becomes more significant, it generates a large spectrum of vibrations, depicted by the amplitude versus frequency characteristics (Fig. 4).

- Rotor electrical faults are represented by the breaking of bars and/or slip rings and higher contact resistance. The frequency spectrum of the current presents harmonic components around the fundamental frequency. The frequency of the harmonic components can be calculated using Eq. (12) [4] (Fig. 5):

$$f_b = \left(\frac{k}{p \cdot (1-s)} \pm s\right) \cdot f_1; \frac{k}{p} = 1, 3, 5\ldots \tag{12}$$

where f_1 is the frequency of the power supply, s the motor slip and p the pole pairs number. The adjacent poles are determined from the ratio k/p and correspond to the oscillation frequencies of the rotor [6]. The spectrum of the current (Fig. 6a)

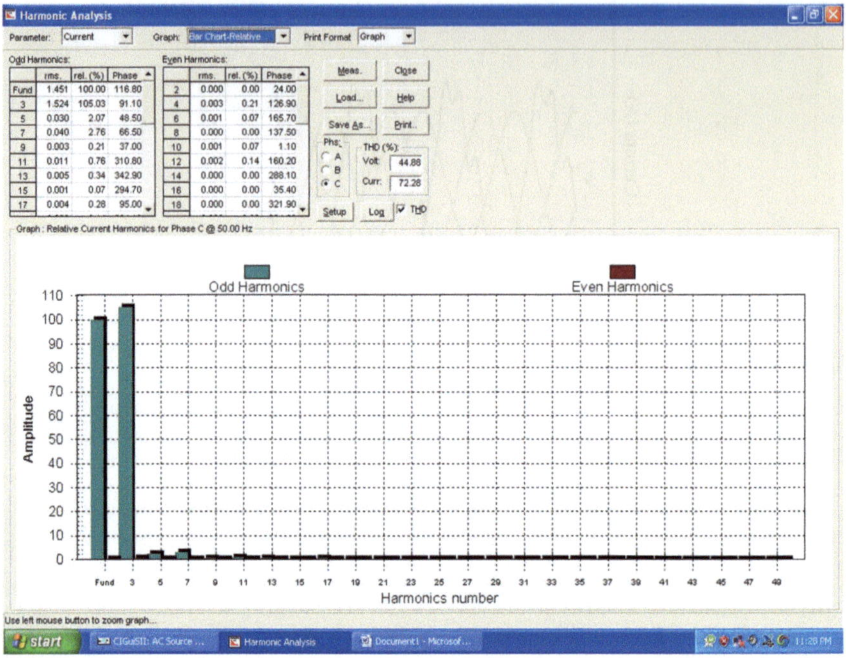

Fig. 4 Motor supplied at the line frequency and the third harmonic component

presents frequency side bands modulated around the oscillating frequency of the rotor (1X).

- Eccentricities with respect to the rotation axis are present both in the case of the rotor and stator. The spectrum of the current differs, depending on the fault location: in stator (static eccentricity) or in rotor (dynamic eccentricity).

Static eccentricity is determined by the modification of the stator axis with respect to the rotation axis and occurs when the ball bearing lager is oval. This type of fault determines in the spectrum of the current distortions whose frequencies are calculated using Eq. (13):

$$f_{static} = \left((k \cdot n) \cdot \left(\frac{1-s}{p} \right) \pm v \right) \cdot f_1 \tag{13}$$

where k is a positive constant, n the slots number and v the harmonic order.

Static eccentricity of the stator leads to the premature aging of the slide bearings and balls damage.

Current spectrum analysis highlights the presence of the side bands ±FL and ±3 FL, around the rotation frequency of the rotor slots (Fig. 6b).

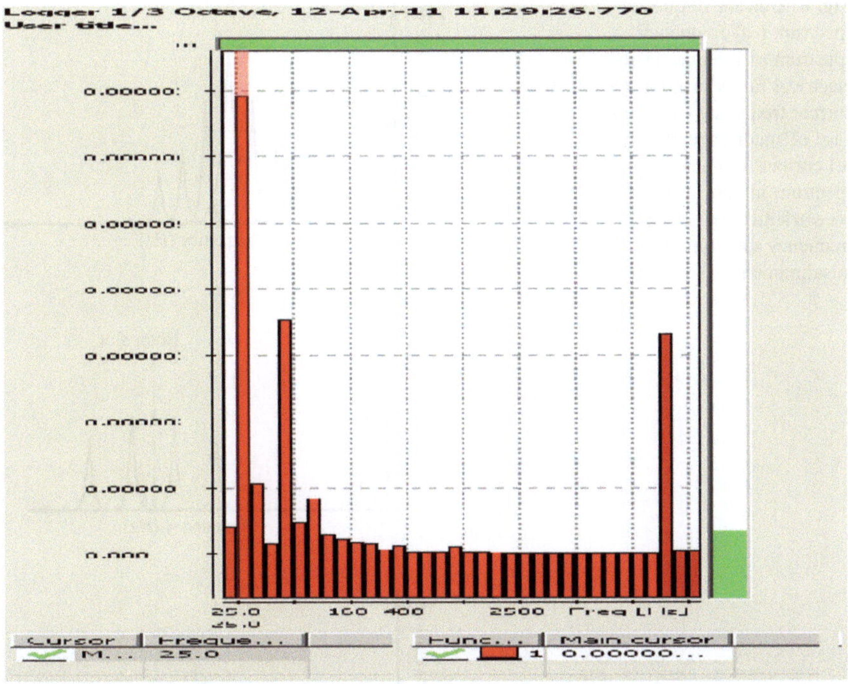

Fig. 5 Vibration spectrum of harmonic current

Dynamic eccentricities [10] occur when the rotor axis does not correspond with the normal rotation axis and are present in the case of bent rotors or in the presence of mechanical resonances. These types of faults determines overheating, ball bearings aging and operation of the machine at critical speeds (see Eqs. (4–6). The current spectrum analysis highlights the presence of the side bands ±FL and ±3 FL, around the rotation frequency of the rotor slots which does not correspond to the rotation frequency RS (Fig. 6c).

- Misalignment between the electric motor and the coupling device determines the presence in the current spectrum of the frequency 1X, different from the rotation frequency (Fig. 6d).

4 Applications and Experimental Results

In the present section are presented two applications of the vibration signature method in determining faults on electrical machines.

Measurements must be properly performed, using an accurate system (accelerometers—vibration analyzer), in order that vibration amplitude versus time and

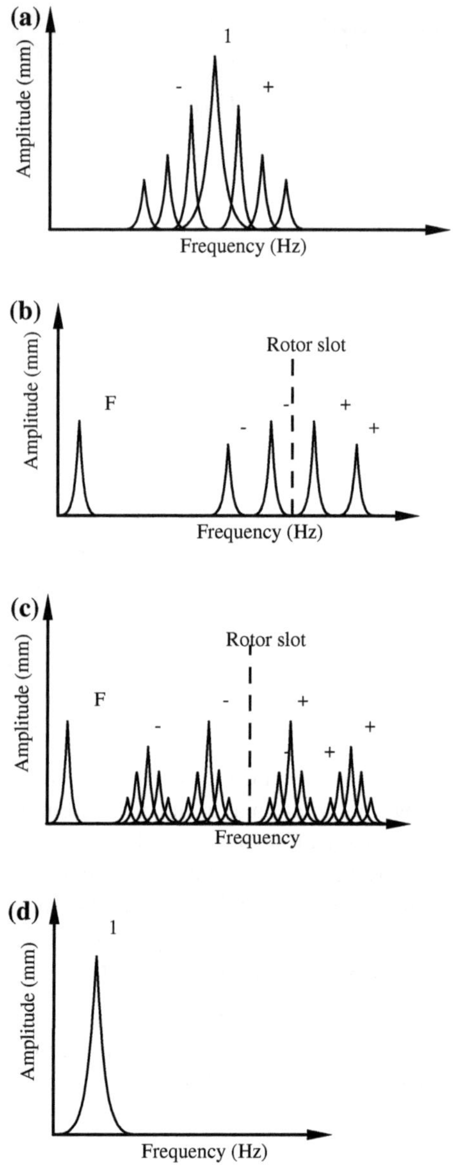

Fig. 6 Current frequency spectrum (**a**) frequency spectrum of the current for electrical faults in rotor, (**b**) current frequency spectrum in case of stator eccentricities, (**c**) current frequency spectrum in case of rotor eccentricities, (**d**) current frequency spectrum in case of misalignment

versus frequency characteristics reflect the actual operation of the electrical machine.

The accelerometer is the primary device used in vibrations monitoring, representing a transducer which converts the acceleration generated by static and dynamic forces in electrical signals. It offers a large voltage excursion, covering the maximum vibration level, usually presenting a sensitivity of 1,000 mV/g. Its sensitivity and frequency range are decisive for the selection. In case of electrical machines monitoring 10,000 Hz frequency range accelerometers should be selected.

A very important issue is to comply with standards from the point of view of mounting.

Accelerometers are connected to vibration analyzers, which can be set to measure vibration amplitudes, velocities and accelerations, along with the analysis versus time or frequency.

In the following applications, mono-channel [27] and three-channel [28] accelerometers were used, conceived for 3D determinations in a three-axis orthogonal system and adapted for three mono-axial accelerometers, placed coplanar on the electrical machine stator.

A first application (Fig. 7) refers to a production line for secure glass [29], where the monitoring system detected an overcome of the levels of the vibration velocity admitted by the standards.

Vibration measurements must be performed in compliance with ISO 10816 Standard [24], which provides vibration amplitude acceptance guidelines for rotating machinery operating from 600 until 12,000 rpm, applied to electric generators, gas or steam turbines, turbo-compressors and vertical, horizontal or inclined shaft fans.

ISO 10816 Standard specifies the RMS vibration velocity limits on a basis of machine horsepower, and covers a frequency range from 10 Hz to 1,000 Hz. The limits in the standard are set in RMS values.

As measuring points, the ISO 10816 Standard guidelines recommend:

- for vertical or inclined mounted machines, areas with maximum value of vibrations must be chosen;

Fig. 7 The main measuring points

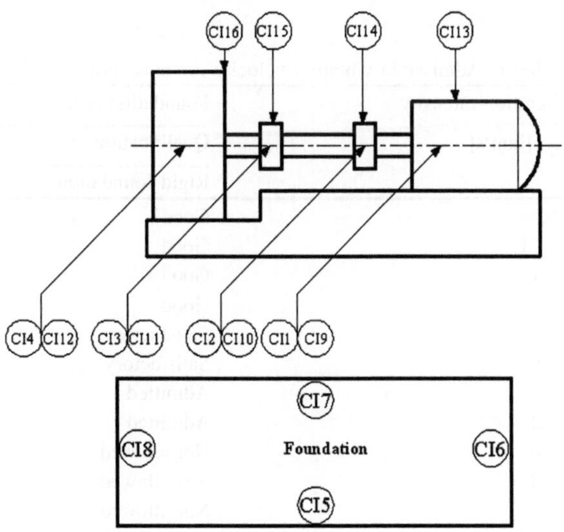

- measurements must be performed on accessible parts of the machines;
- measurements outcomes must have a high level of accuracy, representing the main vibrations and not the local resonation ones.

The ISO 10816 Standard Operation defines also the measurements conditions:

- measurements should be performed only in the steady state regime of rotor operation for constant speed and constant load drives;
- for adjustable speed drives (ASD), measurements must be performed in extreme conditions, covering the operation characteristics;

Moreover, ISO 10816 Standard, differentiates between rigid and elastic building foundations (Table 1), setting admissible vibration velocity limits.

The calculus of the RMS value of the vibration velocity using (Eqs. 8–11) is rather complicated, especially if there are a lot of points in which the FFT (Fast Fourier Transform) must be considered.

A first set of measurements was carried out using mono-axial accelerometers in the case of the driving motor, in horizontal and vertical plane (Fig. 7).

Diagrams corresponding to the stator measuring points: acceleration versus time (point C19, Fig. 8a) and acceleration versus frequency (point C19, Fig. 8b) analyzed using the vibration signature method [26], could not identify the nature and the location of the fault, so a further examination became necessary.

The vibration acceleration in a plane perpendicular to the rotational axis, in three coplanar points displaced first at 120° and then at 90° respectively. Using Eqs. 8–11, the velocity vector of for the two measurements setup can be determined (Fig. 9).

The main drawback of the method used in this application consisted in the use of mono-channel vibration analyzer [27], measurements in the three points being not simultaneous.

Table 1 Admissible vibration velocity limits according to ISO 10816 Standard

Vibration intensity v_{eff} [mm/s]	Foundation type Qualification	
	Rigid foundation	Elastic foundation
0.46	Good	Good
0.71	Good	Good
1.12	Good	Good
1.8	Good	Good
2.8	Satisfactory	Good
4.6	Satisfactory	Satisfactory
7.1	Admitted	Satisfactory
11.2	Admitted	Admitted
18.0	Not allowed	Admitted
28.0	Not allowed	Not allowed
71.0	Not allowed	Not allowed

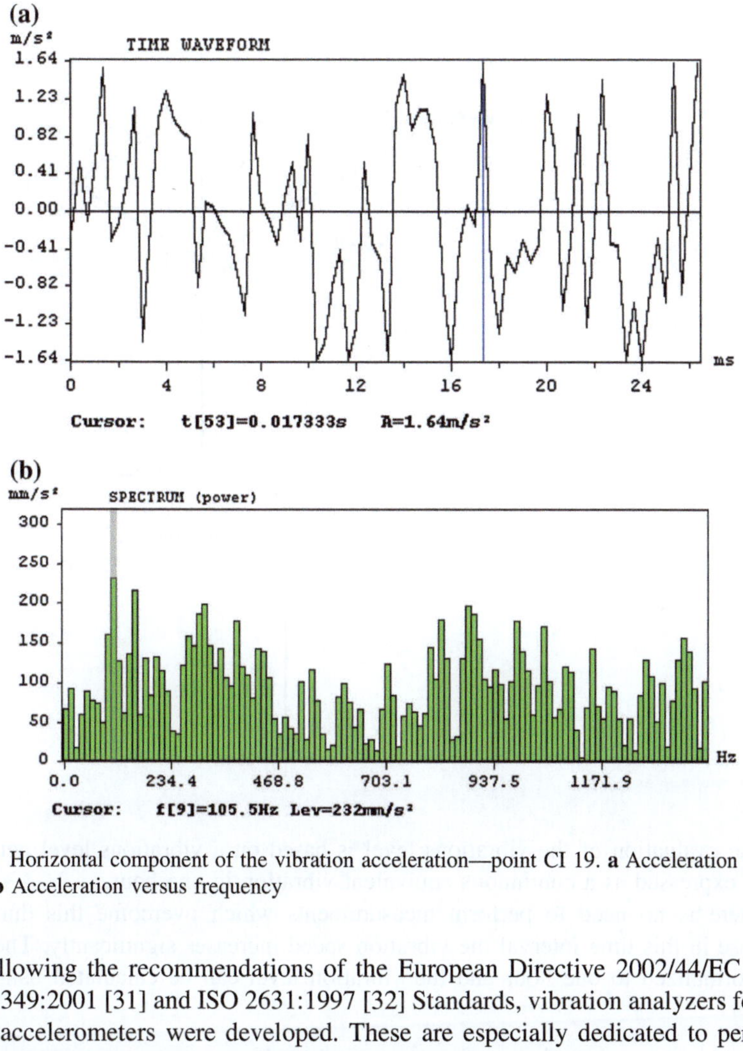

Fig. 8 Horizontal component of the vibration acceleration—point CI 19. **a** Acceleration versus time. **b** Acceleration versus frequency

Following the recommendations of the European Directive 2002/44/EC [30], ISO 5349:2001 [31] and ISO 2631:1997 [32] Standards, vibration analyzers for tri-axial accelerometers were developed. These are especially dedicated to perform analyses of the vibrations transmission from building floors to humans.

Measurements were carried out using such an analyzer, namely SVAN 958 [28], and instead of the tri-axial accelerometer, three mono-axial accelerometers were mounted, in a plane perpendicular to the rotation axis of the electrical motor under diagnosis.

The fault direction on the electric motor's stator was determined, its value provides an indication concerning the faulty area: in the stator or in the rotor area of the electric motor.

The analysis versus time and frequency of the measurements carried out using the software SVAN PC [33] eliminates the previous drawback, because it has the capacity of performing simultaneous measurements and calculus of the vibrations' directions in the three locations, making needless the use of the Eqs. 8–11.

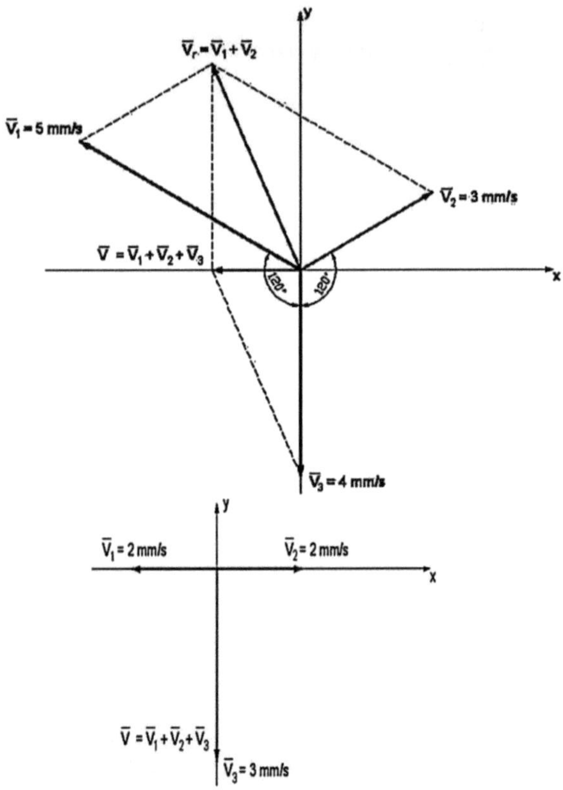

Fig. 9 The fault direction in the two measurement setups

The evaluation of the vibrations level is based on a vibrations level per hour A(1), expressed as a continuous equivalent vibration in one hour.

There is no need to perform measurements which overcome this duration, because in this time interval the vibration speed increases significantly. The data are normalized to one hour and the vibration level can be calculated using Eq. (14):

$$A(1) = v_{ae} \cdot \sqrt{\frac{T_e}{T_0}}, \tag{14}$$

where:

v_{ae} represents the mean value of the weighted vibration speed during the measurements;

T_e represents the total measurement interval;

T_0 the time reference (one hour).

If the measurements are performed using three accelerometers (the analyzer SVAN 958 is designed for tri-axial accelerometers, so its connectors have been adapted for three mono-axial accelerometers), the total vibration speed of the electric motor under diagnosis being calculated using Eq. 15:

$$v_{ve} = \sqrt{v_{w1}^2 + v_{w2}^2 + v_{w3}^2} \tag{15}$$

the indices 1, 2 and 3 corresponding to the three accelerometers respectively.

The presented technique has been used in an application in which the vibration signature for two electric motors has been determined. The first was a faulty one (1.55 kW, 230 V/50 Hz, 1,500 rpm), the second was a normal operating one (2.2 kW, 400 V/50 Hz, 2,880 rpm). The two motors have been used in driving the pumps of the heating central in an office building [25, 29].

5 Conclusions

The vibrations signature analysis determined by the electric motors is quite a difficult task because in real situations the shape of the amplitude versus frequency characteristics (see Figs. 8b, 10b, and 11b) is much more complex than the shape of the ideal ones (see Fig. 1a–e), recommended in literature [26].

The experimental results of the first application highlights a vibration velocity vector determined using three accelerometers, mounted in two different positions on the stator periphery, which presents different magnitude and phase in the two situations depicted.

For a normal operation of a motor, the magnitude of the vector must be as lower as possible, having the same phase displacement or slightly different. However, for an accurate interpretation of the amplitude versus frequency vibrations it is advisable to analyze also the amplitude-frequency characteristics of the stator current, in order to be convinced that current distortions or harmonics do not lead to extra-vibrations.

The result of the analysis obtained with mono-channel vibration analyzers presents the major drawback of the non-simultaneous measurements. The magnitude and the phase displacement between the two mounting positions (Fig. 9) highlights that the fault location is on the rotor, confirmed also by the comparison between the amplitude versus frequency characteristics (see Figs. 1e and 8).

The multi-channel analyzer used in the second application is more straightforward in measuring and interpretation of the measuring results. The amplitude versus frequency characteristics in the case of the normal operating motor highlights that the fundamental component of the frequency is the power frequency (i.e. 50 Hz, in Fig. 10b), while for the faulty one is 1,000 Hz (Fig. 11b).

Another important advantage of the multi-channel analyzers consists in their software, which directly determines the vibration velocity vector, simultaneously for the three accelerometers and for the entire measurement duration.

For the normal operating motor, only a slight variation of the acceleration vector is recorded, its magnitude tending asymptotically to 160 m/s^2 (Fig. 11a).

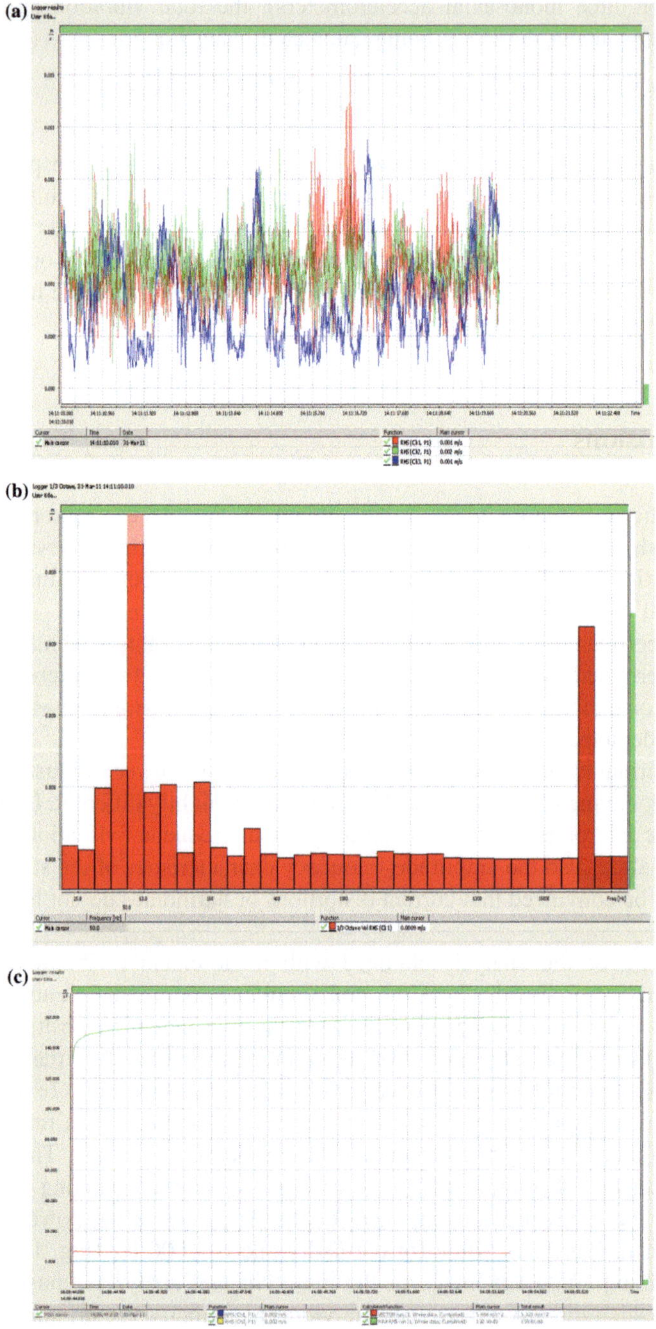

Fig. 10 The normal operating motor. (**a**) Vibration versus time signature. (**b**) Vibration versus frequency signature. (**c**) Vector analysis

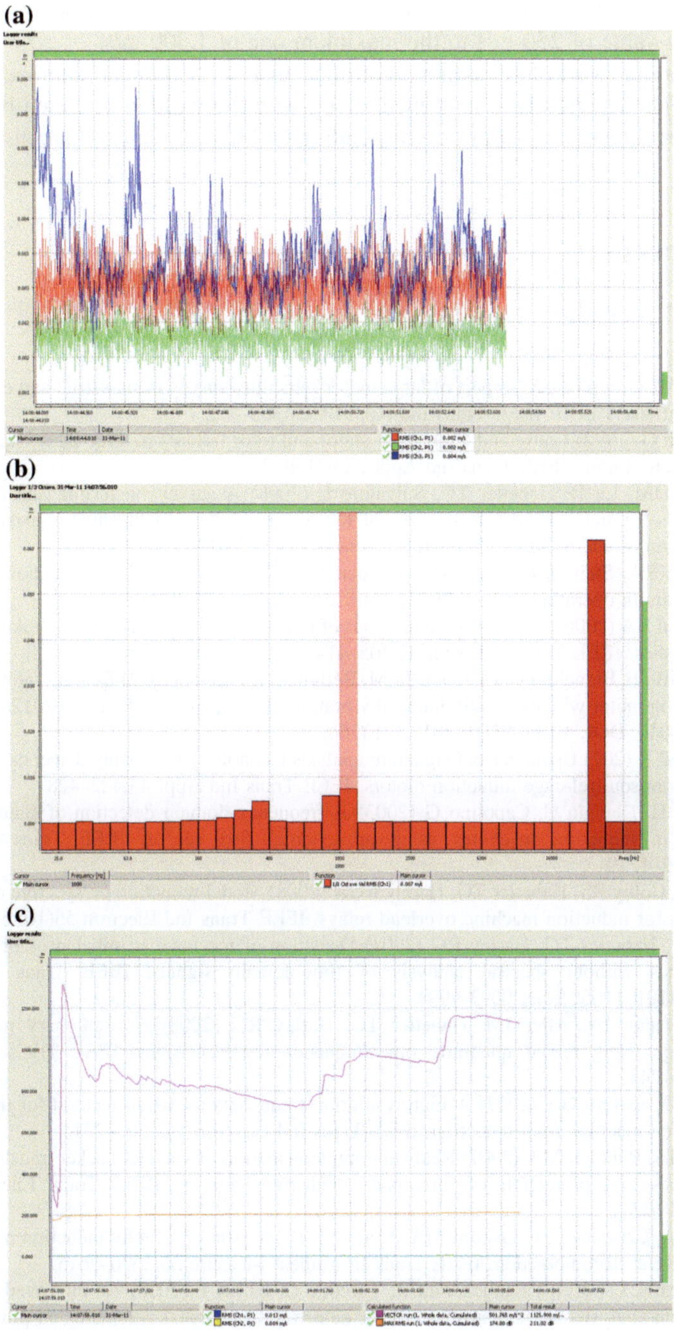

Fig. 11 The faulty motor. (**a**) Vibration versus time signature. (**b**) Vibration versus frequency signature. (**c**) Vector analysis

For the faulty motor, the vibration velocity is changing permanently, from the minimum value of 250 m/s to the maximum one of 1,300 m/s.

As further work, the authors consider that a theoretical approach of the experiments depicted here is needed, along with the buildup of a data base, containing the direction diagrams corresponding to different faulty situations [18, 34].

References

Journal Article

1. M. R. W. Group (1985) Report of large motor reliability survey of industrial and commercial installations. Part II: IEEE Trans Ind Appl IA-21:865–872
2. Bonnett G (1992) Cause and analysis of stator and rotor failures in three-phase squirrel-cage induction motors. IEEE Trans Ind Appl 28:921–937
3. Grubic JM, Lu B, Habetler TG, Salvatore L (2008) A survey on testing and monitoring methods for stator insulation systems of low-voltage induction machines focusing on urn insulation problems. IEEE Trans Ind Electron 55:4127–4136
4. Kliman RA, Stein J, Endicott RD, Madden MW (1998) Non-invasive detection of broken rotor bars in operating motors. IEEE Trans Energy Conv 28:873–879
5. Mirafzal NA (2006) On innovative methods of induction motor interturn and broken-bar fault diagnostics. IEEE Trans Ind Appl 42:405–414
6. Immovilli F, Bianchini C, Cocconcelli M, Bellini A, Rubini R (2012) Bearing fault model for induction motor with externally induced vibration. IEEE Trans Ind Electron 2012(99). ISSN: 0278-0046, DOI: 10.1109/TIE.2012.2213566
7. Wendell R (207) Using current signature analysis technology to reliably detect cage winding defects in squirrel-cage induction motors. IEEE Trans Ind Appl 43:422–428
8. Henao H, Demian C, Capolino G (2003) A frequency-domain detection of stator winding faults in induction machines using an external flux sensor. IEEE Trans Ind Appl 39:1272–1279
9. Gao Z, Colby RS, Habetler TG, Harley RGA (2008) Model reduction perspective on thermal models for induction machine overload relays. IEEE Trans Ind Electron 55(10):3525–3534
10. Kral C, Habetler TG, Harley RG (2004) Detection of mechanical imbalances of induction machines without spectral analysis of time-domain signals. IEEE Trans Ind Appl 40(4):1101–1105. ISSN 0093-9994
11. Rajagopalan JA, Aller JM, Habetler TG, Harley RG (2008) Nonstationary motor fault detection using recent quadratic time-frequency representations. IEEE Trans Ind Appl 44:735–744
12. Kia H, Capolino GA (2007) A high-resolution frequencyestimation method for three-phase induction machine fault detection. IEEE Trans Ind Electron 54:2305–2314
13. Khan TS, Rahman MA (2007) Real-time implementation of wavelet packet transform-based diagnosis and protection of three-phase induction motors. IEEE Trans Energy Conv 22:647–655
14. Ponci A, Cristaldi L, Lazzaroni MALM (2007) Diagnostic of a faulty induction motor drive via wavelet decomposition. IEEE Trans Instrument Measure 56:2606–2615
15. Leonard F, Landry C, Beauchemin R, Turcotte O, Brikci F (2008) An improved algorithm vibration analysis as a diagnostic tool for detecting mechanical anomalies on power circuit breakers. IEEE Trans Power Delivery 23(4): 1986

16. Diallo M, Hamad D, Pierre X (2005) Fault detection and diagnosis in an induction machine drive: a pattern recognition approach based on concordia stator mean current vector. IEEE Trans Energy Conv 20:512–519
17. Boesing M, Schoenen T, Kasper KA, De Donker RW (2010) Vibration synthesis for electrical machines based on force response superposition. IEEE Trans Magnet 46:2986–2989
18. Concari C, Franceschini G, Tassoni C (2011) Toward practical quantification of induction drive mixed eccentricity. IEEE Trans Ind Appl 47(3), 2011, DOI: 10.1109/TIA.2011.2124434

Journal Article Only by DOI

19. Kral C, Habetler TG, Harley RG, Pirker F, Pascoli G, Oberguggenberger H, Fenz CJM (2003) A comparison of rotor fault detection techniques with respect to the assessment of fault severity. In: Symposium on diagnostics for electric machines, power electronics and drives. SDEMPED 2003, pp 265–270. DOI: 10.1109/DEMPED.2003.1234584
20. Frosini L, Bassi E, Fazzi A, Gazzaniga C (2008) Use of the stator current for conditioning monitoring of bearings in induction motors: In: Proceedings of ICEM'08- XVIII international conference on electrical machines, Vilamoura (Algarve), Portugal,from 6–9 Sept 2008. DOI: 10.1109/ICELMACH.2008.4799991
21. Nandi S, Toliyat HA (1999) Condition monitoring and fault diagnosis of electrical machines-a review. IEEE Industrial Applications Society Annual Meeting 1999. DOI: 10.1109/IAS.1999.799956
22. Gaylard A, Meyer A, Landy C (1995) Acoustic evaluation of faults in electrical machines. In: 7th international conference on electrical machines and drives. IEE Conference Publication No. 412, pp 147–150. DOI: 10.1049/cp:19950852

Book

23. Vas P (1993) Parameter estimation, condition monitoring, and diagnosis of electrical machines. Clarendon Press, Oxford

Online Document (no DOI Available)

24. ISO 10816 "Mechanical vibration — Evaluation of machine vibration by measurements on non-rotating parts —Part 3: Industrial machines with nominal power above 15 kW and nominal speeds between 120 and 15, 000 rot/min when measured in situ", www.pruftechnik.com Condition Monitoring
25. Balan H, Vadan I, Stefanescu S, Karaisas P (2008) Vibro-acoustic diagnosis of a ventilator driving system. In: Proceedings of the 2008 international conference on electrical machines, Paper ID 1081; Proceedings of ICEM'08- XVIII international conference on electrical machines, Vilamoura (Algarve), Portugal, 6–9 Sept 2008, on CD, http://www.apdee.org/conferences/icem08
26. Frédéric C (2009) Condition monitoring of induction motors using vibration and electrical signature analysis. In: EE mods conference 14–17 sept 2009, Nantes, France, www.docstoc.com/.../Condition-Monitoring-of-Induction-Motors-Usi
27. ***SVANTEK „SVAN 912 AE four channels sound & vibration analyzer" technical book, www.svantek.com

28. *** Noise and vibration analyzer SVAN 958", Technical Book, www.svantek.com
29. Vibro-acoustic diagnosis for ventilators driving system of UNIGLASS equipment type UGT-2400x4200. Contract 35/2007, www.utcluj.ro
30. *** "Directive 2002/44/EC for occupational vibration exposure", www.ueanet.com/facts2/summ/sum-02-44-en.pdf
31. [ISO 5349:(2001) Measurement and evaluation of human exposure to hand-transmitted vibration", www.freestd.us › Standards Worldwide
32. [*** ISO 8041 (2005) Human response to vibration—Measuring Instrumentation", www.iso.org/iso/catalogue
33. Svan (PC) Software, www.svantek.com
34. Dorrell DG (2011) Sources and characteristics of unbalanced magnetic pull in three-phase cage induction motors with axial-varying rotor eccentricity. In: IEEE Trans Ind Appl 47(1):12–24. DOI: 10.1109/TIA.2010.2090845

Fault Diagnosis in Induction Motor Using Motor's Residual Stator Current Signature Analysis

Khalid Dahi, Soumia Elhani, Said Guedira and Nabil Ngote

Abstract In this paper, we present fault prognosis and diagnosis technique in a three phase asynchronous machine. Based on statistical analysis of scalar indicator resulting from the TSA method (Time Synchronous Averaging), this technique will be dedicated to condition monitoring of machines. In addition, spectral analysis, using the Fast Fourier Transform (FFT) algorithm of the stator- current signature MCSA (Motor Current Signature Analysis), determines their frequency composition, and therefore, allows finding the spectral lines associated to the fault. This work highlights our first results related to the comparison of the spectral representation of the stator current and that of the residual current obtained by the TSA method. We proved the effectiveness of these techniques by simulation and experimental tests made on a wound rotor induction machine. The fault rotor is taken into account by an additional resistance of one of the rotor phases. The results of simulations and experiments underline the practical utility of this method.

Keywords Prognosis · Diagnosis · Asynchronous machine · Fault · MCSA · TSA

K. Dahi (✉) · S. Elhani · S. Guedira · N. Ngote
Research Team: Electromecanical, Diagnosis and Control Research Laboratory: Electrical Engineering, ENSET Rabat Mohammed V Souissi University, Rabat, Morocco
e-mail: khalid.dahi@um5s.net.ma

S. Elhani
e-mail: s.elhani@um5s.net.ma

G. Dalpiaz et al. (eds.), *Advances in Condition Monitoring of Machinery in Non-Stationary Operations*, Lecture Notes in Mechanical Engineering,
DOI: 10.1007/978-3-642-39348-8_54,

1 Introduction

The use of asynchronous machines is mainly due to their robustness, their power and their fabrication cost. Nevertheless, various faults may appear in this type of machines. The appearance of these defects has pushed industrial and scientific communities to find solutions to make these systems more efficient, more competitive and safer within the framework of the conditional preventive maintenance.

Among the recommended methods to detect and characterize defects, we are particularly interested to the analysis of the signature of stator current MCSA (Motor Current Signature Analysis). Its particularity lies in the fact that the current contain more information on almost all defects [1, 3, 9].

Our results cover on the temporal and spectral analysis of the stator current that can extract information about the fault.

However, the stator current signal in the presence of electrical faults (specifically fault of broken bars) shows a behavior related to non-stationary mode and fluctuations in the electrical phase. It has been shown that the formalism of cyclostationarity reflects better the analysis of such signals compared to conventional approaches dedicated to stationary behavior [6].

By the application of this formalism, we conditioned a temporal indicator by using Time Synchronous Averaging method (TSA) [4, 8], whose reliability has been confirmed by our experimental results, this first analysis can concluded the existence of a failure. We remind that conventional diagnostic methods show more appropriate to stationary signals. In order to analyze the electrical current taking account of cyclostationarity we analyzed the stator current residual obtained by synchronous averaging. The experimental results are encouraging.

2 Asynchronous Machine in the Presence of a Rotor Fault

Although the asynchronous machine is deemed to be robust, degraded modes of operation may appear during use. References [5, 7] review a large number of cases of failures and associated detection techniques. In this work we are interested in rotor fault remains ranked among most frequently encountered faults in the machine.

Experimental results of fault rotor

In order to verify experimentally the impact of the presence or absence of fault on the machine; we developed a test bench including a wound rotor induction machine.

2.1 Presentation of the Test Bench

The experimental bench that we used was developed in the laboratory of CPS2I ENIM-Rabat-(National School of Mineral Industry) in the research axis "Diagnosis and monitoring of rotating machinery." This bench has an asynchronous machine power 3 kW and a data acquisition system based on data acquisition card with its software DATA Acquisition (Fig. 1).

2.2 Experimental Results

The rotor fault has been carried out by adding an extra 40 mΩ resistance on one of the rotor phases (i.e. 10 % of the rotor resistance value per phase, $R_r = 0{,}4\Omega$)

From Fig. 2, with the creation of a rotor fault on the bench, we are interested to the stator current and we note well the envelope modulation of the current.

2.3 RMS Indicator

In this section we focus on the signal conditioning elaborated for the development of an indicator for monitoring the induction motor. For that we realize three tests: no loaded motor, 65 %loaded and full loaded motor.

We define a first indicatorK1,such as the RMS of the stator current, according to the relationship:

$$K_1 = \frac{Is_{RMS}(Healthy) - Is_{RMS}(defective)}{Is_{RMS}(Healthy)}$$

The following table summarizes the values for the three tests (Table 1):

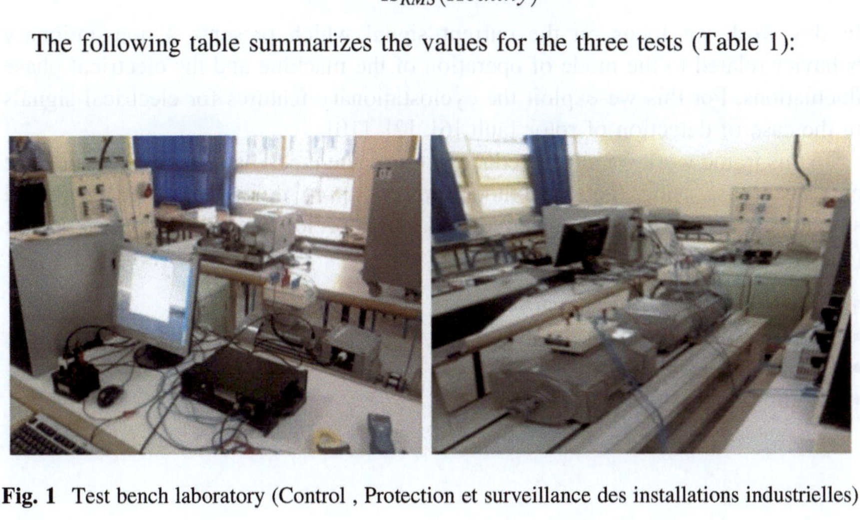

Fig. 1 Test bench laboratory (Control , Protection et surveillance des installations industrielles)

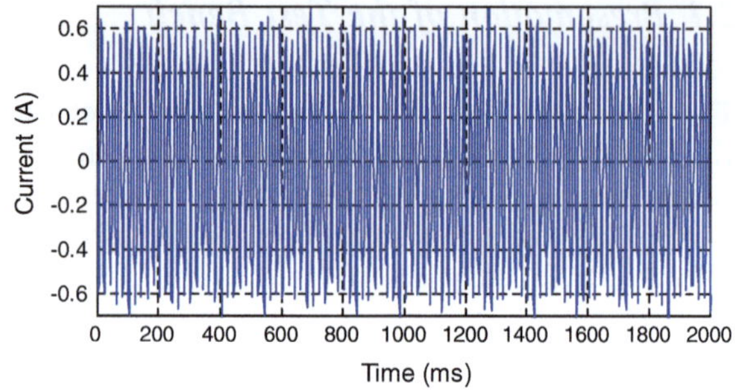

Fig. 2 Experiments with a rotor fault

Table 1 Variation of RMS current with different load levels

	Stator current RMS (A)		Indicator K_1 (%)
	Healthy case	Faulty case	
No-loaded motor	2.88	2.95	2.55
Full-loaded motor	3.78	4.06	7.45

From the table we notice too little variation of the K_1indicator between healthy and defective cases. Therefore, it cannot be used as a sensitive indicator of rotor defect.

For this and with the use of the concept of cyclostationarity and exploitation of the TSA method we develop an algorithm for conditioning a statistical indicator for rotor fault detection.

3 Conditioning of an Indicator with the TSA Method

In this work we focus on the current signal which presents a non-stationary behavior related to the mode of operation of the machine and the electrical phase fluctuations, For this we exploit the cyclostationary features for electrical signals in the case of detection of rotor fault [6], [2], [10].

In this frame work we aim to reduce or eliminate the dynamics of 50 Hz electric currents, thus, using thetas method allowed us to develop an organizational structure of conditioning a sensitive indicator to detect rotor fault, the steps are: [4, 8].

- Synchronization of electrical signals,
- Synchronous Averaging
- Extraction of the residual signal
- Conditioning of the indicator.

3.1 Synchronization of Electrical Signals

In this work, we exploit the first order cyclostationarity of stator current and voltage (Fig. 3a). However, we note a problem of derives cycle of an electrical cycle to another, due to electrical supply fluctuations. The idea will be to decompose the signal into overlapping period and thereafter.

Figure 3b represents a superposition of 500 electrical cycles acquired temporally. The sampling rate is 1 kHz, we have 20 samples per cycle average of 50 Hz (1,000/50 = 20).

To do this,

- Voltage signal is first cut out in slices, each one corresponding to one period (20 ms), and each period containing an integer number of samples N. In our case, the sample rate is 1 kHz, so N = 20 samples per period (20 =1 kHz × 20 ms).
- Then, we estimate the shift between the first period, taken as a reference, and the others. We then shift each period to make it coincide with the first one (reference). If the two periods are already synchronous, the shift is then null. The obtained signal is represented in Fig. 3c

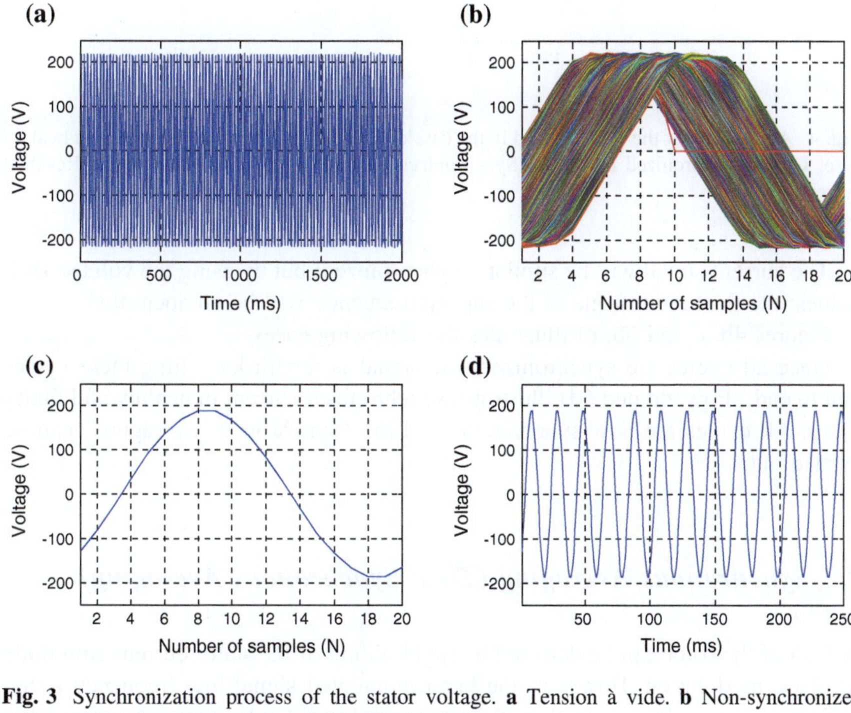

Fig. 3 Synchronization process of the stator voltage. **a** Tension à vide. **b** Non-synchronized voltage. **c** Synchronized voltage. **d** Synchronized voltage

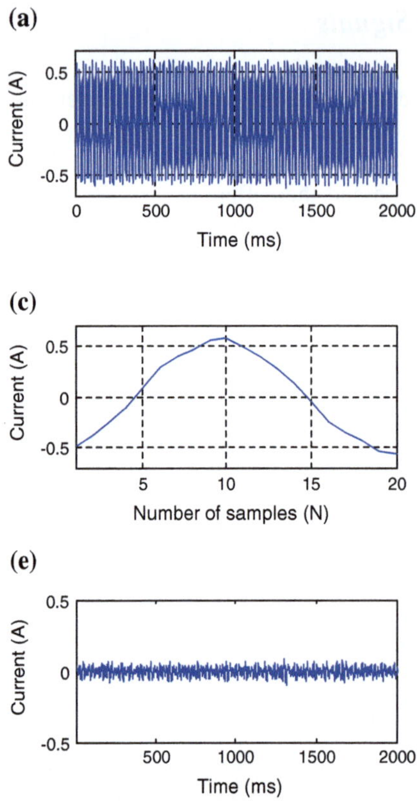

Fig. 4 Application of the TSA method in the case of a healthy machine. **a** Stator current healthy case. **b** Non-synchronized current. **c** Synchronized current. **d** TSA stator current. **e** Résidual current

The current signal will be similarly synchronized; but by using the voltage-shift values: the fluctuations due to the supply frequency will be compensated.

Figures 4b, c and 5b, c) illustrates the following cases.

Once all cycles are synchronized, the signal is rebuilt by setting these cycles end to end. (Figs. 4d and 5d) illustrates synchronized current in healthy and faulty cases. All cycles are now synchronous and the "synchronous averaging" can be carried out.

3.2 *Synchronous Averaging (Time Synchronous Averaging)*

A fault of the rotor can be detected by highlighting of the stator current amplitude or phase modulation. However, the band modulated signal low frequency is too difficult to detect the modulation. To overcome this difficulty, research works have

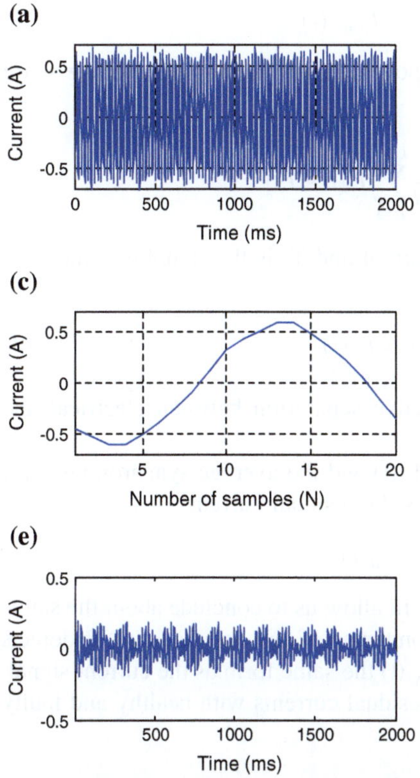

Fig. 5 Application of the TSA method in the case of a defective machine. **a** Stator current. **b** Non-synchronized current. **c** Synchronized current. **d** TSA stator current. **e** Résidual current

used the TSA method [4]. It is a way to reshape the signal before processing. It allows the separation between the excitation sources and therefore, the identification of a fault.

We can decompose the stator current is (t) as follows:

$$I_s(t) = I_{sh}(t) + I_{smec}(t) + b(t) \tag{1}$$

With

I_{sh} The stator-current harmonic component,
I_{smec} The mechanical-structure-related stator current
b (t) the noise

The asynchronous motor monitoring consists of supervising the signal harmonic part. So we have to separate between harmonic frequency (50 Hz) which is related to electrical phenomena and mechanical-structure-related frequency.

For this purpose, we will apply the TSA method to the stator current. In fact, the stator current is the sum of a determinist signal (I_{Sh}) and a random signal (sum of I_{Smec} t and b(t)); whose average value is zero:

$$I_s(t) = I_{sh}(t) + I_{srand}(t) \quad (2)$$

Is_{rand} (t) is the stator current random component. The synchronous average stator current of N samples is done by:

$$Is^N_{moy}(n * Ts) = \frac{1}{N}\sum_{k=1}^{k=N} I^k_s(n * Ts) \quad (3)$$

where: I^k_s is the *k*th synchronized stator current and T_s is the sampling time.

For large value of N, we have:

$$\lim_{N \to \infty} Is^N_{moy}(t) = I_{sh}(t) \quad (4)$$

The average synchronous allows an effective separation between electrical and mechanical components linked.

Subtraction between the stator current I_s (t) and the average synchronous I_{smoy} (t) = I_{sh} (t) (for the high value of N) gives the residual current

$$I_{res}(t) = I_{srand}(t) \quad (5)$$

This is a very interesting property that will allow us to conclude about the status of a mechanical structure related to the monitoring of any defects (such as defects of the rotor.) We note that the current I_{smoy} (t) the same form as the current signal. Figures 4e and 5e show respectively the residual currents with healthy and faulty machine functioning at full load.

3.3 Conditioning of the Indicator

We have seen in paragraph 1.2.2, that an indicator based on the RMS stator current cannot be used as an indicator for fault detection due to the low variation between non load and full load of the machine.

We will try to calculate the RMS residual current obtained after TSA stator current.

The RMS residual current is calculated by the relation

$$I_{res_{RMS}} = \sqrt{\frac{1}{N}\sum_{n=1}^{n=N} I^2_{res}(n * Tech)} \quad (6)$$

where: N is the number of samples (20,000). 20 samples per cycle × 1,000.

We define a second indicator K_2, obtained from the residual RMS stator current according to the equation (Fig. 6):

$$K_2 = \frac{I_{res_{RMS}}(Healthy) - I_{res_{RMS}}(defective)}{I_{res_{RMS}}(Healthy)} \quad (7)$$

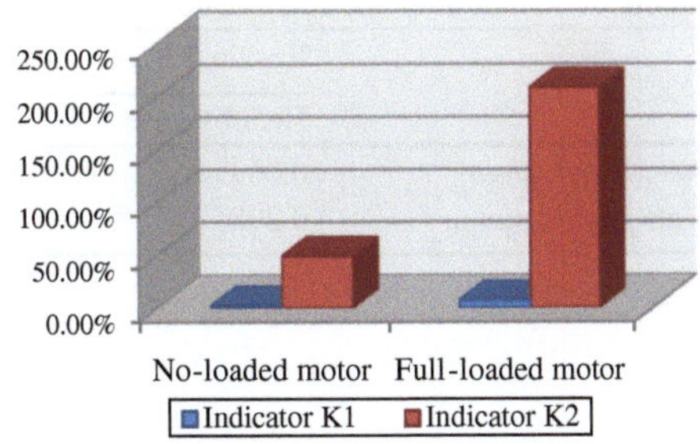

Fig. 6 K1 & K2 evolution according to the motor load

The indicator conditioning K_2, on the other hand, allows easy distinction between healthy and defective cases (Table 2).

4 Comparative Study Between the Signatures of the Line Current and the Residual Stator Current

We have seen that among the best known approaches for the diagnosis of rotor faults in induction machines is based on processing of the stator currents. However, the signature analysis of residual stator current obtained by the TSA method may have advantages, we list below, as the current line, a comparative study between the two analyzes is the subject of this next paragraph.

4.1 Spectral Analysis

4.1.1 Stator Current Analysis

The rotor defect of an asynchronous machine induces changes in the stator currents and therefore leads to the appearance of characteristic harmonics in the signal spectrum. Indeed, the frequency of occurrence of this type of fault is:

$$f_R = f_s(1 \pm 2.k.g) \tag{8}$$

The results of experimentation with a Hanning window are gathered in Fig. 7.

Figure 8 shows the fundamental harmonic of the stator current spectrum of the induction machine operating at full load.

In one hand, we can see an increase in amplitudes of frequency components at $fs - fr$ and $fr + fs$ that are already present in the spectrum of the healthy machine due to the natural eccentricity.

On the other hand, considering the defective rotor, default frequencies are clearly visible in the spectrum; we can easily observe that the harmonics induced by the undulation current are consistent with the generalized formula given by

Table 2 Variation of the RMS residual current with different load levels

	Residual Stator current RMS (A)		Indicator K_2 (%)
	Healthy case	Faulty case	
No-loaded motor	0.069	0.103	49.21
Full-loaded motor	0.298	0.921	209.06

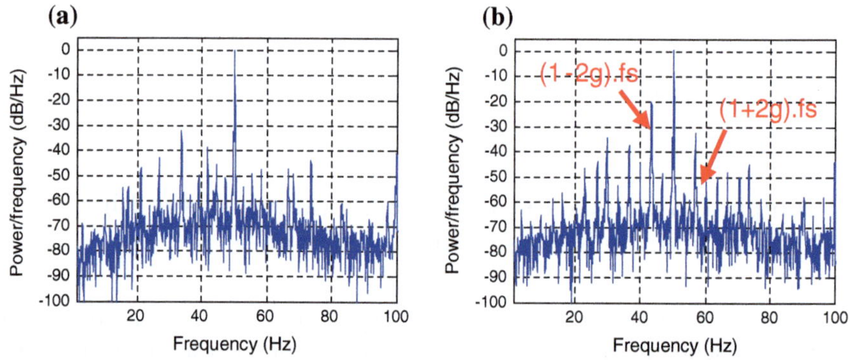

Fig. 7 Spectrum of the stator current at full loaded motor. **a** Healthy. **b** Defective

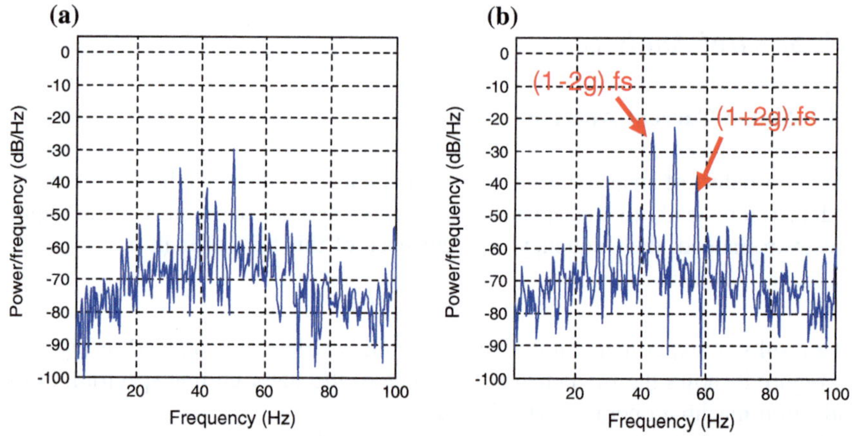

Fig. 8 residual stator current Spectrum obtained after TSA. **a** Full load—healthy, **b** Default

Table 3 The amplitudes associated to the fault characteristic frequency

The amplitudes of frequencies (1 + 2 g)fs			Stator current		Residual current stator	
Load (%)	Slip.	Fault fr (Hz)	Healthy	Defective	Healthy	Defective
0	0.022	52	−33.8	−32.8	−40	−36.42
65	0.044	55.18	−43.23	−34.5	−45.8	−31.57
100	0.068	56.8	−14.9	−31.76	−51.5	−23.54
The amplitudes of frequencies (1 − 2 g)fs			Stator current		Residual current stator	
Load (%)	Slip.	Fault fr (Hz)	Healthy	Defective	Healthy	Defective
0	0.022	47.8	−37.35	−38.13	−38.5	−36.42
65	0.044	45.6	−43.86	−24.63	−43.77	−27.16
100	0.068	43.2	−38.37	−18.54	−44.43	−23.13

$$f_R = (1 \pm 2\ \mathrm{kg})fs.$$

g: is the slip.

4.1.2 Residual Current Analysis

Figure 8 shows the spectrum of the residual current of the machine obtained after application of the TSA method with a Hanning window at full loaded motor, we can clearly see the appearance of additional harmonic components induced by the rotor defect.

And to make a comparison between the two analyzes, we will look in detail the values of the amplitude of the harmonics of the two signals for different loads, from the non loaded motor to the full load. The results obtained are shown in Table 3, and we are interested in amplitudes of frequency around the harmonic $(1 + 2\ g)\ fs$.

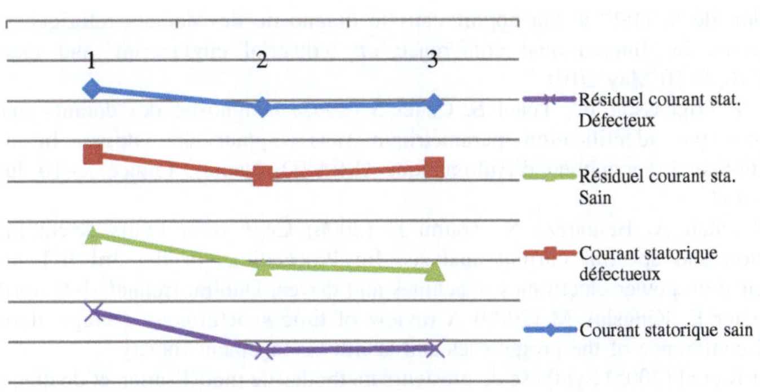

Fig. 9 Evolution of the amplitudes of spectrum function of the load for a healthy and defective case

Figure 9 shows the variation of the harmonic amplitude (1–2 kg*). fs* of the two currents depending on the load in the case of a healthy and defective machine according to the values listed in Table 3.

A simple comparison between the diagnostic analysis of the stator current and the residual obtained by applying the TSA method, shows the change in the evolution of the amplitudes of the second analysis (residual current) between the defective and healthy case (even lower amplitudes), and secondly from the Fig. 8 we can say that the spectral analysis of the residual current improves window, and consequently the good appearance of specific lines of the existence of defects.

5 Conclusion

In this paper, we studied the feasibility of prognosis and diagnosis of rotor faults in asynchronous machine by analyzing the stator currents. For a first step of prognosis, the conditioning of a statistical indicator, by synchronous averaging, for the fault detection is applied to the signal current. The experimental results have allowed us to approve the reliability of the indicator conditioning and that the stator current has a harmonic (1±2 kg) *fs* very sensitive to this type of fault. Spectral analysis is then applied to determine the composition of the current signal frequency and therefore find the harmonics characterizing rotor defects. In this spectral analysis we are interested to the comparison between the spectrum of the stator current and the residual current obtained by the TSA method, and we notice that the last analysis (residual current) improves the windowing and increases the variation in amplitudes between the healthy and defective case.

References

1. Aimer AF, Boudinar AH, Bendiabdellahet A, Mokhtar C (2010) Effet du fenêtrage sur la résolution de la DSP et son apport dans le diagnostic des défauts rotoriques du moteur asynchrone. In: International conference on industrial engineering and manufacturing ICIEM'10, 9–10 May 2010
2. Bachir S, Trigeassou JC, Tnani S, Cauet S (2002) Diagnostic des défauts statoriques et rotoriques par identification paramétrique sans capteur de vitesse. In: Conférence Internationale Francophone d'Automatique CIFA'02, Nantes, France, 8–10 Juillet 2002, pp.447–452
3. Bendiabdellah A, Benouzza N, Toumi D (2006) Cage rotor faults detection by speed estimation and spectral current analysis. In: Proceedings of the 3rd IET international conference on power electronics, machines and drives, Dublin, Ireland, 4–6 April 2006
4. Bechhoefer E, Kingsley M (2009) A review of time synchronous average algorithms. In: Annual conference of the prognostics and health management society
5. Casimir R et al (2005) Synthèse de plusieurs méthodes de modélisation et de diagnostic de la machine asynchrone à cage en présence de défauts. Revue Internationale de Génie Electrique 8(2):287–330

6. Ibrahim A (2009) Contribution au diagnostic de machines électromécaniques: Exploitation des signaux électriques et de la vitesse instantanée. Thèse de Doctorat 2009
7. Nandi S, Toliyat HA (2005) Condition monitoring and fault diagnosis of electrical machines—a review. IEEE Trans Energy Conv 20(4):719–729
8. Ngote N, Guedira S, Cherkaoui M (2010) Conditioning of a statistical indicator for the detection of an asynchronous machine rotor faults
9. Razik H (2002) Le Contenue Spectrale du Courant Absorbé par la Machine Asynchrone en cas de Défaillance, un Etat de l'Art. Revue 3EI, N°29, Juin 2002, pp 48–52
10. Sahraoui M, Zouzou SE, Menacer A, Aboubou ET, Derghal A (2004) Diagnostic des défauts dans les moteurs asynchrones triphasés a cage Partie II : Méthodes dédies a la détection des cassures de barres dans les moteurs asynchrones triphasés a cage. N°05, Juin 2004, pp 57–61

Advanced Data Mining Techniques for Power Performance Verification of an On-Shore Wind Farm

Francesco Castellani, Alberto Garinei, Ludovico Terzi, Davide Astolfi, Michele Moretti and Andrea Lombardi

Abstract The monitoring of wind energy production is fundamental to improve the performances of a wind farm during the operational phase. In order to perform reliable operational analysis, data mining of all available information spreading out from turbine control systems is required. In this work a Supervisory Control and Data Acquisition (SCADA) data analysis was performed on a small wind farm and new post-processing methods are proposed for condition monitoring of the aerogenerators. Indicators are defined to detect the malfunctioning of a wind turbine and to select meaningful data to investigate the causes of the anomalous behaviour of a turbine. The operating state database is used to collect information about the proper power production of a wind turbine, becoming a tool that can be used to verify if the contractual obligations between the original equipment manufacturer and the wind farm operator are met. Results demonstrate that a proper selection of the SCADA data can be very useful to measure the real performances of a wind farm and thus to define optimal repair/replacement and preventive maintenance policies that play a major role in case of energy production.

Keywords Wind energy · Wind turbine condition monitoring · SCADA analysis · Data mining

F. Castellani (✉) · D. Astolfi · M. Moretti
Department of Industrial Engineering, University of Perugia, Perugia, Italy
e-mail: francesco.castellani@unipg.it

A. Garinei
DMII, Università degli Studi Guglielmo Marconi, 00193 Rome, Italy

L. Terzi · A. Lombardi
Sorgenia Green srl, Via Viviani 12 20124 Milan, Italy

G. Dalpiaz et al. (eds.), *Advances in Condition Monitoring of Machinery in Non-Stationary Operations*, Lecture Notes in Mechanical Engineering,
DOI: 10.1007/978-3-642-39348-8_55, © Springer-Verlag Berlin Heidelberg 2014

1 Introduction

The monitoring of wind energy production is fundamental during the operational phase of a wind farm. Usually numerical modelling and experimental campaigns are carried out for wind-resource assessment and power performance prediction of a wind farm prior the installation of the turbines. Even if many techniques were developed for this purpose, disagreement between the predicted and the real production of a wind farm is often remarkable. Thus it is necessary to analyse the operational conditions of a wind farm both to increase the reliability of the assessment techniques and to define optimal repair/replacement and preventive maintenance policies, that play a major role in case of energy production. The Supervisory Control and Data Acquisition (SCADA) system is able to generate and organize a very functional database that can be used to monitor the wind farm and thus to set the optimal configuration for the aerogenerators, in order to maximize the energy production. Starting from the first methods based on the fault detection and analysis, more refined techniques were developed for wind turbines performance assessment [1, 2] and for the prediction and diagnosis of wind turbine faults [3]. In [3] the SCADA data, collected on a 5 min scale, are classified in four groups (wind parameters, energy conversion parameters, vibration parameters, temperature parameters) and are crossed against a database of status codes, entering at given times, ranked in four categories of decreasing fault severity. The power curve is studied both from filtering operational anomalies directly from the SCADA measurements, or crosschecking with the status code database, and a model curve is obtained by training neural networks to interpolate the scattered points. Computational models are developed to predict with certain accuracy a single severe fault occurring one hour before the fault itself appears in the status code report.

In order to increase wind farm efficiency, wind farm power performance tests were defined [4] and wind turbine condition assessment has been performed through the analysis of the power curve [5]: in particular in [6] a systematic analysis of three different operational curves (power curve, rotor curve, blade pitch curve) is set up from SCADA measurements. Reference curves are built, from a 4 year database, for each month independently by removing outliers due to anomalous behaviours: a multivariate outlier detection approach based on Mahanobis distance is used. Therefore skewness and kurtosis of the reference curves are computed and the performance monitoring is based by comparing these moments of the measured operational curves against the reference ones on a 2D-plot.

In [7] the wind control centre performances are analyzed globally and in particular the nature of SCADA analysis is depicted for its increasing role of crosschecking estimated power offer (from forecast) and actual one, measured from historical and meteorological data.

In [8] wind turbine condition monitoring has been developed applying Adaptive Neuro-Fuzzy Interference Systems (ANFIS) to SCADA measurements. A three step strategy has been followed: firstly normal behaviour models are used, by training Neural Networks, in order to monitor and detect anomalies on the relevant

SCADA data. Subsequently occurred anomalies are related to reported faults, and relations are obtained to implement a knowledge database used by the Fuzzy Interference System to output diagnosis.

Modern wind turbines are equipped with a complex monitoring system, so SCADA can provide a very large dataset that requires the development of new post-processing methods. In the present work a SCADA dataset of a wind farm installed in southern Italy is analysed and new post-processing methods are proposed for the monitoring of the aerogenerators. The wind farm is composed of nine wind turbines with a rated power of 2 MW installed on an hillside area with quite gentle slopes. The most important turbine parameters are identified on the SCADA dataset and data mining is performed both for the performance analysis and for the understanding of the wind turbines behaviour during operations.

2 The SCADA Database Analysis

2.1 The Structure of the Database

The SCADA database consists on a series of measurements (e.g. blade pitch, hub, gearbox, inlet temperature, active, reactive and apparent power and so on) performed on a 10 min basis. For each parameter, minimum, maximum, average and standard deviation are recorded. Thus, for each turbine, around 140 data are recorded every 10 min. To understand the real operational conditions of the wind turbines, the analysis was focused on the machine parameters that could be strictly related to the turbine performances. Among the data available from the machine control systems, the following parameters were considered more significant for the present study: active power output (kW), reactive power (kW), inlet temperature (C), nacelle position (°), blade pitch (°), rotor speed (rpm). Measurements from a met mast (wind speed and direction at hub height and hub minus radius) were also used on post-processing to complete the SCADA dataset analysis.

Measured parameters were analysed considering also the control system status code report, that lists the incoming or phasing out of a status. The code report is divided in two groups of data: operating states, which are mutually exclusive, and status codes (error, warning, info and operating state), some of which can coexist at a given time. Thus the complete dataset is composed of the status code database, that is a read only database, and of the SCADA measurement database, that can be cut or modified with respect to appropriate status codes, to obtain a filtered database. The interactions between the databases are shown in Fig. 1.

2.2 The Condition Monitoring Procedures

The first step of post-processing procedures was done developing routines to crosscheck and perform a series of controls on the acquired data. Because of the

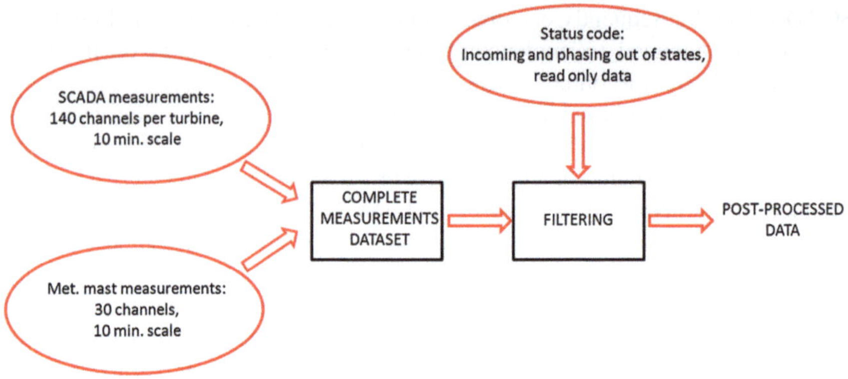

Fig. 1 The available dataset structure

high level of IEC standards, the available data exceeds the typical data used for routine monitoring, so solid mathematical and computational tools are required. A first control was done considering the percentage in which each of the 9 turbines was in every operating state. The status code database coherence was checked summing these percentages and verifying that it was 100, being the operating states mutually exclusive. Then operating states statistics were obtained considering the status of each wind turbine. Considering the whole database (4 month of operational conditions), the Grid Operating State percentage is around 70 %, whereas 20 % of the time the turbines operating state is the automatic start-up, that follows a brake program. Depending on the brake program, the sequence of the other phases that follows the automatic start up may have a different length, but they are of short duration. The graph of Fig. 2 highlights these states and, being a clear indicator of the power availability of each wind turbine, it can be used to detect the malfunctioning of a wind turbine.

Then, for each operating state and status code, the average and standard deviation were calculated on a daily, weekly and monthly basis.

By analysing the operating states of the nine turbines, a malfunctioning was detected for wind turbine T53, as shown in Fig. 3. This has been obtained by an

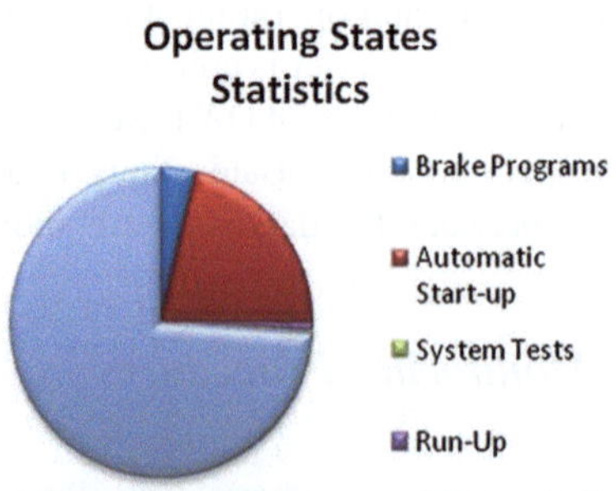

Fig. 2 Operating states statistics for a sample turbine

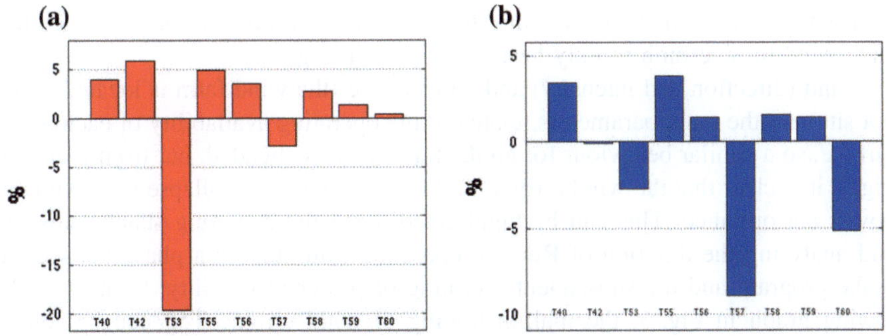

Fig. 3 **a** Grid Operation deviation statistics with respect to a reference turbine (T40) **b** System OK status code deviation statistics with respect to a reference turbine (T40)

automatic routine that computes mean and standard deviation on the whole turbine park of the percentages of each operating states: whereas significant deviations are met from the mean values, the statistics are performed on a shorter scale, weekly or even daily rather than monthly, in order to isolate the anomaly and thus cut a temporal window for which significant crosscheck with SCADA data can be done in order to investigate the causes of the anomaly.

A closer examination can be done by analysing the operating states for each status code. Even if a malfunctioning was detected by the operating states, the percentage of the status code system ok for turbine T53 is comparable with those of the other turbines, as shown in Fig. 3b. This crosscheck between status codes, turbine operating states and SCADA data helps in pointing out the reasons of the anomaly, unveiling if the underproduction comes from environmental condition (strong or rapidly variable wind), or if it comes from network or electrical problems: actually Status Code System OK on indicates that the turbine is potentially in condition of producing power, but if the operating state is not power output production, the anomaly must not come from the machine itself. It might be due, for example, to an electrical problem. Performing also a crosscheck and filtering of the SCADA data helps in verifying diagnosis hypothesis.

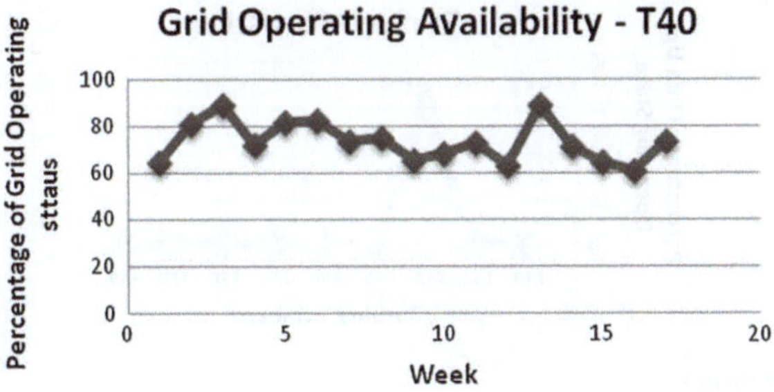

Fig. 4 Grid operating availability for turbine T40

The Grid Operating State and thus the turbine availability was then analysed for each wind turbine on a weekly basis, as shown in Fig. 4.

Wind (direction and intensity) and temperature (the wind farm is located in an hot site) are the main parameters related to the operating availability of each wind turbine, so a similar behaviour for all the turbines was attended, but from graph in Fig. 4 it is clear that the wind turbine T53 has a remarkable collapse in maximum power performance. This can be highlighted using the operating states statistics and analysing the duration of Run Up operating state, that is a phase between a brake program and the subsequent returning of power availability. From the following graph in Fig. 5, the malfunctioning of wind turbine T53 can be easily detected.

Because this malfunctioning may be detected or not depending on the scale of observation (i.e. daily, weekly, monthly, global), this check can be used as an indicator to investigate the causes of the anomalous behaviour of a turbine in a certain period. This can be done by selecting the SCADA measurements corresponding to the Operating State Run Up active and analysing them in detail. This is a clear example of how the Status Code database can be used to select meaningful data from the measurement SCADA database.

The operating state database can also be used to collect information about the proper power production of a wind turbine. The theoretical available power curve (Betz limit) of a wind turbine is:

$$f(v) = \frac{16}{27}\frac{\pi}{8}\rho \cdot d^2 \cdot v^3 \tag{1}$$

.

Where ρ is the density and d is the rotor diameter. If we simply graph the active power measurements of a wind turbine (e.g. wind turbine T40) against the wind speed at the nacelle we obtain the first graph of the following Fig. 6.

A first operation can be the renormalization of the wind speed with the density factor, which is available from the SCADA data. Then, to filter these data, it is necessary to select the operating state significant for the proper power production. The output power can be curtailed due to rapidly changing winds, with the blade

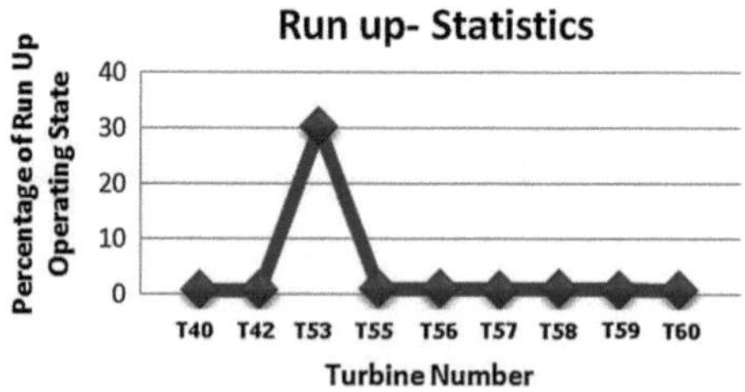

Fig. 5 Run-up statistics

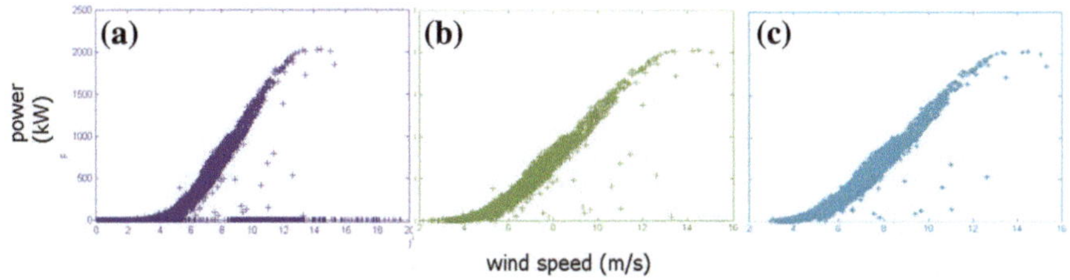

Fig. 6 Unfiltered **a** filtered on power-speed product **b** and double filtered **c** power curve

pitch angle being not fast enough to adjust to the wind direction, and might therefore not be due to a degradation of the wind turbine. A first condition to select significant data is that, for accepted data, the power-speed product is greater than 1. The filtered power curve is the (b) shown in Fig. 6.

An interesting crosscheck can be done by analysing the data excluded by this filter considering the status code database, to inquire whether the turbine is in any problematic operating state, and considering the SCADA measurements.

Then, as a second step, only data acquired when the operating state of maximum power production was on were accepted (Fig. 6c).

Data are now filtered on the operating state, which gives information about the proper power production, and thus they can be used to compare active power measurements against the wind velocity at the nacelle with the theoretical power curve. This tool can be used to verify if the contractual obligations between the original equipment manufacturer and the wind farm operator are met.

2.3 The Study of the Wind Farm Behaviour

An analysis of the behavior of the turbine park has been performed by the point of view of the nacelle response to the wind: discrepancies between the wind direction measured respectively by the anemometer and by the turbine nacelles have been investigated throughout the park and plotted on a three-dimensional graph along all the wind rose.

It is intuitively expected such discrepancy being a function of the distance between the turbine and the anemometer: this effect is clearly visible in the subsequent three-dimensional graph, Fig. 7.

Yet, such amplitude, increases with the nacelle-anemometer distance, is not homogeneously distributed along the wind rose, as might be expected and as is shown in Fig. 8. This should be due to wake effects, which can be further investigated with numerical tools such as Computational Fluid Dynamics (CFD) and the actuator disc model [9].

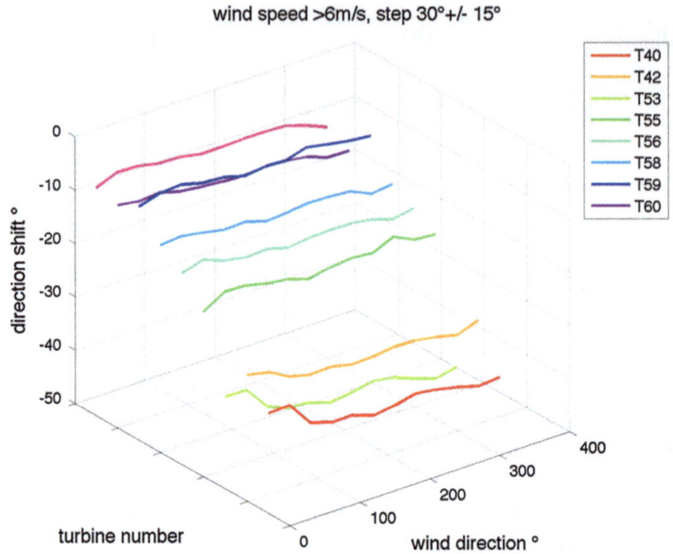

Fig. 7 Anemometer-nacelle wind direction shift calculated referring to the met mast

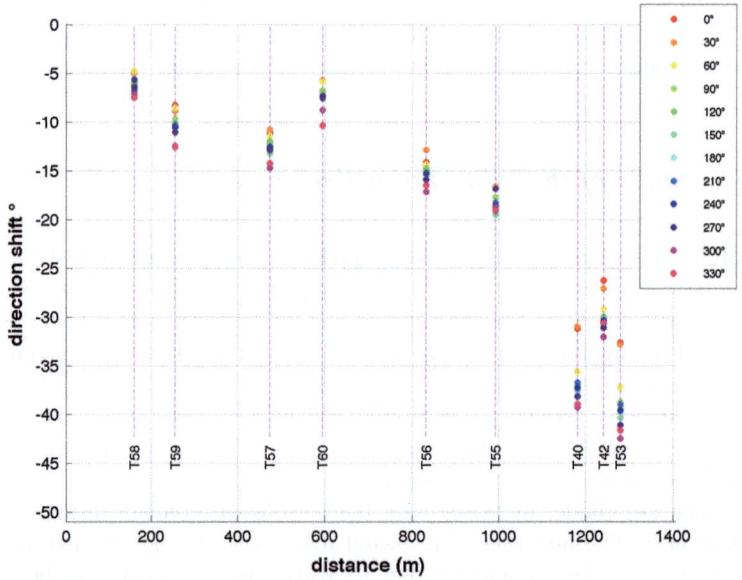

Fig. 8 Anemometer-nacelle wind direction spread vs distance, sector by sector

3 Results and Conclusions

In this work a SCADA data analysis was performed on a wind farm composed of nine wind turbines and new post-processing methods are proposed for the monitoring of the aerogenerators. The most important turbine parameters were identified on the SCADA dataset and data mining was performed both for the performance analysis and for the understanding of the wind turbines behaviour during operations. Then routines were developed to crosscheck and make a series of controls on the acquired data. An automatic routine, providing indicators of the power availability of the wind turbines was defined. Significant deviation from the mean statistics on the whole park, and thus the malfunctioning of wind turbine T53 was detected. The operating states were then used both to confirm the malfunctioning of wind turbine T53 and to select data to be analysed to investigate the causes of its anomalous behaviour. The operating state database was then used to collect information about the proper power production of the wind turbines, to verify if the contractual obligations between the original equipment manufacturer and the wind farm operator are met. Also the measured values can be fundamental to study critical operating situations such as strong wakes loads due to the rotor misalignment. Results demonstrate that a proper selection of the measured parameters, considering the operating and status code dataset, can be very useful to investigate the real performances of a wind farm and to understand the behaviour of each wind turbine. The analysis of the operational conditions of a wind farm can be helpful both to increase the reliability of the assessment techniques and to define optimal repair/replacement and preventive maintenance policies that play a major role in case of energy production.

References

1. Catmull S (2011) Self-organising map based condition monitoring of wind turbines. In Proceedings of EWEA conference, 14–17 Mar 2011, Brussels, Belgium
2. Lapira E, Brisset D, Davari Ardakani H, Siegel D, Lee J (2012) Wind turbine performance assessment using multi-regime modeling approach. Renewable Energy 45:86–95
3. Kusiak A, Li W (2011) The prediction and diagnosis of wind turbine faults. Renewable Energy 36:16–23
4. Carvalho H, Gaião M, Guedes R (2010) Wind farm power performance test, in the scope of the IEC 61400-12.3. In Proceedings of EWEC 2010 European wind energy conference and exhibition, 20–23 Apr 2010, Warsaw, Poland
5. Gill S, Stephen B, Galloway S (2012) Wind turbine condition assessment through power curve copula modeling. IEEE Trans Sustain Energy 3(1):94–101
6. Kusiak A, Verna A (2012) Monitoring wind farms with performance curves. IEEE Transactions Sustain Energy 4(1):192–199
7. Gallardo-Calles J-M, Colmenar-Santos A, Ontanon-Ruiz J, Castro-Gil M (2013) Wind control centres: state of the art. Renewable Energy 51:93–100

8. Schlechtingen M, Ferreira Santos I, Achiche S (2013) Wind turbine condition monitoring based on SCADA data using normal behavior models. Part 1: system description. Appl Soft Comput 13(1):259–270
9. Castellani F, Vignaroli A (2012) An application of the actuator disc model for wind turbine wakes calculations. Appl Energy 101:432. ISSN 0306-2619, doi:10.1016/j.apenergy.2012.04.039

Virtual Assessment of Damage Detection Techniques for Operational Wind Turbine

Emilio Di Lorenzo, Simone Manzato, Bart Peeters and Herman Van der Auweraer

Abstract Operational Modal Analysis (OMA), also known as output-only modal analysis, allows identifying modal parameters only by using the response measurements of the structures in operational conditions when the input forces cannot be measured. These information can then be used to improve numerical models in order to monitor the operating and structural conditions of the system. This is a critical aspect both for condition monitoring and maintenance of large wind turbines, particularly in the off-shore sector where operation and maintenance represent a high percentage of total costs. Although OMA is widely applied, the wind turbine case still remains an open issue. Numerical aeroelastic models could be used, once they have been validated, to introduce virtual damages to the structures in order to analyze the generated data. Results from such models can then be used as baseline to monitor the operating and structural condition of the machine.

Keyword Wind turbine · Damage detection · Operational modal analysis · Fault simulation

1 Introduction

The objective of this paper is to apply advanced Operational Modal Analysis (OMA) techniques to predict in advance failures or damages with a monitoring application based on the analysis of modal parameters and their variations in operating conditions. In this way, the risk to have catastrophic failures and cost associated can be significantly reduced. OMA is a technique that allows extracting the modal parameters from vibration response signals. The main difference compared to the traditional experimental modal analysis is that it does not need the

E. Di Lorenzo · S. Manzato (✉) · B. Peeters · H. Van der Auweraer
LMS International, RTD Test Division, Interleuvenlaan 68, 3001 Leuven, Belgium
e-mail: simone.manzato@imsinth.com

G. Dalpiaz et al. (eds.), *Advances in Condition Monitoring of Machinery in Non-Stationary Operations*, Lecture Notes in Mechanical Engineering,
DOI: 10.1007/978-3-642-39348-8_56, © Springer-Verlag Berlin Heidelberg 2014

measurement of the input forces so that also structures under operating conditions, or in other situations where it is impossible to measure the input forces, can be tested. The information obtained from this analysis can then be used to improve numerical models, to predict the dynamic behavior of new designs, to identify the modal parameters of prototypes and to monitor systems in operating conditions.

The problem associated with the dynamic identification of wind turbines has its roots in 1990 when a special technique known as Natural Excitation Technique (NExT) was developed to estimate modal parameters of wind turbines excited in their operating environment [1]. In the following years, this technique has been applied to other fields such as civil structures as well as automotive and aerospace applications. Although the first application of an Operational Modal Analysis methodology was related to a vertical-axis wind turbine, not many other applications to wind turbines were studied later on. The main reason is the fact that most of the OMA assumptions are violated by operating wind turbines.

With this paper, an advanced OMA technique, the so-called PolyMAX Operational Modal Analysis technique is applied to different set of simulated data obtained introducing virtual damages to the unit under test. The main objective is to predict damages by means of modal parameters variations for condition monitoring and maintenance (O&M) purposes. In Sect. 2 the PolyMAX Operational Modal Analysis technique is briefly presented. The NREL offshore 5-MW baseline wind turbine is sketched out in Sect. 3. Section 4 describes the virtual damages introduced in the wind turbine such as ice on all the blades and unbalanced masses. Finally in Sect. 5 the mentioned OMA technique is applied to the different load cases.

2 PolyMAX Operational Modal Analysis

Operational modal analysis has attracted a significant amount of research interest in the past years. Several operational modal analysis techniques such as Frequency Domain Decomposition (FDD), Stochastic Subspace Identification (SSI) and Operational PolyMAX have been developed and evaluated [2]. In this paper, the PolyMAX method is selected to perform the operational modal analysis. It has been developed as a polyreference version of the least-squares complex frequency-domain (LSCF) estimation method using a so-called right matrix-fraction model. This method, in case of OMA, requires output spectra as primary data. It can be demonstrated that, under the assumption of white noise input, output spectra can be modeled very similarly to FRFs [3].

In case of Experimental Modal Analysis [4], the modal decomposition of an FRF matrix [H(ω)] is:

$$[H(\omega)] = \sum_{i=1}^{N} \frac{\{v_i\}\langle l_i^T\rangle}{j\omega - \lambda_i} + \frac{\{v_i^*\}\langle l_i^H\rangle}{j\omega - \lambda_i^*} \tag{1}$$

where l is the number of outputs; N is the number of modes and half of the system order, * is the complex conjugate operator, ^{H} is the complex conjugate transpose of a matrix, $\{v_i\}$ are the mode shapes, $\langle l_i^T \rangle$are the modal participation factors and λ_i are the poles. The system poles are recurring in complex-conjugate pairs and are related to the eigenfrequencies ω_i and damping ratios ξ_i as:

$$\lambda_i, \bar{\lambda}_i = -\xi_i \omega_i \pm j\sqrt{1-\xi_i^2} \cdot \omega_i \tag{2}$$

Now, the input spectra $[S_{uu}(\omega)]$ and output spectra $[S_{yy}(\omega)]$ of a system represented by the FRF matrix in Eq. (1) are related as:

$$\left[S_{yy}(\omega)\right] = [H(\omega)][S_{uu}(\omega)][H(\omega)]^H \tag{3}$$

In case of operational data, output spectra are the only available information. The deterministic knowledge of the input is replaced by the assumption that the input is white noise, which is characterized by a constant power spectrum and is independent of the frequency. The modal decomposition of the output spectrum matrix can be obtained now by inserting Eq. (1) in Eq. (3) and converting to partial fraction form:

$$\left[S_{yy}(\omega)\right] = \sum_{i=1}^{N} \frac{\{v_i\}\langle g_i \rangle}{j\omega - \lambda_i} + \frac{\{v_i^*\}\langle g_i^* \rangle}{j\omega - \lambda_i^*} + \frac{\{g_i\}\langle v_i \rangle}{-j\omega - \lambda_i} + \frac{\{v_i\}\langle g_i \rangle}{-j\omega - \lambda_i^*} \tag{4}$$

where $\langle g_i \rangle$ are the so-called operational reference factors, which replace the modal participation factors in case of output-only data. Their physical interpretation is less obvious as they are a function of all modal parameters of the system and the constant input spectrum matrix. The main goal of OMA will then be to identify the right-end side of Eq. (4) by using measured output data pre-processed into output spectra.

The PolyMAX algorithm greatly facilitates the operational modal parameter estimation process by producing extremely clear stabilization diagrams, making the pole selection a lot easier by means of estimating unstable poles (i.e. mathematical or noise modes) with negative damping making them relatively easy to separate from the stable poles (i.e. system modes).

The PolyMAX technique has been widely employed for OMA of civil engineering and mechanical structures including bridges, a football stadium [5], a satellite and many others [6]. It has also been computationally optimized to analyze large data sets with a broad frequency band up to high model orders. In the next sections, the Operational PolyMAX method is applied to a 5 MW wind turbine in several conditions.

3 NREL Offshore 5 MW Baseline Wind Turbine Model

In this section the virtual wind turbine model used as test case is described. The NREL offshore 5-MW baseline wind turbine has been developed by the National

Renewable Energy Laboratory (NREL) to support concept studies aimed at assessing offshore wind technology. It is a conventional three-bladed upwind variable-speed variable blade-pitch-to-feather-controlled turbine [7].

The main objective is to analyze the global dynamic behavior of the full-scale turbine; for this reason, the model has been built as simple as possible. The generated model is shown in Fig. 1 and it can be divided in three main components:

- Tower: it is modeled as 5 elastic beam elements with lumped masses and hinged to the ground foundation. The total tower height is 90 m.
- Rotor: in the 3-bladed rotor, each blade is identical and is modeled with 17 sections with specific mass, elastic and aerodynamic properties.
- Drivetrain: the transmission is simplified into a 1 degree-of-freedom system with a gear-ratio of 97 between the Low Speed Shaft (LSS) and the High Speed Shaft (HSS). The generator torque is regulated by the controller model.

The software SAMCEF for Wind Turbines (S4WT) allows the user to define both a structural and an aerodynamic model which are then solved together to obtain the coupled aero-elastic solution [8]. Different parameters can be assigned; in this analysis a turbulent wind has been applied because the interest of the analysis lays in the turbine dynamic response in real conditions. The wind dominant component is in the X direction (in the model, from LSS to HSS), but to have a more realistic response also turbulent fluctuations on the other two directions are included (Table 1).

3.1 Load Cases

The model of the wind turbine is analyzed in different operating conditions and the modal parameters are extracted from the generated acceleration signals by means

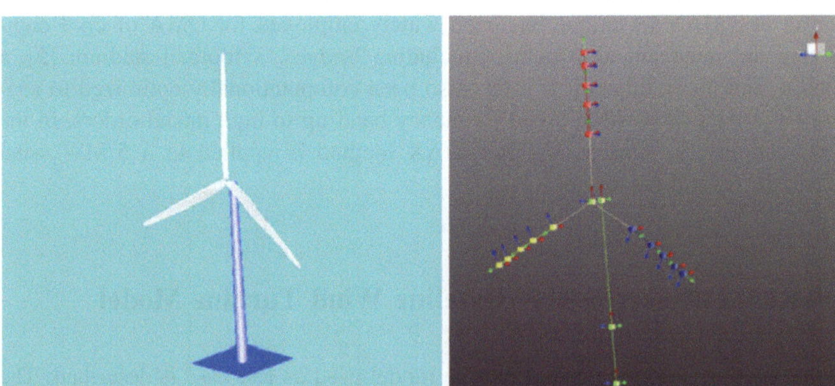

Fig. 1 NREL 5 MW S4WT model (*left*) and Test.Lab geometry (*right*)

Table 1 Wind turbine main parameter

Blade length	61.5 m
Blade overall mass	17,740 kg
Tower height above ground	87.6 m
Tower overall mass	347,460 kg
Hub mass	56,780 kg
Nacelle mass	240,000 kg
Gearbox ratio	97:1

of Operational Modal Analysis. The different models are solved in S4WT and accelerations are computed and exported in LMS Test.Lab for processing and identification [9].

In order to have simulated accelerations that can be considered as accelerations obtained from tri-axial accelerometers mounted on the blades, it is necessary to consider them in the local reference frame in which the X axis is the blade axis (oriented toward the blade tip), the Y axis is aligned with the chord-line and belongs to the blade section plane (oriented toward the leading edge) and the Z axis is normal to the chord line and belongs to the blade section planes. Using this axis configuration, the edge-wise modes are described by bending along the Y axis while flap-wise modes bend the structure along the Z axis. Axial modes along the blade pitch axis can be neglected since they appear at frequencies much higher than those the analysis will focus on in this paper.

The locations selected to measure the accelerations are:

- 3 sensors distributed along the tower
- 1 sensor at the hub center
- 5 sensors per-blade located on the pitch axis.

After analyzing the response of the structure in reference and ideal conditions, different possible damages will be introduced to understand how they affect the measured accelerations. In this preliminary assessment, two main damages are introduced:

- Blade icing
- Mass unbalance on a blade.

The presence of ice on the blades can create excessive turbine vibration and can change the natural frequencies of the blades as well as increase the fatigue loads. It is very important to predict when the icing phenomena occur. Icing has two main effects [10]; on one hand it modifies the blade shape increasing the drag and decreasing the lift and on the other hand the presence of an additional and not uniformly distributed mass could cause unbalancing of the rotor. Effect of ice is not only related to performance issues, but also to safety ones; in fact, during the operational conditions, lumps of ice can detach from the blade and cause damage to people or things.

Within the software S4WT, the conditions “ice formation on all rotor blades” and “ice formation on all rotor blades except one” can be investigated. Calculation is based on the guidelines for certification of wind turbines [11], where it is suggested that the mass distribution (mass/unit length) is assumed at the leading edge of the rotor blade and it increases linearly from zero at the rotor axis to the maximum value μ_{E} at half the radius and then it remains constant up to the outermost radius. The value μ_{E} is calculated as:

$$\mu_E = \rho_E \cdot k \cdot c_{\min} \cdot (c_{\max} + c_{\min}) \tag{5}$$

$$k = 0.00675 + 0.3\exp(-0.32 \cdot R) \tag{6}$$

where $\mu_E$is the mass distribution on the leading edge of the rotor blade at half of the rotor radius, ρ_E is the ice density, R is the rotor radius, c_{max} is the maximum chord length and c_{min} is the length at the blade tip, linearly extrapolated from the blade contour.

The second damage condition which was simulated is the presence of an unbalanced mass that is added to a particular location on the wind turbine. A little cubic mass, which properties can be found in the Table 2, was considered for this preliminary analysis.

Below, the different operating conditions investigated in this paper are summarized:

- Reference parked conditions.
- Parked conditions with ice on all blades.
- Parked conditions on all blades but one.
- Reference power production.
- Power production with ice on all blades.

Table 2 Unbalanced mass properties

Location	Mass	30 kg
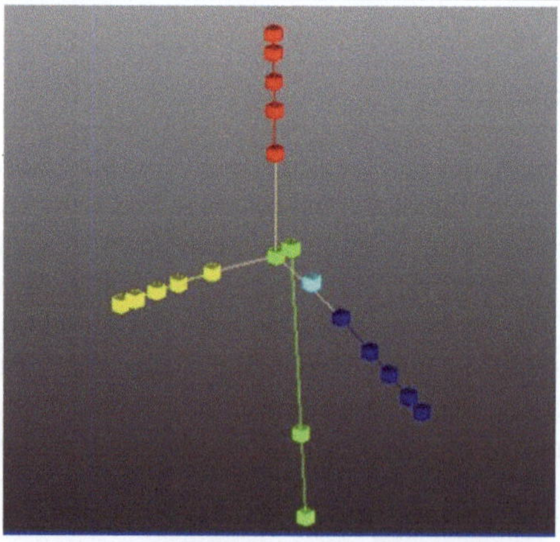	I_{xx} I_{yy} I_{zz}	0.325 kg*m^2 0.325 kg*m^2 0.2 kg*m^2

- Power production with ice on all blades but one.
- Power production with unbalanced mass.

4 Data Analysis

First of all, the wind turbine is analyzed in parked conditions. This load case represents the condition in which the blades are parked and the generator is disconnected; the first seconds are used to place the pitch in its parking position and the rotor at the angle specified as initial rotor angle. At the start time, the rotor is released, but the pitch remains fixed.

To perform Operational Modal Analysis, the point at the tower top is chosen as reference for the correlations and spectra computation, which are then used to extract the modal parameters by means of PolyMAX method [12]. To simplify the entire procedure, the accelerations along the blade axis are neglected. The identified modes are then compared to those one obtained using both the FAST model and the ADAMS model; in FAST the natural frequencies are calculated by performing an eigenanalysis on the first-order matrix created from a linearization analysis, while in ADAMS a command that linearizes the complete model and compute eigenparameters is used.

The results are summarized in Table 3 and the agreement between them is quite good. By using PolyMAX all the first 10 modes can be identified and the biggest differences exist in the predictions of the blades second asymmetric flapwise yaw and pitch modes. "Yaw" and "pitch" mean that these blade asymmetric modes couple with the nacelle-yaw and nacelle-pitching motions, respectively.

The natural frequencies shown in Table 3 are obtained in the standard configuration; they can be compared to those one obtained in other cases such as that one in which the presence of ice is simulated on all the blades or that one in which

Table 3 Numerical modes in parked conditions: STS: side-to-side, FA for-aft; blade modes are described based on their main motion orientation

Mode	Description	Natural frequencies [Hz]		
		Test.Lab	FAST	ADAMS
1	1st tower STS	0.312	0.312	0.319
2	1st tower FA	0.329	0.324	0.316
3	1st flap yaw	0.666	0.666	0.630
4	1st flap pitch	0.675	0.668	0.669
5	1st flap sym	0.720	0.700	0.702
6	1st edge pitch	1.056	1.079	1.074
7	1st edge yaw	1.059	1.089	1.088
8	2nd flap yaw	1.853	1.934	1.651
9	2nd flap pitch	1.888	1.922	1.856
10	2nd flap sym	1.900	2.021	1.960

Table 4 Numerical modes in parked conditions for the different analyzed configurations

Mode	Standard configuration Frequency/damping	Ice on all blades Frequency/damping	Ice on all blades but one Frequency/damping
1st tower FA	0.329 Hz/5.59 %	0.324 Hz/5.13 %	0.324 Hz/5.24 %
1st flap yaw	0.666 Hz/6.20 %	0.588 Hz/7.05 %	0.585 Hz/7.06 %
1st edge yaw	1.059 Hz/0.75 %	0.948 Hz/0.95 %	0.948 Hz/0.99 %
2nd flap yaw	1.853 Hz/3.24 %	1.696 Hz/2.05 %	1.695 Hz/2.33 %

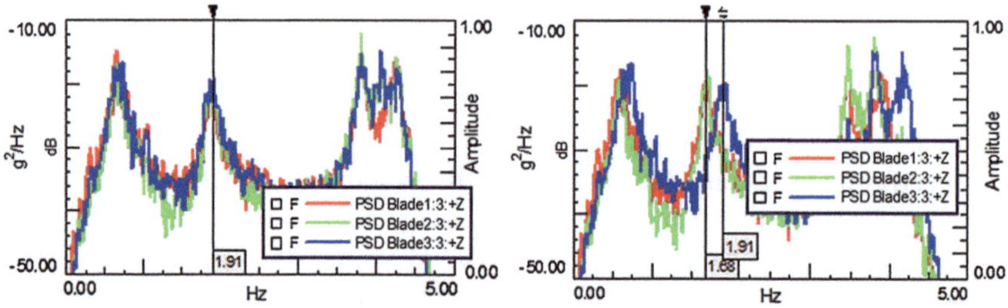

Fig. 2 PSD of acceleration measured on one point on each blade in parked conditions. Standard configuration (*left*) compared to the one with ice on all blades but one (*right*)

the ice is on all the blades but one. In order to perform a better analysis, the point at the blade root is chosen as reference for the correlations and spectra computation and the blade axis accelerations are neglected. The results, in the form of frequencies and damping values, are shown in Table 4 for some of the modes. A frequency shift due to the added ice mass on the blades can be observed and, as expected, increasing the mass decreases the natural frequencies. Not so many considerations can be done regarding the damping values because a global trend cannot be seen moving from one condition to another one, but it should be analyzed mode by mode. Besides, possible effect on damping should be more visible in operating conditions. It should also be mentioned that during ice events, ice accumulates on the rotor blades modifying the blade shape and reducing the aerodynamic efficiency. The amount of lift and the drag variation depends on the quality, shape and position of the ice. In S4WT, the aerodynamic performance modifications introduced by icing are not simulated. The only thing that is currently considered is the mass variation, which leads to changes mostly in natural frequency values and partly on the damping. In real conditions, however, icing will greatly influence both the structural behavior and the power production performance of the machine. In real condition, damping variations, pitch control responses and power production could be also used for detecting icing events, but analyzing the real effects by means of numerical models can be quite difficult [13].

For a qualitative analysis of the signals, the PSD from some sensors is computed and shown in Fig. 2. One point along each one of the three blades at the same distance from the rotor center is taken into account.

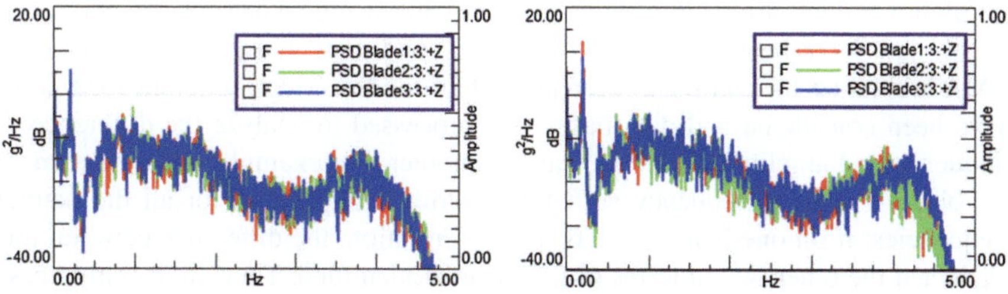

Fig. 3 PSDs from acceleration measured on one point on each blade in operating conditions; ice on all the blades (*left*) compared to ice on all the blades but one (*right*)

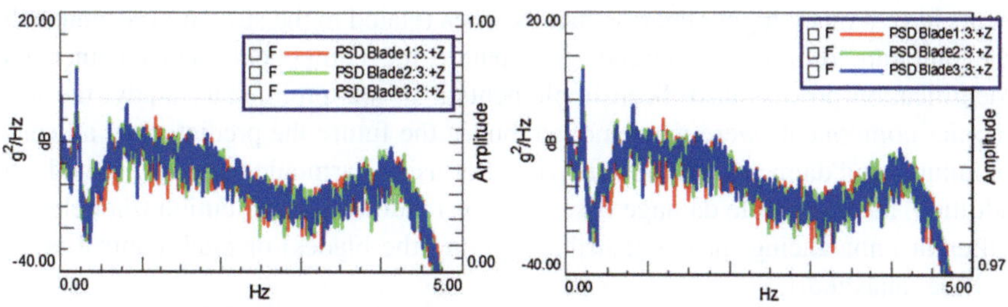

Fig. 4 PSDs from acceleration measured on one point on each blade in operating conditions; standard configuration (*left*) compared to mass unbalance configuration (*right*)

First of all, parked conditions are analyzed. Two different configurations are considered, the standard one and that one with ice formation on all the blades except one. The curves are shown in Fig. 2. Almost all peaks for the two blades with ice in the plot on the right are shifted to lower frequencies, as a consequence of the mass increase.

Figures 3 and 4 show the PSDs in operating conditions for different configurations. The fundamental harmonic frequency is 0.217 Hz because of an averaged rotational speed equal to 13.02 rpm. The PSD amplitude in correspondence of the first harmonic increases for ice configuration in comparison with the standard configuration; the same considerations can be done regarding the mass unbalance configuration, also if the blade mass is quite bigger than the unbalanced mass and the differences between the two curves are not so evident. The other harmonics can also be identified, but the amplitudes are less high than the first one.

5 Conclusions

In this paper, an advanced OMA technique has been applied to a wind turbine to predict in advance failures or damages with a monitoring application based on the analysis of modal parameters and their variations in operating conditions. The simulation model of a wind turbine has been built using an aero-elastic code; first of all, data have been generated for parked conditions and processed using the OMA technique to identify a reference set of modes. Then, different load cases have been considered and data have been processed to analyze the difference in frequency and amplitude from one case to another. For example, adding ice on all the blades causes a frequency shift toward lower frequencies for all the natural frequencies; if on one blade there is no ice formation, the difference between this blade and the other two in terms of PSD can be identified. Then an operating case has been simulated and the data have been processed and analyzed for different conditions, with the icing and with the presence of an unbalanced mass on one of the three blades. In operating conditions, the presence of harmonic components in the signals makes the modal identification process critical because these components have a much higher energy than the ones related to the structural response. In the literature different techniques to separate the components and enhance the identification process have been implemented. In this preliminary study, the harmonics components were not removed, but in the future the prediction in advance of failures and damages will be analyzed after using harmonic removal methods. In addition, more possible damage cases such as random ice distribution (modeled by different unbalancing masses distributed along the blades) or pitch control errors will be analyzed.

Acknowledgments The authors acknowledge the European Commission (EC) for their research grant under the project FP7-PEOPLE-2012-ITN 309395 "MARE-WINT" (new MAterials and Reliability in offshore WINd Turbines technology), see http://marewint.eu/.

References

1. James III GH, Carne TG, Lauffer JP (1993) The natural excitation techniques (NExT) for modal parameters extraction from operating wind turbines. Sandia National Laboratories Technical Report, USA
2. Peeters B, Guillaume P, Van der Auweraer H, Cauberghe B, Verboven P, Leuridan J (2004) Automotive and Aerospace applications of LMS PolyMAX modal parameter estimation method, In Proceedings of IMAC 22, Dearborne, USA
3. Peeters B, Van der Auweraer H (2005) PolyMAX: a revolution in operational modal analysis. In: Proceedings of 1st IOMAC, Copenhagen, Denmark
4. Heylen W, Lemmens S, Sas P (1997) Modal analysis theory and testing, Katholieke Universiteit Leuven, Department Werktuigkunde, Leuven
5. Peeters B, Van der Auweraer H, Vanhollebeke F, Guillaume P (2007) Operational modal analysis for estimating the dynamic properties of a football stadium structure during a football game. Shock Vibr 14(4):283–303
6. Debille J, Pauwels S, Peeters B (2005) The benefits of operational modal analysis of aircraft and spacecraft structures. In: Proceedings of ETTC, Toulouse, France
7. Jonkman J, Buttefield S, Musial W, Scott G (2009) Definition of a 5 MW reference wind turbine for offshore system development. NREL/TP-500-38060-Technical Report, USA
8. Samtech Iberica LMS (2012) Samtech for wind turbines V3.2.01. Online help, Spain

9. Manzato S, Moccia D, Peeters B, Janssens K, White JR (2012) A review of harmonic removal methods for improved operational modal analysis of wind turbines. In: Proceedings of ISMA 2012, Leuven, Belgium
10. Luczak H, Firla M, Sporna D, Manzato S, Peeters B (2011) Structural monitoring of the small wind turbine test stand. In: Proceedings of EURODYN 2011, Leuven, Belgium
11. Germanischer Lloyd Wind Energie GmbH (2010) Guideline for the certification of wind turbines, Edition 2010
12. LMS International (2012) LMS Test.Lab rev. 12A user manual, Belgium
13. Hochart C, Fortin G, Perron J, Ilinca A (2008) Wind turbine performance under icing conditions. Wind Energy 11:319–333

Data-Driven Wind Turbine Power Generation Performance Assessment Using NI LabVIEW's Watchdog® Agent Toolkit

Lodovico Menozzi, Wenyu Zhao and Edzel Lapira

Abstract Power generation performance is a fundamental metric that all wind farm operators use to determine whether expected power throughput is actually being met. IEC 61400-12-1 has been drafted as an exhaustive power performance measurement scheme for wind turbines. The primary weakness of such a standard is the required level of depth of the associated performance tests, which is more than sufficient for operators to use to run daily wind farm activities. In addition, since this IEC test is not really meant for frequent evaluation, it also fails to capture any loss in power generation performance over time. This paper addresses the aforementioned weaknesses of the IEC standard by the application of data-driven approach to model a wind turbine's power curve. A set of measurements during a known good condition is utilized to setup a baseline model. Regular power curve measurements are then compared while taking into account the multi-regime dynamics of the turbine. The approach was implemented using NI LabVIEW's Watchdog Agent® Toolkit and was successfully validated using actual SCADA data collected from an on-shore wind turbine.

Keywords Wind turbine · Power curve · Multi-regime · Prognostics · Health management

L. Menozzi (✉) · E. Lapira
National Instruments, Italy, Centro Dir.Le Milanofiori Nord, Via Del Bosco Rinnovato, 8 20090 Assago, MI, Italy
e-mail: lodovico.menozzi@uc.com

E. Lapira
e-mail: edzel.lapira@uc.edu

W. Zhao
Center for Intelligent Maintenance Systems at University of Cincinnati, 2600 Clifton Avenue, Cincinnati, OH, USA45221-0072,
e-mail: zhaowy@mail.uc.edu

G. Dalpiaz et al. (eds.), *Advances in Condition Monitoring of Machinery in Non-Stationary Operations*, Lecture Notes in Mechanical Engineering,
DOI: 10.1007/978-3-642-39348-8_57,

1 Introduction

The wind power industry is experiencing capacity installations that continuously increase, engineered by advancements in turbine technologies and surging market needs. According to Global Wind Energy Council (GWEC) 2011 Annual Market Update [1], global wind power capacity increased by 18 % and reached 238 GW. While turbine manufacturing and capacity installation have been the main focus of wind power industry, another key issue that has gradually received considerable attention is wind turbine reliability and maintenance. As U.S. Department of Energy reported [2], Operation and Maintenance (O and M) costs possess over 20 % of total life-cycle cost for offshore wind turbines and 15 % for onshore turbines. To monitor and improve the reliability of wind turbines, development of fundamental research and pioneering technology is required to evaluate turbine power performance, detect faults of critical components and predict failure.

To offer a generic methodology to standardize power performance measurement for different participants of the industry, The International Electrotechnical Commission (IEC) has provided a standard IEC 61400-12-1 [3] as a guideline. Participants, including Original Equipment Manufactures (OEM), wind farm operators, service providers, regulators and academic researchers, have adopted the standard [4]. Although recognized as an accurate and comprehensive method, the IEC standard has the following disadvantages, inherently and observed from practices:

1. To achieve the accuracy advocated by the standard, it requires high data fidelity and inherently expensive monitoring routine. The standard is an exhaustive approach that operators hesitate to perform.
2. The approach and metrics defined in the standard do not generate continuous monitoring value for users, mainly because the cost and time incurred with the measurement. For example, the standard advises 180 h of data for evaluation, which usually takes longer to accumulate enough volume of the data. Therefore, it does not provide a deeper insight of turbine performance deterioration.
3. The method used to calculate energy distribution in wind speed bins and annual energy production (AEP) strongly implies that the projection of wind speed will follow a deterministic distribution, and the turbine will be available and operating at a constant performance level for the next year.

To address the issues above, a data-driven approach is applied to model power curve continuously, which is entailed within a two-tier framework that employs Prognostics and Health Management (PHM) techniques for wind turbine monitoring. Prognostics and Health Management (PHM) is an engineering discipline "focusing on detection, prediction, and management of the health and status of complex engineered systems", defined by the International Society of PHM and IEEE Society of Reliability. Having been successfully implemented in industries such as rotary machinery, semiconductor manufacturing and aerospace, data-driven PHM has been proposed and developed to ensure continuous and efficient operation of wind energy assets.

In the domain of wind energy, two types of frequently used data are supervisory control and data acquisition (SCADA) data and condition monitoring system (CMS) data. SCADA system is commonly used by operators to monitor turbine parameters and report alarms. It retains sparse measurements including temperature readings, rotation speed and wind speed, which is used to determine daily site activity [5]. In this paper, selected SCADA variables are used to model the deviation of turbine power performance over time.

On the other hand, CMS data is high-resolution sensor data, which may includes vibration, acoustic emission or oil debris analysis data, and used to decide fault indicator for gearbox, generator and bearings [6, 7]. Critical wind turbine components and their reliabilities have been largely surveyed and studied [9–11]. Integrating component downtime distribution and failure probabilities can generate a criticality analysis chart (Fig. 1).

In this chart, quadrant 4 contains components with low failure probability but require longer lead-time for repair when they fail. These components are identified as critical components for turbine system, and are appropriate for opting predictive maintenance strategy and PHM techniques to monitor component health condition much more closely with CMS data.

This paper presents an enhanced framework of wind turbine monitoring and a case study for turbine power performance monitoring. The reminder of the paper is organized as following: Sect. 2 presents the methodology and analytic framework; Sect. 3 introduces how proposed method for wind turbine power performance monitoring is applied to a real-world wind farm; and Sect. 4 summarizes and concludes the work.

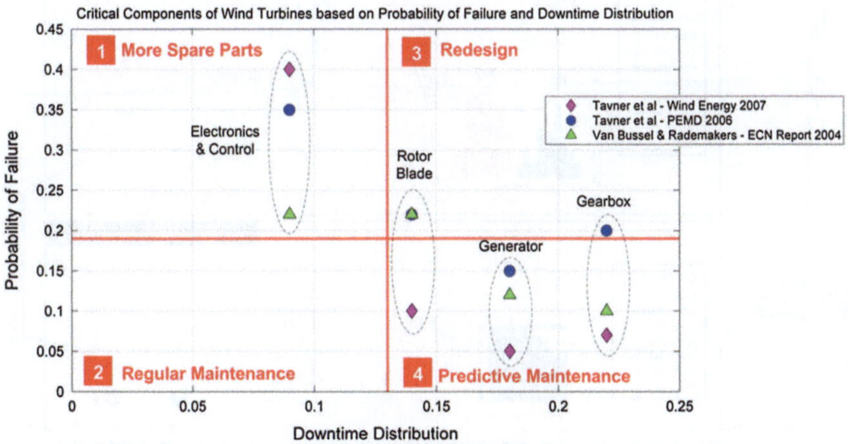

Fig. 1 4-Quadrant criticality analysis for wind turbine components

2 Systematic Methodology

2.1 Overall Framework

A systematic methodology for wind turbine prognostics is proposed and shown in Fig. 2. In this two-tier framework, SCADA data is first used to model turbine overall performance, which is defined as its capability to generate electricity power under varying wind condition. SCADA parameters including output power, wind speed, wind direction and pitch angle are selected to input to a multi-regime model corresponding to turbine's dynamic operating conditions and density function parameters are estimated for each operating regime. In the next step, performance assessment is conducted where current or recent behavior that are represented by the model parameters are compared with normal behavior learned with the same parameters while turbine is known to be new or healthy. A performance indicator frequently generated from the comparison, called Confidence Value (CV) as a Global Health Estimator (GHE) for turbine performance, is then trended over time and predicted with an upper limit R1 and a lower limit R2. The predictions can be converted to forecast when the revenue per unit cycle will drop below a predetermined breakeven level and investigations should be triggered for component Local Damage Estimator (LDE) values.

LDE values are generated from CMS data. Depending on the availability of sensors, the types of data may include vibration, acoustic emission, temperature, and oil debris. Different signal processing tools are used to extract features to

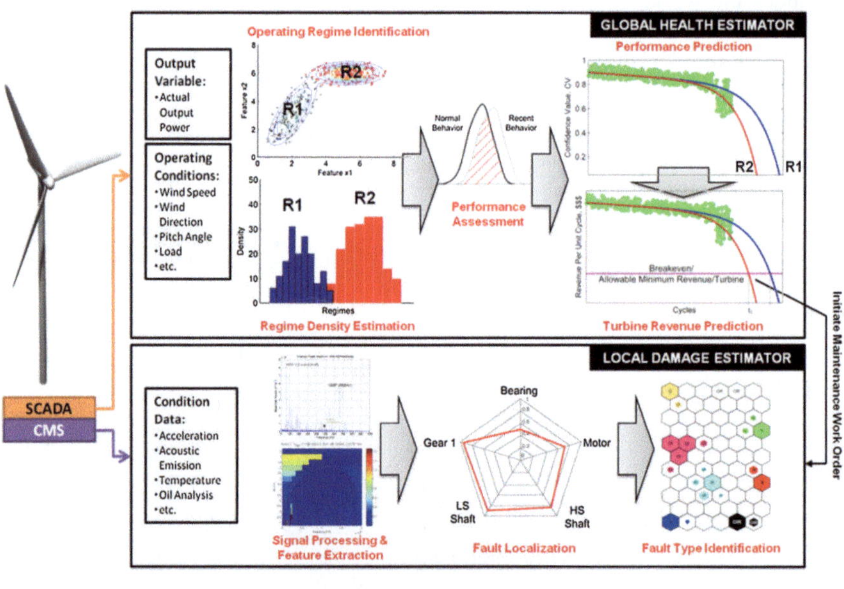

Fig. 2 Systematic methodology for wind turbine prognostics

represent the high-dimension datasets. The features can be used to identify the health degradation status of each instrumented component, and locate component failure that is causing turbine performance reduction. For the located component(s), the specific failure mode can also be identified with diagnosis tools so that correct maintenance action can be suggested.

The proposed framework for wind turbine prognostics can improve existing wind turbine monitoring methods as following:

(1) The highly dynamic environmental condition and operating condition for wind turbines are taken into account with the application of multi-regime modeling method.
(2) Under the framework, various techniques with similar capabilities can be compared and optimized to generate a performance indicator. At the mean time, the indicator is updated frequently and can represent real-time power performance.
(3) A correlation between turbine overall performance and key component defect is investigated, so that the performance metric of CV value can prioritize the effects of degrading components. It allows users to optimize maintenance strategy with a simple yet effective objective.

To aggregate frequently used PHM techniques, the Watchdog Agent® Toolbox is developed as a reconfigurable hardware and software platform for various PHM applications [12]. In National Instruments LabVIEW software, the toolbox includes four categories (as shown in Table 1) of algorithms as Virtual Instruments (VI) for rapid deployment.

The signal processing and feature extraction tools filter, transform and analyze acquired sensor data to extract representative features that are highly related with operation, failure mode or health condition. The feature set then serves as an input to health assessment tools, where pattern recognition and artificial intelligence tools model the similarity between baseline features and features from latest signal, to evaluate overall degradation of the system. Health diagnosis tools, usually

Table 1 Watchdog agent® toolbox algorithms

WATCHDOG AGENT® TOOLS

HEALTH ASSESSMENT		SIGNAL PROCESSING & FEATURE EXTRACTION	
Logistic Regression	Neural Networks	Time Domain Analysis	Wavelet Analysis
Statistical Pattern Recognition	Gaussian Mixture Model	Frequency Domain Analysis	Principal Component Analysis
Self-organizing Maps	Auto-Associative NN	Time-frequency Analysis	Expert Extracted Features
HEALTH DIAGNOSIS		**PERFORMANCE PREDICTION**	
Support Vector Machine	Bayesian Belief Network	Autoregressive Moving Average	Match Matrix
Self-organizing Maps	Hidden Markov Model	Recurrent Neural Network	Trajectory Similarity-based

classification algorithms, can identify the specific impending failure mode at its early stage upon the detection of degradation. Performance prediction tools model the trend of degradation and developing failure mode, to project these indications and estimate the remaining useful life of the system and its components.

For wind turbine PHM, main effort of the estimation of LDE is usually accomplished by selecting suitable tools from numerous algorithms in signal processing category, to extract best features for sensors with different locations and targeted drivetrain components. The computation of GHE, wind turbine's power performance, can be achieved with appropriate health assessment tools.

2.2 Global Health Estimator

To evaluate turbine GHE over time, SCADA data is input to a pre-processing module to be filtered, segmented and normalized. Then parameter selection module determines the relevant variables that will be used to interpret turbine overall performance. As aforementioned that a wind turbine is subjected to dynamic operating conditions, wind turbine data can be represented with as a mixture of distributions through regime partitioning. Several tools can be used here from Watchdog Agent® Toolbox, including Gaussian Mixture Model (GMM), Self-organizing Map (SOM) and Neural Networks. Finally, distance metrics depending on the choice of multi-regime modeling method can be computed to interpret the wind turbine CV through comparing the mixture within similar regime [13] (Fig. 3).

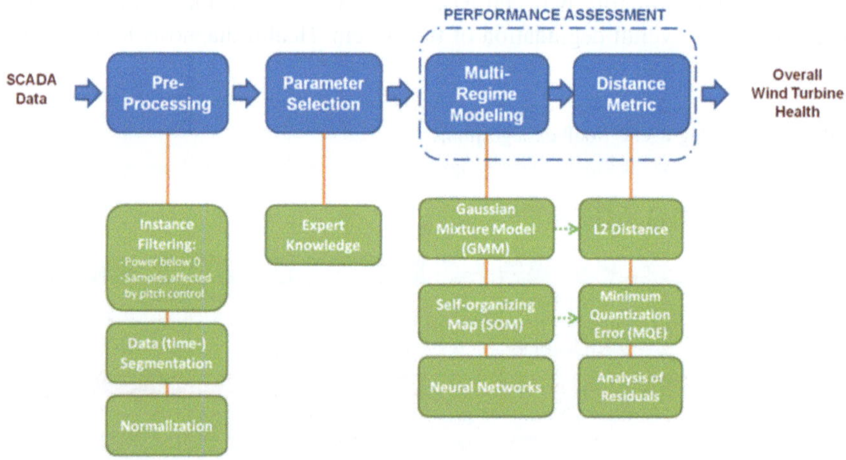

Fig. 3 The overall turbine health represented by GHE

2.3 *Local Damage Estimator*

Data-driven analytical tools from different categories of Watchdog Agent® Toolbox are employed here to obtain LDE values for each component. For generators, bearings and gearbox, signal processing and feature extraction, feature selection, health assessment and health diagnosis tools can be applied to examine the root cause of turbine degradation and health deterioration at component level [13–15] (Fig. 4).

3 Case Study

A SCADA dataset acquired from an onshore large-scale turbine is used to validate the proposed methodology for estimating GHE. The duration of the data is 26 months, during which SCADA module stores the mean, maximum, minimum and standard deviation of all parameters every 10 min. The actual power output is shown in Fig. 5, where three major downtime events are highlighted in grey shadowed areas: (1) Q1-08–Q20-8, (2) Q1-09–Q3-09 and (3) Q4-09–Q1-10.

Data pre-processing methods are used to reject data instances that will eventually not benefit the analysis. First type is when the actual power is less than zero (when turbine is not generating power), which corresponds to when wind speed is below the wind turbine's cut-in speed. The rest of the wind speed range, even

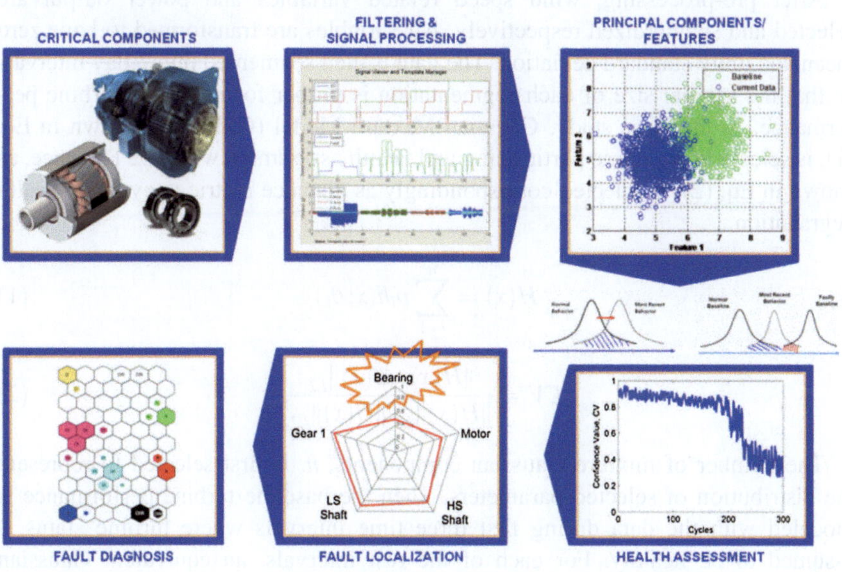

Fig. 4 Systematic approach to estimate LDE for critical turbine components

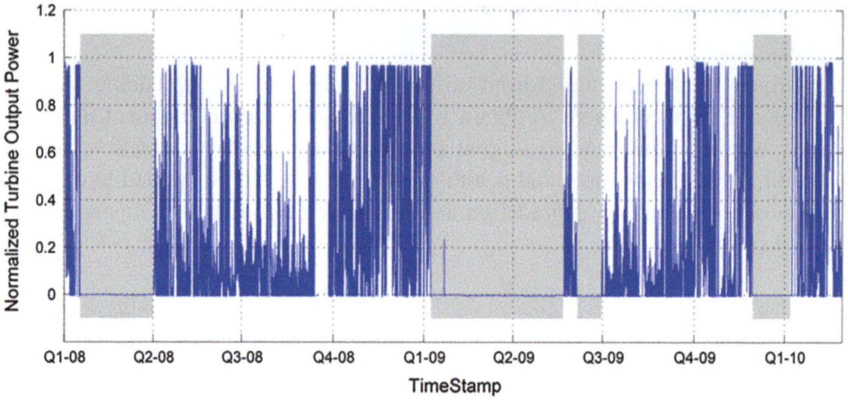

Fig. 5 Actual power generation for the turbine unit

when beyond the turbine's rated cut-out speed, is all kept in this step. Second type of data rejection is based on turbine control mechanism, where power generation is curtailed due to pitch control as counteraction to wind gust rather than turbine's actual degradation. For the wind turbine used in this case study, there is an embedded pitch regulation module designated to adjust the blade's angle of attack with respect to wind direction when high wind speed is observed, so to slow down rotor rotation and limit the drivetrain workload. In the data, power production drops, after high wind speed and rapid change of pitch angle, can be observed. Such instances are rejected to ensure model accuracy.

After pre-processing, wind speed related variables and power output are selected and standardized respectively. All variables are transformed to have zero mean and unity standard deviation. The data is then segmented into 7-day intervals so that the sample size of each segmentation is proper for modeling turbine performance. In this case study, Gaussian Mixture Model (GMM), as shown in Eq. (1), is selected for regime partitioning and health assessment while L2 Distance, as shown in Eq. (2), is selected correspondingly as distance metric to evaluate health degradation.

$$H(x) = \sum_{i=1}^{n} p_i h(x; \theta_i) \tag{1}$$

$$CV = \frac{\|H(x) \cdot G(x)\|_{L2}}{\|H(x)\|_{L2} \cdot \|G(x)\|_{L2}} \tag{2}$$

The number of mixture Gaussian components, *n*, is first selected to represent the distribution of selected parameters. Then the baseline turbine performance is modeled with the data during first three time intervals where turbine status is assumed to be healthy. For each of the rest intervals, an equivalent Gaussian mixture representation is found and CV is computed using L2 distance between baseline and each of the representation [13]. The trend of CV over time shows that the technique captures the gradual degradation of the turbine unit. With appropriate selection and configuration of a prediction algorithm, providing early alarms of ongoing deterioration could prevent major downtime before it happens.

The implementation of the approach in LabVIEW is shown in Fig. 6 through Fig. 7.

As the process of configuring model training is shown in Fig. 6, pre-processed SCADA data variables, including wind speed, power output and their timestamp, are loaded in step 1 and step 2. Step 3 is where the GMM parameters for training a baseline model are configured, including baseline duration, number of clusters and number of iterations. While model training is triggered, the normalized baseline power curve is displayed to provide an evident observation.

Performance evaluation through testing of the GMM model obtained from remaining data is shown in Fig. 8. As continuation from baseline modeling, step 5 is for configuring GMM parameters for testing and step 6 initiates the calculation of CV value based on the configured parameters. Once testing is completed, CV is shown over time on the right hand side of the panel.

Eventually, the progression of turbine unit power performance, shown as CV, together with wind speed and power output is visualized for user (Fig. 9). Furthermore, a scalable monitoring platform for an entire fleet can be constructed by executing a configured and validated algorithm for each individual turbine.

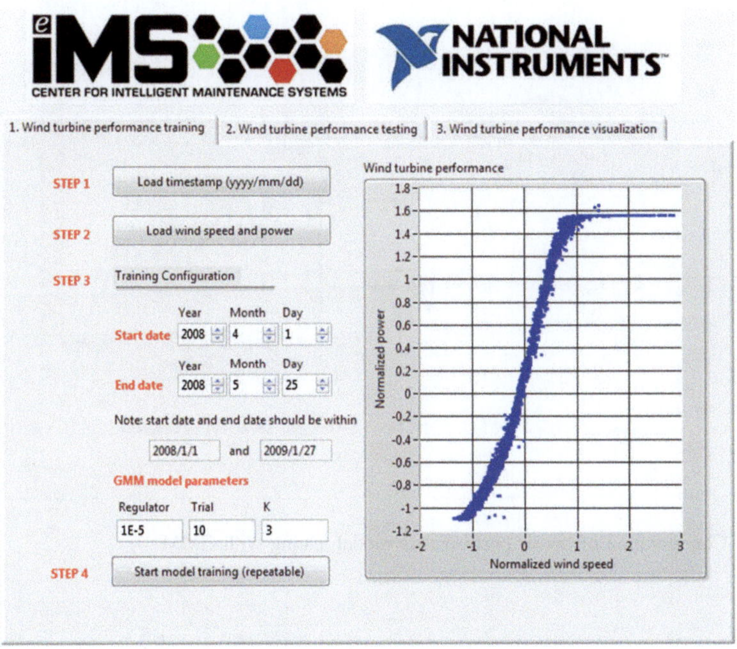

Fig. 6 Configuration of power performance model training with GMM

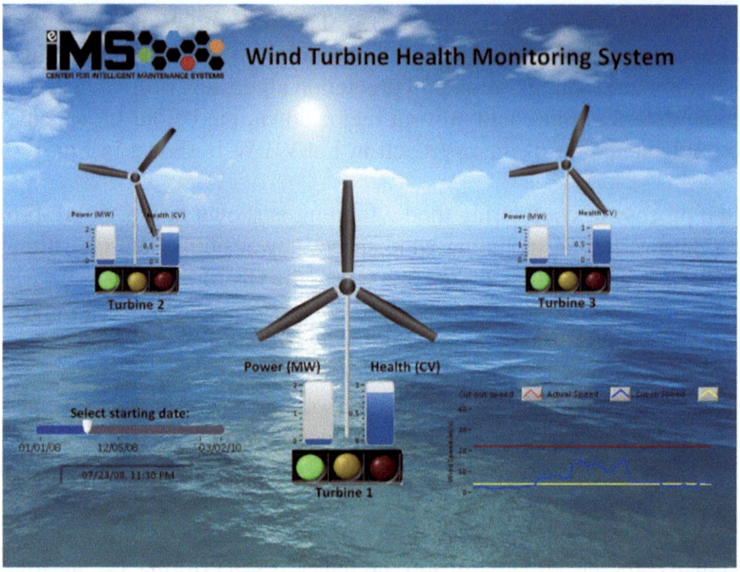

Fig. 7 Demonstration of wind turbine health monitoring system using national instruments LabVIEW [14]

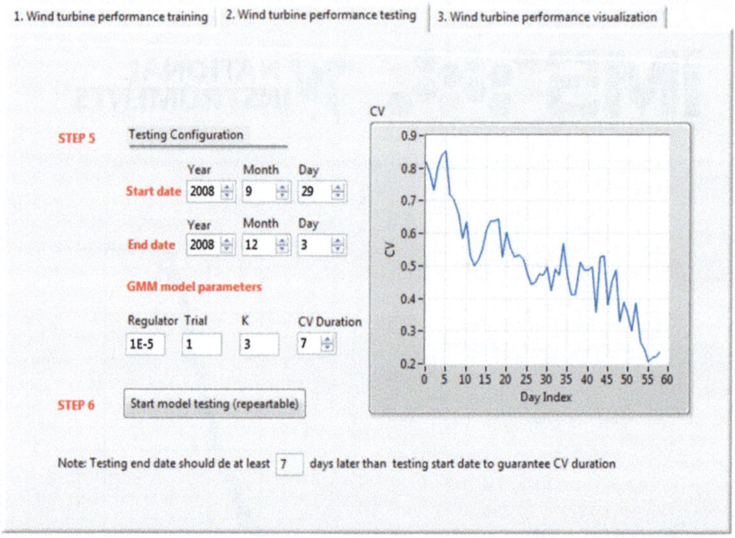

Fig. 8 Configuration of power performance model testing with GMM

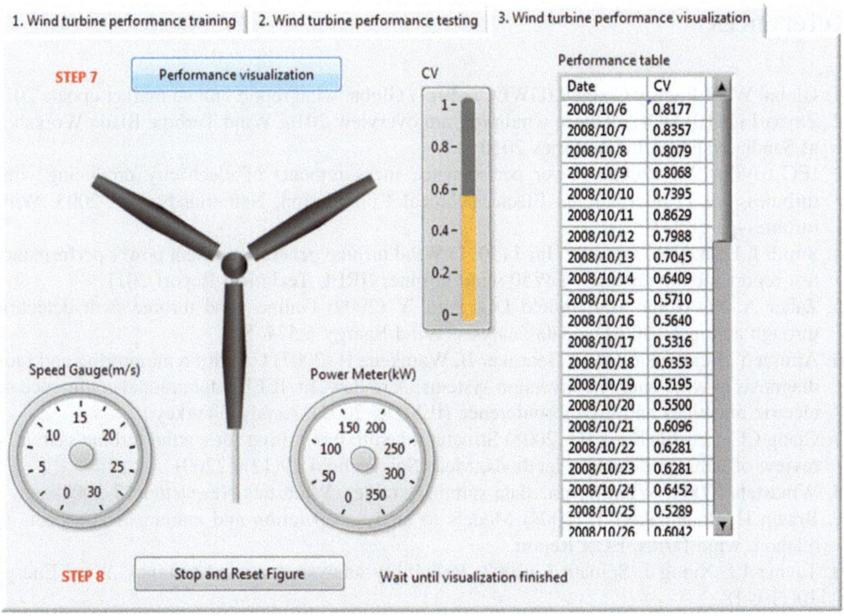

Fig. 9 Visualization of wind turbine power performance

4 Conclusion

This paper presents an advanced framework for wind turbine prognostics and health management. A Global Health Estimator is proposed to enhance current performance testing practices advised by an IEC standard. It is generated with SCADA data to indicate and predict turbine unit overall performance. A Local Damage Estimator is computed to evaluate degradation status of critical turbine drivetrain components, and locate the fault that is the root cause of turbine performance degradation. A case study is presented where Gaussian Mixture Model and L2 Distance are used to compute the GHE. The implementation of GHE computation is presented with Watchdog Toolbox® in National Instruments LabVIEW.

The presented work benefits the wind power industry, as it helps to establish an effective and systematic PHM solution that is capable of dealing with turbines' constantly changing operating conditions and provides predictive maintenance suggestions for wind farm operator. The calculation of GHE utilizes existing SCADA system, provides progressive evaluation of actual power performance and offers opportunities to predict turbine downtime.

References

1. Global Wind Energy Council (GWEC) (2011) Global wind report annual market update 2011
2. Zayas J (2010) DOE offshore wind program overview 2010, Wind Turbine Blade Workshop at Sandia National Laboratories 2010
3. IEC 61400-12-1 (2005) Power performance measurements of electricity producing wind turbines, vol 1. International Electrotechnical Commission, Netherlands, IEC 2005, Wind turbines–part 12–1
4. Smith J, Huskey A, Jager D, Hur J (2011) Wind turbine generator system power performance test report for the entegrity EW50 wind turbine, NREL Technical Report 2011
5. Zaher A, McArthur SDJ, Infield DG, Patel Y (2009) Online wind turbine fault detection through automate SCADA data analysis. Wind Energy 6:574–593
6. Amirat Y, Benbouzid MEH, Bensaker B, Wamkeue R (2007) Condition monitoring and fault diagnosis in wind energy conversion systems: A review. In: IEEE international conference on electric machines and drives conference (IEMDC 2007). Antalya, Turkey
7. Ciang CC, Lee JR, Bang HJ (2008) Structural health monitoring for a wind turbine system: a review of damage detection methods. Meas Sci Technol 19(12):122001
8. Windstats (2004) Wind turbine data summary tables. Windstats Newsletter 17 2004
9. Braam H, Rademakers L (2004) Models to analyse operation and maintenance aspects of offshore wind farms. ECN Report
10. Tavner PJ, Xiang J, Spinato F (2007) Reliability analysis for wind turbines. Wind Energy 10(1):1–18
11. Liao L, Wang H, Lee J (2008) Reconfigurable watchdog agent® for machine health prognostics. Int J COMADEM 11(3):2–15
12. McLachlan GJ (1992) Discriminant analysis and statistical pattern recognition. Wiley, New York
13. Lapira ER, Siegel D, Zhao W, Brisset D, Su J, Wang C, AbuAli M, Lee J (2011) A systematic framework for wind turbine health assessment under dynamic operating conditions, 24th international congress on condition monitoring and diagnostics engineering management (COMDAEM 2011)

ART-2 Artificial Neural Networks Applications for Classification of Vibration Signals and Operational States of Wind Turbines for Intelligent Monitoring

Tomasz Barszcz, Andrzej Bielecki, Mateusz Wójcik and Marzena Bielecka

Abstract In recent years wind energy is the fastest growing branch of the power generation industry. The largest cost for the wind turbine is its maintenance. A common technique to decrease this cost is a remote monitoring based on vibration analysis. Growing number of monitored turbines requires an automated way of support for diagnostic experts. As full fault detection and identification is still a very challenging task, it is necessary to prepare an "early warning" tool, which would focus the attention on cases which are potentially dangerous.

Keywords Wind turbines intelligent monitoring · Early warning · ART neural networks

T. Barszcz
Faculty of Mechanical Engineering and Robotics, Chair of Robotics and Mechatronics, AGH University of Science and Technology, Al. Mickiewicza 30, 30-059 Cracow, Poland
e-mail: tbarszcz@agh.edu.pl

A. Bielecki (✉)
Faculty of Electrical Engineeing, Automation, Computer Science and Biomedical Engineering, Chair of Applied Computer Science, AGH University of Science and Technology, Al. Mickiewicza 30, 30-059 Cracow, Poland
e-mail: azbielecki@gmail.com

M. Wójcik
Faculty of Physics, Astronomy and Applied Computer Science, Department of Computer Design and Graphics, Jagiellonian University, Reymonta 4, 30-059 Cracow, Poland
e-mail: mateusz.wojcik@uj.edu.pl

M. Bielecka
Faculty of Geology, Geophysics and Environmental Protection, Chair of Geoinformatics and Applied Computer Science, AGH University of Science and Technology, Cracow, Poland
e-mail: bielecka@agh.edu.pl

G. Dalpiaz et al. (eds.), *Advances in Condition Monitoring of Machinery in Non-Stationary Operations*, Lecture Notes in Mechanical Engineering,
DOI: 10.1007/978-3-642-39348-8_58,

1 Introduction

In recent years wind energy is the fastest growing branch of the power generation industry. The average yearly growth in the years 1997–2003 achieved 32 % in the United States and 22 % in the European Union [1] and these figures will hold for at least the next decade. The distribution of costs during the life cycle of the unit for wind energy is significantly different from that of traditional, fossil fired units [1]. First of all, initial investment costs are relatively higher, whereas in traditional units cost of fuel plays important role (usually it is the second largest cost). After commissioning, the largest cost for the wind turbine (WT for abbreviation) is maintenance. With proper maintenance policies, wind turbines can achieve the highest level of availability in the power generation sector—even up to 98 %.

Therefore, condition monitoring of wind turbines, including fault diagnostics, in particular at the early stage of a fault occurrence or even participatory actions, is an essential problem in wind turbines engineering in particular [2–4] and in rotating machinery engineering in general [5]. There were several attempts to develop various monitoring tools, in most cases based on various classification methods. Some of them are based on artificial neural networks (ANNs for abbreviation). In most types of ANNs the learning process is unsuitable for cases of continuous machinery intelligence monitoring. This means, among others, that adding a new patterns as inputs requires repetition of the learning process. In ART networks, introduced by Carpenter and Grossberg [6, 7], the learning process is not separated from its operation. Furthermore, ART neural networks are capable to add new states when necessary [8–10]. Therefore, this sort of ANNs was tested as a tool for classification of states in continuous monitoring.

2 The Machine Description

The faults which are sought in wind turbines are primarily of mechanical origin. The wind turbine with the gearbox, which is the most popular type, can be described in the following way. The main rotor with three blades is supported by the main bearing and transmits the torque to the planetary gear. The main rotor is connected to the plate which is the gear input. The planetary gear has three planets, with their shafts attached to the plate. The planets roll over the stationary ring and transmit the torque to the sun. The sun shaft is the output of the planetary gear. The sun drives the two-stage parallel gear which has three shafts: the slow shaft connected to the sun shaft, the intermediate shaft and the fast shaft, which drives the generator. The overall gear ratio is in the range of 1:100. The generator produces alternating current of slightly varying frequency. This current is converted first into direct current power and then into alternating current power of frequency equal to the grid frequency. Electric transformations are performed by the controller at the base of the tower—see Fig. 1.

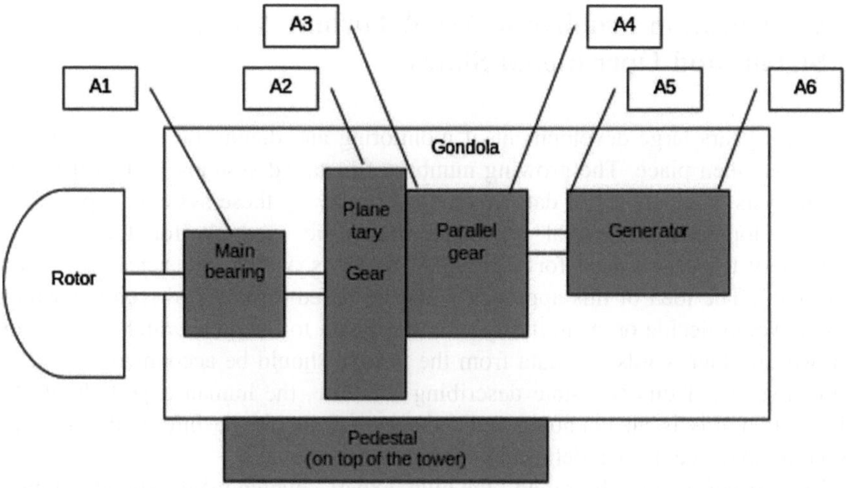

Fig. 1 The mechanical structure of the wind turbine. Location of vibration measurement sensors is shown by An symbols

In the field of vibrodiagnostics, a machine operational state is understood as an accepted range of machine's operational points enabling referential analysis. In practice, machine operating point is defined by values of available measurements of physical quantities such as speed, load, pressure, temperature, etc., usually called "machine process parameters" [11]. Typically, from each vibration record, a set of diagnostic indicators is calculated known as "trends". Each trend point is a combination of representation of true machine technical condition and behaviour, current operating point of the machine measurement error and random factor. In a typical condition monitoring set up, each trend is tracked against a precalculated threshold value. In this case, operational states (shortly called "states") are used for data classification during the data acquisition process. On the basis of these states, data is combined into sets, which are assumed to represent a particular machine. Consequently, the overall number of defined diagnostic indicators and estimators is equal to the number of indicators and estimators multiplied by the number of states. Therefore, from operator's point of view, it is desirable to have as little states as possible. On the other hand, from reliable-diagnostics point of view, in order to minimize the fluctuation of operating points of the machine it is desirable to define ranges of states as low as possible. In this case, the state configuration would result either in (a) single operational state with low permissible fluctuation of operational parameters, (b) a large number of operational state with low permissible fluctuations of operational parameters.

3 Classification Problem of Wind Turbine Vibration Signals and Operational States

In recent years large development of monitoring and diagnostic technologies for WTs has taken place. The growing number of installed systems created the need for analysis of gigabytes of data created every day by these systems. Apart from the development of several advanced diagnostic methods for this type of machinery there is a need for a group of methods, which will act as an "early warning". The idea of this approach could be based on a data driven algorithm, which would decide on a similarity of current data to the data, which are already known. In other words, the data from the turbine should be accounted for one of known states. If this is a state describing a failure, the human expert should be alarmed. If this is an unknown state, the expert should be informed about the situation and asked for a definition of such a new state.

This approach could be called "the blunt expert", which maybe sounds strange, but gives the most important feature of the proposed method. This approach may brake the biggest barrier of application of artificial neural networks (ANNs) in diagnostics, which is availability of significant amount of training data. As in real cases it is not possible to acquire it, it is only possible to train ANNs for a few cases covered by available data.

The problem of classification was investigated by several authors. One of the first works was research by Shuhui et al. [12], who compared classification techniques for the wind curve estimation. This work was often referenced by others, but only multi-layred feed-forward ANNs were studied there. Another important contribution was given by Kim [13], who compared performance of several classification methods. His experiments showed that if the number of independent variables in the system is low, then ANNs perform better than other methods. Again, the investigated network was the multi-layer feed-forward network trained by the back-propagation algorithm.

There are no works, known to us, which would consider application of ART networks for the classification of WT states. There were also works applying ANNs for wind and power generation prediction, but this issue is outside the scope of this paper.

As the ART networks are capable to perform efficient classification and to recognize new states when necessary [6, 7], we performed research of initial classification task. The goal of the experiment was verification of ART classification capabilities with comparison to the human expert. This type of data is acquired in the majority of cases and the successful classifier should create a reasonable number of classes, similar to these by a human expert. This task is the main goal of the following paper.

As such a classification was shown, it is thus possible to filter out states, which are known to be correct. The expert can then focus only on "suspicious" states returned by the algorithm.

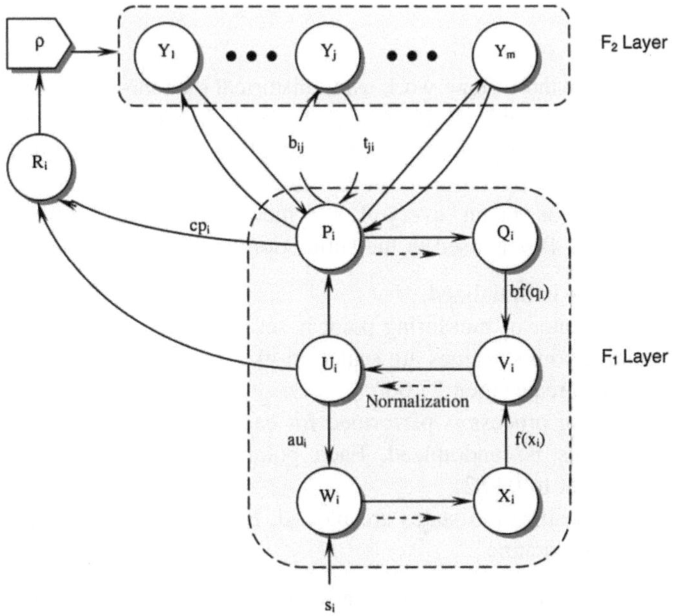

Fig. 2 ART-2 architecture

4 Characteristics of the ART-2 Neural Networks

Let us briefly recall ART-2 neural network properties [7, 14, 15].

The ART-2 is an unsupervised neural network based on adaptive resonance theory (ART). A typical ART-2 architecture, introduced by Carpenter and Grossberg [6, 7], is presented in Fig. 2 (only one unit of each type is shown here). In the attentional sub-system, an input pattern s is first presented to the F_1 layer, which consists of six kinds of units—the W, X, U, V, P and Q cells. It then undergoes a process of activation, including normalization, noise suppression and updating. This results in an output pattern p from the F_1 layer. Responding to this output pattern, an activation is produced across F_2 layer through bottom-up weights b_{ij}. As the F_2 layer is a competitive layer with a winner-takes-all mode, only one stored pattern is a winner. It also represents the best matching pattern for the input pattern at the F_1 layer. Furthermore, the pattern of activation on the F_2 layer brings about an output pattern that is sent back to the F_1 layer via top-down weights t_{ji}. For the orienting sub-system, it contains a reset mechanism R and a vigilance parameter q to check for the similarity between the output pattern from the F_2 layer and the original input pattern from the F_1 layer. If both patterns are concordant, the neural network enters a resonant state where the adaptation of the stored pattern is conducted. Otherwise, the neural network will assign an uncommitted (inhibitory) node on the F_2 layer for this input pattern, and thereafter, learn and transform it into a new stored pattern.

5 Results

A simulation of a wind turbine work using historical data has been done. Changes of operational states and vibration signals were investigated. Historical data contain 27,000 measuring points in time (sampling frequency is 1 per 15 min). Each point has three operational states values and one vibration signal value. Simulation step was taken every 1,000 measuring points. ART-2 network (described in Sect. 4) was used in the simulation in a following way.

1. ART-2 network is initialized.
2. Time t as a pointer of measuring point is set to 1,000.
3. All measuring points' values are scaled to $[0, 1]$.
4. While t is not greater then 27,000.
(a) ART-2 learning process is performed for each measuring point in $[1, t]$. The order of points is randomized. Each point is taken once. The vigilance parameter is set to 0.982.
(b) Figures of classification states are plotted. Each network pattern determines some state of machine.

Figure 3 presents how the ART-2 network has classified states for first 5,000 measuring points. There were 3 dominate states. After 15,000 measuring points 4-th dominate state has joined. That situation is presented on Fig. 4. At the end classification states look like in the Fig. 5.

Table 1 describes how numbers of classified states were changing during simulation process.

Looking at last the column there can be observed first changes of numbers of states. First change has happened after 8,000 measuring points. Next ones after 10,000 and after 14,000 points. Figure 6 presents the vibration signals for all measuring points (ordered by time). After about 14,000 points the biggest damage

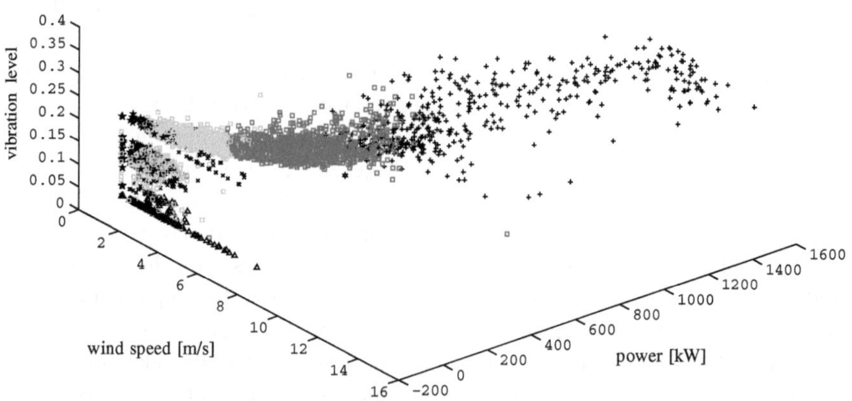

Fig. 3 The operational states (wind, power) and vibration signal classified using ART-2 network using 5,000 measuring points

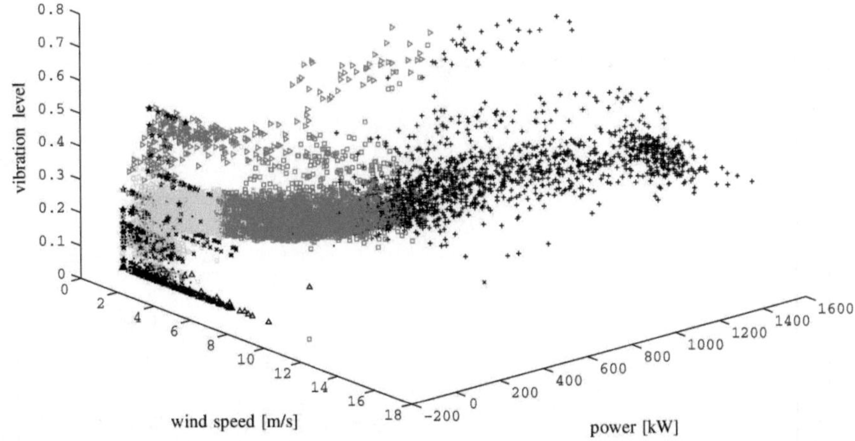

Fig. 4 The operational states (wind, power) and vibration signal classified using ART-2 network using 15,000 measuring points

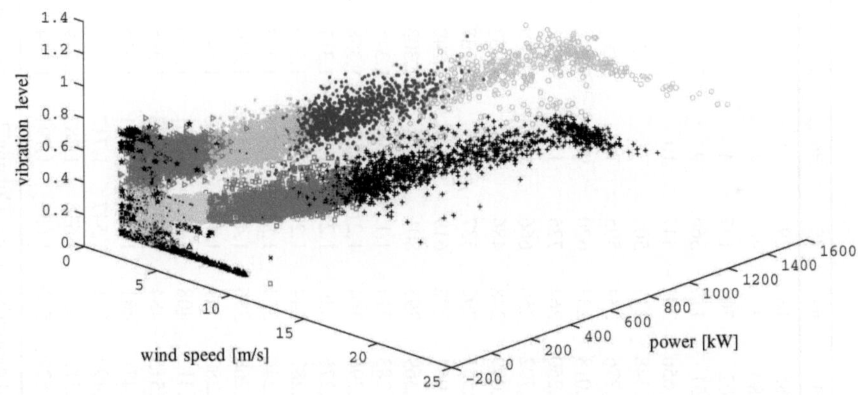

Fig. 5 The operational states (wind, power) and vibration signal classified using ART-2 network using 27,000 measuring points

Table 1 Details of simulation steps

Distance	States	#1	#2	#3	#4	#5	#6	#7	#8	#9	#10	#11	#12	#13	#14	#15	# > 10
1,000	5	677	91	185	18	29											5
2,000	6	1,354	131	380	33	98	4										5
3,000	6	2,033	144	602	70	147	4										5
4,000	7	2,560	184	751	119	366	9	11									6
5,000	7	2,170	210	1,656	208	447	10	299									6
6,000	8	2,269	238	2,348	217	505	7	412	4								6
7,000	8	2,590	257	2,776	234	595	7	535	6								6
8,000	8	3,106	252	3,018	233	620	9	755	7								6
9,000	8	3,531	281	3,250	264	725	12	922	15								8
10,000	9	4,008	256	3,702	256	666	113	977	20	2							8
11,000	10	4,362	240	4,099	248	486	473	1,053	24	14	1						9
12,000	11	4,594	247	4,551	292	537	593	1,137	32	14	1	2					9
13,000	11	4,938	256	4,964	292	619	624	1,249	39	14	2	3					9
14,000	11	5,116	273	5,369	293	801	776	1,303	44	14	4	7					9
15,000	11	5,418	277	5,283	314	1,129	858	1,345	59	14	7	296					10
16,000	11	5,425	280	5,296	384	1,211	972	1,358	129	14	10	921					10
17,000	12	5,515	260	5,276	339	1,217	1,336	1,317	269	12	23	1,433	3				11
18,000	12	5,574	261	5,287	344	1,237	1,346	1,350	274	12	28	2,283	4				11
19,000	12	5,635	279	5,244	354	1,328	1,475	1,353	273	12	35	3,006	6				11
20,000	12	5,633	292	5,268	386	1,440	1,573	1,350	291	12	46	3,700	9				11
21,000	13	5,520	298	5,282	399	1,443	1,610	1,349	288	13	91	3,895	801	11			13
22,000	14	5,299	291	5,412	408	1,432	1,561	1,336	291	12	174	2,977	2,306	497	4		13
23,000	14	5,151	302	5,515	484	1,499	1,643	1,385	315	12	182	3,416	2,546	545	5		13
24,000	14	5,228	314	5,471	481	1,499	1,681	1,365	331	12	200	3,781	2,979	653	5		13
25,000	14	5,233	323	5,499	511	1,547	1,717	1,350	337	12	217	4,283	3,262	700	9		13
26,000	14	5,226	343	5,512	578	1,638	1,782	1,349	340	12	223	4,718	3,554	710	15		14
27,000	15	5,177	346	5,524	572	1,636	1,791	1,394	341	13	229	5,231	3,737	882	30	97	15

Column 1 (Details)—number of processed measuring points. Column 2—number of detected states. Values in column 3–17 represent numbers of measuring point which were classified as a specified state. Last column—number of states with value greater then 10

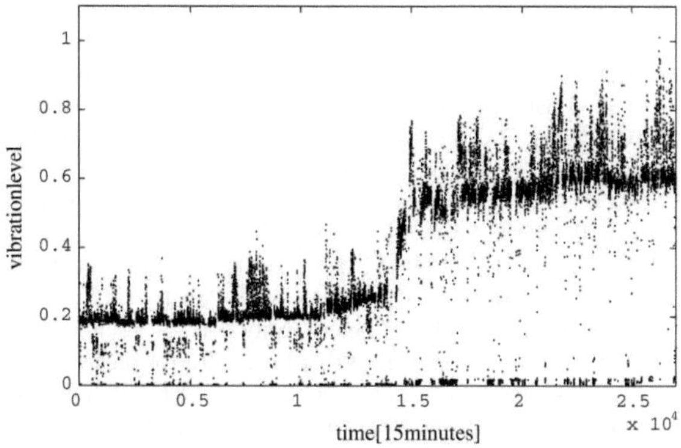

Fig. 6 Vibration signals for all measuring points

happened. There must have been some symptoms of that damage. Mentioned two first changes could be candidates of that symptoms.

6 Concluding Remarks

As it has been mentioned, monitoring is crucial in wind turbines exploitation. On the other hand, there are very few attempts to create system for intelligent monitoring based on artificial intelligence—see [2] and references given there. The experiments described in this paper show that ART ANNs can be effective tool for such task performing—the symptoms of a turbine damage can be detected using ART ANN. It should be stressed however, that the obtained results are preliminary ones—only one vibration channel has been used whereas, usually, a few vibration channels are observed simultaneously—see [16]. The monitoring module based on ANNs is planned to be a module of expert system for intelligent monitoring and fault diagnostics in wind turbines.

The paper was supported by the Polish Ministry of Science and Higher Education under Grant No. N504 147,838.

References

1. Barszcz T, Randall RB (2009) Application of spectral kurtosis for detection of a tooth crack in the planetary gear of a wind turbine. Mech Sys Signal Process 23:1352–1365
2. Hameeda Z, Honga YS, Choa TM, Ahnb SH, Son CK (2009) Condition monitoring and fault detection of wind turbines and related algorithms: a review. Renew Sustain Energy Rev 13:1–39
3. Jabłoński A, Barszcz T, Bielecka M (2011) Automatic validation of vibration signals in wind farm distributed monitoring systems. Measurement 44:1954–1967
4. Kusiak A, Li W (2011) The prediction and diagnosis of wind turbine faults. Renewable Energy 36(2011):16–23

5. Barszcz T, Bielecki A, Romaniuk T (2009) Application of probabilistic neural networks for detection of mechanical faults in electric motors. Electr Rev 8(2009):37–41
6. Carpenter GA, Grossberg S (1987) A massively parallel architecture for a self-organizing neural pattern recognition machine. Comput Vis Graph Image Process 37:54–115
7. Carpenter GA, Grossberg S (1987) ART2: self-organization of stable category recognition codes for analog input pattern. Appl Opt 26:4919–4930
8. Korbicz J, Obuchowicz A, Uciński D. (1994) Artificial neural networks. Foundations and Applications. Academic Press PLJ, Warsaw (in Polish)
9. Rutkowski L. (1996) Neural networks and neurocomputers. Technical University in Czestochowa Press, Czestochowa (in Polish)
10. Tadeusiewicz R (1993) Neural Networks. Academic Press, Warsaw (in Polish)
11. Jabłoński A, Barszcz T (2012) Procedure for data acquisition for machinery working under non-stationary operational conditions, The Ninth International Conference on Condition Monitoring and Machinery Failure Prevention Technologies, 12–14 June 2012, London
12. Shuhui L, Wunsch DC, O'Hair E, Giesselmann MG (2001) Comparative analysis of regression and artificial neural network models for wind turbine power curve estimation. J Sol Energy Eng 123:327–332
13. Kim YS (2010) Performance evaluation for classification methods: a comparative simulation study. Expert Syst Appl 37(3):2292–2306
14. Barszcz T, Bielecki A, Wójcik M (2012) ART-type artificial neural networks applications for classification of operational states in wind turbines. Lecture Notes in Artificial Intelligence 6114, pp 11–18
15. Shieh MD, Yan W, Chen CH (2008) Soliciting customer requirements for product redesign based on picture sorts and ART2 neural network. Expert Syst Appl 34:194–204
16. Barszcz T, Bielecka M, Bielecki A, Wójcik M (2012) Wind speed modelling using Weierstrass function fitted by a genetic algorithm. J Wind Eng Ind Aerodyn 109:68–78

Software Applications for Wind Turbine Vibrations Analysis

I. Cozorici, H. Balan, R. A. Munteanu and P. Karaisas

Abstract In this paper is presented the implementation of a software for the analysis of vibration generated by wind turbines components. This software application was built in LabWiew programming environment and for vibration analysis and fault detection were used techniques as: wavelet analysis, envelope detection, FFT analysis, Cepstrum analysis, Vector RMS analysis. Vibration signal envelope detection was performed using a virtual instrument based on Hilbert transform and rms analysis was performed using virtual instrument "Vector RMS" specifically designed for vibration analysis. The experimental results were obtained by measuring the vibration signals of a bearing with a simulated fault on outer raceway.

Keywords Vibrations · RMS · Hilbert transform

1 Introduction

Vibration signal processing usually is performed by means of analysis in time domain and frequency domain, mainly, due to the simplicity of these methods. In nonstationary signal analysis applications, time domain analysis or frequency

I. Cozorici (✉) · H. Balan · R. A. Munteanu
Technical University of Cluj Napoca 28, Memorandumului Street, Cluj Napoca, Romania
e-mail: Ioan.COZORICI@mas.utcluj.ro

H. Balan
e-mail: horia.balan@eps.utcluj.ro

R. A. Munteanu
e-mail: radu.a.munteanu@ethm.utcluj.ro

P. Karaisas
T. E. I. of Piraeus, 250th Petrou Ralli and Thivon Street, 12244 Egaleo Athens, Greece
e-mail: karaisas@teipir.gr

G. Dalpiaz et al. (eds.), *Advances in Condition Monitoring of Machinery in Non-Stationary Operations*, Lecture Notes in Mechanical Engineering,
DOI: 10.1007/978-3-642-39348-8_59,

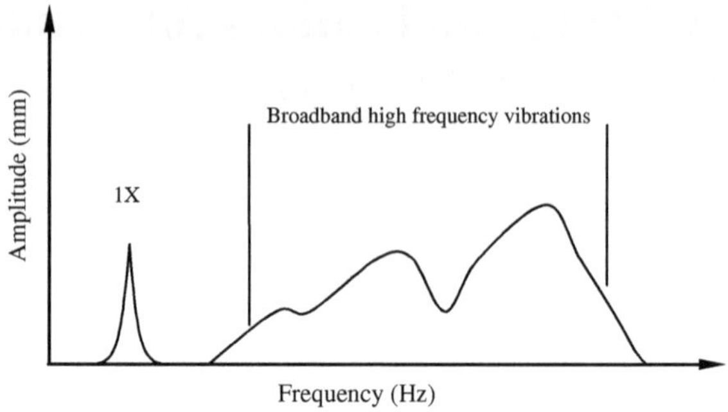

Fig. 1 Amplitude-frequency characteristic of a bearing defect

domain analysis, generates errors, which in some cases are significant. In the case of nonstationary signals, the spectrum changes over time and hence the need for time-frequency analysis.

Signal processing techniques in the time-frequency domain must highlight non-stationarity of the analyzed signal structure by spectro-temporal analysis of energy density, which is the physical significance of the time-frequency representations.

For bearings fault diagnosis, Fig. 1 uses several signal processing techniques, which include: Fourier analysis, Cepstrum analysis, signal analysis methods based on time domain analysis and Wavelet analysis [1, 2, 6–8].

Given that bearings are critical components of the gearboxes which are non-linear operating components of wind turbines, it requires the use of advanced signal processing methods.

Today state of the art in the field of computing, allows the implementation of mathematical algorithms, though known for a long time, have had limited applicability. Among these methods, Hilbert transform is a signal processing technique suitable for the envelope vibration signal analysis.

2 Theoretical Considerations

Hilbert transform is defined in the time domain as the convolution of Hilbert transformation parameter $1/(\pi t)$ with the time domain real vibration signal x (t) [12]:

$$y(t) = H\{x(t)\} = \frac{1}{\pi} \int_{-\infty}^{+\infty} \frac{x(t)}{t - \tau} d(\tau) \tag{1}$$

Knowing the time variation of the signal x(t) and Hilbert transform y(t) it can be introduced the notion of complex analytic signal associated x(t) signal, according to the relationship:

$$z(t) = x(t) + i \cdot y(t), \tag{2}$$

which has the advantage that it can be represented as a function of the instantaneous amplitude and phase:

$$z(t) = A(t) \cdot \exp(i \cdot \varphi(t)), \tag{3}$$

where A(t) is the envelope of z(t) signal, defined by:

$$A(t) = |x(t) + j \cdot y(t)|, \tag{4}$$

and φ(t) is the phase of z(t) complex analytic signal, defined by the relation:

$$\varphi(t) = ctg\frac{y(t)}{x(t)} \tag{5}$$

Square of A(t) function is the expression of instantaneous power signal, depending on time. Hilbert analysis allows to determine the power and instantaneous frequency of a signal.

In [3] the authors proposed a method for detecting bearings faults using vibration signal analysis. The signal is translated into time-frequency domain by the use of Wavelet transform, and then the envelope analysis of the detail coefficients of the high frequency components is performed using the Hilbert transform thus being able to identify the bearing fault characteristic frequencies.

3 Vibrowind Software

The monitoring and diagnosis system GUI software application is realized using Labview programming language.

LabWiew is a visual programming language provided by the National Instruments company [4]. Its name comes from the abbreviation of "Laboratory Virtual Instrument Engineering Workbench".

Thanks to exclusively graphical programming interface, LabWiew is an ideal programming language for the use in process monitoring applications. The library contains several predefined functions, focusing on the following areas:

- data quisition;
- control devices;
- data analysis;
- display and data storage.

LabWiew programs are called virtual instruments (VI) and are based on the concepts of modularity and hierarchy tree. Thanks to the virtual instruments modular nature can be used both as main programs and subroutines in which case they are called "SubVI". These subroutines have the advantage that they can be developed and tested independently.

LabWiew applications have two distinct parts: the front panel and the block diagram. When the program is running these two windows carry out a continuous data exchange between them.

The front panel is the graphical user interface that displays graphical elements as indicators or controls. For the front panel designing are available predefined objects such as buttons, graphics objects, switches, tanks, sliders etc.

For the block diagram implementation are used nodes representing execution elements, terminals representing gates through which data is transferred and wires representing data flow block diagram.

Software "VibroWind" is composed of two modules: data acquisition module (Fig. 2) and data analysis module (Fig. 3). In the window "DATA ACQUISITION" data acquisition is performed, the user being able to select specific data acquisition parameters such as channel acquisition, channel dynamic range, mode and sample rate and also the user can view, both monitored waveforms in real time, for example vibrations from accelerometers (Fig. 2), and histogram of the monitored parameters.

In this window the user can save the data in TDMS file type by selecting the memory address and name in the "TDMS File Path". The user can create a new

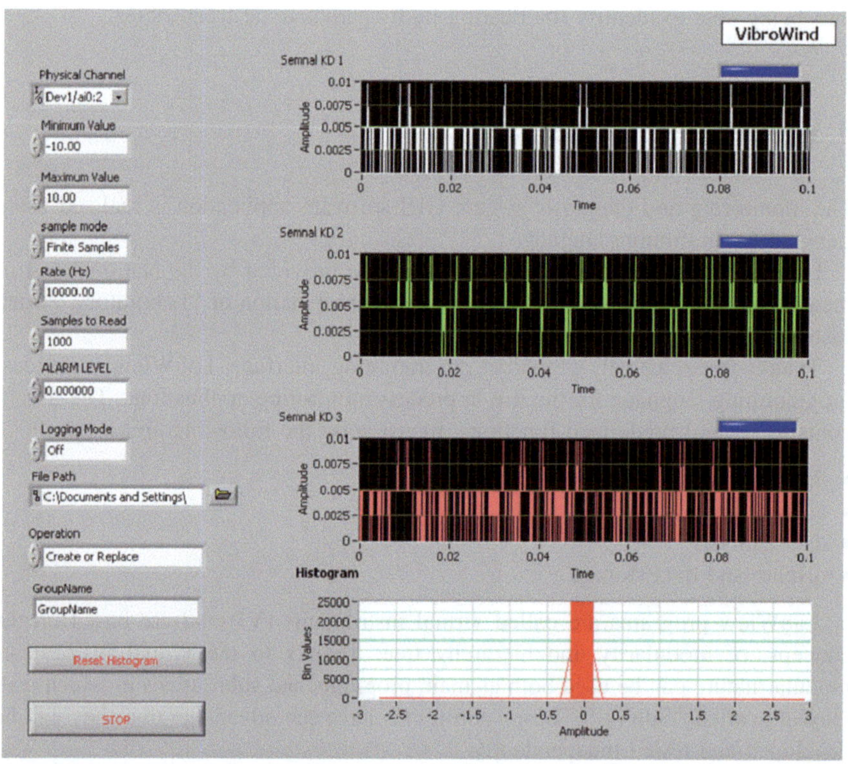

Fig. 2 Data acquisition module GUI

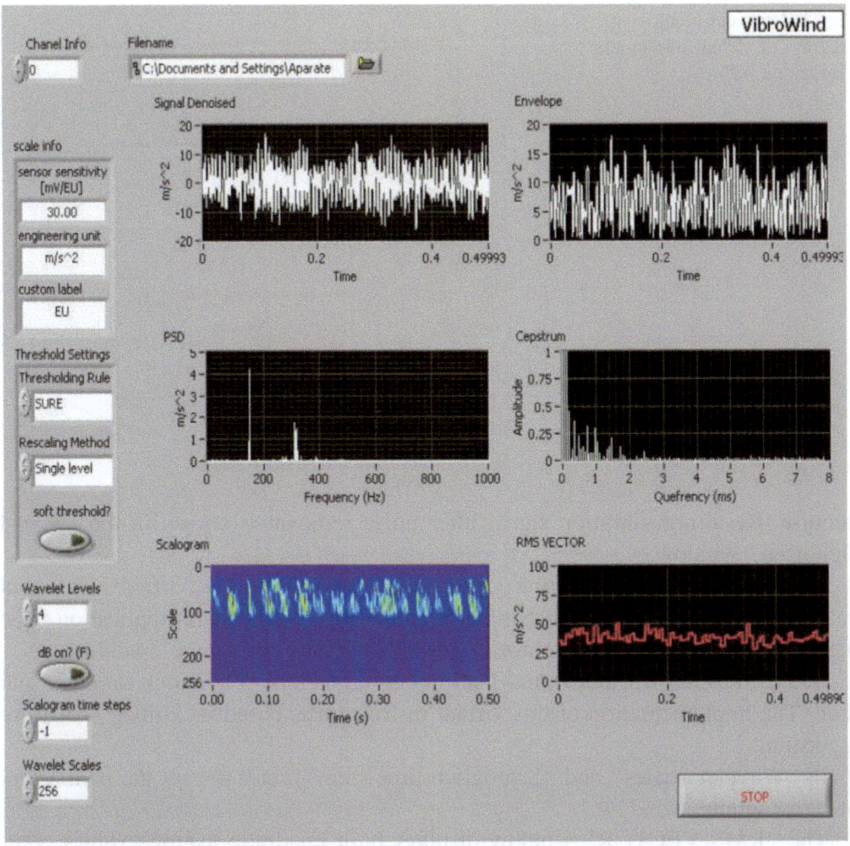

Fig. 3 Data analysis module GUI

file or add data to an existing file. If the user add data to an existing file, the # symbol is added automatically to the group name followed by a number until the user chooses a new group name. The window is provided with a set of three blue LEDs for each acquisition channel, which are designed to alert the operator. These LED turn color to red when the signal amplitude on a particular channel exceeds the value set in the "ALARM LEVEL" field, at which time the operator can proceed to analyze in detail the respective signal.

In the "Data Analysis" window the user can select a file saved in the previous window, or other previously saved file for signal envelope analysis, FFT analysis, Cepstrum analysis and Wavelet analysis. Files which contain saved vibration signals can be selected by specifying the file name and the desired channel.

In this window, in the "File Name" field the user can select the file containing vibration signal. In the "scale info" field it can be selected the vibration sensor sensitivity and measurement unit. In the "threshold settings" field it can be selected the method of Wavelet tresholding implementation for vibration signal

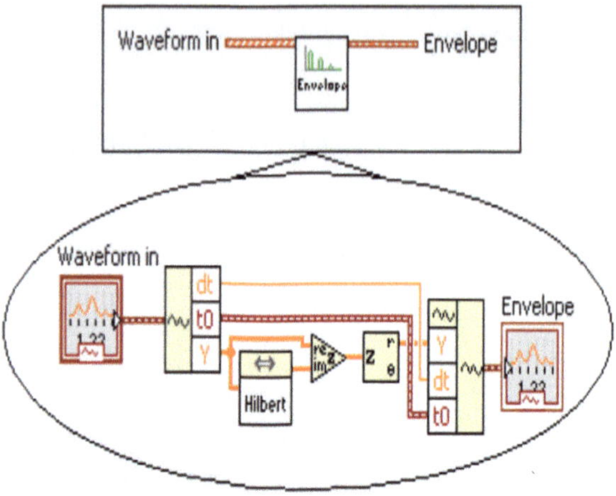

Fig. 4 Icon and block diagram of virtual instrument "Envelope VI"

denoise. Resulting vibration signal after noise removal is shown in the "Signal Denoised" window.

For the analysis of vibration signal envelope was built a virtual instrument called "Envelope VI" which has been embedded in the main application. The detected vibration signal envelope is presented in the "Envelope" window.

In Fig. 4 is shown the virtual instrument "Envelope VI" block diagram and icon. The implementation of this virtual instrument is based on Hilbert transform algorithm.

In "Wevelet Scales" and "Scalogram time steps" fields are set the parameters of "Scalogram" window.

The "RMS VECTOR" window displays built quadratic average values computed for each period of the analyzed signal. This parameter is proposed to be used as indicator of machine condition in the diagnosis of wind turbine structure.

In Fig. 5 is shown the virtual instrument "RMS VECTOR VI" icon and block diagram.

Virtual instrument shown in Fig. 5 uses for the calculation of the RMS value the application "RMS VI" available in Probability and Statistics Labview library.

Figures 6 and 7 shows the block diagram of VibroWind application acquisition module and analysis module.

4 Experimental Results

As shown in Sect. 2, tire vibration signal, defined by the relation (4) is a low frequency signal which follows the rectified input signal peaks.

In the envelope frequency spectrum are found components with frequency equal to the occurrence rate of pulses and amplitude proportional to their energy. The main stages of envelope vibration analysis are filtering, enveloping and

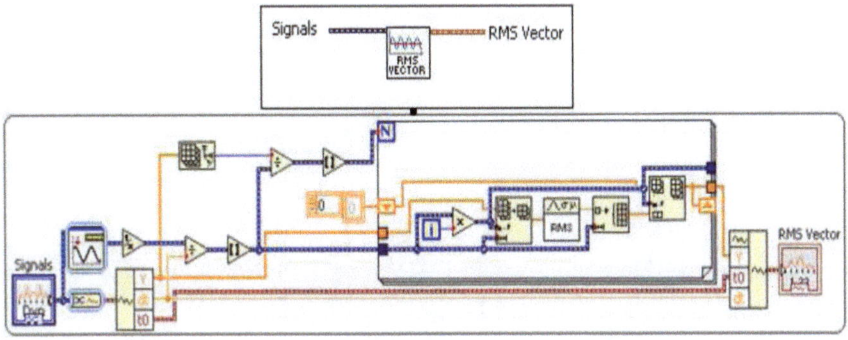

Fig. 5 Icon and block diagram of virtual instrument "RMS Vector VI"

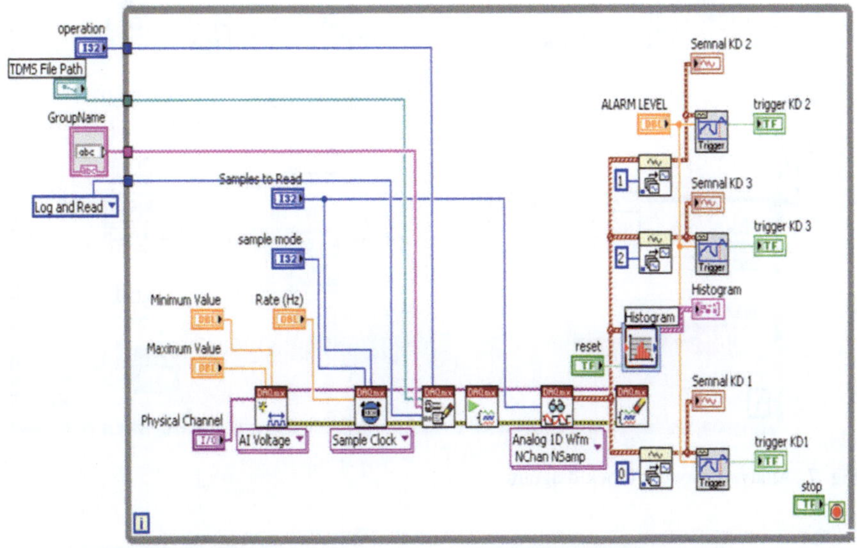

Fig. 6 Acquisition module block diagram

spectral analysis. Filtering must be band-pass type. The effectiveness of this technique is closely related to the selection of filter band frequency [9].

This technique is most often used to diagnose bearings and gearboxes and is based on bearings constructive particularities, emphasizing the presence of the impulses and friction defects at an early stage [10, 11].

In the case study was simulate a fault of a bearing by performing two holes with a diameter of 4 mm on the outer race, Fig. 8.

Virtual instrument "RMS VECTOR" is used to display the waveform constructed from RMS values computed for each period of the analyzed signal. This parameter is proposed to be used as an indicator for diagnosis of bearing condition.

In Fig. 9 are shown the results obtained using "RMS Vector VI".

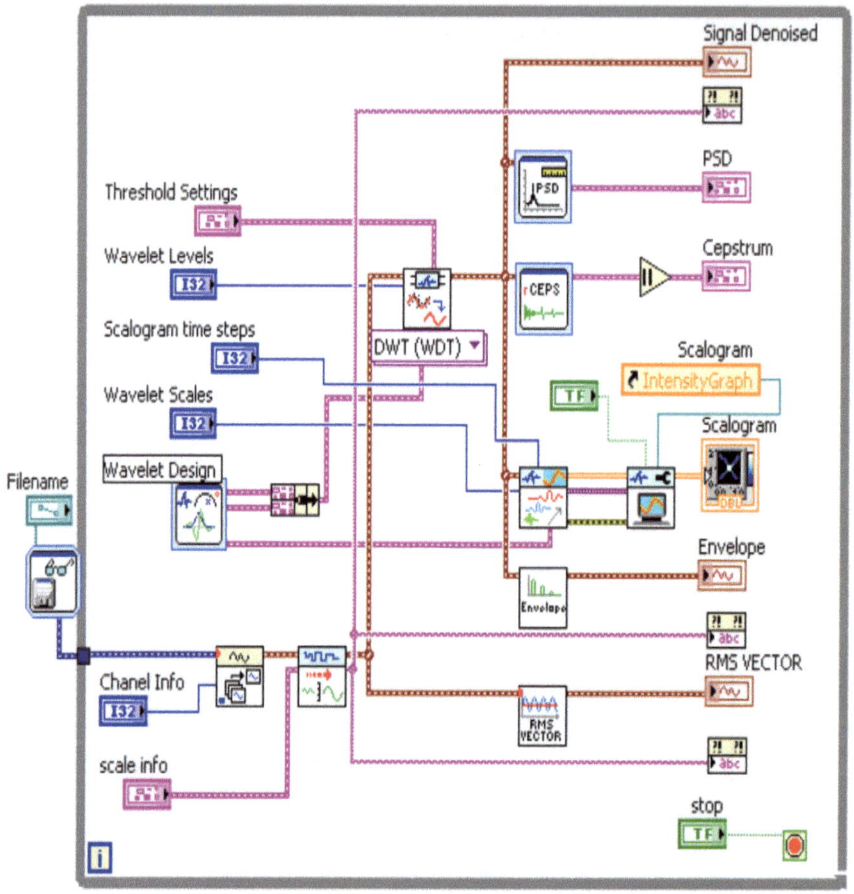

Fig. 7 Analysis module block diagram

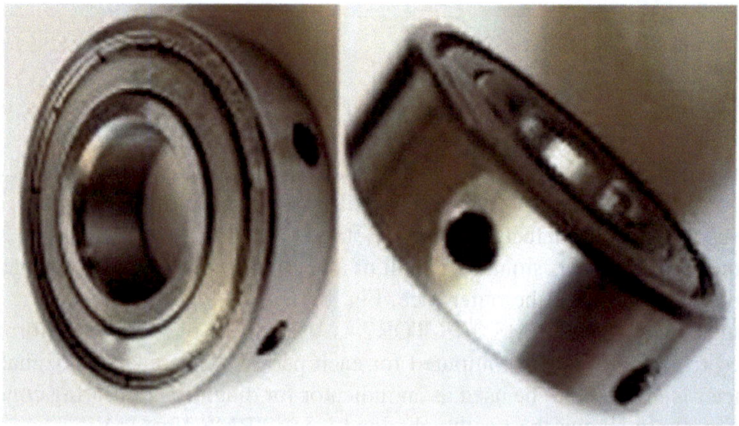

Fig. 8 Bearing with artificial fault

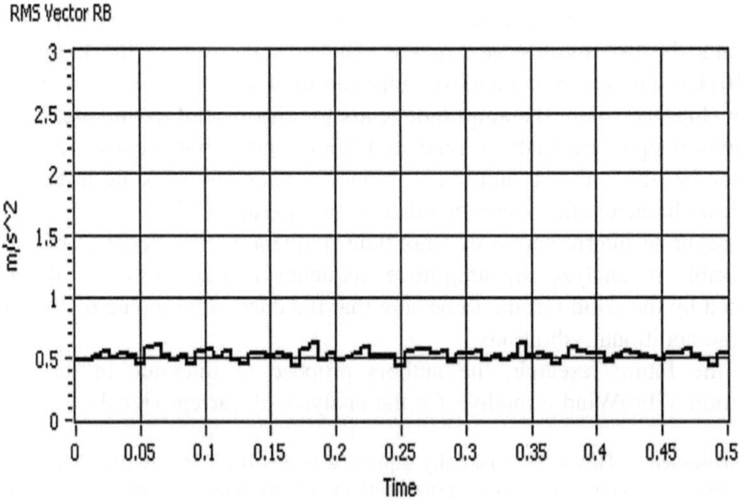

Fig. 9 Vector RMS analysis. Bearing without fault

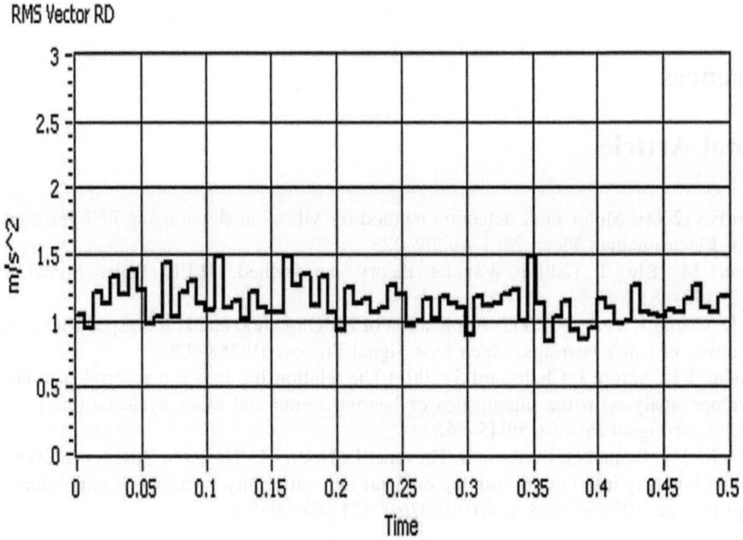

Fig. 10 Vector RMS analysis. Bearing with fault

5 Conclusions

Waveform linearity or lack of is closely related to the variation signal amplitude. So sudden variations in signal amplitude will be characterized by an irregular waveform similar to that in Fig. 10, while a signal whose amplitude variations are less significant will be characterized by an almost linear waveform (Fig. 9). Analyzing the two figures we can see that the waveform of the faulty bearing (Fig. 10) has a much greater and irregular variation than the waveform of the good bearing (Fig. 9), where the amplitude tends to a horizontal asymptote.

Vibration signature analysis generated from wind turbine gearbox is a difficult operation because the real amplitude-frequency characteristics are more complex than ideal characteristics recommended by the literature [5].

For accurate interpretation of amplitude-frequency characteristics of vibration is advisable to analyze the amplitude-frequency characteristics of the current generated by the wind turbine to be sure that the current harmonic distortion does not cause additional vibrations.

For the future research, the authors propose to integrate in the software application VibroWind a module for the analysis of current signals.

Acknowledgments This work is partially supported by PN-II-PT-PCCA-2011-3.2-1696 'Innovative wind energy conversion micro-system with direct-driven electric generator for residential uses'—INNOWECS ANCS Agency Romania.

The authors also gratefully acknowledge the helpful comments and suggestions of the reviewers, which have improved the presentation.

References

Journal Article

1. Lim HS (2006) Motor fault detection method for vibration signal using FFT residuals. Int J Appl Electromagnet Mech 24(3–4):209–223
2. Unser M, Blu T (2003) Wavelet theory demystified. IEEE Trans Signal Process 51(2):470–483
3. Yu D, Cheng J, Yang Y (2005) Application of EMD method and Hilbert spectrum to the fault diagnosis of roller bearings. Mech Syst Signal Process 19:258–270
4. Randall RB, Antoni J, Chobsaard S (2001) The relationship between spectral correlation and envelope analysis in the diagnostics of bearing faults and other cyclostationary machine. Mech Syst Signal Process 5:945–962
5. Gamelin FX, Baquet G, Berthoin S, Thevenet D, Nourry C, Nottin S, Bosquet L (2009) Effect of high intensity intermittent training on heart rate variability in prepubescent children. Eur J Appl Physiol 105:731–738. doi:10.1007/s00421-008-0955-8

Journal Article Only by DOI

6. Niola V, Quaremba G, Avagliano V (2009) Vibration monitoring of gear transmission. In: Proceedings of the 9th wseas international conference on simulation, modelling and optimization, ISBN: 978-960-474-113-7
7. Cocconcelli M, Zimroz R, Rubini R, Bartelmus W (2012) STFT based approach for ball bearing fault detection in a varying speed motor, condition monitoring of machinery in non-stationary operations. Springer, pp 41–50

Book

8. Norton MP, Karczub DG (2003) Fundamentals of noise and vibration analysis for engineers, Cambridge University Press. ISBN 0-521-49913-5, 17 Nov 2003

Online Document (No DOI Available)

9. Crowther A, Eritenel T (2012) Keeping an eye on the health of your wind turbine rotating machinery via condition monitoring will help avoid costly downtime and expensive repairs, wind systems. www.romaxtech.com/wind. Jan 2012
10. Konstantin-Hansen H (2012) Envelope analysis for diagnostics of local faults in rolling element bearings. Appl Note Brüel Kjær, Denmark. www.bksv.com/doc/bo0501.pdf
11. Champavier F (2009) Condition monitoring of induction motors using vibration and electrical signature analysis, EE mods conference, Nantes, France. www.docstoc.com/.../Condition-Monitoring-of-Induction-Motors-Usi. 14–17 Sep 2009

Dissertation

12. Johansson M (1999) The hilbert transform. M.S. Thesis, Vaxjo University

Book

Experimental Characterization of Chatter in Band Sawing

Tilen Thaler, Primož Potočnik and Edvard Govekar

Abstract In the paper the results of the characterization of the chatter phenomenon in the band sawing process are presented. In particular, the influence of the cutting speed and of the distance between the cutting blade supports on chatter characteristics was investigated. In addition to the cutting forces, and emitted sound, the machine vibrations described by the measured acceleration signals were used to characterize the chatter. Based on an analysis of these signals, a hysteresis of the chatter onset and chatter die-out cutting speeds was observed. The observed chatter hysteresis indicates that the chatter onset in band sawing is caused by a Hopf-like bifurcation, and that cutting speed is a promising parameter for chatter control. Additionally a strong effect on chatter characteristics of the distance between the cutting blade supports was experimentally confirmed.

Keywords Band sawing · Chatter · Cutting

1 Introduction

In cutting processes, chatter is a phenomenon that can be described as self-excited, high amplitude vibrations of the cutting tool or workpiece [1]. It is caused by the instability of a nonlinear cutting process, and can have harmful effects on the

T. Thaler (✉)
PETRA Stroji d.o.o, Cesta Andreja Bitenca 68, 1000 Ljubljana, Slovenia
e-mail: tilen.thaler@pe-tra.com

P. Potočnik · E. Govekar
Faculty of Mechanical Engineering, University of Ljubljana,
Aškerčeva 6, 1000 Ljubljana, Slovenia
e-mail: primoz.potocnik@fs.uni-lj.si

E. Govekar
e-mail: edvard.govekar@fs.uni-lj.si

G. Dalpiaz et al. (eds.), *Advances in Condition Monitoring of Machinery in Non-Stationary Operations*, Lecture Notes in Mechanical Engineering,
DOI: 10.1007/978-3-642-39348-8_60,

process performance from the point of view of quality, economy, and ecology [2]. The chatter phenomenon has been investigated intensively in finalizing machining operations such as turning [3, 4], milling [5] and, grinding [6]. Contrary to these machining operations, band sawing is most often used at the beginning of the machining chain, and as such does not influence the final product properties so significantly. However, with the development of expensive and sometimes difficult to cut materials and metal alloys, minimization of waste material for final machining operations and corresponding surface quality are also becoming more and more important.

In band sawing, the negative impact of chatter was first noticed in lumber cutting [7, 8], and was reflected in the so-called wash-boarding [9] of the cut surface. To understand, predict and avoid chatter vibration in band sawing, several mathematical models were derived in order to predict the natural frequencies of the moving continuum representing a band saw blade [7, 8, 10]. Directions for practical solutions take into account control of the cutting speed, avoidance of excitation of cutting blade natural frequencies, and maximizing the tension of the cutting blade [7, 8]. Several theoretical studies have shown that cutting speed and the distance between blade supports are very influential process parameters for the natural frequencies of the band saw blade [7, 8, 10]. However, a description of the experimental characterization of the influence, on chatter characteristics, of cutting speed and of the blade support distance has not been found in the literature. In this paper the results of an experimental characterization of chatter onset, occurring in a band sawing process performed on a horizontal double column band saw, are presented. This characterization was performed based on the acquired signals of the cutting forces, the machine vibration, and the sound emitted during the process. Particular attention was paid to the investigation of the influence, on chatter vibrations, of the cutting speed and the distance between the two blade supports. For this purpose, an experimental system that enables presetting of the blade support distance, controlled variation of the cutting speed, and simultaneous measurements of the cutting forces, machine vibrations and sound signals during the band sawing process was constructed. The experimental system and the performed experiments, including controlled variation of the cutting speed at different blade support distances, are described in the next section. The results of the analysis of selected signals and the corresponding chatter characteristics, such as the chatter onset and chatter die-out cutting speeds, the chatter duration, and the observed hysteresis, as they are related to the distance between the blade supports, are presented in the section preceding the conclusions.

2 Experimental Setup

The experiments were conducted on a double column PE-TRA Toolmaster 300DC band saw of 300 mm maximum cutting width capacity. The cutting width capacity is defined by the distance between the cutting blade supports L_b, i.e. the distance between the two vertical blade supporting arms. This distance L_b can be preset within the range from 230 to 420 mm. A bimetal cutting blade of length 4,150 mm and pitch 2–3 [teeth per inch] was used, and tensioned at approximately 2.0 kN. The detailed characteristic parameters of the cutting blade are given in Table 1. In order to investigate the influence of the cutting speed v_c on the band sawing process, the band saw machine was equipped with a system for controlled variation of the cutting speed v_c.

For the characterization of the band sawing process and the related chatter phenomenon, a three-component Kistler 9257 dynamometer, a three-component PCB piezo accelerometer, and a Brüel and Kjaer microphone were used. The frequency responses of the dynamometer and accelerometer were, in terms of their first natural frequency and according to the manufacturers, 2 and 10 kHz respectively, for the applied type of mounting. The microphone frequency response was ±2 dB in the frequency range 3.15 Hz–20 kHz. Placement of the sensors is shown in Fig. 1. The dynamometer was placed below the workpiece in the vicinity of the cutting blade, in order to measure the cutting force components (F_f, F_c, F_l). The subscripts c, f, and l denote components in the cutting, feed, and lateral directions. In order to measure the machine vibrations (a_f, a_c, a_l) a three component accelerometer was mounted on the left blade supporting arm. The supporting arm is in direct contact with the cutting blade and represents one of the most dynamically exposed parts of the machine structure.

To measure the sound pressure S emitted during the band sawing process the microphone was positioned approximately 32 cm above the workpiece and directed toward the cutting zone. All the sensory data obtained during the band sawing process were acquired by means of a 16-bit resolution A/D data acquisition card, and were transferred into a computer for off-line analysis. The sampling frequencies for the cutting force components (F_f, F_c, F_l), machine vibration (a_f, a_c, a_l), and the sound pressure p signals, were 20, 25.6 and 20 kHz, respectively.

In the band sawing experiments, solid profile workpieces with a width of 100 mm and a height of 60 mm, made of structural steel of type St37 (according to DIN 17100) were used. The main control parameters of the experiments were the cutting speed v_c and the distance between the blade supports L_b. To investigate the influence of the distance L_b on the band sawing process, cutting experiments at predefined values of L_b within the interval [250, 400] mm, with increments of 25 mm, were performed. During the band sawing, the cutting speed v_c was linearly increased, at a preselected value of L_b, from a minimum value of $v_{cmin} = 35$ m/

Table 1 Cutting tool parameters

Tool parameters	
Material	M42
Loop length [mm]	4150
Width/thickness [mm]	34/1.1
Teeth pitch [teeth per inch]	2–3
Rake/clearance angle [°]	10/32

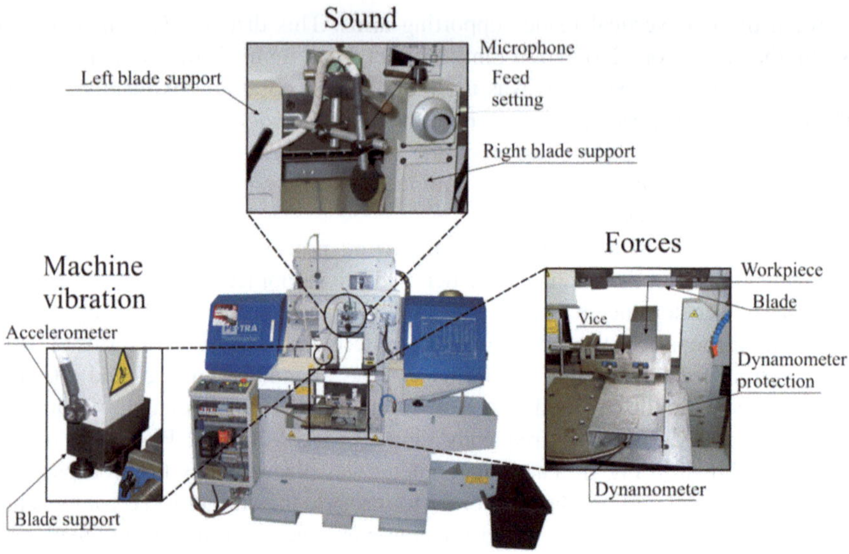

Fig. 1 Experimental setup on an industrial horizontal band saw

min to a maximum cutting speed of $v_{cmax} = 135$ m/min. After this, the cutting speed v_c was linearly decreased back to the minimum cutting speed v_{cmin}. The described variation of the cutting speed was performed within a 60 s time interval, ensuring that the changes occurred slowly and smoothly. The feed speed v_f during the all band sawing experiments was set at a constant value of $v_f = 50 \pm 5$ mm/min.

Examples of the described triangular cutting speed profile $v_c(t)$ and the corresponding feed force F_f, the feed acceleration a_f, and the sound pressure p measured during band sawing of the workpieces at a preset distance between the blade supports of $L_b = 375$ mm are shown in Fig. 2.

Fig. 2 Examples of the cutting speed $v_c(t)$ triangular profile and the corresponding acquired signals of the feed force F_f, the acceleration a_f and the sound pressure p at the blade support distance of $L_b = 375$ mm

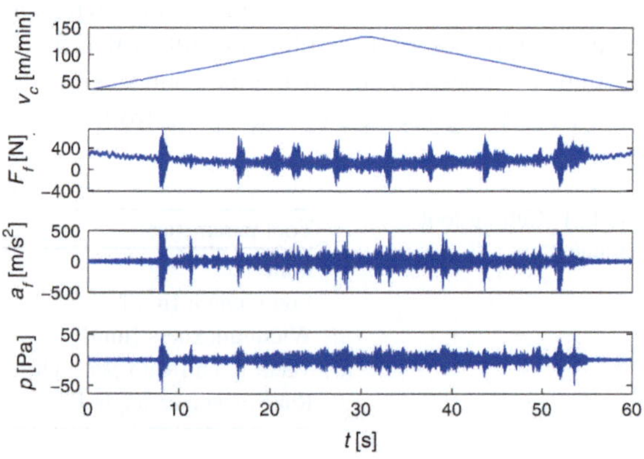

From the time history of the signals it can be seen that, at the beginning and at the end of the workpiece sawing, low amplitude signals of all the variables are typical. With an increase in the cutting speed to $v_c \approx 59.3$ m/min an abrupt onset of large amplitude fluctuations occurs at time $t \approx 7.7$ s. These large amplitude vibrations remain present with a further increase of the cutting velocity to its maximum value of $v_c = 135$ m/min. They die-out during the decrease of the cutting speed to $v_c \approx 49.3$ m/min, at time $t \approx 55.1$ s. The observed onset of large amplitude oscillations at $t \approx 7.7$ s and a cutting velocity of $v_c = 59.3$ m/min is presumably caused by the onset of chatter.

In the cutting force signal, there is, in addition to the observed large amplitude vibration, a time-dependent offset component, which is due to change of the cutting speed and the effects of the workpiece geometry [11].

In general it has been observed that all of the acquired signals appear to be informative with respect to chatter. However, from the point of view of potential practical applications the acceleration signal a_f has several advantages, so this variable was used in the following analysis for the characterization of chatter in band sawing.

3 Analysis and Detection of Chatter Onset

As shown in the previous section, chatter is presumably reflected in high amplitude oscillations of the feed force F_f, the feed acceleration a_f, and the sound pressure p. However, apart from high amplitude vibrations, vibrations with specific pronounced chatter inherent frequencies are characteristic. In order to characterize this property and to overcome the problem of non-stationarity of the signals imposed by the variation of the cutting speed, Short-Time Fourier Transform (STFT) was used to perform further analyses of the acceleration signals a_f. Figure 3 shows examples of spectrograms of the feed acceleration signals a_f acquired during the band sawing process, with variation of the cutting speed v_c performed at three different distances between the blade supports L_b. Horizontal stripes, indicating the excitation of chatter related frequencies, were found to occur in the spectrograms, regardless of the size of the distance between the blade supports L_b, whereas regular chatter-free cutting was characterized by the relatively homogenous, low power frequency pattern in the spectrograms.

From the spectrograms it can be observed that chatter onset takes place at higher and dies out at lower cutting speeds v_c. The observed difference indicates the presence of hysteresis, which is characteristic for the nonlinear chatter phenomenon in cutting, and can be described by a subcritical Hopf bifurcation [1]. It can also be seen that chatter characteristic frequencies depend on the cutting speed v_c. Additionally, the strong effect of the distance between the blade supports L_b on chatter characteristics such as the chatter onset and die-out cutting speeds v_c, the chatter duration, and the length of the hysteresis is evident from the spectrograms. This has also been observed in other theoretical studies [7, 8]. In the following the

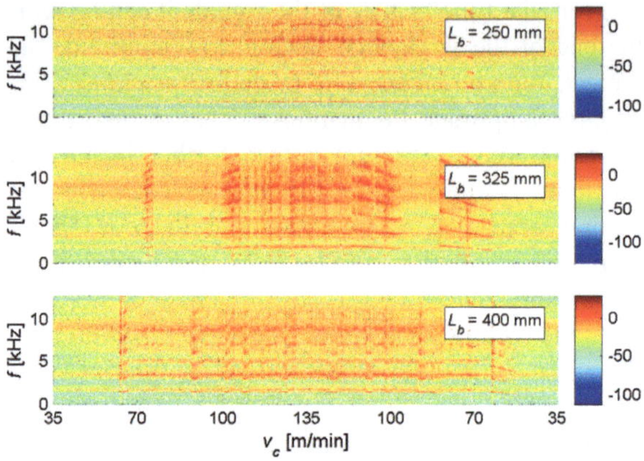

Fig. 3 Spectrograms of the acceleration signal a_f versus cutting speed v_c at three different distances between the blade supports L_b

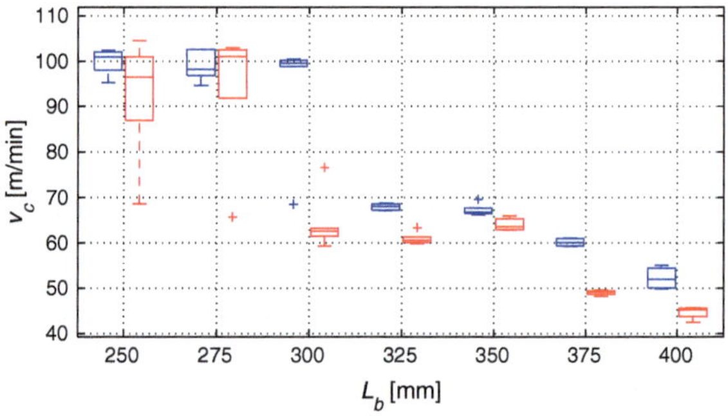

Fig. 4 Box plot of chatter onset (*left*) and chatter die-out cutting speed (*right*) versus the distance between the blade supports L_b

influence of the blade support distance L_b on chatter characteristics is described in more detail.

In order to experimentally characterize the observed chatter dependence on the distance L_b, spectrograms of all the acquired signals were examined, and cutting speed chatter regions were defined based on the frequency content in the spectrograms. Based on the defined chatter regions, the statistical influence of the distance between the blade supports L_b on chatter characteristics such as the chatter onset speed v_{co}, the chatter die-out cutting speed v_{cd}, the chatter duration τ, and the cutting speed hysteresis Δv_c, can be determined as shown in Fig. 4.

Figure 4 shows a box plot diagram of the chatter onset v_{co}, (blue) and the die-out cutting speed v_{cd}, (red) versus the distance between the blade supports L_b. In order to generate the diagram at each investigated support distance L_b a sample of 6 measurements was used. The box plot indicates the median as the central horizontal line, whereas the upper and lower box boundaries are located at the 1st and 3rd quartiles respectively, with whiskers placed at 3-times the standard deviation of the sample, and points (+) outside the whiskers representing outliers. From the box plot (Fig. 4) it can be seen that both the onset chatter and chatter die-out mean cutting speeds are around 100 m/min for the first two shorter distances L_b. At a

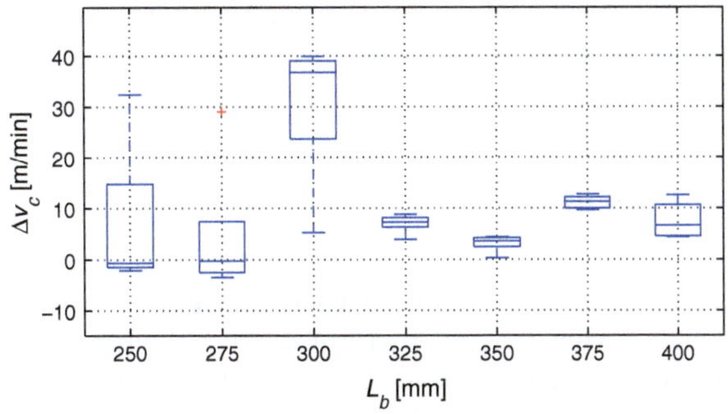

Fig. 5 Length of the cutting speed hysteresis Δv_c versus the blade support distance L_b

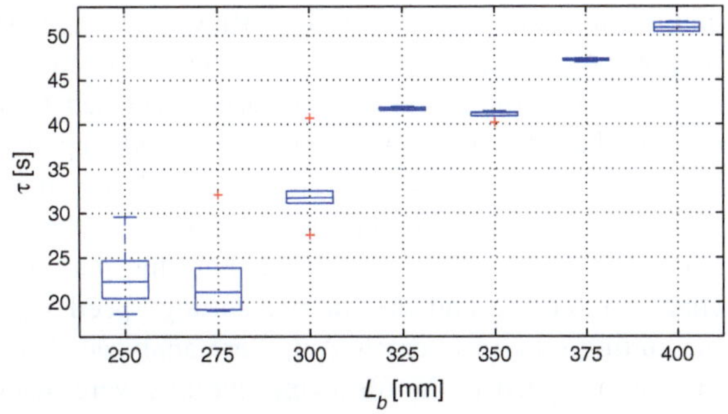

Fig. 6 Duration of chatter τ in dependence on the blade support distance L_b

distance of $L_b = 300$ mm, the die-out chatter cutting speed drops to $v_{cd} = 62.4$ m/min, whereas the chatter onset cutting speed remains equal to $v_{co} = 100$ m/min. With a further increase in the distance to $L_b = 325$ mm the median cutting speed of chatter onset drops to approximately $v_{co} = 68.1$ m/min and further decreases to a value of $v_{co} = 53.4$ m/min at a distance of $L_b = 400$ mm. The chatter onset cutting speeds v_{co} are always slightly higher than the corresponding die-out chatter cutting speeds v_{cd} what indicates the presence of the hysteresis. To characterize the length of the cutting speed hysteresis Δv_c the difference between the chatter onset and chatter die-out cutting speed $\Delta v_c = |v_{co} - v_{cd}|$ was used. The influence of the distance between the blade supports L_b on the length of hysteresis Δv_c is shown in Fig. 5. Using the median values, it can be seen that in the case of smaller distances L_b no hysteresis can be observed.

The largest hysteresis value $\Delta v_c = 34.8$ m/min occurs at a blade support distance of $L_b = 300$ mm, denoting the largest bi-stable region of the band sawing process. With a further increase in the blade support distance L_b the length of the cutting speed hysteresis Δv_c decreases, and scatters around 8 m/min.

A box plot of chatter duration τ versus the distance between the blade supports L_b is presented in Fig. 6. Considering the median values of τ, it can be observed that chatter duration τ, in the case of small distances L_b, is short, and is not affected by the distance L_b. In the case of distances larger than $L_b = 275$ mm, the duration of the chatter τ increased nonlinearly as the blade support distance L_b was increased.

As expected, the longest duration of the chatter $\tau = 51$ s took place in the case of the largest blade support distance $L_b = 400$ mm.

4 Conclusion

In the paper the results of characterization of observed chatter phenomena in a band sawing process performed on a double column horizontal band saw are presented. The characterization was based on the analysis of the acquired cutting force signals, the machine vibrations, and the sound emitted during the sawing process. In particular, the influence, on chatter characteristics, of the cutting speed and of the distance between the cutting blade supports was investigated. For this purpose the cutting speed was increased linearly from $v_{cmin} = 35$ m/min to $v_{cmax} = 135$ m/min, and then linearly decreased back to $v_{cmin} = 35$ m/min. The cutting blade support distance was increased stepwise from 250 to 400 mm, with steps of 25 mm. Although reflected in all the measured signals, the machine vibration as described by acceleration signal in the feed direction was used in this paper for chatter characterization. For this purpose characteristics such as the chatter onset v_{co}, and the die-out cutting speed v_{cd}, the chatter duration τ, and the cutting speed hysteresis Δv_c, were analyzed. With respect to the influence of the cutting speed v_c the following findings were made:

- the specific characteristic frequencies excited by the chatter onset depend on the cutting speed v_c,
- in general, chatter onset takes place at higher cutting speeds v_c, and dies out at lower cutting speeds,
- the observed cutting speed differences Δv_c between the chatter onset and die-out cutting speeds indicate the presence of a hysteresis which is characteristic for the onset of a nonlinear chatter phenomenon, which corresponds to a Hopf-like bifurcation.

Additionally, the strong effect of the distance between the cutting blade supports L_b on chatter characteristics (the chatter onset cutting speed v_{co}, the chatter die-out speed v_{cd}, the chatter duration τ, and the length of the cutting speed hysteresis Δv_c) was confirmed from inspections of the spectrograms of the acceleration signal in the feed direction. Statistically it was determined that:

- the chatter onset cutting speed v_{co} and the corresponding chatter die-out speed v_{cd} decrease nonlinearly as the distance L_b between the cutting blade supports is increased,
- the chatter duration τ also increases nonlinearly as the distance L_b between the cutting blade supports is increased,
- in the case of smaller distances L_b, no cutting speed hysteresis Δv_c was evident. The largest cutting speed hysteresis $\Delta v_c = 37$ m/min was observed at a distance of $L_b = 300$ mm.

The presented results show that the cutting speed v_c and the distance between the blade supports L_b have a significant influence on the chatter phenomenon in band sawing. The defined characteristics (the chatter onset cutting speed v_{co}, the chatter die-out speed v_{cd}, the chatter duration τ, and the length of the cutting speed hysteresis Δv_c) provide basic information which will be useful for the further development and analysis of process stability in terms of stability diagrams. Furthermore, apart from the blade support distance L_b, which is predominantly predefined by the geometry of the workpiece, the cutting speed v_c appears to be a promising parameter for chatter control in band sawing processes.

Acknowledgments The research work described in this paper was partly financed by the European Union, European Social Fund. It was implemented within the framework of the Operational Program for Human Resources Development for the period 2007–2013, Priority axis 1: Promoting entrepreneurship and adaptability, Main type of activity 1.1.: Experts and researchers for competitive enterprises.

The support of the Slovenian Research Agency, the research program P2-0241 Synergetics of Complex Systems and Processes, is also gratefully acknowledged.

The authors also gratefully acknowledge the helpful comments and suggestions of the reviewers, which have improved the presentation.

References

1. Gradišek J, Govekar E, Grabec I (2001) Chatter onset in non-regenerative cutting: a numerical study. J Sound Vib 242:829–838
2. Quintana G, Ciurana J (2011) Chatter in machining processes: a review. Int J Mach Tools Manuf 51:363–376. doi:10.1016/j.ijmachtools.2011.01.001
3. Dombovari Z, Barton DAW, Wilson RE, Stepan G (2011) On the global dynamics of chatter in the orthogonal cutting model. Int J Non-Linear Mech 46:330–338
4. Pan G, Xu H, Kwan CM, Liang C, Haynes L, Geng Z (1996) Modeling and intelligent chatter control strategies for a lathe machine. Control Eng Pract 4:1647–1658
5. Gradišek J et al (2005) On stability prediction for milling. Int J Mach Tools Manuf 45:769–781
6. Govekar E, Baus A, Gradisek J, Klocke F, Grabec I (2002) A new method for chatter detection in grinding. CIRP Ann–Manuf. Technol 51:267–270. doi:10.1016/S0007-8506(07)61514-5
7. Mote CD Jr (1965) A study of band saw vibrations. J Franklin Inst 279:430–444. doi:10.1016/0016-0032(65)90273-5
8. Alspaugh DW (1967) Torsional vibration of a moving band. J Franklin Inst 283:328–338
9. Okai R (2009) Influence of vibration coupling between band saw frame and feed-carriage system on sawdust spillage and surface quality of workpiece during sawing. Eur J Wood Prod 67:189–195. doi:10.1007/s00107-009-0319-z
10. Le-Ngoc L, McCallion H (1999) Self-induced vibration of band saw blades during cutting. Proc Inst Mech Eng 213:371–380. doi:10.1243/0954406991522329
11. Thaler T, Potočnik P, Mužič P, Bric I, Bric R, Govekar E (2012) Characterization of band sawing based on cutting forces. J Mach Eng 12:41–54

